ALGEBRA: GROUPS, RINGS, AND OTHER TOPICS

ALGEBRA: GROUPS, RINGS, AND OTHER TOPICS

Neal H. McCoy

SMITH COLLEGE

Thomas R. Berger

UNIVERSITY OF MINNESOTA

Allyn and Bacon, Inc.

Boston   London   Sydney   Toronto

Portions of this book first appeared in *Fundamentals of Abstract
Algebra,* by Neal H. McCoy, Copyright © 1972 by Allyn and
Bacon, Inc.

Library of Congress Cataloging in Publication Data

McCoy, Neal Henry, 1905–
  Algebra: groups, rings, and other topics.

  Bibliography: p.
  Includes index.
  1. Algebra, Abstract. I. Berger, Thomas R.,
1941–   joint author. II. Title.
QA162.M29   512′.02   76–56743

ISBN 0-205-05699-7

contents

16 SOME ADDITIONAL TOPICS   595

preface

Although substantially different in approach, the content of this book overlaps considerably with the content of *Fundamentals of Abstract Algebra*, Allyn and Bacon, Inc., 1972 (hereafter referred to as FOAA), by one of the present authors. As in FOAA, our goal has been to present abstract algebra in as concrete a setting as possible without sacrifice to the mathematics. To ease the way for students, we have worked from the examples to the theory, often showing how a difficult theorem is proven in the same way as a much easier problem is solved. We have included discussions of groups, rings, fields, and linear algebra; and in each case, we have given sufficient material to allow the instructor considerable freedom of choice of additional topics. To give students a personal and living feeling for mathematics, we have included commentaries and a far-ranging bibliography.

In this present edition, after some necessary concepts have been introduced in the first chapter, we present the theory of groups in careful detail—carrying the subject through quotient groups, homomorphisms, and permutation groups. In order to keep the development concrete, we have focused attention upon only four groups: the additive group of integers, the additive group of least residues modulo n, the group of rigid symmetries of a set of undirected coordinate axes in three dimensions, and the group of rigid symmetries of a cube. By concentrating upon a few nontrivial examples, we are able to develop a deeper and more sophisticated set of examples. Since computation with permutations leads students to understanding via manipulation, emphasis has been placed first upon groups of symmetries. Thus, the students can develop a conceptual framework before they lose themselves in manipulation of permutations. The students (on their own) and the instructor (in class) can develop many further examples through the exercises which occur at the end of each section.

Rings are introduced in Chapter 3, and in obtaining properties of rings, full use is made of the approach and the results already obtained for groups. Numerous examples are presented, which give the student glimpses of some of the more important kinds of rings.

Chapter 4 on integral domains (as in FOAA) presents a characterization of the ring of integers; then follows a chapter on Euclidean domains. The proofs are given in full generality but are cast in the framework of the integers. This keeps the development understandable and concrete, since it concentrates on the only Euclidean domain thoroughly familiar to the students. However, the more general approach is useful in later chapters.

As in FOAA, Chapter 6 discusses elementary properties of fields and, in particular, the rational numbers as the field of quotients of integers. After framing the assumptions of existence and completeness of the real numbers, in Chapter 7 we discuss some consequences of completeness, followed by the algebraic and geometric structure of the field of complex numbers.

Returning to group theory in Chapter 8, we use the methods of actions and orbits to establish the Sylow Theorems. In optional sections we discuss the concept of simple groups (proving that the alternating groups on more than four symbols are simple) and give the structure of the group of units of the integers modulo n. We close with a section in which the classical proof of the Sylow Theorems has been given through a sequence of exercises and hints. Chapter 9 presents a detailed proof of the basis theorem for finite abelian groups (as it appears in FOAA).

Chapter 10 develops in careful detail the properties of polynomial domains, relying on the results of the earlier chapter on Euclidean domains. In Chapter 11 we discuss vector spaces with emphasis on the concepts of basis and dimension. The ideas presented in Chapters 10 and 11 are applied in Chapter 12 to study finite extension fields, where we prove that any two splitting fields of a polynomial are isomorphic. Applications are made to geometric constructions and finite fields.

Linear algebra is developed (as in FOAA) through emphasis on linear systems of equations (Chapter 13). This motivates a discussion of determinants (Chapter 14) and linear transformations (Chapter 15). In Chapter 16 we present some additional topics in such a way that the instructor may assign particular sections as soon as the prerequisites for those sections have been met.

At the end of each chapter there appears a series of commentaries, some of them historical in nature, some of them hinting at related or more general results, and all of them giving references for collateral reading. It is our thought that these commentaries should not be assigned, but that students should be invited to read them at their pleasure. It is hoped that they will be interesting and instructive to eager or questioning students.

Of course, in order to give references for further reading, an expanded bibliography is required. In fact, at the end of the book there is an extensive bibliography of books and articles. A large number of them present material closely related to the topics presented in this book or mentioned in the commentaries. However, some of them are of a more general nature, including essays on the nature of mathematics, foundations of mathematics, fiction in which mathematical concepts appear, and so on. It is our hope that students will be encouraged to look over the bibliography and to browse through the mathematical literature according to their interests.

In the appendix there is a proof of the Fundamental Theorem of Algebra. This theorem, whose proof requires concepts from more than one branch of mathematics (e.g., from algebra and analysis), is frequently not included in text-books. Of course, the material could be assigned at the option of the instructor. However, we think that even if the material is not assigned, students should have a convenient opportunity to see a proof of this rightly famous theorem, thereby gaining a general awareness of how various ideas have produced one of the more transparent proofs of this result.

We are indebted to a number of people for criticisms and suggestions in connection with the preparation of this book. One author's algebra class worked diligently through a preliminary version, offering constructive criticisms and suggestions. Others, including Scot Adams, Gerald Lenz, Philip McCartney, and Ian Richards offered suggestions to improve specific parts of the manuscript. In addition, Robert Gamble, Harvey Wolff, and Charles Pinter read essentially the entire manuscript. We greatly appreciate their numerous suggestions, many of which we have adopted and which have contributed to the accuracy or to the smoothness and consistency of the exposition.

We would also like to thank our wives, Jeanne Berger and Ardis McCoy, for their forbearance, assistance, and encouragement.

Thomas R. Berger
Neal H. McCoy

chapter one

SOME FUNDAMENTAL CONCEPTS

The outstanding characteristic of modern algebra, and indeed also of many other branches of modern mathematics, is its extensive use of what is known as the axiomatic or postulational method. The method itself is not new, since it was used by Euclid (about 300 B.C.) in his construction of geometry as a deductive science. However, in many ways the modern viewpoint is quite different from Euclid's, and the power of the method did not become apparent until this century.

We shall not attempt to give here any description or analysis of the postulational method, but the material of the next few chapters will illustrate the ideas involved. This first chapter will present a few basic concepts to be used repeatedly, and will introduce some convenient notation. Although the reader may have previously met some, or even all, of these concepts, they are so fundamental for our purposes that it seems desirable to start off by presenting them in some detail. Many more illustrations of each concept will appear in later chapters.

1.1 SETS

The concept of *set* (class, collection, aggregate) is fundamental in mathematics as it is in everyday life. A related concept is that of *element* of a set. We make no attempt to define these terms but shall presently give some examples that will illustrate the sense in which they are being used.

First of all, we may say that a set is made up of elements. In order to give an example of a set we need, therefore, to exhibit its elements or to give some rule that will specify its elements. We shall often find it convenient to denote sets by capital letters and elements of sets by lower case letters. If a is an element of the set A, we may indicate this fact by writing $a \in A$ (read, "a is an element of A"). Also, $a \notin A$ will mean that a is not an element of the set A. If both a and b are elements of the set A, we may write $a, b \in A$.

If P is the set of all positive integers, $a \in P$ means merely that a is a positive integer. Certainly, then, it is true that $1 \in P$, $2 \in P$, and so on. If B is the set of all triangles in a given plane, $a \in B$ means that a is one of the triangles of this plane. If C is the set of all books in the Library of Congress, then $a \in C$ means that a is one of these books. We shall presently give other examples of sets.

If $a, b \in A$ and we write $a = b$, it is always to be understood that these are identical elements of A. Otherwise expressed, a and b are merely different symbols designating the same element of A. If $a, b \in A$ and it is not true that $a = b$, we may indicate this fact by writing $a \neq b$ and may say that a and b are *distinct* elements of A.

If A and B are sets with the property that every element of A is also an element of B, we call A a *subset* of B and write $A \subseteq B$ (read, "A is contained in B"). This is also sometimes written in the form $B \supseteq A$. Perhaps we should point out that for every set A it is true that $A \subseteq A$ and hence, according to our definition, one of the subsets of A is A itself. If $A \subseteq B$ and also $B \subseteq A$, then A and B have exactly the same elements and we say that these sets are *equal*, and indicate this by writing $A = B$. If it is not true that $A = B$, we may write $A \neq B$. If $A \subseteq B$ and $A \neq B$, then we say that A is a *proper subset* of B and indicate this fact by the notation $A \subset B$ (read, "A is properly contained in B"). Clearly, $A \subset B$ means that every element of A is an element of B and, moreover, B contains at least one element which is not an element of A.

Sometimes, as has been the case so far, we may specify a set by stating in words just what its elements are. Another way of specifying a set is to exhibit its elements. Thus, $\{x\}$ indicates the set which consists of the single element x, $\{x, y\}$ the set consisting of the two elements x and y, and so on. We may write $A = \{1, 2, 3, 4\}$ to mean that A is the set whose elements are the positive integers 1, 2, 3, and 4. If P is the set of all positive integers, by writing

$$K = \{a \mid a \in P, a \text{ divisible by } 2\},$$

we shall mean that K consists of all elements a having the properties indicated after the vertical bar, that is, a is a positive integer and is divisible by 2. Hence, K is just the set of all *even* positive integers. We may also write

$$K = \{2, 4, 6, 8, \ldots\},$$

the dots indicating that all even positive integers are included in this set. As another example, if

$$D = \{a \mid a \in P, a < 6\},$$

then it is clear that $D = \{1, 2, 3, 4, 5\}$.

Whenever we specify a set by exhibiting its elements, it is to be understood that the indicated elements are distinct. Thus, for example, if we write $B = \{x, y, z\}$, we mean to imply that $x \neq y$, $x \neq z$, and $y \neq z$.

For many purposes, it is convenient to allow for the possibility that a set may have no elements. This unusual set with no elements we shall call the *empty set*. According to the definition of subset given above, the empty set is a subset of every set. Moreover, it is a proper subset of every set except the empty set itself. The empty set is often designated by $\emptyset$, and thus we have $\emptyset \subseteq A$ for every set A.

If A and B are sets, the elements that are in both A and B form a set called the *intersection* of A and B, denoted by $A \cap B$. Of course, if A and B have no elements in common, $A \cap B = \emptyset$. In this latter case we say that A and B are *disjoint* sets.

If A and B are sets, the set consisting of those elements which are elements either of A or of B (or of both) is a set called the *union* of A and B, denoted by $A \cup B$.

As examples of the concepts of intersection and union, let $A = \{1, 2, 3\}$, $B = \{2, 4, 5\}$, and $C = \{1, 3, 6\}$. Then we have

$$A \cap B = \{2\}, A \cap C = \{1, 3\}, B \cap C = \emptyset,$$
$$A \cup B = \{1, 2, 3, 4, 5\}, A \cup C = \{1, 2, 3, 6\},$$

and

$$B \cup C = \{1, 2, 3, 4, 5, 6\}.$$

Although we have defined the intersection and the union of only *two* sets, it is easy to extend these definitions to any number of sets, as follows. The *intersection* of any number of given sets is the set consisting of those elements which are in all the given sets, and the *union* is the set consisting of those elements which are in at least one of the given sets.

If A, B, and C are sets, each of the following is an immediate consequence of the various definitions which we have made:

$A \cap B \subseteq A$ and $A \cap B \subseteq B$.
$A \subseteq A \cup B$ and $B \subseteq A \cup B$.
$A \cap B = A$ if and only if $A \subseteq B$.
$A \cup B = A$ if and only if $B \subseteq A$.
If $B \subseteq C$, then $A \cup B \subseteq A \cup C$ and $A \cap B \subseteq A \cap C$.

In working with sets, so-called Venn diagrams are sometimes used to give a purely symbolic, but convenient, geometric indication of the relationships involved. Suppose, for the moment, that all sets being considered are subsets of some fixed set U. In Figures 1 and 2, the points within the square represent elements of U. If A and B are subsets of U, then the elements of A and B may be represented by the points within indicated circles (or any other closed regions). The intersection and the union of the sets A and B are then represented in an obvious way by the shaded regions in Figures 1 and 2, respectively.

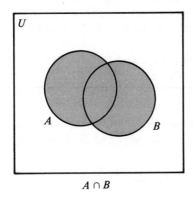

$A \cap B$

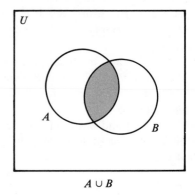

$A \cup B$

FIGURE 1   FIGURE 2

Of course, the use of a Venn diagram is not meant to imply anything about the nature of the sets being considered, whether or not indicated intersections are nonempty, and so on. Moreover, such a diagram cannot in itself constitute a proof of any fact, but it may be quite helpful in suggesting a proof.

Let us make the following remarks by way of emphasis. A problem of frequent occurrence is that of proving the equality of two sets. Suppose that C and D are given sets and it is required to prove that $C = D$. By definition of equality of sets, we need to show that $C \subseteq D$ and that $D \subseteq C$. Sometimes one or both of these conditions follow easily from given facts. If not, the standard procedure is to start with an arbitrary element of C and show that it is an element of D, and then do the same thing with C and D interchanged. When we write "let $x \in C$" or "$x \in C$," we mean that x is to represent a completely arbitrary element of the set C. Hence, to show that $C \subseteq D$, we only need to show that "if $x \in C$, then $x \in D$." Of course, any other symbol could be used in place of x. Let us now give an example by way of illustration.

Example If A, B, and C are sets, prove that

$$A \cup (B \cap C) = (A \cup B) \cap (A \cup C).$$

Solution First, let us take advantage of the opportunity to give another illustration of a Venn diagram. If we think of the meaning of $A \cup (B \cap C)$ as consisting of all elements of A together with all elements that are in both B and C, we see that the set $A \cup (B \cap C)$ may be represented by the shaded portion of the Venn diagram in Figure 3. We leave it to the reader to verify that this same shaded region also represents the set $(A \cup B) \cap (A \cup C)$.

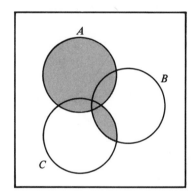

FIGURE 3

We now proceed to give a formal proof of the required formula. Clearly, $B \cap C \subseteq B$, and so $A \cup (B \cap C) \subseteq A \cup B$. Similarly, $B \cap C \subseteq C$, and hence $A \cup (B \cap C) \subseteq A \cup C$. It follows that

$$A \cup (B \cap C) \subseteq (A \cup B) \cap (A \cup C),$$

and we have obtained inclusion one way. To obtain inclusion the other way, let $x \in (A \cup B) \cap (A \cup C)$ and let us proceed to show that $x \in A \cup (B \cap C)$. Now $x \in A \cup B$ and also $x \in A \cup C$. If $x \in A$, then surely $x \in A \cup (B \cap C)$. If $x \notin A$, then $x \in B$ and $x \in C$, so that $x \in B \cap C$, and again we have that $x \in A \cup (B \cap C)$. This shows that $(A \cup B) \cap (A \cup C) \subseteq A \cup (B \cap C)$, and the proof is therefore complete.

At this point we inject another definition. For two sets A and B we shall define the *set difference* $A\backslash B$ to be the set of all elements which are members of the set A but are not members of the set B. In another way, we say this formally as: $A\backslash B = \{x \mid x \in A, x \notin B\}$. Look over at Figure 1: The set $A\backslash B$ is pictured as the white portion inside A where the shaded portion has been removed. The set B has taken a "bite" out of the set A, so to speak. We will rarely use this notation; but when we do, the set B will usually contain some elements which we want removed from the set A.

The final concept to be introduced in this section may be illustrated by the familiar idea of coordinates of a point in a plane. A point is determined by an ordered pair (x, y) of real numbers. The word *ordered* is meant to imply that the order of writing the two numbers x and y is important; that is, that (x, y) is to be considered as a different pair from (y, x) unless, of course, x and y happen to be equal real numbers. If **R** denotes the set of all real numbers, the set of all ordered pairs of elements of **R** is frequently called the *Cartesian product* of **R** by **R** and designated by **R** $\times$ **R**. More generally, if A and B are any sets, the set of all ordered pairs (a, b), where $a \in A$ and $b \in B$, is the *Cartesian product* of A by B, designated by $A \times B$. It may happen, of course, that A and B are identical sets, as in the illustration given above. It is obvious how to define the Cartesian product of more than two sets. Thus, for example, the set $A \times B \times C$ is the set of all ordered triples (a, b, c), where $a \in A$, $b \in B$, and $c \in C$.

As another example of a product set, if $A = \{1, 2, 3\}$, and $B = \{u, v\}$, then

$$A \times B = \{(1,u), (1,v), (2,u), (2,v), (3,u), (3,v)\}.$$

EXERCISES

1. If $A = \{a, b, c\}$, $B = \{c, x, y\}$, and $C = \{x, z\}$, determine each of the following sets: $A \cap B$, $A \cap C$, $B \cap C$, $A \cup B$, $A \cup C$, $B \cup C$, $A \times B$, $A \times C$, $B \times C$.

2. Let P be the set of all positive integers, and define subsets of P as follows:

$$F = \{a \mid a \in P, a < 10\},$$
$$G = \{a \mid a \in P, a > 5\},$$
$$H = \{a \mid a \in P, a \text{ divisible by } 3\}.$$

Determine each of the following sets: $F \cap G$, $F \cap H$, $G \cap H$, $F \cup G$, $F \cup H$, $G \cup H$.

3. Exhibit the four different subsets of a set with two elements. How many subsets does a set with three elements have? A set with four elements?

4. If k is a positive integer, show that a set with $k + 1$ elements has twice as many subsets as a set with k elements. Then find a formula for the number of subsets of a set with n elements, where n is an arbitrary positive integer.

5. If A, B, and C are sets, draw Venn diagrams to illustrate and then give a formal proof that $A \cap (B \cup C) = (A \cap B) \cup (A \cap C)$.

6. If X is a subset of a set U, let us denote by X' the *complement* of X in U, that is, the set of all elements of U which are not in subset X. If A and B are subsets of U, prove each of the following: $(A \cap B)' = A' \cup B'$, $(A \cup B)' = A' \cap B'$.

1.2 MAPPINGS

As a first illustration of the concept to be introduced in this section, let C be the set of all books in the Library of Congress, and P the set of all positive integers. Corresponding to each book there is a unique positive integer; namely, the number of pages in the book. That is, to each element of C there corresponds in this way a unique element of P. This is an example of a mapping of the set C into the set P. As another illustration, let N be the set of all names occurring in a given telephone directory, and L the set of the twenty-six letters of the alphabet. We may then associate with each name the first letter of the surname, and this then defines a mapping of N into L. Additional examples will be given after the following definition.

1.1 **DEFINITION** A *mapping* of a set A into a set B is a correspondence that associates with each element a of A a unique element b of B. The notation $a \rightarrow b$ is sometimes used to indicate that b is the element of B that is associated with the element a of A under a given mapping. We may say that a *maps into* b or that b is the *image* of a under this mapping. We often call the set A the *domain* of a mapping from A into B.

Let us now give an example of a mapping of the set $S = \{1, 2, 3, 4\}$ into the set $T = \{x, y, z\}$. To specify such a mapping, we merely need to select an element of T to be the image of each element of S. Thus

1.2 $$1 \rightarrow x, \quad 2 \rightarrow y, \quad 3 \rightarrow x, \quad 4 \rightarrow y$$

defines a mapping of S into T in which x is the image of 1, y the image of 2, and so on. Note that although every element of S is required to have an image in T, it need not be true that every element of T occurs as the image of at least one element of S.

Before proceeding, let us observe that a *function,* as the term is often used, is just a mapping of the set **R** of all real numbers (or of some subset of **R**) into the same set **R**. For example, the function f defined by $f(x) = x^2 + x + 1$ is the mapping $x \to x^2 + x + 1$ which associates with each real number x the real number $x^2 + x + 1$. In this setting, the mapping is denoted by f and the image of the number x under the mapping f by $f(x)$.

Although we are now concerned with arbitrary sets (not just sets of real numbers), the function notation of the preceding paragraph could be, and frequently is, used for mappings. However, we shall adopt an alternate notation which is also of fairly wide use in algebra and which will have some advantages later on. Mappings will henceforth usually be denoted by Greek letters, such as $\alpha, \beta, \gamma, \ldots$. If α is a mapping of A into B, the image of an element a of A will be denoted by $a\alpha$. Note that α is here written on the right and without parentheses, rather than on the left as in the more familiar function notation $\alpha(a)$. For example, let β be the mapping 1.2 of S into T. Then, instead of writing 1.2 we might just as well write

1.3 $1\beta = x, \quad 2\beta = y, \quad 3\beta = x, \quad 4\beta = y.$

Another mapping γ of S into T is defined by

1.4 $1\gamma = x, \quad 2\gamma = y, \quad 3\gamma = y, \quad 4\gamma = z.$

We shall presently use these mappings to illustrate certain additional concepts.

It is customary to write $\alpha : A \to B$ to indicate that α is a mapping of the set A into the set B. We may also sometimes write $a \to a\alpha$, $a \in A$, to indicate this mapping, it being understood that for each $a \in A$, $a\alpha$ is a uniquely determined element of B. If we have mappings $\alpha : A \to B$ and $\beta : A \to B$, we naturally consider these mappings to be *equal,* and write $\alpha = \beta$, if and only if $a\alpha = a\beta$ for every $a \in A$. Thus, for the mappings $\beta : S \to T$ and $\gamma : S \to T$ exhibited above, we have $\beta \neq \gamma$ since, for example, $3\beta = x$ and $3\gamma = y$.

In order to avoid some trivial special cases, whenever we consider a mapping of a set A into a set B we shall always tacitly assume that the sets A and B are not empty.

In studying mappings it is sometimes suggestive to make use of a geometrical diagram. Figure 4 is supposed to suggest that α is a mapping of A into B and that under this mapping each element a of A has image $a\alpha$ in B.

In the particular mapping $\beta : S \to T$ given by 1.3, the element z of T does not occur as the image of any element of S. However, in the mapping $\gamma : S \to T$,

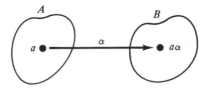

FIGURE 4

defined in 1.4, every element of T is the image of at least one element of S. The language for stating this essential difference between these mappings is given in the following definition.

1.5   DEFINITION A mapping α of A into B is said to be a mapping of A *onto* B if and only if every element of B is the image of at least one element of A under the mapping α.

Thus, γ is a mapping of S onto T, whereas β is not a mapping of S onto T. It is important to observe that "into" is *not* the opposite of "onto." According to our language, every mapping is a mapping of some set into some set. That is, "onto" is a special case of "into," and if $\alpha : A \to B$ is a mapping of A *onto* B, it is perfectly correct to say that it is a mapping of A into B (although this doesn't give the maximum amount of available information).

If $\alpha : A \to B$ is a mapping of A into B, let us denote by $A\alpha$ the set of all elements of B that occur as images of elements of A under the mapping α, that is,

$$A\alpha = \{a\alpha \,|\, a \in A\}.$$

Thus α is an onto mapping if and only if $A\alpha = B$. In any case, an arbitrary mapping $\alpha : A \to B$ may be considered as defining a mapping of A *onto* the subset $A\alpha$ of B. Thus, associated with each mapping is an onto mapping if we suitably restrict the set in which the images lie.

There is one additional concept which plays an important role in the study of mappings. In the mapping $\beta : S \to T$, defined by 1.3, we see that both 1 and 3 have x as image. Similarly, $\gamma : S \to T$, defined by 1.4 is such that both 2 and 3 have y as image. Now let $T = \{x, y, z\}$ as above, and let $U = \{r, s, t, u\}$. Then the mapping $\theta : T \to U$ defined by

1.6 $x\theta = t, \quad y\theta = r, \quad z\theta = u$

is such that every element of U which occurs as an image of some element of T is the image of exactly one element of T. This property has a name which we proceed to introduce.

1.7   DEFINITION A mapping $\alpha : A \to B$ is said to be a *one-one* mapping of A into B if and only if distinct elements of A have distinct images in B or, equivalently, if $a_1, a_2 \in A$ such that $a_1\alpha = a_2\alpha$, then $a_1 = a_2$.

The mapping $\theta : T \to U$ defined by 1.6 is an example of a one-one mapping. Note, however, that it is not an onto mapping. Hence a one-one mapping may or may not be an onto mapping. Clearly, also, an onto mapping need not be a one-one mapping.†

---

† Some other terms will be found in the literature as follows. An onto mapping is also called a *surjection*; a one-one mapping may be called an *injection*. A mapping which is both one-one and onto is also called a *bijection*.

We now give some additional examples to illustrate these concepts.

Example 1   Let C be a nonempty subset of the set D.   The mapping $\phi : C \to D$ defined by $c\phi = c$ for each $c \in C$ is a one-one mapping of C into D. It is an onto mapping if and only if $C = D$.

Example 2   Let $\alpha : A \times B \to A$ be defined by $(a, b)\alpha = a$ for each $(a, b) \in A \times B$.   This is certainly an onto mapping.   However, if $b_1, b_2 \in B$ with $b_1 \neq b_2$, then $(a, b_1)\alpha = (a, b_2)\alpha$ with $(a, b_1) \neq (a, b_2)$, so the mapping is not a one-one mapping.   It will be a one-one mapping if and only if B has exactly one element.   The mapping α of this example is called the *projection* of $A \times B$ onto A.   Similarly, one can define the projection of $A \times B$ onto B.

Example 3   Let $\mathbf{Z}$ be the set of all integers and $\alpha : \mathbf{Z} \to \mathbf{Z}$ be defined by $i\alpha = 2i + 1, i \in \mathbf{Z}$.   In contrast to most of our previous examples, this is an example of a mapping of the set $\mathbf{Z}$ into the same set $\mathbf{Z}$. To determine whether α is an onto mapping, let j be an arbitrary element of $\mathbf{Z}$ and let us find whether j is the image of some element. That is, we need to determine whether there exists an integer i such that $2i + 1 = j$.   Clearly, there will be no such integer i if j is even since $2i + 1$ is odd for every integer i.   Thus, α is not an onto mapping.   Is it a one-one mapping?   To answer this question, suppose that $i_1, i_2 \in \mathbf{Z}$ such that $i_1\alpha = i_2\alpha$, that is, such that $2i_1 + 1 = 2i_2 + 1$. It follows that $i_1 = i_2$, and α is therefore a one-one mapping.

It will probably seem reasonable, and it is indeed true although we shall not discuss this fact here, that if set C has n elements for some positive integer n, then there will exist a one-one mapping of C onto D if and only if D also has n elements.

EXERCISES

1. Let $\mathbf{Z}$ be the set of all integers, and $i \in \mathbf{Z}$.   Determine in each case whether the indicated mapping α of $\mathbf{Z}$ into $\mathbf{Z}$ is an onto mapping and whether it is a one-one mapping.

   (a) $i\alpha = i + 3$,

   (b) $i\alpha = i^2 + i$,

   (c) $i\alpha = i^3$,

   (d) $i\alpha = 2i - 1$,

   (e) $i\alpha = -i + 5$,

   (f) $i\alpha = i - 4$.

2. Let **R** be the set of all real numbers, and $x \in \mathbf{R}$. Determine in each case whether the indicated mapping α of **R** into **R** is an onto mapping and whether it is a one-one mapping.

(a) $x\alpha = 2x + 1$,

(d) $x\alpha = x^3$,

(b) $x\alpha = 1 - x$,

(e) $x\alpha = x^2 + x$,

(c) $x\alpha = x^2$,

(f) $x\alpha = 4x$,

(g) $x\alpha = \begin{cases} x \text{ if } x \text{ is rational,} \\ 2x \text{ if } x \text{ is irrational.} \end{cases}$

3. Let P be the set of all positive integers, and $n \in P$. Determine in each case whether the indicated mapping of P into P is an onto mapping and whether it is a one-one mapping.

(a) $n\alpha = 2n$,

(c) $n\alpha = n^2$,

(b) $n\alpha = n + 1$,

(d) $1\alpha = 1, n\alpha = n - 1$ for $n > 1$.

4. Give an example of a mapping of the set P of all positive integers into the set P such that every element of P is the image of exactly two elements.

5. If **R** is the set of all real numbers, use the fact that every cubic equation with real coefficients has a real root to show that the mapping α of **R** into **R** defined by $x\alpha = x^3 - x, x \in \mathbf{R}$, is a mapping of **R** onto **R**. Is it a one-one mapping?

6. If **R** is the set of all real numbers, why doesn't the formula $x\alpha = 1/x, x \in \mathbf{R}$, define a mapping of **R** into **R**?

7. If $A = \{1, 2, 3\}$ and $B = \{x, y\}$, verify that there exist eight mappings of A into B and nine mappings of B into A. How many mappings of A onto B are there?

8. Let A be a set with m elements, and B a set with n elements (m and n positive integers). Formal proofs are not required, but in each of the following give some indication why you believe your conclusion to be correct.

(a) Determine the number of mappings of A into B.

(b) If $n \geq m$, determine the number of one-one mappings of A into B.

(c) If $m = n$, determine the number of one-one mappings of A onto B.

9. Let R be the set of all real numbers, and let $x \in R$. Use any calculus methods which you know to determine whether each of the following mappings α of R into R is an onto mapping and whether it is a one-one mapping.

(a) $x\alpha = e^x$,

(c) $x\alpha = x + \sin x$,

(b) $x\alpha = \sin x$,

(d) $x\alpha = \dfrac{x}{2} + \sin x$.

1.3 PRODUCTS OF MAPPINGS

Under certain conditions, two or more mappings may be combined in a natural way to form a new mapping. Suppose that A, B, and C are sets and that we have given mappings $\alpha : A \to B$ and $\beta : B \to C$. It is then easy to define in a natural way a mapping of A into C. If $a \in A$, we first take the image $a\alpha$ of a under the mapping α. Now $a\alpha \in B$, so $(a\alpha)\beta$ is a uniquely determined element of C. Thus $a \to (a\alpha)\beta$, $a \in A$, defines a mapping of A into C determined by the given mappings α and β. We denote this mapping by $\alpha\beta$ and call it the *product* (or *composition*) of α by β. More formally, the definition of the mapping $\alpha\beta : A \to C$ is as follows:

1.8 $$a(\alpha\beta) = (a\alpha)\beta, \qquad a \in A.$$

We may point out that, according to the definition just given, $\alpha\beta$ means "first perform α, then perform β." It is here that it makes an essential difference in notation whether we denote the image of a under the mapping α by $a\alpha$ or by the function notation $\alpha(a)$. Had we adopted the latter notation, the mapping which we have denoted by $\alpha\beta$ would map an element a of A into the element $\beta(\alpha(a))$ of C, and it would be natural to denote it by $\beta\alpha$. *Both notations are widely used, and in reading other books the student must be prepared to find either one.*

As a simple illustration of Definition 1.8, let $A = \{1, 2, 3, 4\}$, $B = \{x, y, z\}$, $C = \{r, s\}$, and let $\alpha : A \to B$ and $\beta : B \to C$ be defined, respectively, as follows:

$$1\alpha = y, \quad 2\alpha = x, \quad 3\alpha = x, \quad 4\alpha = z;$$
$$x\beta = s, \quad y\beta = r, \quad z\beta = s.$$

Then the mapping $\alpha\beta : A \to C$ is obtained by the following calculations:

$$1(\alpha\beta) = (1\alpha)\beta = y\beta = r,$$
$$2(\alpha\beta) = (2\alpha)\beta = x\beta = s,$$
$$3(\alpha\beta) = (3\alpha)\beta = x\beta = s,$$
$$4(\alpha\beta) = (4\alpha)\beta = z\beta = s.$$

It should be clear that in defining the product of two mappings a certain condition on the sets involved is necessary. Thus, if α is a mapping of A into B, $\alpha\beta$ is defined only if β is a mapping of the set B into some set.

We now take one more step as follows. Let A, B, C, and D be sets, and suppose that we have mappings $\alpha : A \to B$, $\beta : B \to C$, and $\gamma : C \to D$. Then $\alpha\beta$ is a mapping of A into C, and $(\alpha\beta)\gamma$ is a mapping of A into D. In like manner, $\alpha(\beta\gamma)$ is seen to be a mapping of A into D. It is an important fact that these two mappings are equal; that is, that

1.9 $(\alpha\beta)\gamma = \alpha(\beta\gamma).$

By the definition of equality of mappings, we shall prove 1.9 by verifying that

1.10 $a((\alpha\beta)\gamma) = a(\alpha(\beta\gamma))$

for every element a of A.

First, we observe that by the definition of the product of the mappings $\alpha\beta$ and γ, we have

$$a((\alpha\beta)\gamma) = (a(\alpha\beta))\gamma.$$

Then, by the definition of the product $\alpha\beta$, it follows that

$$(a(\alpha\beta))\gamma = ((a\alpha)\beta)\gamma,$$

and so the left side of 1.10 is equal to $((a\alpha)\beta)\gamma$. In like manner, by applying the definition of the product of α by $\beta\gamma$, and then the definition of the product $\beta\gamma$, we obtain

$$a(\alpha(\beta\gamma)) = (a\alpha)(\beta\gamma) = ((a\alpha)\beta)\gamma.$$

Since both sides of 1.10 are equal to $((a\alpha)\beta)\gamma$, we have proved 1.10 and also 1.9.

The fact that 1.9 holds whenever the indicated products are defined is usually expressed by saying that multiplication of mappings is *associative*. This property of associativity will appear frequently in various settings throughout this book.

For each set A, the *identity mapping on A* will be denoted by ϵ_A. That is, $\epsilon_A : A \to A$ is defined by $a\epsilon_A = a$ for each $a \in A$. If $\alpha : A \to B$, then for $a \in A$ we have $a\alpha = (a\epsilon_A)\alpha = a(\epsilon_A\alpha)$. Thus $\alpha = \epsilon_A\alpha$. Similarly, it may be verified that $\alpha\epsilon_B = \alpha$.

If $\alpha : A \to B$ and $\beta : B \to A$ are such that $\alpha\beta = \epsilon_A$, we say that β is a *right inverse* of α and that α is a *left inverse* of β. The following theorem gives conditions under which a mapping has a right or a left inverse.

1.11 **THEOREM** *Let* $\alpha : A \to B$ *be a given mapping. Then each of the following is true:*

> (*i*) *The mapping* α *has a right inverse if and only if it is a one-one mapping.*
>
> (*ii*) *The mapping* α *has a left inverse if and only if it is an onto mapping.*
>
> (*iii*) *If* α *has a right inverse* $\beta : B \to A$ *and a left inverse* $\gamma : B \to A$, *then* $\beta = \gamma$.

Proof of (i): Suppose, first, that $\beta : B \to A$ is a right inverse of α, so that $\alpha\beta = \epsilon_A$. For each $a \in A$, we have that

$$(a\alpha)\beta = a(\alpha\beta) = a\epsilon_A = a.$$

Hence, if $a_1, a_2 \in A$ such that $a_1\alpha = a_2\alpha$, it follows that $(a_1\alpha)\beta = (a_2\alpha)\beta$, and thus that $a_1 = a_2$. This shows that α is a one-one mapping.

Conversely, let us assume that α is a one-one mapping, and define a mapping $\beta : B \to A$ such that $\alpha\beta = \epsilon_A$. Now $A\alpha$ may well be a proper subset of B, but for each $b \in A\alpha$, the fact that α is one-one assures us that there is exactly *one* element a of A such that $b = a\alpha$. If $b \in A\alpha$, we define $b\beta = a$, where $b = a\alpha$. For each $b \in B$ with b not in $A\alpha$, we choose a fixed element a_1 of A and define $b\beta = a_1$. Then β is defined on all of B and for $a \in A$, we have $a(\alpha\beta) = (a\alpha)\beta = a$. That is, $\alpha\beta = \epsilon_A$ and β is a right inverse of α.

Proof of (ii): Suppose that α has a left inverse $\gamma : B \to A$, so that $\gamma\alpha = \epsilon_B$. Then for each $b \in B$, we see that $(b\gamma)\alpha = b(\gamma\alpha) = b$, and clearly α is an onto mapping.

Conversely, let us assume that $\alpha : A \to B$ is an onto mapping. For each $b \in B$, choose† one element a of A such that $a\alpha = b$ and define $b\gamma = a$. This defines a mapping $\gamma : B \to A$ and clearly $b(\gamma\alpha) = (b\gamma)\alpha = b$ for each $b \in B$. Thus, $\gamma\alpha = \epsilon_B$, and γ is indeed a left inverse of α.

Proof of (iii): Under our assumptions, we have that $\alpha\beta = \epsilon_A$ and that $\gamma\alpha = \epsilon_B$. The reader is asked to justify each step in the following proof of the desired result:

$$\beta = \epsilon_B\beta = (\gamma\alpha)\beta = \gamma(\alpha\beta) = \gamma\epsilon_A = \gamma.$$

---

† The possibility of making such a choice of one element from each of a possibly infinite number of sets involves an interesting and profound assumption called the "Axiom of Choice." We here simply assume that this is permissible. Discussions of this axiom may be found, for example, in Wilder [97], Kurosh [16], and Birkhoff [119] as listed in the bibliography at the end of this book.

   If there exists a mapping $\delta : B \to A$ which is both a right inverse and a left inverse of the mapping $\alpha : A \to B$, δ is said to be an *inverse* of α. A special case of part (iii) of the preceding theorem, applied to the situation in which both β and γ are inverses of α, shows that if a mapping has an inverse, it is *unique*. The unique inverse (if it exists) of a mapping α is usually denoted by α^{-1}. In view of the theorem, we see that a mapping $\alpha : A \to B$ has an inverse if and only if it is a one-one and onto mapping. It is clear that if $\alpha : A \to B$ is a one-one mapping of A onto B, then $\alpha^{-1} : B \to A$ is a one-one mapping of B onto A. The simple relationship between the mappings α and α^{-1} may possibly be suggested by the diagram of Figure 5. The reader should carefully observe that the mapping α^{-1}, as here defined, exists if and only if α is a one-one mapping of A onto B, since in any mapping of B into A *every* element of B must have exactly *one* image in A.

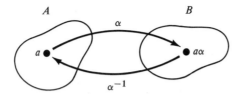

FIGURE 5

   Clearly, a one-one mapping of A onto B may be thought of as a pairing of the elements of A and the elements of B. In view of the mutual relationship between A and B, a one-one mapping of A onto B (or of B onto A) is sometimes called a *one-to-one correspondence between A and B*.

EXERCISES

1. Given mappings $\alpha : A \to B$ and $\beta : B \to C$, prove each of the following:

   (i) If both α and β are one-one mappings, then $\alpha\beta$ is a one-one mapping.

   (ii) If both α and β are onto mappings, then $\alpha\beta$ is an onto mapping.

   (iii) If $\alpha\beta$ is a one-one mapping, then α is a one-one mapping.

   (iv) If $\alpha\beta$ is an onto mapping, then β is an onto mapping.

2. Suppose that $\alpha\beta$ is defined. If α has a right inverse α_1 and β has a right inverse β_1, prove that $\alpha\beta$ has $\beta_1\alpha_1$ as a right inverse.

3. Suppose that $\alpha\beta$ is defined. If α has a left inverse α_2 and β has a left inverse β_2, prove that $\beta_2\alpha_2$ is a left inverse of $\alpha\beta$.

4. If $\alpha\beta$ is defined, and α has the inverse α^{-1}, and β has the inverse β^{-1}, prove that $\beta^{-1}\alpha^{-1}$ is defined and is the inverse of $\alpha\beta$.

5. Let $A = \{1, 2, 3\}$, and define mappings $\alpha : A \rightarrow A$ and $\beta : A \rightarrow A$ as follows:

$$1\alpha = 2, \quad 2\alpha = 1, \quad 3\alpha = 3,$$
$$1\beta = 2, \quad 2\beta = 3, \quad 3\beta = 1.$$

Compute each of the following mappings of A into A: $\alpha\beta$, $\beta\alpha$, α^{-1}, β^{-1}, and $(\alpha\beta)^{-1}$. Verify that

$$(\alpha\beta)^{-1} = \beta^{-1}\alpha^{-1} \neq \alpha^{-1}\beta^{-1}.$$

1.4 EQUIVALENCE RELATIONS

The concept of an equivalence relation, to be defined presently, plays a very important role in modern algebra; accordingly, we present it in this preliminary chapter. Before giving the definition, let us briefly explain what we mean by a relation.

If $\mathbf{Z}$ is the set of all integers, and $i < j$ has the usual meaning for $i, j \in \mathbf{Z}$, then "$<$" is an example of a relation defined on $\mathbf{Z}$. This statement only means that for every ordered pair (i, j) of elements of $\mathbf{Z}$, $i < j$ is either true or false. With the usual meanings of these symbols, "$\leq$" and "$=$" are other relations defined on $\mathbf{Z}$.

In general, let A be a given set. We say that a *relation* "$\sim$" is defined on A if for each ordered pair (a, b) of elements of A, $a \sim b$ is either true or false. We may write $a \nsim b$ to indicate that $a \sim b$ is false.† For the present, we are not concerned with relations in general but only with those relations which have the particular properties stated in the following definition.

1.12 DEFINITION A relation "$\sim$" defined on a set A is called an *equivalence relation* if it has the following three properties, where a, b, and c are arbitrary elements of A:

(1) $a \sim a$ (*reflexive property*).

(2) If $a \sim b$, then $b \sim a$ (*symmetric property*).

(3) If $a \sim b$ and $b \sim c$, then $a \sim c$ (*transitive property*).

---

† Given a relation "$\sim$" on a set A, we may associate with it the uniquely determined subset T of the Cartesian product $A \times A$ consisting of all ordered pairs (a, b) such that $a \sim b$. Conversely, given a subset T of $A \times A$, we may use it to define a relation "$\sim$" on A by agreeing that $a \sim b$ if and only if the ordered pair (a, b) is an element of T. Thus a relation on a set A is uniquely associated with a subset of $A \times A$. For this reason, a relation on a set A is frequently *defined* to be simply a subset of $A \times A$.

If "$\sim$" is an equivalence relation, we may find it convenient to read $a \sim b$ as "a is equivalent to b."

The relation "$<$" on the set $\mathbf{Z}$ of all integers is not an equivalence relation since it has neither the reflexive property nor the symmetric property. The relation "$\leq$" has the reflexive property and the transitive property, but not the symmetric property. Of course, "$=$" is an equivalence relation on $\mathbf{Z}$ (or on any other set), as is perhaps suggested by the word "equivalence."

We shall now give a few examples of equivalence relations, but many more will occur in later chapters of this book. Let T be the set of all triangles in a fixed plane, and let a and b be arbitrary elements of T. Then "$\sim$" is an equivalence relation on T if we agree to define "$\sim$" in any one of the following ways:

   (i) $a \sim b$ to mean "a is congruent to b,"

   (ii) $a \sim b$ to mean "a is similar to b,"

   (iii) $a \sim b$ to mean "a has same area as b,"

   (iv) $a \sim b$ to mean "a has the same perimeter as b."

As another example of an equivalence relation, let $\mathbf{Z}$ be the set of all integers, and let us define $a \equiv b$ to mean that $a - b$ has 3 as a factor; that is, that there exists an integer n such that $a - b = 3n$. It is then readily verified that "$\equiv$" has the three defining properties of an equivalence relation. Furthermore, every integer is equivalent to one of the three integers 0, 1, 2. In this connection, consider the following three subsets of $\mathbf{Z}$:

$$J = \{\ldots, -9, -6, -3, 0, 3, 6, 9, \ldots\},$$
$$K = \{\ldots, -8, -5, -2, 1, 4, 7, 10, \ldots\},$$
$$L = \{\ldots, -7, -4, -1, 2, 5, 8, 11, \ldots\}.$$

It will be observed that every integer is in exactly one of these subsets. In other words, the union of these three subsets is $\mathbf{Z}$ and the intersection of any two of them is the empty set. Moreover, J can be characterized as the set of all elements of $\mathbf{Z}$ that are equivalent to 0 (or to any other element of J), and similar characterizations can be given for K and L. The sets J, K, and L are examples of a concept which we proceed to define.

1.13   DEFINITION   Let A be a set and "$\sim$" an equivalence relation defined on A. If $a \in A$, the subset of A which consists of all elements x of A, such that $x \sim a$ is called an *equivalence set* or *class*. This equivalence set will frequently be denoted by $[a]$.

This definition of the equivalence set $[a]$ can be written formally as follows:

1.14 $$[a] = \{x \mid x \in A, x \sim a\}.$$

In the above example, note that $J = [0]$, $K = [1]$, and $L = [2]$; also that $[0] = [3] = [6]$, and so on. Hence there are just the three different equivalence sets.

To return to the general definition, let us consider a few properties of equivalence sets. First, since $a \sim a$ by the reflexive property of an equivalence relation, we always have $a \in [a]$; that is, $[a]$ is the equivalence set which contains a. This shows that every element of A is in at least one equivalence set. Other important properties of equivalence sets are the following, where a and b are elements of the set A:

1.15

(i) $[a] = [b]$ *if and only if* $a \sim b$,

(ii) *If* $[a] \cap [b] \neq \varnothing$, *then* $[a] = [b]$.

Proof: As a first step in proving 1.15(i), let us assume that $[a] = [b]$ and show that $a \sim b$. It has been pointed out that $a \in [a]$, and hence we have $a \in [b]$. By definition of the equivalence set $[b]$, it follows that $a \sim b$, as we wished to show. Conversely, let us now assume that $a \sim b$. If $x \in [a]$, then $x \sim a$ by definition of $[a]$. Now we have $x \sim a$ and $a \sim b$; so the transitive property of an equivalence relation assures us that $x \sim b$. This then implies that $x \in [b]$, and we have therefore proved that $[a] \subseteq [b]$. We leave as an exercise the similar proof that $[b] \subseteq [a]$, from which we conclude that $[a] = [b]$, as desired.

We now prove 1.15(ii). Since $[a] \cap [b] \neq \varnothing$, there exists at least one element s of A such that $s \in [a]$ and also $s \in [b]$. It follows that $s \sim a$ and $s \sim b$. (Why?) By the symmetric property of an equivalence relation, we have $a \sim s$. Since $a \sim s$ and $s \sim b$, the transitive property assures us that $a \sim b$. The fact that $[a] = [b]$ then follows at once from 1.15(i).

A collection of subsets of a set A is often called a *partition* of A if A is the union of these subsets and any two of the subsets have empty intersection. In view of 1.15(ii), together with the fact that every element of A is in some equivalence set, we see that the different equivalence sets relative to an equivalence relation defined on A form a partition of A. Conversely, it may be verified that a given partition of A determines an equivalence relation on A if we *define* elements in the same subset to be equivalent.

1.5 OPERATIONS

There is one other term that we wish to introduce in this preliminary chapter. First, we consider a familiar concept as follows. Let $\mathbf{Z}$ be the set of all integers.

Associated with each ordered pair (i, j) of elements of $\mathbf{Z}$ there is a uniquely determined element $i + j$ of $\mathbf{Z}$. Accordingly, we say that addition, denoted by "$+$," is an operation on $\mathbf{Z}$. More precisely, we may call it a *binary* operation to emphasize that it is defined for each ordered *pair* of elements of $\mathbf{Z}$. The general definition is as follows.

1.16   DEFINITION†   Let A be a given set. A *binary operation* "$\circ$" on A is a correspondence that associates with each ordered pair (a, b) of elements of A a uniquely determined element $a \circ b$ of A.

Later on we shall seldom have occasion to use any unfamiliar symbol to denote an operation. For the most part, we shall find it convenient to call an operation "addition" or "multiplication," and to use the familiar notations, $a + b$ and $a \cdot b$ (or simply ab). However, for the moment we continue to use the symbol "$\circ$" for a binary operation on a set A. We may emphasize that saying that "$\circ$" is a binary operation on A asserts two conditions upon A, namely: $a \circ b$ is uniquely determined (i.e., has only one value), and $a \circ b$ is an element of A. This latter fact, that $a \circ b$ is an element of A, is often expressed by saying that "$\circ$" has *closure*, or that A is closed under (or with respect to) "$\circ$." Some important concepts are introduced in the following definition.

1.17   DEFINITION   Let "$\circ$" be a binary operation defined on the set A. Then

   (i) The operation "$\circ$" is said to be a *commutative* operation, if and only if $a \circ b = b \circ a$ for all $a, b \in A$.

   (ii) The operation "$\circ$" is said to be an *associative* operation if and only if $(a \circ b) \circ c = a \circ (b \circ c)$, for all $a, b, c \in A$.

   (iii) An element e of A is said to be an *identity* for the operation "$\circ$" if and only if $a \circ e = e \circ a = a$ for every $a \in A$.

As examples of these concepts, let us again consider the set $\mathbf{Z}$ of all integers. For the present, we assume as known the familiar properties of addition and multiplication on $\mathbf{Z}$; in particular, that they are both commutative and associative. Moreover, since $a + 0 = 0 + a = a$ for every $a \in \mathbf{Z}$, we see that 0 is an identity for addition; and clearly 1 is an identity for multiplication.

On the same set $\mathbf{Z}$, let us define $a \circ b = a - b$. Since $3 \circ 2 = 1$ and $2 \circ 3 = -1$, we see that this operation is not commutative. Note that just *one* instance in which $a \circ b \neq b \circ a$ implies that the operation is not commutative.

---

† Expressed more formally, this definition merely asserts that a binary operation "$\circ$" on A is a mapping of $A \times A$ into A. The image of an element (a, b) of $A \times A$ under this mapping is then denoted by $a \circ b$.

The reader may verify that neither is this operation associative. Does there exist an identity for this operation? Since $a \circ 0 = a - 0 = a$ for every integer a, it might appear at first glance that 0 is an identity. However, $0 \circ a = -a$ and the definition of an identity is not met.

EXERCISES

1. If "$\sim$" is an equivalence relation on a set A, carefully prove each of the following:

   (i) If $a, b \in A$ such that $a \nsim b$, then $[a] \cap [b] = \varnothing$.

   (ii) If $a, b, c, d \in A$ such that $c \in [a]$, $d \in [b]$ and $[a] \neq [b]$, then $c \nsim d$.

2. If a and b are integers, let us define $a \equiv b$ to mean that $a - b$ has 5 as a factor. Verify that "$\equiv$" is an equivalence relation on the set $\mathbf{Z}$ of all integers, and exhibit all the different equivalence sets.

3. In each of the following, a relation "$\sim$" is defined on the set $\mathbf{Z}$ of integers. Show whether "$\sim$" is or is not an equivalence relation. Where possible, give the equivalence sets. For $a, b \in \mathbf{Z}$, we set $a \sim b$ if and only if:

   (i) Either $a = b$ or $a + b = 17t$ for some $t \in \mathbf{Z}$.

   (ii) $(-1)^a = (-1)^b$.

   (iii) $a < bn$ for some integer n.

   (iv) $a - b = abt$ for some integer t.

   (v) $a^2 + 2a = b^2 + 2b$.

4. Suppose that $\alpha : A \to B$ is a mapping of a set A onto a set B. For $a, b \in A$ let $a \sim b$ if and only if $a\alpha = b\alpha$. Prove that "$\sim$" is an equivalence relation on A, and describe the equivalence sets.

5. If $A = \{x, y\}$ and we define $x \sim x$, $x \nsim y$, $y \sim x$, $y \nsim y$, verify that the relation "$\sim$" has exactly two of the three defining properties of an equivalence relation.

6. Find an example of a relation on some set which is reflexive and symmetric, but not transitive.

7. In each of the following, "∘" is the specified binary operation on the set **Z** of integers. Determine in each case whether the operation is commutative, whether it is associative, and whether there is an identity for the operation.

   (i) $a \circ b = b$,

   (ii) $a \circ b = a + b + ab$,

   (iii) $a \circ b$ is the larger of a and b,

   (iv) $a \circ b = 2a + 2b$,

   (v) $a \circ b = a + b - 1$,

   (vi) $a \circ b = a + ab$.

8. Suppose that "∼" is an equivalence relation upon a set A. Then "∼" is symmetric and transitive. What is wrong with the following "proof" of the reflexive property: (i) $x \sim y$, so by symmetry $y \sim x$; (ii) $x \sim y$ and $y \sim x$, so by transitivity $x \sim x$; (iii) conclude that "∼" is reflexive.

1.6   SOME PROPERTIES OF INTEGERS

Consider the set of whole counting numbers $\{1, 2, 3, \ldots\}$. This set is also called the set of *positive integers*. If we add *zero* and the *negative* integers to this set, we obtain the set of all *integers* which we shall call $\mathbf{Z} = \{\ldots, -2, -1, 0, 1, 2, \ldots\}$ (from the German word "Zahlen" meaning numbers). We are all familiar with certain properties of addition, multiplication, and order for integers, so that initially we shall draw upon our experience with integers. Strictly speaking, we wish to develop algebra rigorously. That is, except for certain axioms, we wish to prove all facts before we use them. In this instance, however, we shall defer a careful study of properties of integers until later chapters. In the meantime, we shall need to know certain properties of integers. This section will concentrate upon setting these properties down and illustrating their uses. Be forewarned that we shall prove very little now; but also, be assured that proofs will be given later.

   To begin, we state a fundamental property (axiom or postulate) concerning sets of positive integers.

1.18   WELL-ORDERING PRINCIPLE   *If S is a nonempty set of positive integers, then S contains a least element, that is, there is a positive integer n in S so that n is less than or equal to every integer in S (i.e., $n \leq x$ for all $x \in S$).*

The Well-Ordering Principle is eminently reasonable. To see this, note that the set S does contain positive integers since it is nonempty. Choose one of these, m, from S. Now count upward from 1 to m: $1, 2, 3, \ldots, m - 1, m$. From the resulting list cross off every number which does not lie in S. Observe that the very first number, n, not crossed off must certainly be the least integer in S, for every integer in S which is greater than m is certainly greater than n, and all other numbers from S appear on our list and are greater than or equal to n.

The argument just given is not a proof of the Well-Ordering Principle. We cannot prove the Well-Ordering Principle since it is one of our axioms or postulates. The problem with the above argument is that we cannot be sure that there are not other positive integers less than one or between successive integers in our list (for example, between 2 and 3). This fact seems intuitively obvious, but "intuitively obvious" is not good enough. In fact, we can use the Well-Ordering Principle to prove the very points we have assumed; for example, there are no positive integers less than one. The argument above was given only to show that the truth of the Well-Ordering Principle is quite understandable.

Example   Let us next illustrate how the Well-Ordering Principle is used. Possibly we are familiar with the formula

1.19
$$1 + 2 + 3 + \cdots + n = \frac{n(n + 1)}{2}$$

where n is a positive integer.

This formula tells us that $1 = 1(1 + 1)/2$ when $n = 1$, $1 + 2 = 2(2 + 1)/2$ when $n = 2$, $1 + 2 + 3 = 3(3 + 1)/2$ when $n = 3$, and so on. We may prove the truth of this equality using the Well-Ordering Principle. We proceed by contradiction, that is, we assume the formula is incorrect for some value of n, say a positive integer n', and seek a contradiction to this assumption. Assemble into a set S all positive integers n such that the formula of 1.19 is incorrect for that n. Since n' is in S by our assumption, S is not empty. By the Well-Ordering Principle, S, being a nonempty set of positive integers, contains a least integer m. Since m is in S, the formula 1.19 is incorrect for $n = m$. Observe that if $n = 1$ then $1 = 1(1 + 1)/2$; so the formula is correct for $n = 1$, in particular, $m \neq 1$. We conclude that $m - 1$ is also a positive integer. Further, $m - 1$ is not in S since m is the least positive integer in S, so that the formula is correct for $n = m - 1$. In other words,

$$1 + 2 + \cdots + (m - 1) = \frac{(m - 1)[(m - 1) + 1]}{2} = \frac{(m - 1)m}{2}.$$

We may now write

$$1 + 2 + \cdots + m = [1 + 2 + \cdots + (m - 1)] + m$$

$$= \frac{(m - 1)m}{2} + m$$

$$= \frac{2m + (m - 1)m}{2}$$

$$= \frac{m(m + 1)}{2}.$$

But this is exactly Equation 1.19 with $n = m$. However, we have assumed that the formula does not hold for m, and we have the desired contradiction. Our original assumption was that Equation 1.19 was incorrect for some value of n. This assumption must be wrong, that is, Equation 1.19 holds for all values of n.

Proofs using the Well-Ordering Principle all follow this basic pattern: A particular property fails, by assumption, to hold for an integer n but holds for all smaller integers; this fact is used to obtain a contradiction. This example may appear to be "disguised mathematical induction," and this is almost correct. However, we have taken the Well-Ordering Principle as an axiom, and later on we shall prove what we may call the Induction Principle, making essential use of the Well-Ordering Principle. We shall discuss all this more completely and clearly in Chapter 4, at which point we shall also discuss some of the finer points which have been ignored here.

Next let us consider division of integers. We have all learned methods for long division; for example, let us divide 164 by 12:

$$
\begin{array}{r}
13 \\
12\overline{)164} \\
12 \\
\hline
44 \\
36 \\
\hline
8
\end{array}
$$

We see that the quotient is 13, and the remainder is 8. Written another way, this says that

$$164 = 13 \cdot 12 + 8$$

and the remainder is equal to or between 0 and 11. We have learned the following facts about long division. The remainder r is always such that $0 \leq r \leq n - 1$ where n is the divisor; and subject to this condition, the quotient and remainder are unique. In other words, there is only one "correct answer" in a long division problem.

With a little ingenuity, we can devise a method for dividing into negative integers and still obtain a nonnegative remainder. For example, how do we divide 12 into -164? From above, we conclude that

$$-164 = -(13 \cdot 12 + 8) = (-13) \cdot 12 + (-8).$$

Adding and subtracting 12 gives

$$-164 = [(-13) \cdot 12 + (-12)] + [(-8) + 12]$$
$$= (-14) \cdot 12 + 4.$$

The quotient is -14 and the remainder is 4. Again the remainder is such that $0 \leq 4 \leq 11$.

We formulate the ideas illustrated here as a theorem.

1.20 THEOREM (*Division Algorithm*) *If a, b are integers with $b > 0$, then there exist unique integers q (called the quotient) and r (called the remainder) such that*

$$a = bq + r \qquad 0 \leq r < b.$$

As illustrated above, we use long division to determine the numbers q and r. When we were young we worked a large number of examples of "long division." We convinced ourselves that "long division works." But we certainly have not worked all possible examples; so, we do not know for sure that the Division Algorithm always works. In Chapter 5 we shall show that the Division Algorithm holds as soon as certain rather more fundamental properties hold. The interested reader may turn to Theorem 5.5 of Section 5.1 and read the proof following it with profit.

The Division Algorithm is very useful. Again consider 12 and 164. The greatest positive integer which divides both 12 and 164 with zero remainder is 4. We can find this number experimentally. The number 4 is called the *greatest common divisor* (g.c.d.) of 12 and 164. There is a purely mechanical method for arriving at the greatest common divisor called the *Euclidean Algorithm*, and it involves successive applications of the Division Algorithm. We start with the results of the first division:

$$164 = 13 \cdot 12 + 8.$$

Next divide the original divisor 12 by the remainder 8, thus obtaining

$$12 = 1 \cdot 8 + 4.$$

Finally, divide the preceding divisor, 8, by the preceding remainder, 4,

$$8 = 2 \cdot 4 + 0.$$

We stop here because the remainder is now 0. We obtain this string of equations using the Division Algorithm, and we stop when the last remainder is 0. The next to the last remainder is 4, the greatest common divisor of 12 and 164.

By reversing the steps, we may find integers x and y so that

$$12 \cdot x + 164 \cdot y = 4.$$

Starting with the next to the last equation, we have

$$4 = 12 - 1 \cdot 8.$$

From the first equation,

$$8 = 164 - 13 \cdot 12.$$

Plugging in for 8 gives

$$
\begin{aligned}
4 &= 12 - 1 \cdot (164 - 13 \cdot 12) \\
&= (12 + 13 \cdot 12) - 1 \cdot 164 \\
&= 14 \cdot 12 + 164 \cdot (-1).
\end{aligned}
$$

Here we have illustrated a rather important result called Euclid's Lemma.

1.21 THEOREM *If a and b are nonzero integers and d is the greatest common divisor of a and b, then there are integers x and y so that*

$$ax + by = d.$$

Further, an integer e divides d with zero remainder if and only if e divides both a and b with zero remainder.

The proofs which will justify these examples and theorems are given in Chapter 5.† We turn our attention now to factorizations of integers. If the greatest common divisor of two integers a and b is one, then we say that a and b are *relatively prime*. If $p > 1$ is a positive integer and none of the integers $2, 3, \ldots, p - 1$ divides p with 0 remainder (i.e., all pairs $2, p; 3, p; \ldots; p - 1, p$ are relatively prime) then we call p a *prime*. We also call $-p$ a prime. For example, $-2, -3, 5, 7$, and -11 are all primes.

---

† A discussion of the Euclidean Algorithm follows the proof of Theorem 5.18 and may be read now. In that section, read $a\delta$ to mean $|a|$, and $a \,|\, b$ to mean $b = ac$, that is, $a \,|\, b$ means a is a factor of b. Theorem 1.21 is a special case of Theorem 5.18 of Section 5.3, whose proof involves quite a bit of unfamiliar terminology.

We have learned that we can often factor positive integers, for example,

$$720 = 10 \cdot 72 = 15 \cdot 48.$$

We have also been told that if we continue factoring, the process stops:

$$720 = 2 \cdot 2 \cdot 2 \cdot 2 \cdot 3 \cdot 3 \cdot 5.$$

We probably observed that the process stops precisely when all factors of 720 are primes. Further, no matter how we start factoring (with $10 \cdot 72$ or $15 \cdot 48$), the final product of primes is unique except for the ordering of the positive prime factors, e.g.,

$$720 = 3 \cdot 2 \cdot 2 \cdot 3 \cdot 2 \cdot 2 \cdot 5.$$

This factorization property of integers is so important that it is called the Fundamental Theorem of Arithmetic.

1.22   FUNDAMENTAL THEOREM OF ARITHMETIC *Every positive integer $a > 1$ is a prime or can be expressed as a product of positive primes in one and only one way (except for the order of the factors).*

Proof: This theorem is so very important that we sketch a proof now, different from that given in the text (Theorem 5.26 of Section 5.4). We proceed by way of contradiction and assume that the theorem is false for some positive integer $n > 1$. The Well-Ordering Principle assures us that there is a least positive integer m for which the theorem is false, in particular, m is not a prime. Now $m = ab$ is the product of two lesser integers a and b to which the theorem applies, that is, a and b have factorizations into primes, so that their product m also has a prime factorization. Incidentally, if we assume that a is a prime, then the preceding argument shows that any prime factor of m appears in some prime factorization of m into a product of primes. In other words, if a number has only one prime factorization and we know a prime factor of that number, then the prime must appear in the factorization. (In applying the theorem we could have concluded that each of a and b has only one prime factorization, up to ordering of prime factors. Why can we *not* conclude from this that m has only one prime factorization?)

Since the theorem is false, it must be that m has at least two prime factorizations:

$$m = p_1 p_2 p_3 \cdots p_r$$

and

$$m = q_1 q_2 q_3 \cdots q_s$$

where the p's and q's are positive primes. There is no harm in arranging the primes in order so that $p_1 \leq p_2 \leq p_3 \leq \cdots \leq p_r$ and $q_1 \leq q_2 \leq q_3 \leq \cdots \leq q_s$. By interchanging the p's and q's, if necessary, we may also assume that $p_1 \leq q_1$. If $p_1 = q_1$, then we may cancel p_1 ($=q_1$) from both factorizations obtaining

$$a = q_2 q_3 \cdots q_s = p_2 p_3 \cdots p_r$$

as two different factorizations of a lesser integer than m into primes. Since the theorem applies to a, these two factorizations of a must agree, and this certainly cannot be the case, so that $p_1 < q_1$.

We now have $p_1 q_1 < q_1 q_1 \leq q_1 q_2 \leq m$, so that $m - p_1 q_1$ is a positive integer. But

$$m - p_1 q_1 = p_1(p_2 p_3 \cdots p_r - q_1),$$

so that p_1 is a prime factor of $m - p_1 q_1$. A similar argument shows that q_1 is also a prime factor of $m - p_1 q_1$. Since $m - p_1 q_1$ is less than m, the theorem applies to $m - p_1 q_1$, that is, $m - p_1 q_1$ has *only one* factorization into a product of positive primes. Since p_1 and q_1 are different prime factors of $m - p_1 q_1$, they must both appear in this prime factorization. Therefore,

$$m - p_1 q_1 = p_1 q_1 t,$$

where t is the product of all primes other than p_1 and q_1 in the factorization of $m - p_1 q_1$. Rewriting this equation gives

$$m = p_1 q_1 (1 + t).$$

From a preceding equation,

$$m = q_1 q_2 q_3 \cdots q_s.$$

Cancelling q_1 from these equations gives

$$a = q_2 q_3 \cdots q_s = p_1(1 + t).$$

We have already concluded that a has only one prime factorization (now we know this factorization involves q's and p_1 as factors). In particular, p_1 must equal one of the q's. Unfortunately, $p_1 < q_1 < q_i$ for $i = 2, 3, \ldots, s$. This contradiction shows that there are no exceptions to the theorem, completing the sketch.

In this section it has been our main aim to state and illustrate certain facts about integers. We have generally avoided proofs (the proofs are all given in later chapters). From time to time we shall need these facts, but for now, a certain competency is needed in manipulating integers. The exercises below will help to provide that competency. If you are worried about why these computational techniques work and why these theorems hold, you might want to thumb through Chapters 4 and 5. Be assured, however, that when we reach that point we will prove all the details which have only been illustrated here.

EXERCISES

1. We list the following pairs (a, b): $(7,11)$, $(24,15)$, $(38,62)$, $(129, -264)$, $(57, -93)$. Using the Euclidean Algorithm, find the greatest common divisor of a and b.

2. For each pair of integers (a, b) in Exercise 1, let d be the greatest common divisor of a and b. Find integers x and y so that $ax + by = d$.

3. For each pair of integers (a, b) in Exercise 1, let d be the greatest common divisor of a and b. Reduce the fractions ab/d. The *least common multiple* of a and b is the least positive integer m which is divisible by both a and b with zero remainder. Find the least common multiple of a and b for the pairs (a, b) of Exercise 1. Can you guess a theorem we might want to prove later?

4. Factor each of the following integers into a product of primes: 97, 168, 187, 360, 899, 16,637, 70,560.

5. Using the Well-Ordering Principle, prove that

$$1 + 3 + 5 + \cdots + (2n - 1) = n^2$$

is correct for all positive integers n.

6. Consider the following proposition: Every positive integer is interesting. A proof might go as follows. Suppose there are uninteresting positive integers. Let S be the set of all these. Now $S \neq \varnothing$. By the Well-Ordering Principle there is a least positive integer n which is in S. That is, n is the smallest uninteresting number. So n has a very unique property. Isn't that interesting? Comment.

7. Assume that you have a row of balls lined up on a table. At the left end of the row is a green ball, and immediately to the right of every green ball in the row is another green ball. Use the Well-Ordering Principle to prove that all the balls in the row are green.

8. Use the Well-Ordering Principle to prove that the following formula is correct for all positive integers n:

$$1^2 + 2^2 + \cdots + n^2 = \frac{n(n + 1)(2n + 1)}{6}.$$

 COMMENTARY

1 The dictionary tells us that *abstraction* is the process of stripping an idea of its concrete accompaniments, and that a *concrete* thing exists in real or material form. The difficulty in interpreting these definitions lies in understanding the meaning of the word "real." Many areas of mathematics that are real or concrete for us were difficult abstractions in the past; for example, topics that confused the best minds of the fourteenth century are routinely taught in high schools today. Further, ideas that are abstract when we first encounter them may become concrete as our familiarity with them increases. In any case, the material in this text is considered abstract by the usual reader. Assimilating abstractions is often extremely difficult; therefore, a few comments and hints may be of some value.

The story is often told that when Ptolemy asked Euclid (c. 300 B.C.) for an easy way to learn geometry, the latter replied that there is no royal road to geometry. More recently, Poincaré (1854–1912), while allowing that some of his greatest inspirations occurred in strange places (such as while riding trams), underscored the point that these moments only came after periods of intense hard work. Eminent mathematicians have said again and again over the ages that mastery of mathematics is difficult. It appears that brilliance in mathematics is a matter of the rate rather than the ease with which new ideas are learned. Those who wish to learn mathematics must be willing to work very hard at the task. However, unlike many subjects, in mathematics there are objective measures we can apply to determine how much effort we have invested: how many and how difficult are the problems we have solved? Remember that a problem has not been solved until a complete and careful solution has been written down. Even the best mathematicians have "found" trivial solutions to problems, only to have them evaporate when put to the test of careful exposition. If the assigned exercises are either too difficult or too easy, then the texts described in section 3 of this commentary should provide more appropriate ones. The important point is to solve a sufficient quantity of problems of sufficient quality to guarantee that the material has been learned.

The evolution of newly created abstractions into old concrete ideas takes place very slowly. It is little wonder then that a student may encounter difficulties when asked to assimilate in one year something that took centuries to evolve. Many times, the original concrete thing that preceded the abstraction has been lost or is not quite appropriate for presentation now. Consequently, one encounters rigorously defined concepts stripped of all conception; to all appearances they are suffering from *rigor mortis*. In fact, students often point to a vicious circle, namely: "I cannot understand, hence I cannot learn. I cannot learn, therefore I cannot understand." In such circumstances, memory can be a most powerful weapon. Memorize verbatim (not approximately, but exactly) the offending definition or theorem, and then find a few exercises that can be solved with memorized facts alone. It is surprising how quickly memorized sentences change into knowledge. Within limits, this learning technique can be useful, especially when applied to a few definitions related to new and difficult abstract concepts.

Sometimes even memory seems to fail us, for all exercises appear too difficult for our newly memorized facts. Often it is neither that our knowledge fails us nor that the exercises are too difficult, but rather that we do not understand what is needed for a solution. The chosen exercise is a nonproblem in the sense that it is not clear what is being asked for or what an answer might be. At times like these it is a good idea to seek outside help: a fellow student, your teacher or tutor (who is being paid partially to help you at such times), or possibly a book that either solves (e.g., textbooks listed in section 3 of this commentary) or describes how to solve [130] problems.

Mathematicians study mathematics because, in spite of the hard work, it is exciting. It is hoped that readers of this book will feel some of that excitement—and that is much easier if the invested effort is not overwhelming.

2   Usually a section of the commentary is reserved for collateral reading and reference. Since we have not begun a study of "real" algebra as yet, most references lead us away from algebra. For example, set theory has been a nonalgebraic subject studied since its first real creation by Cantor (1845–1918). We have probably all been told, at one time or another, that π and $\sqrt{2}$ are not rational numbers. The number π has the additional property that it is a transcendental number, that is, π is not the root of any algebraic equation with rational coefficients. Euclid (c. 300 B.C.) gives a proof of the irrationality of $\sqrt{2}$ in his *Elements*; but it was not until 1882 that Lindemann (1852–1939) showed that π is transcendental [133]. Actually, it is very difficult to show that a particular number is transcendental; the first such proof was given by Liouville (1809–1882) in 1844 when he showed that a specially concocted number is transcendental. On the other hand, in 1874 Cantor showed that the set of all transcendental numbers is very much "larger" than the set of all roots of algebraic equations with rational coefficients. In particular, Cantor showed, using his set theory, not only that transcendental numbers exist, but also that there are very many such numbers . . . without producing a single specific example of one [89]!

For his ideas, Cantor was both vilified and praised; one of his most ardent enemies was the very famous Kronecker (1823–1891) who personally blocked the appointment of Cantor to a prestigious position at Berlin. However, Hilbert (1862–1943) said, "No one shall drive us out of the paradise which Cantor has created for us." An interesting account of Cantor's life is contained in a very good book by Bell [135]. There are very many sources which discuss the theory of sets; among them are Halmos [91], a textbook; Kamke [92], a readable account of Cantor's set theory; and Breuer [88], a very elementary book on set theory.

3   Most people intrigued by mathematics have a natural interest in its history, its philosophy, and its pastimes. There are some very good histories of mathematics.

Two of the best are Bell [134] and Kline [144]; but these often assume a good deal of mathematical sophistication on the part of the reader. Among the very good elementary histories are Boyer [136] and Eves [141]. Mathematics has always been a favorite stopping place for philosophers. Illustrious names ranging from Zeno (c. 450 B.C.) to Russell (1872–1970) appear on the list of mathematical philosophers. A very small sample of books on the philosophy of mathematics might include Frege [90], Hadamard [168]; Körner [173], Quine [94], Russell [179], and Wittgenstein [98]. Stories about mathematics and mathematicians, and games and pastimes, possibly make up the most universally intriguing aspect of mathematics to the uninitiated. Possibly the best collections of anecdotes, stories, and one-liners on mathematics are Moritz [174] and Eves [156–158]. Some of the best fiction on the subject is compiled in Fadiman [159,160]. Of the various collected pieces [172,175], possibly *The World of Mathematics* is the best; edited by J. R. Newman, these four volumes are a gold mine of pleasurable reading. For the puzzler, the very fine *Scientific American* collections edited by Martin Gardner are available [162,163,165–167]. In addition, there are books compiled by some of the most famous puzzlers [164,176,177]. There are also excellent books which attempt to explain and/or describe many of the most popular mathematical games and pastimes [154,180]. The various books listed here contain bibliographies, and the interested reader can pursue the thread into the literature upon mathematics via these. Each of the successive commentaries will also contain additional references on various topics. We have by no means been exhaustive even in the lists of this commentary. More references are included in the bibliography under the headings *History* and *Recreational Mathematics*.

It is sometimes helpful to refer to other algebra texts for coverage of some topics. In the *General Algebra* section of the bibliography we have listed a few of the better algebra texts. From this list we have selected a few which might be found in most libraries and given annotations below.

Birkhoff, Garrett, and Saunders MacLane [3]
> Numbers. Groups. Rings. Fields. Linear algebra. Galois theory. Clearly written. Many examples. Good source for problems.

Bourbaki, N. [4]
> Abstract. Modern outlook. Graduate level. Comprehensive coverage.

Burton, David M. [5]
> Numbers. Groups. Rings. Fields. Linear algebra. Clearly written. Many problems.

Dean, Richard A. [6]
> Numbers. Groups. Rings. Fields. Galois theory. Clear. Concise. Slightly higher level.

Fraleigh, John B. [9]
> Groups. Rings. Fields. Galois theory. Clearly written. Some proofs omitted. Many easier problems. Strong on group theory.

Goldstein, Larry J. [11]
> Numbers. Groups. Rings. Fields. Galois theory. Historical approach. Some nice number theory. Well written. Slightly higher level.

Herstein, I. N. [12]

> Numbers. Groups. Rings. Fields. Galois theory. Linear algebra. Transcendence of e. Excellent on ring theory and Sylow theorems. Difficult text. Many challenging problems.

Jacobson, Nathan [14]

> Graduate level. Volume I: General algebra. Volume II: Very extensive high-level coverage of linear algebra. Volume III: Difficult and broad coverage of field and Galois theories.

Lang, Serge [17]

> Graduate level. Broad coverage. Modern viewpoint.

MacLane, Saunders and Garrett Birkhoff [20]

> Advanced level. Numbers. Groups. Rings. Fields. Linear algebra. Category theoretic approach.

Mostow, George D., Joseph H. Sampson, and Jean-Pierre Meyer [21]

> Numbers. Groups. Rings. Fields. Linear algebra. Clearly written. Very extensive coverage. A good source on linear algebra.

van der Waerden, B. L. [24]

> Graduate level. A classic text. Still one of the best. Excellent coverage. Volume I: General algebra through Galois theory. Volume II: More advanced material on rings, fields, and algebras.

chapter two

GROUPS:
AN INTRODUCTION

In this chapter we begin our study of algebra by starting with algebraic systems called groups. In a certain sense groups are the building blocks of algebra, since most other systems contain groups. Thus it is logical to start here. All properties that are used to define a group are suggested by simple properties of the integers; we begin by pointing out some of these properties. These same properties are shared by certain motions of familiar geometric figures. We shall look at these too. The first section is therefore of a preliminary nature and is merely intended to furnish a partial motivation of the material to follow.

This chapter is just an introduction to the theory of groups. Additional topics on group theory will be presented in later chapters. After defining the word "group" and covering certain preliminaries, the main purpose of this chapter is to introduce the very important concepts of homomorphism and isomorphism of groups. These notions will be defined and discussed in Sections 2.5, 2.9, and 2.10. As a secondary purpose we shall discuss certain very important kinds of groups, namely, those called "cyclic groups" and those called "permutation groups."

2.1 INTEGERS AND GEOMETRIC FIGURES

The simplest numbers are the numbers $1, 2, 3, \ldots$ used in counting. These are called the *natural numbers*, or *positive integers*. Addition has a simple interpretation if we consider a natural number as indicating the number of elements in a set. For example, suppose we have two piles of stones, the first one containing m stones, and the second one n stones. If the stones of the first pile are placed on the second pile, there results a pile of $n + m$ stones. If, instead, the stones of the second pile are placed on the first pile, we get a pile of $m + n$ stones. It thus seems quite obvious that

$$m + n = n + m$$

for every choice of m and n as natural numbers, i.e., addition of natural numbers is commutative. This property of the natural numbers is an example of what is sometimes called a *law* or *formal property*. Another example is the associative law of addition.

   Historically, the natural numbers were no doubt used centuries before there was any consideration of their formal properties. However, in modern algebra it is precisely such formal properties that are of central interest. Some of the reasons for this changed viewpoint will become evident as we pursue our study of modern algebra.

   Of course, if m and n are natural numbers, there need not be a natural number x such that $m + x = n$; for example, $5 + x = 3$ has no natural number solution x. In order to be able to solve all equations of this kind, we need to have available the negative integers and zero in addition to the positive integers. The properties with which we shall be concerned in this section are suggested by well-known properties of addition in the set of all integers (positive, negative, and zero). In a sense, these properties are just the ones needed in order to be able to solve equations such as $5 + x = 3$. We shall use the word "group" in the following examples. This word will be defined formally in the next section. In each example we introduce a set which has a single binary operation having certain properties, namely, it is associative, and has identities and inverses. Informally, a set with such an operation is called a *group*.

   Example 1 (*Additive group of integers*) We let $\mathbf{Z}^+$ denote the set of integers with the operation of addition. The superscript $+$ is to remind us that we are ignoring the operation of multiplication. Addition has the following properties for arbitrary integers a, b, and c:

   (1) $(a + b) + c = a + (b + c)$ (*associative law*).
   (2) $a + 0 = 0 + a = a$ (*existence of an identity*).
   (3) $a + (-a) = (-a) + a = 0$ (*existence of inverses*).
   (4) $a + b = b + a$ (*commutative law*).

These are properties which are familiar to us and clearly hold for integers. Let us construct another different system of numbers starting from $\mathbf{Z}^+$.

Example 2 (*Additive group of least residues modulo n*) We shall proceed to explain what we have in mind. Fix a positive integer n, for example, $n = 6$. Consider the integers $0, 1, \ldots, n - 1$, which we call *least residues* modulo n. In our example of $n = 6$ the numbers would be $0, 1, 2, \ldots, 5$. We call this set $\mathscr{Z}_n{}^+$ (or $\mathscr{Z}_6{}^+$). We wish to define an operation $a \circ b$ upon the set $\mathscr{Z}_n{}^+$. The Division Algorithm tells us that if a is any integer, then there are unique integers q and r so that $a = qn + r$ and $0 \leq r < n$. For example, if $a = 47$, then $47 = 7 \cdot 6 + 5$, where $0 \leq 5 < 6$. For two elements $a, b \in \mathscr{Z}_n{}^+$ there exist unique integers q and r so that $a + b = qn + r$, where $0 \leq r < n$. We define $a \oplus b = r$, and thus have an operation upon $\mathscr{Z}_n{}^+$. In our example for $n = 6$, if $a = 5$ and $b = 3$, then $5 + 3 = 1 \cdot 6 + 2$ so that $5 \oplus 3 = 2$.

The associative law is valid for this operation $\oplus$, and the demonstration goes as follows. Suppose a, b, and c are elements of $\mathscr{Z}_n{}^+$. We wish to argue that

$$(a \oplus b) \oplus c = a \oplus (b \oplus c).$$

Using the Division Algorithm several times, we obtain the equations

$$
\begin{aligned}
a + b &= q_1 n + r_1, & 0 \leq r_1 < n. \\
r_1 + c &= q_2 n + r_2, & 0 \leq r_2 < n. \\
b + c &= q_3 n + r_3, & 0 \leq r_3 < n. \\
a + r_3 &= q_4 n + r_4, & 0 \leq r_4 < n.
\end{aligned}
$$

From these equations we obtain the equations

$$
\begin{aligned}
(a \oplus b) \oplus c &= r_1 \oplus c = r_2, \\
a \oplus (b \oplus c) &= a \oplus r_3 = r_4.
\end{aligned}
$$

We must prove that $r_2 = r_4$, which we do by combining our equations above.

$$
\begin{aligned}
(a + b) + c &= (q_1 n + r_1) + c = q_1 n + (r_1 + c) \\
&= q_1 n + (q_2 n + r_2) = (q_1 + q_2)n + r_2, \\
a + (b + c) &= a + (q_3 n + r_3) = q_3 n + (a + r_3) \\
&= q_3 n + (q_4 n + r_4) = (q_3 + q_4)n + r_4.
\end{aligned}
$$

Let $q' = q_1 + q_2$ and $q'' = q_3 + q_4$.

$$
\begin{aligned}
(a + b) + c &= q'n + r_2, \\
a + (b + c) &= q''n + r_4.
\end{aligned}
$$

Since addition of integers is associative, the equality

$$(a + b) + c = a + (b + c)$$

must hold. Now both r_2 and r_4 lie between 0 and $n - 1$, and the Division Algorithm tells us there is only one value q and one value r so that

$$a + (b + c) = qn + r.$$

Since q' or q'' will work for q and since r_2 or r_4 will work for r, we must have

$$q' = q = q''$$

and

$$r_2 = r = r_4.$$

We have proven that

$$(a \oplus b) \oplus c = a \oplus (b \oplus c).$$

The following properties hold for $\mathscr{Z}_n{}^+$, the set of least residues modulo n with the new "addition": $\oplus$. Suppose that a, b, and c are arbitrary elements of $\mathscr{Z}_n{}^+$.

(1) $(a \oplus b) \oplus c = a \oplus (b \oplus c)$ (*associative law*).
(2) $a \oplus 0 = 0 \oplus a = a$ (*existence of an identity*).
(3) For some $d \in \mathscr{Z}_n{}^+, a \oplus d = d \oplus a = 0$ (*existence of inverses*).
(4) $a \oplus b = b \oplus a$ (*commutative law*).

Observe that if $0 < a < n$, then $0 < n - a < n$ also. Therefore, in (3) we may let $d = 0$ if $a = 0$ and $d = n - a$ if $a \neq 0$. The verification of these other properties is left to the exercises.

The four listed properties for $\mathbf{Z}^+$ are the same as the four listed properties of $\mathscr{Z}_n{}^+$. As we shall see, property (4) is somewhat special. The next two examples are of a very different kind, since they involve the symmetrical shape of a geometric object. Neither example has property (4).

Example 3 (*Group of symmetries of x,y,z-axes in space*) Imagine an x,y,z-axis system with origin O. We imagine each axis as continuing indefinitely in opposite directions, so that Figure 1 only shows a segment of each axis. View the axes as having no markings upon them, that is, each axis continues indefinitely in two directions and is unoriented. There is no "positive" or "negative" axis; and there are no coordinates on the axes. Think of the axis system as

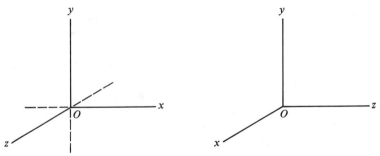

FIGURE 1   FIGURE 2

occupying a hole in space, noting the positions of the axes. We shall form a set of elements called *symmetries* or *motions*. A symmetry may be performed by lifting the axes out of their hole in space, moving the rigid configuration of axes around, and replacing them in their hole in space. Imagine that the axes started in the orientation of Figure 1 and finished in the orientation of Figure 2. We may call this symmetry σ. The *change of position* is the element σ. The starting position of Figure 1 is, alone, unimportant; the final position of Figure 2 alone, is unimportant; and how we perform this change is not important. Only the change itself is important. For example, σ may be performed by looking down the vertical axis and rotating around that axis by 90° clockwise. The axes continue indefinitely in two directions and are not oriented. So, this rotation moves the z-axis into the x-axis position, and the x-axis into the z-axis position. Rotating the opposite or counterclockwise direction by 90° results in exactly the same change from Figure 1 to Figure 2. Therefore, σ may be performed in at least two different ways. It should also be noted that σ is independent of the starting configuration, for rotating 90° about the vertical axis will change Figure 3 into Figure 4. The element or symmetry does not depend

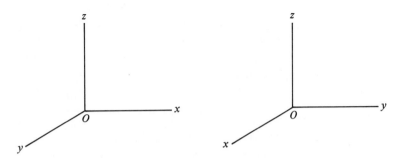

FIGURE 3   FIGURE 4

upon the initial or final configurations alone; it is completely described by the change from a *given initial position* to a *given final position*. This point is worth repeating. If we specify a starting position, then the symmetry σ is uniquely specified by the final position resulting from applying σ to a given starting position. If a symmetry τ causes Figure 1 to change into Figure 3, then since σ and τ are described by the same initial and different final configurations, σ and τ are different symmetries of the axes.

We shall denote the set of all symmetries of the x,y,z-axes by A. Now we may define a product operation $\sigma\tau$ upon A as follows: First perform σ and then perform τ; $\sigma\tau$ is described by the total change effected by these two successive changes. Starting with Figure 1, σ will change the axes to Figure 2; then τ will change this configuration (Figure 2) to Figure 5. In other words, $\sigma\tau$ will cause the configuration of Figure 1 to change to Figure 5.

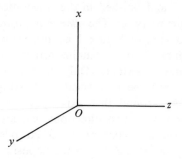

FIGURE 5

We have defined an operation of multiplication upon the set A of all symmetries of our x,y,z-axis system. We shall now argue that the following properties hold for any three symmetries σ, τ, $\rho \in A$.

(1) $(\sigma\tau)\rho = \sigma(\tau\rho)$ *(associative law)*.

(2) There is a symmetry $\iota \in A$ (which may be described as "no change") such that
    $\iota\sigma = \sigma\iota = \sigma$.   *(existence of an identity)*.

(3) For each σ there is an element η so that
    $\sigma\eta = \eta\sigma = \iota$.   *(existence of inverses)*.

Let us denote positions of our axes by capital letters. For example, we could label the configuration of Figure 1 as configuration X.

We prove that property (2) holds for A. Suppose σ changes configuration X into some other configuration Y. The symmetry ι

which causes no change will change X into X and Y into Y. Therefore, $\iota\sigma$ makes the changes $X \xrightarrow{\iota} X \xrightarrow{\sigma} Y$. The symmetry $\sigma\iota$ makes the changes $X \xrightarrow{\sigma} Y \xrightarrow{\iota} Y$. Both of these products cause the change $X \xrightarrow{\sigma} Y$. Therefore, $\iota\sigma = \sigma\iota = \sigma$.

Next we argue that property (3) holds. Suppose σ changes position X to position Y. We may define a symmetry η which changes position Y into position X. Then $\sigma\eta$ causes the change $X \xrightarrow{\sigma} Y \xrightarrow{\eta} X$, or $X \xrightarrow{\sigma\eta} X$. In other words, $\sigma\eta = \iota$. It seems clear that whatever it is that σ does, η undoes. Or stating this another way, whatever η does, σ undoes. That is, $\eta\sigma$ causes the changes $Y \xrightarrow{\eta} X \xrightarrow{\sigma} Y$, or $Y \xrightarrow{\eta\sigma} Y$. We conclude that $\eta\sigma = \iota$. This proves that property (3) holds.

Finally we prove that property (1) holds. Suppose σ, τ, ρ cause the following changes: $W \xrightarrow{\sigma} X$, $X \xrightarrow{\tau} Y$, $Y \xrightarrow{\rho} Z$. Then $\sigma\tau$ will cause $W \xrightarrow{\sigma\tau} Y$, so that $(\sigma\tau)\rho$ causes $W \xrightarrow{(\sigma\tau)\rho} Z$. Similarly $\tau\rho$ causes $X \xrightarrow{\tau\rho} Z$, so that $\sigma(\tau\rho)$ causes $W \xrightarrow{\sigma(\tau\rho)} Z$. Since $\sigma(\tau\rho)$ and $(\sigma\tau)\rho$ cause the same changes from position W, they are equal. That is, $(\sigma\tau)\rho = \sigma(\tau\rho)$, or property (1) holds.

Consider the symmetry σ which changes Figure 1 into Figure 2; and the symmetry τ which changes Figure 1 into Figure 3. We saw that $\sigma\tau$ changes Figure 1 into Figure 5. The element σ can be described by a rotation of 90° around the vertical axis, so that it will change Figure 3 into Figure 4. The end result of applying $\tau\sigma$ to Figure 1 is Figure 4. The end result of applying $\sigma\tau$ to Figure 1 is Figure 5. We conclude that $\sigma\tau \neq \tau\sigma$. The system A of symmetries does not have a property like property (4) of Examples 1 and 2. However, it does have properties (1) to (3) which correspond respectively to (1) to (3) in Examples 1 and 2.

Before describing another example, let us make a few observations about this example.

The arguments in the proofs we gave did not depend upon the fact that our geometric system was the axis system. This means that our argument should work for the set of all symmetries of any geometric object. In other words, every geometric object has a set of symmetries and these symmetries obey properties (1) to (3) for the product operation. For purposes of illustration, let us look at yet another example.

Example 4 (*Group of symmetries of a cube*)　This example will play a large role in illustrating the concepts of group theory. A cube marked as in Figure 6 will be of great assistance in visualizing this example. On page 647 there is a marked fold-out cube. The cube you need can be obtained by cutting and folding the cube in the book

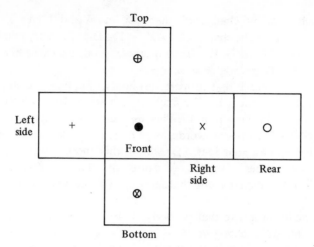

FIGURE 6

or marking a block of wood appropriately. It is important to label the cube exactly as the one in the book is marked since all our illustrations will use this particular marking. Position the cube flat on the table in front of you, and turn a face toward yourself. Observe that the configuration of faces is uniquely determined by (1) the face toward you and (2) the right-hand face. Therefore, the position of the cube can be shown by giving only the front face and the right-hand face. Figure 7 gives a possible position for the cube. In the future, all diagrams will picture cube positions in this way.

Imagine that the cube occupies a hole in space. A *motion* or *symmetry*† of the cube, as for the axes, is a *change* of position.

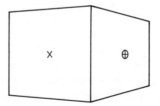

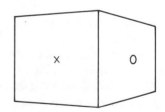

FIGURE 7   FIGURE 8

† Many authors call the "symmetries" defined here by the name "rigid symmetries"; they include other new "symmetries" which we disallow. These additional "symmetries" are obtained as reflections. After reading this example, put a mirror in front of your cube facing toward it. Copy down the positions of all faces as you see them in the mirror; and call this configuration of the faces Y. If the cube sits in position X, then the change from X to Y is called a reflection. No amount of rolling or bouncing of the cube will change the cube from X to Y. It literally must be cut apart and reassembled to change from configuration X to configuration Y. For obvious reasons, we ignore these "symmetries" called reflections.

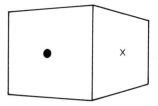

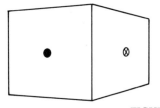

FIGURE 9   FIGURE 10

For example, note the starting position of the cube; lift the cube out of its hole in space; roll it around; and then replace it in its hole in space. Note the new position of the faces. Our cube could have started as in Figure 7 and finished as in Figure 8. The motion or symmetry σ which causes this change of position would also cause the change from Figure 9 to Figure 10. Incidentally, we may describe σ as: "rotate the cube counterclockwise 90°," or equally, "rotate the cube 270° clockwise."

As for the symmetries of the x,y,z-axes, the change from the initial to the final position describes σ. How this change is effected is unimportant. For two symmetries σ and τ, suppose τ changes Figure 10 to Figure 11. So, if we first perform σ and then τ, we obtain $\sigma\tau$ which changes Figure 9 to Figure 11. We shall denote the set of all symmetries of the cube by C. For the set of symmetries C of the cube and this operation of multiplication the following properties hold. Suppose σ, τ, $\rho \in C$ are arbitrary.

(1) $(\sigma\tau)\rho = \sigma(\tau\rho)$ (*associative law*).

(2) There is a motion $\iota \in C$ so that
  $\iota\sigma = \sigma\iota = \sigma$ (*existence of an identity*).

(3) For each σ there is an η so that
  $\sigma\eta = \eta\sigma = \iota$ (*existence of inverses*).

It will be left as an exercise to prove that these properties hold. The following observations are helpful: (a) Any symmetry is uniquely determined by an arbitrary starting position X and the final position

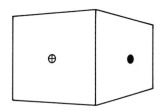

FIGURE 11

Y into which *X* is carried by the symmetry. (b) The proofs in Example 3 may be read verbatim with the cube in mind.

Can you find symmetries σ and τ of the cube so that $\sigma\tau \neq \tau\sigma$? If you can, then a property analogous to property (4) in Examples 1 and 2 does not hold for the symmetries *C* of a cube.

Let us now turn our imaginations loose. A geometric object is made up of vertices (points), edges (line segments), and faces (plane regions). Other geometric objects ought also to have symmetries. These symmetries also ought to satisfy properties (1) to (3) of Example 4 when we view "multiplication" as one symmetry followed by another. Wad up a piece of paper. That is a "geometric figure." What is its set of symmetries? What is the set of symmetries of a crystal ball?

These examples are all instances of an algebraic system called a *group*. In the next section we shall give the definition for a group. We shall also derive some simple computational properties of groups.

EXERCISES

1. Verify that properties (2) to (4) hold in Example 2.

2. Show that properties (1) to (3) hold in Example 4.

3. Cut an equilateral triangle from a piece of paper. Describe the symmetries of the triangle. For the operation composition of symmetries show that properties (1) to (3), but not (4), of Example 1 hold for this set of symmetries.

4. Consider a regular polygon of *n* (> 2) sides. Describe the symmetries of the regular *n*-gon. Which properties hold for this set of symmetries?

5. A regular solid called a tetrahedron can be assembled from four identical equilateral triangles. Discuss the set of symmetries of the regular tetrahedron.

6. Each symmetry in *A* (of Example 3) may be described as a rotation about some axis through *O*. There are six symmetries in *A*. Find them and describe each as a rotation about some axis.

7. Each symmetry in *C* (of Example 4) may be described as a rotation about some axis through the center of the cube. There are 24 symmetries in *C*. Find them and describe them as rotations about some axis.

8. Recall the equilateral triangle of Exercise 3. The vertices (points) of the triangle are held at fixed distances and relationships by the edges and face of the triangle. Imagine we have the three points fixed in space. We remove the face and edges of the triangle leaving only the three points; that is, the points occupy three "holes" in space, and a symmetry of these points will redistribute them in the three holes. Describe the set of symmetries of these detached three points. Is this set of symmetries "the same" as the set of symmetries of the triangle? That is, does each symmetry of the points correspond uniquely to a symmetry of the triangle?

9. Repeat Exercise 8 using a square in place of the triangle.

10. Notice that $\mathscr{L}_{12}{}^{+}$ of Example 2 consists of the set

$$\{1, 1 \oplus 1, (1 \oplus 1) \oplus 1, ((1 \oplus 1) \oplus 1) \oplus 1, \ldots\}.$$

Show that $\mathscr{L}_{12}{}^{+}$ is not equal to the set $\{3, 3 \oplus 3, \ldots\}$. For which least residues m modulo 12 is $\mathscr{L}_{12}{}^{+} = \{m, m \oplus m, \ldots\}$?

11. The set $\mathbf{Z}^{+}$ of Example 1 is equal to the set

$$\{0, \pm 1, \pm(1 + 1), \pm((1 + 1) + 1), \ldots\}.$$

For which integers m is $\mathbf{Z}^{+} = \{0, \pm m, \pm(m + m), \ldots\}$? Show that if n is any integer then the set $I = \{0, \pm n, \pm(n + n), \ldots\}$ satisfies properties (1) to (4) of Example 1 under the operation of addition.

12. A dodecahedron is a regular solid assembled from 12 identical pentagons. Repeat Exercises 4 and 7 for the dodecahedron.

13. There are five regular solids. We have mentioned the tetrahedron, cube, and dodecahedron. Find the others and discuss their sets of symmetries.

2.2 DEFINITION OF A GROUP

The concepts to be presented in this section are of fundamental importance, although a full realization of their generality will probably not become apparent until we have made a number of applications of these ideas in later chapters. We shall now develop some of the most elementary properties of algebraic systems called groups.

We begin with a set G on which is defined a binary operation usually called "multiplication" and for which we use the familiar notation. Accordingly, if $a, b \in G$, then ab (or $a \cdot b$) is a uniquely determined element of the set G. We now assume the following properties or laws in which a, b, and c are arbitrary elements, distinct or identical, of G.

G_1: $(ab)c = a(bc)$ (*associative law*).

G_2:   There exists an element e of G such
that $ea = ae = a$ for every element a
of G (*existence of an identity*).

G_3:   If $a \in G$, there exists an element x of
G such that $ax = xa = e$ (*existence of inverses*).

Under these conditions G is said to be a *group*. Let us repeat the definition in the following formal way.

2.1   **DEFINITION**   If G is a nonempty set on which is defined a binary operation, usually called "multiplication," such that properties G_1 to G_3 hold, we say that G is a *group* (with respect to this definition of multiplication).

Let us make a few remarks about the defining properties of a group. First, we emphasize that the examples of the previous section show us we should not think of the elements of a group as necessarily being *numbers* (or even symmetries of some geometric object). Further, multiplication is not assumed to have any properties other than those specified. The existence of other properties is left open. For example, the group of integers under addition also obeys the commutative law, but the group of symmetries of the cube does not. The element e, whose existence is asserted in G_2, is called an identity element for the operation of multiplication in Section 1.5. However, we shall now call it an identity of the group. We do not assume there is only *one* identity e, but later on we shall prove this to be true. As suggested by the indicated name of the third property, the element x whose existence is asserted in G_3 is called an inverse of the element a. Note that *every* element in a group has an inverse. We do not assume that the inverse of a is unique, but eventually we shall give a proof of this fact.

Examples 1 to 4 of Section 2.1 are all instances of groups. In each case a set G ($= \mathbf{Z}^+, \mathscr{L}_n{}^+, A, C$) is given and, respectively, an operation ($+, \oplus, \cdot, \cdot$) upon G is given which satisfies G_1 to G_3.

Let us consider another example of a group.

Example   Let T be the set $\{e, a, b, c\}$ with an operation of multiplication defined by Table 1.

Table 1

$(\cdot)$	e	a	b	c
e	e	a	b	c
a	a	e	c	b
b	b	c	e	a
c	c	b	a	e

This table is used as follows. To compute the product ab, we look down the first column under $(\cdot)$ until we find a. We look across the top row to the right of $(\cdot)$ until we find b. The product ab is then listed in the table as that entry lying in the row labeled a and column labeled b. In this case, $ab = c$. Notice also that $ba = c$.

The properties G_2 and G_3 are easily checked. Property G_1 is somewhat tedious to check but it also holds. The only meanings we can obviously attach to the elements of T are the properties of multiplication given by the table. If the elements of a group are given by symbols with no special additional meaning, and the multiplication is given by some table, as above, we call such a group an *abstract group*. This is in contrast to the more concrete groups of Examples 1 to 4 of Section 2.1. For example, when we think of the integers, we do not usually limit ourselves to properties of addition. Likewise, C, the group of symmetries of a cube, conjures up a cube and other geometric ideas. Of course, we may write a multiplication table for these groups, and then ignore the meaning of the symbols for the elements of these groups. In this fashion any group may be viewed as an abstract group.

If we are given the multiplication table for a group, then properties G_1 to G_3 must hold. Is there any way we can see properties just by glancing at the table? Property G_1 is difficult to check since, for example, to check it for a table of only four elements would require an enormous amount of calculation. Property G_2 is somewhat easier. We must look for an element (in this case e) which heads a row, and this row repeats the column headings. We look also for a column headed by e which repeats the row headings. If there is such an element, that element satisfies G_2. We may relabel this element e and call it an identity. If each row and column has e as an entry, then property G_3 holds. We may arrange the rows and columns of our table so that the rows appear in the same order as the columns. This is the case for Table 1. Observe that the table is symmetrical about the diagonal of e's. This means that $xy = yx$ for all elements x and y of T. In other words, the commutative law holds for the group T.

Generally speaking, a table is the worst way to "know" a group. Even a group with only 10 elements has a table with 100 entries. We once memorized two tables: addition of integers $0, 1, \ldots, 9$ and multiplication of integers $1, \ldots, 10$. But it took years of practice to gain familiarity with these tables. An abstract group of 100 elements might seem hopeless. Consequently, we shall discover other ways to "know" groups (even abstract groups) as we go along. There may be better ways to "know" a group than a multiplication table. For example, the abstract group T of Table 1 is completely determined (up to names of elements and the name of the operation) by the two properties: (1) T contains four elements, and (2) $xx = e$ for all $x \in T$. These two properties contain all the information of that 4×4 table. This fact is left to the problems at the end of this section.

All of our examples with the exception of Examples 3 and 4 of Section 2.1 also satisfy the commutative law. That is, they are abelian groups according to the following definition.

2.2   **DEFINITION**   If in a group G with operation multiplication $ab = ba$ for all $a, b \in G$, G is said to be an *abelian group* (or commutative group).

The term "abelian group" is most commonly used for this concept. The name is derived from that of Niels Henrik Abel (1802–1829), a now famous mathematician who lived and died in poverty, and who did fundamental work in both algebra and analysis. His work has stood as an inspiration for many later mathematicians.

We could give the group operation any name we pleased. It is most often called multiplication, but there are many cases where another name is preferred. For example, addition, as we know it, is always commutative. So, when we are talking about abelian groups, we shall often call the operation "addition" and denote it by "+." We shall never use addition as the operation of a nonabelian group. That is, whenever addition is used as the operation, we shall always assume, whether or not it is explicitly mentioned, that the group is abelian.

When the operation of a group is called multiplication, the group may be either abelian or nonabelian. Accordingly, when we come to prove a property of arbitrary groups, we shall think of the operation as multiplication, and use the implied notation. In particular, we shall show that a group element a has a unique inverse. Then the inverse of a will be denoted by a^{-1}. We shall usually let e be the identity of the group, and reserve the symbol "1" for the smallest positive integer.

Perhaps we should emphasize that writing ab for $a, b \in G$ and calling it multiplication does not necessarily mean it is "ordinary" multiplication (of numbers, for example). It is just an operation having *only* the properties required in the definition of a group; other properties may hold in special cases, but these additional properties depend upon the special case and *not* the definition.

Now let G be an arbitrary group with operation multiplication. Since the word "multiplication" means nothing special here, the properties of a group are not obvious, except for those actually used in the definition. We now consider some properties of multiplication in a group.

2.3   THEOREM   *The following hold in every group G:*

(*i*) *The identity of G is unique.*

(*ii*) *If $a \in G$, a has unique inverse a^{-1}. Further, $(a^{-1})^{-1} = a$.*

(*iii*) *(Left Cancellation) If a, b, $c \in G$ such that $ab = ac$, then $b = c$.*

(*iv*) *(Right Cancellation) If a, b, $c \in G$ such that $ba = ca$, then $b = c$.*

(*v*) *If $a, b \in G$, there exists a unique element x of G such that $ax = b$ and a unique element y of G such that $ya = b$. In fact, $x = a^{-1}b$ and $y = ba^{-1}$.*

(*vi*) *The inverse of a product is the product of the inverses in the reverse order; that is, if $a, b \in G$, then $(ab)^{-1} = b^{-1}a^{-1}$.*

Proof: We have used the word "unique" here. We mean "one and only one" when we say "unique." Consider (i). Property G_2 tells us that there is at least one identity $e \in G$ satisfying G_2. Suppose that $e' \in G$ is also an identity of G. Applying G_2 $(a = ae)$ to the element $e' = a$ and the identity e, we obtain

$$e' = e'e.$$

Now using the identity e' and the element $e = a$ in G_2 $(a = e'a)$, we have

$$e = e'e.$$

Thus

$$e' = e'e = e$$

so that $e' = e$. In other words, e' could not be different from e. There is only one e in G satisfying G_2.

In view of (i), we are justified in speaking of *the* identity of a group. An element which is not the identity may naturally be called a *nonidentity* element.

We next prove that (iii) holds. We therefore assume that

2.4 $ab = ac.$

By G_3 there is an element x such that

2.5 $xa = e.$

Multiplying both sides of Equation 2.4 by x and observing that multiplication is an operation yields

2.6
$$x(ab) = x(ac).$$

But

$$\begin{aligned} x(ab) &= (xa)b & \text{(\textit{associative law})} \\ &= eb & \text{(\textit{Equation 2.5})} \\ &= b & \text{(\textit{existence of identity}).} \end{aligned}$$

Similarly,

$$x(ac) = (xa)c = ec = c.$$

From these calculations and Equation 2.6, we see that $b = c$, as we wished to show.

Using the above proof as a guide, write down a proof that property (iv) holds.

Although the definition of e requires that $ae = a$ and $ea = a$ for all $a \in G$, we can now observe that the identity of a group is completely determined by any *one* element and one of the above equations. By this statement we mean that *if d is some element of G and either $dx = d$ or $xd = d$, then we must have $x = e$.* Of course, this is an immediate consequence of (iii) and (iv). For suppose $dx = d$. Then $de = d$ and $dx = d$ imply that $de = dx$, from which (iii) tells us that $e = x$.

Similarly, (ii) follows from (iii) and (iv). Suppose both $ax = e$ and $ay = e$. Then $ax = ay$; so by (iii), $x = y$, proving that a has only one inverse.

Since each element a of G has exactly one inverse, we shall find it convenient to denote this inverse by a^{-1}. It may be helpful to have in mind a verbal definition of a^{-1} as follows: "a^{-1} is the element of G which when multiplied by a gives e." That is, if $ax = e$ (or equally well, $xa = e$), it follows that $x = a^{-1}$.

Since $a^{-1}a = e$, we see also that a solves $a^{-1}x = e$, and consequently, that a is the inverse of a^{-1}, that is, that $(a^{-1})^{-1} = a$, completing the proof of (ii).

Now we consider (v). It is easy to verify that $x = a^{-1}b$ is a solution to $ax = b$. Suppose also that $az = b$ for some $z \in G$. Then $ax = az$. Applying (iii), we conclude that $x = z$. In particular, $x = a^{-1}b$ is the unique solution to the equation $ax = b$. It might be worth writing out a detailed proof of the part of (v) which refers to the equation $ya = b$.

Finally, we consider part (vi). Let us compute the following product:

$$(ab)(b^{-1}a^{-1}) = ((ab)b^{-1})a^{-1} \qquad \text{(\textit{associative law})}$$
$$= (a(bb^{-1}))a^{-1} \qquad \text{(\textit{associative law})}$$
$$= (ae)a^{-1} \qquad \text{(\textit{definition of inverse})}$$
$$= aa^{-1} \qquad \text{(\textit{definition of identity})}$$
$$= e \qquad \text{(\textit{definition of inverse})}$$

Also, by the definition of inverse,

$$(ab)(ab)^{-1} = e.$$

Therefore

$$(ab)(ab)^{-1} = (ab)(b^{-1}a^{-1}).$$

Applying left cancellation to ab, we obtain

$$(ab)^{-1} = b^{-1}a^{-1}.$$

This completes the proof of part (vi) and the theorem.

Looking back, we see that the definition of a group implies certain properties for multiplication of elements of a group. That is, the properties of Theorem 2.3 hold for all groups, including those of Examples 1 to 4 of Section 2.1.

EXERCISES

1. Complete the proof of Theorem 2.3 by proving (iv) and the remaining part of (v).

2. With the indicated operation, which of the following are groups?

   (a) The set of integers which are multiples of 3 under addition.

   (b) The set $\{0, 2, 4, 6, 8\}$ of least residues modulo 10 with the operation $\oplus$.

   (c) The set $\{0, 3, 6, 9\}$ of least residues modulo 10 with the operation $\oplus$.

   (d) The set of all symmetries of a cube which can be described as rotations about a vertical axis passing through the center of the cube and perpendicular to a face with the operation composition of symmetries.

(e) The set of all integers with the operation $\circ$ defined as follows:
$a \circ b = a + b + 1$.

(f) The set of all integers with the operation $\circ$ defined as follows:
$a \circ b = a - b$.

3. Prove that $(ab)(ab) = (aa)(bb)$ for all choices of a and b as elements of a group G if and only if G is an abelian group.

4. Consider the set of all mappings θ_b which map integers to integers and which are given by a formula $x\theta_b = x + b$ for some integer b. Define $\theta_b \circ \theta_c$ to be a composition of mappings. With the operation $\circ$, is this set a group? Is it commutative?

5. Suppose that $G = \{e, a, b, c\}$ is a group with identity e containing four elements. Assume that $xx = e$ for all choices of x in G. Show that G is the group T of the example in this section by writing down a multiplication table for G.

6. Assume that G is a set with an operation called multiplication which satisfies the following three properties:

G'_1: $(ab)c = a(bc)$ for every $a, b, c \in G$.

G'_2: There exists an element e of G so that for every $a \in G$, $ae = a$.

G'_3: If $a \in G$, then there is a $b \in G$ such that $ab = e$.

(a) If $ab = e$, then show that $ba = e$.

(b) Show that $ea = a$ for all $a \in G$.

(c) Show that G is a group.

7. The group $\mathscr{Z}_3{}^+$ contains three elements. Write a table for $\mathscr{Z}_3{}^+$. Show that if G is any group containing exactly three elements, then by renaming the elements of G, a table for G will be identical to that of $\mathscr{Z}_3{}^+$.

8. Suppose that G is a set with an associative binary operation which we call "multiplication." Assume that for any $a, b \in G$ there are unique elements $x, y \in G$ so that $ax = b$ and $ya = b$. Show that G is a group.

9. Suppose that G is a finite set with an associative binary operation which we call "multiplication." Assume that for any $a, b, c \in G$ such that either $ac = bc$ or $ca = cb$, then $a = b$. Prove that G is a group.

10. Give an example of a nonabelian group G such that $(ab)(ab) = (ba)(ba)$ for all $a, b \in G$. *Hint:* Examine the group of symmetries of a square.

11. Show that the positive rational numbers (fractions) form a group with the operation of ordinary multiplication of numbers.

2.3 GENERAL PRODUCTS

We all know that addition of integers is an associative operation. We only add two integers at a time, but we often tend to ignore parentheses telling us how additions take place. We write

$$27 + 32 + 19 + 21,$$

instead of

$$((27 + 32) + 19) + 21.$$

We pass off our omission of parentheses with the comment, "The associative law holds." The associative law only discusses the sum or product of three elements. We must demonstrate that we may forget parentheses in longer sums or products.

Suppose G is a group and $a_1, a_2, \ldots, a_n$ are elements of G (possibly not all different). Clearly $a_1 a_2$ is an element of G. We may define $a_1 a_2 a_3$ by setting it equal to $(a_1 a_2) a_3$. The associative law tells us that

2.7　　　　　　　　　　　　$(a_1 a_2) a_3 = a_1 (a_2 a_3).$

Thus $a_1 a_2 a_3$ is really independent of the way parentheses might be introduced to indicate the manner of association of the elements.

Now that we have defined a product of three elements of a group, let us define a product of four elements as follows:

2.8　　　　　　　　　　　　$a_1 a_2 a_3 a_4 = (a_1 a_2 a_3) a_4.$

The associative law then tells us that

$$a_1 a_2 a_3 a_4 = ((a_1 a_2) a_3) a_4 = (a_1 a_2)(a_3 a_4)$$

and also that

$$a_1 a_2 a_3 a_4 = (a_1 (a_2 a_3)) a_4 = a_1 ((a_2 a_3) a_4) = a_1 (a_2 a_3 a_4).$$

These calculations verify that the product of four elements is also independent of the way in which the elements may be associated.

It seems fairly clear that similar statements hold for products of more than four elements. A general proof can be given using the Well-Ordering Principle of the positive integers. We shall use this property to prove these ideas about associativity. First we formulate a statement which generalizes what we have said about products of three and four elements.

In 2.7 and 2.8 we defined the product of three or four elements of a group. In general, if k is a positive integer such that

$$a_1 a_2 \cdots a_k$$

has been defined, we define

2.9 $$a_1 a_2 \cdots a_k a_{k+1} = (a_1 a_2 \cdots a_k) a_{k+1}.$$

It should appear, then, that this gives us a definition of a product of any number n of elements of a group. Such a definition is called a *recursive* definition; and definitions of this kind will be considered in more detail in a later chapter.

We now state a theorem that generalizes the ideas of associativity which we have discussed.

2.10 **THEOREM** *Let n be an arbitrary positive integer, and let $a_1, a_2, \ldots, a_n$ be elements of a group G. The generalized associative law holds in G. That is, for each positive integer r such that $1 \le r < n$, we have*

2.11 $$(a_1 a_2 \cdots a_r)(a_{r+1} \cdots a_n) = a_1 a_2 \cdots a_n.$$

Proof†: If $n = 3$ or 4, we have verified this theorem in 2.7 and 2.8.

We prove the theorem by contradiction. That is, we assume the theorem is false and show that this assumption leads to a contradiction. In particular, our assumption must be wrong if it leads to a contradiction. If our assumption is wrong, the right conclusion is that the theorem is true.

Assume the theorem is false. If it is false, there must be an instance where 2.11 is wrong, and in particular, there is some positive integer n and some positive integer r such that $1 \le r < n$ for which 2.11 does not hold. (For example, it might be that $n = 107$, $r = 93$.) Let S be the set of positive integers n' for which there is a counterexample to 2.11, that is, there is a positive integer r' so that $1 \le r' < n'$ and there are some elements $a_1, \ldots, a_{n'}$ from G so that 2.11 does not hold for this r' and for these elements. Notice

_____

† The reader might want to skip the proof here upon first reading.

that $n \in S$ so that S contains some integers. By the Well-Ordering Principle of sets of positive integers, S contains a least positive integer n''. For this n'', which we now call n, there is some r'', which we call r, such that $1 \le r < n$, and for some elements $a_1, \ldots, a_n$ from G, 2.11 does not hold. That is (using n, r for n'', r''),

2.12
$$(a_1 \cdots a_r)(a_{r+1} \cdots a_n) \ne a_1 \cdots a_n.$$

If $r = n - 1$, then

$$(a_1 \cdots a_{n-1})a_n \ne a_1 \cdots a_n$$

which contradicts

$$(a_1 \cdots a_{n-1})a_n = a_1 \cdots a_n,$$

as assured by Definition 2.9. We must therefore have $r < n - 1$. By Definition 2.9, we have

$$a_{r+1} \cdots a_n = (a_{r+1} \cdots a_{n-1})a_n.$$

So we may write

2.13
$$(a_1 \cdots a_r)(a_{r+1} \cdots a_n) = (a_1 \cdots a_r)((a_{r+1} \cdots a_{n-1})a_n).$$

Letting

$$a = a_1 \cdots a_r,$$
$$b = a_{r+1} \cdots a_{n-1},$$

and

$$c = a_n,$$

the associative law

$$(ab)c = a(bc)$$

tells us that

2.14
$$(a_1 \cdots a_r)((a_{r+1} \cdots a_{n-1})a_n) = ((a_1 \cdots a_r)(a_{r+1} \cdots a_{n-1}))a_n.$$

Look at the expression

$$(a_1 \cdots a_r)(a_{r+1} \cdots a_{n-1}).$$

Since n was the least integer in S, $n - 1$ is not in S. The associative law holds, so that $n > 3$. That is, $n - 1$ is positive. Now S was the set of all positive integers n for which 2.11 was false in some way. In particular, $n - 1 \notin S$ means

2.15
$$(a_1 \cdots a_r)(a_{r+1} \cdots a_{n-1}) = a_1 \cdots a_{n-1}.$$

Substituting 2.15 in 2.14 gives

2.16
$$(a_1 \cdots a_r)((a_{r+1} \cdots a_{n-1})a_n) = (a_1 \cdots a_{n-1})a_n.$$

Combining 2.16 with 2.13, we get

$$(a_1 \cdots a_r)(a_{r+1} \cdots a_n) = (a_1 \cdots a_{n-1})a_n.$$

Looking back at Definition 2.9, we must have

2.17
$$(a_1 \cdots a_r)(a_{r+1} \cdots a_n) = (a_1 \cdots a_{n-1})a_n$$
$$= a_1 \cdots a_n.$$

Comparing 2.12 with 2.17, we see we have a contradiction. Our assumption that the theorem is false leads to a contradiction; therefore, the theorem is true.

Our proof actually contains a logical gap. That was in Definition 2.9 where we used the definition for the product of n elements. However, we do not know that 2.9 has defined such a product. Later we shall show that such *recursive definitions* really are definitions.

A remark should be made concerning the generalized associative law. In the proof we used only the property that the product is an associative operation. In particular, our proof shows that the generalized associative law will hold for any binary associative operation upon a set G. It is not necessary that G actually be a group. Properties G_2 and G_3 of a group do not enter into the proof.

In abelian groups there is also a law concerning commutativity. Using the generalized associative law, we may omit parentheses in the conclusion of the following statement.

2.18   THEOREM   *Let n be a positive integer; let $i_1, \ldots, i_n$ be the integers $1, 2, \ldots, n$ in some ordering, and $a_1, \ldots, a_n \in G$ where G is an abelian group. The generalized commutative law holds for G. That is,*

2.19
$$a_{i_1} a_{i_2} \cdots a_{i_n} = a_1 a_2 \cdots a_n.$$

The proof of this theorem is similar to the proof of the generalized associative law. We shall use the generalized associative law in the course of the proof. First let us illustrate the theorem when $n = 3$, and $i_1 = 3$, $i_2 = 1$, $i_3 = 2$. Here 2.19 states that

$$a_3 a_1 a_2 = a_1 a_2 a_3.$$

This special case may be readily verified as follows:

$$a_3 a_1 a_2 = (a_3 a_1) a_2 \qquad \textit{(definition)}$$
$$= a_3 (a_1 a_2) \qquad \textit{(associative law)}$$
$$= (a_1 a_2) a_3 \qquad \textit{(commutative law)}$$
$$= a_1 a_2 a_3 \qquad \textit{(definition)}.$$

Proof†:   To prove the theorem, we suppose it is false.   Let S be the set of all positive integers $n > 2$ such that there is a counterexample to 2.19.   That is, if $n \in S$, then there exists an ordering $i_1, \ldots, i_n$ of $1, \ldots, n$ and elements $a_1, \cdots, a_n \in G$ so that

2.20
$$a_{i_1} \cdots a_{i_n} \neq a_1 \cdots a_n.$$

Since the theorem is false, there is a counterexample for some n. This particular n lies in S, so that the set S is certainly not empty. By the Well-Ordering Principle of the integers, S contains a least element which we will also call n, and for this n there is some counterexample as in 2.20.

Let us denote i_n by j, so that a_j is the last element in the product occurring on the left in 2.20, and let us consider the product

$$a_{i_1} \cdots a_{i_{n-1}}.$$

Then $i_1, \ldots, i_{n-1}$ is a reordering of $1, 2, \ldots, j - 1, j + 1, \ldots, n$. Here we have $n - 1$ integers, and it will be convenient to introduce a change of notation which will produce a corresponding reordering of the integers $1, 2, \ldots, n - 1$.   Let us therefore define integers i'_s as follows: Set $i'_s = i_s$ for $i_s = 1, 2, \ldots, j - 1$.   Define $i'_s = i_s - 1$ for $i_s = j + 1, \ldots, n$.   Then $i'_1, \ldots, i'_{n-1}$ is a reordering of $1, \ldots, n - 1$.   Set $b_{i'_j} = a_{i_j}$.   We are now ready to apply the generalized commutative law.   Since n is the least integer in S, $n - 1$ is not contained in our set S, that is, there is no counterexample to our theorem for $n - 1$.   The commutative law holds so that $n > 2$, and therefore $n - 1$ is positive.   In particular, the generalized commutative law applies to $b_{i'_1} \cdots b_{i'_{n-1}}$.   We obtain

2.21
$$b_{i'_1} \cdots b_{i'_{n-1}} = b_1 \cdots b_{n-1}.$$

Plugging the a_i's back into 2.21, we obtain

2.22
$$a_{i_1} \cdots a_{i_{n-1}} = a_1 \cdots a_{j-1} a_{j+1} \cdots a_n.$$

---

† The reader may wish to defer this proof until later.

We now have the following equalities:

2.23 $\quad a_{i_1} \cdots a_{i_n} = (a_{i_1} \cdots a_{i_{n-1}})a_{i_n}$ (*definition*)

$\qquad\qquad = (a_1 \cdots a_{j-1}a_{j+1} \cdots a_n)a_j$ (*2.22*)

$\qquad\qquad = (a_1 \cdots a_{j-1})(a_{j+1} \cdots a_n a_j)$ (*generalized associative law*).

Look at the product $a_{j+1} \cdots a_n a_j$.

2.24 $\qquad a_{j+1} \cdots a_n a_j = (a_{j+1} \cdots a_n)a_j$ (*definition*)

$\qquad\qquad\qquad = a_j(a_{j+1} \cdots a_n)$ (*commutative law*)

$\qquad\qquad\qquad = a_j a_{j+1} \cdots a_n$ (*generalized associative law*).

Using 2.24 in 2.23, we obtain

2.25 $\qquad\qquad a_{i_1} \cdots a_{i_n} = (a_1 \cdots a_{j-1})(a_j \cdots a_n)$

$\qquad\qquad\qquad = a_1 \cdots a_n$ (*generalized associative law*).

Comparing 2.20 and 2.25, we have a contradiction, so that our assumption that the theorem has a counterexample is false. The theorem must be true.

Our proof used the generalized associative law and the commutative law. We saw in Theorem 2.10 that the associative law guarantees the truth of the generalized associative law. This means our proof will work if the associative and commutative laws are valid. In particular, the generalized commutative law holds for any binary operation which is both associative and commutative. The set G need not be a group with the given operation.

The generalized associative law tells us that in any product we may insert parentheses to indicate association in any way we wish without changing the value of the product. The generalized commutative law tells us that in an abelian group we may order the terms of a product in any way without changing the value of the product. Accordingly we shall omit parentheses in products unless we wish to emphasize a part of the product. Further, in an abelian group we will freely reorder the terms of a product to fit our immediate needs. Some special cases of these theorems will appear in the exercises, and they should help to clarify the proofs.

We next observe that integral exponents may be defined in any group G in the usual way. If a is an arbitrary element of G, we may define $a^0 = e$, $a^1 = a$, $a^2 = a \cdot a$. More generally, if k is a positive integer such that a^k has been defined, we define $a^{k+1} = a^k \cdot a$. If k is a negative integer, then $-k$ is positive so that we define $a^k = (a^{-1})^{(-k)}$ in this case. The following familiar laws of exponents now hold, where m and n are arbitrary integers.

2.26
$$\text{(i)} \quad a^m \cdot a^n = a^{m+n},$$

$$\text{(ii)} \quad (a^m)^n = a^{mn}.$$

Suppose, now, that a, $b \in G$. Then $(ab)^2 = (ab)(ab)$, and if $ba \neq ab$, $(ab)^2$ may not be equal to a^2b^2. However, if $ba = ab$, it does follow that $(ab)^2 = a(ba)b = a^2b^2$. In general, it is not difficult to show that if G is an abelian group and m is any integer, then

2.27 $$(ab)^m = a^m \cdot b^m.$$

Recall that in an abelian group we will often use "+" (addition) to denote the group operation, so that the laws of exponentiation change into laws about multiples. Our definitions become: $0a = e$, $1a = a$, $2a = a + a$, and for positive k such that ka is defined, $(k + 1)a = (ka) + a$. For negative integers k, we set $ka = (-k)(-a)$. Now our laws of exponents may be rewritten in the form of laws of multiples. Let m and n be arbitrary integers. Then

2.28

$$\text{(i)} \quad (ma) + (na) = (m + n)a,$$

$$\text{(ii)} \quad n(ma) = (nm)a,$$

$$\text{(iii)} \quad (ma) + (mb) = m(a + b).$$

There is a little ambiguity in these laws, since we use "+" to denote the operation of addition of integers and addition of group elements. These two "additions" may be very different, but there is no problem here as long as we do not try to add an integer to a group element.

Suppose m is a positive integer. It should perhaps be emphasized that a^m is a convenient way of writing the product $a \cdot a \cdots a$ (m factors) and ma is a convenient way of writing the sum $a + a + \cdots + a$ (m terms). The expression ma is not a product of elements of a group G.

The laws of exponents and multiples given in 2.26 to 2.28 may be proven using the Well-Ordering Principle of the integers, and in a later chapter we will discuss these laws in more detail. Unless explicitly stated to the contrary, as in some problems and a later chapter where we give proofs of some of these laws, we shall henceforth assume the truth of 2.26 to 2.28. In particular, we will often write a^3 in place of $a \cdot a \cdot a$ or $3a$ in place of $a + a + a$.

In this section it may appear that we have belabored the obvious. This might be true if our group G were the one of Example 1 (the integers $\mathbf{Z}^+$) or possibly even Example 2 (residues mod n, $\mathscr{Z}_n{}^+$) of Section 2.1. If we reread the formulas thinking about Example 3 of Section 2.1 (rigid symmetries A of axes), the formulas of this section do not seem at all obvious. The lesson in all this is that no matter how obvious a property of an algebraic system may seem, it may only be obvious because we are thinking about a particular example which is not at all general. Such thinking is useful for guessing, but it can lead to mistakes. We must check every step we make against the properties we *know* to hold.

EXERCISES

1. Using only the associative law [i.e. $a(bc) = (ab)c$] verify that for elements a_1, a_2, a_3, a_4, a_5 of a group G, $a_1a_2a_3a_4a_5$ equals each of the following:

    (a) $a_1(a_2(a_3(a_4a_5)))$,

    (b) $a_1(a_2((a_3a_4)a_5))$,

    (c) $(a_1((a_2a_3)a_4))a_5$.

2. Using only the generalized associative law and the commutative law, verify that for elements a_1, a_2, a_3, a_4, a_5 of an abelian group G, $a_1a_2a_3a_4a_5$ equals each of the following:

    (a) $a_1a_2a_5a_3a_4$,

    (b) $a_5a_4a_3a_2a_1$,

    (c) $a_5a_1a_4a_2a_3$.

3. Using the associative and commutative laws, show that for a_1, a_2, a_3, a_4, a_5, elements of an abelian group G,

$$a_5(a_4(a_3(a_2a_1))) = a_1a_2a_3a_4a_5.$$

4. Suppose that G is a group in which $a^3b^3 = (ab)^3$, $a^4b^4 = (ab)^4$, and $a^5b^5 = (ab)^5$ for all choices of a and b in G. Prove that G is abelian.

5. Show that if $a^2 = e$ for all choices of a in a group G, then G is abelian.

6. Suppose that G is a group in which $a^2 = e$ for all choices of a from G. Further, we assume that G contains n distinct elements where $3 \leq n \leq 7$. Show that $n = 4$ and consequently if $G = \{e, a, b, c\}$, then G is the group T of Table 1 in Section 2.2. *Hint:* First show that G contains a set of elements $T = \{e, a, b, c\}$ which form the group of Table 1 in Section 2.2 for the multiplication of G. Then prove that $T = G$.

7. Assume that G is a group with identity e containing more than one element. For $a, b \in G$ define $[a, b] = a^{-1}b^{-1}ab$. Suppose that $[[a,b],c] = e$ for all choices of $a, b, c \in G$. Prove that there is some element $a \in G$ ($a \neq e$) which commutes with all elements of G.

8. Show that 2.26(i) and 2.28(i) follow from 2.11.

9. Give an example of a set G with a binary operation (which we call multiplication) such that $(aa)a \neq a(aa)$ for some $a \in G$.

10. Suppose that a group G contains exactly five distinct elements. Prove that G is abelian.

2.4 SUBGROUPS AND DIRECT PRODUCTS

We now introduce the concept of a subgroup of a group.

> **Example** Look again at the group of Example 4 of Section 2.1: C, the rigid symmetries of a cube. Place the cube in its hole in space in front of you on the table. Consider all motions obtained by rotating the cube, so that the top is moved away from you and the bottom toward you. Call this set of rotations D. There are exactly four faces which can be toward you after performing a rotation from D, so that D contains four elements. If we draw a horizontal axis through the center of the cube running from left to right, and if we orient this axis by putting arrows on it pointing to the right, then D is the set of all elements from C which take this oriented axis into itself. Each element of D can be described as a rotation about this axis. That is, D is the subset of C "fixing" or "leaving invariant" the oriented axis. Equivalently, D may be described as the set of all symmetries from C which keep the left face of the cube on the left side and the right face of the cube on the right side. It is not difficult to verify that D is a group since the group operation of D is identical to that of C.

This example motivates the following definition.

2.29   DEFINITION   A nonempty set H of elements of a group G is called a *subgroup* of G if H is itself a group with respect to the operation already defined on G.

Clearly if e is the identity of G then both G and $\{e\}$ are subgroups of G, and we call these two subgroups the *trivial* subgroups. Any other subgroup is called a *proper* subgroup. That is, in our example above, D is a proper subgroup of C. Incidentally, if $G = \{e\}$ we call G a *trivial group*.

Working directly from the definition of a group, it is often somewhat complicated to verify that a particular subset H is a subgroup of the group G. We may simplify matters a little by proving the following theorem.

2.30   THEOREM   (*a*) *A nonempty subset K of a group G is a subgroup if and only if the following two conditions are satisfied:*

(i) *If $a, b \in K$, then $ab \in K$.*

(ii) *If $a \in K$, then $a^{-1} \in K$.*

(*b*) *If K contains only a finite number of elements, condition (ii) is implied by condition (i).*

In view of this theorem, we see that a nonempty set of elements of a group G having a finite number of elements is a subgroup of G if and only if the set is closed under the operation of G. However, for infinite groups, condition (ii) is not a consequence of condition (i). As an example, the set of positive integers in the group $\mathbf{Z}^+$ is a nonempty subset which satisfies condition (i) but not condition (ii).

Proof: We start by proving that if K is a subgroup then conditions (i) and (ii) are satisfied. If $a, b \in K$, then $ab \in K$ since K is a subgroup. Further, by the "existence of inverse" property G_3 of a group, any element a in K must have an inverse a^{-1} in K. Both conditions hold.

Now assume conditions (i) and (ii) hold. We will show that K is a subgroup. Suppose a, b, c are elements of K. By property G_1 of the group G,

$$(ab)c = a(bc).$$

We conclude that K satisfies the associative law.

By condition (ii), since a is an element of K, a^{-1} is an element of K. Applying condition (i) to a and a^{-1}, we conclude that $aa^{-1} = e$ is an element of K, so that K has the identity property G_2. That is, there is an e' (namely e itself) in K so that $be' = e'b = b$ for all $b \in K$ (in fact for all $b \in G$). In this argument we used property (ii) to tell us that if $a \in K$ then $a^{-1} \in K$. The inverse property G_3 is now an immediate consequence. That is, for any $a \in K$ there is $a^{-1} \in K$ so that $a^{-1}a = aa^{-1} = e$. We conclude that K is a subgroup.

Suppose now that K contains a finite number of elements. We may label them as $a_1, a_2, \ldots, a_n$. We prove that condition (i) implies condition (ii) in these circumstances. Consider the products $a_1a_j, a_2a_j, a_3a_j, \ldots, a_na_j$. By condition (i) these are all elements of K. Suppose that two of these products are equal, that is, $a_sa_j = a_ta_j$. The cancellation laws for G then tell us that $a_s = a_t$ or $s = t$. In particular, all the products $a_1a_j, a_2a_j, a_3a_j, \ldots, a_na_j$ are different. Since there are n of these products, they must be $a_1, a_2, a_3, \ldots, a_n$ in some reordering. Therefore, the element a_j occurs somewhere in our list of products. Suppose that $a_ia_j = a_j$, so that we have $a_ia_j = a_j = ea_j$. The cancellation laws tell us that $a_i = e$. Now the element $a_i = e$ appears in our list of products. It is, for example, a_ma_j, that is, $e = a_i = a_ma_j$. We may write $a_j^{-1}a_j = e = a_ma_j$. The cancellation laws tell us that $a_j^{-1} = a_m$, and we conclude that the element $a_j^{-1} \in K$. This argument works for each j, implying that if $a \in K$, then $a^{-1} \in K$. Condition (i) does imply condition (ii) when K is a finite set.

We conclude this section by introducing one additional concept. Suppose G and H are groups, distinct or identical. It is possible to construct a new group from G and H called the *direct product* of G and H and written $G \times H$. Let us consider the Cartesian product $G \times H$ whose elements are the ordered pairs (a, b), $a \in G$, $b \in H$. On this set we define multiplication as follows:

2.31 $$(a_1, b_1)(a_2, b_2) = (a_1a_2, b_1b_2).$$

It is understood, of course, that $a_1, a_2 \in G$ and that $b_1, b_2 \in H$. Moreover, although the same notation is used for multiplication in both groups, a_1a_2 is the product of a_1 and a_2 in G and b_1b_2 is the product of b_1 and b_2 in H. We leave as an exercise the proof that with respect to the operation in 2.31, the set $G \times H$ becomes a group.

What conditions on G and H will assure us that $G \times H$ is commutative? If both G and H are abelian, then for $a_1, a_2 \in G$ and $b_1, b_2 \in H$ we have

$$(a_1, b_1)(a_2, b_2) = (a_1a_2, b_1b_2) = (a_2a_1, b_2b_1) = (a_2, b_2)(a_1, b_1),$$

so that $G \times H$ is abelian in this case. When G and H are abelian and when the operations of both G and H are written additively, we call $G \times H$ the *direct sum* of G and H and write $G \oplus H$ to denote $G \times H$ in this case. The operation in $G \oplus H$ is:

$$(a_1, b_1) + (a_2, b_2) = (a_1 + a_2, b_1 + b_2).$$

We shall return to consider direct products of groups in the next section.

EXERCISES

1. If H_1 and H_2 are subgroups of a group G, prove that $H_1 \cap H_2$ is a subgroup of G. Generalize this result by proving that the intersection of any number of subgroups of G is a subgroup of G.

2. Find all subgroups of each of the following groups:

   (a) The additive group $\mathscr{Z}_7{}^+$ of least residues modulo 7.

   (b) The additive group $\mathscr{Z}_{30}{}^+$ of least residues modulo 30.

   (c) The group A of symmetries of the x,y,z-axes.

   (d) The group C of symmetries of the cube.

   Hint: These latter groups may be described as all symmetries leaving something invariant. Determine the "somethings."

3. If G and H are groups, prove that their direct product $G \times H$ is a group.

4. If G and H are groups, give necessary and sufficient conditions on G and H so that their direct product $G \times H$ is abelian.

5. Consider the direct sum $\mathscr{Z}_5{}^+ \oplus \mathscr{Z}_3{}^+$. Prove that there is an element a in this group so that $\{a, a \oplus a, a \oplus a \oplus a, \ldots\}$ is the set of all elements in this group.

6. Label the elements of $\mathscr{Z}_2{}^+ \oplus \mathscr{Z}_2{}^+$ as e, a, b, c, where $e = (0,0)$. Make a table for the resulting group. Compare the table to Table 1 of the example in Section 2.2. We never did verify that T is a group. Without now verifying $G_1 - G_3$ of 2.1, explain why T must be a group.

7. Let P^* be a regular pentagon, and S^* a square cut from a stiff piece of paper. Push P^* and S^* onto a pencil by piercing the center of P^* and S^*. Let P be the group of symmetries obtained by rotating P^* around the pencil, and let S be the group of symmetries obtained by rotating S^* around the pencil. Next consider the configuration consisting of both the pentagon and the square in their positions on the pencil. We also have a group G obtained by rotating either P^* or S^* or both around the pencil. Show that G is "just" the direct product $P \times S$.

8. Let a be a fixed element of a group G. Let $C_G(a) = \{x \mid x \in G, ax = xa\}$. Prove that $C_G(a)$ is a subgroup of G which contains a. This subgroup is called the *centralizer* of a in G.

9. Let H be a subgroup of a group G which contains only a finite number of elements. Let $N_G(H) = \{x \mid x \in G, xyx^{-1} \in H \text{ for all } y \in H\}$. Prove that $N_G(H)$ is a subgroup of G which contains H. This subgroup is called the *normalizer* of H in G.

10. Suppose that G is a group and H and K are subgroups. Set $HK = \{ab \mid a \in H, b \in K\}$. Prove that HK is a subgroup if and only if $HK = KH$.

2.5 HOMOMORPHISMS AND ISOMORPHISMS

In this section we shall discuss a concept which is among the most important in all of algebra. The notion will occur again and again in our examples, our proofs, and our theorems. We introduce the concept of group homomorphisms.

> **Example 1** Consider the two groups of Examples 3 and 4 of Section 2.1. Imagine our x,y,z-axis system is fixed with O at the center of the cube and with the axes perpendicular to the cube faces. We imagine that the axes are oriented as in Figure 12. Notice that any symmetry of the cube causes a symmetry of the axes. For example, if the symmetry σ of the cube changes Figure 12 into Figure 13, then σ causes a corresponding symmetry σ' of the axes also shown in Figures 12 and 13. Let us define a mapping ψ from the symmetries C of the cube to the symmetries A of the axes. If $\sigma \in C$ is a symmetry of the cube and causes the cube in position X to change to position Y, then σ causes a corresponding change of the axes from

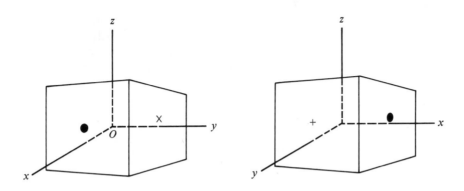

FIGURE 12   FIGURE 13

a position X' to a position Y'. This change is a symmetry σ' of the axes. We set $\sigma\psi = \sigma'$. That is, σ in C causes the change $\sigma\psi = \sigma'$ in A. If τ changes the cube from position Y to position Z, then $\tau' = \tau\psi$ will cause a corresponding change in the axes from position Y' to a position Z'. Since $\sigma\tau$ causes the change of the cube from X to Z via Y, $(\sigma\tau)'$ will cause a change of the axes from position X' to position Z' via Y'. Thus $\sigma'\tau' = (\sigma\tau)'$. We may state this principle in the form $(\sigma\tau)\psi = (\sigma\psi)(\tau\psi)$, so that ψ is an example of a mapping satisfying the following definition.

2.32   **DEFINITION**   Assume that G and H are groups and $\theta : G \to H$ is a mapping such that

2.33 $$(ab)\theta = (a\theta)(b\theta)$$

for all a, b elements of G. We call θ a *homomorphism* of G into H. If there exists a homomorphism of G *onto* H, we may say that G is *homomorphic* to H or that H is a *homomorphic image* of G.

The mapping ψ is an example of a homomorphism of C into A. The property 2.33 tells us that the group operation is *preserved under the mapping* θ. We ought to note that Equation 2.33 also states that $(a\theta)(b\theta) = (ab)\theta$.

We have introduced conventions for writing the group operations. If G is an additive abelian group and H is a multiplicative group, 2.33 would read "$(a + b)\theta = (a\theta)(b\theta)$." If G is multiplicative and H is additive abelian, then 2.33 would read "$(ab)\theta = (a\theta) + (b\theta)$." If both G and H are additive abelian groups, then 2.33 reads "$(a + b)\theta = (a\theta) + (b\theta)$."

Let us look at another example.

Example 2   Recall the groups $\mathbf{Z}^+$ of Example 1 and $\mathscr{Z}_n{}^+$ of Example 2 of Section 2.1. We shall now define a rather important homomorphism ϕ of the group $\mathbf{Z}^+$ into the group $\mathscr{Z}_n{}^+$. If a is any integer, by the Division Algorithm, integers q and r exist uniquely so $a = qn + r$, $0 \le r < n$. We set $a\phi = r$ which is an element of $\mathscr{Z}_n{}^+$. Suppose we have the following equations given by the Division Algorithm:

$$a = q_1 n + r_1, \quad 0 \le r_1 < n;$$
$$b = q_2 n + r_2, \quad 0 \le r_2 < n;$$
$$r_1 + r_2 = q_3 n + r_3, \quad 0 \le r_3 < n$$

where a and b are integers. Then

$$a + b = (q_1 + q_2 + q_3)n + r_3, \quad 0 \le r_3 < n.$$

We now have

$$(a\phi) \oplus (b\phi) = r_1 \oplus r_2 = r_3 = (a + b)\phi.$$

The mapping $\phi : \mathbf{Z}^+ \to \mathcal{Z}_n^+$ is another example of a homomorphism.

Homomorphisms come in many forms. We shall give another example now.

Example 3 Suppose G and H are two groups, possibly identical. Recall that $G \times H$, the direct product of G and H, is a group with multiplication given by

$$(a_1, b_1)(a_2, b_2) = (a_1a_2, b_1b_2)$$

where $a_1, a_2 \in G$ and $b_1, b_2 \in H$. We let $\phi : G \times H \to G$ be the *projection* of the set $G \times H$ onto the set G, as defined in Example 2 of Section 1.2. Then ϕ is a mapping of the group $G \times H$ into the group G defined by

$$(a, b)\phi = a, \qquad (a, b) \in G \times H.$$

For elements $(a_1, b_1), (a_2, b_2) \in G \times H$ we may compute

$$((a_1, b_1)(a_2, b_2))\phi = (a_1a_2, b_1b_2)\phi = a_1a_2 = (a_1, b_1)\phi(a_2, b_2)\phi.$$

This sequence of equalities demonstrates that ϕ is a homomorphism of $G \times H$ onto G. There is also a similar projection homomorphism of $G \times H$ onto H.

Let us consider yet another example of a different type.

Example 4 We have a cube C^* whose symmetry group C is described in Example 4 of Section 2.1. Consider another cube D^* whose markings are the same as those on C^*. The symmetry group D of D^* is also described by Example 4 of Section 2.1. Consider the symmetry described by the change from Figure 14 to Figure 15.

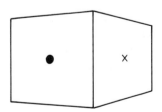

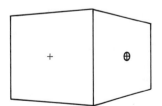

FIGURE 14   FIGURE 15

If we view this change as taking place for cube C^*, then we obtain a symmetry τ in the group C. On the other hand, we may view this as a change of position of the second cube D^*. In this way, we obtain an element τ' in the group D. If we consider the change of position described from Figure 12 to Figure 13, in a similar fashion we obtain elements σ in C and σ' in D. The products $\sigma\tau$ and $\sigma'\tau'$ exist in their respective groups. In fact, $\sigma\tau$ can be described by the change from Figure 12 to Figure 16 on the cube C^*. [It would be a very good idea to verify this point. Observe that τ (or τ') can be described by a certain rotation about one of the diagonals of the cube.] Also, $\sigma'\tau'$ is described by the same change on the cube D^*.

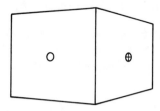

FIGURE 16

We define a mapping θ of C into D by setting $\sigma\theta = \sigma'$ for all choices of σ. In other words, if σ is a motion of C^*, then σ' is the "same" or "corresponding" motion of D^*. This "sameness" or "corresponding" quality allows us to draw the following immediate conclusions about θ. (1) As a mapping, θ is both one-one and onto. It gives a one-one correspondence between C and D. (2) θ is a homomorphism in that for any $\sigma, \tau \in C$ $(\sigma\tau)\theta = (\sigma\theta)(\tau\theta)$. We place "same" in our remarks in quotation marks because C and D are obviously *not* the same: a symmetry from C moves cube C^*; and a symmetry from D moves D^*. In fact, our use of the word "same" to define θ is backwards. It is the properties of θ which we mean by "same," so that this word ought to be reserved as a synonym for "identical." We introduce another word for this notion of "sameness," and this example suggests the following definition.

2.34 **DEFINITION** If G and H are groups and $\theta : G \to H$ is a homomorphism of G into H which is also a one-one mapping, then we call θ an *isomorphism* of G into H. Furthermore, if θ maps G onto H, then we say that G is *isomorphic* to H. If G is isomorphic to H, we sometimes call H an *isomorphic image* of G.

In the example we have been discussing, we may say that C is isomorphic to D. This example suggests that isomorphic groups may be considered as differing only in the notation (and meanings given that notation) used to indicate the elements of the groups. In other words, if we made group multiplication tables for the two groups, then the two resulting abstract groups would differ only in the names of their elements. Accordingly, isomorphic groups are sometimes said to be *abstractly identical*.

We may emphasize that, for the sake of generality, we have not required a homomorphism or isomorphism to be an onto mapping. However, whenever we say that G is homomorphic (isomorphic) to H or that H is a homomorphic (isomorphic) image of G, we do mean to imply that it is an onto mapping. Although the more general concept is useful in certain parts of the theory, we shall seldom have occasion to refer to homomorphisms (isomorphisms) that are not onto mappings. The reader should watch carefully to see whether the word "onto" is used.†

It may be worth observing that there always exists a *trivial* homomorphism of any group G into any group H. We have only to define the image of every element of G to be the identity element of H. Of course, this does not assert that H is a homomorphic image of G.

Another point is worth noting. Suppose that $\theta : G \to H$ is a homomorphism of the group G into the group H so that θ is a mapping of the set G into the set H. If K is a subgroup of G, then simply by restricting the domain of θ from G to K we obtain a mapping $\theta' : K \to H$ satisfying Equation 2.33, where $k\theta' = k\theta$ for $k \in K$. In other words, $\theta' : K \to H$ is a homomorphism of K into H. The point here is that we can obtain new homomorphisms from old ones just by restricting the old ones to subgroups.

If the mapping $\theta : G \to H$ is an isomorphism of the group G *onto* the group H, it may be verified that the mapping $\theta^{-1} : H \to G$, as defined in Section 1.2, is an isomorphism of H onto G. Accordingly, if G is isomorphic to H, then H is isomorphic to G and we may sometimes simply say that G and H are *isomorphic* groups.

The symbol $\cong$ is widely used for "is isomorphic to." Thus, if G and H are groups and we write $G \cong H$, it means that there exists an isomorphic mapping of G *onto* H. Using this notation, we have just shown in the preceding paragraph that if $G \cong H$, then $H \cong G$. We may emphasize that the homomorphic mapping of G into H whose existence is asserted by writing $G \cong H$ is both a *one-one* and an *onto* mapping.

The most elementary properties of homomorphisms are stated in the following theorem. It should be kept in mind that an isomorphism is a special case of a homomorphism, so that isomorphisms certainly have the stated properties.

---

† A homomorphism that is a one-one mapping is also called a *monomorphism*. Likewise, a homomorphism that is an onto mapping is sometimes called an *epimorphism*. In this new language an *isomorphism* onto a group is both a monomorphism and an epimorphism, and thus means the same as our symbol $\cong$ to be introduced presently.

2.35 **THEOREM** *Let* $\theta: G \to H$ *be a homomorphism of the group* G *into the group* H. *Then each of the following is true:*

(*i*) *If* e *is the identity of* G, *then* $e\theta$ *is the identity of* H.

(*ii*) *If* $a \in G$, *then* $(a^{-1})\theta = (a\theta)^{-1}$.

(*iii*) *If* G *is an abelian group and* θ *is an onto mapping, then* H *is an abelian group.*

Proof of (i): Let e' be the identity element of H. We must prove that $e' = e\theta$. The cancellation laws may be used to prove this as follows. We have, by the existence of an identity,

$$(e\theta)e' = e\theta = (ee)\theta = (e\theta)(e\theta).$$

Cancelling $e\theta$ on the left, we obtain

$$e' = e\theta.$$

The proof of the other parts of the theorem are left as an exercise.

The reader may sometimes have to make a suitable modification in notation. For example, if the operation in G is multiplication and that in H is addition, property (ii) of the preceding theorem should be interpreted as stating that $(a^{-1})\theta = -(a\theta)$, since $-(a\theta)$ is the inverse of $a\theta$ in H.

The concept introduced in the following definition plays an important role in the study of homomorphisms of groups.

2.36 **DEFINITION** Let $\theta: G \to H$ be a homomorphism of the group G into the group H. The set of all elements of G which map into the identity of H is called the *kernel* of the homomorphism θ, and may be denoted by ker θ.

In Example 1 above, we may describe ker ψ as the set of all symmetries of the cube which do not alter the positions of the x,y,z-axes. Recall Figure 12. The z-axis may be identified as that axis passing through the top and bottom faces of the cube. A symmetry which preserves the z-axis will keep the top and bottom faces on the top and bottom, respectively, or will interchange them. Similar statements may be made for the y- and x-axes and the faces they pass through. In other words, ker ψ is the set of symmetries of the cube which interchange or leave on their respective sides opposite pairs of faces. For example, if ρ is described by rotating from Figure 12, 180° clockwise about the x-axis as viewed in Figure 12, then $\rho\psi$ will be the identity transformation of the axes. Suppose μ is a symmetry of the cube which also preserves the positions of the axes. Clearly ρ^{-1} and $\rho\mu$ will both preserve the positions of the axes. In other words, Theorem 2.30 tells us that ker ψ is a subgroup of the group of symmetries of the cube.

Now look at Example 2. Notice that $a\phi = 0$ if and only if $a = qn$ is a multiple of n. If we have two multiples of n, say $q_1 n$ and $q_2 n$, then their sum, $q_1 n + q_2 n = (q_1 + q_2)n$, is again a multiple of n. Similarly if qn is a multiple of n, then $-qn = (-q)n$ is a multiple of n. Again Theorem 2.30 tells us that ker ϕ is a subgroup of $\mathbf{Z}^+$. It is the subgroup of multiples of n.

The next theorem will show that these facts are no accident.

2.37   **THEOREM**   *If* $\theta : G \to H$ *is a homomorphism, then* ker θ *is a subgroup of* G. *Moreover, if* e *is the identity of* G, ker $\theta = \{e\}$ *if and only if* θ *is an isomorphism.*

Proof: Let $K = \ker \theta$. By Theorem 2.35(i), the identity of H is $e\theta$. If $a, b \in K$, then $(ab)\theta = (a\theta)(b\theta) = (e\theta)(e\theta) = e\theta$, and thus K is closed with respect to multiplication. Moreover, if $a \in K$, by Theorem 2.35(ii) we have that $(a^{-1}\theta) = (a\theta)^{-1} = (e\theta)^{-1} = e\theta$, and hence $a^{-1} \in K$. Theorem 2.30 then shows that K is a subgroup of G. To prove the second statement of the theorem, suppose first that θ is an isomorphism. Since $e \in \ker \theta$, and the mapping is one-one, this assures us that no other element of G has an image the identity $e\theta$ of H, and it follows that ker $\theta = \{e\}$. Conversely, let us assume that θ is a homomorphism of G into H such that ker $\theta = \{e\}$. Suppose that $a, b \in G$ such that $a\theta = b\theta$. It follows that

$$(ab^{-1})\theta = (a\theta)(b^{-1}\theta) = (a\theta)(b\theta)^{-1} = e\theta,$$

and hence that $ab^{-1} \in K$. Thus, $ab^{-1} = e$ and $a = b$. This shows that θ is a one-one mapping and is therefore an isomorphism. This completes the proof of the theorem.

The result just established is often useful in the following way. To show that a mapping of a group G into a group H is an isomorphism, we may first show that it is a homomorphism and then verify that its kernel consists of the identity only.

We shall close this section by proving a rather useful fact about homomorphisms. Suppose that $\theta : G \to H$ is a homomorphism of the group G into the group H. If S is a subset of G then we let

$$S\theta = \{a\theta \mid a \in S\},$$

that is, $S\theta$ is the image of the set S under the mapping θ. This notation was introduced in Section 1.2.

2.38   **THEOREM** (*Universal Mapping Property*)   *Assume that* $\theta : G \to H$ *is a homomorphism of a group* G *into a group* H, *and* $\phi : G \to K$ *is a* *homomorphism of G onto a group K. If* ker $\theta \supseteq$ ker ϕ *then there is a unique homomorphism* $\psi: K \rightarrow H$ *of K into H such that* $\theta = \phi\psi$, *i.e.,*

$$a\theta = a(\phi\psi) \quad \text{for all } a \in G.$$

Further, (ker θ)ϕ = ker ψ.

This theorem becomes clearer if we draw a picture:

The theorem asserts that if we have the two homomorphisms θ and ϕ of the picture so that ker $\theta \supseteq$ ker ϕ, then the homomorphism ψ of the picture exists and is uniquely determined by the condition that $\phi\psi = \theta$. Observe that while θ may be into H, ϕ must map G onto K.

Proof: In order to prove this theorem, we must do five things. First we must prove that a mapping ψ exists. Second we must show that ψ is a homomorphism. Third we show that $\phi\psi = \theta$. Fourth we must prove that ψ is unique. And finally we prove that (ker θ)ϕ = ker ψ.

Let us define the mapping ψ. Since the mapping ϕ is onto K, for each $a \in K$ we may choose an $a' \in G$ such that $a'\phi = a$.† Set $a\psi = a'\theta$. Is ψ a mapping? For each $a \in K$ we can find an $a' \in G$ such that $a'\phi = a$. It might be that there is another choice $a'' \in G$ so that $a''\phi = a$. We want ψ to be a mapping, and so we must show that $a''\theta = a'\theta$. Since G is a group, Theorem 2.3(v) tells us that there is an element $b' \in G$ such that $a'b' = a''$. Applying ϕ, we obtain

$$a(b'\phi) = (a'\phi)(b'\phi) = (a'b')\phi = a''\phi = a = ae,$$

where e is the identity element of K. Cancelling on the left, we conclude that $b'\phi = e$ or $b' \in$ ker ϕ. Recall that by hypothesis, ker $\phi \subseteq$ ker θ, so that $b' \in$ ker θ or $b'\theta = e_0$ where e_0 is the identity of H. We use θ to compute

$$a''\theta = (a'b')\theta = (a'\theta)(b'\theta) = (a'\theta)e_0 = a'\theta.$$

---

† If K contains a very large infinite number of elements, this choice of a' depends upon the Axiom of Choice. We have mentioned this axiom in a footnote of Section 1.3, and we will return to it in Chapter 16.

In other words, if $a''\phi = a = a'\phi$, then $a''\theta = a'\theta$. This proves that the element $a'\theta$ is uniquely determined by a, and therefore, the equation $a\psi = a'\theta$ defines a mapping.

Next we must prove that the mapping $\psi: K \to H$ is a homomorphism. Assume that $a, b \in K$ and that $a', b' \in G$ are chosen so that $a'\phi = a$ and $b'\phi = b$. Then $(a'b')\phi = (a'\phi)(b'\phi) = ab$. We now have

$$(ab)\psi = (a'b')\theta = (a'\theta)(b'\theta) = (a\psi)(b\psi).$$

Therefore, ψ is a homomorphism.

We now argue that $\phi\psi = \theta$. Suppose $a' \in G$. Now ψ was defined so that $(a'\phi)\psi = a'\theta$. In other words, $a'(\phi\psi) = a'\theta$ for all $a' \in G$. We have shown that ψ is a homomorphism of K into H such that $\phi\psi = \theta$. The first three steps of our proof are complete.

Fourth, we must show that ψ is unique. In other words, if $\psi': K \to H$ is a homomorphism such that $\phi\psi' = \theta$, then $\psi' = \psi$. Suppose $a \in K$. Since ϕ maps G onto K, there is an $a' \in G$ so that $a'\phi = a$. Computing again:

$$a\psi' = (a'\phi)\psi' = a'(\phi\psi') = a'\theta = a\psi.$$

Since this equality holds for all $a \in K$, we must have $\psi' = \psi$. Therefore, ψ is unique.

To finish the proof we must prove the final statement: $(\ker \theta)\phi = \ker \psi$. Suppose $a \in (\ker \theta)\phi$. Then there is an $a' \in \ker \theta$ so that $a = a'\phi$. In other words, $e_0 = a'\theta = (a'\phi)\psi = a\psi$. Therefore, a lies in $\ker \psi$, and this shows that $(\ker \theta)\phi \subseteq \ker \psi$. To obtain the inclusion the other way, let $b \in \ker \psi$, and let $b' \in G$ be chosen so that $b'\phi = b$. Then

$$e_0 = b\psi = (b'\phi)\psi = b'(\phi\psi) = b'\theta.$$

Therefore, $b' \in \ker \theta$ so that $b \in (\ker \theta)\phi$. The set $\ker \psi$ is contained in $(\ker \theta)\phi$. We conclude that $\ker \psi = (\ker \theta)\phi$.

We shall use this theorem to prove other theorems about homomorphisms. As we shall see later in the text, this theorem is not "just a theorem of group theory"; it is very general and true, in a certain form, for other algebraic systems. This theorem is used in the following way. We are asked to prove that a group H is an isomorphic (homomorphic) image of a group K. Instead of manufacturing the mapping $\psi: K \to H$ directly, we use mappings $\theta: G \to H$ and $\phi: G \to K$ of a group G which are onto homomorphisms and which we already know for some reason. We do this in such a way that the theorem applies, telling us that ψ exists.

EXERCISES

1. Let $\theta : G \to H$ be an isomorphism of G onto H; that is, $G \cong H$. Prove that $\theta^{-1} : H \to G$ is a mapping and, in fact, that it is an isomorphism of H onto G.

2. Prove that isomorphism ("$\cong$") is an equivalence relation for groups.

3. (a) If G is an abelian group having n elements, then define a mapping $\theta : G \to G$ by $a\theta = a^k$ for all $a \in G$, where k is some fixed integer. Prove that θ is a homomorphism.

   (b) If $G = \mathscr{Z}_7{}^+$ and $k = 3$, then describe ker θ.

   (c) If $G = \mathscr{Z}_{30}{}^+$ and $k = 6$, then describe ker θ.

   (d) If $G = \mathscr{Z}_{30}{}^+$ and $k = 14$, then describe ker θ.

   (e) If k is a divisor of n, then describe ker θ of part (a).

4. Prove (ii) and (iii) of Theorem 2.35.

5. Let $K = $ ker ψ, where ψ is given by Example 1 of this section. Prove that K is isomorphic to the group T of the example in Section 2.2. Using Exercise 8 of Section 2.4, show that K is isomorphic to $\mathscr{Z}_2{}^+ \oplus \mathscr{Z}_2{}^+$.

6. Show that the group D of the example in Section 2.4 is isomorphic to $\mathscr{Z}_4{}^+$.

7. Show that the homomorphism ψ of Example 1 of this section is onto. That is, A is a homomorphic image of C.

8. Show that $\mathscr{Z}_3{}^+ \oplus \mathscr{Z}_5{}^+ \cong \mathscr{Z}_{15}{}^+$. Show that $\mathscr{Z}_2{}^+ \oplus \mathscr{Z}_2{}^+$ is not isomorphic to $\mathscr{Z}_4{}^+$. Would you be willing to conjecture when $\mathscr{Z}_n{}^+ \oplus \mathscr{Z}_m{}^+ \cong \mathscr{Z}_{nm}{}^+$? Is $\mathscr{Z}_3{}^+ \oplus \mathscr{Z}_6{}^+$ isomorphic to $\mathscr{Z}_{18}{}^+$?

9. Prove that G is abelian if and only if the mapping $\theta : G \to G$ defined by $a\theta = a^{-1}$ for $a \in G$ is a homomorphism.

10. If $\theta : G \to H$ and $\phi : H \to K$ are group homomorphisms, show that $\theta\phi$ also is a homomorphism.

11. Let G be a group and H merely a set on which a binary operation is defined. If there exists a mapping $\theta : G \to H$ of G onto H which preserves the operation, prove that H is a group with respect to the given operation and that θ is a homomorphism of the group G onto the group H. (This fact is usually expressed by saying that a homomorphic image of a group is a group.)

12. If $\theta : G \to H$ is a homomorphism of G into H and $a \in G$, prove that $a^k\theta = (a\theta)^k$ for every integer k. *Hint:* Use the Well-Ordering Principle to prove this for positive integers k.

13. Recall Exercise 5 of Section 2.1. Show that there is a nontrivial homomorphism of the group of symmetries of a regular tetrahedron into the group A of symmetries of the x,y,z-axes. *Hint:* Connect the midpoints of opposite edges of the tetrahedron. Try these lines as axes.

2.6 CYCLIC GROUPS

If a is an element of an arbitrary group G, then since G is closed with respect to the operation (which we will consider to be multiplication), we see that $a^k \in G$ for every positive integer k. Moreover, a^0 is the identity e of G by definition, and a^{-k} is the inverse of a^k. It follows easily that the set $\{a^k \,|\, k \in \mathbf{Z}\}$ is a subgroup of G. We are particularly interested in the case in which this subgroup turns out to be all of G. Accordingly, let us make the following definition.

2.39   **DEFINITION**   If the group G contains an element a such that $G = \{a^k \,|\, k \in \mathbf{Z}\}$, we say that G is a *cyclic group* and that G is *generated by* a or that a is a *generator of G.*

Since $a^i \cdot a^j = a^j \cdot a^i$ for $i, j \in \mathbf{Z}$, we see that a cyclic group is necessarily abelian.

Whether or not a group G is cyclic, if $a \in G$, the subgroup $\{a^k \,|\, k \in \mathbf{Z}\}$ of G is a cyclic group which we naturally call the *subgroup of G generated by a.*

Let us now give some examples of cyclic groups.

Example 1   Recall the additive group of integers $\mathbf{Z}^+$ (of Example 1 in Section 2.1). When the operation is addition, the analogue of a^k, used above, is ka. The integer 1 is a generator of this group since every element is of the form $k \cdot 1$ for some integer k. In this case we have chosen $a = 1$ as a generator. We could equally well have chosen $a = -1$ as our generator since every integer is of the form $k \cdot (-1)$ for some integer k. We could not have chosen $a = 3$ as a generator since there is no integer k such that $k \cdot 3 = 2$. In particular, the choice of a may not be free; we may not be able to choose any a as a generator. Further, the choice of a may not be unique; we may have several possible choices for a which can serve as generators.

Example 2 Recall the additive group $\mathscr{Z}_n{}^+$ of least residues modulo n (of Example 2 in Section 2.1). For this example let $n = 6$. This group is cyclic with 1 as a generator. Namely, $0 = 0 \cdot 1$, $1 = 1 \cdot 1$, $2 = 2 \cdot 1 = 1 \oplus 1$, $3 = 3 \cdot 1 = 1 \oplus 1 \oplus 1$, etc. The residue 2 has the multiples 2, 4, 0, so that 2 is not a generator of $\mathscr{Z}_6{}^+$, even though it does generate a cyclic subgroup of $\mathscr{Z}_6{}^+$. On the other hand, 5 is a generator.

Example 3 Recall the Example of Section 2.4. We considered all symmetries D of the cube C obtained by rotating the cube around the y-axis in Figure 12. There are four such rotations possible. They are generated by a rotation through $90°$ moving the top of the cube away and the bottom toward us.

We are now ready to give another definition as follows:

2.40 **DEFINITION** (i) If a group G has n elements, where n is a positive integer, G is said to have a *finite order* or, more precisely, to have order n. If there exists no such positive integer, G is said to have *infinite order*.

(ii) The *order of an element* a of a group G is the order of the cyclic subgroup of G generated by a.

In the language here introduced, we may say that the additive group of integers $\mathbf{Z}$ has infinite order, the additive group of least residues modulo n has order n, and the subgroup D of C has order 4. All these examples, except the first, are groups of finite order.

Let us look for a moment at the computation of the order of a symmetry group.

Example 4 Consider the group A of symmetries of the x,y,z-axes (Example 3 of Section 2.1). If we specify a starting position, say Figure 17(1), then each element σ of A is uniquely determined by

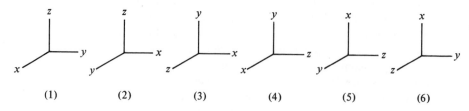

(1)   (2)   (3)   (4)   (5)   (6)

FIGURE 1

the final position of the axes after applying σ to the position of Figure 17(1). In other words, the order of A is equal to the number of allowable positions of the axes. We count these as follows. There are three possible choices for a vertical axis. Once we have chosen a vertical axis, there are two remaining choices for the right-left axis. Once we have chosen the vertical and right-left axes, the position of the axes is determined. Therefore, the order of A is $3 \cdot 2 = 6$. In Figure 17 we have drawn all six positions for the axes. We may write the elements of A as follows. Let σ_i move Figure 17(1) to Figure 17(i) for $i = 1, 2, \ldots, 6$. Then $A = \{\sigma_1, \sigma_2, \ldots, \sigma_6\}$. Each σ_i is describable geometrically, for example, σ_2 is obtained by a rotation around the vertical axis of $90°$ (in either direction). On the other hand, we may view σ_3 as follows: in Figure 17(1) we may imagine an axis running up, out, and to the right from O; we may imagine this axis as making equal angles with the x,y,z-axes; and looking down this axis and rotating $120°$ counterclockwise changes Figure 17(1) into Figure 17(3) and gives σ_3. Note that σ_3 will change Figure 17(2) into Figure 17(6).

Example 5 Recall the group C of symmetries of a cube given in Example 4 of Section 2.1. The order of this group is the same as the number of allowable positions of the cube. (Why?) A position is determined by (1) the front face of the cube and (2) the right-hand side of the cube. There are six faces which may appear on the front, and once the front face is chosen, there are four adjacent faces which may appear on the right-hand side. In particular, C has order $6 \cdot 4 = 24$.

Until now we have been proving theorems to aid our computations, and using examples to illustrate the ideas involved. Sometimes the ideas are difficult, but the examples help to clarify the situation, but we have no assurance that the examples are typical. There may be other examples which appear very strange. The next theorem is very important in that it is typical of the kind of theorem algebraists would like to have for all algebraic systems. We will *classify* all cyclic groups by showing that any cyclic group G is isomorphic to an example we know about. In other words, using an isomorphism, we may replace G by a concrete example in order to discover algebraic properties of G. Such *classification* theorems allow us to "think of G concretely."

2.41 THEOREM (*i*) *Every cyclic group of infinite order is isomorphic to the additive group of integers* $\mathbf{Z}^+$.

(*ii*) *Every cyclic group of order* m *is isomorphic to the additive group of least residues modulo* m, $\mathscr{Z}_m{}^+$.

We shall use the Well-Ordering Principle of the integers, the Universal Mapping Property, the Division Algorithm, and Theorem 2.37 in the proof of this theorem. The proof illustrates the use of some of our new mathematical ideas.

Let us assume that G is a cyclic group with a generator a, that is, $G = \{a^k \,|\, k \in \mathbf{Z}\}$. Define the following mapping $\theta : \mathbf{Z}^+ \to G$ given by

$$k\theta = a^k, \qquad k \in \mathbf{Z}^+.$$

Clearly θ is an onto mapping. If $i, j \in \mathbf{Z}^+$, then

$$(i + j)\theta = a^{i+j} = a^i \cdot a^j = (i\theta)(j\theta),$$

and hence θ is a homomorphism of the additive group of integers onto the group G. Let $K = \ker \theta$ be the kernel of θ, that is, the set of all integers i such that $i\theta = e$, where e is the identity element of G. The following lemma is helpful in determining the integers which lie in K.

2.42   LEMMA   *Let $\theta : \mathbf{Z}^+ \to H$ be a homomorphism of the group $\mathbf{Z}^+$ into a group H. Then there is a unique nonnegative integer n such that $\ker \theta$ is the set $\{q \cdot n \,|\, q \in \mathbf{Z}\}$ of all multiples of n.*

Proof: We recall that $\ker \theta = K$ is a subgroup of $\mathbf{Z}^+$. Thus $0 \in K$ is the identity of K. If $K = \{0\}$, then $K = \{q \cdot 0 \,|\, q \in \mathbf{Z}\}$. We may therefore assume that K contains numbers other than just zero. If $m \in K$, $m \neq 0$, then both m and $-m$ are in K since K is a subgroup of $\mathbf{Z}^+$. One of the numbers m or $-m$ is positive, and therefore K contains positive integers. Let S be the set of all positive integers contained in K so that S is a nonempty set. By the Well-Ordering Principle, S contains a smallest positive integer n. We shall argue that $K = \{q \cdot n \,|\, q \in \mathbf{Z}\}$ for this integer n. Since K is a subgroup, the set $\{q \cdot n \,|\, q \in \mathbf{Z}\}$ is contained in K. We must show that K is contained in this set.

Let $b \in K$. Using the Division Algorithm, we have unique integers q and r so that

$$b = qn + r, \qquad 0 \leq r < n.$$

Since both b and qn lie in K, and since K is a subgroup,

$$b - qn = b + (-qn) = r$$

lies in K. If r is positive, then r is a positive integer in K and, therefore, lies in S. But $r < n$ and n is the smallest integer in S so that

r does not lie in S and thus r is not positive. Since $r \geq 0$, we must have $r = 0$ or, equivalently,

$$b = qn.$$

We conclude that $K = \{qn \mid q \in \mathbf{Z}\}$.

We must now prove that n is unique. Suppose that $K = \{qm \mid q \in \mathbf{Z}\}$ is the set of multiples of m for a nonnegative integer m. If $n = 0$, then $K = \{0\}$ so that $m = 0$ also. Suppose $n \neq 0$. The preceding argument applied to m in place of n shows that $m \neq 0$. Now $m \in K = \{qn \mid q \notin \mathbf{Z}\}$. Thus $m = xn$ for some integer x. Since both m and n are positive, x is positive; thus $m \geq n$. Reverse the roles of m and n and repeat this argument to obtain $n \geq m$. We conclude that $m = n$, in other words, n is unique. We have proven the lemma.

We apply this lemma in our proof of the theorem by observing that $K = \ker \theta = \{qn \mid q \in \mathbf{Z}\}$ is the set of multiples of a unique nonnegative integer n. If $n = 0$, then $K = \{0\}$. In this case, Theorem 2.37 tells us that θ is an isomorphism. Since θ is onto, we may write:

2.43   If $n = 0$, then $\mathbf{Z}^+ \cong G$ and G has infinite order.

Next suppose that n is a positive integer. Recall the homomorphism $\phi : \mathbf{Z}^+ \to \mathscr{L}_n{}^+$ of Example 2 of Section 2.5. If $b \in \mathbf{Z}^+$, then the Division Algorithm tells us there are unique integers q and r so that

$$b = qn + r, \quad 0 \leq r < n.$$

We set $b\phi = r$. In the argument preceding Theorem 2.37 we showed that $\ker \phi = \{qn \mid q \in \mathbf{Z}\}$, in particular, $\ker \theta = \ker \phi$.

We may draw a diagram of homomorphisms so that $\ker \theta \supseteq \ker \phi$:

The Universal Mapping Property applies to tell us there is a unique homomorphism $\psi : \mathscr{L}_n{}^+ \to G$ so that $\theta = \phi\psi$. Further, $\ker \psi = (\ker \theta)\phi = (\ker \phi)\phi = \{0\}$.

Since $\ker \psi = \{0\}$, Theorem 2.37 tells us that ψ is an isomorphism. Since θ maps $\mathbf{Z}^+$ onto G and since $\theta = \phi\psi$, the mapping ψ takes $\mathscr{L}_n{}^+$ onto G. The mapping ψ is a one-one correspondence between $\mathscr{L}_n{}^+$ and G so that G has order n. We may now write

2.44   If $n > 0$, then $\mathscr{L}_n{}^+ \cong G$ and G has order n.

The statements 2.43 and 2.44 combine to prove (i) and (ii) of Theorem 2.41. For example, to prove (i), 2.43 tells us that $n \neq 0$; then 2.44 completes the proof.

Notice that this theorem proves that two cyclic groups are isomorphic if and only if they have the same order.

Let us continue working within the context of the preceding proof. For the integers $0, 1, \ldots, n - 1$, we obtain the distinct residues $0 = 0\phi; 1 = 1\phi, \ldots, n - 1 = (n - 1)\phi$. For these residues we must have $i\psi = a^i$. Since ψ is a one-one correspondence, $e = a^0, a^1, \ldots, a^{n-1}$ are all the n distinct elements of G. Suppose that b is any other integer. Then $b = qn + r$, $0 \leq r < n$, for integers q and r by the Division Algorithm. Therefore

$$a^b = b\theta = (b\phi)\psi = r\psi = a^r.$$

From this we may conclude that $a^i = a^j$ if and only if i and j leave the same remainder upon division by n; that is,

$$i = q_1 n + r,$$
$$j = q_2 n + r, \qquad 0 \leq r < n.$$

This happens if and only if $i - j$ is divisible by n, so that we have almost proven the following theorem.

2.45   THEOREM *An element a of a group H has order t if and only if t is the smallest positive integer such that $a^t = e$, where e is the identity of H. If no such positive integer exists, then a has infinite order. Further, if t exists, then for integers $i, j \in \mathbf{Z}$, $a^i = a^j$ if and only if $i - j$ is divisible by t.*

Proof: Let G be the group generated by a. In the terminology of Theorem 2.41, if $n = 0$, then $\mathbf{Z}^+ \cong G$ so that a has infinite order. Suppose $n > 0$. Now $a^i = a^j$ if and only if $i - j$ is divisible by n. In particular, $a^t = e = a^0$ if and only if n divides t. Since $a^n = n\theta = n(\phi\psi) = 0\psi = a^0 = e$, and t is minimal such that $a^t = e$, we conclude that n divides t with 0 remainder and $t \leq n$, so that $n = t$. That is, a has order t. This completes the proof of the theorem.

This theorem characterizes the order of an element a in a group G. In fact, in order to compute the order of a given element, it is usually simpler to apply this theorem than to use the definition. This theorem tells us that if G is cyclic of order n and a is a generator, then

2.46 $a^0 = e, a^1, a^2, \ldots, a^{n-1}$

are all distinct and are all the elements of G. We illustrate this theorem in the following example.

Example 6 Consider the element 5 of $\mathscr{Z}_{15}{}^+$. The multiples of 5 are $1 \cdot 5 = 5$, $2 \cdot 5 = 5 \oplus 5 = 10$, $3 \cdot 5 = 5 \oplus 5 \oplus 5 = 0$, so that 3 is the smallest positive integer t such that $t \cdot 5 = 0$. By the theorem, in $\mathscr{Z}_{15}{}^+$, 5 has order 3.

The isomorphism ψ of Theorem 2.41 was constructed in such a way that $1\psi = a$ for a fixed generator a of G and $1 \in \mathscr{Z}_n{}^+$ (or $1 \in \mathbf{Z}^+$ if G has infinite order). Since a cyclic group may have more than one generator, there may be different isomorphisms between two isomorphic groups. In particular, a **group** may be isomorphic to itself in more than one way.

Example 7 As an example, let us show that $\mathscr{Z}_6{}^+$ has two isomorphisms with itself. The identity map is clearly one such isomorphism. We now make use of the fact that 5, as well as 1, is a generator of $\mathscr{Z}_6{}^+$. Pictured are the mappings defined in the proof of Theorem 2.41 applied to this special case where $G = \mathscr{Z}_6{}^+$, $n = 6$, $a = 5$:

That is, ϕ is the mapping of Example 2 of Section 2.5, just as in the proof of Theorem 2.41. Also, in additive notation we have θ defined for integers k by the formula

$$k\theta = k \cdot 5.$$

Therefore, the isomorphism ψ exists by the Universal Mapping Property. If $k = 6q + r$, $0 \le r < 6$, then

$$r \cdot 5 = r\psi = (k\phi)\psi = k \cdot 5.$$

In other words, the mapping

$$r\psi = r \cdot 5$$

is an isomorphism of $\mathscr{Z}_6{}^+$ onto $\mathscr{Z}_6{}^+$. Since $1\psi = 1 \cdot 5 = 5$, ψ is certainly not the identity isomorphism. In particular, two groups which are isomorphic may be isomorphic in more than one way.

Our final theorem about cyclic groups is the following.

2.47 THEOREM *Every subgroup and every homomorphic image of a cyclic group G is itself a cyclic group.*

We shall leave the proof for homomorphic images to the exercises.

Proof: Suppose that G is generated by a, and let H be a subgroup of G. Let m be the smallest positive integer such that $a^m \in H$. We shall show that H is a cyclic group generated by a^m. Since $H \subseteq G$, any element of H is of the form a^k for some integer k. By the Division Algorithm, we may write $k = qm + r$, where $0 \leq r < m$. Hence,

$$a^k = a^{qm+r} = (a^m)^q \cdot a^r,$$

and from this it follows that

$$a^r = (a^m)^{-q} \cdot a^k.$$

Since $a^m \in H$ and $a^k \in H$, this equation implies also that $a^r \in H$. In view of the choice of m as the smallest positive integer such that $a^m \in H$, and since $r < m$, we must have $r = 0$. We conclude that $k = qm$, and hence that every element a^k of H is of the form $(a^m)^q$ for some integer q. This shows that H is a cyclic group generated by a^m.

As an almost immediate consequence of the *proof* of the preceding theorem, we obtain the following result.

2.48   COROLLARY   *If a cyclic group G has finite order n and is generated by a, every subgroup H of G is generated by an element of the form a^m, where m is a divisor of n.*

Since Theorem 2.45 shows that $a^n = e$, and $e \in H$, we apply the above argument with $k = n$ and obtain $n = qm$. Hence, m is a divisor of n.

Of course, by a simple change in notation, these results apply equally well to the case in which the operation is addition. As an illustration of the preceding corollary, let us find all subgroups of the additive group $\mathscr{Z}_{14}$. This is a cyclic group of order 14 generated by the element 1; hence the only subgroups are the cyclic subgroups generated by 1, 2, 7, and 0. The subgroup generated by 0 consists only of the identity 0. The subgroup generated by 2 has order 7 and the subgroup generated by 7 has order 2.

EXERCISES

1. Prove that a homomorphic image of a cyclic group is a cyclic group.

2. Find the order of each element of $\mathscr{Z}_{12}{}^+$.

3. Find the order of each element of $\mathscr{Z}_n{}^+$ where $n > 1$ is an integer.

4. Find the order of each element of the group A of symmetries of the x,y,z-axes.

5. Find the order of each element of the group C of symmetries of a cube.

6. For which integers $n > 1$ does $\mathscr{Z}_n{}^+$ have no proper subgroup?

7. Prove: If G is cyclic of order n, and k divides n, then G contains exactly one subgroup of order k.

8. Let G be a cyclic group of order n with a generator a. Prove that a^k generates G if and only if k and n are relatively prime.

9. Prove that if a and b are elements of a group, then ab and ba have the same order.

10. A dodecahedron is assembled from 12 identical pentagons. Does the group of symmetries of the regular dodecahedron contain a subgroup isomorphic to (a) the group of symmetries A of the x,y,z-axes; or (b) the group of symmetries C of the cube? *Hint to (b):* Use Exercise 5. Compute the order of the group of the dodecahedron. Find all elements as rotations about appropriate axes. Compute the possible orders for elements.

11. Suppose an abelian group G contains an element a of order k and an element b of order l, where k and l are relatively prime. Show that ab has order kl. *Hint:* If $(ab)^t = e$, raise both sides to the k power and conclude that t must be divisible by l. Similarly show that t must be divisible by k. Finally show that the smallest integer divisible by both k and l is kl.

12. Continuing Example 7, find all isomorphisms $\phi : \mathscr{Z}_6{}^+ \to \mathscr{Z}_6{}^+$. Show that composition of mappings is a binary operation upon the set of isomorphisms and that with this operation, the set is a group.

13. Repeat Exercise 12 for the following groups: (a) $\mathscr{Z}_7{}^+$, (b) $\mathscr{Z}_8{}^+$, (c) $\mathscr{Z}_{15}{}^+$.

2.7 COSETS AND LAGRANGE'S THEOREM

Let G be an arbitrary group, and H a subgroup of G. If $a \in G$, we shall designate by aH the *set* of all elements of G of the form ah, where $h \in H$. That is, $aH = \{ah \mid h \in H\}$.

2.49   **DEFINITION**   If H is a subgroup of the group G and $a \in G$, we call aH a *coset* of H (in G).†

Since $eH = H$, we see that H is itself a coset. Moreover, since $e \in H$, it is clear that $a \in aH$.

The following lemma will be very useful in studying cosets.

2.50   **LEMMA**   *If H is a subgroup of the group G and $a, b \in G$, then each of the following is true:*

    (*i*) *If $aH \cap bH \neq \varnothing$, then $aH = bH$.*

    (*ii*) *$aH = bH$ if and only if $a \in bH$.*

[handwritten: $h = a^{-1}bh_1$ $H = a^{-1}b$ $hh_1^{-1} = a^{-1}b$ $ah = bh_1$]

Proof of (i):   Suppose that aH and bH have at least one element in common. Thus, there exist $h_1, h_2 \in H$ such that $ah_1 = bh_2$. Then $a = bh_2h_1^{-1}$ and any element ah of aH can be expressed in the form $bh_2h_1^{-1}h$. Since $h_2h_1^{-1}h \in H$, it follows that $ah \in bH$. We have therefore shown that $aH \subseteq bH$. In a similar way we can show that $bH \subseteq aH$, and therefore we conclude that $aH = bH$. One way of stating the property we have just proved is to say that two cosets either coincide or have no element in common.

Proof of (ii):   Since $a \in aH$, it is obvious that if $aH = bH$, then $a \in bH$. Conversely, suppose that $a \in bH$. Then $a \in aH \cap bH$, and part (i) of the lemma implies at once that $aH = bH$.

In view of this lemma, we see that every element of G is in exactly one coset of H in G, that is, that the distinct cosets of H in G form a partition of G.

Example 1   Let us return to the group A of Example 3 of Section 2.1. The symmetries of the x,y,z-axes are described and named in Example 4 of Section 2.6. Consider the set of all symmetries H which fix the vertical axis. From Figure 17 we see that $H = \{\sigma_1, \sigma_2\}$, where H is a subgroup of G. Using Figure 17, we may compute the cosets of H in A.

$$\sigma_1 H = \{\sigma_1, \sigma_2\}, \qquad \sigma_2 H = \{\sigma_2, \sigma_1\},$$
$$\sigma_3 H = \{\sigma_3, \sigma_4\}, \qquad \sigma_4 H = \{\sigma_4, \sigma_3\},$$
$$\sigma_5 H = \{\sigma_5, \sigma_6\}, \qquad \sigma_6 H = \{\sigma_6, \sigma_5\}.$$

Every element appears in exactly one distinct coset and the size (number of elements) of each coset is equal to the order of H. There

---

† More precisely, we have here defined a *left* coset, and one can similarly define a right coset Ha. However, in accordance with the definition just given we shall in this section use the word *coset* to mean *left coset*.

are $3 = \frac{6}{2}$ distinct cosets where 6 is the order of A and 2 is the order of H. As an illustration of 2.50(ii), we may observe that $\sigma_5 H = \sigma_6 H$ since $\sigma_5 \in \sigma_6 H$, but that $\sigma_4 H \neq \sigma_5 H$ since $\sigma_4 \notin \sigma_5 H$.

So far we have used multiplication as the operation but, as usual, it is easy to make the necessary modifications if the operation is addition.

Example 2   Let G be the group $\mathscr{Z}_{12}{}^+$ of residues modulo 12 (Example 2 of Section 2.1). Let H be the cyclic subgroup generated by 3; $H = \{0, 3, 6, 9\}$. It may be verified that the *different* cosets of H in G are the following:

$$0 \oplus H = \{0, 3, 6, 9\},$$
$$1 \oplus H = \{1, 4, 7, 10\},$$
$$2 \oplus H = \{2, 5, 8, 11\}.$$

Again every element appears in exactly one coset, and the size of each coset is equal to the order of H. There are $3 = \frac{12}{4}$ different cosets where 12 is the order of G and 4 is the order of H.

Let us make the following definition.

2.51   DEFINITION   If the group G has finite order and H is a subgroup of G, the number of distinct cosets of H in G is called the *index* of H in G.

Although we have been using *coset* to mean *left coset*, we should perhaps point out that Exercise 9 below shows that there are the same number of right cosets as left cosets of H in G. Accordingly, in the definition just given it does not matter whether we think of left cosets or right cosets.

We shall next prove the following theorem, called Lagrange's Theorem, which is of fundamental importance in the study of groups of finite order.

2.52   THEOREM (*Lagrange's Theorem*)   *Suppose that the group G has order n. If H is a subgroup of G of order m and of index k, then $n = km$. In particular, both the order and the index of H are divisors of the order of G.*

Proof: We first observe that every coset of H has exactly m elements. For if $a \in G$ and $h_1, h_2 \in H$, then $ah_1 = ah_2$ if and only if $h_1 = h_2$. Hence an arbitrary coset aH has the same number of elements as H, namely, m.

We have already shown that the distinct cosets of H in G form a partition of G. Since there are k distinct cosets and each of them contains m elements, G must contain km elements. This proves that $n = km$, and completes the proof.

There are some interesting consequences of the fact that the order of a subgroup of a finite group is a divisor of the order of the group. First of all, since the order of an element of a group is the order of the cyclic subgroup generated by that element, we have at once the following corollary.

2.53 COROLLARY *The order of an element of a group of finite order is a divisor of the order of the group.*

If the order of a group is a prime p, then every element of the group, other than the identity, must have order p. This yields the next result as follows.

2.54 COROLLARY *A group of order p, where p is a prime, is a cyclic group. Moreover, every element except the identity is a generator of the group.*

If the group G has order n, and the element a of G has order m, then, by Corollary 2.53, we have $n = mk$ for some integer k. By Theorem 2.45, we know that $a^m = e$, and hence $a^n = (a^m)^k = e^k = e$. We have established the following corollary.

2.55 COROLLARY *If a is an element of a group of order n, then $a^n = e$.*

EXERCISES

1. Recall Example 4 of Section 2.6. Let K be the cyclic subgroup generated by σ_3. Exhibit all the distinct cosets of K in A, the group of symmetries of the x,y,z-axes.

2. Recall Example 2 of Section 2.5. Let K be the kernel of ϕ. Show that for $0 \le j < n$, $j + K$ is precisely the set of elements a of $\mathbf{Z}^+$ for which $a\phi = j$.

3. Recall Example 1 of Section 2.5. Let K be the kernel of ψ. Exhibit all the distinct cosets of K in C, the group of symmetries of a cube. Show that for $\sigma \in C$, σK is precisely the set of all elements τ of C such that $\tau\psi = \sigma\psi$, that is, $\tau\psi = \sigma'$.

4. Exhibit all cosets of the subgroup D in C of the Example of Section 2.4.

5. Exhibit all cosets of the subgroup $\{0, 4, 8, 12, 16\}$ of the additive group $\mathcal{Z}_{20}{}^+$.

6. Prove that a group of order $n > 1$ has a proper subgroup if and only if n is not a prime.

7. In Exercise 2 show that $j + K = K + j$ for all $0 \le j < n$. In Exercise 3 show that $\sigma K = K\sigma$ for all $\sigma \in C$. In Exercise 4 show that for some $\sigma \in C$, $\sigma D \ne D\sigma$. Sometimes right and left cosets are the same; sometimes they are not.

8. Let H be a subgroup of a group G. If $a, b \in G$, let $a \sim b$ mean that $b^{-1}a \in H$. Show that "$\sim$" is an equivalence relation defined upon G. If $[a]$ is the equivalence set which contains a, show that $[a] = aH$ and therefore the cosets of H in G are the equivalence sets relative to this equivalence relation.

9. Let H be a subgroup of a group G and define a mapping β of the set of left cosets of H into the set of right cosets of H as follows: $(aH)\beta = Ha^{-1}$. Prove that β is a well-defined mapping (i.e., β is, in fact, a mapping) and that it is a one-one mapping of the set of all left cosets of H onto the set of all right cosets of H in G.

10. A regular tetrahedron is assembled from four identical equilateral triangles. Let T be the group of symmetries of a tetrahedron, so that T has order 12. The converse to Lagrange's Theorem would say that for every divisor of the group order there is a subgroup of that order. Show this is false; T has no subgroup of order 6.

11. Assume that H and K are subgroups of a finite group G. Set $HK = \{ab \mid a \in H, b \in K\}$. Show that if H contains n elements, K contains m elements, and $H \cap K$ contains l elements, then HK contains nm/l elements.

2.8   PERMUTATION GROUPS AND THE SYMMETRIC GROUP

We shall now introduce a new class of groups called permutation groups. Since such groups can be viewed as geometric symmetry groups, we will start with this point of view.

Fix a positive integer n (for example, $n = 4$). Imagine n points arranged evenly in a row. To help visualize this idea, we can arrange, say, four small objects into a row upon a sheet of paper, realizing of course that we think of each object as being only a point. Imagine that our row runs from left to right, and number the positions of the points $1, 2, \ldots, n$ (where $n = 4$ in our example) running from left to right. We wish to investigate the geometric symmetry group of the n points. A symmetry, recall, is an allowable change of position of a geometric figure. In this case, any change which puts the n points into the positions $1, 2, \ldots, n$ would be a symmetry, that is, any rearrangement of the n points would be a symmetry. For example, for $n = 4$ we have numbered the points and positions below. The arrow indicates a change σ from configuration (a) to configuration (b).

Points   ① ② ③ ④   ② ③ ④ ①

$\xrightarrow[\sigma]{\text{Change}}$

Positions   1   2   3   4   1   2   3   4

(a)   (b)

2.56

③ ① ② ④   ① ② ④ ③

$\xrightarrow[\sigma]{}$

1   2   3   4   1   2   3   4

(c)   (d)

Looking at the change from (a) to (b), we see that σ may be described as moving the point in position 1 to position 4, the point in position 2 to position 1, and so on. Thus, the same symmetry would also cause the change from (c) to (d) above. We may describe σ crudely as "shift one position to the left." The left-hand point cannot "shift to the left" so it just moves to the other end of the row. Recalling our discussion in Examples 3 and 4 of Section 2.1, we conclude that the collection of all symmetries of n points in a row is a symmetry group which we call $\mathscr{S}_n$. The group $\mathscr{S}_n$ is a particular instance of a general class of groups called *permutation groups*, and we shall now define the more general concept.

Let A be a set, for example, the set of n points in a row which we have just discussed. Associated with the set A is the family S of all *one-one* mappings of A *onto* itself. If we have two mappings $\alpha, \beta \in S$ then the *product* (or *composite*) $\alpha\beta$ is defined in Section 1.3 by the equation

2.57 $a(\alpha\beta) = (a\alpha)\beta$ for all $a \in A$ and all $\alpha, \beta \in S$.

Since α and β both map the set A onto itself, $\alpha\beta$ does also; in addition, since both mappings are one-one, the product is too. By Equation 1.9 and Theorem 1.11 of Section 1.3, the set S with the operation given by 2.57 is a group.

2.58 **DEFINITION** Let S be the set of all one-one mappings of a set A onto itself. A mapping in S is called a *permutation* of the set A. We call S the *symmetric group* on the set A.

From our observations above, we have the following theorem.

2.59 **THEOREM** *The symmetric group S on a set A is a group with multiplication given by 2.57.*

In the particular case where A is the set $\{1, 2, 3, \ldots, n\}$ of all positive integers from 1 to n, we denote the symmetric group on A by S_n. It is the group S_n which will be our major concern in this section. To give some idea as to the structure of this group, let us indicate why it is isomorphic to the geometric symmetry group $\mathscr{S}_n$ of n points in a line. Our sketch will not depend upon the fact that S is a group; and since the image of a group under a homomorphism is a group (Exercise 11 of Section 2.5), we may also use the fact that $\mathscr{S}_n$ is a group as an alternate way to conclude that S_n is a group.

We may view a symmetry $\rho \in \mathscr{S}_n$ of the n points as changing the positions of the n points. Since we may think of each position as being named by one of the numbers $1, 2, 3, \ldots, n$, ρ causes a one-one mapping ρ' of the set $\{1, 2, 3, \ldots, n\}$ onto itself. In our particular example with σ of 2.56, σ' is the mapping given by

$$1\sigma' = 4, \quad 2\sigma' = 1, \quad 3\sigma' = 2, \quad 4\sigma' = 3.$$

In other words, each symmetry ρ of $\mathscr{S}_n$ gives rise to a permutation ρ' in S_n, and we shall denote the mapping sending ρ to ρ' by $\phi : \mathscr{S}_n \to S_n$.

We may reverse this whole process, so that if ρ' is any permutation of S_n, then ρ' gives rise to a symmetry ρ in $\mathscr{S}_n$. To illustrate the reverse process, consider the permutation τ' of S_4 given by

$$1\tau' = 2, \quad 2\tau' = 3, \quad 3\tau' = 1, \quad 4\tau' = 4.$$

The symmetry τ described by τ' will carry (a) of 2.56 into (c) of 2.56. It is evident that $\tau\phi = \tau'$ since this reverse correspondence is just the inverse mapping of ϕ. In other words, ϕ is a one-one mapping of $\mathscr{S}_n$ onto S_n.

We shall now indicate why ϕ is a homomorphism by showing that $(\tau\sigma)\phi = (\tau\sigma)' = \tau'\sigma' = (\tau\phi)(\sigma\phi)$. Since τ changes 2.56 from (a) to (c) and σ changes 2.56 from (c) to (d), $\tau\sigma$ must change 2.56 from (a) to (d). Therefore, $(\tau\sigma)'$ is given by

$$1(\tau\sigma)' = 1, \quad 2(\tau\sigma)' = 2, \quad 3(\tau\sigma)' = 4, \quad 4(\tau\sigma)' = 3.$$

From 2.57 we may compute the product $\tau'\sigma'$ so that

$$(1\tau')\sigma' = 2\sigma' = 1, \qquad (2\tau')\sigma' = 3\sigma' = 2,$$
$$(3\tau')\sigma' = 1\sigma' = 4, \qquad (4\tau')\sigma' = 4\sigma' = 3.$$

We conclude that $(\tau\sigma)' = \tau'\sigma'$, so that $(\tau\sigma)\phi = (\tau\phi)(\sigma\phi)$. The computations we have given here work in general, that is, ϕ satisfies the condition:

2.60 $(\rho\nu)\phi = (\rho\phi)(\nu\phi)$ for any $\rho, \nu \in \mathscr{S}_n$.

Evidently, ϕ is an isomorphism of $\mathscr{S}_n$ onto S_n.

If A is the set $\{a_1, a_2, a_3, \ldots, a_n\}$ of n elements, then in our preceding discussion we could label the positions of the points by using a_1 instead of 1, a_2 instead of 2, and so on. In this case, the mapping ϕ would be an isomorphism of $\mathcal{S}_n$ onto the symmetric group S on the set A. Since both S and S_n are isomorphic to $\mathcal{S}_n$, we conclude that S is isomorphic to S_n; that is, *any two symmetric groups upon sets of the same size are isomorphic.*

We may use the group $\mathcal{S}_n$ to compute the order of symmetric groups. The order of any geometric symmetry group is equal to the total number of allowable configurations. In the case of the cube, this was 24, and for the axes it was 6. For n points in a row we may compute the number of allowable configurations as follows: we put any one of the n points in position 1; after we have done this, we choose any one of the $n - 1$ remaining points to place in position 2, that is, the first two choices may be made in $n(n - 1)$ ways since any one of the first n points may be followed by any one of the $n - 1$ remaining points; continuing in this way, we find that there are $n \cdot (n - 1) \cdot (n - 2) \cdots 3 \cdot 2 \cdot 1$ ways to arrange the points in a row. This latter number is usually denoted by $n!$ and called "*n factorial.*" We have shown that $\mathcal{S}_n$ (hence also S_n since ϕ is a one-one mapping) has order $n!$. Our discussion has outlined a method for proving the following theorem.

2.61 **THEOREM** *If A is a set of n elements for a positive integer n, then the symmetric group S of all permutations of A is a group of order $n!$ (n factorial). The symmetric group S_n upon the set $\{1, 2, 3, \ldots, n\}$ is isomorphic to S.*

Since $\mathcal{S}_n$ is isomorphic to S, we may view S as the geometric symmetry group of n points in a line as follows. Suppose that $\theta : S \to \mathcal{S}_n$ is an isomorphism of S onto $\mathcal{S}_n$. If H is a subgroup of S, then $\mathcal{H} = \{a\theta \mid a \in H\} = H\theta$ is a subgroup of $\mathcal{S}_n$ isomorphic to H. In particular, via θ, any subgroup of S may be viewed as a subgroup of a geometric symmetry group. Suppose that G is a group of order n and $\psi : G \to S$ is an isomorphism of G into the group of permutations of the set A. Applying $\psi\theta$ to G, we obtain an isomorphism of G into $\mathcal{S}_n$; in other words, via $\psi\theta$, we may view G as a geometric symmetry group. That is, if the mapping ψ always exists, then our examples of geometric symmetry groups are not really so special; every group of order n could then be viewed, via some isomorphism, as a subgroup of the geometric symmetry group of n points in a row. This remarkable fact is known as Cayley's Theorem since it was first proven by the famous English mathematician, Arthur Cayley (1821–1895).

2.62 **THEOREM** (*Cayley's Theorem*) *Every finite group G is isomorphic to a group of permutations.*

Proof: In order to prove this result, we need first to determine the *set*, some of whose permutations we shall associate with the elements of the given group G. We make what is perhaps the most obvious choice, namely, the set of elements of G itself. Moreover, we shall let G denote both the group and the set of its elements as suits our convenience. Actually, the desired permutations will be obtained by multiplication by the elements of the group. More precisely, let us first observe that if $a \in G$, then

$$\{xa \mid x \in G\} = G.$$

Therefore the mapping $\theta_a : G \to G$ defined by $x\theta_a = xa$, $x \in G$, is a mapping of G onto G and it is clearly also a one-one mapping; hence θ_a is a permutation of the set G associated with the element a of the group G.

Now let us set

$$H = \{\theta_a \mid a \in G\},$$

that is, H is the set of all permutations of the type introduced above. Since for $x \in G$, we have

$$x(\theta_a \cdot \theta_b) = (x\theta_a)\theta_b = (x\theta_a)b = xab = x\theta_{ab},$$

we see that

2.63
$$\theta_a \cdot \theta_b = \theta_{ab},$$

and H is therefore closed with respect to multiplication. Moreover, H has identity θ_e, where e is the identity of G; and 2.63 shows that $\theta_{a^{-1}}$ is the inverse of θ_a. Hence H is a subgroup of the group of all permutations of the set G.

We now assert that the mapping $\alpha : G \to H$ defined by $a\alpha = \theta_a$, $a \in G$, is an isomorphism of G onto H. It is clearly an onto mapping. Moreover, it is a homomorphism since by 2.63,

$$(ab)\alpha = \theta_{ab} = \theta_a \cdot \theta_b = (a\alpha)(b\alpha).$$

There remains only to prove that α is a one-one mapping. However, if $\theta_a = \theta_b$, it follows that $xa = xb$ for every element x of G, and clearly we must have $a = b$. Thus α is indeed a one-one mapping, and this completes the proof.

In view of this theorem, in order to prove a theorem for arbitrary groups, it is sufficient to prove it for groups of permutations. Although the subject of group theory is sometimes approached through the study of permutation groups, we shall not limit ourselves to this point of view.

For the rest of this section we shall study permutation groups. In particular, we shall study the symmetric group S_n upon the set $A = \{1, 2, \ldots, n\}$. From here until the end of this section the word "*permutation*" will mean an element of S_n for some positive integer n, and we shall sometimes find it convenient to refer to the elements of A as "symbols."

We shall first study permutations of the particular type described in the following definition.

2.64   **DEFINITION**   A permutation α of S_n is said to be a *cycle of length* k if there exist distinct elements $a_1, a_2, \ldots, a_k$ $(k \geq 1)$ of A such that

$$a_1\alpha = a_2, \quad a_2\alpha = a_3, \quad \ldots, \quad a_{k-1}\alpha = a_k, \quad a_k\alpha = a_1,$$

and $i\alpha = i$ for each element i of A other than $a_1, a_2, \ldots, a_k$. The cycle α may be designated by $(a_1a_2 \cdots a_k)$.

It will be observed that a cycle of length 1 is necessarily the identity permutation. It sometimes simplifies statements to consider the identity permutation as a cycle, but we shall usually be interested in cycles of length greater than 1.

As an example of a cycle, suppose that β is the element of S_6 defined by

$$1\beta = 3, \quad 3\beta = 2, \quad 2\beta = 5, \quad 5\beta = 6, \quad 6\beta = 1, \quad 4\beta = 4.$$

Then β is a cycle of length 5, and we may write $\beta = (13256)$. In a cycle such as (13256) the symbols appearing are permuted cyclically; that is, each symbol written down maps into the next one, except that the last maps into the first. A symbol which is not written down, such as 4 in this example, is assumed to map into itself. There are other ways of writing the cycle defined above. For example, $\beta = (32561) = (25613)$, and so on.

Recall the element σ which changes 2.56 from (a) to (b). The permutation $\sigma' \in S_4$ may be written $\sigma' = (1432)$. The element τ which changes (a) to (c) in 2.56 has an associated permutation $\tau' = (123)$. The product $\tau'\sigma' = (34)$.

It is not true that all permutations are cycles, for example, if $\nu' \in S_4$ and $\nu' = (12)$, then $\nu'(\tau'\sigma') = (12)(34)$ cannot be written as a single cycle.

There is another notation for permutations which is in common usage. Imagine $\delta \in S_n$ is a permutation. We may associate with δ the symbol

$$\delta = \begin{pmatrix} 1 & 2 & \cdots & n \\ 1\delta & 2\delta & \cdots & n\delta \end{pmatrix}.$$

Below each i we place $i\delta$, the number to which i is sent by δ. For the permutations $\beta \in S_6$, τ', $\sigma' \in S_4$ we have symbols as follows:

$$\beta = \begin{pmatrix} 1 & 2 & 3 & 4 & 5 & 6 \\ 3 & 5 & 2 & 4 & 6 & 1 \end{pmatrix},$$

$$\tau' = \begin{pmatrix} 1 & 2 & 3 & 4 \\ 2 & 3 & 1 & 4 \end{pmatrix},$$

$$\sigma' = \begin{pmatrix} 1 & 2 & 3 & 4 \\ 4 & 1 & 2 & 3 \end{pmatrix},$$

$$\tau'\sigma' = \begin{pmatrix} 1 & 2 & 3 & 4 \\ 1 & 2 & 4 & 3 \end{pmatrix}.$$

In this form we may multiply $\tau'\sigma'$ by reading the diagrams as illustrated.

$$\tau'\sigma' = \begin{pmatrix} 1 & 2 & 3 & 4 \\ 2 & 3 & 1 & 4 \end{pmatrix} \begin{pmatrix} 1 & 2 & 3 & 4 \\ 4 & 1 & 2 & 3 \end{pmatrix}.$$

The image of 2 under $\tau'\sigma'$ is found by reading down under the 2 to 3 in τ', then over to the 3 in σ' and down to the 2. So $2(\tau'\sigma') = 2$.

As further illustrations of all the various notations used, let us consider elements of S_6 and verify that

$$(1345)(146) = \begin{pmatrix} 1 & 2 & 3 & 4 & 5 & 6 \\ 3 & 2 & 6 & 5 & 4 & 1 \end{pmatrix}.$$

In the first cycle factor, 1 maps into 3, and in the second cycle factor, 3 is unchanged; hence in the product, 1 maps into 3. The symbol 2 does not appear in either factor, and hence 2 maps into 2. In the left cycle factor, 3 maps into 4, and then in the second cycle factor, 4 maps into 6; hence in the product, 3 maps into 6. Similarly, the other verifications are easily made.

Now let α be the cycle $(a_1 a_2 \cdots a_k)$ of S_n of length k, and let us consider the powers of α. Under the mapping α^2, we see that a_1 maps into a_3 (if $k \geq 3$), for

$$a_1 \alpha^2 = (a_1 \alpha)\alpha = a_2 \alpha = a_3.$$

Similarly, under the mapping α^3, a_1 maps into a_4 (if $k \geq 4$), and so on. Continuing, we find that $a_1 \alpha^k = a_1$. Since we could just as well write $\alpha = (a_2 a_3 \cdots a_k a_1)$, a similar argument shows that $a_2 \alpha^k = a_2$ and, in general, that $a_i \alpha^k = a_i$ for $i = 1, 2, \ldots, k$. It follows that $\alpha^k = \epsilon$, the identity permutation, and, moreover, k is the smallest power of α which is equal to ϵ. The following result then follows immediately from Theorem 2.45.

2.65   THEOREM   *A cycle of length k has order k.*

Two cycles $(a_1 a_2 \cdots a_k)$ and $(b_1 b_2 \cdots b_l)$ of S_n are said to be *disjoint* if the sets $\{a_1, a_2, \ldots, a_k\}$ and $\{b_1, b_2, \ldots, b_l\}$ have no elements in common. A set of more than two cycles is said to be disjoint if each pair of them is disjoint. The next result shows why cycles play an important role in the study of permutations.

2.66   THEOREM   *Every element γ of S_n that is not itself a cycle is expressible as a product of disjoint cycles.*

Before considering the proof, let us look at an example. Suppose that

$$\gamma = \begin{pmatrix} 1 & 2 & 3 & 4 & 5 & 6 \\ 3 & 1 & 4 & 2 & 6 & 5 \end{pmatrix},$$

and let us start with any symbol which does not map into itself, for example, the symbol 1. We see that $1\gamma = 3$, $3\gamma = 4$, $4\gamma = 2$, and $2\gamma = 1$. Now take any symbol which has not yet been used and which does not map into itself, for example, 5. Then $5\gamma = 6$ and $6\gamma = 5$. It is then almost obvious that $\gamma = (1342)(56)$.

The proof in the general case follows the same pattern as in this example. Since the identity permutation is a cycle (of length 1), we assume that γ is not the identity. Start with any symbol a_1 such that $a_1\gamma \neq a_1$, and suppose that $a_1\gamma = a_2$, $a_2\gamma = a_3$, $a_3\gamma = a_4$, and so on until we come to the point where, say, $a_k\gamma$ equals some one of the symbols $a_1, a_2, \ldots, a_{k-1}$ already used. Then we must have $a_k\gamma = a_1$ since every other one of these symbols is already known to be the image of some symbol under the mapping γ. Thus γ has the same effect on the symbols $a_1, a_2, \ldots, a_k$ as the cycle $(a_1 a_2 \cdots a_k)$, and also effects a permutation of the remaining symbols (if any). If b_1 is a symbol other than $a_1, a_2, \ldots, a_k$ and $b_1\gamma \neq b_1$, we proceed as above and obtain a cycle $(b_1 b_2 \cdots b_l)$. Now if all symbols that do not map into themselves have been used, we have

$$\gamma = (a_1 a_2 \cdots a_k)(b_1 b_2 \cdots b_l).$$

If there is another symbol c_1 such that $c_1\gamma \neq c_1$, we can similarly obtain another cycle. Evidently, the process can be continued to obtain the desired result. A complete proof can be given by using Well-Ordering of the positive integers.

EXERCISES

1. In each of the following, γ is an element of S_7. Express it as a product of disjoint cycles.

(a) $1\gamma = 3$, $2\gamma = 4$, $3\gamma = 1$, $4\gamma = 7$, $5\gamma = 5$, $6\gamma = 6$, $7\gamma = 2$.

(b) $1\gamma = 5$, $2\gamma = 3$, $3\gamma = 4$, $4\gamma = 7$, $5\gamma = 6$, $6\gamma = 1$, $7\gamma = 2$.

(c) $\gamma = \begin{pmatrix} 1 & 2 & 3 & 4 & 5 & 6 & 7 \\ 3 & 4 & 1 & 2 & 6 & 7 & 5 \end{pmatrix}$.

(d) $\gamma = \begin{pmatrix} 1 & 2 & 3 & 4 & 5 & 6 & 7 \\ 2 & 3 & 1 & 5 & 4 & 7 & 6 \end{pmatrix}$.

2. Express each of the following elements of S_7 as a product of disjoint cycles:

   (a) (123)(16543),

   (b) (213456)(172),

   (c) (4215)(3426)(5671),

   (d) (1234)(124)(3127)(56).

3. In the following, α and β are the given permutations of the set $A = \{1, 2, 3, 4, 5\}$. Compute, in each case, $\alpha\beta$, $\beta\alpha$, α^2, and β^2.

   (a) $1\alpha = 2,\quad 2\alpha = 1,\quad 3\alpha = 3,\quad 4\alpha = 5,\quad 5\alpha = 4;$
       $1\beta = 1,\quad 2\beta = 4,\quad 3\beta = 2,\quad 4\beta = 3,\quad 5\beta = 5.$

   (b) $1\alpha = 4,\quad 2\alpha = 3,\quad 3\alpha = 5,\quad 4\alpha = 1,\quad 5\alpha = 2;$
       $1\beta = 2,\quad 2\beta = 3,\quad 3\beta = 1,\quad 4\beta = 4,\quad 5\beta = 5.$

   (c) $1\alpha = 2,\quad 2\alpha = 1,\quad 3\alpha = 4,\quad 4\alpha = 5,\quad 5\alpha = 3;$
       $1\beta = 2,\quad 2\beta = 3,\quad 3\beta = 4,\quad 4\beta = 5,\quad 5\beta = 1.$

   (d) $\alpha = \begin{pmatrix} 1 & 2 & 3 & 4 & 5 \\ 5 & 4 & 3 & 1 & 2 \end{pmatrix}$, $\beta = \begin{pmatrix} 1 & 2 & 3 & 4 & 5 \\ 3 & 2 & 1 & 5 & 4 \end{pmatrix}$.

   (e) $\alpha = \begin{pmatrix} 1 & 2 & 3 & 4 & 5 \\ 1 & 3 & 2 & 5 & 4 \end{pmatrix}$, $\beta = \begin{pmatrix} 1 & 2 & 3 & 4 & 5 \\ 2 & 3 & 1 & 4 & 5 \end{pmatrix}$.

   (f) $\alpha = \begin{pmatrix} 1 & 2 & 3 & 4 & 5 \\ 5 & 4 & 3 & 2 & 1 \end{pmatrix}$, $\beta = \begin{pmatrix} 1 & 2 & 3 & 4 & 5 \\ 5 & 4 & 2 & 1 & 3 \end{pmatrix}$.

4. Find all subgroups of the symmetric group S_3.

5. Prove that the mapping ϕ of 2.60 is an isomorphism of $\mathscr{S}_n$ onto S_n. In this way, give a complete proof of Theorem 2.61.

The cyclic structure of permutations is an important tool in the study of groups of permutations, and we shall put some of our notation and definitions to work. The cycles of length 2 are of special interest, so that we make the following definition.

2.67   DEFINITION   A cycle of length 2 is called a *transposition*.

A transposition (ij) merely interchanges the symbols i and j, and leaves the other symbols unchanged. Since $(ij)(ij) = \epsilon$, it follows that a *transposition is its own inverse*.

It is quite easy to show that every cycle of length more than 2 can be expressed as a product of transpositions. In fact, this result follows from the observation that

$$(a_1 a_2 \cdots a_k) = (a_1 a_k)(a_2 a_k) \cdots (a_{k-1} a_k),$$

which can be verified by direct calculation. In view of Theorem 2.66, it follows immediately that *every* permutation can be expressed as a product of transpositions. However, it is easy to verify that there is more than one way to express a permutation as such a product. As examples, we see that

$$(1234) = (14)(24)(34) = (32)(12)(14) = (13)(24)(34)(12)(24),$$
$$(123)(14) = (12)(13)(14) = (14)(24)(34) = (14)(24)(34)(23)(23),$$

and so on. Since $(ij)(ij) = \epsilon$, we can insert as many such pairs of identical transpositions as we wish. Clearly, then, a permutation can be expressed as a product of transpositions in many different ways.

The following theorem, of which the first statement has already been proved, is one of the principal theorems about permutations.

2.68   THEOREM   *Every permutation α can be expressed as a product of transpositions. Moreover, if α can be expressed as a product of r transpositions and also as a product of s transpositions, then either r and s are both even or they are both odd.*

Suppose that α is a permutation of the set $A = \{1, 2, \ldots, n\}$. Suppose, further, that

2.69 $\alpha = \beta_1 \beta_2 \cdots \beta_r = \gamma_1 \gamma_2 \cdots \gamma_s,$

where each β and each γ is a transposition. To establish the theorem, we need to prove that r and s are both even or that they are both odd. There are ways to prove this fact by calculating entirely with permutations, but we proceed to give a well-known proof which is simpler in its details but which involves the introduction of a certain "counting device" which has no inherent connection with the permutations themselves. Let $x_1, x_2, \ldots, x_n$ be independent symbols (or variables, if you wish) and let P denote the polynomial† with integral coefficients defined as follows:

2.70
$$P = \prod_{i<j} (x_i - x_j),$$

it being understood that this stands for the product of all expressions of the form $x_i - x_j$, where i and j take values from 1 to n, with $i < j$. We now define

2.71
$$P\alpha = \prod_{i<j} (x_{i\alpha} - x_{j\alpha}),$$

that is, $P\alpha$ is the polynomial obtained by performing the permutation α on the *subscripts* of the symbols $x_1, x_2, \ldots, x_n$.

As an illustration of this notation, if $n = 4$, we have

$$P = (x_1 - x_2)(x_1 - x_3)(x_1 - x_4)(x_2 - x_3)(x_2 - x_4)(x_3 - x_4).$$

Moreover, if

$$\alpha = \begin{pmatrix} 1 & 2 & 3 & 4 \\ 4 & 1 & 2 & 3 \end{pmatrix},$$

we find that

$$P\alpha = (x_4 - x_1)(x_4 - x_2)(x_4 - x_3)(x_1 - x_2)(x_1 - x_3)(x_2 - x_3).$$

and it is easily verified that $P\alpha = -P$. In general, it is fairly clear that always $P\alpha = \pm P$, with the sign depending in some way on the permutation α. We next prove the following lemma.

2.72   LEMMA   *If $\delta = (kl)$ is a transposition, then $P\delta = -P$.*

Proof: Now $k \neq l$, and there is no loss of generality in assuming that $k < l$. Hence, one of the factors in P is $x_k - x_l$ and in $P\delta$ the corresponding factor is $x_l - x_k$; that is, this factor is just changed in sign under the mapping δ on the subscripts. Any factor of P of the form $x_i - x_j$, where neither i nor j is equal to k or l, is clearly unchanged under the mapping δ. All other factors of P can be paired to form products of the form $\pm(x_i - x_k)(x_i - x_l)$, with the sign determined by the relative magnitudes of i, k, and l. But since the effect of δ is just to interchange x_k and x_l, any such product is

---

† Logically, this proof should be deferred until after polynomials have been studied in some detail. However, it seems preferable to insert it here with the expectation that it will be convincing, even though use is made of a few simple properties of polynomials which will not be established until later. The polynomial P is sometimes called the Van der Monde polynomial after a mathematician who discovered some of its applications.

unchanged. Hence, the only effect of δ is to change the sign of P, and the lemma is established.

The proof of the theorem now follows easily. Since, by 2.69, $P\alpha$ can be computed by performing in turn the r transpositions $\beta_1, \beta_2, \ldots, \beta_r$, and by the lemma each of these merely changes the sign of P, it follows that $P\alpha = (-1)^r P$. In like manner, using the fact that $\alpha = \gamma_1\gamma_2 \cdots \gamma_s$, we see that also $P\alpha = (-1)^s P$. Hence, we must have $(-1)^r P = (-1)^s P$ from which it follows that $(-1)^r = (-1)^s$. This implies that r and s are both even or they are both odd, and the proof is complete.

2.73 DEFINITION A permutation is called an *even* permutation or an *odd* permutation, respectively, if it can be expressed as a product of an even or an odd number of transpositions.

If the permutation α can be expressed as a product of k transpositions, and the permutation β can be expressed as a product of l transpositions, it is obvious that $\alpha\beta$ can be expressed as a product of $k + l$ transpositions. It follows that the product of two even, or of two odd, permutations is an even permutation, whereas the product of an odd permutation and an even permutation is an odd permutation.

Another observation of some importance is the following. Suppose that α is a product of k transpositions, say $\alpha = \alpha_1\alpha_2 \cdots \alpha_k$. Then, since a transposition is its own inverse, it is easy to see that $\alpha^{-1} = \alpha_k\alpha_{k-1} \cdots \alpha_1$. It follows that α^{-1} is an even permutation if and only if α is an even permutation. We shall conclude our study of permutation groups by proving the following theorem.

2.74 THEOREM *The set A_n of all even permutations of the symmetric group S_n is a subgroup of S_n of order $n!/2$.*

Proof: The fact that A_n is a subgroup of S_n follows at once from the preceding remarks and Theorem 2.30. This subgroup A_n of S_n is usually called the *alternating group* on n symbols.

Let us now consider the order of A_n. If β is a fixed odd permutation, all the elements of the coset βA_n are odd permutations since the product of an odd permutation by an even permutation is necessarily an odd permutation. We proceed to show that *all* odd permutations of S_n are in the coset βA_n. If γ is an arbitrary odd permutation, we may write $\gamma = \beta(\beta^{-1}\gamma)$, and $\beta^{-1}\gamma$ is an even permutation since β^{-1} and γ are both odd. It follows that $\beta^{-1}\gamma \in A_n$, and hence that $\gamma \in \beta A_n$. We have shown that the coset βA_n consists of *all* the odd permutations, and hence that there are just the two cosets A_n and βA_n of A_n in S_n. Since these cosets have the same number of elements and S_n has order $n!$, it follows that the alternating group A_n has order $n!/2$.

There is yet another way to view this particular theorem. Theorem 2.68 tells us that to each permutation is attached a sign. If α can be expressed as the product of r transpositions then the sign of α would be $(-1)^r$, leading us to the following definition.

2.75   DEFINITION   If α is a permutation of a finite set, we define

$$\text{sign } \alpha = \begin{cases} +1 & \text{if } \alpha \text{ is an even permutation,} \\ -1 & \text{if } \alpha \text{ is an odd permutation.} \end{cases}$$

If a permutation β can be expressed as the product of s transpositions, then $\text{sign } \beta = (-1)^s$. Obviously, $\alpha\beta$ can be expressed as the product of $r + s$ transpositions, so that

2.76 $$\text{sign } \alpha\beta = (-1)^{r+s} = (-1)^r(-1)^s = (\text{sign } \alpha)(\text{sign } \beta).$$

So "sign" is a mapping of S_n into the multiplicative group $\{+1, -1\}$ of integers, which is a homomorphism. The kernel of "sign" is A_n, and the set of elements α such that $\text{sign } \alpha = -1$ is the coset βA_n, where β is an odd permutation. Theorem 2.37 tells us that A_n is a subgroup of S_n, where A_n has two cosets in S_n (one "belonging" to $+1$ and the other to -1). Lagrange's Theorem, 2.52, tells us that $n!$ equals the order of S_n which is two times the order of A_n, and thus A_n has order $n!/2$.

Incidentally, since "sign" is a homomorphism onto the multiplicative set $\{+1, -1\}$, taking "multiplication" in S_n to multiplication of integers, we conclude that $\{+1, -1\}$ is indeed a group with the group operation being multiplication of integers. (See Exercise 11 of Section 2.5.)

EXERCISES

1. Verify that a cycle of length k is an even or an odd permutation according as k is odd or even, respectively.

2. Prove that every even permutation is a cycle of length 3 or can be expressed as a product of cycles of length 3. *Hint:* $(12)(13) = (123)$, and $(12)(34) = (134)(321)$.

3. Exhibit the elements of the alternating group A_3 and of the alternating group A_4.

4. Let G be a subgroup of the symmetric group S_n and suppose that G contains at least one odd permutation. By a suitable modification of the proof of Theorem 2.74, prove that the set of all even permutations in G is a subgroup of G (try using the "sign" mapping), and then prove that G contains the same number of odd permutations as of even permutations.

5. Recall Example 3 of Section 2.1. By numbering the positions of the axes, prove that A is isomorphic to S_3.

6. Show that the group of symmetries of an equilateral triangle is isomorphic to S_3.

7. Recall Example 4 of Section 2.1. A diagonal of the cube passes through the center and two vertices of the cube. There are four diagonals of a cube. Show that C is isomorphic to S_4.

8. Exhibit explicitly a homomorphism of S_4 *onto* S_3. *Hint:* See Exercises 5 and 6 and Example 1 of Section 2.5.

9. Recall Example 4 of Section 2.1. Show that there are isomorphisms of the group C into the following symmetric groups: S_4, S_6, S_8, S_{12}, S_{24}. Show that each isomorphism θ may be chosen so that $C\theta$ is a *transitive* group, that is, if $k = 4, 6, 8, 12, 24$ and $\theta: C \to S_k$, then for each j, $1 \le j \le k$, there is an $a \in C$ so that if $a\theta = \alpha$ then $1\alpha = j$. *Hint:* Consider axes, or faces, or vertices, etc.

10. A regular tetrahedron is assembled from four identical equilateral triangles. Show that the group of symmetries of the regular tetrahedron is isomorphic to a subgroup of C. Then, using Exercise 6, show that this subgroup is isomorphic to A_4. *Hint:* See Exercise 12 of Section 2.5.

11. Show that the group of symmetries S of a square is of order 8. By placing the square in the center of a cube parallel to a face, argue that S is isomorphic to a subgroup of C of Example 4 in Section 2.1. Argue then that S is isomorphic to a subgroup of S_4 (see Exercise 6). Now argue directly that S is isomorphic to a subgroup of S_4 by numbering the sides of the square.

12. Find all subgroups of order 8 in S_4. *Hint:* Find all subgroups of C of order 8. Describe them geometrically.

13. Number the vertices of a regular n-gon (polygon of n sides) consecutively clockwise: $1, 2, \ldots, n$. Consider the group of rotations of the regular n-gon obtained by rotating the n-gon about an axis through its center perpendicular to the n-gon face. Show that the resulting cyclic group is isomorphic to the cyclic group in S_n generated by the cycle $(123 \cdots n)$. Consider the cyclic group T generated by (13425) in S_5. Show that for some numbering of the vertices of a regular pentagon we obtain an isomorphism of the rotations of this 5-gon as above with the cyclic group T.

14. A regular dodecahedron is assembled from 12 identical pentagons. There are 30 edges on the dodecahedron. By connecting the midpoints of opposite edges, we get $\frac{30}{2} = 15$ pairs of edges paired by "connecting lines." Look at one of these connecting lines. There are exactly two other connecting lines perpendicular to the chosen one. The three perpendicular connecting lines may be thought of as x,y,z-axes. Any x,y,z-axes chosen in this way connect $3 \cdot 2 = 6$ edges. Thus there are exactly $\frac{30}{6} = 5$ different x,y,z-axes systems in the dodecahedron in this fashion. Number these systems 1, 2, 3, 4, 5. Show that there is an isomorphism of the group D of symmetries of the dodecahedron *onto* A_5, the alternating group on five symbols.

15. Suppose that A is a set which is not finite. Let S be the set of all permutations of A. That is, S is the set of all one-one mappings of A onto itself. Prove that S is a group. Prove that for each positive integer n, S_n is isomorphic to a subgroup of S.

16. The converse of Lagrange's Theorem is false. Show that A_4 has no subgroup of order 6. (See Exercise 9.)

2.9   NORMAL SUBGROUPS AND QUOTIENT GROUPS

We begin this section with an example.

> **Example 1**   Recall the groups of Examples 3 and 4 of Section 2.1. These were the group of symmetries A of the x,y,z-axes and the group of symmetries C of the cube. Recall Example 1 of Section 2.5; there we embedded the axes with O at the center of the cube and introduced a homomorphism ψ of C onto A. Recall that ker ψ was the subgroup of the cube, all of whose elements induced the identity motion of the axes. Let us call this subgroup $K = \ker \psi$. It contains the identity motion of the cube. It also contains three other elements. These can be described respectively as rotations about the x-, y-, or z-axes of 180° (clockwise or counterclockwise), that is, K has order 4. Let $\sigma \in C$ be any symmetry of the cube. To

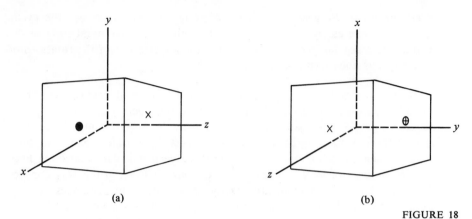

(a)   (b)

FIGURE 18

be specific, let us think of σ as being given by a rotation about a
diagonal of the cube as described in Figure 18 from (a) to (b). Let
τ be any element of K. What is the effect of $\sigma\tau\sigma^{-1}$ upon the axes?
Now σ causes the axes to rotate from (a) to (b) in Figure 18. The
element τ may then cause the cube position to change, but since it
is in K it will not cause the axes to change. In other words, $\sigma\tau$
changes the axes from (a) to (b) in Figure 18. But now σ^{-1} causes
the axes to change from (b) to (a), that is $\sigma\tau\sigma^{-1}$ causes no change
in the axes even though it may cause a change in the cube position.
We conclude that $\sigma\tau\sigma^{-1}$ is also in K. What we have argued in particular must hold in general. The point is: For any element τ in
K and any element σ in C, $\sigma\tau\sigma^{-1}$ is again in K. To see this, it might
be worth experimenting with a cube using several different choices
for σ.

It may not be obvious but this property says something about cosets of K.
Consider the cosets σK and $K\sigma$. We have just shown that for any $\tau \in K$, $\sigma\tau\sigma^{-1}$
is in K. In other words, $\sigma\tau\sigma^{-1} = \rho$, an element of K. But then $\sigma\tau = \rho\sigma$, $\sigma\tau$ is
in σK, and $\rho\sigma$ is in $K\sigma$. This little argument may be used to show that $\sigma K = K\sigma$.
Let us outline what we have shown: (1) $K = \ker \psi$, the kernel of a homo-
morphism, has the property that $K = \sigma K\sigma^{-1} = \{\sigma\tau\sigma^{-1} \mid \tau \in K, \sigma \in C\}$; (2) a
group K with this latter property has the further property that $\sigma K = \{\sigma\tau \mid \tau \in K\} =$
$K\sigma = \{\tau\sigma \mid \tau \in K\}$ for all $\sigma \in C$, that is, any right coset of K is equal to a left coset
of K. Actually, as we shall see, (3) if K has the property that every right coset
is also a left coset, then K is also the kernel of some homomorphism. Our task
in this section will be to straighten out these ideas and close the gap at (3) mentioned above. First let us recall some terminology and define the ideas illustrated.
If H is a subgroup of the group G and $a \in G$, the set $aH = \{ah \mid h \in H\}$,
which in Section 2.7 was called a coset, we shall for the present call a *left* coset

of H in G. Similarly, the set $Ha = \{ha \mid h \in H\}$ is called a *right* coset of H in G. It need not be true that a left coset aH is equal to the right coset Ha. However, we shall be interested in subgroups which do have this property, and we therefore introduce the following definition.

2.77 DEFINITION A subgroup K of a group G is said to be a *normal* (or *invariant*) subgroup of G if and only if $aK = Ka$ for every element a of G.

We may emphasize that this definition does not state that necessarily $ak = ka$ for each $a \in G$ and $k \in K$; it merely states that the *sets* aK and Ka coincide. In particular, if $a \in G$ and $k \in K$, there must exist an element k_1 of K (not necessarily the same element k) such that $ak = k_1a$.

There is another equivalent way of characterizing a normal subgroup as follows: If we define

$$a^{-1}Ka = \{a^{-1}xa \mid x \in K\},$$

then K is a normal subgroup of G if and only if $a^{-1}Ka = K$ for every $a \in G$.

Clearly, every subgroup of an abelian group is a normal subgroup. As an example of a subgroup which is not normal, consider Example 1 of Section 2.7 where we have given the cosets of the group $H = \{\sigma_1, \sigma_2\}$ in the group A of symmetries of the x,y,z-axes. It may be verified there that $\sigma_3H = \{\sigma_3, \sigma_4\}$, whereas using Figure 17 of Example 4 of Section 2.6, we may show that $H\sigma_3 = \{\sigma_3, \sigma_6\}$, and hence that $\sigma_3H \neq H\sigma_3$. So, H is not a normal subgroup of A, however, A does have a proper normal subgroup $N = \{\sigma_1, \sigma_3, \sigma_5\}$. The reader should verify this by direct calculation.

In nonabelian groups, subgroups are more apt to be nonnormal than normal, that is, normality is the exceptional situation. We do have a large store of examples, however, of normal subgroups, namely, the alternating subgroup of the symmetric group on n symbols. We shall now prove this.

Example 2 A_n is a normal subgroup of S_n for $n > 1$. If α is an even permutation, then $\alpha \in A_n$ and $\alpha A_n = A_n\alpha = A_n$, that is, $\alpha \in A_n =$ ker (sign), the kernel of the sign homomorphism. If α is an odd permutation, the proof of Theorem 2.74 shows that αA_n is the set of all odd permutations of S_n. In particular, αA_n is the set of all $\beta \in S_n$ such that

$$\text{sign } \beta = -1.$$

A similar argument will show also that $A_n\alpha$ is the set of all odd permutations of S_n. Hence, for every $\alpha \in S_n$, we have $\alpha A_n = A_n\alpha$, and A_n is therefore a normal subgroup of S_n.

We may see this fact in another form which is equivalent. Using the "sign" mapping for any $\tau \in A_n$ and $\sigma \in S_n$, we have

$$
\begin{aligned}
\text{sign } (\sigma\tau\sigma^{-1}) &= (\text{sign } \sigma)(\text{sign } \tau)(\text{sign } \sigma^{-1}) \\
&= (\text{sign } \sigma)^2 \\
&= +1.
\end{aligned}
$$

Therefore $\sigma\tau\sigma^{-1} \in \text{ker (sign)} = A_n$. So $\sigma A_n \sigma^{-1} = A_n$ for all $\sigma \in S_n$. This also shows that A_n is normal in S_n.

Now let G be an arbitrary group, and K a normal subgroup of G. Since K is normal, we need not distinguish between left cosets and right cosets; so we shall again simply call them cosets and write them as left cosets. On the set of all cosets of K in G we propose to define an operation of multiplication which will make this set into a group. Accordingly, let us define

2.78 $\qquad\qquad (aK)(bK) = (ab)K, \qquad a, b \in G.$

In order to verify that this does define an operation on the set of all cosets, we need to show that multiplication is well-defined by this equation. That is, we must show that if $aK = a_1K$ and $bK = b_1K$, then $(ab)K = (a_1b_1)K$. By 2.50(ii), this fact can be established by showing that if $a \in a_1K$ and $b \in b_1K$, then $ab \in (a_1b_1)K$. Suppose then that $a = a_1k$, and $b = b_1k'$, where $k, k' \in K$. Thus $ab = a_1kb_1k'$, and since K is a normal subgroup of G, there exists $k'' \in K$ such that $kb_1 = b_1k''$. Hence, $ab = a_1b_1k''k'$ and it follows that $ab \in (a_1b_1)K$, as we wished to show. This proves that 2.78 does indeed define an operation of multiplication on the set of all cosets of K in G, and we proceed to prove the following theorem.

2.79 THEOREM *Let K be a normal subgroup of the group G. With respect to the multiplication 2.78 of cosets, the set of all cosets of K in G is a group, usually called the* quotient group *of G by K and denoted by G/K. Moreover, the mapping $\theta: G \rightarrow G/K$ defined by $a\theta = aK$, $a \in G$, is a homomorphism of G onto G/K, with kernel K.*

Proof: The associative law in G/K is an almost immediate consequence of the associative law in G, and we leave this part of the proof to the reader. Now, if e is the identity of G, then since by 2.78,

$$(aK)(eK) = (eK)(aK) = aK, \qquad a \in G,$$

we see that $eK = K$ is the identity of G/K. Finally, 2.78 implies that

$$(aK)(a^{-1}K) = (aa^{-1})K = eK = K$$

and, similarly, $(a^{-1}K)(aK) = K$. Hence $a^{-1}K$ is the inverse of aK, and we have proved that G/K is a group. (Cf. Exercise 11 of Section 2.5.) Furthermore, the definition of the mapping θ shows that it is a mapping of G onto G/K, and the definition of multiplication of cosets shows that

$$(ab)\theta = (ab)K = (aK)(bK) = (a\theta)(b\theta), \qquad a, b \in G,$$

and hence that θ is a homomorphism. Finally, since K is the identity of G/K, an element a of G is in ker θ if and only if $aK = K$, that is, if and only if $a \in K$. This completes the proof of the theorem.

We may remark that since the elements of the quotient group G/K are the distinct cosets of K in G, if G has finite order, the order of the group G/K is the index of K in G. In fact, by Theorem 2.52, we see that the order of G/K is a divisor of the order of G. More precisely, if G has finite order, we have

$$\text{Order of } G/K = \frac{\text{Order of } G}{\text{Order of } K}.$$

In the above, we have used multiplication as the operation in G. If G is abelian and the operation is considered to be addition, it is important to keep in mind that a coset is of the form $a + K$, and the multiplication 2.78 of cosets is replaced by addition of cosets defined as follows:

2.80 $(a + K) + (b + K) = (a + b) + K, \qquad a, b \in G.$

In this case, the identity of G is called the "zero" as usual, and the zero of the quotient group G/K is the coset $K = 0 + K$.

We have shown that if K is a normal subgroup of an arbitrary group G, then there exists a homomorphism of G, with kernel K, onto the quotient group G/K. We shall next prove that "essentially" all homomorphisms of G are of this type. More precisely, we shall show that the kernel of every homomorphism of G is a normal subgroup of G and that every homomorphic image of G is isomorphic to a quotient group G/K for some choice of the normal subgroup K. This is the content of the following theorem.

2.81 FUNDAMENTAL THEOREM ON GROUP HOMOMORPH-ISMS *Let $\phi: G \to H$ be a homomorphism of the group G onto the group H with kernel K. Then K is a normal subgroup of G, and $H \cong G/K$. More precisely, the mapping $\alpha: G/K \to H$ defined by*

2.82 $(aK)\alpha = a\phi, \qquad a \in G,$

is an isomorphism of G/K onto H.

Proof: Let us first show that the kernel K of a homomorphism $\phi: G \to H$ is necessarily a normal subgroup of G. We have already proved in Theorem 2.37 that K must be a subgroup, so there only remains to prove that it is normal. If $a \in G$ and $k \in K$, then

$$(aka^{-1})\phi = (a\phi)(k\phi)(a\phi)^{-1}.$$

But if e is the identity of G, $e\phi$ is the identity of H and $k\phi = e\phi$ by definition of ker ϕ. Thus

$$(aka^{-1})\phi = (a\phi)(k\phi)(a\phi)^{-1} = (a\phi)(e\phi)(a\phi)^{-1}$$
$$= (a\phi)(a\phi)^{-1} = e\phi.$$

and $aka^{-1} \in$ ker $\phi = K$. Hence $aka^{-1} = k_1$ for some element k_1 of K. It follows that $ak = k_1 a$ and this shows that $aK \subseteq Ka$. In like manner, it can be shown that $Ka \subseteq aK$, so that $aK = Ka$ and K is indeed a normal subgroup. Thus we can now speak of the quotient group G/K.

Before proving the rest of the theorem, its meaning may perhaps be clarified by reference to the diagram below. Here ϕ is the given homomorphism of G onto H, $K = $ ker ϕ, and

$$G \xrightarrow{\phi} H$$
$$\theta \searrow \quad \nearrow \alpha$$
$$G/K$$

$\theta: G \to G/K$ is the homomorphism of G onto G/K defined by $a\theta = aK$, $a \in G$, as in the preceding theorem. Our present theorem may then be interpreted as stating that $\phi = \theta\alpha$ for some homomorphism $\alpha: G/K \to H$, that is, that $a\phi = (a\theta)\alpha = (aK)\alpha$ for each $a \in G$. Otherwise expressed, an element a of G has the same image in H no matter which of the two paths from G to H is taken.

The diagram and the fact that we want $\phi = \theta\alpha$ should call to mind the Universal Mapping Property, Theorem 2.38. By Theorem 2.79 we know that ker $\theta = K = $ ker ϕ; in particular, the Universal Mapping Property applies here to tell us that not only does the homomorphism α exist, but also that it is uniquely determined by the condition that $\phi = \theta\alpha$ and that ker $\alpha = ($ker $\phi)\theta = ($ker $\theta)\theta = eK$. In particular, α is an isomorphism of G/K into H by Theorem 2.37. Since ϕ maps G onto H, and since $\phi = \theta\alpha$, we conclude that α is an isomorphism of G/K onto H. This proves the Theorem, i.e. $G/K \cong H$. Incidentally, if $aK \in G/K$, then $a\phi = a(\theta\alpha) = (a\theta)\alpha = (aK)\alpha$ showing that 2.82 is correct.

We shall prove one more isomorphism theorem in this section, whose proof will once again require the use of the Universal Mapping Property.

2.83   THEOREM   *If H and K are normal subgroups of G such that*
$H \subseteq K$, then K/H is a normal subgroup of G/H and

$$(G/H)/(K/H) \cong G/K.$$

This theorem is easy to remember.   It says we may "cancel denominators,"
even though this is not at all what is really happening.

Proof: Theorem 2.79 tells us there are two homomorphisms of G,
namely, $\theta: G \to G/K$ given by $a\theta = aK$ for $a \in G$, and $\phi: G \to G/H$
given by $a\phi = aH$ for $a \in G$.   Notice that Theorem 2.79 tells us
what the kernels of ϕ and θ are: $H = \ker \phi \subseteq K = \ker \theta$, so that
$\ker \phi \subseteq \ker \theta$.   We now refer to the following diagram.

The hypotheses of the Universal Mapping Property, Theorem
2.38, are satisfied, so that a homomorphism $\psi: G/H \to G/K$ exists
such that $\theta = \phi\psi$ and $\ker \psi = (\ker \theta)\phi$.   But the kernel of θ is K
so that $\ker \psi = (\ker \theta)\phi = K\phi = K/H$.   Since θ maps G onto G/K
and since $\theta = \phi\psi$, ψ maps G/H onto G/K.   In particular, G/K is a
homomorphic image of G/H under ψ, and the kernel of ψ is K/H.
By Theorem 2.81, the Fundamental Theorem on Group Homomorphisms, we conclude that K/H is a normal subgroup of G/H
and that

$$G/K \cong (G/H)/(K/H).$$

This completes the proof of the theorem.

Example 3   We shall now discuss a group $\mathbf{Z}_n{}^+$ called the group of
integers modulo n.   Fix a positive integer n, and recall the group
$\mathscr{L}_n{}^+ = \{0, 1, \ldots, n - 1\}$ of least residues modulo n.   In Example 2
of Section 2.5 we introduced a homomorphism $\phi: \mathbf{Z}^+ \to \mathscr{L}_n{}^+$ of
$\mathbf{Z}^+$ onto $\mathscr{L}_n{}^+$, and we showed there that the kernel of ϕ consists of
the set $\{qn \mid q \in \mathbf{Z}\}$ of all multiples of n.   There is a notation for this
set of multiples of n, namely $(n) = \{qn \mid q \in \mathbf{Z}\}$, which is in such
common usage that we shall adopt it now.   That is, (n) denotes the
set of all integral multiples of the integer n.
   The Fundamental Theorem of Group Homomorphisms tells us

that, since $\mathscr{Z}_n{}^+$ is a homomorphic image of $\mathbf{Z}^+$ under the mapping ϕ, $\mathscr{Z}_n{}^+$ is isomorphic to a factor group of $\mathbf{Z}^+$; namely,

$$\mathbf{Z}^+/(n) \cong \mathscr{Z}_n{}^+.$$

The cosets of $\mathbf{Z}^+/(n)$ are the sets $r + (n) = r + \{qn \,|\, q \in \mathbf{Z}\} = \{qn + r \,|\, q \in \mathbf{Z}\}$, where $0 \leq r < n$. That is, two integers r and r' such that $0 \leq r < n$ and $0 \leq r' < n$ belong to the same coset if and only if $r = r'$. Using the Division Algorithm, two integers a and b will belong to the same coset of $\mathbf{Z}^+/(n)$ if and only if they both leave the same remainder upon division by n, that is, $a = qn + r$ and $b = q'n + r$, where $0 \leq r < n$. This latter circumstance occurs if and only if $a - b$ is a multiple of n, that is, $a - b = (qn + r) - (q'n + r) = (q - q')n$. We paraphrase these three conditions as (1) "same coset," (2) "same remainder," and (3) "difference is a multiple of n." For example, when $n = 6$, if $a = 33$ and $b = 257$, then a and b belong to different cosets in $\mathbf{Z}^+/(6)$ because they leave different remainders upon division by 6, that is, $33 = 5 \cdot 6 + 3$ and $257 = 42 \cdot 6 + 5$. On the other hand, 47 and 395 belong to the same coset since their difference is a multiple of 6, that is, $395 - 47 = 58 \cdot 6$.

The additive group $\mathbf{Z}^+/(n)$ of cosets will play a large role in our studies of algebra. We shall denote it by $\mathbf{Z}_n{}^+$ and call it the *additive group of integers modulo n*. We even introduce a special notation to designate its elements, i.e., its cosets. The coset $a + (n) = a + \{qn \,|\, q \in \mathbf{Z}\} = \{qn + a \,|\, q \in \mathbf{Z}\}$ will be denoted by the symbol $[a]$. In our example for $n = 6$ above then, we observe that Lemma 2.50 tells us that $[33] \neq [257]$ but that $[47] = [395]$. Further,

$$[33] = \{\ldots, -9, -3, 3, 9, 15, \ldots, 27, 33, 39, \ldots\}$$
$$= \{6q + 3 \,|\, q \in \mathbf{Z}\} = 33 + (6) = 3 + (6).$$

Addition of cosets, as given by 2.80, tells us that for any two integers a and b,

2.84
$$[a] + [b] = [a + b].$$

For this group operation, $[0]$ is the identity and $-[a] = [-a]$ is the inverse of $[a]$. In our present notation $[0]$ and (n) are the same sets.

Let us make an observation, and then view this whole quotient group from another angle. The homomorphism $\phi \colon \mathbf{Z}^+ \to \mathscr{Z}_n{}^+$ sends an integer j to a least residue r, where r is uniquely determined by the Division Algorithm: $j = qn + r$, $0 \leq r < n$. By the Fundamental Theorem of Group Homomorphisms, there is an isomorphism $\psi \colon \mathscr{Z}_n{}^+ \to \mathbf{Z}_n{}^+$ which will assign the least residue r to the coset $[r]$. The composite mapping $\theta : \mathbf{Z}^+ \to \mathbf{Z}_n{}^+$ given by $j\theta = j(\phi\psi) = r\psi = [r] = [qn + r] = [j]$ assigns an integer j to the coset $[j]$ to which it belongs. This mapping θ is a homomorphism so that

$$(i + j)\theta = [i + j] = [i] + [j] = (i\theta) + (j\theta).$$

We shall now alter our viewpoint slightly. Lemma 2.50 tells us that the cosets of $\mathbf{Z}_n{}^+$ partition the set of integers $\mathbf{Z}$. Whenever we have a partition of a set, we have an equivalence relation upon this set. We shall denote this equivalence relation by "$\equiv$" and call it *congruence modulo n*. That is, we write $a \equiv b$ (mod n) if and only if a and b belong to the same coset in $\mathbf{Z}_n{}^+$, the group of integers modulo n. We have derived two other equivalent ways of defining congruence. Namely, a and b are congruent if and only if they leave the same remainder upon division by n. Or, a and b are congruent if and only if $a - b$ is a multiple of n. In the example for $n = 6$ we would write $47 \equiv 395$ (mod 6) and $33 \not\equiv 257$ (mod 6). Why can we write $231 \equiv 525$ (mod 6)? (Give two reasons.) Continuing this one step further, 2.84 and the observations about θ above tell us there is an addition upon the equivalence classes (cosets). In the case for $n = 6$ we may write $47 + 231 \equiv 395 + 525$ (mod 6). More generally, if $a \equiv b$ (mod n) and $c \equiv d$ (mod n), then $a + c \equiv b + d$ (mod n), since $[a] = [b]$, $[c] = [d]$, and therefore $[a + c] = [a] + [c] = [b] + [d] = [b + d]$. It is often conceptually easier to work with integers and congruence modulo n than it is to work with the cosets of $\mathbf{Z}_n{}^+$, even though each mirrors the other. We shall use both points of view.

Recall that $\psi : \mathscr{Z}_n{}^+ \to \mathbf{Z}_n{}^+$ is an isomorphism such that $r\psi = [r]$. In particular, in each coset there is a unique integer r such that $0 \leq r < n$. If $[a]$ is a coset, then $a = qn + r$ where $0 \leq r < n$, and therefore $[a] = [r]$. This unique integer r is a distinguished coset representative. It now becomes clear from whence the group $\mathscr{Z}_n{}^+$ comes. From each coset the distinguished representative is chosen. This representative is called a *least residue modulo n*. If r and r' are two least residues, then we need only define addition $r \oplus r'$ of least residues so that (1) $r \oplus r'$ is a least residue and (2) $[r] + [r'] = [r \oplus r']$. In this way, the operation in $\mathscr{Z}_n{}^+$ will mirror that in $\mathbf{Z}_n{}^+$. In fact, this is how $\mathscr{Z}_n{}^+$ in Example 2 of Section 2.1 was actually constructed.

We shall return to the integers modulo n many times as an example. In other words, this example is of great importance in understanding many facts about algebra. For instance, to restate Theorem 2.41(ii), we may say that if G is a cyclic group of order n, then G is isomorphic to the group $\mathbf{Z}_n{}^+$ of integers modulo n.

Summing up, we have shown that the following statements are all equivalent for a fixed positive integer n and arbitrary integers a, b, c:

(a) $[a] + [b] = [c]$.

(b) $a + b \equiv c$ (mod n).

(c) $a + b - c$ is a multiple of n.

(d) $(a + b)$ and c leave the same remainder upon division by n.

With $n = 6$, $a = 33$, $b = 29$, $c = 14$, these statements become:

(a) $[33] + [29] = [14]$.

(b) $33 + 29 \equiv 14 \pmod 6$.

(c) $33 + 29 - 14 = 48 = 8 \cdot 6$.

(d) $33 + 29 = 62 = 10 \cdot 6 + 2$ and $14 = 2 \cdot 6 + 2$.

EXERCISES

1. Prove that any subgroup of an abelian group is normal.

2. Consider the group of symmetries of a regular pentagon. Find all normal subgroups of this group.

3. Suppose that H is a subgroup of a group G and a is an element of G. Define a mapping θ of H into G by setting $h\theta = aha^{-1}$. Prove that θ is an isomorphism. In particular, $H\theta = aHa^{-1} = \{aha^{-1} \mid h \in H\}$ is a subgroup of G isomorphic to H.

4. Prove that the intersection of two or more normal subgroups of a group G is a normal subgroup of G.

5. Consider the homomorphism ϕ of Example 2 of Section 2.5. This homomorphism maps $\mathbf{Z}^+$ onto $\mathscr{Z}_{20}^+$ when $n = 20$. In particular, we obtain a homomorphism of $\mathbf{Z}^+$ onto $\mathbf{Z}_{20}^+$. Consider the subgroup

$$H = \{[0], [5], [10], [15]\}$$

where $[j]$ is the coset for $j \in \mathbf{Z}^+$ in $\mathbf{Z}_{20}^+$. Show that H is a normal subgroup of $\mathbf{Z}_{20}^+$ and that $\mathbf{Z}_5^+ \cong \mathbf{Z}_{20}^+/H$.

6. Prove: If there exist exactly two left cosets (or right cosets) of a subgroup H in a group G, then H is necessarily a normal subgroup of G.

7. If n is a positive integer and a and b are integers, then find all solutions x to $a + x \equiv b \pmod n$.

8. Let n be a positive integer, and a, b, c, and d be integers. Using only the fact that $a \equiv b \pmod n$ if and only if $a - b$ is a multiple of n, prove that if $a \equiv b \pmod n$ and $c \equiv d \pmod n$, then $a + c \equiv b + d \pmod n$.

9. Suppose G is a group of order $2m$, where m is an odd number. We shall eventually show that G contains an element a of order 2. Assuming that G contains an element of order 2, prove that G contains a normal subgroup of order m. *Hint:* Recall Cayley's Theorem 2.62, and Exercise 4 at the end of Section 2.8.

10. Prove Theorem 2.79 by first observing that θ is a homomorphism and then applying Exercise 11 of Section 2.5.

11. Suppose that n is a positive integer. Prove that for integers a, b, c, if $a + c \equiv b + c \pmod{n}$, then $a \equiv b \pmod{n}$. *Hint:* See Theorem 2.3.

12. Prove: If n is a positive integer, then $[a]$ is a generator of $\mathbf{Z}_n{}^+$ if and only if there exists an integer b so that $ab \equiv 1 \pmod{n}$.

13. Suppose that H and K are normal subgroups of G. Let $\theta: G \to G/H$ and $\phi: G \to G/K$ be homomorphisms given as in Theorem 2.79. Define a mapping $\psi: G \to (G/H) \times (G/K)$ by setting $a\psi = (a\theta, a\phi)$. Show that ψ is a homomorphism of G into the direct product $(G/H) \times (G/K)$. By Theorem 2.81 we know that ker ψ is a normal subgroup of G. What subgroup is ker ψ?

14. Let H be any subgroup of a group G. Let $B = \{Ha \,|\, a \in G\}$ be the collection of right cosets of H in G. For each $b \in G$ define a mapping $\theta_b: B \to B$ defined by $(Ha)\theta_b = H(ab)$. Show that the mappings θ_b are well-defined (i.e., they are really mappings). For $b \in G$, set $b\phi = \theta_b$.

   (a) Prove that ϕ is a homomorphism of G into the symmetric group S on the set B.

   (b) What is the kernel of ϕ?

   (c) Show that if H has index 3 in G, then G contains a normal subgroup K of index 3 or index 6.

15. Find all normal subgroups of the group of the cube. *Hint:* $C \cong S_4$ and $A \cong S_3$. Any proper normal subgroup of C must contain ker ψ, where ψ is given in Example 1 of Section 2.5.

2.10 HOMOMORPHISMS AND SUBGROUPS

In this section we shall present a few of the relationships which hold between subgroups of a given group G and subgroups of a homomorphic image of G.

First, suppose that $\theta: G \to H$ is merely a *mapping* of a *set* G into a *set* H. A little later, we shall be primarily interested in the case in which G and H are groups and θ is a homomorphism of G onto H.

We recall that if A is a subset of G, we have introduced the notation $A\theta$ for the set of images of elements of A under the mapping θ. That is,

2.85
$$A\theta = \{a\theta \mid a \in A\}.$$

It will now be convenient to introduce a little additional notation. If U is a subset of H, let us denote by $U\theta^{-1}$ the *inverse image* of U, that is, the set of all elements of G whose images under the mapping θ lie in the subset U. Expressed formally,

2.86
$$U\theta^{-1} = \{x \mid x \in G, x\theta \in U\}.$$

This use of θ^{-1} should not be confused with the inverse mapping introduced in Section 1.3. Note that as here defined, θ^{-1} is not a mapping of H into G, but it does map *subsets* of H into *subsets* of G, and it is a convenient concept for our present purposes.

> **Example** Consider the group C of symmetries of the cube, the group A of symmetries of the x,y,z-axes, and the homomorphism $\psi: C \to A$ of Example 1 in Section 2.5. Let B be the subgroup of symmetries of the cube obtained by rotations of the cube about a vertical (that is, y) axis. From Figure 18(a) and Figure 17 we see that
>
> $$U = B\psi = \{\sigma_1, \sigma_6\}.$$
>
> On the other hand, $U\psi^{-1}$ contains not only B, but other elements as well. If K is the kernel of ψ then $U\psi^{-1}$ contains all elements in the set $U_0 = \{\beta\kappa \mid \beta \in B, \kappa \in K\}$ since $(\beta\kappa)\psi = (\beta\psi)(\kappa\psi) = (\beta\psi)\iota_0 = \beta\psi \in B\psi$, where ι_0 is the identity of A. In particular, if we connect the midpoints of the right-hand front vertical edge and the left-hand rear vertical edge of the cube, then $U\psi^{-1}$ contains a rotation of $180°$ (either way) about this axis but B does not. This particular element is obtained as $\beta\kappa$ for $\beta \in B$ and $\kappa \in K$. Can you find β and κ so that $\beta\kappa$ is this element?

The following facts, which we shall number for easy reference, are almost immediate consequences of the definitions. We shall list their proofs as an exercise below.

2.87   *If A is a subset of G, then $A \subseteq (A\theta)\theta^{-1}$.*

2.88 *If A is a subset of G and θ is a one-one mapping, then $A = (A\theta)\theta^{-1}$.*

2.89 *If U is a subset of H, then $(U\theta^{-1})\theta \subseteq U$.*

2.90 *If U is a subset of H and θ is an onto mapping, then $U = (U\theta^{-1})\theta$.*

Using the notation just presented, we now state the following theorem.

2.91 **THEOREM** (*Correspondence Theorem*) *Let $\theta: G \to H$ be a homomorphism of the group G onto the group H, and let $K = \ker \theta$. Then each of the following is true:*

(i) *If A is a subgroup (normal subgroup) of G, then $A\theta$ is a subgroup (normal subgroup) of H.*

(ii) *If A is a subgroup of G which contains K, then $A = (A\theta)\theta^{-1}$.*

(iii) *If U is a subgroup (normal subgroup) of H, then $U\theta^{-1}$ is a subgroup (normal subgroup) of G which contains K.*

In (i) and (iii) it is to be understood that "normal subgroup" may be substituted for "subgroup" both times it appears in the statement.

Proof of (i): Let A be a subgroup of G. To prove that $A\theta$ is a subgroup of H, we need to show that it is closed with respect to multiplication and with respect to taking inverses. Let $a_1\theta$ and $a_2\theta$ be elements of $A\theta$, a_1 and a_2 being elements of A. Then $(a_1\theta)(a_2\theta) = (a_1a_2)\theta \in A\theta$, since $a_1a_2 \in A$. Moreover, since A is a subgroup of G and therefore $a_1^{-1} \in A$, $(a_1\theta)^{-1} = a_1^{-1}\theta \in A\theta$. This proves that $A\theta$ is a subgroup of H.

Now suppose that A is a normal subgroup of G. An appropriate way of expressing this fact is to say that $x^{-1}Ax = A$ for each $x \in G$. Consider $h^{-1}(A\theta)h$, where h is an arbitrary element of H. Since θ is an onto mapping, there exists an element a in G such that $a\theta = h$. Then

$$h^{-1}(A\theta)h = (a\theta)^{-1}(A\theta)(a\theta) = (a^{-1}Aa)\theta = A\theta,$$

since $a^{-1}Aa = A$ by normality of A. Thus $h^{-1}(A\theta)h = A\theta$ for each h in H, and therefore $A\theta$ is a normal subgroup of H.

Proof of (ii): From 2.87, we know that $A \subseteq (A\theta)\theta^{-1}$. If $c \in (A\theta)\theta^{-1}$, then $c\theta \in A\theta$. Thus, there exists $a \in A$ such that $c\theta = a\theta$. It follows that

$$(ca^{-1})\theta = (c\theta)(a^{-1}\theta) = (c\theta)(a\theta)^{-1} = (c\theta)(c\theta)^{-1},$$

and this is the identity of H. Accordingly, $ca^{-1} \in K = \ker \theta$. Since we are assuming that $K \subseteq A$, we conclude that $ca^{-1} \in A$. Since $a^{-1} \in A$, we see that $c \in A$. This shows that $(A\theta)\theta^{-1} \subseteq A$, completing the proof that $A = (A\theta)\theta^{-1}$.

Proof of (iii): Let U be a subgroup of H and suppose that $x, y \in U\theta^{-1}$. This means that $x\theta \in U$ and $y\theta \in U$. Since U is a subgroup of H, $(x\theta)(y\theta) = (xy)\theta \in U$. Accordingly, $xy \in U\theta^{-1}$, and $U\theta^{-1}$ is closed under multiplication. It is also closed under taking inverses. For if $x \in U\theta^{-1}$, then $x\theta \in U$ and $x^{-1}\theta = (x\theta)^{-1} \in U$. In turn, this implies that $x^{-1} \in U\theta^{-1}$. Hence $U\theta^{-1}$ is a subgroup of G, and it clearly contains K since U must contain the identity of H.

Now suppose that U is a normal subgroup of H, and let us show that $U\theta^{-1}$ is a normal subgroup of G. Let y be an arbitrary element of G. Then, using 2.90 and the fact that U is a normal subgroup of H, we have

$$[y^{-1}(U\theta^{-1})y]\theta = y^{-1}\theta[(U\theta^{-1})\theta]y\theta$$
$$= (y\theta)^{-1}U(y\theta) = U.$$

Therefore $y^{-1}(U\theta^{-1})y \subseteq U\theta^{-1}$. We shall now show inclusion the other way. Let $a \in U\theta^{-1}$, so that $a\theta \in U$. If we set $t = yay^{-1}$, then $a = y^{-1}ty$. But $t\theta = (y\theta)(a\theta)(y\theta)^{-1} \in U$ since $a\theta \in U$ and U is normal. It follows that $t \in U\theta^{-1}$ and therefore $a \in y^{-1}(U\theta)^{-1}y$. This shows that $U\theta^{-1} \subseteq y^{-1}(U\theta^{-1})y$, and we conclude that for every y in G, $y^{-1}(U\theta^{-1})y = U\theta^{-1}$, and $U\theta^{-1}$ is therefore a normal subgroup of G, completing the proof.

We now apply the theorem just established to the following situation. Let K be a normal subgroup of G and let $\theta: G \to G/K$ be the homomorphism defined (as in Theorem 2.81) by $a\theta = aK$, $a \in G$. Then $\ker \theta = K$, and if A is a subgroup of G, $A\theta$ consists of those cosets of K in G which contain an element of A. On the other hand, if U is a subgroup of G/K, $U\theta^{-1}$ consists of all those elements of G appearing in any one of the cosets which make up the subgroup U of G/K. In other words, $U\theta^{-1}$ is the union of those cosets which are elements of the subgroup U of G/K. Accordingly, we have the following immediate consequences of the preceding theorem.

2.92 COROLLARY *Let K be a normal subgroup of the group G. Then*

    (i) If A is a subgroup (normal subgroup) of G, the cosets of K in G which contain an element of A are the elements of a subgroup (normal subgroup) of the quotient group G/K.

    (ii) If U is a subgroup (normal subgroup) of the quotient group G/K,

the union of the cosets appearing as elements of U is a subgroup
(normal subgroup) of G.

Suppose now that G is a finite group. In particular, suppose that the subgroup U of G/K has order r and that the subgroup K of G has order s. Then the subgroup of G whose existence is asserted in part (ii) of the preceding corollary has order rs since it consists of elements of G occurring in r different cosets, each of which contains s elements. In particular, we have the following result.

2.93   COROLLARY   *If K is a normal subgroup of G of order s and the quotient group G/K contains a subgroup of order r, then the group G contains a subgroup of order rs.*

We may use the Correspondence Theorem, 2.91, to prove another isomorphism theorem.

2.94   THEOREM   *Let H and K be subgroups of a group G, with K a normal subgroup of G. Then*

(1) $HK = \{hk \mid h \in H, k \in K\}$ is a subgroup of G.

(2) K is a normal subgroup of HK.

(3) $H \cap K$ is a normal subgroup of H, and

(4) $HK/K \cong H/(H \cap K)$.

The proof of this theorem involves a clever weaving together of some of our more important theorems in this chapter. The following diagram should help in remembering this theorem:

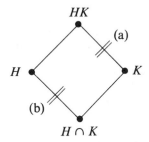

The vertices of the square may be thought of as subgroups as labeled. An edge means the vertex (subgroup) below is contained in the one above. When factor groups exist, an edge may be identified with a factor group; for example, HK/K is represented by edge (a) and $H/(H \cap K)$ by edge (b). Using "$=$" on edges (a) and (b), we mean to say that edge (a) is the "same as" (isomorphic to) edge (b). We shall call a diagram like the one above a *containment diagram*.

Proof: Since K is a normal subgroup of G, Theorem 2.79 tells us there is a homomorphism θ of G onto G/K given by $a\theta = aK$ for $a \in G$. The Correspondence Theorem, 2.91, tells us that both $H\theta$ and $(H\theta)\theta^{-1}$ are groups. In particular, $(H\theta)\theta^{-1}$ is a subgroup of G. Since $H\theta$ consists of all cosets hK where $h \in H$, by Corollary 2.92, $(H\theta)\theta^{-1}$ is the union of all elements in the cosets hK, $h \in H$. That is, $(H\theta)\theta^{-1} = \{hk \mid h \in H, k \in K\} = HK$. We conclude that HK is a subgroup of G.

Notice that θ (when we restrict its domain to HK) is a homomorphism of HK into G/K. The image of HK is $(HK)\theta = H\theta = HK/K$ by Corollary 2.92. Thus θ is a homomorphism of HK onto HK/K.

Restrict the domain of θ even further to H. Then since $H\theta = HK/K$, we conclude that θ is a homomorphism of H onto HK/K. What is the kernel of θ with domain H? It will consist of all elements in the kernel of θ (without restricted domain, i.e., all elements of K) which lie in H. That is, θ with domain H has kernel $H \cap K$. By the Fundamental Theorem on Group Homomorphisms, 2.81, we conclude that $H \cap K$ is a normal subgroup of H and that $H/(H \cap K) \cong HK/K$. The proof of the theorem is complete.

EXERCISES

1. Prove 2.87 to 2.90.

2. In the notation of Theorem 2.91, prove that the mapping $A \to A\theta$ defines a one-one mapping of the set of subgroups (normal subgroups) of G which contain K onto the set of all subgroups (normal subgroups) of H.

3. If K is given as a normal subgroup of G, what conclusions can you draw from the preceding exercise about the subgroups of G which contain K and the subgroups of the quotient group G/K?

4. Recall Example 1 of Section 2.5. Find a subgroup H of C which is isomorphic to A. *Hint:* Consider the symmetries which preserve a single diagonal of the cube. Verify that $H\psi = A$, and that therefore $H(\ker \psi) = C$. Show that $H \cap \ker \psi = \{e\}$.

5. In Exercise 4 let L be the subgroup of symmetries of the cube obtained by rotations about a fixed line connecting the midpoints of two opposite edges of the cube. What are $L\psi$ and $(L\psi)\psi^{-1}$? Show that $(L\psi)\psi^{-1}$ has order 8. Prove that this group is isomorphic to the group of symmetries of a square.

6. In Exercise 4 let M be the subgroup of symmetries of the cube obtained by rotations about a fixed diagonal of the cube. What are $M\psi$ and $(M\psi)\psi^{-1}$? Show that $(M\psi)\psi^{-1}$ is isomorphic to A_4 and hence has order 12. Show that $(M\psi)\psi^{-1}$ has no subgroup of order 6.

7. Let P_n^* be a regular polygon with n vertices, and P_n the group of all symmetries of P_n^*. If $m > 1$ and s are integers so that $ms = n$, then m is a factor of n. For each factor m of n show that there is a homomorphism of P_n onto P_m. In this way determine all normal subgroups of P_n as kernels of appropriate homomorphisms.

8. Prove part (4) of Theorem 2.94 by displaying an explicit isomorphism of HK/K onto $H/(H \cap K)$.

2.11 MORE GROUPS OF SYMMETRIES (OPTIONAL)

Up to this point all our geometric groups of symmetries have been either the symmetries of some regular Euclidean figure or the symmetries of a set of disconnected points. Let us now introduce geometric figures which lie somewhere in between. We shall discuss figures called graphs. Our object here is the study of groups, so that we will not study many properties of graphs. But graph theory (the study of graphs and their properties) is a growing branch of mathematics with many applications inside and outside mathematics.

> **Example 1** A cube is made up of vertices, edges, and faces. The vertices are connected by edges. Suppose that we forget about the faces and imagine the edges as being rubber bands of incredible strength and stretch. In Figure 19 are two sketches of how this

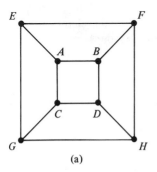

(a)

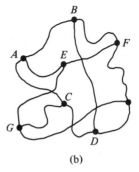

(b)

FIGURE 19

emaciated cube might look if dropped. So far as this geometric object is concerned, the only important things are (1) the vertices and (2) the connections between the vertices. The set of vertices is $\mathscr{V} = \{A, B, C, D, E, F, G, H\}$. The set of edges is $\mathscr{E} = \{\{A,B\}, \{A,E\}, \{A,C\}, \{B,D\}, \{B,F\}, \{C,D\}, \{C,G\}, \{D,H\}, \{E,G\}, \{E,F\}, \{F,H\}, \{G,H\}\}$. Observe that it is only necessary to tell which vertices are connected to one another. Thus $\{A,B\}$ being an edge tells us that this edge connects the vertices A and B.

Example 2 The set of edges and vertices determines the configuration. For example, suppose we have a vertex set $\mathscr{V} = \{A, B, C, D\}$ and an edge set $\mathscr{E} = \{\{A,B\}, \{A,C\}, \{B,D\}\}$. Two different pictures of this configuration are given in Figure 20. Again only the vertices and connections between them are important.

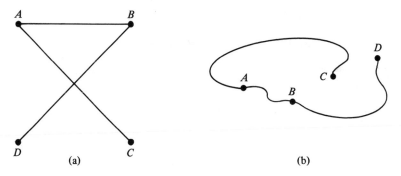

(a)   (b)

FIGURE 20

These two examples are instances of the following definition.

2.95 DEFINITION A *graph* Γ consists of two sets $\mathscr{V}$ and $\mathscr{E}$, where (i) $\mathscr{V}$ is nonempty and (ii) the members of $\mathscr{E}$ are two-element subsets of $\mathscr{V}$. We call the elements of $\mathscr{V}$ *vertices*, and the elements of $\mathscr{E}$ *edges* of the graph. The elements of $\mathscr{E}$ are of the form $\{A,B\} \in \mathscr{E}$ where A and B are distinct elements of $\mathscr{V}$. The set $\mathscr{E}$ may be empty or nonempty.

In Example 2 we saw that a graph may be visualized as a picture consisting of points with connections made between these points. In Example 1 we saw that by labeling points we could obtain a graph by taking the labels to be the set of vertices and the edges to be the sets of vertices connected to each other by a segment passing between two points without passing through a third labeled point. In other words, a graph Γ is a "geometric object."

As such, graphs should have groups of symmetries. A symmetry would be a shuffling of the points which preserves connections by edges. For example, in Figure 19(a) a symmetry might move vertex E to vertex F. Since an edge connects E to A, vertex A must move to vertex E, B, or H, the vertices connected to F. Vertex A could not move to vertex C since E and A would go to F and C, respectively. That is, they would be connected by an edge before but not after. This change would not be a symmetry since it does not preserve connection by edges.

Example 3 If we label the vertices of a cube A, B, C, D, E, F, G, H to correspond to those connected by edges in Figure 19, then each symmetry of the cube will cause a symmetry of the graph of Figure 19. Consider the mapping given by $A\phi = E$, $B\phi = F$, $C\phi = G$, $D\phi = H$, $E\phi = A$, $F\phi = B$, $G\phi = C$, $H\phi = D$. Then ϕ is a symmetry of the graph of Figure 19 but is not a symmetry of the cube. Graphs exhibit a "looseness" not found in rigid objects.

We may view symmetries of a graph as permutations of the vertices. That is, ϕ above is the permutation

$$\phi = \begin{pmatrix} A & B & C & D & E & F & G & H \\ E & F & G & H & A & B & C & D \end{pmatrix}$$
$$= (AE)(BF)(CG)(DH).$$

In particular, since ϕ preserves edges, for any edge $X = \{W, U\}$, the set $Y = \{W\phi, U\phi\}$ must also be an edge. This observation allows us to give a specific definition of a symmetry of a graph.

2.96 DEFINITION Suppose that Γ is a graph with edge set $\mathscr{E}$ and vertex set $\mathscr{V}$. We call a permutation ϕ from the symmetric group on $\mathscr{V}$ a *graph automorphism* or *graph symmetry* if whenever $\{A, B\} \in \mathscr{E}$ is an edge of Γ then $\{A\phi, B\phi\} \in \mathscr{E}$ is also an edge of Γ.

The group of symmetries of a graph, then, is a subgroup of the symmetric group of the set of vertices of the graph. We have already seen one graph in Section 2.8, as we proceed to point out in the following example.

Example 4 Suppose Γ consists of vertices $\{1, 2, \ldots, n\}$ for some positive integer n and edges $\mathscr{E} = \varnothing$, the empty set. We may envision this graph as n points in a line with no connections. In Section 2.8 we saw that the group of symmetries of this graph is S_n, the symmetric group on $\mathscr{V}$. This observation prompts part of the following definition.

2.97 **DEFINITION** A graph Γ is called *totally disconnected* if its edge set is empty. A graph Γ is called *complete* if its edge set contains all possible edges. If A and B are any two distinct vertices, then A and B are joined by an edge, that is, $\{A, B\}$ is an edge of Γ.

In Example 4 we saw that a graph with n vertices has the symmetric group on n symbols as its symmetry group. Actually, we may prove the following theorem.

2.98 **THEOREM** *A graph Γ with vertex set $\mathscr{V}$ and edge set $\mathscr{E}$ has the symmetric group on $\mathscr{V}$ as its group of symmetries if and only if Γ is either complete or totally disconnected.*

Proof: We will leave to the exercises the proof that the complete graph with vertex set $\mathscr{V}$ has the symmetric group on $\mathscr{V}$ as its symmetry group.

Suppose Γ has the symmetric group on $\mathscr{V}$ as its group of symmetries. Suppose that Γ is neither complete nor totally disconnected. We wish to obtain a contradiction. Since Γ is neither complete nor totally disconnected, there must be vertices $A, B, C, D \in \mathscr{V}$ so that $\{A,B\}$ is an edge but $\{C,D\}$ is not an edge of Γ. Arranging the labeling correctly, we may assume that either A, B, C, D are four distinct vertices or $A = C$, B, D are three distinct vertices. In any case,

$$\phi = \begin{pmatrix} A & C & B & D & a_1 & a_2 & \cdots \\ C & A & D & B & a_1 & a_2 & \cdots \end{pmatrix}$$

where $\mathscr{V} = \{A, B, C, D, a_1, a_2, \ldots\}$ and ϕ is a permutation of $\mathscr{V}$. The symbols $a_1, a_2, \ldots$ are used to mean that $(b)\phi = b$ for all vertices not among A, B, C, D. Now ϕ is a symmetry of Γ so that $\{A\phi, B\phi\} = \{C,D\}$ must be an edge of Γ. But $\{C,D\}$ is not an edge. This contradiction completes the proof of the theorem.

There are certain visual advantages to thinking of symmetries. For most considerations it is probably easier to think about the group C of the cube than it is to think about the symmetric group S_4 even though these two groups are isomorphic (see Exercise 6 at the end of Section 2.8). Until now we have started with geometric configurations and studied their groups. Cayley's Theorem helps us to reverse this process. Using his theorem, 2.62, and Theorem 2.98, we see that any finite group is isomorphic to a subgroup of symmetries of a complete or totally disconnected graph. Such a graph is not of too much use; either there are too few (none) or too many (all) connections. There are other ways to construct graphs associated with given groups. Using such associations, we are able to more easily "see" properties of a given group. Let us consider another example.

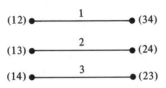

FIGURE 21

Example 5 Let S_4 be the symmetric group on the set $\{1,2,3,4\}$. Let $\mathscr{V} = \{(12), (13), (14), (23), (24), (34)\}$ be the set of all transpositions in S_4. We form edges in the following way. We consider $\{\alpha,\beta\}$ to be an edge if and only if $\alpha \neq \beta$; $\alpha, \beta \in \mathscr{V}$ and $\alpha\beta = \beta\alpha$. That is, $\{\alpha,\beta\}$ is an edge if α and β are distinct commuting transpositions. We obtain a graph Γ whose edge set is $\mathscr{E} = \{\{(12),(34)\}, \{(13),(24)\}, \{(14),(23)\}\}$. This graph is pictured in Figure 21. For any permutation x in S_4 define a mapping $\phi_x \colon \mathscr{V} \to \mathscr{V}$ by setting

$$\nu\phi_x = x^{-1}\nu x$$

where $\nu \in \mathscr{V}$ is a vertex. In Exercise 4 it will be shown that $x^{-1}\nu x$ is again a transposition, so that ϕ_x is a mapping of $\mathscr{V}$ to $\mathscr{V}$. In Exercise 5 it will be shown that $\phi_x\phi_y = \phi_{xy}$ under composition of mappings. This has two consequences. First, $\phi_{x^{-1}}$ is the inverse mapping to ϕ_x so that ϕ_x is both one-one and onto; that is, ϕ_x is a permutation of the set $\mathscr{V}$. Second, the mapping

$$x\theta = \phi_x$$

is a homomorphism of S_4 into the symmetric group of $\mathscr{V}$.

Now suppose that $\{\mu,\nu\}$ is an edge. That is, $\mu\nu = \nu\mu$. Then $\mu\phi_x = x^{-1}\mu x$ and $\nu\phi_x = x^{-1}\nu x$ so that $(x^{-1}\mu x)(x^{-1}\nu x) = x^{-1}\mu\nu x = x^{-1}\nu\mu x = (x^{-1}\nu x)(x^{-1}\mu x)$. Since the vertices satisfy $(\mu\phi_x)(\nu\phi_x) = (\nu\phi_x)(\mu\phi_x)$, $\{\mu\phi_x, \nu\phi_x\}$ is an edge. In other words, θ is a homomorphism of S_4 into the group of symmetries of the graph Γ. Starting with S_4, we have constructed a graph Γ such that a homomorphic image of S_4 is a subgroup of the symmetry group of Γ.

In Exercise 6 it is shown that θ is actually an isomorphism. The image $S_4\theta$ now acts to permute the edges of Γ by sending an edge $\{\mu,\nu\}$ into an edge $\{\mu\phi_x, \nu\phi_x\} = \{\mu(x\theta), \nu(x\theta)\}$ for $x \in S_4$, so that we may view $S_4\theta$ as a permutation group on the edges $\{1,2,3\}$ of the graph Γ as pictured in Figure 21. That is, we have a homomorphism θ' of S_4 into the symmetric group S_3 upon the set $\{1,2,3\}$. Notice that $x = (12)$ has the property that $(12)\phi_x = (12)$, $(13)\phi_x = (23)$, $(14)\phi_x = (24)$. That is, ϕ_x fixes edge 1 and interchanges edge 2 and edge 3. As a permutation of $\{1,2,3\}$ then ϕ_x gives rise to (23). In a similar fashion we may show that $x = (13)$ gives rise to (13) as a permutation of $\{1,2,3\}$ and $x = (14)$ gives rise to (12) as a permutation of $\{1,2,3\}$. That is, the image of θ' contains (12), (23), and (13). These are all the transpositions of S_3. By Theorem 2.68 we know that θ' is a mapping of S_4 onto S_3.

This graphical construction obtained from S_4 makes it easy to see that there is a homomorphism of S_4 onto S_3. Starting from a group we constructed a graph. We could then geometrically see other properties of the group which were not obvious to begin with. The particular method used in this example has recently been applied by group theorists to advance our understanding of group theory.

Let us now describe how this example might work in general. Let G be a group. If $a \in G$, then any element $b^{-1}ab$ for $b \in G$ is called a *conjugate* of a. If a and c are conjugate in G, we often write $a \sim c$, and the relation "$\sim$" is an equivalence relation (see Exercise 7). The equivalence classes of $\sim$ are called the *conjugacy classes* of G. For example, the transpositions in S_n form a conjugacy class of S_n (see Exercise 4).

2.99 DEFINITION Suppose that $K_1, K_2, \ldots, K_t$ are some of the conjugacy classes of G. We define a graph Γ on the set of vertices $\mathscr{V} = K_1 \cup K_2 \cup \cdots \cup K_t$. If $a, b \in \mathscr{V}$, then a, b are elements of G. As such, we call $\{a,b\}$ an edge of Γ if and only if $a \neq b$ and $ab = ba$, that is, a and b commute with one another. In this way we obtain a graph Γ on the vertices $\mathscr{V}$ with edge set $\mathscr{E}$. Thus, the elements of $\mathscr{V}$ are elements of G and, in particular, have the property that if $v \in \mathscr{V}$ and $a \in G$, then $a^{-1}va \in \mathscr{V}$. Moreover, if $v \in K_i$, then $a^{-1}va \in K_i$.

By itself, this graph is of little use. We want to show that there is a homomorphism of G into the group of symmetries of Γ.

2.100 THEOREM *Suppose* $\mathscr{V} = K_1 \cup \cdots \cup K_t$, *where* $K_1, \ldots, K_t$ *are conjugacy classes of* G. *If* $a \in G$ *is a fixed element of* G, *and* $v \in \mathscr{V}$ *varies over the elements of* $\mathscr{V}$, *then setting* $v\phi_a = a^{-1}va$ *defines a permutation* ϕ_a *of the set* $\mathscr{V}$. *The mapping defined by*

2.101 $$a\theta = \phi_a$$

is a homomorphism of G *into the symmetric group upon* $\mathscr{V}$. *For each* $a \in G$, ϕ_a *is a symmetry of the graph given by Definition 2.99. In other words,* θ *is a homomorphism of* G *into the group of symmetries of the graph* Γ.

The proof of this theorem is left to the exercises. Actually, Example 5 is just an instance of this theorem. The elements ϕ_x of that example are the elements ϕ_a of the theorem. The homomorphism θ of the example is given by 2.101.

Of special interest to modern finite group theorists are the graphs obtained when all the elements of all the conjugacy classes K_i have order 2. Further, the group G is assumed to have no normal subgroups other than the trivial ones, $\{e\}$ and G. Such a group G is called a *simple* group. In particular, if G is a non-abelian simple group, then θ of 2.101 is an isomorphism. That is, G is isomorphic to a subgroup of a symmetry group of the graph Γ. Since group theorists are interested in describing the class of finite simple groups, this geometric construction has proven to be of some value.

EXERCISES

1. Compute the order of the group of symmetries of the graph shown in Figure 19.

2. Compute the order and explicitly exhibit the elements of the group of symmetries of the graph shown in Figure 20.

3. Complete the proof of Theorem 2.98 by showing that a complete graph of n vertices has the symmetric group upon n symbols as its symmetry group.

4. Suppose that $x = (a_1 \cdots a_{j})(a_{j_1+1} \cdots a_{j_2}) \cdots (a_{j_{t-1}+1} \cdots a_{j_t})$, $j_t = n$, is an element of S_n written as a product of disjoint cycles. Suppose that

$$y = \begin{pmatrix} a_1 & a_2 & \cdots & a_n \\ b_1 & b_2 & \cdots & b_n \end{pmatrix}$$

is another element of S_n. Show that $y^{-1}xy$ is equal to $(b_1 \cdots b_{j_1})(b_{j_1+1} \cdots b_{j_2})$ $\cdots (b_{j_{t-1}+1} \cdots b_{j_t})$ written as a disjoint union of cycles. This supplies a missing step in Example 5. That is, if x is a transposition, then $y^{-1}xy$ is also a transposition.

5. Suppose that $K_1, \ldots, K_t$ are conjugacy classes of a group G. For $x \in G$ and $v \in K_1 \cup \cdots \cup K_t$ show that $v\phi_x = x^{-1}vx$ defines a permutation ϕ_x of $K_1 \cup \cdots \cup K_t$. Further show for $y \in G$ that $\phi_x \phi_y = \phi_{xy}$. Consequently prove that the mapping θ defined by $x\theta = \phi_x$ is a homomorphism of G into the symmetric group upon $K_1 \cup \cdots \cup K_t$. This exercise supplies a missing step in Example 5 and partially proves Theorem 2.100.

6. Consider the graph Γ and the homomorphism θ of Example 5. Argue that

θ is an isomorphism by proving that ker $\theta = \{e\}$. This exercise supplies a missing step in Example 5.

7. Prove that $\sim$ (is conjugate to) is an equivalence relation upon the elements of G.

8. After completing Exercise 5, complete the proof of Theorem 2.100 by showing that the elements $a\theta$ for $a \in G$ preserve connections. That is, $G\theta$ is a subgroup of the group of symmetries of Γ.

9. Write down all elements of order 3 in S_3. Form the graph Γ for this set of vertices as in Definition 2.99. Show that there is a homomorphism of S_3 onto S_2. Compare this with the "sign" mapping of S_3.

10. Write down all elements of order 2 in S_4. Form the graph Γ for this set of vertices as in Definition 2.99. From this graph, show that there is a homomorphism of S_4 onto S_3.

11. What is the group of symmetries of each of the following graphs?

(a)   (b)   (c)   (d)

12. Let Γ be a graph with vertex set $\mathscr{V}$ and edge set $\mathscr{E} \neq \varnothing$. Define a new thing Γ^*. Let $\mathscr{V}^* = \mathscr{E}$. Define $\mathscr{E}^*$ to be the set of all subsets $\{A,B\}$ of $\mathscr{V}^*$ such that $A \neq B$, and $A \cap B \neq \varnothing$. Show that Γ^* is a graph. Show that there is a homomorphism of the group of symmetries of Γ into the group of symmetries of Γ^*.

13. For each graph Γ of Exercise 11 sketch the graph Γ^*. Also sketch the graph Γ^{**} when possible.

14. Show that the symmetry ϕ of Example 3 is not an element of C, the group of symmetries of a cube. *Comment:* Many authors allow symmetries to involve "reflections." For these people ϕ would be a symmetry of the cube. A reflection is just that; look in the mirror, label the faces of the cube as you see them. No symmetry will give that labeling of the cube.

15. Let $\mathscr{V}$ be the set of elements of order 5 in S_5. Construct the graph Γ of Definition 2.99 upon $\mathscr{V}$. Show that there is an isomorphism θ of S_5 into S_6. Show that $S_5\theta$ is a transitive subgroup of S_6. That is, if j is any integer in $\{1,2,3,4,5,6\}$, then there is an $\alpha \in S_5$ such that $1(\alpha\theta) = j$.

COMMENTARY

1 Where did group theory come from? Where is it going? Like most branches of mathematics, group theory evolved from several folds in the fabric of mathematics. The ideas of transformation groups evolved slowly in geometry; in analysis certain group theoretic ideas started appearing around 1800. But group theory actually sprang to life with the study of algebraic equations.

In the sixteenth and seventeenth centuries methods for finding the roots of quadratic, cubic, and quartic equations were found. The actual history of these discoveries is quite fascinating, covers a considerable span of time ranging from 1500 to 1600, and involves intrigue and vituperation [137,147,151]. Naturally, mathematicians wanted to discover how to find the roots of quintic and higher degree equations. Some very great mathematicians tried their hands at this puzzle; among them are Euler (1707–1783) and Lagrange (1736–1813). Lagrange knew that groups had something to do with equations; he saw that because the structure of the permutation groups S_2, S_3, S_4 was particularly "tame" that second-, third-, and fourth-degree equations could be "solved." He also saw that the group S_5 was, in some way, not quite so "tame"; but he could find no criterion which would directly connect this with the solvability of fifth-degree equations [53]. It was left to Abel (1802–1829) to show once and for all that one could not, in general, give an algebraic solution to any quintic equation; as put today, the general quintic equation is not solvable by radicals [148].

During this time Galois (1811–1832) was trying to discover a necessary and sufficient condition for an nth-degree equation to be solvable by radicals. This he succeeded in doing before he reached the age of 21. His life is interesting, first because he was ignored by mathematicians until well after his death, and second, because he was, as we say today, a political radical, involved with the aftermath of the French Revolution. He was killed at age 21 in a duel of which he did not want to be a party, and about which little is known but much is conjectured. An exciting fictional account of Galois' life has been written by Leopold Infeld [143] in which an appendix discusses the factual basis for the novel.

Galois showed that to each algebraic equation of degree n is attached a group G of permutations upon the roots of the equation. The equation may be solved or is solvable by radicals precisely when the group G is solvable; that is, exactly when G has a sequence of subgroups $H_1, \ldots, H_t$ such that $G = H_1$, $H_i \supseteq H_{i+1}$, $H_t = \{e\}$, and such that H_{i+1} is a normal subgroup of H_i with the quotient group H_i/H_{i+1} being cyclic of prime order. As we shall see in Chapter 8, A_5, the alternating group on five symbols, contains no nontrivial normal subgroup; therefore, if A_5 is the group associated with some equation, then that equation is not solvable by radicals [54,55,152].

Since Galois' discoveries, group theory has come to influence many parts of mathematics. In the late nineteenth century, the famous mathematician Klein (1849–1929), through his Erlanger Program, attempted to describe all geometries by their groups of symmetries [153]. Since that time, groups have played a role in all the major branches of mathematics. During the past 20 years, finite group theory has been an especially active branch of mathematical research. To get some idea of current directions, you might want to page ahead and read the Commentary of Chapter 8 and the first few paragraphs of Section 8.3 which discuss some reasons for current interest in those groups called "simple groups."

2   The exercises of Section 2.1 indicated that every symmetry of a regular polyhedron could be given as a rotation about some axis. How can one see this in general? First notice that we may either circumscribe or inscribe a regular polyhedron in a sphere. In particular, the symmetries of the polyhedron are just symmetries of the sphere. Is it possible to obtain every rigid symmetry of the sphere as a rotation about some axis? If so, then this will imply the corresponding result for any regular polygon or polyhedron inscribed in the sphere. We shall show that every symmetry of the sphere can be obtained by a rotation about some axis. We shall do this by arguing that any symmetry has a fixed axis. Why does this prove the statement about rotations?

We shall find the fixed axis as the intersection of two planes. In order to show this, we must use reflections. Pass a plane P through the center of a sphere S. Erect a perpendicular from each point Q on the sphere to the plane P. If we continue the perpendicular through the plane, it will meet the sphere on the opposite side in a point Q'. A reflection ρ is obtained by moving every point Q to its "reflected position" Q'. The sphere intersects the plane P in a great circle C and none of the points on C will move in this exchange. So any reflection ρ of the sphere will be associated with a particular plane P passing through the center of the sphere. Reflections "mess up" the sphere in that they will change clockwise oriented circles (looking from the outside at the sphere) drawn on the sphere into counterclockwise oriented circles. But two successive reflections put things back in order. In other words, as we have defined them, two successive reflections applied to the sphere give a rigid symmetry of the sphere.

The converse is also true; that is, any rigid symmetry may be obtained as two successive reflections. To see this, we must mark the sphere so that we may note

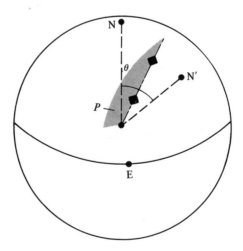

FIGURE 22

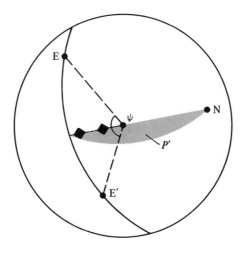

FIGURE 23

changes of position accurately. Imagine that the sphere is a globe with a marked north pole N and a 0° point E marked on the equator. Note that if we fix the center point of the sphere in space, then it is sufficient to know the positions of N and E in order to know completely the position of the sphere. Let τ be a symmetry of the sphere, and suppose that τ changes the sphere from the position in Figure 22 to the position in Figure 24. For purposes of illustration we shall assume that both N and E have moved, and that neither point has moved to an opposite position on the sphere; you may supply the argument for these special cases. Notice that N of Figure 24 is marked upon Figure 22 as N'. Connect both N and N' to the center of the sphere so that the two lines determine a plane P_0. Bisect the angle determined in this plane by N, N' and the center of the sphere. Erect a plane P on this bisector which is perpendicular

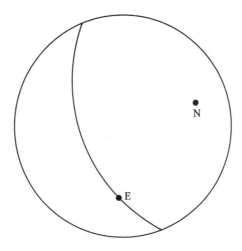

FIGURE 24

to the plane P_0. If we perform the reflection ρ in this plane P, note that N and N′ will switch positions. The reflection ρ will take Figure 22 into Figure 23. Now N is in the right position, but E is wrong. Observe that E is still on the equator, and if we mark E of Figure 24 upon Figure 23 as E′, then E′ is also upon the equator.

Again connect E and E′ of Figure 23 to the center of the sphere to give a plane P_0' and bisect the resulting angle. This time, when we erect a plane P' perpendicular to the plane P_0' containing the bisector, we obtain a plane which, because E and E′ are on the equator, also contains the point N. The reflection ρ' in the plane P' fixes N and switches E and E′. Now we have $\tau = \rho\rho'$; but even more, the points along the intersection of P with P' remain fixed under ρ and ρ'; therefore also under $\tau = \rho\rho'$. If the intersection of P and P' is a line l, then l is precisely the axis fixed by τ, and τ is obtained as a rotation about l!

There is a very general theorem about the groups of rotations of a sphere which can have finite order [7,39]. The following is a complete list of all finite rotation groups of a sphere:

(1) The group of those rotations of a regular n-sided polygon inscribed in a great circle which do *not* "turn the polygon upside down" (cyclic of order n)

(2) The group of *all* symmetries of the polygon of (1) (called a dihedral group—it has $2n$ elements and is nonabelian)

(3) The group of a regular inscribed tetrahedron (isomorphic to A_4)

(4) The group of a regular inscribed cube or octahedron (isomorphic to S_4)

(5) The group of a regular inscribed dodecahedron or icosahedron (isomorphic to A_5)

3   There are a very large number of books on group theory; the Bibliography lists a few of these. A few suggestions might help you in your search for more reading. Most textbooks do not put heavy emphasis upon geometric symmetry groups; groups of permutations serve as the primary examples of finite nonabelian groups. There are several very good general algebra books which cover material similar to that of this chapter [5,6,9,11]; in particular, Birkhoff and MacLane [3] is considered a classic, and Herstein [12] is considered a "modern classic." On a more comprehensive level are the textbooks of group theory. Lederman [35] and Macdonald [36] can be considered reasonably elementary textbooks; on the other hand, Curtis and Reiner [121], Gorenstein [29], and Huppert [32] can be considered standard references for group theorists. The books by Kaplansky [33] and Fuchs [28] concentrate upon abelian groups, while that by Kurosh [34] has become a standard reference on infinite groups. On a more recreational level are the excellent books by Weyl [185], Grossman and Magnus [30], Hilbert and Cohn-Vossen [105], and Courant and Robbins [120]; each of these contains interesting sidelights on group theory. If your library is a comprehensive one, you may want to find the collected works of G. A. Miller [127]. At the beginning and end of some volumes are comments on the history of group theory. Professor Miller was a serious student of history and wrote several articles on various aspects of the history of group theory. By browsing through the table of contents of his collected works, you will find some interesting historical reading.

chapter three

RINGS

In the previous chapter we studied groups that are algebraic systems with a single binary operation. In this chapter we will begin a study of certain algebraic systems with two related binary operations called rings. The properties which hold for a ring are all suggested by familiar properties of the integers. In fact, throughout algebra, the integers play such an important role that we shall aim our study, for the next few chapters, at determining properties of the integers. Besides the importance of the integers, there is another reason for concentrating upon this particular algebraic system. Many important algebraic systems resemble the integers very closely. Just to mention one example, recall your studies of polynomials. If you learned how to divide one polynomial into another, you probably recall that this long division process had many properties in common with long division for integers. It is the similarity of integers and polynomials which leads to the similarity of division methods for the two systems. It is the similarity of the integers with other rings, along with their familiarity, which makes them such an important example.

In studying rings we shall not forget completely about groups. Every ring has a group associated with it, and many times, more than one group. So, from time to time we will refer back to results in Chapter 2 to assist us in this chapter.

The first section of this chapter gives the definition of a ring. We then proceed in the second section to some examples used to illustrate the concepts. In the final two sections we shall carry over to rings the various theorems on homomorphisms and isomorphisms proved in Sections 2.5, 2.9, and 2.10.

3.1 DEFINITION OF A RING

We begin with a nonempty set R on which there are defined two binary operations, which we shall call "addition" and "multiplication," and for which we shall use the familiar notation. Accordingly, if $a, b \in R$, then $a + b$ and ab (or $a \cdot b$) are uniquely determined elements of the set R. We now assume the following properties or laws, in which a, b, and c are arbitrary elements, distinct or identical, of R.

R_1: $a + b = b + a$ *(commutative law of addition)*.
R_2: $(a + b) + c = a + (b + c)$ *(associative law of addition)*.
R_3: There exists an element 0 of R such that
        $a + 0 = a$ for every element a of R *(existence of a zero)*.
R_4: If $a \in R$, there exists $x \in R$ such that
        $a + x = 0$ *(existence of additive inverses)*.
R_5: $(ab)c = a(bc)$ *(associative law of multiplication)*.
R_6: $a(b + c) = ab + ac, (b + c)a = ba + ca$ *(distributive laws)*.

Under all these conditions R is said to be a *ring*. Let us repeat this definition in the following formal way.

3.1   DEFINITION If R is a nonempty set on which there are defined binary operations of addition and multiplication such that properties R_1 to R_6 hold, we say that R is a *ring* (with respect to these definitions of addition and multiplication).

Let us make a few remarks about the defining properties of a ring. First, we may emphasize that we should not think of the elements of a ring as necessarily being *numbers*. Moreover, addition and multiplication are not assumed to have any properties other than those specified. The element "0," whose existence is asserted in R_3, and which we call a *zero*, is actually an identity for the operation of addition since, by R_1, $0 + a = a + 0$ and therefore we also have $0 + a = a$. In addition, the element x whose existence is given by R_4 must satisfy the equation $x + a = a + x = 0$ by R_1. From these observations we see that properties R_1 to R_4 imply that the set R with the operation of addition is an abelian group. We shall have occasion to refer to this group, so that we call the set R with the operation of addition the *additive group of R* and denote it by R^+. The sign "+" is to remind us that addition is the only operation being considered. We may carry over directly certain properties of addition described in the previous chapter. One outcome of all this is that the additive inverse of a, namely x, whose existence is asserted by R_4, is unique. We shall denote the additive inverse of a by $-a$, and shall also often write $b + (-a)$ in the form $b - a$.

3.2　THEOREM　*Suppose that R is a ring.*

(i) *The zero of R, whose existence is asserted by* R_3 *is unique.*

(ii) *(Cancellation Laws of Addition) If a, b, and c are elements of R then the following are true:*

(a) *If* $a + b = a + c$, *then* $b = c$.

(b) *If* $b + a = c + a$, *then* $b = c$.

(iii) *The additive inverse of an element a of R is unique. Further,* $-(-a) = a$.

(iv) *If a and b are elements of R, then*

$$-(a + b) = -a - b.$$

(v) *If a and b are elements of R, the equation* $a + x = b$ *has, in R, the unique solution* $x = b - a$.

(vi) *The generalized associative and commutative laws of addition hold in R. (See 2.11 and 2.19.)*

Parts (i) to (v) follow directly from Theorem 2.3. Part (vi) follows from Theorems 2.10 and 2.18. The results of these theorems must be translated from multiplicative notation into additive notation.

All the properties used to define a ring are certainly familiar properties of the integers. Hence, with the usual definitions of addition and multiplication, the set of all integers is a ring. This ring is denoted by **Z**. For this ring, the zero whose existence is asserted in R_3 is the familiar number zero.

Now let E be the set of all *even* integers (positive, negative, and zero). Using addition and multiplication as already defined in **Z**, we see that the sum of two elements of E is an element of E, and similarly for the product of two elements. Hence, the operations of addition and multiplication, originally defined on the larger set **Z**, are also operations *on the set E*. This fact, as pointed out in Section 1.5, is expressed by saying that E is *closed* under these operations. It is easy to verify that E is itself a ring.

If all elements of a ring S are contained in a ring R, it is natural to call S a *subring* of R. It is understood that addition and multiplication of elements of S are to coincide with addition and multiplication of these elements considered as elements of the larger ring R. Naturally, a set S of elements of R cannot possibly be a subring of R unless S is closed under the operations of addition and multiplication on R since, otherwise, we would not have operations *on the set S*. We see that E, as defined above, is a subring of the ring **Z**. However, the set of all odd integers cannot be a subring of **Z** since this set is not closed under addition; that is, the sum of two odd integers is not always (in fact, is never) an odd integer.

It is important to observe that the definition of a ring does not require that the operation of multiplication be commutative. However, we shall frequently want to consider this property, so let us give it a number as follows:

R_7: If $a, b \in R$, then $ab = ba$ (*commutative law of multiplication*).

A ring which has property R_7 is called a *commutative ring*. If R_7 does not hold, that is, if there exist at least two elements c and d of R such that $cd \neq dc$, then R is said to be a *noncommutative* ring.

We may also point out that in a ring there need not be an identity for the operation of multiplication. If in a ring R there exists an identity for multiplication, we shall usually call it a *unity* of R and say that R is a *ring with unity*. For convenience of reference, let us give this property a number as follows:

R_8: There exists a nonzero element e of R such that $ea = ae = a$ for every element a of R (*existence of a unity*).

We may emphasize that a ring need not have either of the properties R_7 or R_8. However, many of the rings that we shall study in detail will have both of these properties. The ring Z is an example of a commutative ring with unity, whereas the ring E of all even integers is a commutative ring without a unity. A few cases of noncommutative rings will occur among the examples of the next section. Naturally, they will have to be quite different from the familiar number systems.

Let us look for a moment at multiplicative properties of R.

3.3 THEOREM *A ring can have at most one unity.*

The proof here is just like the uniqueness proof for the identity in a group. We leave the proof to the exercises.

If a ring has a unity e, then there may be elements a, b of the ring so that $ab = ba = e$. This situation is important enough to warrant a definition.

3.4 DEFINITION Let a be an element of a ring R with unity e. If there exists an element s of R such that $as = sa = e$, then s is called a *multiplicative inverse* of a. If a has a multiplicative inverse in R, then a is called a *unit* of R.

Observe that if a is a unit and $as = sa = e$, then s is also a unit with a multiplicative inverse a. A warning is needed at this point: the words "unit" and "unity" are different; they should not be confused. A ring can have at most one unity. If that unity is e, then $ee = e$, so that e has a multiplicative inverse, namely e itself. We conclude that e is a unit of R, but there may very well be other units in R *besides the unity*. For example, in the ring Z of integers, -1 is a unit but not a unity.

We may now prove the following theorem.

3.5 THEOREM *If R is a ring with unity e, then the set U of all units in R is a group with the group operation being multiplication in R. We call U the* group of units *of R.*

Proof: Let *a* be an element of *U*. Since $e \in U$ is the unity of *R*, we conclude that

$$ae = ea = a.$$

This statement is property G_2 for a group; that is, *U* contains an identity element. Since *a* is a unit, it has a multiplicative inverse *s* in *R*. But *a* is a multiplicative inverse of *s*, and thus *s* is also a unit and in *U*. We now have

$$as = sa = e.$$

This is property G_3 of a group; each element of *U* has a multiplicative inverse. Suppose *b* is another unit from *U* with a multiplicative inverse *t* from *U*. Since $bt = tb = e$, we must have

$$(ab)(ts) = a(b(ts)) = a((bt)s) = a(es) = as = e,$$

and, similarly,

$$(ts)(ab) = ((ts)a)b = (t(sa))b = (te)b = tb = e.$$

We conclude that *ab* is a unit in *U*. Another way of putting this is to say that multiplication is a binary operation upon *U*, or *U* is closed under multiplication. Since *U* is closed, the associative law of multiplication, R_5, is just property G_1 of a group; therefore we conclude that *U* is a group.

Combining this theorem with Theorem 2.3(ii), we have the following corollary.

3.6 COROLLARY *If R is a ring with unity and a is a unit in R, then the multiplicative inverse of a, denoted by a^{-1}, is unique. Further, $(a^{-1})^{-1} = a$.*

We turn now to the connection between multiplication and addition. The sole connection is supplied by property R_6, the distributive laws. Property R_8 does mix an additive element, zero, with a multiplicative element, unity, but only to the extent that zero and unity must be different elements. No other property supplies any connection whatsoever between multiplication and addition. Apparently, then, the distributive laws are very important. In fact, they are the reason the following theorem holds.

3.7 **THEOREM** *For each element a of a ring R, we have*

$$a \cdot 0 = 0 \cdot a = 0.$$

Proof: Since $a + 0 = a$, it follows that

$$a(a + 0) = a \cdot a.$$

But, by one of the distributive laws,

$$a(a + 0) = a \cdot a + a \cdot 0.$$

Hence,

$$a \cdot a + a \cdot 0 = a \cdot a.$$

Now we know that $a \cdot a + 0 = a \cdot a$ and, by Theorem 3.2(ii), we conclude that $a \cdot 0 = 0$.

In case R is a commutative ring, it follows from what we have just proved that also $0 \cdot a = 0$. If R is not commutative, a proof that $0 \cdot a = 0$ can easily be given using the other one of the distributive laws. This proof will be left as an exercise.

The following can now be verified in turn for arbitrary elements a, b, and c of a ring:

(i)	$a(-b) = -(ab)$,
(ii)	$(-a)b = -(ab)$,
(iii)	$(-a)(-b) = ab$,
(iv)	$a(b - c) = ab - (ac)$,
(v)	$(b - c)a = ba - (ca)$.

3.8

The proof of (i) goes as follows. We have

$$a(b + (-b)) = a \cdot 0 = 0.$$

However, by one of the distributive laws, we know that

$$a(b + (-b)) = ab + a(-b).$$

Hence,

$$ab + a(-b) = 0.$$

But since ab has a unique additive inverse $-(ab)$, it follows that $a(-b) = -(ab)$. The proofs of the other parts of 3.8 will be left as exercises.

In view of 3.8(i) and (ii), we see that

$$-(ab) = (-a)b = a(-b).$$

Accordingly, in later sections we shall usually write simply $-ab$ for any one of these equal expressions.

Recall Theorems 2.10 and 2.18. Following each theorem we remarked upon which properties were needed in a proof: for the generalized associative law we needed the associative law, R_5; for the generalized commutative law we needed both the associative law, R_5, and the commutative law, R_7. We may therefore state the following theorem.

3.9   **THEOREM**   *The generalized associative law (2.11) holds for multiplication in a ring. The generalized commutative law (2.19) holds for multiplication in a commutative ring.*

Actually, as far as the arithmetic of rings is concerned, we may carry over directly the results of Section 2.3. We may add to the list there slightly. That is, if $a, b, a_1, a_2, \ldots, a_k$ are elements of a ring R and m and n are integers, then the following equations hold.

(i) $a^m \cdot a^n = a^{m+n}, m, n > 0$.

(ii) $(a^m)^n = a^{mn}, m, n > 0$.

(iii) If R is commutative then

$$(ab)^m = a^m \cdot b^m, m, n > 0.$$

(iv) $(ma) + (na) = (m + n)a$.

3.10   (v) $n(ma) = (nm)a$.

(vi) $(ma) + (mb) = m(a + b)$.

(vii) *(Generalized Distributive Laws)*

$$a(a_1 + \cdots + a_k) = (aa_1) + (aa_2) + \cdots + (aa_k)$$

and

$$(a_1 + \cdots + a_k)a = (a_1a) + (a_2a) + \cdots + (a_ka).$$

(viii) $m(ab) = (ma)b = a(mb)$.

Just as for 2.26 to 2.28, the proof of equations like those above will be discussed in Section 4.3.

Let us now make a few remarks about the concept of *subring*. If S is a subring of the ring R, then not only is S a subset of R but the operations of addition and multiplication on S coincide with those operations as defined in the ring R. In particular, it follows that the zero of R is also the zero of S and, moreover, the additive inverse of an element of the subring S is identical with the additive inverse of this element considered as an element of R. (Why?) The following theorem, whose proof will be left as an exercise, furnishes a convenient way to determine whether a set of elements of R is actually a subring of R.

3.11 THEOREM *Let R be a ring and S a nonempty subset of the set R. Then S is a subring of R if and only if the following conditions hold:*

(i) *S is closed under the operations of addition and multiplication defined on R.*

(ii) *If $a \in S$, then $-a \in S$.*

The proof can be made to depend upon Theorem 2.30. Comparing this theorem with Theorem 2.30, we see that S is a subring of R if and only if (i) S with the operation of addition (i.e., S^+) is a subgroup of R with the operation of addition (i.e., R^+) and (ii) S is closed under the operation multiplication in R.

EXERCISES

1. Prove Theorem 3.3.

2. Prove that in a noncommutative ring R, $0 \cdot a = 0$ for any element a in R. (This exercise completes the proof of Theorem 3.7.)

3. Prove that the equations (ii) to (v) of 3.8 are all correct for a ring R.

4. Prove Theorem 3.11.

5. Give an example of a ring with exactly one element.

6. Property R_8 tells us that the unity of a ring is nonzero. Suppose that R is a ring with an element e such that $ea = ae = a$ for all elements a of R. Prove that if R contains more than one element, then e is a unity of R.

7. Let R be an arbitrary ring and $a \in R$. Prove that the set $\{x \mid x \in R, ax = 0\}$ is a subring S of R. Show that for any $y \in R$ and $z \in S$ that $zy \in S$.

8. Suppose that $a \cdot a = a^2 = a$ for all elements a in a ring R. Prove that $a + a = 2a = 0$ for all $a \in R$. Then prove that R is necessarily a commutative ring. A ring in which $a^2 = a$ for all elements a is called a *Boolean ring*.

9. Assume that G is an additive abelian group. For any elements a and b of G, set $ab = 0$. This defines a multiplication upon G. Is G a ring with the operations of addition and multiplication?

10. If S and T are subrings of a ring R, show that $S \cap T$ is a subring of R. More generally, show that the intersection of a collection of subrings of R is a subring of R.

11. Suppose that R is a ring with the following two properties:

(1) $a + a + a = 3a = 0$ for all $a \in R$.

(2) $a \cdot a \cdot a = a^3 = a$ for all $a \in R$.

Show that R is a commutative ring.

3.2 EXAMPLES OF RINGS

In order to give an example of a ring R, it is necessary to specify the elements of R and to define the operations of addition and multiplication on R so that Properties R_1 to R_6 hold. The ring $\mathbf{Z}$ of integers has been mentioned as a well-known example of a ring. Other examples are the ring of all real numbers and the ring of all complex numbers, with the usual definitions of addition and multiplication. It will be recalled that the *rational numbers* are those numbers which can be expressed in the form m/n, where m and n are integers with $n \neq 0$. With respect to the familiar definitions of addition and multiplication of rational numbers, the set of all rational numbers is also a ring. Clearly, the ring $\mathbf{Z}$ is a subring of the ring of all rational numbers; the ring of all rational numbers is a subring of the ring of all real numbers; and the ring of all real numbers is a subring of the ring of all complex numbers. All these number systems will be considered in detail in later chapters.

We proceed to give some other, less familiar, examples of rings. For the most part, we shall not write out the verifications of the properties R_1 to R_6. Some of these verifications will be required in the next list of exercises. The purpose of these examples is to clarify the concept of a ring and to show that there are rings of many different kinds.

> **Example 1** Let S be the set of all real numbers of the form $x + y\sqrt{2}$, where $x, y \in \mathbf{Z}$, with addition and multiplication defined in the usual way. It may be verified that S is closed under these operations.

Actually, S is a commutative ring with unity. Of course, it is a sub-ring of the ring of all real numbers.

Example 2 Let T be the set of all real numbers of the form $u + v\sqrt[3]{2} + w\sqrt[3]{4}$, where u, v, and w are rational numbers. Using the usual definitions of addition and multiplication, T is a commutative ring with unity.

Example 3 *The ring of integers modulo n.* Recall the group $\mathbf{Z}_n{}^+$ of Example 3 in Section 2.9 of integers modulo n, where $n \geq 1$ is a fixed integer. It consisted of the cosets of the set of all multiples of n in $\mathbf{Z}$. That is, the coset $[k]$ is equal to $\{k + qn \mid q \in \mathbf{Z}\}$ for an integer k. Let $\mathbf{Z}_n$ denote the collection of cosets in $\mathbf{Z}_n{}^+$. From $\mathbf{Z}_n{}^+$ we already have an addition on $\mathbf{Z}_n$ defined by (Equation 2.84 of Section 2.9)

3.12
$$[k] + [j] = [k + j]$$

for integers k and j. Since $\mathbf{Z}_n{}^+$ is an abelian group, properties R_1 to R_4 hold for this operation of addition.

We shall show that a multiplication upon $\mathbf{Z}_n$ is defined by setting

3.13
$$[k][j] = [kj]$$

for integers k and j. We will prove that with respect to these two operations $\mathbf{Z}_n$ is a ring called the *ring of integers modulo n*. Actually, $\mathbf{Z}_n$ is a commutative ring with unity if $n \geq 2$. In order to demonstrate these facts, we must show that 3.13 does define a binary operation upon $\mathbf{Z}_n$ and then show that R_5 to R_8 hold.

To show that Equation 3.13 defines an operation, we must show that if $[k'] = [k]$ and $[j'] = [j]$ then $[k'j'] = [kj]$. Accordingly, assume we have k', k, j', and j as above. Since $[k'] = [k]$, k' is in the coset $[k] = \{k + qn \mid q \in \mathbf{Z}\}$. In other words, $k' = k + q'n$ for some value of $q' \in \mathbf{Z}$, and similarly $j' = j + q''n$ for some value of $q'' \in \mathbf{Z}$. Multiplying k' and j', we obtain

$$\begin{aligned}
k'j' &= (k + q'n)(j + q''n) \\
&= kj + kq''n + q'nj + q'nq''n \\
&= kj + (kq'' + q'j + q'nq'')n.
\end{aligned}$$

This tells us that $k'j'$ is in the coset $[kj]$. By Lemma 2.50 we conclude that $[k'j'] = [kj]$, and, therefore, an operation is defined by 3.13.

The properties R_5 to R_8 now follow directly from their counterparts for $\mathbf{Z}$. We shall leave this verification to the exercises.

If we let $n = 6$, then $[4][5] = [20] = [2]$ since $20 = 3 \cdot 6 + 2$, and similarly $[2][3] = [0]$. The unity of $\mathbf{Z}_6$ is $[1]$. The ring $\mathbf{Z}_6$ is an example of a ring with exactly six elements.

Recall that in Example 3 of Section 2.9 we introduced the concept of congruence modulo n; that is, two integers a and b are congruent modulo n, written $a \equiv b \pmod{n}$, if $[a] = [b]$ in $\mathbf{Z}_n{}^+$, or equivalently in $\mathbf{Z}_n$. The identity 3.12 can be interpreted to say that if $a \equiv b \pmod{n}$ and $c \equiv d \pmod{n}$, then $a + c \equiv b + d \pmod{n}$, and, in addition, 3.13 says that $ac \equiv bd \pmod{n}$. For example, with $n = 6$, if we have $a = 11$, $b = -13$, $c = 3$, $d = 33$, then $11 \equiv -13 \pmod 6$ since $11 - (-13) = 24 = 4 \cdot 6$ and $3 \equiv 33 \pmod 6$ since $3 - 33 = -30 = (-5) \cdot 6$. Therefore $11 \cdot 3 \equiv (-13) \cdot 33 \pmod 6$. Checking this directly, we find that $(11 \cdot 3) - ((-13) \cdot 33) = 33 + 429 = 462 = 77 \cdot 6$, proving that $11 \cdot 3$ is congruent to $(-13) \cdot 33$ modulo 6. The foregoing congruences could have been stated by saying that $[11] = [-13]$ and $[3] = [33]$ in $\mathbf{Z}_6$ so that $[11][3] = [-13][33]$. A congruence class, for example, $[3] = \{\ldots, -9, -3, 3, 9, 15, \ldots\}$, is conceptually more difficult to think about than are integers together with the concept of congruence, so it is often easier to think of the integers 3 or 15 in place of the congruence sets $[3]$ or $[15]$ and to think of $[3] = [15]$ as $3 \equiv 15 \pmod 6$.

Example 4 Let C be the set of all functions which are continuous on the closed interval $0 \le x \le 1$, with the usual definitions of addition and multiplication of functions. Since a sum or product of two continuous functions is a continuous function, C is closed under these operations. It can be shown that C is a ring. What is the zero of C? Does it have a unity?

There are many ways in which to construct new rings from old ones. The following example is an especially important one.

Example 5 Let W be the set of all symbols of the form

$$\begin{bmatrix} a & b \\ c & d \end{bmatrix},$$

where a, b, c, and d are arbitrary elements of $\mathbf{Z}$. Our definitions of addition and multiplication are as follows:

$$\begin{bmatrix} a & b \\ c & d \end{bmatrix} + \begin{bmatrix} e & f \\ g & h \end{bmatrix} = \begin{bmatrix} a + e & b + f \\ c + g & d + h \end{bmatrix},$$

$$\begin{bmatrix} a & b \\ c & d \end{bmatrix} \cdot \begin{bmatrix} e & f \\ g & h \end{bmatrix} = \begin{bmatrix} ae + bg & af + bh \\ ce + dg & cf + dh \end{bmatrix}.$$

With respect to these definitions of addition and multiplication, W is a ring. It is called the *ring of all matrices of order 2 over the integers.* The reader may verify, by examples, that the commutative law of multiplication does not hold and hence that W is a non-commutative ring.

We may point out that the elements of W are quadruples of elements of $\mathbf{Z}$, and could just as well have been written in the form (a, b, c, d). However, the above notation is more convenient and is the traditional one.

If we modify this example by letting a, b, c and d be rational (or real, or complex) numbers instead of integers, we obtain *the ring of all matrices of order 2 over the rational (or real or complex) numbers (each of which is also noncommutative).* More general matrices will be considered in a later chapter.

For later reference we give another example of a noncommutative ring.

Example 6   Let L be the set $\mathbf{Z} \times \mathbf{Z} \times \mathbf{Z}$. That is, L is the set of all ordered triples (a, b, c), where $a, b, c \in \mathbf{Z}$. We make the following definitions:

$$(a, b, c) + (d, e, f) = (a + d, b + e, c + f),$$
$$(a, b, c)(d, e, f) = (ad, bd + ce, cf).$$

To avoid any possible confusion, we may again state that we consider two elements of a set to be equal only if they are identical. Hence, if (a, b, c) and (d, e, f) are elements of L, then $(a, b, c) = (d, e, f)$ means that $a = d$, $b = e$, and $c = f$.

It is easy to verify that $(0, 0, 0)$ is the zero of the ring L, and that $(1, 0, 1)$ is a unity. This is another noncommutative ring since, for example,

$$(0, 1, 0)(1, 0, 0) = (0, 1, 0),$$

whereas

$$(1, 0, 0)(0, 1, 0) = (0, 0, 0).$$

Let us verify one of the distributive laws for this ring. If (a, b, c), (d, e, f), and (g, h, i) are elements of L, let us show that

$$(a, b, c)((d, e, f) + (g, h, i)) = (a, b, c)(d, e, f) + (a, b, c)(g, h, i).$$

The equality of these expressions is a consequence of the following simple calculations:

$(a, b, c)((d, e, f) + (g, h, i))$
$$= (a, b, c)(d + g, e + h, f + i)$$
$$= (a(d + g), b(d + g) + c(e + h), c(f + i)),$$

and

$$(a, b, c)(d, e, f) + (a, b, c)(g, h, i)$$
$$= (ad, bd + ce, cf) + (ag, bg + ch, ci)$$
$$= (ad + ag, (bd + ce) + (bg + ch), cf + ci).$$

Example 7 Let $R = \{u, v, w, x\}$; that is, R consists of just these four elements. We define addition and multiplication in R by means of the following tables.

(+)	u	v	w	x		(·)	u	v	w	x
u	u	v	w	x		u	u	u	u	u
v	v	u	x	w		v	u	v	w	x
w	w	x	u	v		w	u	w	x	v
x	x	w	v	u		x	u	x	v	w

These tables are read just as Table 1 of Section 2.2; for example, $v + x = w$, $w + w = u$, $vw = w$, and $xx = w$. From the addition table, it is seen that the zero of the ring R is the element u; and from the multiplication table it follows that v is the unity. The reader may verify that this is a commutative ring.

We may emphasize that in this example, as in others in which addition and multiplication of more than two elements are defined by tables, it would be exceedingly tedious to verify the associative and distributive laws. Of course, the tables have not been written down at random but have been obtained by methods not yet available to the student. At present, the associative and distributive laws will have to be taken on faith but there is no real difficulty in verifying the other defining properties of a ring.

Recall that in Section 2.2 we remarked that a table was a poor way to "know" a group. Since a ring involves two tables, understandably, tables are even a worse way in which to "know" a ring. In the example above, the ring R can be described as the commutative ring with unity containing exactly four elements, such that every nonzero element of R has a multiplicative inverse. Contrast this ring R with the ring $\mathbf{Z}_4$, of integers modulo 4, which also has exactly four elements, not all of whose nonzero elements have multiplicative inverses.

Example 8 This final example is of a type quite different from any of the previous examples. Let A be a given set, and let R be the set of *all* subsets of A, including the empty set and the entire set A.

We shall now denote elements of R by lower case letters—even though they are sets of elements of A.

Our definitions of addition and multiplication are as follows. If $a, b \in R$, $a + b$ is the set of all elements of A that are in subset a or in subset b, *but not in both*. Also, we define $ab = a \cap b$, the intersection of a and b; in other words, it is the set of elements in both a and b. We may observe that $a + b$ is not, in general, the union of the sets a and b, but it will be this union whenever $a \cap b$ is the empty set. In the Venn diagram shown in Figure 1, region 1 represents those elements of A which are in neither subset a nor subset b, region 2, those elements of A which are in a but not in b, and so on. Hence ab is represented by region 4 and $a + b$ by regions 2 and 3.

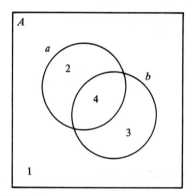

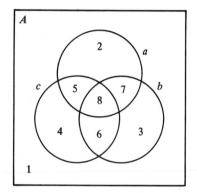

FIGURE 1

FIGURE 2

We now assert that with the above definitions of addition and multiplication, R is a commutative ring with unity.

The commutative laws of addition and multiplication are obvious, as is also the associative law of multiplication. Let us briefly consider the associative law of addition, and let a, b, and c be arbitrary elements of R. In Figure 2, $a + b$ is represented by regions 2, 3, 5, and 6. Since c is made up of regions 4, 5, 6, and 8, it follows that $(a + b) + c$ is represented by regions 2, 3, 4, and 8. This pictorial representation suggests that $(a + b) + c$ consists of those elements of A which are in exactly one of the subsets a, b, and c; together with those which are in all three. To complete the verification of the associative law of addition by means of Venn diagrams, we need to characterize the set $a + (b + c)$. We omit the details, but it is not difficult to verify that we again get the set represented by regions 2, 3, 4, and 8. Hence, $(a + b) + c = a + (b + c)$, as we wished to show. In an exercise below the reader is asked to consider how one could turn this geometrical argument into a formal proof.

If we denote the empty set by "0," it follows that $a + 0 = a$, and the empty set is the zero of the ring R. Moreover, the subset of A consisting of A itself is the unity of the ring. (Why?) If $a \in R$, it is interesting to observe that $a + a = 0$, and thus a is its own additive inverse. Another unusual property of this ring is that $a \cdot a = a$ for every element a of R. We shall refer to this ring as the *ring of all subsets of the set A*.

We conclude this section not by giving still another example of a ring but by presenting a simple, but quite useful, way to construct new rings from given rings. Suppose that R and S are rings, distinct or identical, and let us consider the Cartesian product $R \times S$ whose elements are the ordered pairs (r, s), $r \in R$, $s \in S$. On this set $R \times S$, we define addition and multiplication as follows:

$$(r_1, s_1) + (r_2, s_2) = (r_1 + r_2, s_1 + s_2),$$
$$(r_1, s_1)(r_2, s_2) = (r_1 r_2, s_1 s_2).$$

It is understood, of course, that $r_1, r_2 \in R$ and that $s_1, s_2 \in S$. Moreover, although the same symbol for addition is used in both rings, $r_1 + r_2$ is the sum of r_1 and r_2 in the ring R, and $s_1 + s_2$ is the sum of s_1 and s_2 in the ring S (and similarly for products). We leave as an exercise the proof that with respect to the above definitions the set $R \times S$ becomes a ring. In spite of the "product" notation for the set, this ring is usually called the *direct sum* of the rings R and S, and frequently denoted by $R \oplus S$. What conditions on R and S will assure us that $R \oplus S$ is commutative? That it has a unity?

Observe the similarity between the direct sum of two rings and the direct product (or sum) of two groups defined in Section 2.4. In fact, in the instance described above, the ring $R \oplus S$ contains the direct sum $R^+ \oplus S^+$ of the two additive groups R^+ and S^+. We may go even one step further to sketch the following theorem.

3.14 THEOREM *If R and S are both rings with unity, and if U and V are the groups of units for R and S, respectively, then the direct product $U \times V$ is the group of units of the direct sum ring $R \oplus S$.*

Proof: Let e be the unity of R and e' the unity of S. It is straightforward to verify that (e, e') is the unity of $R \oplus S$. We now know that $R \oplus S$ has a group of units W. The proof can be completed by showing that $W = U \times V$, that is, if $(a, b) \in R \oplus S$, then (a, b) has a multiplicative inverse in $R \oplus S$ if and only if a and b have multiplicative inverses in R and S, respectively.

EXERCISES

In these exercises, it is to be assumed that the real numbers (in particular, the rational numbers and the integers) have all the familiar properties which are freely used in elementary algebra.

1. Which of the following are rings with respect to the usual definitions of addition and multiplication? In this exercise, the ring of all even integers is denoted by E.

   (a) The set of all positive integers.

   (b) The set of all integers (positive, negative, and zero) that are divisible by 3.

   (c) The set of all real numbers of the form $x + y\sqrt{2}$, where $x, y \in E$.

   (d) The set of all real numbers of the form $x + y\sqrt[3]{2}$, where $x, y \in \mathbf{Z}$.

   (e) The set of all real numbers of the form $x + y\sqrt[3]{2} + z\sqrt[3]{4}$, where $x, y, z \in \mathbf{Z}$.

   (f) The set of all real numbers of the form $x + y\sqrt{3}$, where $x \in E$ and $y \in \mathbf{Z}$.

   (g) The set of all rational numbers that can be expressed in the form m/n where $m \in \mathbf{Z}$ and n is a positive odd integer.

2. What is the additive inverse of each element of the ring R of Example 7?

3. Verify that the subset $S = \{u, v\}$ of the ring R of Example 7 is a subring of R. Show that, except for the notation employed, this is the ring $\mathbf{Z}_2$.

4. For the ring R of Example 7, use the tables to verify each of the following:

$$(u + v) + w = u + (v + w),$$
$$(v + w) + x = v + (w + x),$$
$$w(v + x) = wv + wx,$$
$$(w + v)x = wx + vx,$$
$$(xv)w = x(vw).$$

5. For the ring L of Example 6, verify the other distributive law and the associative law of multiplication.

6. For the ring W of Example 5, verify the associative law of multiplication and the distributive laws. What is the zero of this ring? Verify that

$$\begin{bmatrix} 1 & 0 \\ 0 & 1 \end{bmatrix}$$

is a unity of W. Give examples to show that W is a noncommutative ring.

7. For the ring R of Example 8, consider how a formal proof of the associative law of addition could be given without use of Venn diagrams, and write out at least a part of the proof.

8. For the ring R of Example 8, use Venn diagrams to verify that if $a, b, c \in R$, then $a(b + c) = ab + ac$. How do you know without further calculation that the other distributive law must also hold?

9. Using the notation for complements as in Exercise 6 of Section 1.1, show that the addition used in Example 8 could have been defined as follows: $a + b = (a \cap b)' \cap (a \cup b)$.

10. In Example 8, why would we not obtain a ring if we were to define $ab = a \cap b$ and $a + b = a \cup b$?

11. On the set $S = \mathbf{Z} \times \mathbf{Z}$, let us define addition and multiplication as follows:

$$(a, b) + (c, d) = (a + c, b + d),$$
$$(a, b)(c, d) = (ac + 2bd, ad + bc).$$

Prove that S is a commutative ring with unity.

12. Show that the set C of Example 4 is a ring. Consider the subset $\{f \mid f \in C, f(\tfrac{1}{2}) = 0\}$ of all functions which vanish at $\tfrac{1}{2}$. Is this subset a subring of C?

13. Verify that properties R_5 to R_7 hold for the set $\mathbf{Z}_n$ of Example 3. If $n > 1$ show that R_8 holds. Why does R_8 not hold for $n = 1$?

14. If R and S are rings, give a detailed proof that the direct sum $R \oplus S$ is a ring.

15. What is the group of units in $\mathbf{Z}_{15}$? In $\mathbf{Z}_3 \oplus \mathbf{Z}_5$? Are these two groups isomorphic?

16. Assume that S is a ring with unity containing exactly four elements, and every nonzero element of S has a multiplicative inverse. Show that except

for the names of elements, S is the ring R of Example 7. *Hint:* Show that the group U of units of S is cyclic of order 3. From this make a multiplication table for S. Show that $2a = 0$ for every a in S. Now construct an addition table for S.

17. If a and b are any integers, let us give the following new definitions of "addition" and "multiplication," indicated respectively by "$\oplus$" and "$\odot$":

$$a \oplus b = a + b - 1, \qquad a \odot b = a + b - ab.$$

Verify that with respect to these definitions of "addition" and "multiplication," the set of all integers is a commutative ring with unity. What is the zero of this ring?

18. Complete the proof of Theorem 3.14.

19. Let R be an arbitrary ring and consider matrices

$$\begin{bmatrix} a & b \\ c & d \end{bmatrix},$$

where $a, b, c, d \in R$. If addition and multiplication are defined as in Example 5, prove that we obtain a ring. This ring is called the *ring of all matrices of order two over the ring R*.

20. Prove that if a ring R contains elements s and t such that $st \neq 0$, then the ring of all matrices of order two over R is a noncommutative ring.

21. Assume that $n = lm$ for integers l and m such that $l > 1$ and $m > 1$. Show that if we denote the elements of $Z_n{}^+$ by $[a]$ and those of $Z_m{}^+$ by $\{a\}$, then for any $[a] \in Z_n{}^+$, $[a]\phi = \{a\}$ defines a homomorphism $\phi: Z_n{}^+ \to Z_m{}^+$ of $Z_n{}^+$ onto $Z_m{}^+$. Use this to prove that for integers a and b, if $a \equiv b \pmod{n}$, then $a \equiv b \pmod{m}$.

22. Assume that m, l are integers greater than one, and a and b are arbitrary integers. Prove that if $al \equiv bl \pmod{ml}$, then $a \equiv b \pmod{m}$.

23. Suppose that $n > 1$ is an integer. Discuss the group of units in Z_n. Recall Corollary 2.55 of Section 2.7. Prove that if p is a prime, then $[a]^{p-1} = [1]$ for every integer a which does not have p as a factor. Use this fact to prove: (i) *Fermat's Little Theorem:* If a is an integer and p is a prime, then $a^p \equiv a \pmod{p}$; and (ii) if b is not divisible by p and $ab \equiv cb \pmod{p}$ for integers a and c, then $a \equiv c \pmod{p}$.

3.3 HOMOMORPHISMS AND ISOMORPHISMS

The concepts of homomorphism and isomorphism apply to rings as well as groups. In fact, many of the theorems we proved in Sections 2.5, 2.9, and 2.10 carry over directly to rings without modification. We shall work now in close analogy with those sections of Chapter 2. After reading this section, it should be quite apparent that homomorphisms play a very large role in algebra. It cannot be stressed strongly enough that the concepts of homomorphism and isomorphism are among the most important in all of algebra.

As in Section 2.5, we start with an example to illustrate the concept of homomorphism.

Example 1 Let $\mathbf{Z}$ be the ring of integers, and $\mathbf{Z}_n$ the ring of integers modulo n for a positive integer n. (See Example 3 of Section 3.2.) We define a mapping $\phi: \mathbf{Z} \to \mathbf{Z}_n$ by setting

$$3.15 \qquad k\phi = [k]$$

for $k \in \mathbf{Z}$ and $[k] = \{k + qn \mid q \in \mathbf{Z}\} \in \mathbf{Z}_n$.

From Example 2 of Section 2.5 and Example 3 of Section 2.9, we know that ϕ is a homomorphism of the *group* $\mathbf{Z}^+$ onto the *group* $\mathbf{Z}_n{}^+$. In other words,

$$3.16 \qquad (k + j)\phi = (k\phi) + (j\phi).$$

In any case, this identity is easily verified directly. For multiplication we have, by Equation 3.13,

$$3.17 \qquad (kj)\phi = [kj] = [k][j] = (k\phi)(j\phi).$$

From 3.13 and 3.14 we see that ϕ is a mapping of $\mathbf{Z}$ onto $\mathbf{Z}_n$ which preserves both addition and multiplication.

This example motivates the following definition.

3.18 DEFINITION A mapping $\theta: R \to S$ of a ring R into a ring S is called a *homomorphism* if and only if the operations of addition and multiplication are preserved under θ, that is, if and only if for arbitrary elements a, b of R, the following hold:

$$3.19 \qquad (a + b)\theta = a\theta + b\theta,$$

and

$$3.20 \qquad (ab)\theta = (a\theta)(b\theta).$$

If there exists a homomorphism of R *onto* S, we may say that R is *homomorphic* to S or that S is a *homomorphic image* of R.

In a fashion similar to Section 2.5 we define isomorphism.

3.21 DEFINITION A homomorphism which is a one-one mapping is called an *isomorphism*. If there exists an isomorphism of R onto S, we may say that R is *isomorphic* to S or that S is an *isomorphic image* of R.

Of course, just as with groups, if θ is an isomorphism of R onto S, then θ^{-1} is an isomorphism of S onto R, and we may say that S and R are isomorphic. We may write $R \cong S$ to indicate that R is isomorphic to S. As with groups, the relation "is isomorphic to," denoted by "$\cong$", is an equivalence relation upon rings.

Let us look at more examples.

Example 2 Let R and S be arbitrary rings, and let $\phi: R \times S \to R$ be the *projection* of the set $R \times S$ onto R, as defined in Example 2 of Section 1.2. Since the set $R \times S$ becomes a ring $R \oplus S$ under natural definitions of addition and multiplication, ϕ may be considered as a mapping of the ring $R \oplus S$ onto the ring R defined by

$$(r, s)\phi = r, \qquad (r, s) \in R \oplus S.$$

We assert that the operations of addition and multiplication are preserved under the mapping ϕ. That this is true for addition is a consequence of the following simple calculations:

$$[(r_1, s_1) + (r_2, s_2)]\phi = (r_1 + r_2, s_1 + s_2)\phi = r_1 + r_2$$
$$= (r_1, s_1)\phi + (r_2, s_2)\phi.$$

A similar calculation will verify that multiplication also is preserved under the mapping ϕ. Hence ϕ is a homomorphism of $R \oplus S$ onto R.

Example 3 Let $\mathbf{Z}_2$ be the ring of integers modulo 2, and R the ring of Example 8 in Section 3.2 where $A = \{x\}$ is a one-element set. Then $R = \{\varnothing, \{x\}\}$. Define $\theta: \mathbf{Z}_2 \to R$ by setting

$$[0]\theta = \varnothing$$

and

$$[1]\theta = \{x\}.$$

In the exercises it will be verified that θ is an isomorphism.

Example 4 Let W_0 be the set of all matrices of order 2 over the integers of the form

$$\begin{bmatrix} a & 0 \\ b & c \end{bmatrix}.$$

In the exercises we will show that W_0 is a subring of the ring W of Example 5 in Section 3.2. Recall the ring L of Example 6 in Section 3.2. We define a mapping $\phi: W_0 \to L$ by setting

$$\begin{bmatrix} a & 0 \\ b & c \end{bmatrix}\phi = (a, b, c).$$

That ϕ is a homomorphism follows from the calculations below.

$$\left(\begin{bmatrix} a & 0 \\ b & c \end{bmatrix} + \begin{bmatrix} d & 0 \\ e & f \end{bmatrix}\right)\phi = \begin{bmatrix} a+d & 0 \\ b+e & c+f \end{bmatrix}\phi$$

$$= (a+d, b+e, c+f)$$

$$= (a, b, c) + (d, e, f)$$

$$= \begin{bmatrix} a & 0 \\ b & c \end{bmatrix}\phi + \begin{bmatrix} d & 0 \\ e & f \end{bmatrix}\phi.$$

$$\left(\begin{bmatrix} a & 0 \\ b & c \end{bmatrix}\begin{bmatrix} d & 0 \\ e & f \end{bmatrix}\right)\phi = \begin{bmatrix} ad & 0 \\ bc+ce & cf \end{bmatrix}\phi$$

$$= (ad, bd+ce, cf)$$

$$= (a, b, c)(d, e, f)$$

$$= \left(\begin{bmatrix} a & 0 \\ b & c \end{bmatrix}\phi\right)\left(\begin{bmatrix} d & 0 \\ e & f \end{bmatrix}\phi\right).$$

In the exercises we will show that ϕ is both one-one and onto so that W_0 is isomorphic to L.

In analogy with Theorem 2.35 we have the following.

3.22 THEOREM *Let $\theta: R \to S$ be a homomorphism of the ring R into the ring S, and $S \neq \{0\}$. Then each of the following is true:*

(i) *If 0 is the zero of R, then 0θ is the zero of S.*

(ii) *If $a \in R$ then $(-a)\theta = -(a\theta)$.*

(iii) *If R has unity e and θ is an onto mapping, then S has $e\theta$ as a unity.*

(iv) *If R has a unity and U is the group of units of R, then $U\theta = \{a\theta \mid a \in U\}$ is a multiplicative group in S. In particular, if θ is an onto mapping, then $U\theta$ is a subgroup of the group of units of S, and if $a \in U$, then $(a^{-1})\theta = (a\theta)^{-1}$.*

(v) *If R is a commutative ring and θ is an onto mapping, then S is also a commutative ring.*

Parts (i), (ii), and (iv) follow directly from Theorem 2.35. Other parts of the theorem will be proved in the exercises.

Continuing our analogy with groups, we define the kernel of a homomorphism.

3.23 DEFINITION Let $\theta: R \to S$ be a homomorphism of the ring R into the ring S. The set of all elements of R which map into the zero of S is called the *kernel* of the homomorphism θ, and may be denoted by ker θ.

From 3.19 we see that θ is a homomorphism of the additive *group R^+* into the additive *group S^+*. The kernel of θ is precisely the kernel of the group homomorphism $\theta: R^+ \to S^+$. In other words, if we disregard multiplication and consider only the additive groups of rings, then ring homomorphisms become group homomorphisms and kernels stay kernels.

This observation is of some help in computing kernels. In Example 1 above, ker ϕ is the kernel of the group homomorphism $\phi: \mathbf{Z}^+ \to \mathbf{Z}_n{}^+$. In Section 2.5 we observed that ker ϕ was the set of all multiples of n, i.e., the set $(n) = [0] = \{qn \mid q \in \mathbf{Z}\}$.

In Example 2, ker ϕ is the set $\{(0, s) \mid s \in S\}$. This is easily verified since $(r, s)\phi = r = 0$ if $(r, s) \in$ ker ϕ.

In both these examples it is easy to show that ker ϕ is a subring.

3.24 THEOREM *If $\theta: R \to S$ is a ring homomorphism, then ker θ is a subring of R. If $a \in$ ker θ and $b \in R$, then both ab and ba lie in ker θ. Furthermore, θ is an isomorphism if and only if ker $\theta = \{0\}$.*

Proof: We already know by Theorem 2.37 that $(\text{ker } \theta)^+$ is a subgroup of R^+. By Theorem 3.11 it is sufficient to show that if a and b lie in ker θ, then ab lies in ker θ. This will certainly follow if we prove the second statement of the theorem. Accordingly, assume that $a \in$ ker θ and $b \in R$. Now $(ab)\theta = (a\theta)(b\theta) = 0(b\theta) = 0$ so that $ab \in$ ker θ. A similar argument shows that $ba \in$ ker θ.

We may conclude the proof using Theorem 2.37 again. The mapping $\theta: R^+ \to S^+$ is one-one (or an isomorphism of groups) if and only if ker $\theta = \{0\}$. This fact implies the final statement of the theorem. The proof of the theorem is complete.

Our final theorem of this section will be the Universal Mapping Property for ring homomorphisms.

3.25   THEOREM (*Universal Mapping Property*)   *Assume that* $\theta: R \to S$ *is a homomorphism of a ring R into a ring S, and* $\phi: R \to T$ *is a homomorphism of R onto a ring T. If* $\ker \theta \supseteq \ker \phi$ *then there is a unique homomorphism* $\psi: T \to S$ *of T into S such that* $\theta = \phi\psi$, *i.e.*

$$a\theta = a(\phi\psi) \qquad \text{for all } a \in R.$$

Further, $(\ker \theta)\phi = \ker \psi$.

Proof: From Theorem 2.38 we recall that this theorem has a diagram:

Actually, by Theorem 2.38 we do know that there is a unique *group* homomorphism ψ so that $\theta = \phi\psi$ and $(\ker \theta)\phi = \ker \psi$, where the groups in question are the additive groups R^+, S^+, and T^+ of R, S, and T, respectively.

So, to complete the proof, we need only show that ψ preserves multiplication. Accordingly, fix a and b in T. Because ϕ is onto, we may choose a' and b' in R so that $a'\phi = a$ and $b'\phi = b$. Since ϕ is a ring homomorphism, $(a'b')\phi = (a'\phi)(b'\phi) = ab$. Therefore,

$$(ab)\psi = ((a'b')\phi)\psi = (a'b')(\phi\psi) = (a'b')\theta = (a'\theta)(b'\theta)$$
$$= (a'\phi\psi)(b'\phi\psi) = (a\psi)(b\psi).$$

This concludes the proof of the Universal Mapping Property for rings.

Example 5   Let us turn once again to Example 1. We shall relate it to Example 3 of Section 2.9. In the latter example we observed that the cosets $[k]$ in $\mathbf{Z}_n$ formed a partition of $\mathbf{Z}$. This partition gives rise to an equivalence relation "$\equiv$," called congruence. That is, $a \equiv b \pmod{n}$ if and only if $[a] = [b]$, or equivalently if and only if $a - b$ is divisible by n. Since $[a] + [b] = [a + b]$ and $[a][b] = [ab]$ by Example 1, we have the following congruences: if $n > 1$ is an integer and a, b, c, d are any integers such that $a \equiv b \pmod{n}$ and $c \equiv d \pmod{n}$, then

3.26 $a + c \equiv b + d \pmod{n}$ and $ac \equiv bd \pmod{n}$.

With $n = 72$, $a = 36$, $b = 108$, $c = 29$, $d = -115$, since both $108 - 36 = 72$ and $29 - (-115) = 144$ are divisible by 72, we have $36 \equiv 108 \pmod{72}$ and $29 \equiv -115 \pmod{72}$. Consequently,

$$36 + 29 = 65 \equiv -7 = 108 + (-115) \pmod{72}$$

and

$$36 \cdot 29 = 944 \equiv -12390 = 108 \cdot (-115) \pmod{72}$$

as may be checked directly.

Let us apply these ideas to show that any integer of the form $a^2 + b^2$ (where a and b are integers) will not be of the form $4n + 3$. We shall work modulo 4. Any integer will be congruent to one of 0, 1, 2, 3 modulo 4. We must show that $a^2 + b^2$ is not congruent to 3 modulo 4. First observe that squaring 0, 1, 2, 3 tells us that any square is congruent to 0 or 1 modulo 4. Next, if we add two squares, the result must be congruent to $0 = 0 + 0$, $1 = 0 + 1$, or $2 = 1 + 1$ modulo 4. In other words, $a^2 + b^2 \equiv 0, 1, 2 \pmod 4$. We conclude that for any two integers a and b, $a^2 + b^2$ is never equal to $4n + 3$ for any value of n.

In this illustration, we have used the fact that there is a ring homomorphism $\phi: \mathbf{Z} \to \mathbf{Z}_4$. We could have argued that for no elements $[a]$ and $[b]$ is there a solution to $[a]^2 + [b]^2 = [3]$ in $\mathbf{Z}_4$. Instead of doing this, we worked with elements $a \in [a]$ and $b \in [b]$ showing that $a^2 + b^2 \equiv 3 \pmod 4$ has no solution, but this just translates into the statement that $[a]^2 + [b]^2 = [3]$ has no solution in $\mathbf{Z}_4$. Therefore, congruence modulo 4 is a disguised way of computing in $\mathbf{Z}_4$.

EXERCISES

1. Show that in $\mathbf{Z}_5$ there are no solutions to $[a]^2 = [2]$ or $[3]$. From this, argue that there are no integers a such that a^2 is of the form $5n + 2$ or $5n + 3$.

2. Show that there are no integers a such that $a^3 \equiv 2, 3, 4, 5 \pmod 7$. Then argue that for every integer a, a^3 is of the form $7n$, $7n + 1$, or $7n + 6$.

3. Verify that θ of Example 3 is an isomorphism.

4. Show that W_0 of Example 4 is a subring of W.

5. In Example 4 prove that the mapping ϕ is both one-one and onto. That is, show that $W_0 \cong L$.

6. Complete the proof of Theorem 3.22.

7. Use the Universal Mapping Property to prove there is a homomorphism of $\mathbf{Z}_{24}$ onto $\mathbf{Z}_6$. What is the kernel of this homomorphism? More generally, prove that there is a homomorphism of $\mathbf{Z}_{nm}$ onto $\mathbf{Z}_n$, where n and m are positive integers. What is the kernel in this more general situation?

8. If R and S are any two rings, then verify that the set of all elements of $R \oplus S$ of the form $(r, 0)$, $r \in R$, is a subring of $R \oplus S$ isomorphic to R.

9. If L is the ring of Example 6 of Section 3.2, show that the mapping $\theta: L \to Z$ defined by $(a, b, c)\theta = a$ is a homomorphism of L onto Z. What is the kernel of θ?

10. Prove that $\mathbf{Z}_3 \oplus \mathbf{Z}_5 \cong \mathbf{Z}_{15}$.

11. Show that the group of units of the ring $\mathbf{Z}_{13}$ is isomorphic to the additive group $\mathbf{Z}_{12}{}^+$.

12. For an integer $n > 1$ show that there is a ring $\mathscr{Z}_n$ of least residues modulo n of which $\mathscr{Z}_n{}^+$ is the additive subgroup. Show that the map $b\phi = [b]$ is an isomorphism of $\mathscr{Z}_n$ onto $\mathbf{Z}_n$.

13. Show that a homomorphic image of a ring is a ring. That is, if R is a ring, S is a set with binary operations of addition and multiplication, and $\theta: R \to S$ is a mapping of R *onto* S satisfying 3.19 and 3.20, then S is a ring.

14. If R and S are rings, prove that $R \oplus S \cong S \oplus R$.

15. Let $\theta: R \to S$ be a homomorphism of a ring R into a ring S. Show that the mapping ψ defined by

$$\begin{bmatrix} a & b \\ c & d \end{bmatrix}\psi = \begin{bmatrix} a\theta & b\theta \\ c\theta & d\theta \end{bmatrix}$$

is a homomorphism of the ring of matrices of order 2 over R into the ring of matrices of order 2 over S. (Recall Exercise 19 of Section 3.2.) What is the kernel of ψ if I is the kernel of θ?

16. Let R be a ring without unity. On the set $R \times \mathbf{Z}$ let us define addition and multiplication as follows:

$$(a, i) + (b, j) = (a + b, i + j),$$
$$(a, i)(b, j) = (ab + ja + ib, ij).$$

Prove that with respect to this addition and multiplication $R \times \mathbf{Z}$ is a ring with unity and that this ring contains a subring which is isomorphic to R.

3.4 HOMOMORPHISMS, IDEALS, AND SUBRINGS

We continue in this section to parallel the developments given in Chapter 2. Recall that if $\phi: G \to H$ is a homomorphism of groups then ker ϕ is a normal subgroup of G and the image of G in H is isomorphic to the quotient group $G/\mathrm{ker}\ \phi$. A similar situation holds for rings. In studying rings, instead of "normal subring" we use a different word, "ideal."

3.27   **DEFINITION**   A subring I of a ring R is called an *ideal* of R if for any elements a of R and b of I, both ab and ba are elements of I.†

By Theorem 3.24, if $\theta: R \to S$ is a homomorphism of rings, then ker θ is an ideal in R. We shall show that the converse statement is also true. We do this by proving analogues to Theorems 2.79 and 2.81.

We have already seen several examples of ideals in rings. These are the kernels of the homomorphisms given in Section 3.3. For example, consider the ring homomorphism $\phi: \mathbf{Z} \to \mathbf{Z}_n$ of Example 1 in Section 3.3. The kernel of ϕ is the set $(n) = \{qn \,|\, q \in \mathbf{Z}\}$ of all multiples of the fixed integer n. Clearly this set is an abelian group under addition of integers. Further, if qn is a multiple of n and r is any integer, then $r(qn) = (rq)n$ is also a multiple of n. The set of multiples of n is, therefore, an ideal in the ring of integers.

We next construct *quotient rings* in much the same way that we constructed quotient groups in Section 2.9. Let R be a fixed ring and I an ideal in R. Associated with R is its additive group R^+. Associated with I is its additive group I^+ which is a subgroup of R^+. Since R^+ is an abelian group, I^+ is a normal subgroup of R^+. Consequently, by Theorem 2.79, the quotient group R^+/I^+ of cosets of I^+ in R^+ exists with addition given by the formula 2.78 which, when rewritten additively, becomes

3.28 $$(a + I) + (b + I) = (a + b) + I, \qquad a, b \in R.$$

We now wish to define a multiplication upon the cosets of I in R.

Recall the group $\mathbf{Z}_n{}^+$ of integers modulo n defined in Example 3 of Section 2.9. In Example 3 of Section 3.2 we showed how to define a multiplication in $\mathbf{Z}_n{}^+$ via formula 3.13. Thus $\mathbf{Z}_n{}^+$ is a ring: $\mathbf{Z}_n$. The additive group $\mathbf{Z}_n{}^+$ is $\mathbf{Z}^+/(n)^+$, where (n) is the set $\{qn \,|\, q \in \mathbf{Z}\}$ of all multiples of n. We have already observed that (n) is an ideal in $\mathbf{Z}$. As a consequence, the multiplication given

---

† A useful concept in the study of noncommutative rings is that of a right (left) ideal. A subring S of a ring R is called a *right ideal* of R if for any $a \in S$ and $b \in R$, $ab \in S$; that is, right multiplication of elements of S by elements of R preserves S. Analogously, we may define left ideals of R. An ideal is both a right ideal and a left ideal, and in a commutative ring right (or left) ideals are just ideals. In general, right (left) ideals do not arise as kernels of ring homomorphisms. Accordingly, they fit more nicely into the study of modules (see Section 16.3) than into our present discussion.

by $[k][j] = [kj]$ defines a product. That is, in terms of cosets, $(k + (n))(j + (n)) = kj + (n)$. We shall mimic this definition of multiplication for our arbitrary ring R and ideal I, by defining

3.29 $$(a + I)(b + I) = (ab) + I, \qquad a, b \in R.$$

We will show that this defines a product and that 3.28 and 3.29 together give us a ring called the *quotient ring of R by I*, denoted by R/I.

Let us argue that 3.29 defines a product. Suppose that $a + I = a' + I$ and $b + I = b' + I$. Thus there are elements x and y of I such that $a = a' + x$ and $b = b' + y$. It follows that

$$\begin{aligned} ab &= (a' + x)(b' + y) \\ &= a'b' + a'y + xb' + xy \\ &= a'b' + z, \end{aligned}$$

where $z = a'y + xb' + xy$. Each of the elements $a'y$, xb', and xy lies in I since x and y do and since I is an ideal. We conclude that z lies in I. Our calculations and Lemma 2.50 together show that $(ab) + I = (a'b') + I$. Therefore, 3.29 does define a multiplication upon the cosets of I in R.

We have the following analogue to Theorem 2.79 for rings.

3.30 THEOREM *Let I be an ideal in the ring R. With respect to addition and multiplication of cosets defined by 3.28 and 3.29, the set of all cosets of I in R is a ring, usually called the* quotient ring of R by I, *and denoted by R/I. Moreover, the mapping $\theta: R \to R/I$ defined by $a\theta = a + I$, $a \in R$, is a homomorphism of R onto the ring R/I with kernel I.*

Proof: We observe that the mapping θ is a homomorphism of R into R/I in that θ preserves addition and multiplication because of the equations of 3.28 and 3.29. Clearly θ is an onto mapping. The kernel of θ is the set of all $a \in R$ such that $a\theta = a + I = 0 + I = I$, but $a + I = I$ if and only if $a \in I$. We conclude that the kernel of θ is I. The fact that R/I is a ring now follows directly from Exercise 13 of Section 3.3, that is, a homomorphic image of a ring is again a ring.

Because rings admit a quotient structure, as groups do, rings also have homomorphism theorems essentially identical to those for groups.

3.31 FUNDAMENTAL THEOREM ON RING HOMOMORPHISMS
Let $\phi: R \to S$ be a homomorphism of the ring R onto the ring S with

kernel I. Then I is an ideal of R; and $S \cong R/I$. More precisely, the mapping $\alpha: R/I \to S$ defined by

$$(a + I)\alpha = a\phi$$

is an isomorphism of R/I onto S.

Proof: Observe that I is an ideal by Theorem 3.24. If we consider only the additive groups of R, I, and S, respectively, then the theorem is an immediate consequence of Theorem 2.81. To complete the proof, we need only show that the mapping α preserves multiplication. That is, for $a + I$ and $b + I$ in R/I we have

$$((a + I)(b + I))\alpha = ((ab) + I)\alpha = (ab)\phi$$
$$= (a\phi)(b\phi) = ((a + I)\alpha)((b + I)\alpha).$$

This completes the proof of the theorem.

An analogue to Theorem 2.83 is the following.

3.32 **THEOREM** *If I and J are ideals of a ring R such that $I \subseteq J$, then J/I is an ideal of R/I and*

$$(R/I)/(J/I) \cong R/J.$$

Similarly we obtain the following Correspondence Theorem. Recall the meaning of the notation θ^{-1} appearing in Section 2.9; we shall again use that notation.

3.33 **THEOREM** *Let $\theta: R \to S$ be a homomorphism of the ring R onto the ring S, and let $I = \ker \theta$. Then each of the following is true:*

(*i*) *If J is a subring (ideal) of R, then $J\theta$ is a subring (ideal) of S.*

(*ii*) *If J is a subring of R which contains I, then $J = (J\theta)\theta^{-1}$.*

(*iii*) *If K is a subring (ideal) of S, then $K\theta^{-1}$ is a subring (ideal) of R which contains I.*

Using this theorem we may prove the following analog to Theorem 2.94.

3.34 **THEOREM** *Let I and J be subrings of a ring R with J an ideal of R. Then*

(*i*) *$I + J = \{a + b \,|\, a \in I, b \in J\}$ is a subring of R.*

(*ii*) *J is an ideal of $I + J$.*

(*iii*) *$I \cap J$ is an ideal of I.*

(*iv*) *$(I + J)/J \cong I/(I \cap J)$.*

The proofs to the preceding three theorems are left to the exercises.

It should be apparent by now that the homomorphism theorems proved in Chapter 2 are not theorems just of group theory, but are theorems of all of algebra. As such, they receive extensive use by algebraists and are among the most important theorems at this level.

EXERCISES

1. Describe all ideals in $\mathbf{Z}$.

2. Describe, using Theorem 3.31, "all" homomorphisms of $\mathbf{Z}$.

3. Prove Theorem 3.32.

4. Prove Theorem 3.33.

5. Prove Theorem 3.34.

6. Suppose that R is a finite commutative ring with unity e. If R^+ is a cyclic group generated by e, prove that R is isomorphic to $\mathbf{Z}_n$ for some n. Show that if R is not finite, then R is isomorphic to $\mathbf{Z}$. *Comment:* This result might be interpreted as the ring theoretic version of Theorem 2.41.

7. If A and B are ideals in a ring R, prove that $A \cap B$ is an ideal in R. Generalize this to any number of ideals.

8. An element a of a ring R is said to be *nilpotent* if $a^n = 0$ for some positive integer n (which may depend upon a). Prove that the set of all nilpotent elements in a commutative ring R is an ideal N of R. Prove that $0 + N$ is the only nilpotent element in R/N.

9. Assume that R is a commutative ring with unity, e. Suppose that a is a nilpotent element of R. Prove that $e + a$ is in the group of units U of R. [*Hint:* Recall that $1/(1 + x) = 1 - x + x^2 - x^3 + \cdots$.] Suppose that b is a unit of R. Is ba nilpotent? Is $b + a$ a unit?

10. For which values of n (a positive integer) will $\mathbf{Z}_n$ contain nilpotent elements?

11. Assume that in Exercise 8, $R = \mathbf{Z}_n$. Is $\mathbf{Z}_n/N$ isomorphic to $\mathbf{Z}_m$ for some m?

12. Prove that if I is a proper ideal of $\mathbf{Z}_n$, then $\mathbf{Z}_n/I$ is isomorphic to $\mathbf{Z}_m$ for some divisor m of n. Show that for each divisor m of n, $\mathbf{Z}_m$ is a homomorphic image of $\mathbf{Z}_n$.

13. If R is a ring and I, J are two ideals in R, then prove that their sum

$$I + J = \{x + y \,|\, x \in I, y \in J\}$$

is an ideal in R.

 # COMMENTARY

1 Throughout the history of mathematics, the ring of integers has played a very large role. The study of properties of the positive integers is called "number theory" (or sometimes "the higher arithmetic"). The subject has been a source for many important problems in all branches of mathematics, and as new results develop within mathematics these often are drawn into number theory to further enhance our knowledge. In one sense, ring theory may owe part of its existence to a very famous problem of number theory called "Fermat's Last Theorem." Certainly it was this famous "theorem" that led to the invention of ideals.

Fermat (1601–1665) served much of his life as a lawyer in the parliament of Toulouse, France, but studied mathematics as a pastime. He sometimes took notes upon his number theoretical observations while reading a translation by Bachet of *Diophantus' Arithmetica*, for many of his results are jotted in the margin of his copy of *Diophantus*. Something is known about the techniques used by Fermat, and most of his "marginal" theorems can be proved by his methods. However, one of his notes is a puzzle to this day. He writes that he has discovered a remarkable theorem whose proof is too long to put in the margin. That "theorem" can be stated simply as follows: If $n > 2$ is an integer, then there exist no positive integers x, y, z such that $x^n + y^n = z^n$. This "theorem" is now known generally as "Fermat's Last Theorem." A huge amount of effort has gone into attempts to prove it. In 1920 the famous number theorist L. J. Mordell (1888–1972) gave a very exciting set of three lectures on attempts to prove the "Last Theorem," and fortunately these lectures are written down [123].

In any case, the "Last Theorem" was a significant impetus to the development of a very important branch of number theory called "algebraic number theory." It was discovered that the integers might be widened to larger rings called "algebraic integers"; in these larger rings the "Last Theorem" still has meaning. For many of these rings it was possible to prove the theorem, and therefore, for a very large number of values for n, it is known that $x^n + y^n = z^n$ has no integral solutions. Unfortunately, there are still an infinite number of values for n about which nothing is known [83].

In 1843, Kummer (1810–1893) extended the notion of integer and gave a "proof" of the Last Theorem. Dirichlet (1805–1859) found that Kummer had made an incorrect assumption about factorizations of numbers, so that Kummer again worked very hard to correct his "proof." In order to rectify the erroneous factorizations, Kummer introduced the concept of "ideal numbers," which allowed him to complete his proof in certain special cases. These "ideal numbers" were not numbers but corresponded rather to sets of numbers. Dedekind (1831–1916) used this latter observation to invent the notion of an ideal, and it was at this point that ring theory was born in the guise of algebraic number theory. Kronecker (1823–1891) gave the name "order" to a ring of algebraic integers, a name used today in this same way. The word "ring" seems to have come later from Hilbert (1862–1943) and his axiomatic approach to mathematics [82,ii].

The study of rings of algebraic integers has turned out to be very fruitful. The Fundamental Theorem of Arithmetic tells us that, except for order, every positive integer $n > 1$ may be factored uniquely into a product of primes. For more general rings, unique factorization may not be true; but if one introduces certain notions of "ideal" numbers, unique factorization is retrieved. That is, ideals are expressed uniquely as "products" of other ideals in certain rings of algebraic integers. These ideas led in two directions: to the discovery of ideals and to a better understanding of what conditions lead to theorems of unique factorization.

Here then is an example of a problem which led to new mathematics: the creation of ideals and the birth of ring theory. The new mathematics only partially solved the problem for which it was created, but it turned out to be so important that this new mathematics has solved many problems different from the one which helped create it.

2   In the text it is intimated that one should think of an ideal as being the kernel of some homomorphism. Actually, ideals were studied and defined before homomorphisms, so how did the emphasis switch from the ideals to the homomorphisms? A twentieth century German mathematician was very instrumental in this change of viewpoint. Her view of algebra has been so fruitful that her hand is felt almost everywhere.

Emmy Noether's (1882–1935) effect upon algebra is difficult to measure directly since she did not publish a large number of papers and only a few of these exhibit her unifying ideas. Van der Waerden's (1903–   ) book [24] upon algebra set down in 1931 the modern view of algebra; with modifications to take into account some recent mathematics, this book remains one of the standard graduate algebra texts. In his introduction, Van der Waerden credits courses by E. Artin (1898–1962) and E. Noether for a great part of the text. There is a very definite shift in viewpoint from earlier texts to this one.

In these times of women's liberation, it is difficult to realize that even though women's rights are not fully recognized today, 50 years ago they were essentially nonexistent, even in an intellectual subject like mathematics. Emmy Noether was hindered at every turn in her attempts to study mathematics. Even for so obviously great a thinker, she could not find a good position in her own country; finally, due to the rise of the Nazis and her circumstances, she was forced to take a position at a small American college (Bryn Mawr).

The story of Emmy Noether and her influence upon algebra can be partially found in an exciting book by Constance Reid on the life of David Hilbert [150]. Other references upon her life include [146,149,193,w]. The last article by Kimberling contains references to other articles about Emmy Noether.

3   In casting around for books on elementary ring theory, you might try Refs. 3,5,6,11,12. Of the more specialized books, Herstein [45] gives a very good survey of noncommutative ring theory, and Kaplansky [47] contains some of the most recent material on commutative rings. At a lower level, but still substantial are the books on commutative rings [48] and on noncommutative rings [49]. On the recreational side are many books of elementary number theory. For example, Weiss [25] makes a point of using algebra to develop the theory. The four volume set, *The World of Mathematics* [175], is a gold mine of information on recreational mathematics. An entire chapter in the first volume is devoted to stories about number theory; and the article in Volume 1 on Cayley and Sylvester gives the story of invariant theory, a branch of mathematics which was a precursor to modern commutative ring theory.

        The historical threads running through ring theory are far more complex than those in group theory. Number theory and analysis, of course, developed the rings which are our number systems. It was not until after 1900 that a subject of ring theory itself began to evolve. And even today there is quite a clear division between those who study commutative and those who study noncommutative rings. The great triumphs of noncommutative ring theory are the theorems of Wedderburn (1882–1948) [24,49] and Goldie [45]. Commutative ring theory is much broader as it encompasses our usual number systems and polynomials. Further, in analysis, there are rings of functions of many kinds. Among these are the rings of continuous functions [44] and Banach algebras [114]. Consequently, it is very difficult to point to any particular achievement. On the other hand, certain abstract developments stand above others in importance. And one of the most significant developments, following a thread started by Hilbert (1862–1943), has been the view of ring theory expounded by Emmy Noether. In fact, that class of rings which she and others have so carefully probed are today called Noetherian rings. Included among these very important rings are the integers and the rings of polynomials [48,50].

chapter four

INTEGRAL DOMAINS

The properties which we used to define a ring were suggested by simple properties of the integers. However, since we have had numerous examples of commutative rings with unity that bear little resemblance to the ring of integers, it is clear that the system of integers must have some other properties in addition to those which make it a commutative ring with unity. Accordingly, in order to specify in some sense *all* the properties of the ring of integers, we need to consider some properties not mentioned in the previous chapter. In the present chapter we proceed to restrict the rings studied and, eventually, shall have enough properties listed that, in a sense to be described precisely later on, the *only* system which has all these properties is the ring of integers. We may then say that we have obtained a characterization of the ring of integers.

One of the properties that we shall require in characterizing the ring of integers is a property which leads in a natural way to the method of proof by mathematical induction. Accordingly, we shall introduce this important method of proof and use it to establish a few of the results that were stated without proof in Section 2.3.

We shall conclude the chapter with a few remarks about an alternate method of approaching the study of the integers in which all the familiar properties are derived from a few simple properties of the *positive* integers only.

4.1 DEFINITION OF INTEGRAL DOMAIN

We have proved that if 0 is the zero of a ring R, then $a \cdot 0 = 0 \cdot a = 0$ for every element a of R. Of course, this is a familiar property of our elementary number systems. However, in some of the rings previously mentioned there exist elements c and d, both of which are different from zero, such that $cd = 0$. For example, in the ring $\mathbf{Z}_6$ we have $[2][3] = [0]$. As another example, consider the ring of all subsets of a given set (Example 8 of Section 3.2). The empty set is the zero of this ring and, by the definition of multiplication in this ring, if c and d are subsets whose intersection is the empty set, then $cd = 0$. In discussing elements of the type just mentioned, it will be convenient to make the following definition.

4.1   **DEFINITION**   An element a of a ring R is said to be a *divisor of zero in R* if there exists a *nonzero* element c of R such that $ac = 0$ or a *nonzero* element d of R such that $da = 0$.

It is trivial that the zero of a ring R is a divisor of zero (provided R has more than one element and therefore has a nonzero element to play the role of c or d in the above definition). The elementary number systems have no divisors of zero except the zero or, as we shall say, have no nonzero divisors of zero. An alternate way of stating that a ring R has no nonzero divisors of zero is to say that it has the following property:

If $r, s \in R$ such that $rs = 0$, then $r = 0$ or $s = 0$.

We next prove the following simple result.

4.2   **THEOREM** (*Cancellation Laws of Multiplication*)   *If a is not a divisor of zero in a ring R, then each of the following holds:*

(*i*) *If $b, c \in R$ such that $ab = ac$, then $b = c$.*
(*ii*) *If $b, c \in R$ such that $ba = ca$, then $b = c$.*

Proof: Let us prove part (i) of this theorem. If $ab = ac$, it follows that $a(b - c) = 0$. Then, since a is not a divisor of zero, we must have $b - c = 0$ or $b = c$. Of course, part (ii) follows by a similar argument.

It is important to keep in mind that the cancellation laws of multiplication hold *only if a is not a divisor of zero.*

In most of this chapter we shall be studying rings without nonzero divisors of zero. In such a ring the cancellation laws of multiplication as stated in Theorem 4.2 always hold provided only that $a \neq 0$. Moreover, in order to restrict ourselves for the present to rings more like the ring of integers, we shall also re-quire our rings to be commutative and to have a unity. The next definition gives a convenient way to refer to rings having all of these properties.

4.3   **DEFINITION**†   A ring D with more than one element is called an *integral domain* if it is commutative, has a unity, and has no nonzero divisors of zero.

The most familiar examples of integral domains are the ring of integers, the ring of real numbers and the ring of complex numbers. The reader may verify that the rings of Examples 1, 2, and 7 of Section 3.2 are integral domains, whereas the rings of Examples 4, 5, 6, and 8 are not integral domains.

4.2 ORDERED INTEGRAL DOMAINS

One important property of the integers that has not been mentioned so far is that they can be *ordered*. If we think of the integers as being exhibited in the following way

$$\ldots, -4, -3, -2, -1, 0, 1, 2, 3, 4, \ldots,$$

and a and b are integers, we say that "*a* is greater than *b*" if a occurs to the right of b in the above scheme. It is clear that "*a* is greater than *b*" means merely that $a - b$ is a positive integer. This observation suggests that the concept of "order" can be defined in terms of the concept of "positive." We therefore make the following definition.

4.4   **DEFINITION**   An integral domain D is said to be an *ordered integral domain* if D contains a subset D^p with the following proper-ties:

(i) If $a, b \in D^p$, then $a + b \in D^p$ (*closed under addition*).

(ii) If $a, b \in D^p$, then $ab \in D^p$ (*closed under multiplication*).

(iii) For each element a of D exactly *one* of the following holds:
$a = 0$, $a \in D^p$, $-a \in D^p$ (*trichotomy law*).

The elements of D^p are called the *positive* elements of D. The non-zero elements of D that are not in D^p are called the *negative* elements of D.

---

† In more recent literature, an integral domain is often just called a *domain*. In older literature, one is often called a *domain of integrity*. Presumably, nonzero divisors of zero have no integrity whatsoever.

In view of the definition of an integral domain, the cancellation laws of multiplication (as stated in Theorem 4.2) are always valid in an integral domain so long as $a \neq 0$.

We may emphasize that D^p is just the notation used to designate a particular subset of an ordered integral domain D. No significance is to be attached to the use of the symbol "p" in this connection.

Obviously, the set $\mathbf{Z}^p$ of positive integers has the properties required of D^p in the above definition, and hence $\mathbf{Z}$ is an ordered integral domain. However, there are other ordered integral domains such as, for example, the integral domain of all rational numbers or the integral domain of all real numbers. However, not all integral domains are ordered integral domains. For example, we shall prove later on that the integral domain of all complex numbers has no subset with the three properties listed in the preceding definition, and therefore this integral domain is not ordered. See also Exercise 14 at the end of this section.

Now let D be any ordered integral domain, and let D^p be the set of positive elements of D, that is, the set having the three properties stated in the preceding definition. If $c, d \in D$, we *define* $c > d$ (or $d < c$) to mean that $c - d \in D^p$. Then it is clear that $a > 0$ means that $a \in D^p$, that is, that a is a positive element of D. Similarly, $a < 0$ means that $-a \in D^p$ or that a is a negative element of D. The three properties of Definition 4.4 can then be restated in the following form:

(i) If $a > 0$ and $b > 0$, then $a + b > 0$.

4.5   (ii) If $a > 0$ and $b > 0$, then $ab > 0$.

(iii) If $a \in D$, then exactly one of the following holds:

$$a = 0, \quad a > 0, \quad a < 0.$$

It is now not difficult to verify the following additional properties of inequalities:

(i) If $a > b$, then $a + c > b + c$ for every $c \in D$.

(ii) If $a > b$ and $c > 0$, then $ac > bc$.

4.6   (iii) If $a > b$ and $c < 0$, then $ac < bc$.

(iv) If $a > b$ and $b > c$, then $a > c$.

(v) If $a \neq 0$, then $a^2 > 0$.

The proof of the first of these is as follows. If $a > b$, we have $a - b > 0$. However, $a + c - (b + c) = a - b$ and we see at once that $a + c - (b + c) > 0$, that is, that $a + c > b + c$.

Let us now prove 4.6(v). If $a \neq 0$, then by the form 4.5(iii) of the trichotomy law, either $a > 0$ or $-a > 0$. If $a > 0$, it follows from 4.5(ii) that

$a^2 > 0$. If $-a > 0$, the same argument shows that $(-a)^2 > 0$. Since, by 3.8(iii), $(-a)^2 = a^2$, it follows again that $a^2 > 0$.

Proofs of the other parts of 4.6 will be left as exercises.

It is obvious that one can define $a \geq b$ (or $b \leq a$) to mean that either $a = b$ or $a > b$, without specifying which. We shall henceforth use this notation whenever convenient to do so. If $a \geq 0$, it is sometimes convenient to say that a is *nonnegative*. By writing $a < b < c$, we shall mean that $a < b$ and also that $b < c$.

In any ordered integral domain it is possible to introduce the concept of absolute value in the usual way as follows.

4.7   DEFINITION   Let D be any ordered integral domain and $a \in D$. The *absolute value* of a, written as $|a|$, is defined as follows:

(i) If $a \geq 0$, then $|a| = a$.

(ii) If $a < 0$, then $|a| = -a$.

From this definition it follows that $|0| = 0$ and that if $a \neq 0$, then $|a| > 0$.

EXERCISES

1. If a is a divisor of zero in a commutative ring R, show that ar also is a divisor of zero for every element r of R.

2. Let N be the set of all elements of an arbitrary ring R which are *not* divisors of zero. Prove that N is closed under multiplication, and verify by an example that N need not be closed under addition.

3. Prove that if a has a multiplicative inverse a^{-1} in a ring R, then a is not a divisor of zero in R.

4. Verify that each of the following is a divisor of zero in the ring of all matrices of order 2 over $\mathbf{Z}$:

$$\begin{bmatrix} 0 & 1 \\ 0 & 0 \end{bmatrix}, \quad \begin{bmatrix} 1 & 2 \\ 0 & 0 \end{bmatrix}, \quad \begin{bmatrix} 1 & 2 \\ 2 & 4 \end{bmatrix}.$$

5. Prove 4.6(ii), (iii), (iv).

In Exercises 6 to 12, the letters a, b, c, and d represent elements of an ordered integral domain.

6. Prove that if $a > b$, then $-a < -b$.

7. Prove that if $a > b$ and $c > d$, then $a + c > b + d$.

8. Prove that if a, b, c, and d are all positive with $a > b$ and $c > d$, then $ac > bd$.

9. Prove that if $a > 0$ and $ab > ac$, then $b > c$.

10. Prove that $|ab| = |a| \cdot |b|$.

11. Prove that $-|a| \le a \le |a|$.

12. Prove that $|a + b| \le |a| + |b|$.

13. Prove: There cannot be a greatest element in an ordered integral domain D (that is, for each $d \in D$ there exists $c \in D$ such that $c > d$).

14. Use the result of the preceding exercise to give a convincing argument (a formal proof is not required) why an integral domain with a finite number of elements cannot be an ordered integral domain.

15. If property 4.2(i) holds for every nonzero element a of a ring R, prove that R has no nonzero divisor of zero.

16. If property 4.2(i) holds for every nonzero element a of a ring R, prove that property 4.2(ii) also holds for every nonzero element a of R.

17. Prove that in a Boolean ring, as defined in Exercise 9 of Section 3.1, every nonzero element except the unity (if it has a unity) is a divisor of zero.

18. Prove that an isomorphic image of an integral domain is an integral domain.

19. Show that a homomorphic image of an integral domain need not be an integral domain.

4.3 WELL-ORDERING AND MATHEMATICAL INDUCTION

We need one further condition to characterize the ring of integers among the ordered integral domains. Let us first make the following general definition.

4.8   DEFINITION   A set S of elements of an ordered integral domain is said to be *well-ordered* if each nonempty subset U of S contains a least element, that is, if for each nonempty subset U of S there exists an element a of U such that $a \leq x$ for every element x of U.

It is apparent that the set of all positive integers is well-ordered, and we shall presently find that this property is precisely what distinguishes the ring of integers from other ordered integral domains. The rational numbers will be considered in detail later on in this book, but we may observe now that the set of positive rationals is not well-ordered. In fact, the set of all positive rational numbers has no least element. For if r is any positive rational number, then $r/2$ is also a positive rational number and $r/2 < r$. Hence there can be no least positive rational number.

Let us pause to clarify our point of view about the ring of integers. We have not *proved* any property of the integers, instead we have from time to time merely assumed that they have certain properties. We are now able to state precisely as follows just what properties of the integers we do wish to consider as known. *We assume that the ring of integers is an ordered integral domain in which the set of positive elements is well-ordered.* Accordingly, when we shall henceforth speak of a proof of any property of the integers we shall mean a proof based on this assumption only. In particular, we have assumed that the Well-Ordering Principle, 1.18 of Section 1.6, holds for the ring of integers. The theorem of the next section will indicate why no other properties are required.

The following theorem, which is the basis of proofs by mathematical induction, is just as "obvious" as the fact that the set of positive integers is well-ordered. However, in accordance with our chosen point of view, we shall give a proof of this result.

4.9   THEOREM   *Let K be a set of positive integers with the following two properties:*

(i) $1 \in K$.

(ii) If k is any positive integer such that $k \in K$, then also $k + 1 \in K$.

Then K consists of the set of all positive integers.

Proof: To prove this theorem, let us assume that there is a positive integer not in K, and obtain a contradiction. Let U be the set of all positive integers not in K and therefore, by our assumption, U is not empty. Then, by the well-ordering property, U must contain a least element m. Since, by (i), we have $1 \in K$, clearly $m \neq 1$ and it follows that $m > 1$ and therefore $m - 1 > 0$. Moreover, $m - 1 \in K$ since m was chosen to be the least element of U. Now, by (ii) with $k = m - 1$, we see that $m \in K$. But $m \in U$, and we have obtained the desired contradiction. The proof is therefore complete.

The most frequent application of Theorem 4.9 is to a proof of the following kind. Suppose that there is associated with each positive integer n a *statement* (or proposition) S_n, which is either true or false, and suppose we wish to prove that the statement S_n is true for every positive integer n. Let K be the set of all positive integers n such that S_n is a true statement. If we can show that $1 \in K$, and that whenever $k \in K$, then also $k + 1 \in K$, it will follow from Theorem 4.9 that K is the set of all positive integers. Since $n \in K$ merely means that S_n is true, we may reformulate these remarks in the following convenient form.

4.10   INDUCTION PRINCIPLE   *Suppose that there is associated with each positive integer n a statement S_n. Then S_n is true for every positive integer n provided the following hold:*

(*i*) *S_1 is true.*

(*ii*) *If k is any positive integer such that S_k is true, then also S_{k+1} is true.*

A proof making use of the Induction Principle (or of Theorem 4.9) is usually called a proof by induction or a proof by mathematical induction.

We may remark that there is another useful form of the Induction Principle in which condition (ii) is replaced by a somewhat different condition. (See Exercise 9 at the end of this section.)

Example As a first illustration of the language and notation just introduced, we consider a simple example from elementary algebra. It would be a very good idea to compare this proof with the proof of 1.19 in Section 1.6. There are similarities in both the statements and the proofs. If n is a positive integer, let S_n be the statement that

$$2 + 4 + 6 + \cdots + 2n = n(n + 1),$$

it being understood that the left side is the sum of the first n positive even integers. We now prove that S_n is true for every positive integer n, by verifying (i) and (ii) of 4.10. Clearly, S_1 is true since S_1 merely states that $2 = 1 \cdot 2$. Suppose, now, that k is any positive integer such that S_k is true, that is, such that the following is true:

$$2 + 4 + 6 + \cdots + 2k = k(k + 1).$$

Then, by adding the next even integer, $2(k + 1)$, to both sides we obtain

$$\begin{aligned} 2 + 4 + 6 + \cdots + 2k + 2(k + 1) &= k(k + 1) + 2(k + 1) \\ &= (k + 1)(k + 2). \end{aligned}$$

However, this calculation shows that S_{k+1} is true, and hence we have verified both (i) and (ii) of 4.10. The Induction Principle then assures us that S_n is true for every positive integer n.

We now consider again part of the material of Section 2.3, and we first illustrate by a simple example how a recursive definition really involves the Induction Principle. The recursive definition of a^n, which was given earlier, may be stated for rings in the following formal way.

4.11 DEFINITION If a is an element of a ring R, we define $a^1 = a$. Moreover, if k is a positive integer such that a^k is defined, we define $a^{k+1} = a^k \cdot a$.

Now let S_n be the statement, "a^n is defined by 4.11." The Induction Principle then shows that S_n is true for every positive integer n, that is, that a^n is defined by 4.11 for every positive integer n.

Let us now prove 2.26(i) that if m and n are arbitrary positive integers, then

4.12
$$a^m \cdot a^n = a^{m+n}.$$

Let S_n be the statement that for the positive integer n, 4.12 is true for *every* positive integer m. Then, by definition of a^{m+1}, we see that $a^m \cdot a^1 = a^{m+1}$, and hence S_1 is true. Let us now assume that k is a positive integer such that S_k is true, that is, such that

4.13
$$a^m \cdot a^k = a^{m+k}$$

for every positive integer m. Then

$$
\begin{aligned}
a^m \cdot a^{k+1} &= a^m \cdot a^k \cdot a && \textit{(by definition of } a^{k+1}\textit{)}, \\
&= a^{m+k} \cdot a && \textit{(by 4.13)}, \\
&= a^{m+k+1} && \textit{(by definition of } a^{(m+k)+1}\textit{)}.
\end{aligned}
$$

We have now shown that S_{k+1} is true, and the Induction Principle then assures us that S_n is true for every positive integer n. Thus we have given a formal proof of the very familiar law of exponents stated in 4.12. In the above proof we have tacitly made use of the associative law of multiplication.

In a similar manner the other results that were stated in Sections 2.3 and 3.1 can be established by induction. Some of them are listed in the following set of exercises.

EXERCISES

1. Prove the generalized distributive law 3.10(vii):

$$b(a_1 + a_2 + \cdots + a_n) = ba_1 + ba_2 + \cdots + ba_n.$$

2. Prove 2.26(ii) that for arbitrary positive integers m and n,

$$(a^m)^n = a^{mn}.$$

3. If a and b are elements of an abelian group, prove 2.27 that $(ab)^m = a^m b^m$ for every positive integer m.

4. If n is a positive integer and $a_1, a_2, \ldots, a_n$ are elements of an integral domain such that $a_1 a_2 \cdots a_n = 0$, show that at least one of the a's is zero.

5. If $\theta : R \to S$ is a homomorphism of the ring R into the ring S and $a \in R$, prove that $a^n \theta = (a\theta)^n$ for every positive integer n.

6. Prove (3.10(vi)) that if a and b are elements of a ring, for every integer m (positive, negative, or zero),

$$m(a + b) = ma + mb.$$

7. Prove (3.10(viii)) that if a and b are elements of a ring, for every integer m,

$$m(ab) = (ma)b = a(mb).$$

8. Prove (3.10(iv)) that if a is an element of a ring, for all integers m and n,

$$ma + na = (m + n)a.$$

Hint: Make a number of cases as follows: either m or n is zero; both m and n are positive; one of m, n is positive and the other negative; both m and n are negative.

9. Use the fact that the set of positive integers is well-ordered to prove the following alternate form of the Induction Principle:

Suppose that there is associated with each positive integer n a statement S_n. Then S_n is true for every positive integer n provided the following hold:

(i) S_1 is true.

(ii) If k is a positive integer such that S_i is true for every positive integer $i < k$, then also S_k is true.

10. Prove that the products $a_1 a_2 \cdots a_n$ are actually defined by 2.9.

11. All billiard balls are the same color. The proof is by induction upon the size n of a set S_n of billiard balls. If $n = 1$ then $S_1 = \{0\}$ contains one billiard ball, so that the statement is true if $n = 1$. Assume any set S_n of n billiard balls contains only billiard balls of one color. Consider a set of $n + 1$ billiard balls.

$$S_{n+1} = \{(0[000 \cdots 000)0]\}$$

Within ( ) and [ ] are grouped n billiard balls which all must be the same color by our induction hypothesis. Note that the balls between [ ) are the same color. So S_{n+1} contains balls of only one color. We now apply induction to conclude that all billiard balls are the same color. Comment.

4.4 A CHARACTERIZATION OF THE RING OF INTEGERS

The purpose of this section is to prove the following theorem.

4.14 THEOREM   *Let both D and D' be ordered integral domains in which the set of positive elements is well-ordered. Then D and D' are isomorphic.*

Since we are assuming that **Z** is an ordered integral domain in which the set of positive elements is well-ordered, this theorem will show that **Z** is the *only* ring with these properties (if we do not consider isomorphic rings as "different" rings).

As a first step in the proof, we shall prove a lemma. In the statement of this lemma, and henceforth whenever it is convenient to do so, we shall make use of the notation introduced in Definition 4.4 and let $\mathbf{Z}^p$ denote the set of all positive integers.

4.15 LEMMA   *Let D be an ordered integral domain in which the set D^p of positive elements is well-ordered. If e is the unity of D, then*

$$D^p = \{me \mid m \in \mathbf{Z}^p\},$$

and

$$D = \{ne \mid n \in \mathbf{Z}\}.$$

Moreover, if $n_1, n_2 \in \mathbf{Z}$ such that $n_1 e = n_2 e$, then $n_1 = n_2$.

Proof: We recall that $a \in D^p$, can also be expressed by writing $a > 0$. It is clear that $e > 0$ since $e^2 = e$, and $e^2 > 0$ by 4.6(v). For each positive integer n, let S_n be the statement that $ne > 0$. Since, by definition, $1e = e$, we have just verified the truth of S_1. If, now, k is a positive integer such that S_k is true, it follows from 4.5(i) that $(k + 1)e = ke + e > 0$, and therefore that S_{k+1} is true. We have therefore proved by mathematical induction that $me > 0$ for every positive integer m. That is, $me \in D^p$ for every positive integer m. We now proceed to show that all elements of D^p are of this form.

First, since D^p is well-ordered, D^p itself has a least element. Actually, e is this least element. For suppose that c is the least element of D^p, and that $0 < c < e$. It follows by 4.6(ii) that $0 < c^2 < c$, since $ce = c$. Hence $c^2 \in D^p$ and $c^2 < c$. However, this violates the assumption that c is the least element of D^p, and it follows that the least element of D^p is the unity e.

We can now complete the proof that every element of D^p is of the form me for some positive integer m. Suppose that this is false, and let U be the nonempty set of elements of D^p that are not of this form. Then U must have a least element, say d. We have proved that e is the least element of D^p, and hence we must have $d > e$, or $d - e > 0$. Hence, $d - e \in D^p$ and, since $e > 0$, it follows that $d - e < d$ and hence that $d - e \notin U$; therefore $d - e = m_1 e$ for some positive integer m_1. It then follows that $d = e + m_1 e = (1 + m_1)e$, and $1 + m_1$ is a positive integer. But d, being an element of U, is *not* of this form, and we have a contradiction. It follows that U must be the empty set, that is, that every element of D^p is of the required form.

It is now easy to complete the proof of the first statement of the lemma. If $a \in D$, and $a \notin D^p$, then 4.4(iii) implies that $a = 0$ or $-a \in D^p$. If $a = 0$, then $a = 0 \cdot e$. If $-a \in D^p$, then by what we have just proved, $-a = m_2 e$ for some positive integer m_2. It follows that $a = (-m_2)e$, and so every element of D is of the form ne, where n is an integer (positive, negative, or zero).

Now suppose that $n_1, n_2 \in \mathbf{Z}$ such that $n_1 e = n_2 e$. If $n_1 \neq n_2$, we can assume that the notation is so chosen that $n_1 > n_2$. It follows that $n_1 - n_2 > 0$ and, by the part of the lemma already proved, $(n_1 - n_2)e \in D^p$. Hence $(n_1 - n_2)e \neq 0$, or $n_1 e \neq n_2 e$. Thus the assumption that $n_1 \neq n_2$ leads to a contradiction, and we conclude that $n_1 = n_2$. This completes the proof of the lemma.

Incidentally, in the proof of this lemma we have proven the fact that e is the smallest positive element of D. In $\mathbf{Z}$ this says 1 is the least positive integer, a fine point mentioned in Section 1.6.

It is now easy to prove Theorem 4.14. If e and e' are the respective unities of D and of D', the lemma shows that

$$D = \{ne \mid n \in \mathbf{Z}\}$$

and

$$D' = \{ne' \mid n \in \mathbf{Z}\}.$$

Moreover, the last statement of the lemma asserts that the elements of D are *uniquely* expressible in the form $ne, n \in \mathbf{Z}$. Of course, the elements of D' are likewise uniquely expressible in the form $ne', n \in \mathbf{Z}$.

We now assert that the mapping $\theta \colon D \to D'$ defined by

$$(ne)\theta = ne', \qquad n \in \mathbf{Z},$$

is the desired isomorphism of D onto D'. By the uniqueness property just obtained, θ is a one-one mapping of D onto D'. Moreover, under this mapping we have

$$(n_1 e + n_2 e)\theta = [(n_1 + n_2)e]\theta = (n_1 + n_2)e' = n_1 e' + n_2 e'$$
$$= (n_1 e)\theta + (n_2 e)\theta$$

and

$$[(n_1 e)(n_2 e)]\theta = [(n_1 n_2)e]\theta = (n_1 n_2)e' = (n_1 e')(n_2 e')$$
$$= [(n_1 e)\theta][(n_2 e)\theta].$$

Hence, addition and multiplication are preserved and we indeed have an isomorphism. This completes the proof of the theorem.

4.5 THE PEANO AXIOMS (OPTIONAL)

So far, we have merely *assumed* that the system of all integers has the properties of an ordered integral domain in which the set of positive elements is well-ordered. In this section we shall briefly indicate how it is possible to assume as a starting point only a few simple properties of the natural numbers (positive integers) and then to *prove* all the other properties that are required. This program was first carried out by the Italian mathematician, G. Peano, and the simple properties with which we start are therefore called Peano's Axioms. If we denote by N the set of all natural numbers, these axioms are often stated as follows.†

---

† In more modern language, Axioms 2 to 4 simply assert that there exists a one-one mapping of N into N with the property that the element 1 does not occur as an image. Using Axiom 5, it is easy to show that every other element of N is an image.

Axiom 1 $1 \in N$.

Axiom 2 To each element m of N there corresponds a unique element m' of N called the *successor* of m.

Axiom 3 For each $m \in N$ we have $m' \neq 1$. (That is, 1 is not the successor of any natural number.)

Axiom 4 If $m, n \in N$ such that $m' = n'$, then $m = n$.

Axiom 5 Let K be a set of elements of N. Then $K = N$ provided the following two conditions are satisfied:

 (i) $1 \in K$.

(ii) If $k \in K$, then $k' \in K$.

This last axiom is essentially our Theorem 4.9 and is the basis of proofs by mathematical induction. In this approach to the study of the natural numbers it is taken as one of the defining properties or axioms.

Using only these five simple axioms, it is possible to *define* addition and multiplication on N and then to *prove* that N has all the properties of an integral domain except that it does not have a zero and its elements do not have additive inverses. We proceed to give the definitions of addition and multiplication, but shall not carry out the rest of the program. The details can be found in many algebra texts.

The definition of addition is as follows.

4.16 **DEFINITION** Let m be an arbitrary element of N. First, we define $m + 1 = m'$. Moreover, if $k \in N$ such that $m + k$ is defined, we define $m + k' = (m + k)'$.

By Axiom 5, it follows that the set of all elements n of N such that $m + n$ is defined by 4.16 is the set of *all* elements of N. In other words, an operation of addition is now defined on N.

The operation of multiplication is defined in a similar way as follows.

4.17 **DEFINITION** Let m be an arbitrary element of N. First, we define $m \cdot 1 = m$. Moreover, if $k \in N$ such that $m \cdot k$ is defined, we define $m \cdot k' = m \cdot k + m$.

It is also possible to define an order relation on N as follows. If $m, n \in N$, we define $m > n$ to mean that there exists an element k of N such that $m = n + k$. It can then be proved that ">" has all the properties that we would expect it to have when applied to the positive elements of an ordered integral domain. Moreover, the set N is well-ordered according to our Definition 4.8 of this concept.

Up to this point we have outlined a program for using Peano's Axioms to establish all the familiar properties of the natural numbers or positive integers. In order to obtain the *ring* of all integers, we still have to introduce into the system the negative integers and zero. This can be done by a method quite similar to that which we shall use in Chapter 6 to construct the rational numbers from the integers. Accordingly, we postpone any further discussion of this program until Section 6.7 at the end of Chapter 6.

 # COMMENTARY

1 The history of the integers as an algebraic system is quite interesting. It leads us to the philosophical question: what is a number? Our current notions about number were worked out during the nineteenth century. Kronecker's (1823–1891) view was that "God gave us the integers, the rest is the work of man." By this he meant that the only valid number systems are those built up from the integers in a finite number of steps by *finite* processes. This restriction rules out the familiar real numbers which cannot be so constructed. Opposing this were the views of Dedekind (1831–1916) and Cantor (1845–1918) who admitted both *infinite* sets of numbers and *nonfinite* operations upon these sets. Toward the end of the century, efforts were underway to reduce the concept of number to even more fundamental ideas. As we have seen, Peano (1858–1932) set down a very primitive set of axioms for the nonnegative integers.

The goal still remained to reduce the concept of number to a logical one. This program was undertaken jointly by Bertrand Russell (1872–1970) and Alfred North Whitehead (1861–1947). The basic idea is quite simple: The number 3 is the size of a set of three elements. In fact, it seems apparent that it is the only property common to all sets of three elements; so we might as well define 3 to be the collection of all such sets. Of course, our statement is somewhat circular because we have used "3" to define "three." But a difficulty like this can be circumvented by hard work. There are worse problems; we have to be very careful about our use of the word "all." In fact, "the set of all sets" is a meaningless phrase.

Suppose that the set S of all sets really is a set. We divide it into two disjoint subsets: A, the set of all sets which contain themselves as elements; and B, the set of all sets which do not contain themselves as elements. The set B is in S, so that it is an element of A or of B. If B is in A, then B is in itself, B. If B is in B, then B is an element of itself and lies in A. Since A and B are disjoint, B is in both and neither of A and B. This situation is called "Russell's Paradox," and it arises because we have tried to treat S as a set.

Needless to say, Russell and Whitehead had a very difficult time. The monument which was their effort is entombed forever as *Principia Mathematica* [96]. Russell has written an expository book [95] which expounds their point of view in *Principia*. One rather important point came out of their considerations. If one is using a symbolic language to describe mathematics, then it is possible to discuss that language in English (or many other languages). In particular, it might be that we can know the truth of a theorem in English which has no proof in the symbolic language.

Certainly then, we would want to have a *complete* symbolic language in the sense that every "true" theorem which can be stated in the language actually has a proof in that language. Kurt Gödel (1906–   ) showed in 1931 that this was too much to hope for. In fact, in any usual symbolic language, in which whole numbers and their arithmetic can be described, there must be unprovable but "true" theorems [93].

The mathematics of the number concept which has evolved since 1850 has given much fuel to the philosophers. We have already mentioned books [90,98,173,179] which discuss some of these ideas. Along a much more historical line are the excellent book by Dantzig [139], and a chapter in the book by Kramer [146].

2   Whole numbers have been around a long time, but as a part of mathematics their entry upon the Western scene in the form we know them today is relatively recent, dating from about 1200 or so [141,144]. Actually, around that time it was the practical people (merchants and farmers) who developed much about our numbers. Don't be mistaken, the Greeks did some very fine arithmetic, but not as we do it today. For example, Euclid (c. 300 B.C.) in his *Elements* proves that for numbers a and b that $(a + b)^2 = a^2 + 2ab + b^2$. What he really says in Book II, Proposition 4, is that, "If a straight line segment is cut into two smaller segments then the area of a square whose edge is the whole segment is equal to the sum of the areas of the two squares whose edges are respectively the two smaller segments plus twice the area of the rectangle whose length and width are the two smaller segments [104]." Counters were also displayed in geometric arrays to indicate whole numbers. To the Greeks and many followers, geometry was the proper domain for what we today call arithmetic.

The story of the entry of numbers and algebra (as we know this word from our high school mathematics) into mathematics is an interesting one. Arithmetic and number names as we know them evolved in India and the Middle East. For centuries such numbers were used for reckoning and keeping accounts on trade. In the Eastern world, a symbolic algebra slowly evolved. As Europe emerged from the Dark Ages, explorers such as Marco Polo brought the enlightenment of the East to the West. Arithmetic became an indispensable tool to European merchants and money lenders. The first important book on reckoning published by a European and attempting to explain the Arabic methods was probably the *Liber Abaci* of Fibonacci (1170–1250) which appeared in 1202. Fibonacci himself was a product of the merchant class in Pisa [151].

3   There is no shortage of books on our number systems. A textbook which develops the properties of our number system in a careful way is Olmsted [129]. Most textbooks along this line take their cue from the thoroughly rigorous account given in Landau [124]. A book by Waismann [184] takes a much more philosophical route. Here, as in Landau, he starts with the Peano Axioms for our number system. But the intent is not a rigorous development, but rather a philosophical discussion of various mathematical methods and their meanings. Other good books include Dubisch [8] and the appendix of Ref. 15.

chapter five

EUCLIDEAN DOMAINS

In this chapter we shall be interested in properties (usually called arithmetic properties) of integral domains, and we shall work with the integers **Z** as our analogy. We have obtained a characterization of the ring of integers much as we obtained a characterization of cyclic groups. This characterization required not only the properties associated with an integral domain but also certain additional properties of order. However, one need not require all the order properties which hold for the integers to encounter interesting rings. In fact, many of the properties we shall study now hold not only for the ring **Z** but for a larger class of integral domains which have proved to be important. In giving illustrative examples, we shall make use of the ring of integers and our familiar decimal notation, but most of the proofs will be based only upon the properties used to distinguish certain integral domains. These defining properties will depend heavily upon the fact that the set of positive integers is well-ordered. In particular, mathematical induction will play a central role in many of the proofs, although we shall frequently omit some of the details. At certain points our results are of real interest only for the domain of integers; consequently, we shall sometimes restrict our attention to the particular ring **Z**.

Our examples use only the ring of integers to illustrate the concept of a Euclidean domain. We leave to Chapter 10 the development of our other major example of Euclidean domains (rings of polynomials), since a careful development of rings of polynomials would lead us too far astray at this point. For purposes of illustration, however, we may convince ourselves that the set F of all real valued polynomial functions on the closed interval $0 \leq x \leq 1$ is a subring of the ring C of Example 4 in Section 3.2. This ring F is, in fact, an integral domain with many properties akin to the ring **Z** of integers. Long division of polynomials is strikingly similar to long division of integers. In a certain sense,

Euclidean domains are precisely those domains which do have a "long division," and, as such, F is also a Euclidean domain. A small amount of material from Chapters 6 and 7 is used in the preliminary sections of Chapter 10 but, even so, the interested reader might want to compare developments in Chapter 10 with the properties to be discussed in this chapter.

5.1 DIVISORS AND THE DIVISION ALGORITHM

We begin with the following familiar definition.

5.1   DEFINITION   Suppose that D is an integral domain. If $a, d \in D$ with $d \neq 0$, d is said to be a *divisor* (or *factor*) of a if there exists an element a_1 of D such that $a = a_1 d$. If d is a divisor of a, we say also that *d divides a* or that *a is divisible* by d.

For example, 2 is a divisor of 6 in $\mathbf{Z}$ but 4 is not. We could just as well allow d to be zero in the above definition, but this case is unimportant and it is convenient to exclude it.

We shall often write $d \mid a$ to indicate that d divides a. We now list as follows a number of simple facts involving the concept of divisor.

5.2   THEOREM   *Assume that a, b, and d are elements of an integral domain d.   Suppose that c is a unit of D.*

   (i) *If $d \mid a$ and $a \mid b$, then $d \mid b$.*

   (ii) *$d \mid a$ if and only if $d \mid (ca)$.*

   (iii) *$d \mid a$ if and only if $(cd) \mid a$.*

   (iv) *$c \mid a$ for every $a \in D$.*

   (v) *$d \mid 0$ for every nonzero $d \in D$.*

   (vi) *If $d \mid c$, then d is a unit of D.*

   (vii) *If $a \mid b$ and $b \mid a$, then $a = cb$ for some unit c of D.*

   (viii) *If $d \mid a$ and $d \mid b$, then $d \mid (ax + by)$ for arbitrary x and y of D.*

We shall prove a few of these statements and leave the proof of the others as an exercise.

Proof of 5.2(i):   If $d \mid a$, then there is an element a_1 of D so that $a = a_1 d$.   Similarly, if $a \mid b$, there is an element b_1 of D such that $b = b_1 a$.   We now have

$$b = b_1 a = b_1(a_1 d) = (b_1 a_1)d.$$

Therefore, $d \mid b$.

Proof of 5.2(vii):   Assume that $a \mid b$ and $b \mid a$.   There are elements a_1 and b_1 such that $b = b_1 a$ and $a = a_1 b$.   Suppose that e is the unity of D.   Then $eb = b = b_1 a = b_1(a_1 b) = (b_1 a_1)b$.   Since $b \mid a$, b is nonzero by Definition 5.1.   Cancelling b on the right, we obtain $e = b_1 a_1 = a_1 b_1$.   Therefore, both b_1 and a_1 have multiplicative inverses.   In other words, a_1 and b_1 are units of D.   We have shown that $a = a_1 b$ for a unit a_1 of D.

Proof of 5.2(viii):   Since $d \mid a$ and $d \mid b$, there exist elements a_1 and b_1 such that $a = a_1 d$ and $b = b_1 d$.   Hence if $x, y \in D$, by the distributive law we have $ax + by = (a_1 x + b_1 y)d$, and therefore, $d \mid (ax + by)$.

For the ring of integers $\mathbf{Z}$, we may add the following additional properties to the list in 5.2.

5.3   THEOREM   *Assume that d and a are integers.*

(*i*) *If $a \neq 0$ and $d \mid a$ then $|d| \leq |a|$.   Moreover, if $a \neq 0$, $d \mid a$, and $d \neq \pm a$, then $|d| < |a|$.*

(*ii*) *The only units of $\mathbf{Z}$ are $+1$ and -1.*

Proof: Let us first point out that one part of the proof of Lemma 4.15 implies the not surprising fact that the unity 1 of $\mathbf{Z}$ is the smallest positive integer.   Now if $d \mid a$, there exists an integer a_1 such that $a = a_1 d$, and $a \neq 0$ implies that $a_1 \neq 0$ and $d \neq 0$.   Hence $|a_1| \geq 1$ and $|a_1| \cdot |d| \geq |d|$.   However, using the well-known result of Exercise 10 of Section 4.2, it follows that $|a| = |a_1 d| = |a_1| \cdot |d| \geq |d|$.   This completes the proof of the first statement of 5.3(i).   The second statement follows from the part just proved and the observation that $|d| = |a|$ if and only if $d = \pm a$.

Next we prove part (ii).   If c is a unit of $\mathbf{Z}$ and c_1 is its multiplicative inverse, then $c_1 c = 1$ so that $c \mid 1$.   From part (i) we have $|c| \leq 1$.   But $c \neq 0$ so that $|c| > 0$.   Since c is an integer, $|c| = 1$, and we conclude that $c = \pm 1$.   The proof of the theorem is complete.

The following concept is an important one in studying divisibility properties of elements in integral domains.

5.4   DEFINITION†   A nonzero element p of an integral domain D which is not a unit is called a *prime* if its only divisors are c and cp for units c of D.

For p an integer other than ± 1 or 0, p is a prime if p is divisible only by ± 1 and $\pm p$. If $n > 1$ is an integer, it follows from this definition that n is *not* a prime if and only if there exist positive integers n_1 and n_2 with $1 < n_1 < n$ and $1 < n_2 < n$, such that $n = n_1 n_2$.

It is obvious that for any unit c of D, cp is a prime if and only if p is a prime. The first few positive prime integers are listed below:

$$2, 3, 5, 7, 11, 13, 17, 19, \ldots.$$

One of the principal reasons for the importance of primes is that for an integer n other than 0, 1, and -1, n is either a prime or can be expressed as a product of primes. This is sometimes taken for granted in arithmetic, and no doubt seems almost obvious. However, the fact that any integer can be so expressed (in only one way, in a sense to be made more precise later) is not trivial to prove and is so important that it is often called the "Fundamental Theorem of Arithmetic." This theorem is also of great importance for systems other than the ring of integers. In a later section of this chapter, we shall prove this theorem for a certain class of integral domains which includes the ring of integers.

We proceed to a consideration of the Division Algorithm stated as Theorem 1.20 but not yet proven. (Compare this theorem with Theorem 10.16 of Section 10.3.)

5.5   DIVISION ALGORITHM   *If* $a, b \in \mathbf{Z}$ *with* $b > 0$, *there exist unique integers q and r such that*

5.6 $a = qb + r, \qquad 0 \le r < b.$

Before giving a detailed proof, we can make the existence of q and r appear plausible by use of a geometric argument. Consider a coordinate line with the multiples of b marked off as in Figure 1. Then if a is marked off, it either falls on a multiple of b, say qb, or it falls between two successive multiples of b, say qb and $(q + 1)b$. (In the figure, $q = -4$.) In either case, there exists an integer q such that $qb \le a < (q + 1)b$. If we set $r = a - qb$, then $a = qb + r$ and it is clear that $0 \le r < b$.

---

† Technically, in 5.4 we have defined an *irreducible* element p of D. A *prime* p is an element which satisfies the condition that if $p \mid ab$, then $p \mid a$ or $p \mid b$ (see Lemma 5.23). In a Euclidean domain, as defined in 5.8, the words "irreducible" and "prime" mean the same thing. Because of this, even though slightly incorrect, we shall use the terminology above.

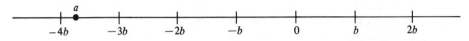

FIGURE 1

Proof: Let us now give a proof which does not make use of geometric intuition. An outline of the proof is as follows. Since 5.6 can be written in the form $r = a - qb$, we consider those nonnegative integers of the form $a - xb$, where $x \in \mathbf{Z}$, and shall show that one of them is necessarily less than b. This one we shall then identify with the integer r, whose existence we wish to establish. In order to carry out the details, let S be the set of integers defined as follows:

$$S = \{a - xb \mid x \in \mathbf{Z}, a - xb \geq 0\}.$$

First, we show that the set S is not empty. Now $b \geq 1$, since b is assumed to be a positive integer. It follows that $|a| \cdot b \geq |a|$, and hence that $a + |a| \cdot b \geq a + |a| \geq 0$. Hence, by using $x = -|a|$, we see that S contains the integer $a + |a| \cdot b$, and is therefore not empty. If $0 \in S$, clearly 0 is the least element of S. If $0 \notin S$, S is a nonempty set of positive integers and therefore has a least element, since the set of positive integers is well-ordered. Hence, in either case, S has a least element, say r. There must then exist an integer q such that $a - qb = r$. We therefore have $a = qb + r$, where $0 \leq r$, and we proceed to show that also $r < b$. Let us suppose, on the contrary, that $r \geq b$. Then $r - b \geq 0$ and, since $r - b = a - (q + 1)b$, we see that $r - b \in S$. But since $b > 0$, $r - b < r$ and we have a contradiction of the fact that r is the least element of S. Hence, $r < b$, and this completes the proof that there exist integers q and r satisfying 5.6. It remains to be proved that they are unique.

Suppose that q and r satisfy 5.6, and that also

$$a = q_1 b + r_1, \qquad 0 \leq r_1 < b.$$

Then $qb + r = q_1 b + r_1$, and it follows that

5.7
$$b(q - q_1) = r_1 - r.$$

Using the fact that the absolute value of a product is equal to the product of the absolute values, and that $|b| = b$, we obtain

$$b \cdot |q - q_1| = |r_1 - r|.$$

But, since $0 \leq r < b$ and $0 \leq r_1 < b$, we must have $|r_1 - r| < b$ and it follows that $b \cdot |q - q_1| < b$. Since $|q - q_1|$ is a nonnegative integer, this implies that $|q - q_1| = 0$, that is, that $q = q_1$; and 5.7 then shows that also $r = r_1$. We have therefore proved the uniqueness of the integers q and r satisfying 5.6.

The unique integers q and r which satisfy 5.6 are called, respectively, the *quotient* and the *remainder* in the division of a by b. It is important to observe that a is divisible by b if and only if the remainder in the division of a by b is zero.

In a numerical case, at least if $a > 0$, the actual calculation of q and r can be carried out by the familiar process of long division. The method is easily adapted also to the case in which a is negative.

As mentioned earlier, there is a process of long division for polynomials. This process also leads to a quotient and a remainder. We are not prepared now to discuss polynomials in any detail, but we want to keep in mind that there are other algebraic systems which have a "Division Algorithm." We shall give such systems a name now.

5.8   **DEFINITION**   An integral domain D is called a *Euclidean domain* if there is a mapping $\delta: D \to Z$ of D into the set of nonnegative integers which satisfies the following two properties:

(i) If a and b are nonzero elements of D, then $(ab)\delta \geq a\delta$.

(ii) If a and b are nonzero elements of D, then there exist elements q and r of D so that $a = bq + r$ and $r\delta < b\delta$.

Part (i) of this definition tells us that if $a, b \in D$ and $(ab)\delta < a\delta$, then $ab = 0$. Together, Proposition 5.3 and Theorem 5.5 (actually Exercise 7 below is also needed) tells us that if we set $a\delta = |a|$ for an integer a, then $\mathbf{Z}$ is a Euclidean domain. Property (ii) of the definition tells us that a Euclidean domain is a domain which has a "Division Algorithm." Accordingly, we shall often refer to (ii) as the "Division Algorithm." As remarked above, there are Euclidean domains which are not isomorphic to the ring of integers. In Chapter 10 we shall study one such example in some detail.

The absolute value mapping of the integers has the following values: $|0| = 0$, $|\pm 1| = 1$, $|n| > 1$ if n is not $0, \pm 1$. That is, the smallest value belongs to 0, the next smallest value belongs to the units, and all other values are larger than these two. This situation holds in any Euclidean domain.

5.9   **THEOREM**   *Assume that D is a Euclidean domain with unity e. Set $n = 0\delta$ and $m = e\delta$. Then for $a, b \in D$:*

(i) $n < m$.

(ii) *If $a \neq 0$ then $a\delta \geq m$, in particular, $a\delta = n$ if and only if $a = 0$.*

(iii) *If $a \neq 0$ then $a\delta = (ab)\delta$ if and only if b is a unit.*

(iv) *a is a unit if and only if $a\delta = m$.*

Proof: Assume first that a is nonzero. Then by (i) of Definition 5.8, $a\delta = (ea)\delta \geq e\delta = m$. This proves the first part of (ii). By (ii) of Definition 5.8 there are elements q and r so that $a = eq + r$, where $r\delta < e\delta = m$. By (ii) we conclude that $r = 0$ and $n < m$. This proves (i) and the second part of (ii).

For part (iii) assume that $a \neq 0$ and $a\delta = (ab)\delta$. By (ii) of Definition 5.8 there are elements q and r so that $a = (ab)q + r$, where $r\delta < (ab)\delta = a\delta$. Writing this another way gives $r = a - (ab)q = a(e - bq)$. Therefore, $r\delta = (a(e - bq))\delta \geq a\delta$ unless $a(e - bq) = 0$. Since $r\delta < a\delta$, we conclude that $a(e - bq) = 0$. But D is a domain, so that $e - bq = 0$ or $bq = e$. We have shown that b is a unit. Next assume that b is a unit with multiplicative inverse, c. Then $(ab)\delta \geq a\delta$ and $a\delta = (ae)\delta = (abc)\delta \geq (ab)\delta$ by (i) of Definition 5.8. We conclude that $a\delta = (ab)\delta$, completing the proof of (iii).

If we set $e = a$ in part (iii), we obtain part (iv). The proof of the theorem is now complete.

One consequence of part (iii) of this theorem is that if a, b are nonzero nonunits of D, then $a\delta < (ab)\delta$. This fact is obvious for the ring of integers when $a\delta = |a|$. The strict inequality given above will allow us to prove theorems by using the Induction Principle as given in Exercise 9 of Section 4.3.

EXERCISES

1. Prove 5.2(ii) to (vi).

2. Show that if $x = y + z$ and d is a divisor of any two of the integers x, y, and z, it is also a divisor of the third.

3. Let b and m be positive integers. If q is the quotient and r is the remainder when the integer a is divided by b, show that q is the quotient and mr is the remainder when ma is divided by mb.

4. If p and q are positive primes such that $p \mid q$, show that $p = q$.

5. If n is a positive integer and $p_1, p_2, \ldots, p_n$ are distinct positive primes, show that the integer $(p_1 p_2 \cdots p_n) + 1$ is divisible by none of these primes.

6. For each of the following pairs of integers find the quotient and the remainder in the division of the first integer by the second, and verify Equation 5.6:

    (i) 1251, 78   (iii) 4357, 418

    (ii) 31, 158   (iv) -168, 15.

7. Prove the following generalized form of the Division Algorithm. If $a, b \in \mathbf{Z}$ with $b \neq 0$, there exist unique integers q and r such that $a = qb + r$, $0 \leq r < |b|$. *Hint:* Make use of the case already proved in which b was assumed to be positive.

8. Give a formal proof that if $i \in \mathbf{Z}$, the largest integer which is less than i is $i - 1$, that is, that there exists no integer j such that $i - 1 < j < i$.

9. Let a, b, and c be integers with $b > 0$ and $c > 0$. If q is the quotient when a is divided by b and q' is the quotient when q is divided by c, prove that q' is the quotient when a is divided by bc.

10. Let $\mathbf{Q}$ be the ring of rational numbers. Set $0\delta = 0$ and $a\delta = 1$ if $a \neq 0$. Show that $\mathbf{Q}$ is a Euclidean domain. Can you invent a mapping δ so that the ring R of Example 7 of Section 3.2 is a Euclidean domain?

5.2 DIFFERENT BASES (OPTIONAL)

In this section we make use of the Division Algorithm to prove a result for the ring of integers $\mathbf{Z}$ which, although independent of the rest of this book, is of some interest in itself.

It is customary to use the integer 10 as the base of our number system. By this we mean that when we write 4371, for example, it is understood to stand for $4 \cdot 10^3 + 3 \cdot 10^2 + 7 \cdot 10 + 1$. The numbers 4, 3, 7, and 1 are called the *digits* of this number. The possible digits of a number are then the ten integers 0, 1, 2,..., 9. Actually, any positive integer greater than 1 can be used as a base in the way in which we ordinarily use 10. By this statement we mean the following.

5.10   **THEOREM**   *Let b be a positive integer greater than 1. Then every positive integer a can be expressed uniquely in the form*

5.11 $$a = r_m b^m + r_{m-1}b^{m-1} + \cdots + r_1 b + r_0,$$

where m is a nonnegative integer and the r's are integers such that

$$0 < r_m < b \quad and \quad 0 \leq r_i < b \quad for \ i = 0, 1, \ldots, m - 1.$$

If 5.11 holds, it will be convenient to say that the right side of 5.11 is "a representation of a, using the base b." We shall first sketch a proof that every positive integer has such a representation, and then prove the uniqueness. In the proof we shall use the form of the Induction Principle given in Exercise 9 of Section 4.3.

Proof: If $a < b$, then 5.11 holds with $m = 0$ and $r_0 = a$; hence every positive integer less than b has a representation, using the base b. In particular, the integer 1 has such a representation. To complete the proof by induction, let us assume that every positive integer less than a has a representation, and show that a must then have a representation. By the remarks above, the case in which $a < b$ has already been disposed of, so we may assume that $a \geq b$. By the Division Algorithm, we have that

5.12
$$a = qb + r, \qquad 0 \leq r < b,$$

with $q > 0$ since $a \geq b$, and clearly $q < a$ since $b > 1$. Therefore, by our assumption, q has a representation, using the base b. That is, we may write

$$q = s_k b^k + \cdots + s_1 b + s_0,$$

where k is a nonnegative integer, $0 < s_k < b$, and $0 \leq s_i < b$ for $i = 0, 1, \ldots, k - 1$. If we substitute this expression for q in Equation 5.12, we obtain

$$a = s_k b^{k+1} + \cdots + s_1 b^2 + s_0 b + r,$$

and it may be verified that this is a representation of the required form (with the m of 5.11 being $k + 1$). The Induction Principle thus shows that every positive integer has a representation, using the base b.

Let us now establish the *uniqueness* of the representation of a, using the base b. If $a < b$, a representation 5.11 must reduce merely to $a = r_0$ (that is, $m = 0$), and it is clear that there is no other representation. Accordingly, let us suppose that $a \geq b$ and for the purposes of our induction proof let us now assume that for every positive integer less than a there is a unique representation, using the base b. If 5.11 is one such representation for a, we may write

$$a = (r_m b^{m-1} + \cdots + r_1)b + r_0,$$

and since $0 \leq r_0 < b$, we see that r_0 is the remainder and that $r_m b^{m-1} + \cdots + r_1$ is the quotient in the division of a by b. Moreover, the quotient is greater than zero since $a \geq b$. Suppose, now, that in addition to 5.11 we have the following representation of a, using the base b:

$$a = t_n b^n + \cdots + t_1 b + t_0,$$

with the appropriate restrictions on these various numbers. Then, by the same argument as applied above to 5.11, we see that t_0 is the remainder and $t_n b^{n-1} + \cdots + t_1$ is the quotient in the division of a by b. However, in the Division Algorithm the quotient and the remainder are unique. Hence $r_0 = t_0$; also we have

$$r_m b^{m-1} + \cdots + r_1 = t_n b^{n-1} + \cdots + t_1,$$

and the two sides of this equation give representations, using the base b, of a positive integer less than a. But, by our assumption, it follows that these representations must be identical. Hence we conclude that $m = n$ and that $r_i = t_i$ for $i = 1, 2, \ldots, m$. Since we already know that $r_0 = t_0$, we see that our two representations of a are identical, and the proof of uniqueness is completed by an application of the Induction Principle.

Just as we omit the powers of 10 in the usual notation, we may specify a number a, using the base b, by giving in order the "digits" $r_m, r_{m-1}, \ldots, r_1, r_0$. In order to indicate the base being used, let us specify the number a, given by 5.11, by writing $(r_m r_{m-1} \cdots r_1 r_0)_b$. If no base is indicated, it will be understood that the base is 10. For example, $(3214)_5$ really means

$$3 \cdot 5^3 + 2 \cdot 5^2 + 1 \cdot 5 + 4,$$

and it is easily verified that $(3214)_5 = 434$. The proof of the uniqueness part of the above theorem suggests an easy way to obtain the representation of a given number a, using the base b. That is, r_0 is the remainder in the division of a by b, r_1 is the remainder in the division of the preceding quotient by b, and so on. We exhibit the calculations for 434, using the base 5:

$$
\begin{array}{r|rl}
5 & 434 & \\
\hline
5 & 86 & 4 \\
\hline
5 & 17 & 1 \\
\hline
 & 3 & 2 \\
\end{array}
$$

Here the remainders in the successive divisions are written off to the right, and the divisions are carried out until the last quotient is less than 5. We conclude, therefore, that $434 = (3214)_5$, which agrees with our previous calculations.

It is possible to carry out all the usual operations of arithmetic using entirely some fixed base other than 10. As an illustration, let us use base 5. In order to add or multiply any two numbers, we need only learn addition and multiplication tables for the integers less than 5. In designating an integer less than 5, it is not necessary to indicate whether the base is 5 or 10. However, we have $5 = (10)_5$, $3 + 4 = (12)_5$, $3 \cdot 4 = (22)_5$, and so on. The reader may verify the addition and multiplication given below, following the usual procedure of arithmetic but using base 5 throughout.

$$(3204)_5$$
$$(23)_5$$

$(3142)_5$	$(20122)_5$
$(1224)_5$	$(11413)_5$
$(4421)_5$	$(134302)_5$
Addition	*Multiplication*

Of course, any other base can be used just as well as 5. However, the only base other than 10 that is in use to any extent is the base 2, and this system is called the *binary* system. The possible "digits" in the binary system are just 0 and 1, and expressing a number in this system involves expressing it as a sum of *different* powers of 2. For example, $(1011)_2 = 2^3 + 2 + 1$. The binary system is a most convenient one for use with many of the modern high-speed computing machines. Some of these machines are so constructed that information can be fed into the machine in the usual decimal system. The machine then expresses the given numbers in the binary system, carries out the calculations in the binary system, changes the results back into the decimal system, and automatically prints the answers.

EXERCISES

1. Write each of the following numbers using the base 5 and also using the base 2:

$$24, \quad 116, \quad 412, \quad 3141, \quad 2384.$$

2. Carry out the following additions using the indicated base:

$(1130)_5$	$(2143)_5$	$(101101)_2$
$(432)_5$	$(1434)_5$	$(11011)_2$

3. Carry out the following multiplications using the indicated base:

$$
\begin{array}{ccc}
(143)_5 & (4312)_5 & (10101)_2 \\
(244)_5 & (324)_5 & (1101)_2 \\
\hline
\end{array}
$$

4. Prove that every positive integer a can be expressed in the form

$$3^m + a_{m-1}3^{m-1} + \cdots + a_1 3 + a_0$$

where m is a nonnegative integer and each of the integers $a_{m-1}, \ldots, a_1, a_0$ has the value 0, 1, or -1.

5.3 GREATEST COMMON DIVISOR

Recall the definition of an ideal I in a ring R. It is a subring such that ab and ba are in I if b is in I and a is in R. Since an integral domain is commutative, we will have $ab = ba$, so that it is only necessary in an integral domain to know that either ab or ba is in I. We wish to investigate the ideals in a Euclidean domain D where the results depend upon the Division Algorithm. It is not too surprising then that the ideals of the ring of integers are much like the ideals of other Euclidean domains in many ways.

To get started, we introduce some notation.

5.13 DEFINITION Suppose D is an integral domain and $a_1, \ldots, a_m$ are elements of D for some positive integer m. If $s_1, \ldots, s_m$ are elements of D, we shall call $s_1 a_1 + \cdots + s_m a_m$ a *linear combination* of $a_1, \ldots, a_m$. We shall denote the set of all such linear combinations by $(a_1, \ldots, a_m)$. In the special case that $m = 1$, the set (a_1) is precisely the set of all elements of D divisible by a_1. We shall call the set (a_1) *principal*. In any case, we call the elements $a_1, \ldots, a_m$ *generators* of the set $(a_1, \ldots, a_m)$.

We have already used (e.g., in Example 3 of Section 2.9 and Example 3 of Section 3.2) the notation (n) to denote the set $\{qn \mid q \in \mathbf{Z}\}$ of multiples of an integer n in $\mathbf{Z}$.

The notation $(a_1, \ldots, a_m)$ is used in several ways in this text. The context will make it clear when we mean the set of all linear combinations of the elements $a_1, \ldots, a_m$. The next theorem is true in any commutative ring with unity; but our main concern is with integral domains so that we state it for this type of ring.

5.14   THEOREM   *If D is an integral domain and $a_1, \ldots, a_m \in D$, then $(a_1, \ldots, a_m)$ is an ideal of D.*

Proof: Set $I = (a_1, \ldots, a_m)$. We must show that if $a, b \in I$ and $c \in R$, then $a - b$ and ac are elements of I. If $a = s_1 a_1 + \cdots + s_m a_m$ and $b = r_1 a_1 + \cdots + r_m a_m$ are linear combinations in D, then $a - b = (s_1 - r_1)a_1 + \cdots + (s_m - r_m)a_m$ is a linear combination. Further, $ac = (s_1 c)a_1 + \cdots + (s_m c)a_m$ is also a linear combination. Therefore, both $a - b$ and ac lie in I. By Theorem 3.11 and Definition 3.27, we conclude that I is an ideal of D.

In the integers, the ideal $(2,3)$ is all of **Z**. In fact, $1 = 1 \cdot 3 + (-1) \cdot 2$ so that $a = a \cdot 3 + (-a) \cdot 2$ for $a \in \mathbf{Z}$ lies in the ideal $(2,3)$. In particular, $(2,3) = (1) = \mathbf{Z}$. In view of Definition 5.13, the ideal $(2,3)$ is an example of a principal ideal, since it can be expressed in the form (1). The remarkable fact here is that all ideals of **Z** are principal ideals. This property holds not only for the ring of integers **Z** but also for all Euclidean domains.

5.15   THEOREM   *If D is a Euclidean domain, then every ideal in D is principal. In fact, if $I = (a)$ is an ideal in D for $a \neq 0$, then $a\delta$ is the smallest value of $b\delta$ for any nonzero $b \in I$. In particular, every ideal of **Z** is of the form (n) for some integer n.*

Proof: Let I be an ideal in D. If $I = \{0\}$, then $I = (0)$. We may therefore assume that I contains nonzero elements. We let S be the set of all integers $a\delta$ for $a \neq 0$ and $a \in I$. By Theorem 5.9 we know that $0 \leq 0\delta < e\delta \leq a\delta$, where e is the unity of D. In particular, S is a nonempty set of positive integers. Since the set of positive integers is well-ordered, S contains a least positive integer n. We may choose an element a from I so that $a\delta = n$. Notice that if D is the ring of integers, then $|a| = n$, so that $n = \pm a$. In this case, we may take a to be n itself, that is, a is the smallest positive integer in I.

Returning once again to the general situation, we wish to show that $I = (a)$, that is, for any $b \in I$, $b = aq$ for some $q \in D$. Clearly every element of the form aq lies in I; so that $I \supseteq (a)$. To show the reverse containment, let $b \in I$. By (ii) of Definition 5.8, that is, by the Division Algorithm in D, there are elements q and r so that

$$b = aq + r \qquad \text{and} \qquad r\delta < a\delta.$$

But I is an ideal, and both a and b lie in I. Therefore, $r = b - aq$ lies in I. If $r \neq 0$, then $r\delta > 0$ and $r\delta \in S$. But $a\delta$ is the smallest integer in S and $r\delta$ is smaller yet. We conclude that $r = 0$, or that $b = aq$, and therefore, $I = (a)$. This completes the proof of the theorem.

The proof contains some hidden information. Suppose that $a \neq 0$ and $I = (a) = (b)$. Then $b\delta \in S$ so that $b\delta \geq a\delta$. But $(b) = (a)$ so that $a = bc$ for some c in D. Therefore $a\delta = (bc)\delta \geq b\delta \geq a\delta$. This implies the equality $(bc)\delta = b\delta$. By Theorem 5.9(iii), c must be a unit. We have the following corollary to the proof of the theorem.

5.16 COROLLARY *If a and b are nonzero elements of a Euclidean domain D, then $(a) = (b)$ if and only if there is a unit c so that $a = bc$.*

Proof: We have shown that if $(a) = (b)$, then $a = bc$ for a unit c. If c is a unit such that $a = bc$, then $ac^{-1} = b$. Therefore $(a) = (b)$, completing the proof of the corollary.

This theorem has another very important corollary, so important, in fact, that it has received the name "Euclid's Lemma" since it appears in disguised form in Euclid's *Elements*. First we must state a definition.

5.17 DEFINITION If a and b are nonzero elements of an integral domain D, then a *greatest common divisor* (g.c.d.) of a and b is an element d with the following two properties:

(i) $d \mid a$ and $d \mid b$.

(ii) If c is any element such that $c \mid a$ and $c \mid b$ then $c \mid d$.

As an example of this definition, it is straightforward to verify for integers 15 and 10 that 5 is a g.c.d.

We shall presently show that in a Euclidean domain, not only do two nonzero elements have a g.c.d., but also that if d_1 and d_2 are both g.c.d.'s of a and b, then there is a unit c so that $d_1 = cd_2$. In the integers, the only units are ± 1; therefore, if nonzero integers a and b have a g.c.d. d, then $\pm d$ are the only g.c.d.'s of a and b. Exactly one of $\pm d$ is positive. Consequently, for integers we will stipulate that the g.c.d. of a and b *must be positive*. A consequence of this is that the g.c.d. of two nonzero integers a and b is *unique*.

We prove Euclid's Lemma next.

5.18 EUCLID'S LEMMA *Assume that a and b are nonzero elements of a Euclidean domain D. An element d of D is a g.c.d. of a and b if and only if the ideals (a, b) and (d) are equal. From this we may conclude:*

(i) A g.c.d. of a and b exists.

(ii) If d_1 and d_2 are both g.c.d.'s of a and b, then there is a unit c so that $d_1 = cd_2$.

(iii) There are elements x and y of D so that $ax + by = d$ where d is a g.c.d. of a and b.

(iv) $d\delta$ is the smallest value of $(ax_1 + by_1)\delta$ for any x_1 and y_1 of D such that $ax_1 + by_1$ is nonzero.

Proof: Assume first that $(a, b) = (d)$. Then d is a divisor of every element in $(d) = (a, b)$. Since $a = ea + 0b$ and $b = 0a + eb$, where e is the unity of D, both a and b lie in (a, b). Now we know that $d\,|\,a$ and $d\,|\,b$. Suppose that $c\,|\,a$ and $c\,|\,b$. By Proposition 5.2(viii), $c\,|\,(ax + by)$ for every x and y in D, that is, every element of (a, b) lies in (c); equivalently, $(a, b) = (d) \subseteq (c)$. In particular, d lies in (c) so that there is an element w for which $d = cw$. We have shown that $c\,|\,d$. Therefore, d is a g.c.d. of a and b.

Assume that d is a g.c.d. of a and b. Since $d\,|\,a$ and $d\,|\,b$, our argument above for (c) and (a, b) shows that $(a, b) \subseteq (d)$. By Theorem 5.15, $(a, b) = (d_1)$ for some element d_1 of D. Thus $d_1\,|\,a$ and $d_1\,|\,b$. By the definition of a g.c.d., we conclude that $d_1\,|\,d$ or that $(d_1) = (a, b) \supseteq (d)$; therefore, $(d) = (d_1) = (a, b)$.

Part (i) follows now from Theorem 5.15 and the proof above. Part (ii) follows from Corollary 5.16. Part (iii) follows from the equality $(a, b) = (d)$ and the definition of (a, b). Finally, part (iv) follows from the last part of Theorem 5.15. The proof of 5.18 is now complete.

An integral domain in which every ideal is a principal ideal is called a *principal ideal domain*, naturally. The important fact demonstrated by Theorem 5.15 is that every Euclidean domain is also a principal ideal domain. Examples are difficult to examine, but not all principal ideal domains are Euclidean domains. In particular, the subring of all complex numbers of the form $a + b(1 + \sqrt{-19})/2$, where $a, b \in \mathbf{Z}$, is a principal ideal domain but is not a Euclidean domain. At the end of this chapter is a reference to an article which proves this fact. Ultimately, we wish to show that nonzero nonunits of a Euclidean domain factor into primes in an essentially unique way. This fact is true also for principal ideal domains, but the proof is more involved than we want to get at this point. Further, the principal ideal domains which we shall encounter are all Euclidean domains. Therefore, the more general theorem is of no immediate use to us (see Section 16.2).

Although the preceding lemma establishes the existence of g.c.d.'s, its proof does not suggest a method for computing the g.c.d. of two nonzero elements. We next present a procedure, known as the *Euclidean Algorithm*, which will be useful in this connection. This process was sketched for integers in Section 1.6. We shall now see why it works.

Consider two nonzero elements a and b of a Euclidean domain. By the Division Algorithm we may write

$$a = qb + r, \qquad 0 \leq r\delta < b\delta.$$

If $r = 0$, then $b \mid a$ and the g.c.d. of a and b is b; henceforth we assume that $r \neq 0$. We now divide b by r using the Division Algorithm, getting

$$b = q_1 r + r_1, \qquad 0 \leq r_1 \delta < r\delta.$$

If $r_1 \neq 0$, we divide r by r_1, and obtain

$$r = q_2 r_1 + r_2, \qquad 0 \leq r_2 \delta < r_1 \delta$$

and repeat this process. Since $r\delta > r_1 \delta > r_2 \delta > \cdots$, and all these δ-values are nonnegative integers, we must eventually get the smallest possible δ-value. That is, we must eventually obtain a zero remainder.

If r_{k+1} is the first zero remainder, we then have the following system of equations:

5.19
$$
\begin{aligned}
a &= qb + r, & 0 &< r\delta \ \ < b\delta \\
b &= q_1 r + r_1, & 0 &< r_1 \delta < r\delta \\
r &= q_2 r_1 + r_2, & 0 &< r_2 \delta < r_1 \delta \\
r_1 &= q_3 r_2 + r_3, & 0 &< r_3 \delta < r_2 \delta \\
&\ \cdots\cdots\cdots & &\ \ \cdots\cdots\cdots \\
r_{k-2} &= q_k r_{k-1} + r_k, & 0 &< r_k \delta < r_{k-1} \delta \\
r_{k-1} &= q_{k+1} r_k.
\end{aligned}
$$

We now assert that r_k (the last nonzero remainder) is the g.c.d. of a and b. To establish this fact, we need to verify the two properties (i) and (ii) of Definition 5.17. First, let us show that r_k is a common divisor of a and b. We do so by starting with the last of Equations 5.19 and working back to the first as follows. It is clear from the last equation that r_k is a divisor of r_{k-1}. Since now r_k is a common divisor of r_k and r_{k-1}, the next-to-last equation shows that it is also a divisor of r_{k-2}. Proceeding in this way, when we get to the second equation we will know that r_k is a common divisor of r_1 and r, and hence is also a divisor of b. The first equation then shows that r_k is also a divisor of a. Hence, r_k is a common divisor of a and b, and 5.17(i) is established. To establish 5.17(ii), let c be any common divisor of a and b. We now use Equations 5.19 in the other order. The first equation shows that c is a divisor of r, the next that it is a divisor of r_1, and so on. Eventually, we find that it is a divisor of r_k, and the proof is complete.

Example Let us now give a numerical example. Suppose that we desire to compute the g.c.d. of the integers 26 and 382. By ordinary division we find that Equations 5.19 take the following form:

$$382 = 14 \cdot 26 + 18,$$
$$26 = 1 \cdot 18 + 8,$$
$$18 = 2 \cdot 8 + 2,$$
$$8 = 4 \cdot 2.$$

In this case, 2 is the g.c.d. since it is the last nonzero remainder.

Not only is the Euclidean Algorithm useful in computing the g.c.d. of two elements, but it is also useful in expressing the g.c.d. of two integers in the form given by (iii) of Euclid's Lemma. That is, a g.c.d. of a and b is a linear combination $ax + by$ of a and b. We show how to find such a linear combination now.

Each of the remainders in Equations 5.19 can be expressed, in turn, as a linear combination of a and b. From the first of Equations 5.19, we see that

5.20 $$r = a - qb,$$

and hence r is a linear combination of a and b. Substituting this expression for r in the second equation, and solving for r_1, we get

5.21 $$r_1 = b - q_1(a - qb) = (1 + q_1q)b - q_1a,$$

and hence r_1 is a linear combination of a and b. Now substituting from 5.20 and 5.21 into the third of Equations 5.19, we obtain

$$r_2 = r - q_2r_1 = (a - qb) - q_2[(1 + q_1q)b - q_1a]$$
$$= (1 + q_2q_1)a - (q + q_2 + q_2q_1q)b,$$

so that r_2 is a linear combination of a and b. By continuing in this way, we see that each remainder and, in particular, the g.c.d. r_k is expressible as a linear combination of a and b (see Exercise 8 below). We are not here interested in general formulas which express these remainders as linear combinations of a and b since, in a numerical case, it is easy to compute, in turn, each of these linear combinations.

Example As an example, let us carry out the calculations for the case in which $a = 382$ and $b = 26$. The Euclidean Algorithm has been applied above to these two integers to find that their g.c.d. is 2. Let us now use the equations previously exhibited to express each of the remainders as a linear combination of 382 and 26. For simplicity, we shall write a in place of 382 and b in place of 26. The calculations are as follows:

$$18 = a - 14b,$$
$$8 = b - 18 = b - (a - 14b) = 15b - a,$$
$$2 = 18 - 2 \cdot 8 = a - 14b - 2(15b - a) = 3a - 44b.$$

Hence,

$$2 = 3 \cdot 382 - 44 \cdot 26,$$

and we have expressed the g.c.d. of 382 and 26 as a linear combination of these two integers.

We shall sometimes find it convenient to let (a, b) designate the g.c.d. of two integers a and b. Thus, for example, we have that $(382, 26) = 2$. There can be no possible confusion with other uses of the number pair notation since the context will make it clear that we are considering the g.c.d. of two integers and not, for example, the coordinates of a point in the plane, or the ideal generated by the g.c.d.

We shall frequently need to refer to a pair of elements with e as their g.c.d. Accordingly, it is convenient to make the following definition.

5.22   DEFINITION   The elements a and b of the Euclidean domain D are said to be *relatively prime* if and only if a g.c.d. of a and b is e, that is, if and only if $(a, b) = e$, the unity of D.

This definition tells us that a and b are relatively prime if and only if the ideal (a, b) is equal to the ideal $(e) = D$.

Observe that a g.c.d. of $a, b \in D$ is denoted by $(a, b) = d$. By Euclid's Lemma, 5.18, the ideal $(a, b) = (d)$. In other words, the notation (a, b) for a g.c.d. is already suggested by the notation for the ideal (a, b).

EXERCISES

Unless otherwise specified, the letters represent arbitrary nonzero integers.

1. Find the g.c.d. of each of the following pairs of integers and express it as a linear combination of the two integers: (i) 52 and 38, (ii) 81 and 110, (iii) 320 and 112, (iv) 7469 and 2387, (v) 10,672 and -4147.

2. Show that a and b are relatively prime if and only if 1 is expressible as a linear combination of a and b.

3. If $d = (a, b)$ and $a = a_1 d$, $b = b_1 d$, show that $(a_1, b_1) = 1$.

4. If m is a positive integer, show that $(ma, mb) = m(a, b)$.

5. Show each of the following:

   (i) If p is a positive prime and a is a nonzero integer, then either $(a, p) = 1$ or $(a, p) = p$.

   (ii) If p and q are distinct positive primes, then 1 is expressible as a linear combination of p and q.

6. If $x = yz + t$, prove that $(x, z) = (z, t)$.

7. Prove that $(a, bc) = 1$ if and only if $(a, b) = 1$ and $(a, c) = 1$.

8. Write out a formal proof that every remainder in Equations 5.19 is expressible as a linear combination of a and b. *Hint:* Assume that this is false, and obtain a contradiction.

9. If a, b, and n are given, prove that n is expressible as a linear combination of a and b if and only if $(a, b)|n$.

10. (i) Define the g.c.d. of *three* nonzero integers.

   (ii) Establish the existence of the g.c.d. of three integers by proving a result analogous to Euclid's Lemma 5.18.

   (iii) If d is the g.c.d. of a, b, and c, show that $d = ((a, b), c) = ((a, c), b) = (a, (b, c))$.

   (iv) Let d be the g.c.d. of a, b, and c. If $a = a_1 d$, $b = b_1 d$, and $c = c_1 d$, show that 1 is the g.c.d. of the three integers a_1, b_1, and c_1.

5.4 THE FUNDAMENTAL THEOREM

The principal theorem to be proved in this section has to do with the factorizations of an integer into a product of primes. We begin with the following important preliminary result.

5.23 **LEMMA**† *If a and b are nonzero elements of a Euclidean domain D and p is a prime such that $p \mid ab$, then $p \mid a$ or $p \mid b$.*

Proof: To prove this lemma, let us suppose that p does not divide a, and show that $p \mid b$. Since p is a prime which is not a divisor of a, the definition of a prime implies that $(a, p) = e$. Then Euclid's Lemma shows that there exist elements x and y such that $e = ax + py$. Multiplying by b, we obtain

$$b = abx + bpy.$$

Since we are given that $p \mid ab$, clearly p divides the right member of this equation, and therefore divides b.

It is almost obvious that the preceding lemma can be generalized to apply to a product of more than two elements. For future reference we now state this more general result.

5.24 **LEMMA** *Let p be a prime and m an arbitrary positive integer. If $a_1, a_2, \ldots, a_m$ are nonzero elements of a Euclidean domain D such that $p \mid (a_1 a_2 \cdots a_m)$, then $p \mid a_i$ for at least one i, $1 \leq i \leq m$.*

This lemma is easily established by induction, and the proof will be left as an exercise. The case in which $m = 2$ is covered by the preceding lemma.

It is easy to verify, for example, that $60 = 2 \cdot 2 \cdot 3 \cdot 5$, and hence that 60 can be expressed as a product of positive primes. We could also write $60 = 2 \cdot 5 \cdot 3 \cdot 2$, but we shall not consider these two factorizations as essentially different since they differ only in the order in which the prime factors are written down. With this understanding, it is true that 60 has only one factorization into a product of positive primes. We could also have factored 60 as $2 \cdot (-2) \cdot 3 \cdot (-5)$. That is, we can multiply the primes by units (in this case ± 1) to alter the appearance of the factors. Somehow, even doing this does not change the essence of the factorization. Let us formulate these ideas.

We say that a and b are *equal up to units*‡ if there is some unit c in D so that $a = cb$. Since $a = cb$ implies that $c^{-1}a = b$, Corollary 5.16 tells us that a and b are equal up to units if and only if a and b generate the same principal ideal, that is, $(a) = (b)$. In the integers, if $a = 3$, then b must be either 3 or -3 if a and b are equal up to units. An element a is *determined up to units* if we know what value a must equal up to units. The numbers in the set $\{3, -3\}$ are each

---

† Historically, this lemma is called Euclid's Lemma, since it is what Euclid proved. However, the proof depends upon the more fundamental result 5.18(iii) so that we slide the name back to this more useful lemma.

‡ Elements equal up to units are usually called *associates*.

determined up to units. We may now say that the prime factors in a factorization of 60 into primes are uniquely determined up to ordering and up to units.

We now state the following important theorem.

5.25   **THEOREM** (*Unique Factorization*)   *Assume that a is a nonzero nonunit in a Euclidean domain D. Then a is either a prime or we may factor a into a product of primes. The prime factors in such a product are uniquely determined up to units (except for the order of the factors).*

In the case of the ring of integers, if we consider only positive integers, we may stipulate that all the prime factors be positive. Clearly any positive integer, determined up to units, is uniquely determined since the only units are ± 1. Unique factorization is much easier to state for the ring of integers. In this case, the theorem is so important that it is called the Fundamental Theorem of Arithmetic. We shall state it separately below, understanding that it is a special case of unique factorization. This theorem was given in Section 1.6.

5.26   **THEOREM** (*Fundamental Theorem of Arithmetic*)   *Every positive integer $a > 1$ is either a prime or can be expressed as a product of positive primes in exactly one way (except for the order of the factors).*

In general, if p is a prime, then p is also considered to be a "product" of primes. This means we could have omitted "is either a prime or" from the statements of these theorems. This interpretation is somewhat confusing so that we have given the clearer statement, even though it is redundant.

> **Proof of Theorem 5.25:**   In proving Theorem 5.25, we shall assume that it is false. That is, for some nonzero nonunit a of D, either a cannot be factored into a product of primes or, if it can be so factored, such a factorization is not unique in the sense of the theorem. We shall assemble together all elements of this "bad" kind into a set K. In particular, a is in K. Let L be the set $\{b\delta \mid b \in K\}$ of all δ-values as b ranges through K. Since every element of K is a nonzero nonunit, Theorem 5.9(i) and (ii) tells us that L is a set of positive integers. The set of positive integers is well-ordered, so that L contains a least positive integer n. For this value of n we may choose $b \in K$ so that $b\delta = n$.
>
> Either b cannot be expressed as a product of primes or such a product does not have the correct uniqueness properties. Assume the former happens. From this assumption we shall derive our first contradiction.
>
> Since $b \in K$, it certainly is not a prime. In other words, we may factor b as $b = c_1 c_2$, where each of c_1 and c_2 is a nonzero nonunit.

By Theorem 5.9(iii) and Definition 5.8(i), we must have both $c_1\delta < b\delta$ and $c_2\delta < b\delta$. By the minimality of $b\delta = n$, we conclude that neither c_1 nor c_2 lies in K. In other words, each of c_1 and c_2 can be expressed as a product of primes. Since $b = c_1c_2$, b itself is a product of primes. This contradicts our assumption.

We are left with only the possibility that b has at least two "non-unique" factorizations into primes. In other words, we may write

$$b = p_1p_2\cdots p_r$$

and

$$b = q_1q_2\cdots q_s,$$

where the elements p_i and q_j are all primes. Since q_1 is a divisor of b, we know that $q_1|(p_1p_2\cdots p_r)$. By Lemma 5.24, we must have $q_1|p_i$ for some i. The definition of a prime, 5.4, indicates that $p_i = c_1q_1$ for a unit c_1 of D. We now set

$$d = q_2\cdots q_s$$

so that

$$b = q_1d$$
$$= (c_1{}^{-1}p_1)p_2\cdots p_{i-1}q_1p_{i+1}\cdots p_r.$$

Using the cancellation laws,

5.27
$$d = (c_1{}^{-1}p_1)p_2\cdots p_{i-1}p_{i+1}\cdots p_r$$
$$d = q_2\cdots q_s.$$

Our assumptions tell us that both q_1 and d are nonzero nonunits so that $d\delta < b\delta$. Further, $c_1{}^{-1}p_1$ is a prime. Since $d \notin K$ and since 5.27 gives two factorizations of d, there must be an ordering $i_2, \ldots, i_r$ of $1, 2, \ldots, i-1, i+1, \ldots, r$ and units $c_2, \ldots, c_r$ so that $s = r$ and $q_j = c_jp_{i_j}$ if $i_j \neq 1$ and $q_j = c_jc_1p_1$ if $i_j = 1$. In other words, $r = s$ and except for ordering, the primes $p_1, \ldots, p_r$ (or $q_1, \ldots, q_s$) are uniquely determined up to units. This contradicts the fact that $b \in K$. We conclude that the theorem could not be false. This completes the proof of Theorem 5.25.

Let us once again consider the example of the integers. Of course, the primes occurring in a factorization of an integer into prime factors need not all be distinct. By combining the equal primes, we see that every integer $a > 1$ can be expressed uniquely in the form

5.28
$$a = p_1{}^{n_1}p_2{}^{n_2}\cdots p_k{}^{n_k},$$

where the p's are distinct positive primes and each of $n_1, n_2, \ldots, n_k$ is a positive integer. The right side of 5.28 may conveniently be called the *standard form* of the integer a. As an example, $2^2 \cdot 3 \cdot 5$ is the standard form of the integer 60.

An integral domain in which every nonzero nonunit factors uniquely into primes (as described by the conclusion of Theorem 5.25) is called a *unique factorization domain*. We have shown that every Euclidean domain is a unique factorization domain; and have remarked that every principal ideal domain is also a unique factorization domain. But there are unique factorization domains which are neither Euclidean nor principal ideal domains. The easiest example of such a domain is a polynomial domain in two indeterminates over a field. These concepts have not been discussed as yet so that an example will have to wait until Chapter 10.

Not all integral domains are unique factorization domains. For example, the subring of all complex numbers of the form $a + b\sqrt{-5}$, where $a, b \in \mathbf{Z}$ is an integral domain in which 6 factors into at least two possible products of irreducible elements, namely $2 \cdot 3$ and $(1 - \sqrt{-5})(1 + \sqrt{-5})$.

5.5 SOME APPLICATIONS OF THE FUNDAMENTAL THEOREM

The following results may be proved for an arbitrary Euclidean domain; however, the results become somewhat complicated to state because of the condition "up to units" in the Unique Factorization Theorem, 5.25. Therefore, we shall confine most of our considerations to the ring of integers. If a and c are positive integers and c is a divisor of a, then $a = cd$ for some positive integer d. If c and d are expressed as products of prime factors, then clearly a is a product of all prime factors of c times all prime factors of d. Moreover, the Fundamental Theorem then states that this gives the unique factorization of a as a product of prime factors. It follows that the only possible prime factors of c (or of d) are the primes that are factors of a. If then a is expressed in the standard form 5.28, any divisor c of a is necessarily of the form

$$c = p_1^{m_1} p_2^{m_2} \cdots p_k^{m_k},$$

where $0 \leq m_i \leq n_i$ $(i = 1, 2, \ldots, k)$. Conversely, any integer c of this form is clearly a divisor of a.

It is now easy to obtain the g.c.d. (a, b) of two integers a and b if both a and b are expressed in standard form. Clearly, (a, b) is the product of those primes which are factors of both a and b, each such prime occurring to the smaller of the two powers to which it occurs in a and in b. For example, $60 = 2^2 \cdot 3 \cdot 5$ and $252 = 2^2 \cdot 3^2 \cdot 7$. It follows that $(60, 252) = 2^2 \cdot 3$.

We have previously had a method for computing the g.c.d. of two integers by the use of Euclid's Algorithm—a method which does not involve finding any prime factors of the given integers. From a computational point of view, the previous method may involve much less work than the present one since it may be exceedingly difficult to find the prime factors of fairly large numbers, and therefore difficult to express them in standard form.

We shall not have much occasion to use the concept we now define, but we include it here for the sake of completeness.

5.29 **DEFINITION** The *least common multiple* (l.c.m.) of two nonzero integers a and b is the positive integer m with the following two properties:

(i) $a|m$ and $b|m$.

(ii) If c is an integer such that $a|c$ and $b|c$, then $m|c$.

It is easy to verify the *uniqueness* of the l.c.m., and its existence may be established in various ways. (See Exercises 3 and 5 below.)

Next we give a formal proof of the following well-known result. We shall show that there do not exist nonzero integers a and b such that

5.30 $$a^2 = 2b^2.$$

Proof: Let us suppose, on the contrary, that there do exist such integers, which we may obviously assume to be positive. If $d = (a, b)$, by Exercise 3 of the preceding set, we have $a = da_1$, $b = db_1$, where $(a_1, b_1) = 1$. Substituting in 5.30, and dividing by d^2, we find that

$$a_1{}^2 = 2b_1{}^2.$$

This equation implies that $2|a_1{}^2$ and Lemma 5.23 (or the Fundamental Theorem) shows that $2|a_1$. But then $4|a_1{}^2$ and therefore $4|2b_1{}^2$. This implies that $2|b_1{}^2$ and we must have $2|b_1$. Thus 2 is a common divisor of a_1 and b_1. We have therefore obtained a contradiction of the fact that $(a_1, b_1) = 1$, and hence there can be no nonzero integers satisfying 5.30. The proof is therefore complete.

Another, perhaps more familiar, way of stating the result just proved is to say that $\sqrt{2}$ is not a rational number; that is, it is not expressible in the form a/b, where a and b are integers.

Let us turn our attention now to a Euclidean domain D. Suppose that I is an ideal in D. We wish to know when the factor ring of cosets D/I defined in Section 3.4 is an integral domain. In Theorem 5.15 we proved that I is a principal ideal; that is, $I = (a)$ for some element a of D. If D is the ring $\mathbf{Z}$ of integers and a is a positive integer n, then D/I is the ring $\mathbf{Z}/(n)$ which is $\mathbf{Z}_n$, the ring of integers modulo n. It would be good to keep this example in mind.

We shall assume that a is neither zero nor a unit; that is, the ideal I is neither (0) nor D. We let e denote the unity of D. Consider the case where a is not a prime in D. That is, $a = bc$ where b and c are nonzero nonunits. From Theorem 5.9(iii) and Definition 5.8(i), we must have both $b\delta < a\delta$ and $c\delta < a\delta$. By Theorem 5.15, $a\delta$ is minimal for nonzero elements of I. Therefore, neither b nor c lie in I. The elements $b + I$ and $c + I$ are nonzero in the quotient ring D/I. If we multiply them, then by 3.29 we obtain $(b + I)(c + I) = (bc) + I = a + I = 0 + I$ since $a \in I$. This is the zero element of D/I. Hence the element $b + I$ is a nonzero divisor of zero in the ring D/I, and we conclude that if D/I is an integral domain, then a must be a prime.

If $D = \mathbf{Z}$ and $a = 6$ so that $I = (6)$, then $6 = 2 \cdot 3$. The cosets $2 + I$ and $3 + I$ are elements of the ring $\mathbf{Z}_6$ of integers modulo 6: they are $[2]$ and $[3]$. The product is $[2][3] = [6] = [0]$; therefore $[2]$ is a divisor of zero in $\mathbf{Z}_6$. This example illustrates the argument just given.

5.31   THEOREM   *If D is a Euclidean domain, I an ideal of D not equal to (0) or D, and a a generator of I, then D/I is an integral domain if and only if a is a prime. In this case, every nonzero element of D/I has a multiplicative inverse.*

We have already shown that if D/I is an integral domain, then a is a prime.

Suppose for a moment that $a = 5$. An integer lying between 0 and 5, say 3, must be relatively prime to 5. As in Euclid's Lemma, 5.18, we have integers, in this case -1 and 2, so that

$$2 \cdot 3 - 1 \cdot 5 = 1.$$

In $\mathbf{Z}_5$, since $[-5] = [0]$, we have

$$[2] \cdot [3] = [2 \cdot 3] + [0] = [6] + [-5]$$
$$= [6 - 5] = [1].$$

The element $[3]$ has a multiplicative inverse $[2]$.

Proof: This argument generalizes to prove that if a is a prime, then every nonzero element of D/I has a multiplicative inverse. Choose $b + I$ nonzero in D/I. Since b is not in $I = (a)$, b is not divisible by a. The g.c.d. of b and a cannot be ca for a unit c. Therefore, the g.c.d. of b and a is a unit which may be taken to be unity e since

a is a prime. By Euclid's Lemma, 5.18, there are elements r and s of D so that

$$ra + sb = e.$$

We use 3.28 and 3.29 to compute (recall that $e + I$ is the unity of D/I), $e + I = (ra + sb) + I = [(ra) + I] + [(sb) + I]$. Since ra lies in the ideal I, the coset $(ra) + I$ equals the coset $0 + I$. This latter coset is the zero of D/I. Consequently,

$$e + I = (sb) + I = (s + I)(b + I).$$

Therefore, $s + I$ is a multiplicative inverse to $b + I$ in D/I. We conclude that every nonzero element of D/I has a multiplicative inverse.

It is now easy to show that if a is a prime, then D/I is an integral domain. Suppose that $(b + I)(c + I) = 0 + I$, the zero of D/I. Assume that $b + I$ is not $0 + I$. Then $b + I$ has a multiplicative inverse which we may call $s + I$. If we multiply zero by anything, we get zero. Consequently,

$$0 + I = (s + I)(0 + I) = (s + I)(b + I)(c + I)$$
$$= (e + I)(c + I) = c + I.$$

That is, $c + I$ is the zero in D/I. Hence the product of any two nonzero elements of D/I must again be nonzero, which means that D/I is an integral domain. The proof of the theorem is complete.

This theorem tells us that such rings as $\mathbf{Z}_7$ or $\mathbf{Z}_{11}$ are integral domains, but such rings as $\mathbf{Z}_9$ or $\mathbf{Z}_{15}$ are not.

Let us return once again to the ring of integers $\mathbf{Z}$. In order to factor a number $n > 1$ into a product of primes, we must find the prime divisors of n. One method of doing this is to test various primes to see if they divide n. To carry out such testing, we need a table of prime numbers. Over 2000 years ago Eratosthenes suggested his *sieve* for finding prime numbers. Since that time certain special methods have been discovered which help to test the primality of particular integers; but no better method has been found for making tables of successive primes. The basic method is pictured below.

$$
\begin{array}{cccccccccc}
1 & 2 & 3 & \not4 & 5 & \not6 & 7 & \not8 & \not9 & \not{10} \\
11 & \not{12} & 13 & \not{14} & \not{15} & \not{16} & 17 & \not{18} & 19 & \not{20} \\
\not{21} & \not{22} & 23 & \not{24} & \not{25} & \not{26} & \not{27} & \not{28} & 29 & \not{30}
\end{array}
$$

Suppose we want to tabulate all primes less than 30. We list all integers less than 30. Starting with 2 and excepting 2, we cross off all integers divisible by 2. After 2, the very next number not crossed off must be a prime (in fact, 3, 5, 7 are all primes), so starting with it, namely 3, we except it and cross off all multiples of it. Passing to the next number not crossed off, namely 5, we repeat this process again. If we try again with 7, we see that there are no numbers to cross off. Let us stop here.

Suppose n has proper factors, c_1 and c_2 so that $n = c_1 c_2$, where $1 < c_1 < n$ and $1 < c_2 < n$. Since $c_1 c_2 = n$, clearly $c_1 \leq \sqrt{n}$ or $c_2 \leq \sqrt{n}$. Assume $c_1 \leq \sqrt{n}$, so that c_1 must have a prime divisor p such that $p \leq \sqrt{n}$. In the sieve we shall reach p by the time we have worked our way up to the largest prime q such that $q \leq \sqrt{n}$. In other words, if we stop sieving at a prime q, then all numbers in our table between 1 and q^2 will be primes. Since $\sqrt{30} < 7$, we know our table from 1 to 30 lists primes in that range since $q = 5$ is the largest prime such that $q \leq \sqrt{30}$.

The *Sieve of Eratosthenes* is very old, but since his time only minor improvements have been made on his method for making tables of primes.

EXERCISES

1. Express 120 and 4851 in standard form, and find their g.c.d. and their l.c.m.

2. Do the same for 970 and 3201.

3. Explain how one can find the l.c.m. of any two integers if their standard forms are known.

4. (i) For the ring $\mathbf{Z}$ of integers, using a method similar to that used in the proof of Lemma 5.23, show that if a is a divisor of bc and $(a, b) = 1$, then a is a divisor of c.

   (ii) Prove the same result by use of the Fundamental Theorem.

5. For the ring $\mathbf{Z}$ of integers, if a and b are positive and $a = a_1 d$ and $b = b_1 d$, where $d = (a, b)$, show that the l.c.m. of a and b is $a_1 b_1 d$.

6. Show that a positive integer $a > 1$ is a perfect square (that is, is the square of an integer) if and only if in the standard form of a all the exponents are even integers.

7. Show that if b and c are positive integers such that bc is a perfect square and $(b, c) = 1$, then both b and c are perfect squares.

8. Prove that there do not exist nonzero integers a and b such that $a^2 = 3b^2$.

9. If n is a positive integer which is not a perfect square, prove that there do not exist nonzero integers a and b such that $a^2 = nb^2$.

10. For each positive integer n, show that there are more than n positive primes. *Hint:* Use the result of Exercise 5, Section 5.1.

11. Prove Lemma 5.24.

12. Is the integral domain $\mathbf{Z}_p$ (p a prime) an ordered integral domain?

13. Prove that $[a]$ is in the group of units of $\mathbf{Z}_n$ if and only if a and n are relatively prime.

14. Write a multiplication table for $\mathbf{Z}_5$. Show that each nonzero element has an inverse. Use this table to solve the following pair of simultaneous equations:

$$[3]x + [2]y = [1],$$
$$[2]x + [4]y = [2].$$

15. Suppose that p is a prime and $n > 1$ an integer. If a is any integer, show that $[1 + ap]$ is a unit in $\mathbf{Z}_{p^n}$. Show that if p is odd, then $[1 + p]$ has order p^{n-1} in the group of units of $\mathbf{Z}_{p^n}$.

5.6 PYTHAGOREAN TRIPLES (OPTIONAL)

If x, y, and z are *positive* integers such that

5.32 $$x^2 + y^2 = z^2,$$

we shall call the ordered triple (x, y, z) a *Pythagorean triple*. Clearly, (x, y, z) is a Pythagorean triple if and only if there exist right triangles whose sides have respective lengths x, y, and z units. Well-known examples of Pythagorean triples are $(3,4,5)$, $(6,8,10)$, and $(5,12,13)$. In this section we shall determine all Pythagorean triples.

First, we observe that we can limit our problem somewhat. If (x, y, z) is a Pythagorean triple, then so is (kx, ky, kz) for every positive integer k. Conversely, let (x, y, z) be a Pythagorean triple and suppose that d is a common divisor of x, y, and z. If we write $x = x_1 d$, $y = y_1 d$, and $z = z_1 d$, we can cancel d^2 from each term of the equation

$$(x_1 d)^2 + (y_1 d)^2 = (z_1 d)^2,$$

and find that (x_1, y_1, z_1) is also a Pythagorean triple. If it happens that d is the g.c.d. of the *three* integers x, y, and z (see Exercise 10, Section 5.3), then x_1, y_1, and z_1 have 1 as their g.c.d. Let us say that a Pythagorean triple (a, b, c) is a *primitive* Pythagorean triple if a, b, and c have 1 as their g.c.d. Then the observations that we have just made assure us that *every* Pythagorean triple is of the form (ra, rb, rc), where (a, b, c) is a primitive Pythagorean triple and r is a positive integer. Our general problem is therefore reduced to the problem of finding all primitive Pythagorean triples.

Now let (x, y, z) be a primitive Pythagorean triple. It is easy to see that *each pair* of the numbers x, y, and z must be relatively prime. If, on the contrary, two of these numbers were not relatively prime, they would have a common prime factor p. Then Equation 5.32 would show that p is also a factor of the third, which would contradict the assumption that (x, y, z) is primitive. As a special case of what we have just proved, we see that x and y cannot both be even. We next show that, also, x and y cannot both be odd. If they were both odd, we could write $x = 2m + 1$ and $y = 2n + 1$, where m and n are properly chosen integers. But then we would have

$$z^2 = x^2 + y^2 = (2m + 1)^2 + (2n + 1)^2$$
$$= 2(2m^2 + 2n^2 + 2m + 2n + 1).$$

Since the second factor in this last expression is odd, we see that z^2 would be divisible by 2 but not by 4, and this is clearly impossible. It follows that x and y cannot both be odd. We have therefore proved that one of the integers x and y must be even and the other odd. It is trivial that (a, b, c) is a primitive Pythagorean triple if and only if (b, a, c) is also, and there will be no real loss of generality if we now limit ourselves to the study of primitive Pythagorean triples (x, y, z) in which x is even, and therefore y is odd.

We are now ready to prove the following theorem.

5.33 THEOREM *If (x, y, z) is a primitive Pythagorean triple in which x is even, then*

5.34 $$x = 2uv, \quad y = u^2 - v^2, \quad z = u^2 + v^2,$$

where u and v are positive integers satisfying the following three conditions:

(i) u and v are relatively prime.

(ii) $u > v$.

(iii) One of u, v is even and the other is odd.

Conversely, if u and v are any positive integers satisfying these three conditions and x, y, and z are determined by formulas 5.34, then (x, y, z) is a primitive Pythagorean triple in which x is even.

Proof: To prove the first part of the theorem, let (x, y, z) be a primitive Pythagorean triple in which x is even. We have proved that no two of x, y, and z can be even; hence y and z are both odd. This implies that $z + y$ and $z - y$ are both even; that is, that there exist positive integers r and s such that

5.35 $$x + y = 2r, \quad z - y = 2s.$$

From these, it is easy to verify that

5.36 $$z = r + s, \quad y = r - s.$$

Now r and s must be relatively prime since any common factor of r and s would be a common factor of the relatively prime integers z and y. Moreover, since

$$x^2 = z^2 - y^2 = (z + y)(z - y),$$

it follows from Equations 5.35 that

5.37 $$x^2 = 4rs.$$

Since x is even, $x = 2t$ for some integer t, and the preceding equation shows that

5.38 $$t^2 = rs.$$

Since r and s are relatively prime, Exercise 7 of the preceding set implies that both r and s are perfect squares. That is, there exist positive integers u and v such that

5.39 $$r = u^2, \quad s = v^2.$$

Then Equations 5.36 and 5.37 show that

5.40 $$x = 2uv, \quad y = u^2 - v^2, \quad z = u^2 + v^2,$$

and formulas 5.34 are satisfied. There remains only to prove that u and v have the required properties. Since r and s are relatively prime, it follows from 5.39 that u and v are relatively prime. Next, we see that $u > v$ since $y > 0$. We already know that u and v cannot both be even inasmuch as they are relatively prime. Finally, from 5.40 it follows that they cannot both be odd since otherwise y (and z also) would be even, whereas we know that it is odd. This completes the proof of the first part of the theorem.

To prove the second part, suppose that u and v are any positive integers satisfying conditions (i), (ii), and (iii); and let x, y, and z be defined by formulas 5.34. Clearly x, y, and z are all positive, and it is easy to verify that

$$(2uv)^2 + (u^2 - v^2)^2 = (u^2 + v^2)^2,$$

and hence that (x, y, z) is a Pythagorean triple. We shall prove that it is necessarily primitive by showing that y and z are relatively prime. By condition (iii), $u^2 - v^2$ and $u^2 + v^2$ are both odd, that is, y and z are both odd. If y and z were not relatively prime, they would have a common prime factor $p \neq 2$. But since $z + y = 2u^2$ and $z - y = 2v^2$, it would follow that p is also a common factor of u and v. However, it is given that u and v are relatively prime, and therefore y and z can have no common prime factor. Hence, (x, y, z) is a primitive Pythagorean triple. It is obvious that x is even, and the proof is therefore complete.

It follows from the theorem that there are infinitely many primitive Pythagorean triples. The triple (4,3,5) is obtained by setting $u = 2$, $v = 1$ in 5.34; the triple (12,5,13) by choosing $u = 3$, $v = 2$; the triple (8,15,17) by choosing $u = 4$, $v = 1$; and so on.

5.7 HOMOMORPHISMS OF EUCLIDEAN DOMAINS (OPTIONAL)

Assume that D is a Euclidean domain and I is an ideal of D. In Section 3.4 we showed how to construct the quotient ring D/I of cosets of I in D. In this section we will discuss direct sum decompositions of D/I.

Example 1 In order to illustrate what we have in mind, consider the quotient ring $\mathbf{Z}/(15) = \mathbf{Z}_{15}$ of integers modulo 15, and the two additional rings $\mathbf{Z}_3$ and $\mathbf{Z}_5$ of integers modulo 3 and 5 respectively. Recall that the direct sum of $\mathbf{Z}_3$ and $\mathbf{Z}_5$ is $\mathbf{Z}_3 \oplus \mathbf{Z}_5$, the set of all ordered pairs (a, b) where $a \in \mathbf{Z}_3$ and $b \in \mathbf{Z}_5$. We may let $\{n\}$ denote the coset $n + (3)$ in $\mathbf{Z}_3$ and $\langle n \rangle$ denote the coset $n + (5)$ in $\mathbf{Z}_5$ for an integer n. The mapping $\theta : \mathbf{Z} \to \mathbf{Z}_3 \oplus \mathbf{Z}_5$ defined by

$$n\theta = (\{n\}, \langle n \rangle)$$

is a homomorphism of $\mathbf{Z}$ into $\mathbf{Z}_3 \oplus \mathbf{Z}_5$. (This is a particular case of Exercise 7 of this section.) What is the kernel of this mapping θ? In the first component of $(\{n\}, \langle n \rangle)$, $\{n\} = \{0\}$ if and only if $n +$

(3) = (3), or $n \in$ (3). Similarly, the second component $\langle n \rangle = \langle 0 \rangle$ if and only if $n + (5) = (5)$, or $n \in (5)$. In other words, $n\theta$ is zero if and only if $n \in (3) \cap (5)$. The ideal (3) is the set of all multiples of 3, and (5) is the set of all multiples of 5; therefore, n is a multiple of both 3 and 5 if and only if n is a multiple of $15 = 3 \cdot 5$. That is, the kernel of θ is the set (15) of all multiples of 15.

We have a homomorphism $\phi: \mathbf{Z} \to \mathbf{Z}_{15}$ given by $n\phi = [n] = n + (15)$:

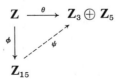

The kernel of ϕ is also (15). We may now apply the Universal Mapping Property, 3.25, to obtain a homomorphism $\psi: \mathbf{Z}_{15} \to \mathbf{Z}_3 \oplus \mathbf{Z}_5$ whose kernel is $(\ker \theta)\phi = (15)\phi = ([0])$. Thus ψ is an isomorphism of $\mathbf{Z}_{15}$ into $\mathbf{Z}_3 \oplus \mathbf{Z}_5$. Actually, ψ is onto $\mathbf{Z}_3 \oplus \mathbf{Z}_5$. To see this, notice that ψ is a one-one mapping from $\mathbf{Z}_{15}$ to $\mathbf{Z}_3 \oplus \mathbf{Z}_5$. But both $\mathbf{Z}_{15}$ and $\mathbf{Z}_3 \oplus \mathbf{Z}_5$ contain exactly 15 elements, so ψ is actually an isomorphism of $\mathbf{Z}_{15}$ onto $\mathbf{Z}_3 \oplus \mathbf{Z}_5$. We have proven that

$$\mathbf{Z}_{15} \cong \mathbf{Z}_3 \oplus \mathbf{Z}_5.$$

We may extend this result to any Euclidean domain as the following theorem shows. In fact, you will notice that the proof is identical to the one we outlined above in Example 1.

5.41   THEOREM   *Assume that $a = bc$ where b and c are relatively prime nonzero nonunits in a Euclidean domain D. The mapping $\psi: D/(a) \to D/(b) \oplus D/(c)$ defined by*

5.42 $$(d + (a))\psi = (d + (b), d + (c))$$

is an isomorphism of $D/(a)$ onto $D/(b) \oplus D/(c)$.

For the ring of integers, for example, this theorem says that $\mathbf{Z}_{30}$ and $\mathbf{Z}_6 \oplus \mathbf{Z}_5$ are isomorphic rings. Or, as we showed in Example 1, this theorem proves that $\mathbf{Z}_{15} \cong \mathbf{Z}_3 \oplus \mathbf{Z}_5$.

Proof: Recall that there are natural homomorphisms $\phi: D \to D/(a)$, $\phi_1: D \to D/(b)$, and $\phi_2: D \to D/(c)$ described in Theorem 3.25. In particular, $d\phi = d + (a)$. For any element $d \in D$ we define

5.43 $$d\theta = (d\phi_1, d\phi_2).$$

In the exercises we will show that θ is a homomorphism of D into $D/(b) \oplus D/(c)$.

We have the following diagram:

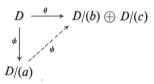

The object is to show that a homomorphism ψ exists such that $\theta =: \phi\psi$. This should recall the Universal Mapping Property, 3.21, so that we now verify that the hypotheses of that theorem hold. The mapping ϕ is clearly onto $D/(a)$, and the kernel of ϕ is (a). We know that a is divisible by both b and c so that both $a\phi_1$ and $a\phi_2$ are zero. Therefore, for any element da of the ideal (a), $(da)\theta$ is zero. We have shown that $\ker \phi = (a) \subseteq \ker \theta$.

Actually, $\ker \theta = (a)$ as we will now show. We shall use a generalization of Lemma 5.23 given as Exercise 4 of Section 5.5. Suppose d lies in $\ker \theta$; then d also lies in both $\ker \phi_1 = (b)$ and $\ker \phi_2 = (c)$. The ideal (b) is the set of all elements of D divisible by b, and (c) is the set of all elements of D divisible by c. Thus both b and c divide d. Choose f so that $d = bf$. Since c divides bf, and b and c are relatively prime, the generalization of Lemma 5.23 tells us that $c \mid f$. Choose g so that $f = cg$. Then $d = fb = (cg)b = (bc)g = ag$. We have shown that d lies in $(a) = \ker \phi$. We conclude that

$$\ker \phi = (a) = \ker \theta.$$

The Universal Mapping Property, 3.25, applies to tell us there is a unique mapping ψ with kernel equal to $(\ker \theta)\phi$ such that $\phi\psi = \theta$. The kernel of ψ is $(\ker \theta)\phi = (a)\phi = (0 + (a))$, that is, ψ is an isomorphism. If $d + (a)$ lies in $D/(a)$ then

$$(d + (a))\psi = d\phi\psi = d\theta = (d + (b), d + (c)),$$

as dictated by 5.42.

To complete the proof we must show that ψ is an onto mapping, and we shall use Euclid's Lemma to do this. This part of the proof is important since it exhibits a method for solving problems. Let us denote cosets in the following ways. We shall set $d + (a) = [d]$, $d + (b) = \{d\}$, and $d + (c) = \langle d \rangle$. Now we must show that for arbitrary elements d, f of D there is an element g of D so that

$[g]\psi = (\{d\}, \langle f \rangle)$. By Euclid's Lemma there are elements r and s of D so that

$$rb + sc = e$$

where e is the unity of D since b and c are relatively prime. We set $g = frb + dsc$. Evaluating ψ at $[g]$, we obtain

$$[g]\psi = g\phi\psi = g\theta = (g\phi_1, g\phi_2).$$

We shall show that $g\phi_1 = \{d\}$, and an analogous argument will then prove that $g\phi_2 = \langle f \rangle$. Since frb is in (b), $(frb)\phi_1 = \{0\}$, is zero. Similarly $(drb)\phi_1 = \{0\}$. Computing $g\phi_1$ and using the fact that ϕ_1 is a homomorphism we obtain the following equalities.

$$\begin{aligned}
g\phi_1 &= (frb + dsc)\phi_1 = (frb)\phi_1 + (dsc)\phi_1 \\
&= \{0\} + (dsc)\phi_1 = (drb)\phi_1 + (dsc)\phi_1 \\
&= (drb + dsc)\phi_1 = (d(rb + sc))\phi_1 \\
&= (de)\phi_1 = d\phi_1 \ = \{d\}.
\end{aligned}$$

Since $g\phi_2 = \langle f \rangle$ also, we have

$$[g]\psi = (\{d\}, \langle f \rangle)$$

completing the proof that ψ is an onto mapping. The proof of the theorem is now complete.

If D is the ring $\mathbf{Z}$ of integers, then this last part of the argument may be shortened as follows. In $\mathbf{Z}$ we may assume that a, b, and c are positive integers greater than 1. We showed that the mapping $\psi: \mathbf{Z}/(a) \to \mathbf{Z}/(b) \oplus \mathbf{Z}/(c)$ is an isomorphism. Actually, the ring $\mathbf{Z}/(n) = \mathbf{Z}_n$, the ring of integers modulo n. So we wish to show that the one-one mapping $\psi: \mathbf{Z}_a \to \mathbf{Z}_b \oplus \mathbf{Z}_c$ is onto. The ring $\mathbf{Z}_a$ contains exactly a elements, and the image $\mathbf{Z}_a\psi$ contains exactly a elements also. But the ring $\mathbf{Z}_b \oplus \mathbf{Z}_c$ contains $bc = a$ elements. Therefore $\mathbf{Z}_a\psi = \mathbf{Z}_b \oplus \mathbf{Z}_c$ and ψ is an onto mapping. This particular conclusion to the proof of the theorem does not tell us how to find g so that $[g]\theta = (\{d\}, \langle f \rangle)$, whereas the earlier version does.

We may state a generalization to this theorem, but first we must define the direct sum of more than two rings. If R and S are rings, then $R \oplus S$ was defined in Section 3.2 to be the set of all ordered pairs (a, b) where $a \in R$ and $b \in S$. For rings $R_1, R_2, \ldots, R_m$, where $m > 1$ is an integer, we set $R_1 \oplus R_2 \oplus \cdots \oplus R_m$ equal to the set of all ordered m-tuples $(a_1, a_2, \ldots, a_m)$ where $a_i \in R_i$. Just as in the case for two rings, we define sums and products component by component, i.e.,

$$(a_1, a_2, \ldots, a_m) + (b_1, b_2, \ldots, b_m) = (a_1 + b_1, a_2 + b_2, \ldots, a_m + b_m)$$

and

$$(a_1, a_2, \ldots, a_m)(b_1, b_2, \ldots, b_m) = (a_1 b_1, a_2 b_2, \ldots, a_m b_m)$$

where $a_i, b_i \in R_i$. Of course, we call $R_1 \oplus R_2 \oplus \cdots \oplus R_m$ the *direct sum* of the rings $R_1, R_2, \ldots, R_m$. The previous theorem extends to a direct sum of several rings as the next theorem shows.

5.44   GENERALIZED CHINESE REMAINDER THEOREM   *Assume that* $a = a_1 a_2 \cdots a_m$ *for a positive integer* m *where* $a_1, a_2, \ldots, a_m$ *are pairwise relatively prime nonzero nonunits in a Euclidean domain* D. *The mapping* $\psi\colon D/(a) \to D/(a_1) \oplus \cdots \oplus D/(a_m)$ *defined by*

5.45
$$(d + (a))\psi = (d + (a_1), \ldots, d + (a_m))$$

is an isomorphism of $D/(a)$ *onto* $D/(a_1) \oplus \cdots \oplus D/(a_m)$.

We may set $b = a_1$ and $c = a_2 a_3 \cdots a_m$. By Theorem 5.41 $\psi_1\colon D/(a) \to D/(b) \oplus D/(c)$ is an isomorphism onto the direct sum. Since $c = a_2 a_3 \cdots a_m$ an induction argument will now complete the proof of the theorem. We leave the details of such an argument to the exercises, but we do indicate that our Induction Assumption upon m allows us to conclude that

$$D/(c) \cong D/(a_2) \oplus D/(a_3) \oplus \cdots \oplus D/(a_m).$$

Let us interpret these theorems in the setting of the integers. It may not be obvious, but the last theorem tells us about simultaneous solution of certain congruences. Such problems are very old, and examples of such problems, with solutions, appear on Chinese artifacts dating from several thousand years B.C. For this reason, the last theorem and the theorem we shall now prove are called the *Chinese Remainder Theorems*.

5.46   CHINESE REMAINDER THEOREM   *Assume that for an integer* $m > 1$, $n_1, \ldots, n_m$ *are pairwise relatively prime integers greater than* 1. *If* $b_1, \ldots, b_m$ *are any integers whatsoever then there exists an integer* x *which solves the following simultaneous congruences:*

5.47
$$
\begin{aligned}
x &\equiv b_1 \ (\mathrm{mod}\ n_1), \\
x &\equiv b_2 \ (\mathrm{mod}\ n_2), \\
&\ \vdots \\
x &\equiv b_m \ (\mathrm{mod}\ n_m).
\end{aligned}
$$

Further, if $n = n_1 n_2 \cdots n_m$ *then an integer* y *is a solution to the simultaneous congruences 5.47 if and only if there is an integer* c *so that* $y = x + cn$.

Proof: Recall that $a \equiv b \pmod{k}$ means $[a] = [b]$ in $\mathbf{Z}_k$; or equivalently, $a - b$ is divisible by k. Let us denote by $[a]_i$ the congruence class [or coset $a + (n_i)$] of a in $\mathbf{Z}_{n_i}$; and by $[a]$ the congruence class of a in $\mathbf{Z}_n$. Further, we shall set $D_i = \mathbf{Z}_{n_i}$. The Generalized Chinese Remainder Theorem tells us that ψ of 5.45 is an isomorphism of $\mathbf{Z}_n$ onto $D_1 \oplus \cdots \oplus D_m$. In particular, there is an integer x so that $[x]\psi = ([x]_1, \ldots, [x]_m) = ([b_1]_1, \ldots, [b_m]_m)$. Therefore $[x]_i = [b_i]_i$ for all $i = 1, 2, \ldots, m$; and x solves the congruences 5.47.

Suppose y solves the congruences 5.47. Then $[y]_i = [b_i]_i$ for each i so that $[y]\psi = [x]\psi$ or $[y - x]$ is in the kernel of ψ. The kernel of ψ is (n) so that $y = x + cn$ for some integer c.

Finally, since n is congruent to zero for each n_i, $y = x + cn \equiv x \pmod{n_i}$ for each i. We conclude that if x solves 5.47 then y does also. The theorem follows.

Example 2 We shall illustrate this theorem by solving the simultaneous congruences:

5.48
$$x \equiv 2 \pmod{7},$$
$$x \equiv 5 \pmod{15}.$$

First we need integers r and s so that

$$15r + 7s = 1.$$

Such integers may be found using the Euclidean Algorithm. We have here:

$$15 \cdot 1 - 2 \cdot 7 = 1.$$

The reason we need such a linear combination is clear from the conclusion of the proof of Theorem 5.41 where we found r and s in D so that $rb + sc = e$. Recall that we set $g = frb + dsc$. In this present case, $g = x$, $f = 5$, and $d = 2$. We then set $x = 5 \cdot (-2) \cdot 7 + 2 \cdot 15 = -40$. From the final step in the proof of Theorem 5.41 we know x solves the congruences 5.48. Further, any other solution has the form $-40 + c \cdot 7 \cdot 15$. In particular, $y = -40 + 7 \cdot 15 = 65$ solves these congruences.

The Chinese Remainder Theorem is actually a very useful tool in number theory and algebra. It finds applications in linear algebra (to be discussed in later chapters) as well as in ring theory.

EXERCISES

1. What are all the solutions to $x \equiv 4 \pmod{15}$, $x \equiv 2 \pmod 4$?

2. Solve the following simultaneous congruences: $x \equiv 21 \pmod 5$, $x \equiv 35 \pmod 3$, $x \equiv 15 \pmod 7$.

3. Solve the simultaneous congruences: $x \equiv 1 \pmod{14}$, $x \equiv 13 \pmod{15}$, and $x \equiv 7 \pmod{11}$.

4. Can you find a solution to $x \equiv 6 \pmod 9$ and $x \equiv 15 \pmod{21}$? What about $x \equiv 3 \pmod 9$ and $x \equiv 15 \pmod{21}$? In this case, what is the problem?

5. Consider the mapping given by 5.42 where $b = 3$ and $c = 5$. Which element of $\mathbf{Z}_{15}$ is mapped to $(2 + (3), 3 + (5))$ by ψ. More generally, compute an inverse mapping to ψ.

6. In Theorem 5.41 let $D = \mathbf{Z}$, $a = 90$, $b = 15$, and $c = 18$. Notice that these choices of a, b, c do not satisfy all the hypotheses of this theorem.

   (a) Is ψ of 5.42 a homomorphism in this case?

   (b) If ψ is a homomorphism, is it an isomorphism?

   (c) If ψ is a homomorphism, is ψ onto $\mathbf{Z}_b \oplus \mathbf{Z}_c$?

7. Prove that 5.43 defines a homomorphism.

8. Prove Theorem 5.44.

9. Find a solution to

$$[6]x + [4]y = [13]$$
$$[4]x + [2]y = [4]$$

   in $\mathbf{Z}_{35}$. *Hint:* Solve the equation in each of $\mathbf{Z}_5$ and $\mathbf{Z}_7$, and use Theorem 5.41.

10. If $n = p_1{}^{e_1}p_2{}^{e_2}\cdots p_s{}^{e_s}$ is a prime factorization of n written in standard form, show that the group of units of $\mathbf{Z}_n$ is isomorphic to the direct product of the groups of units of $\mathbf{Z}_{p_i{}^{e_i}}$, $i = 1, \ldots, s$. *Hint:* See Theorem 3.14.

COMMENTARY

1 The Fundamental Theorem of Arithmetic tells us that every positive integer $n > 1$ must factor uniquely into primes. But what are the primes? We probably have all been told that there are an infinite number of primes. Euclid (c. 300 B.C.) gave a proof based upon the fact that if n is any positive whole number then none of the numbers $1, 2, 3, \ldots, n$ divide $1 + (1 \cdot 2 \cdot 3 \cdots n)$. Even more to the point, is there a formula which gives only primes? The history of prime valued formulas is long and interesting, and there have been many attempts at such formulas. Probably the most famous observation along this line was made by Fermat (1601–1665) when he observed that the numbers $2^{2^0} + 1 = 3$, $2^{2^1} + 1 = 5$, $2^{2^2} + 1 = 17$, $2^{2^3} + 1 = 257$, and $2^{2^4} + 1 = 65,537$ are all primes. He wondered if maybe $2^{2^n} + 1$ might be a prime for all positive integral values of n. About 70 years after his death, Euler (1707–1783) showed that $2^{2^5} + 1 = 641 \times 6,700,417$ ending the chances that $2^{2^n} + 1$ is always a prime. Subsequently, it has been shown that $2^{2^n} + 1$ is not a prime for quite a few values of n [70]. Actually, the largest known prime of the form $2^{2^n} + 1$ is 65,537 so that it is not even known if there are infinitely many "Fermat primes." The Fermat primes, even though little is known about them, play a significant role in mathematics. Gauss (1777–1855) showed that a regular polygon with a prime number of sides can be constructed by straight edge and compass if and only if the prime is a Fermat prime. In 1956 two group theorists, P. Hall and G. Higman, showed that Fermat primes were intimately connected with pathological examples in the theory of groups. They also showed that another kind of prime, a "Mersenne prime," is important in the study of finite groups.

 Mersenne (1588–1648) did not actually give a formula for primes; rather, he felt that if one already had an odd prime p, then the numbers $2^p - 1$ might be interesting. In fact, he gave a list of 11 primes $p \le 257$ which he asserted were the only ones for which $2^p - 1$ is prime. Actually, his assertion contained numbers which are not primes and missed ones which are primes. One of the very largest known primes is the Mersenne prime $2^p - 1$, where $p = 11213$. This fact was discovered using a computer at the University of Illinois; and for many years the mathematics department there used the postmark: "$2^{11213} - 1$ is a prime [141,146]."

 Formulas which give us just primes do exist, but most of these are totally unsatisfactory. They either involve knowing the primes first, or they involve calculations akin to determining the primes [76]. A major breakthrough has occurred in the past few years. In 1970, Matiyasevič showed that there is no algorithm which will determine the integral solutions to any given polynomial equation with integer coefficients [i]. Because of these investigations, he was able to give a polynomial in 31 variables with integer coefficients such that if we substitute positive integers in for the variables, and the value of the polynomial is positive, then it is prime. Further, we may obtain every prime for some positive values of the variables [189]. The trouble with Matiyasevič's remarkable polynomials is that for most positive integral values of the variables, the polynomial is negative valued. Nonetheless, his feat signals a major advance over previous attempts at prime value formulas.

2   As we have mentioned, if $p_1, p_2, \ldots, p_t$ are distinct primes then $N = (p_1 \cdot p_2 \cdots p_t) + 1$ is not divisible by any of the primes p_i. But N is divisible by some prime number, p_{t+1}. In other words, the set of primes is not finite. Being infinite, this set is distributed along the number line in some fashion. If we are given a very large integer B, then approximately how many primes are there between 1 and B? Several guesses were made, most notably by Euler (1707–1783), Legendre (1752–1833), and Gauss (1777–1855). They guessed that there should be approximately $B/(\ln B)$ primes between 1 and B. In fact, if $\pi(B)$ is the actual number of such primes, then it should be that $\lim_{B \to \infty} (\pi(B) \ln B)/B = 1$. This latter statement is called the *prime number theorem* [z]. A vast amount of effort went into proving this theorem, but it was not until 1896 that Hadamard (1865–1963) and de la Vallée-Poussin (1866–1962) independently discovered proofs of the theorem using the theory of functions of a complex variable [78,r]. The prime number theorem tells us that the primes, in the large, are fairly regularly distributed through the positive numbers.

However, it is not difficult to show that there are often gigantic intervals between two successive primes. For example, consider a large integer M. Both $(M + 1)!$ and T are divisible by T for $2 \leq T \leq M + 1$ so that $(M + 1)! + 2$, $(M + 1)! + 3, \ldots, (M + 1)! + M, (M + 1)! + (M + 1)$ constitute M consecutive composite (nonprime) integers. (Of course, $(M + 1)!$ is very large compared to M, so that the existence of such gaps between primes does not in any way contradict the overall regularity mentioned above.)

After the prime number theorem, the average gap between primes, $p_{n+1} - p_n$, is around $\ln p_n$, however, larger gaps occur. The best result to date is due to Rankin [191]: for some constant $c > 0$, and for infinitely many n,

$$p_{n+1} - p_n > c \, \frac{(\ln p_n)(\ln \ln p_n)(\ln \ln \ln \ln p_n)}{(\ln \ln \ln p_n)^2}.$$

Shortly before Rankin's theorem came out, Erdos proved a weaker result; his result was like Rankin's, except that the term $(\ln \ln \ln \ln p_n)$ was missing. This type of analysis has been called the "log log school" of number theory.

There is an unproved conjecture called the *Conjecture on Twin Primes* which says that for any positive integer B (no matter how large) there is a prime $p > B$ so that both p and $p + 2$ are prime: the pair $p, p + 2$ are called twin primes. For example, the following are some twin primes: (i) 3, 5; (ii) 29, 31; (iii) 1277, 1279. From any table of primes, it is easy to find such twins by inspection. They seem to be quite common, even though no one has yet been able to prove the validity of the conjecture [70,183].

The moral of all this seems to be that in toto, primes occur quite regularly, but in small sequences (i.e., relatively short sequences of consecutive integers compared to the size of the integers) they distribute themselves very badly. Not very much is known as yet about this subject.

3   Besides geometry, the theory of numbers is probably the oldest branch of mathematics with an unbroken history. The interested reader might want to pursue it on a leisurely level [70,72,82,178,183], or on a deeper, more substantial level [71,76,

79,85,86]. Incidentally, the topics of this chapter are also discussed in other textbooks on algebra [3,5,6,9,11,12]. The account by Ore [82] is highly recommended. Also see the articles [192,g]. In [190] Motzkin shows that $\sqrt{-19}$ is related to a principal ideal domain which is not a Euclidean domain. For proofs of the Chinese Remainder Theorem, see [25,74,79].

chapter six

FIELDS AND THE RATIONAL NUMBERS

The ring $\mathbf{Z}$ of integers has the property, which is not true of the natural numbers (positive integers) alone, that every equation of the form $a + x = b$, where $a, b \in \mathbf{Z}$, has a solution x in $\mathbf{Z}$. In fact, one of the principal reasons for the introduction of the negative integers and zero is to assure us that every such equation is solvable. That is, the negative integers are introduced to assure that $\mathbf{Z}^+$ is a group. However, in $\mathbf{Z}$, an equation of the form $ax = b$ is solvable if and only if a is a divisor of b. Clearly, in order that an equation of this form, where $a, b \in \mathbf{Z}$ and $a \neq 0$, always have a solution, we need to have available the rational numbers as well as the integers. More generally, we need all the rational numbers to assure us that the group of units is exactly the set of nonzero numbers. Later on in this chapter we shall show how to extend the ring $\mathbf{Z}$ of integers to the larger system of rational numbers. In this process we shall not use any previous knowledge of the rational number system, except perhaps to motivate the procedure used, but shall carry out the construction using only the properties of the integers which have already been given. Before presenting this construction we shall introduce and discuss the important concept of a *field*.

In a final optional section we shall briefly indicate how the ring of integers can be constructed from the system of natural numbers. This material is presented here because the method closely parallels that by which the rational numbers are constructed from the integers.

6.1 FIELDS

Let us make the following definition.

6.1   **DEFINITION**   A commutative ring F with more than one element and having a unity is said to be a *field* if every nonzero element of F has a multiplicative inverse in F, that is, $F \backslash \{0\}$ is the group of units of F.

In view of Corollary 3.6, we know that every nonzero element of a field has a *unique* multiplicative inverse. As indicated in Section 3.1, we may denote the multiplicative inverse of a nonzero element r of a field F by r^{-1}. If 1 is the unity of F, r^{-1} is therefore the unique element of F such that

6.2 $$r \cdot r^{-1} = r^{-1} \cdot r = 1.$$

However, the commutative law of multiplication is required to hold in a field, and we shall henceforth use it without explicit mention. In particular, we may consider that r^{-1} is defined by the single equation $r \cdot r^{-1} = 1$.

We shall now prove the following result. (Compare this with the proof of Theorem 5.31.)

6.3   **THEOREM**   *A field is necessarily an integral domain.*

> **Proof:**   Suppose that r and s are elements of a field F such that $rs = 0$. If $r \neq 0$, r has a multiplicative inverse r^{-1} in F and it follows that
>
> $$r^{-1}(rs) = (r^{-1}r)s = 1 \cdot s = s.$$
>
> But also,
>
> $$r^{-1}(rs) = r^{-1} \cdot 0 = 0.$$
>
> Hence $s = 0$, and we have shown that $r = 0$ or $s = 0$. This proves that F has no nonzero divisors of zero and F therefore satisfies Definition 4.3 of an integral domain.

Since $F \backslash \{0\}$ is the group of units of a field F, an immediate consequence of Theorem 2.3 of Section 2.2 is the following.

6.4   **THEOREM**   *If r and s are elements of a field F and $r \neq 0$, there exists a unique element y of F such that $ry = s$. Moreover, $y = r^{-1} \cdot s$.*

Let us now give a few examples of fields. It is to be understood that the usual definitions of addition and multiplication are implied.

Example 1 The set of all rational numbers; that is, all numbers of the form a/b, where $a, b \in \mathbf{Z}$ with $b \neq 0$.

Example 2 The set of all real numbers of the form $x + y\sqrt{2}$, where x and y are rational numbers. What is the multiplicative inverse of each nonzero element?

Example 3 The set of all real numbers of the form $u + v\sqrt{3}$, where u and v are elements of the field of the preceding example. It is true that every nonzero element has a multiplicative inverse in this set, but we shall not here write out a proof of this fact.

The following theorem gives some other examples of fields of quite a different type.

6.5 **THEOREM** *If p is a prime, the ring $\mathbf{Z}_p$ of integers modulo p is a field.*

This theorem is implied by Theorem 5.31, however, we repeat the proof of this important theorem.

Proof: We already know that $\mathbf{Z}_p$ is a commutative ring with unity and, using the notation of Section 2.9, the unity is [1]. Let [r] be any nonzero element of $\mathbf{Z}_p$. In order to show that we have a field, we need to show that there exists an element [x] of $\mathbf{Z}_p$ such that $[r] \cdot [x] = [1]$. The fact that $[r] \neq [0]$ implies that $r \not\equiv 0 \pmod{p}$, that is, that r is not divisible by p. Since p is a prime, it follows that r and p are relatively prime. Theorem 5.18 then assures us that there exist integers x, y such that $rx + py = 1$. This implies that $rx \equiv 1 \pmod{p}$, and hence that $[r] \cdot [x] = [1]$, as required.

We may remark that if n is not a prime, we know by Theorem 5.31 that the ring $\mathbf{Z}_n$ is not even an integral domain, and certainly then is not a field.

A field $\mathbf{Z}_p$ differs from the usual fields of elementary algebra in that it has only a finite number of elements. However, in this as in any field we can always perform the so-called rational operations of addition, multiplication, subtraction, and division (except by zero). We are here using the familiar word "subtraction" to mean addition of additive inverse, and "division" to mean multiplication by the multiplicative inverse. We may emphasize that division by zero is not defined in any field F since $0 \cdot x \neq 1$ for every x in F, and therefore 0 cannot have a multiplicative inverse.

We have required that a field contain at least two elements, and we may now observe that there does exist a field having exactly two elements, namely, the field $\mathbf{Z}_2$.

In view of Theorems 5.31 and 6.5, we know that if the ring $\mathbf{Z}_n$ is an integral domain, it is actually a field. On the other hand, the ring $\mathbf{Z}$ of integers is an integral domain which is not a field. In this connection, the following theorem may be of some interest.

6.6   THEOREM   *An integral domain S with a finite number of elements is necessarily a field.*

> **Proof:**   Suppose that S has exactly n distinct nonzero elements a_1, $a_2, \ldots, a_n$ and, for later convenience, let us agree that a_1 is the unity of S. Now suppose that a_k is an arbitrary nonzero element of S and let us show that it has a multiplicative inverse in S. Consider the set $A = \{a_k a_1, a_k a_2, \ldots, a_k a_n\}$ consisting of the product of the n nonzero elements of S by the element a_k. Since $a_k \neq 0$, the cancellation law of multiplication shows that $a_k a_i = a_k a_j$ only if $a_i = a_j$. Hence no two of the indicated elements of A can be equal. That is, the elements of A are n distinct elements of S and therefore are *all* the nonzero elements of S. Since one of them is the unity a_1, there must exist some element a_l of S such that $a_k a_l = a_1$, and a_l is the multiplicative inverse of a_k. This argument shows that every nonzero element of S has a multiplicative inverse in S and therefore that S is a field.

Incidentally, the proof of this theorem furnishes an alternate proof of Theorem 6.5 since we already know by Theorem 5.31 that $\mathbf{Z}_p$ is an integral domain (with a finite number of elements) if p is a prime.

The argument used here is the same as the one used to prove part of Theorem 2.30 of Section 2.4; that is, to show that a subset K of a finite group G is a subgroup, we need only prove that K is closed under multiplication of elements.

EXERCISES

1. Find the multiplicative inverse of each nonzero element of each of the following fields: $\mathbf{Z}_5, \mathbf{Z}_7, \mathbf{Z}_{11}, \mathbf{Z}_{13}, \mathbf{Z}_{17}$.

2. Find the multiplicative inverse of each of the following elements of the field $\mathbf{Z}_{1847}$: [12], [35], [416], [800].

3. Show that there exists a finite field which is not isomorphic to Z_p for any prime p.   *Hint:* Consider Example 7 of Section 3.2.

4. Determine at exactly what point the proof of Theorem 6.6 makes use of the assumption that S has a finite number of elements.   *Hint:* Try to apply the proof to the integral domain Z and see where it breaks down.

5. Prove: If R is a commutative ring with more than one element and with the property that for $a, b \in R$, $a \neq 0$, there exists $x \in R$ such that $ax = b$, then R is a field.

6. Let P be the set of all positive real numbers, and q a fixed positive real number not equal to 1.   If $a, b \in P$, we define operations of addition "$\oplus$" and multiplication "$\odot$" on P as follows:

$$a \oplus b = ab \text{ (ordinary multiplication)}, \qquad a \odot b = a^{\log_q b}.$$

Assuming as known all the familiar elementary properties of logarithms, prove that P is a field with respect to these definitions of addition and multiplication.

7. If F is a field, then describe all ideals of F.

6.2 THE CHARACTERISTIC

Although we are now primarily interested in fields, the concept to be introduced in this section applies to any ring and we therefore give the definition in its general form.   We recall that if a is an element of a ring and n is a positive integer, we have given in Section 2.3 a recursive definition of na.   We now make the following definition.

6.7   DEFINITION   Let R be a ring.   If there exists a positive integer n such that $na = 0$ for every element a of R, the smallest such positive integer n is called the *characteristic* of R.   If no such positive integer exists, R is said to have *characteristic zero*.

All the familiar number systems of elementary algebra certainly have characteristic zero.   However, let us consider, for example, the ring Z_4 of integers modulo 4.   If $[r]$ is any element of this ring, then $2[r] = [r] + [r] = [2r]$ and, generally, if k is a positive integer, $k[r] = [kr]$.   The smallest positive integer k

such that $[kr] = [0]$ for every element $[r]$ of $\mathbf{Z}_4$ is clearly 4, so $\mathbf{Z}_4$ has characteristic 4. In general, the ring $\mathbf{Z}_n$ has characteristic n.

The definition of the characteristic of a ring makes an assertion about *every* element of the ring. However, in an important special case, the following theorem shows that the characteristic is determined by some one particular element.

6.8   **THEOREM** *Let R be a ring with a unity e. If there exists a positive integer n such that $ne = 0$, then the smallest such positive integer is the characteristic of R. If no such positive integer exists, then R has characteristic zero.*

Proof: If n is the smallest positive integer such that $ne = 0$, the characteristic of R certainly cannot be a positive integer less than n. Moreover, if $a \in R$, then

$$na = (na)e = (ne)a = 0a = 0,$$

so that $na = 0$ for every element a of R; hence R has characteristic n. The last sentence of the theorem is an immediate consequence of the definition of characteristic zero.

We know that the ring $\mathbf{Z}_n$ is a field if and only if n is a prime. Hence the characteristic of every field that has been mentioned so far is either zero or a prime. In fact, we shall now prove that this is always true for every integral domain and certainly then for every field.

6.9   **THEOREM** *Every integral domain D has characteristic zero or a prime.*

Proof: To prove this theorem, suppose that D has characteristic $n > 0$, and that n is *not* a prime. Then $n = n_1 n_2$, where $1 < n_1 < n$, and $1 < n_2 < n$. If e is the unity of D, we have $ne = 0$ and therefore $(n_1 n_2)e = 0$. However, this implies that $(n_1 e)(n_2 e) = 0$ and, by the definition of an integral domain, it follows that $n_1 e = 0$ or $n_2 e = 0$. But if $n_1 e = 0$, the preceding theorem shows that D cannot have characteristic $n > n_1$, hence $n_1 e \neq 0$. Similarly, $n_2 e \neq 0$, and we have a contradiction of the assumption that n is not a prime. The proof is therefore complete.

We next consider the following result.

6.10   **THEOREM** *A commutative ring with unity of characteristic zero has a subring which is isomorphic to $\mathbf{Z}$. A commutative ring with unity of characteristic n has a subring which is isomorphic to $\mathbf{Z}_n$.*

We shall present some of the steps in the proof of this theorem and leave some of the details to the reader.

Proof: Let R be a commutative ring with unity e, and let us set

$$Ze = \{ne \mid n \in Z\}.$$

Then Ze is a subring of R, and it is this subring which we consider.

First, suppose that R has characteristic zero, and let $\theta: Z \to Ze$ be the mapping defined by $n\theta = ne$, $n \in Z$. Clearly, θ is an onto mapping; so let us verify that it is a one-one mapping. If $n_1e = n_2e$ with $n_1, n_2 \in Z$, the notation can be so chosen that $n_1 \geq n_2$. Then $(n_1 - n_2)e = 0$ and, by Theorem 6.8, we cannot have $n_1 - n_2 > 0$ since this would violate the assumption that R has characteristic zero. Accordingly, we conclude that $n_1 = n_2$ and θ is indeed a one-one mapping. It is then easily verified that addition and multiplication are preserved under the mapping θ, and hence that θ is an isomorphism of Z onto Ze.

Next suppose that R has characteristic n and let us, as usual, denote the equivalence set modulo n which contains the integer m by $[m]$. Let $\phi: Z_n \to Ze$ be the mapping defined by $[m]\phi = me$, $m \in Z$. We leave it to the reader to verify that this defines a mapping. Actually, if we use the mapping $\theta: Z \to Ze$ by setting $m\theta = me$, then the mapping ϕ can be shown to exist by invoking the Universal Mapping Property, 3.25. However, let us show that ϕ is a one-one mapping. Suppose that $n_1, n_2 \in Z$ such that $n_1e = n_2e$, or $(n_1 - n_2)e = 0$. By the Division Algorithm,

$$n_1 - n_2 = qn + r, \qquad 0 \leq r < n.$$

Thus

$$(n_1 - n_2)e = (qn + r)e = qne + re = re.$$

This shows that $re = 0$ and, by Theorem 6.8, it cannot be true that r is a positive integer less than n. Hence we must have $r = 0$ and therefore $n_1 - n_2 = qn$. It follows that $[n_1] = [n_2]$ and ϕ is thus a one-one mapping. We leave it to the reader to verify that ϕ is the desired isomorphism of Z_n onto the subring Ze of R.

The literature on the theory of fields is quite extensive. In much of this work the concept of the characteristic of a field plays an essential role. It frequently happens, for example, that although a theorem may be true for every field, different proofs have to be given for the case in which the characteristic is zero and that in which it is a prime. Later on in this book we shall present additional examples of fields.

6.3 SOME FAMILIAR NOTATION

Let F be a field with unity 1, and t a nonzero element of F. We have introduced the symbol t^{-1} to designate the multiplicative inverse of t, and have found that if $s \in F$, the unique element x of F such that $tx = s$ is given by $x = t^{-1}s$. In accordance with familiar usage, we shall also designate this element $t^{-1}s$ by $\dfrac{s}{t}$ or by s/t. In particular, we have $t^{-1} = 1/t$.

Suppose, now, that v is also a nonzero element of F. Since $(tv)(v^{-1}t^{-1}) = 1$, the multiplicative inverse of tv is $v^{-1}t^{-1}$, that is

6.11
$$(tv)^{-1} = v^{-1}t^{-1}.$$

It is now easy to see that

6.12
$$\frac{sv}{tv} = \frac{s}{t}.$$

This follows by the following calculation, making use of 6.11:

$$\frac{sv}{tv} = (tv)^{-1}(sv) = v^{-1}t^{-1}sv = t^{-1}s = \frac{s}{t}.$$

As a generalization of 6.12, let s and u be arbitrary elements of F, and t and v arbitrary nonzero elements of F. Then we assert that

6.13
$$\frac{s}{t} = \frac{u}{v} \quad \text{if and only if} \quad sv = tu.$$

Suppose, first, that $s/t = u/v$, that is, that $t^{-1}s = v^{-1}u$. Multiplication by tv yields $sv = tu$. Conversely, if $sv = tu$, multiplication by $t^{-1}v^{-1}$ shows that $t^{-1}s = v^{-1}u$ or, otherwise expressed, that $s/t = u/v$.

The following are also easy to establish and will be left as exercises:

6.14

$$\text{(i)} \quad \frac{s}{t} + \frac{u}{v} = \frac{sv + tu}{tv},$$

$$\text{(ii)} \quad \frac{s}{t} \cdot \frac{u}{v} = \frac{su}{tv}.$$

Now a few remarks about exponents. If t is a nonzero element of F, we have a definition of t^{-1}; and if n is any positive integer, we now define t^{-n} to be $(t^{-1})^n$; also we define $t^0 = 1$. Under these definitions, the following laws of exponents hold for every choice of m and n as arbitrary integers (positive, negative, or zero), it being understood that t and v are arbitrary nonzero elements of F:

6.15

$$t^m \cdot t^n = t^{m+n},$$

$$\frac{t^m}{t^n} = t^{m-n},$$

$$(t^m)^n = t^{mn},$$

$$(tv)^m = t^m \cdot v^m,$$

$$\left(\frac{t}{v}\right)^m = \frac{t^m}{v^m}.$$

Of course, these are generalizations of the laws 3.10(i) to (iii), which hold for a commutative ring. Complete proofs of 6.15 can be given by mathematical induction.

EXERCISES

1. Let R be a ring with a finite number of elements, and r a nonzero element of R.

  (i) Show that there must exist a positive integer m such that $mr = 0$.

  (ii) Show that R cannot have characteristic zero.

2. Let a be a fixed nonzero element of an integral domain D such that $ma = 0$ for some positive integer m. Prove that the smallest such positive integer is the characteristic of D (that is, it is independent of the particular nonzero element a which is chosen). Show, by an example, that this result is not necessarily true for a *ring*.

3. Given the characteristics of rings R and S, what can you say about the characteristic of the direct sum $R \oplus S$?

4. Prove 6.14(i), (ii).

5. If s, t, u, and v are elements of a field F, prove (without using the laws of exponents) each of the following in which it is assumed that the necessary elements are different from zero:

  (i) $(t^{-1})^{-1} = t$,

  (ii) $(-t)^{-1} = -(t^{-1})$,

  (iii) $\left(\frac{s}{t}\right)^{-1} = \frac{t}{s}$,

(iv) $\dfrac{s/t}{u/v} = \dfrac{vs}{ut}$,

(vi) $\dfrac{s}{t} + \dfrac{u}{t} = \dfrac{s+u}{t}$,

(v) $-\dfrac{s}{t} = \dfrac{(-s)}{t} = \dfrac{s}{(-t)}$,

(vii) $\dfrac{s}{t} - \dfrac{u}{v} = \dfrac{sv - tu}{tv}$.

6. Let c and d be distinct elements of a field F. If new operations of addition and multiplication are defined on F as follows:

$$x \oplus y = x + y - c, \quad x \odot y = c + \frac{(x-c)(y-c)}{d-c},$$

prove that one obtains a field F'. What are the zero and the unity of F'?

7. If F and F' are as in the preceding exercise, prove that the mapping $\theta: F \to F'$ defined by $x\theta = (d-c)x + c$, $x \in F$, is an isomorphism of F onto F'.

8. If a and b are elements of a commutative ring with characteristic the prime p, prove that $(a+b)^p = a^p + b^p$. Then generalize this result by proving that for every positive integer n, $(a+b)^{p^n} = a^{p^n} + b^{p^n}$.

6.4 THE FIELD OF RATIONAL NUMBERS

We now change our point of view as follows. Instead of studying properties of a given field, let us see how we can start with the integral domain **Z** of the integers and *construct* a field which contains **Z**. This is our first example of an important algebraic problem which may be stated in a general way as follows: Given an algebraic system U which does not have some specified property, to construct a larger system V which contains U and which does have the property in question. Naturally, this is not always possible, but it is in a number of interesting cases. At present, we start with the integral domain **Z** in which not every nonzero element has a multiplicative inverse, and shall construct a larger system—the field of rational numbers—which contains **Z** and in which every nonzero element necessarily has a multiplicative inverse. In this construction, our previous knowledge of the rational numbers will certainly be useful in suggesting procedure, but will be used in no other way.†

---

† The process described here is a special case of an important technique called localization. In particular, we will "localize at (0)."

Let S denote the set of all ordered pairs (a, b), where $a, b \in \mathbf{Z}$ and $b \neq 0$, that is,

$$S = \{(a, b) \mid a, b \in \mathbf{Z}, b \neq 0\}.$$

What we are going to do will be *suggested* by thinking of (a, b) as the familiar a/b, but we use an unfamiliar notation in order to clarify the logical procedure and to avoid using any property until we have actually proved it. If (a, b) and (c, d) are elements of S, we define $(a, b) \sim (c, d)$ to mean that $ad = bc$. Actually, "$\sim$" is an equivalence relation defined on S. The reflexive and symmetric properties are obviously true, and we now prove the transitive property. Suppose that $(a, b) \sim (c, d)$ and $(c, d) \sim (e, f)$, and let us show that $(a, b) \sim (e, f)$. Since $(a, b) \sim (c, d)$, we have $ad = bc$; and, similarly, we have $cf = de$. Multiplication of these equations by f and by b, respectively, yields $adf = bcf$ and $bcf = bde$. Hence $adf = bde$ and, since $d \neq 0$, it follows that $af = be$, that is, that $(a, b) \sim (e, f)$.

Now that we have an equivalence relation "$\sim$" defined on S, we follow a procedure somewhat like that previously used in obtaining the ring of integers modulo n. That is, we shall consider equivalence sets relative to "$\sim$," and give appropriate definitions of addition and multiplication of these sets.

If $(a, b) \in S$, according to our previous usage the equivalence set containing (a, b) would be designated by $[(a, b)]$. However, we shall now use the simpler notation $[a, b]$ to designate this equivalence set. In the sequel it is important to keep in mind that $[a, b] = [a_1, b_1]$ if and only if $(a, b) \sim (a_1, b_1)$, that is, if and only if $ab_1 = ba_1$. Of course, this is just the general property 1.15(i) of equivalence sets as applied in this particular case. The equivalence set $[a, b]$ may therefore be expressed as follows:

6.16 $[a, b] = \{(x, y) \mid (x, y) \in S, xb = ya\}.$

We now define addition and multiplication of equivalence sets as follows:

6.17 $[a, b] + [c, d] = [ad + bc, bd],$

and

6.18 $[a, b] \cdot [c, d] = [ac, bd].$

First, we observe that since (a, b) and (c, d) are elements of S, we have $b \neq 0$ and $d \neq 0$. Hence, $bd \neq 0$, so that in fact $(ad + bc, bd)$ and (ac, bd) are elements of S and the right sides of 6.17 and 6.18 are equivalence sets.

Now, just as in the case of integers modulo n, we must show that addition and multiplication of equivalence sets are well-defined by 6.17 and 6.18. Suppose, then, that

6.19 $[a, b] = [a_1, b_1]$ and $[c, d] = [c_1, d_1].$

In order to show that addition of equivalence sets is well-defined by 6.17 we must show that necessarily

$$[a, b] + [c, d] = [a_1, b_1] + [c_1, d_1];$$

that is, that

6.20 $$[ad + bc, bd] = [a_1d_1 + b_1c_1, b_1d_1].$$

From 6.19, we have that $ab_1 = ba_1$, and that $cd_1 = dc_1$. If we multiply the first of these equations by dd_1, the second by bb_1, and add the corresponding members, it follows that

$$(ad + bc)b_1d_1 = bd(a_1d_1 + b_1c_1).$$

However, this implies 6.20, and therefore addition of equivalence sets is well-defined by 6.17. The proof that multiplication is well-defined by 6.18 will be left as an exercise.

We may now state the following theorem.

6.21 **THEOREM** *Let* $\mathbf{Q}$ *denote the set of all equivalence sets of S relative to the equivalence relation "$\sim$." Then with respect to the operations of addition and multiplication on* $\mathbf{Q}$ *defined by 6.17 and 6.18,* $\mathbf{Q}$ *is a field. Moreover, the set of all elements of* $\mathbf{Q}$ *of the form* $[a, 1]$, $a \in \mathbf{Z}$, *is a subring* $\mathbf{Z}'$ *of* $\mathbf{Q}$; *and the mapping* $\theta : \mathbf{Z} \to \mathbf{Z}'$ *defined by*

$$a\theta = [a, 1], \qquad a \in \mathbf{Z},$$

is an isomorphism of $\mathbf{Z}$ *onto* $\mathbf{Z}'$.

Proof: The commutative laws of addition and multiplication, as well as the associative law of multiplication, are almost obvious, and we omit the details. The associative law of addition may be verified by the following straightforward calculation. Let $[a, b]$, $[c, d]$, and $[e, f]$ be elements of $\mathbf{Q}$. Then

$$([a, b] + [c, d]) + [e, f] = [ad + bc, bd] + [e, f]$$
$$= [adf + bcf + bde, bdf],$$

and

$$[a, b] + ([c, d] + [e, f]) = [a, b] + [cf + de, df]$$
$$= [adf + bcf + bde, bdf],$$

and we therefore have

$$([a, b] + [c, d]) + [e, f] = [a, b] + ([c, d] + [e, f]).$$

Since $[0, 1] + [a, b] = [a, b]$, and $[1, 1] \cdot [a, b] = [a, b]$, it follows that $[0, 1]$ is the zero and $[1, 1]$ the unity of $\mathbf{Q}$. However, if d is a nonzero integer, we have $[d, d] = [1, 1]$ and, similarly, $[0, 1] = [0, d]$. Hence the unity is $[d, d]$ and the zero is $[0, d]$ for *any* non-zero integer d. We may also observe that $[a, b] = [0, 1]$ if and only if $a = 0$, and to say, therefore, that $[a, b]$ is a nonzero element of **Q** is to say that $a \neq 0$.

Since $[a, b] + [-a, b] = [0, b^2]$, and $[0, b^2]$ is the zero of **Q**, it follows that the additive inverse of $[a, b]$ is $[-a, b]$, that is, we have $-[a, b] = [-a, b]$, and each element of **Q** has an additive inverse.

One of the distributive laws is a consequence of the following calculations in which, at one point, we make use of the fact that $[b, b]$ is the unity of **Q**:

$$[a, b]([c, d] + [e, f]) = [a, b] \cdot [cf + de, df]$$
$$= [acf + ade, bdf],$$
$$[a, b] \cdot [c, d] + [a, b] \cdot [e, f] = [ac, bd] + [ae, bf]$$
$$= [acbf + bdae, b^2df]$$
$$= [acf + ade, bdf] \cdot [b, b]$$
$$= [acf + ade, bdf].$$

The other distributive law is an immediate consequence of this one since multiplication is commutative.

Up to this point we have proved that **Q** is a commutative ring with unity. To prove that **Q** is a field, there remains only to show that every nonzero element of **Q** has a multiplicative inverse in **Q**. If $[a, b]$ is a nonzero element of **Q**, then $a \neq 0$ as well as $b \neq 0$, and it is clear that $[b, a] \in$ **Q**. Moreover,

$$[a, b] \cdot [b, a] = [ab, ab] = [1, 1],$$

and the multiplicative inverse of $[a, b]$ is $[b, a]$. That is, if $[a, b]$ is a nonzero element of **Q**, then $[a, b]^{-1} = [b, a]$. This completes the proof that **Q** is a field.

Now let **Z'** be the set of elements of **Q** of the form $[a, 1]$, $a \in$ **Z**, and consider the mapping $\theta: \mathbf{Z} \to \mathbf{Z'}$ defined by $a\theta = [a, 1]$, $a \in$ **Z**. This is clearly an onto mapping and it is also a one-one mapping since $[a, 1] = [b, 1]$ implies that $a = b$. Moreover, for $a, b \in$ **Z**, we have

$$(a + b)\theta = [a + b, 1] = [a, 1] + [b, 1] = a\theta + b\theta,$$

and

$$(ab)\theta = [ab, 1] = [a, 1] \cdot [b, 1] = (a\theta)(b\theta).$$

Thus θ is an isomorphism of $\mathbf{Z}$ onto $\mathbf{Z}'$, and the theorem is established.

Since the subring $\mathbf{Z}'$ of $\mathbf{Q}$ is isomorphic to $\mathbf{Z}$, we shall henceforth find it convenient to identify $\mathbf{Z}'$ with $\mathbf{Z}$ and, as a matter of notation, write simply a to designate the element $[a, 1]$ of $\mathbf{Q}$. We may then consider that the field $\mathbf{Q}$ actually contains the ring $\mathbf{Z}$ of integers.

As a further simplification of notation, let us observe that

$$[a, b] = [a, 1] \cdot [1, b] = [a, 1] \cdot [b, 1]^{-1},$$

and hence we are justified in writing $a \cdot b^{-1}$ or a/b for the element $[a, b]$ of $\mathbf{Q}$. Now that we have justified our familiar notation, we shall henceforth call an element of $\mathbf{Q}$ a *rational number* and the field $\mathbf{Q}$ the *field of rational numbers*. All of the notation of the preceding section naturally applies to the field $\mathbf{Q}$. Throughout the rest of this book, $\mathbf{Q}$ will consistently be used to designate the field of rational numbers.

In the notation which we have finally introduced, the field $\mathbf{Q}$ consists of all numbers of the form a/b, where a and b are integers with $b \neq 0$, addition and multiplication being defined in the usual way (6.17, 6.18).

Let us emphasize the meaning of the notation we have introduced by considering, for example, the rational number $1/2$. We are writing $1/2$ for the equivalence set $[1, 2]$ used above. Now $[1, 2] = [c, d]$ if and only if $d = 2c$, so we see that $1/2$ represents the equivalence set consisting of all ordered pairs of the form $(c, 2c)$, where c is a nonzero integer. Moreover, for example, $1/2 = 3/6$ simply because, by our definition of equivalence, $(1, 2) \sim (3, 6)$ and therefore $[1, 2] = [3, 6]$.

Since $(-a)/b = a/(-b)$, we see that every rational number can be written in the form c/d, where $d > 0$. Moreover, if the integers c and d have a common nonzero factor k, so that $c = c_1 k$ and $d = d_1 k$, then $c/d = c_1/d_1$. It follows that every rational number r can be written uniquely in the form a/b, where a and b are relatively prime integers with $b > 0$. If r is expressed in this form, it is sometimes said that r is expressed *in lowest terms*.

Before proceeding to establish a few properties of the field $\mathbf{Q}$ of rational numbers, let us point out that in the above construction of the field $\mathbf{Q}$, the *only* properties of the integers which were used are those that imply that $\mathbf{Z}$ is an integral domain. Accordingly, by exactly the same construction we could start with an arbitrary integral domain D and obtain the *field of quotients* of D whose elements are expressible in the form ab^{-1}, where $a, b \in D$ with $b \neq 0$. In this terminology, the field $\mathbf{Q}$ of rational numbers is the field of quotients of $\mathbf{Z}$.

One should be careful not to mix "ring of quotients" with "quotient ring." The phrases are different and have *very* different meanings which can be seen by comparing the definition of the previous paragraph with Theorem 3.30 of Section 3.4.

6.5 A FEW PROPERTIES OF THE FIELD OF RATIONAL NUMBERS

We have defined in Section 4.2 what we mean by an ordered integral domain. Since a field is necessarily an integral domain, by an *ordered field* we shall naturally mean a field which is an ordered integral domain. We shall now prove the following result.

6.22 **THEOREM** *Let $\mathbf{Q}^p$ denote the set of all rational numbers a/b, where a and b are integers such that $ab > 0$. Then $\mathbf{Q}^p$ has the Properties 4.4 which define an ordered integral domain, and therefore the field $\mathbf{Q}$ is an ordered field whose positive elements are the elements of $\mathbf{Q}^p$.*

We may point out that when we write $ab > 0$, we mean that ab is a positive *integer* and we are only making use of the fact that $\mathbf{Z}$ is an ordered integral domain.

Proof: First, we need to show that the definition of an element of $\mathbf{Q}^p$ does not depend upon the particular representation of a rational number. That is, we need to show that if $a/b = c/d$ and $ab > 0$, then also $cd > 0$. This follows from the observation that $a/b = c/d$ means that $ad = bc$, and $ab > 0$ implies that either a and b are both positive or they are both negative. The same must therefore be true of c and d; hence also $cd > 0$.

Now let us show [4.4(i)] that the set $\mathbf{Q}^p$ is closed under addition. Let a/b and c/d be elements of $\mathbf{Q}^p$, and therefore $ab > 0$ and $cd > 0$. Then

$$\frac{a}{b} + \frac{c}{d} = \frac{ad + bc}{bd},$$

and we wish to show that

$$(ad + bc)bd = abd^2 + cdb^2 > 0.$$

However, this inequality follows easily from the following known inequalities: $ab > 0$, $cd > 0$, $b^2 > 0$, and $d^2 > 0$.

It is trivial that $\mathbf{Q}^p$ is closed under multiplication [4.4(ii)]. Moreover, if a/b is a nonzero rational number, then either $ab > 0$ or $ab < 0$. It follows that for every rational number a/b, exactly one of the following holds [4.4(iii)]:

$$\frac{a}{b} = 0, \qquad \frac{a}{b} > 0, \qquad -\frac{a}{b} > 0.$$

Hence $\mathbf{Q}^p$ has the three required properties, and the field $\mathbf{Q}$ of rational numbers is ordered.

It will be observed that what we have done is to make use of the known ordering of the integers to establish an ordering of the rational numbers. Inasmuch as we have identified the integer a with the rational number $a/1$, it is clear that a is a positive integer if and only if a is a positive rational number. In other words, our ordering of the rational numbers is an *extension* of the previous ordering of the integers.

In view of Theorem 6.22, we can introduce inequalities involving rational numbers in the usual way. That is, if $r, s \in \mathbf{Q}$, we write $r > s$ (or $s < r$) to mean that $r - s \in \mathbf{Q}^p$, and so on. We now have available all the usual properties (4.6) of inequalities for rational numbers. In the future we shall make use of these properties without specific reference.

The following is a significant property of the rational numbers.

6.23 THEOREM *Between any two distinct rational numbers there is another rational number.*

Proof: Suppose that $r, s \in \mathbf{Q}$ with $r < s$. The theorem will be established by showing that

$$r < \frac{r + s}{2} < s,$$

and hence that $(r + s)/2$ is a rational number between r and s. Since $r < s$, we have $r + r < r + s$, or $2r < r + s$. Now multiplying this last inequality by the positive rational number $1/2$, we see that $r < (r + s)/2$. In a similar manner, it can be shown that $(r + s)/2 < s$, and we omit the details.

The property of the rational numbers stated in the preceding theorem is often expressed by saying that the rational numbers are *dense*. We shall now prove in the following theorem another simple, but important, property of the rational numbers.

6.24 THEOREM (*Archimedean Property*) *If r and s are any positive rational numbers, there exists a positive integer n such that $nr > s$.*

Proof: Let $r = a/b$, $s = c/d$, where a, b, c, and d are positive integers. If n is a positive integer, then $n(a/b) > c/d$ if and only if $n(ad) > bc$. We now assert that this last inequality is necessarily satisfied if we choose $n = 2bc$. For $ad \geq 1$, and therefore $2ad > 1$. Multiplying this inequality by the positive integer bc shows that $2adbc > bc$. Hence, $n = 2bc$ certainly satisfies our requirement. Of course, we do not mean to imply that this is necessarily the smallest possible choice of n.

6.6 SUBFIELDS AND EXTENSIONS

Let us make the following convenient definition.

6.25 DEFINITION A subring F' of a field F which is itself a field is called a *subfield* of F. If F' is a subfield of F, F is frequently called an *extension* of F'.

Although we are here primarily interested in fields, we shall first prove the following fairly general result.

6.26 THEOREM *Let D be an integral domain with unity e, and suppose that R is a subring of D having more than one element. Then, if R has a unity, $e \in R$ and e is the unity of R.*

Proof: Let f be the unity of R and let us prove that $f = e$. Clearly,

$$f(fe - e) = f^2 e - fe = fe - fe = 0.$$

However, $f \neq 0$ since R has more than one element, and therefore f is not a divisor of zero in the integral domain D. Hence we must have $fe - e = 0$, or $fe = e$. But e is the unity of D and $f \in D$, so $fe = f$ and we conclude that $f = e$, as we wished to show.

Since a field F is an integral domain and a subfield of F necessarily has a unity and has more than one element, we have the following special case of the result just obtained.

6.27 COROLLARY *The unity of a field F is also the unity of each subfield of F.*

We may point out that this corollary and Theorem 6.8 show that a field and all of its subfields have the same characteristic.

Since $\mathbf{Z}_p$ is a field for each prime p, the first statement of the following theorem has already been established as a part of Theorem 6.10.

6.28 THEOREM *A field of characteristic p contains a subfield which is isomorphic to the field $\mathbf{Z}_p$. A field of characteristic zero contains a subfield which is isomorphic to the field $\mathbf{Q}$.*

Proof: To prove the second statement of the theorem, let F be a field of characteristic zero and with e as unity. We already know by Theorem 6.10 that the subring $\mathbf{Z}e$ of F is isomorphic to $\mathbf{Z}$. It is perhaps then not surprising that F contains a subfield which is isomorphic to the field $\mathbf{Q}$ of quotients of $\mathbf{Z}$. However, we shall present some of the steps in the proof of this fact.

Since F has characteristic zero, if n is a nonzero integer, $ne \neq 0$ and therefore ne has a multiplicative inverse $(ne)^{-1}$ in F. Let $\mathbf{Q}'$ be the set of all elements of F expressible in the form $(me)(ne)^{-1}$, where $m, n \in \mathbf{Z}$ with $n \neq 0$. We leave it to the reader to verify that $\mathbf{Q}'$ is a subfield of F. (Of course, it is the field of quotients of the integral domain $\mathbf{Z}e$.) We proceed to show that there exists an isomorphism of $\mathbf{Q}$ onto $\mathbf{Q}'$. To this end, we start by defining a mapping $\alpha: \mathbf{Q} \to \mathbf{Q}'$ as follows:

$$(mn^{-1})\alpha = (me)(ne)^{-1}, \qquad m, n \in \mathbf{Z}; n \neq 0.$$

Since elements of $\mathbf{Q}$ are not uniquely expressible in the form mn^{-1}, we must first show that α is in fact a well-defined mapping of $\mathbf{Q}$ into $\mathbf{Q}'$. Suppose that m, n, k, and l are integers with $n \neq 0$ and $l \neq 0$ such that in $\mathbf{Q}$, $mn^{-1} = kl^{-1}$. Then $ml = nk$ and therefore in $\mathbf{Q}'$, $(me)(le) = (ne)(ke)$. It follows that $(me)(ne)^{-1} = (ke)(le)^{-1}$ and the mapping α is indeed well-defined. To verify that it is a one-one mapping, suppose that $(mn^{-1})\alpha = (kl^{-1})\alpha$, that is, that $(me)(ne)^{-1} = (ke)(le)^{-1}$. From this equation it follows that $(ml - nk)e = 0$. Then, since F has characteristic zero, we must have $ml - nk = 0$ or $mn^{-1} = kl^{-1}$. Thus α is a one-one mapping. It is clearly an onto mapping and we leave as an exercise the verification that the operations of addition and multiplication are preserved, and hence that α is the desired isomorphism.

In view of this theorem, we may observe that if we do not consider isomorphic fields as "different," *every* field is an extension of the rational field $\mathbf{Q}$ or of one of the fields $\mathbf{Z}_p$ for some prime p. The study of extensions of a given field is an important part of the general theory of fields, and we shall return to a further consideration of this topic in Chapter 12.

EXERCISES

1. Prove that multiplication of equivalence sets is well-defined by 6.18.

2. Go through the proof of Theorem 6.21 and verify that all the steps can be carried out with an arbitrary integral domain D in place of $\mathbf{Z}$, thus obtaining the field of quotients of D. In this construction, why cannot an arbitrary commutative ring be used in place of $\mathbf{Z}$?

3. Complete the proof of Theorem 6.23, and state what properties of inequalities have been used.

4. If u and v are positive rational numbers with $u < v$, show that $1/u > 1/v$.

5. If $r, s \in \mathbf{Q}$ with $r < s$, and $u, v \in \mathbf{Q}^p$, show that

$$r < \frac{ur + vs}{u + v} < s.$$

6. If $r, s \in \mathbf{Q}$ with $r < s$, and n is an arbitrary positive integer, show that there exist rational numbers $t_1, t_2, \ldots, t_n$ such that

$$r < t_1 < t_2 < \cdots < t_n < s.$$

7. Prove that addition and multiplication are preserved under the mapping α defined in the proof of Theorem 6.28.

8. Prove that a subring R (with more than one element) of a field F is a subfield of F if and only if the multiplicative inverse in F of each nonzero element of R is an element of R.

9. Let D and D' be integral domains with respective fields of quotients F and F'. If $\theta: D \to D'$ is an isomorphism of D onto D', prove that the mapping $\alpha: F \to F'$ defined by

$$(ab^{-1})\alpha = (a\theta)(b\theta)^{-1}, \qquad a, b \in D; b \neq 0,$$

is well-defined and is an isomorphism of F onto F'.

6.7 CONSTRUCTION OF THE INTEGERS FROM THE NATURAL NUMBERS (OPTIONAL)

We indicated in Section 4.5 how all the familiar properties of the natural numbers, that is, the positive integers, can be obtained from a few simple axioms. Let N be the system of all natural numbers. In this system we have operations of addition and multiplication, and all the properties of an integral domain hold except that there is no zero, and elements do not have additive inverses. In this section we shall outline a procedure by which we can start with N and *construct* the ring $\mathbf{Z}$ of all integers. The method closely parallels that by which we have constructed the rational numbers from the ring of integers. We may emphasize that we now assume as known only the properties of the natural numbers.

Let T be the set of all ordered pairs (a, b) of elements of N. Our procedure will be *suggested* by thinking of (a, b) as meaning $a - b$, but we must so formulate our statements that only natural numbers are involved. If (a, b) and (c, d) are elements of T, we shall write $(a, b) \sim (c, d)$ to mean that $a + d = b + c$. It is easy to verify that "$\sim$" is an equivalence relation on T. One way to characterize the equivalence set $[a, b]$ which contains (a, b) is as follows:

$$[a, b] = \{(x, y) \mid x, y \in N, x + b = y + a\}.$$

Now let $\mathbf{Z}$ be the set of all such equivalence sets, and let us make the following definitions:

6.29
$$[a, b] + [c, d] = [a + c, b + d],$$

and

6.30
$$[a, b] \cdot [c, d] = [ac + bd, ad + bc].$$

It can be shown that addition and multiplication are well-defined, and hence that we have operations of addition and multiplication defined on $\mathbf{Z}$. The following theorem can now be established.

6.31 THEOREM *With respect to the Definitions 6.29 and 6.30 of addition and multiplication, $\mathbf{Z}$ is an integral domain and, by a suitable change of notation, we may consider that $\mathbf{Z}$ contains the set N of natural numbers. If we now define the set $\mathbf{Z}^p$ of positive elements of $\mathbf{Z}$ to be the set N, then $\mathbf{Z}$ is an ordered integral domain in which the set of positive elements is well-ordered.*

We shall make a few remarks about the proof of this theorem, but shall not write out all the details. The zero of $\mathbf{Z}$ is $[c, c]$ for an arbitrary natural number c. The additive inverse of $[a, b]$ is $[b, a]$, that is, $-[a, b] = [b, a]$.

Let N' be the set of all elements of $\mathbf{Z}$ of the form $[x + 1, 1]$, $x \in N$. Then the mapping $\theta \colon N \to N'$ defined by

$$x\theta = [x + 1, 1] \qquad x \in N,$$

is a one-one mapping of N onto N' and, moreover, addition and multiplication are preserved under this mapping. Hence, as a matter of notation, let us identify

N' with N; that is, let us write x in place of $[x + 1, 1]$ so that $\mathbf{Z}$ now actually contains N. If $[a, b] \in \mathbf{Z}$ it is easy to verify that

$$[a, b] = [a + 1, 1] + [1, b + 1]$$
$$= [a + 1, 1] - [b + 1, 1]$$
$$= a - b.$$

We have therefore justified writing $a - b$ in place of $[a, b]$.

If $c, d \in N$, we defined $c > d$ in Section 4.5 to mean that there exists a natural number f such that $c = d + f$. Since $[a, b] = a - b$, we see that $[a, b]$ is an element of N if $a > b$, and that $-[a, b]$ is an element of N if $b > a$. The elements of $\mathbf{Z}$ therefore consist of the natural numbers, the additive inverses of the natural numbers, and zero. Of course, the integral domain $\mathbf{Z}$ is called the *ring of integers*.

If we set $\mathbf{Z}^p = N$, then $\mathbf{Z}^p$ has the properties (4.4) which make $\mathbf{Z}$ an ordered integral domain. Finally, then, since the set N is well-ordered, we have that $\mathbf{Z}$ is an ordered integral domain in which the set of positive elements is well-ordered. Our viewpoint in this book has been to *assume* that the ring of integers has all the properties implied in this statement. However, we have now indicated how this result can be proved by starting only with the Peano Axioms for the natural numbers.

 # COMMENTARY

1 What is good mathematics? How does one recognize great mathematics? Probably the best answer is that when good mathematics is encountered, it can be felt and it can be seen. Good mathematics is beautiful and moving, very much like good art, music, and literature. After he lost what he felt was his creative ability and his feeling of value as a mathematician, G. H. Hardy (1877–1947) tried to put into words what it was in mathematics that drew him into the subject [169]. He is overly severe on applied mathematics, and is quite incorrect about the applications of some of his own creations. But his overall description of mathematics rings true for many who enjoy mathematics.

Great mathematics cannot be defined, but we can make some observations about it. There is an enduring quality to it. Each succeeding generation of mathematicians comes back to relearn it. There is an element of surprise to it; at the same time it verifies our intuition. A great problem of mathematics is one that arises naturally; many mathematicians pose the problem. It is not easily solved. Through the years, many solutions are attempted, and new mathematics is created in these attempts. When a solution finally comes, it is surprising in its twists. The solution may involve intricate, involved, and deep computations, but to anyone who has thought deeply about the problem, the solution feels right. As time passes, new solutions replace old ones, and the intricate, clever solutions are replaced by ones of striking simplicity and beauty. Even though the original problem is solved, it does not lose its appeal. Either the problem itself or various attempts to solve it send ripples in many directions through mathematics. So above all, great mathematics is classical in the sense that it is time tested. It rests upon a long history of similar preceding mathematics.

As an example of this, let us consider one of the greatest of algebraic theories: Galois Theory. The solution of equations dates back many years. Babylonian tablets as old as 2000 B.C. indicate that even at that time certain problems were posed which involved the solution of simple algebraic equations. The Greeks knew how to find roots of quadratic equations by geometric construction. During the fifteenth century there was a great onslaught on the solution of cubic and quartic equations. Over this long period, various solutions to fourth- and lower-degree equations led to great improvements in notations and in ideas about numbers. The final attack upon solution to fifth- and higher-degree equations by algebraic means occurred in the early 1800s when Abel (1802–1829) first showed that the general quintic has no "algebraic" solution, and then when Galois (1811–1832) gave necessary and sufficient conditions for an nth-degree equation to have an "algebraic" solution.

The problem originated in attempts to solve very specific and simple equations. It grew into the general one of finding all roots of an arbitrary nth-degree equation. The final solution is hinted at in works of Cardan (1501–1576), Lagrange (1736–1813), Ruffini (1765–1822), and Abel. But this solution is surprising, beautiful, and certainly not obvious from work done prior to Galois. Galois' methods led to the creation of finite group theory and field theory as subjects. Each of these subjects has grown into a major branch of algebra. In other words, Galois not only solved a problem reaching back to antiquity, but his methods also have had a strong influence on the directions algebra has taken since his time.

We may contrast this to a not insignificant amount of mathematics of a very different character which is created in each generation. In some new areas of mathematics, a "classical" theorem is one which is only two years old. The theorem solves a problem which is only four years old. The branches of this avant-garde mathematics are honestly created in attempts to solve classical problems. The creators make enough new mathematics to meet the needs of classical problems and pass on to consideration of other deep and difficult problems. But once a field is open to be tilled, any number will rush in to homestead the land. The subject is developed quickly and cheaply in all and aimless directions. If the new subject survives and becomes a substantial branch of mathematics, then a certain amount of aimless flogging is helpful. It defines the important problems, and reveals the meaningless alleys. For example, in finite group theory, around 1900 it became clear that, as things stood, it was of no profit to describe all finite groups of order 1, 2, 3, 4, 5, ..., 99, 100, ..., ad nauseam, one order at a time. And yet, many mathematicians were involved in this fruitless effort. Similarly, in invariant theory, many mathematicians were involved in the computation of invariants, turning out papers which never have and probably never will be read by more than three persons. Cheap mathematics (some of this is created at great expense of effort) is created in every generation. We post-Sputnik mathematicians may be as guilty as any generation, with our great pressure to create more scientists and mathematicians.

As appreciators of mathematics, we should develop taste, discarding that which is ugly and remembering that which is elegant. But we must remember that the best mathematics is also hard. The best mathematicians compute (but not aimlessly), and the best problems involve a balance between abstract and computational reasoning in their solution.

2   Fields come in many shapes and sizes, and are fundamental in most branches of mathematics. We shall discuss one particular example which arises in geometry and combinatorics. Imagine for a moment the Cartesian product $\mathbf{R} \times \mathbf{R}$ of the real field with itself. To each ordered pair $(a, b) \in \mathbf{R} \times \mathbf{R}$ we associate a unique point in the Cartesian plane. The lines of the plane are the sets $\{(x, y) \mid y = mx + b, x \in \mathbf{R}\}$, $\{(a, y) \mid y \in \mathbf{R}\}$ where $a, m, b \in \mathbf{R}$ are fixed. In the usual algebraic way we may associate various subsets of $\mathbf{R} \times \mathbf{R}$ with their geometric counterpart in the plane.

Let us try this same technique with a field other than the real field. For example, let us consider the field $\mathbf{Z}_p$ of integers modulo p for a prime p. What we are about to describe holds for any field, but we illustrate the construction only for $\mathbf{Z}_2$. Let us call the set $\mathbf{Z}_p \times \mathbf{Z}_p = \{(a, b) \mid a, b \in \mathbf{Z}_p\}$ our set of *points*. For fixed $a, m, b \in \mathbf{Z}_p$ we call the sets $\{(x, y) \mid y = mx + b, x \in \mathbf{Z}_p\}$ and $\{(a, y) \mid y \in \mathbf{Z}_p\}$ *lines*. In fact, we just carry over our geometric names for the corresponding algebraic sets associated with the Cartesian plane. For example, $\{(a, b) \mid a^2 + b^2 = [1]\}$ would be a circle. Corresponding to these sets we may draw a "Cartesian plane." That is, we use dots to represent points $(a, b) \in \mathbf{Z}_p \times \mathbf{Z}_p$ and curves to represent lines. In Figure 1 we have drawn the "plane" with its "lines" for $\mathbf{Z}_2$. The plane we have constructed is called the *affine* plane coordinatized by $\mathbf{Z}_2$.

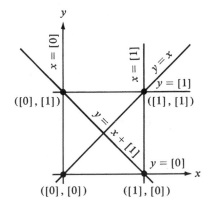

FIGURE 1

Let us go one step further and supply more points and one more line to our plane. It is almost true that any two lines in the plane must intersect, the exception being two parallel lines. I suppose we have all looked down a long straight railway track, or, at one time or another, said, "Parallel lines meet at infinity." Let us now cause this very thing to happen for the Cartesian plane. We shall, for lack of a better word, say that the vertical lines $\{(a, y) \mid y \in \mathbf{R}\}$ have slope infinity, ∞. By fiat, then we say that all lines with a given slope m meet at a point (m) "at infinity." We stipulate

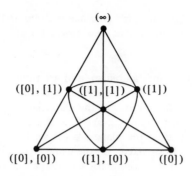

FIGURE 2

that all these "points at infinity" (m) are different for distinct values of m, $m \in$ $\mathbf{R} \cup \{\infty\}$; the set of these points forms a line, called the "line at infinity." Of course, these points and this line cannot be drawn on the "finite" Cartesian plane, but we can imagine them to lying "just outside" of this plane.

We may also carry out such a construction for any field. The situation is easier with $\mathbf{Z}_p$, since we may actually draw a picture of our affine plane along with its line at infinity. After bending things a bit, for $\mathbf{Z}_2$, we arrive at Figure 2. Let us make some observations about properties (i.e., axioms) which hold for such a "bent" plane constructed over a field $\mathbf{Z}_p$ for a prime p.

A1   *Any two lines intersect in exactly one point.*

If the lines are not parallel, we can solve the equations for the point of intersection. If they are parallel, they have the same slope and meet at that slope point on the line at infinity.

A2   *Any two points determine a unique line.*

If both points are "finite" (i.e., not on the line at infinity) then we can determine the equation for the line with the help of familiar slope intercept equations. If "points at infinity" are involved we still can settle this, as a little fiddling with Figure 2, or with the slope-intercept form $y = mx + b$ for the equation of a line, will show.

A3   *There are four points, no three of which are on the same line.*

For any prime p, the choices in $\mathbf{Z}_p$ given by ([0], [0]), ([0], [1]), ([1], [0]), and ([1], [1]) will obviously always suffice. Similar choices can be made for other fields.

What we have here is a geometrical configuration. Because it satisfies A1 to A3 it is called a *projective plane*. Our argument shows that there is a projective plane belonging to any (finite) field.

Observe that for $\mathbf{Z}_p$ (i) each line will contain $p + 1$ points, (ii) each point has $p + 1$ lines passing through it, (iii) there are $p^2 + p + 1$ points, and (iv) there are $p^2 + p + 1$ lines. The number p is called the *order* of the projective plane. Every finite projective plane has an order and satisfies the properties (i) to (iv) with respect to this order. We have just shown that there is a projective plane of order p for every prime p; namely the one coordinatized by the field $\mathbf{Z}_p$.

If n is composite, a little experimentation will show that $\mathbf{Z}_n$ does *not* coordinatize a plane; for our construction we need a field. On the other hand, there are projective planes (i.e., systems satisfying the axioms A1 to A3 above) which do not arise for a field as we have described. In a projective plane every line has $n + 1$ points, where n is the order of the plane. For all known planes, n is a prime power; further, at least one projective plane is known to exist for each prime power order. It is believed that the order of a finite projective plane must be a prime power. It has been shown that there is no plane of order 6. But for the very next candidate, 10, the existence of a plane of that order is completely unknown. Research is currently being carried on in this area.

Since it is a geometric configuration, the plane we have constructed over the field $\mathbf{Z}_p$ for a prime p does have a symmetry group. In our particular case with the affine plane of Figure 1 over the field $\mathbf{Z}_2$, the group of all symmetries of the plane which fix the point ([0], [0]) is isomorphic to S_3. Can you prove this?

In axiom A3, if we take the four points, we may choose four pairs of them so that we obtain four lines no three of which meet in a single point. So A3 may be replaced by:

A3′ *There are four lines, no three of which meet in a single point.*

Because of the equivalence of A3 and A3′, we see that if we interchange the words "point" and "line" in the axioms A1 to A3, we obtain three new axioms which are also the axioms of a projective plane! This switching is called *duality*. It tells us that for any theorem we prove about points and lines, we may prove another theorem just by interchanging the words "point" and "line" wherever they occur in the statement of the theorem and its proof. Obviously, the notion of duality is important and useful.

Dorwart [102] has written a delightful book on projective geometry. Other books include Refs. [99,101,106].

3   In this chapter we have only touched the surface of a very extensive branch of algebra, the theory of fields. Material parallel to that of this chapter may be found in Refs. [3,5,6,9,11,12,18,21,25]. Landau [124] and Omsted [129] are excellent for coverage of the rational numbers. For some of the more advanced topics, Artin [52], Postnikov [55], and Zariski and Samuel [50] may be consulted.

chapter seven

REAL AND
COMPLEX NUMBERS

In the preceding chapter we gave a detailed construction of the field of rational numbers, starting with the integral domain of integers. In this chapter we shall be concerned with extensions of the field of rational numbers to the field of real numbers and of the field of real numbers to the field of complex numbers. There are different methods of carrying out the first of these extensions but any one of them involves rather long and detailed calculations. Accordingly, instead of presenting the details, we shall merely state the existence of a certain extension of the field of rational numbers which we shall call the field of real numbers, and briefly discuss a few properties of this field. It is quite easy to construct the field of complex numbers from the field of real numbers and we shall carry out this construction and establish a number of fundamental properties of the complex numbers.

7.1 THE FIELD OF REAL NUMBERS

The rational numbers are sufficient for use in all simple applications of mathematics to physical problems. For example, measurements are usually given to a certain number of decimal places, and any finite decimal is a rational number. However, from a theoretical point of view, the system of rational numbers is entirely inadequate. The Pythagoreans made this discovery about 500 B.C. and were profoundly shocked by it. Consider, for example, an isosceles right triangle whose legs are 1 unit in length. Then, by the Pythagorean theorem, the hypotenuse has length $\sqrt{2}$; and from this geometrical consideration it appears that there must exist a "number" $\sqrt{2}$, although we have shown in Section 5.5 that it cannot be a rational number.

    The inherent difficulty in extending the field of rational numbers to the field of real numbers is perhaps indicated by the fact that a satisfactory theory of the real numbers was not obtained until the latter half of the nineteenth century. Although other men also made contributions to the theory, it is usually attributed to the German mathematicians Dedekind (1831–1916) and Cantor (1845–1918). We shall not present here the work of either of these men but shall presently state without proof the fundamental theorem which each of them essentially proved and by quite different methods. In order to do this, we must first make a few preliminary definitions.

    So far, the only ordered field which we have studied is the field $\mathbf{Q}$ of rational numbers. However, for the moment, suppose that F is an arbitrary ordered field and let us make the following definition.

7.1   **DEFINITION**   Let S be a set of elements of an ordered field F. If there exists an element b of F such that $x \leq b$ for every element x of S, then b is called an *upper bound* of the set S in F.

As an example, the set $S_1 = \{1/2, 1, 2\}$ of elements of $\mathbf{Q}$ has an upper bound 2. Also, 5/2 is an upper bound of this set, as is 117, and so on. Thus, if a set has an upper bound, it has many upper bounds. Clearly, the set $\mathbf{Z}^p$ of all positive integers does not have an upper bound in $\mathbf{Q}$. As another example, consider the set

$$S_2 = \{a \mid a \in \mathbf{Q}, a > 0, a^2 < 2\}.$$

Then S_2 has upper bounds in $\mathbf{Q}$, one of them being 3.

7.2   **DEFINITION**   Let S be a set of elements of an ordered field F. If there exists an upper bound c of S in F such that no smaller element of F is an upper bound of S, then c is called the *least upper bound* (l.u.b.) of S in F.

It follows from this definition that if a set S has a l.u.b., it is unique. More-over, if c is the l.u.b. of the set S in F and $d \in F$ such that $d < c$, then there must exist an element s of S such that $s > d$, since, otherwise, d would be an upper bound of S less than the least upper bound.

For the set S_1 exhibited above, the element 2 of S_1 is clearly the l.u.b. of S_1 in $\mathbf{Q}$. However, for the set S_2 the situation is not quite so obvious. Although we shall not give the details, it is true and should perhaps not be surprising that there exists no rational number which is the l.u.b. of the set S_2. That is, if $c \in \mathbf{Q}$ is an upper bound of S_2, there exists $d \in \mathbf{Q}$ such that $d < c$ is also an upper bound of S_2. Therefore S_2 has no l.u.b. in $\mathbf{Q}$. Thus we have an example of a set of elements of $\mathbf{Q}$ which has upper bounds in $\mathbf{Q}$ but no l.u.b. in $\mathbf{Q}$. In the field of real numbers, whose existence is asserted in the next theorem, this situation cannot arise. In fact, it is the existence of an ordered field with this property which may be considered to be the principal contribution of Dedekind and Cantor to the subject. Let us state this result as the following theorem.

7.3   **THEOREM**   *There exists a field* $\mathbf{R}$, *called the field of real numbers with the following properties:*

(*i*) $\mathbf{R}$ *is an extension of the field* $\mathbf{Q}$ *of rational numbers. Moreover,* $\mathbf{R}$ *is an ordered field and* $\mathbf{Q}^p \subset \mathbf{R}^p$.

(*ii*) *If* S *is a nonempty set of elements of* $\mathbf{R}$ *which has an upper bound in* $\mathbf{R}$, *it has a l.u.b. in* $\mathbf{R}$.

The elements of $\mathbf{R}$ are called *real numbers*. An element of $\mathbf{R}$ which is not an element of $\mathbf{Q}$ is called an *irrational* number. Considered as elements of $\mathbf{R}$, the set S_2 defined above has a l.u.b. [by 7.3(ii)] and this l.u.b. we may *define* to be the number $\sqrt{2}$. Clearly, $\sqrt{2}$ is an irrational number and it is the fact that this number is not an element of $\mathbf{Q}$ which prevents S_2 from having a l.u.b. *in the field* $\mathbf{Q}$.

The fact that $\mathbf{Q}^p \subset \mathbf{R}^p$ is sometimes expressed by saying that the ordering of $\mathbf{R}$ is an extension of the ordering of $\mathbf{Q}$. That is, a rational number is a positive rational number if and only if, considered as a real number, it is a positive real number. A similar situation arose when we passed from the integers to the rational numbers.

7.2 SOME PROPERTIES OF THE FIELD OF REAL NUMBERS

In this section we shall prove two fundamental properties of real numbers and state one additional property without proof. Of course, our proofs will be based on the assumed properties (i) and (ii) of Theorem 7.3.

Throughout the rest of this book we shall continue to denote the field of real numbers by $\mathbf{R}$ and the set of positive real numbers by $\mathbf{R}^p$.

7.4   **THEOREM** (*Archimedean Property*) *If* $a, b \in \mathbf{R}^p$, *there exists a positive integer* n *such that* $na > b$.

Proof: Let us assume that $ka \leq b$ for every positive integer k, and seek a contradiction. Another way of stating this assumption is to assert that b is an upper bound of the set $S = \{ka \mid k \in \mathbf{Z}^p\}$. Since this set has an upper bound, by 7.3(ii) it has a l.u.b., say c. Now $c - a < c$ and therefore $c - a$ is not an upper bound of the set S. This implies that there exists an element la of S, $l \in \mathbf{Z}^p$, such that $la > c - a$. It follows that $(l + 1)a > c$ and since $(l + 1)a \in S$, we have a contradiction of the fact that c is the l.u.b. of the set S. The proof is therefore complete.

It was shown in Section 6.5 that between any two distinct rational numbers there is another rational number. A generalization of this result is given in the following theorem.

7.5   **THEOREM** *If* $a, b \in \mathbf{R}$ *with* $a < b$, *there exists a rational number* m/n ($m, n \in \mathbf{Z}$) *such that*

$$a < \frac{m}{n} < b.$$

Proof: For simplicity, we shall assume that $a > 0$ and leave the rest of the proof as an exercise.

Since $b - a > 0$, by the preceding theorem there exists $n \in \mathbf{Z}^p$ such that $n(b - a) > 1$. Let n be some such fixed integer. Again applying the preceding theorem to the real numbers 1 and na, there exists $m \in \mathbf{Z}^p$ such that $m > na$, and let m be the *least* positive integer with this property. Now $m > na$ implies that $a < m/n$ and we proceed to complete the proof by showing that also $m/n < b$ or, equivalently, that $m < nb$. Suppose that $m \geq nb$. Since $n(b - a) > 1$, we have $m \geq nb > na + 1$. Thus $m > 1$ and $(m - 1) \in \mathbf{Z}^p$ such that $(m - 1) > na$. Since $m - 1 < m$, this violates our choice of m as the least positive integer which is greater than na. Our assumption that $m \geq nb$ has led to a contradiction, and we conclude that $m < nb$. This completes the proof.

In particular, this theorem tells us that between any two irrational numbers, there is a rational number. It is also true that between any two rational numbers there is an irrational number. (See Exercise 2 below.) Thus the rational and irrational numbers are very closely intertwined.

Although it is true that all the properties of the real numbers can be established using only the properties (i) and (ii) of Theorem 7.3, we shall give no further proofs in this book. However, let us conclude this brief discussion of the real numbers by stating without proof the following familiar and important result.

7.6   THEOREM *For each positive real number a and each positive integer n, there exists exactly one positive real number x such that*
$$x^n = a.$$

The real number x whose existence is asserted by this theorem may be called the *principal* nth root of a and designated by the familiar notation $a^{1/n}$ or by $\sqrt[n]{a}$.

EXERCISES

1. Define lower bound and greatest lower bound of a set of elements of an ordered field. Prove that if a nonempty set of elements of **R** has a lower bound, it has a greatest lower bound in **R**. *Hint:* Consider the additive inverses of elements of the set.

2. Prove that if $a, b \in \mathbf{R}$ with $a < b$, then
$$a < a + \frac{b - a}{\sqrt{2}} < b.$$

   Hence prove that between any two distinct rational numbers there is an irrational number.

3. Complete the proof of Theorem 7.5 by showing that the stated result holds also for the case in which $a \leq 0$.

4. Let S_1 and S_2 be nonempty sets of real numbers having, respectively, b_1 and b_2 as least upper bounds. If $S_3 = \{s_1 + s_2 | s_1 \in S_1, s_2 \in S_2\}$, prove that $b_1 + b_2$ is the least upper bound of the set S_3.

5. Prove Theorem 7.6. *Hint:* See the set S_2 of Section 7.1.

7.3 THE FIELD OF COMPLEX NUMBERS

In order to construct the field of complex numbers, we begin by considering ordered pairs (a, b) of *real* numbers. Our definitions of addition and multiplication will be motivated by the formal properties of expressions of the form $a + bi$, where $i^2 = -1$. However, we are not justified in assuming that there *is* a "number" whose square is -1 until we have constructed a field which has an element with this property. Accordingly, as in the case of the construction of the rational numbers, we begin with an unfamiliar notation in order to avoid using any property until we have established it. We may remind the reader that the equal sign is being used in the sense of identity, that is, $(a, b) = (c, d)$ means that $a = c$ and $b = d$.

We proceed to prove the following theorem, which establishes the existence of the field we shall presently call the field of complex numbers.

7.7　　**THEOREM**　*Let* **C** *be the set of all ordered pairs* (a, b) *of elements of the field* **R** *of real numbers, and let us define operations of addition and multiplication on* **C** *as follows:*

7.8
$$(a, b) + (c, d) = (a + c, b + d),$$

and

7.9
$$(a, b)(c, d) = (ac - bd, ad + bc).$$

Then **C** *is a field with respect to these definitions of addition and multiplication. Moreover, the set of all elements of* **C** *of the form* $(a, 0)$, $a \in$ **R**, *is a subfield of* **C** *which is isomorphic to the field* **R**.

The required properties of addition are almost obvious. From 7.8, it follows that addition is commutative and associative, that $(0, 0)$ is the zero of **C**, and that the additive inverse of (a, b) is $(-a, -b)$.

The associative law of multiplication is a consequence of the following straightforward calculations:

$$((a, b)(c, d))(e, f) = (ac - bd, ad + bc)(e, f)$$
$$= (ace - bde - adf - bcf, acf - bdf + ade + bce),$$
$$(a, b)((c, d)(e, f)) = (a, b)(ce - df, cf + de)$$
$$= (ace - adf - bcf - bde, acf + ade + bce - bdf),$$

and these turn out to be equal elements of **C**.

Next, let us verify one of the distributive laws as follows:

$$(a, b)((c, d) + (e, f)) = (a, b)(c + e, d + f)$$
$$= (ac + ae - bd - bf, ad + af + bc + be),$$
$$(a, b)(c, d) + (a, b)(e, f) = (ac - bd, ad + bc) + (ae - bf, af + be)$$
$$= (ac - bd + ae - bf, ad + bc + af + be),$$

and again we have equal elements of $\mathbf{C}$. The other distributive law follows from this one as soon as we show that multiplication is commutative, and the commutativity of multiplication follows easily from 7.9. For, by interchanging (a, b) and (c, d) in 7.9, we see that

$$(c, d)(a, b) = (ca - db, cb + da),$$

and the right side of this equation is equal to the right side of 7.9. Hence,

$$(a, b)(c, d) = (c, d)(a, b).$$

We have now proved that $\mathbf{C}$ is a commutative ring, and it is easily verified that it has the unity $(1, 0)$. To show that $\mathbf{C}$ is a field, we need only show that each nonzero element (a, b) of $\mathbf{C}$ has a multiplicative inverse in $\mathbf{C}$. Since the zero is $(0, 0)$, to say that (a, b) is not the zero of $\mathbf{C}$ is to say that a and b are not both equal to zero. Since a is an element of the ordered field $\mathbf{R}$, we know that if $a \neq 0$, then $a^2 > 0$. Similarly, if $b \neq 0$, we have $b^2 > 0$. It follows that if (a, b) is not the zero of $\mathbf{C}$, then necessarily $a^2 + b^2 > 0$ and, in particular, $a^2 + b^2 \neq 0$. Hence,

$$\left(\frac{a}{a^2 + b^2}, \frac{-b}{a^2 + b^2} \right)$$

is an element of $\mathbf{C}$ and it may be verified by direct calculation (using 7.9) that

$$(a, b)\left(\frac{a}{a^2 + b^2}, \frac{-b}{a^2 + b^2} \right) = (1, 0).$$

We have therefore shown that every nonzero element of $\mathbf{C}$ has a multiplicative inverse in $\mathbf{C}$, and hence we have proved that $\mathbf{C}$ is a field.

To complete the proof of the theorem, let $\mathbf{R}'$ be the set of all elements of $\mathbf{C}$ of the form $(a, 0)$, $a \in \mathbf{R}$. Then the mapping $\theta: \mathbf{R}' \to \mathbf{R}$ defined by $(a, 0)\theta = a$, $a \in \mathbf{R}$, is a one-one mapping of $\mathbf{R}'$ onto $\mathbf{R}$. Moreover,

$$[(a, 0) + (b, 0)]\theta = (a + b, 0)\theta = a + b = (a, 0)\theta + (b, 0)\theta,$$

and

$$[(a, 0)(b, 0)]\theta = (ab, 0)\theta = ab = [(a, 0)\theta][(b, 0)\theta].$$

Hence, the operations of addition and multiplication are preserved under this mapping, and the mapping therefore defines an isomorphism of $\mathbf{R}'$ onto $\mathbf{R}$. This completes the proof of the theorem.

An element of the field which we have constructed is called a *complex number*, and $\mathbf{C}$ is called the *field of complex numbers*.

We shall henceforth adopt a more familiar notation by identifying $\mathbf{R}'$ with $\mathbf{R}$, that is, we shall write a in place of $(a, 0)$, and consider that the field $\mathbf{C}$ of complex numbers actually contains the field $\mathbf{R}$ of real numbers. Also, for simplicity of notation, as well as for historical reasons, we shall use the symbol i to designate the particular element $(0, 1)$ of $\mathbf{C}$. Since $(0, 1)^2 = (-1, 0)$, in our new notation we have $i^2 = -1$. Now it is easily verified that

$$(a, 0) + (b, 0)(0, 1) = (a, b)$$

and, using the notation we have introduced, it follows that $a + bi = (a, b)$. Accordingly, in the future we shall write $a + bi$ in place of (a, b). In this notation, the product 7.9 of two elements of $\mathbf{C}$ may be expressed in the following form:

7.10 $(a + bi)(c + di) = ac - bd + (ad + bc)i.$

Of course, the right side of 7.10 may be obtained from the left by multiplying out with the aid of the usual distributive, associative, and commutative laws, and replacing i^2 by -1.

We have now extended the field of real numbers to the field of complex numbers. It should be pointed out, however, that one familiar property of the field of rational numbers and of the field of real numbers does not carry over to the field of complex numbers.

7.11   THEOREM   *The field $\mathbf{C}$ of complex numbers is not an ordered field.*

> **Proof:**   By this statement we mean that there does not exist any set $\mathbf{C}^p$ of elements of $\mathbf{C}$ having the properties (4.4) required for $\mathbf{C}$ to be an ordered field. This fact is a consequence of the following observations. If $\mathbf{C}$ were ordered, 4.6(v) would show that the square of every nonzero element would be positive; in particular, both i^2 and 1 would be positive. Then -1 would be negative, and we have a contradiction since $i^2 = -1$.

The fact that $\mathbf{C}$ is not ordered means that inequalities cannot be used between complex numbers. In other words, it is meaningless to speak of one complex number as being greater or less than another.

Throughout the rest of this book we shall continue to denote the field of complex numbers by $\mathbf{C}$.

7.4 THE CONJUGATE OF A COMPLEX NUMBER

Let us make the following definition.

7.12   DEFINITION   If $u = a + bi \in \mathbf{C}$, we define the *conjugate* of u to be the element u^* of $\mathbf{C}$ given by: $u^* = a - bi$.†

As examples, we have $(1 + 7i)^* = 1 - 7i$, $(2 - 2i)^* = 2 + 2i$, $4^* = 4$, and so on.

Now the mapping $\alpha\colon \mathbf{C} \to \mathbf{C}$ defined by $u\alpha = u^*$, $u \in \mathbf{C}$, is a one-one mapping of $\mathbf{C}$ onto $\mathbf{C}$. We proceed to show that the operations of addition and multiplication are preserved under this mapping. Let $u = a + bi$ and $v = c + di$ be elements of $\mathbf{C}$. Then

$$(u + v)\alpha = (u + v)^* = [(a + c) + (b + d)i]^* = a + c - (b + d)i$$
$$= (a - bi) + (c - di) = u^* + v^* = u\alpha + v\alpha,$$

and

$$(uv)\alpha = (uv)^* = [ac - bd + (ad + bc)i]^* = ac - bd - (ad + bc)i$$
$$= (a - bi)(c - di) = u^*v^* = (u\alpha)(v\alpha).$$

Since the operations of addition and multiplication are preserved under the mapping α, it follows that this is an isomorphism of the field $\mathbf{C}$ onto itself. Such an isomorphism is frequently called an *automorphism*.

In working with complex numbers, the concept of conjugate plays an important role. A number of simple, but significant, properties are presented in Exercise 2 below.

EXERCISES

1. Find the multiplicative inverse of the nonzero element (a, b) of $\mathbf{C}$ by assuming that r and s are real numbers such that $(a, b)(r, s) = (1, 0)$, and solving for r and s.

---

† Historically, the usual notation for the conjugate of a complex number z is $\bar{z}$ instead of z^*. The present notation has been adopted only because it makes it easier to print such expressions as the conjugate of the sum of two or more complex numbers.

2. Prove each of the following:

    (i) If $u \in C$, then $uu^* \in R$ and $u + u^* \in R$; moreover, if $u \neq 0$, then $uu^* > 0$.

    (ii) If $u \in C$, then $(u^*)^* = u$.

    (iii) If $u \in C$ and $u \neq 0$, then $(u^{-1})^* = (u^*)^{-1}$.

    (iv) If $u \in C$, then $u = u^*$ if and only if $u \in R$.

    (v) If $u \in C$ and n is a positive integer, then $(u^n)^* = (u^*)^n$.

3. If $\alpha: C \to C$ is the isomorphism defined above, part (iv) of the preceding exercise shows that $u\alpha = u$ if and only if $u \in R$. Prove that if $\phi: C \to C$ is an isomorphism of C onto C with the property that $a\phi = a$ for every $a \in R$, then $\phi = \alpha$ or ϕ is the identity mapping on C. *Hint:* Consider the possibilities for $i\phi$.

4. Let S be an arbitrary ring and T the set of all ordered pairs (a, b) of elements of S. If addition and multiplication are defined by 7.8 and 7.9, respectively, verify each of the following:

    (i) T is a ring.

    (ii) T is a commutative ring if and only if S is a commutative ring.

    (iii) T has a unity if and only if S has a unity.

5. In the notation of the preceding exercise, let S be the ring Z_2 of integers modulo 2. Exhibit addition and multiplication tables for the corresponding ring T. Is T a field in this case? Is it an integral domain? Is it isomorphic to any of the rings with four elements given in Chapter 3?

6. Repeat Exercise 5 with Z_3 in place of Z_2.

7.5 GEOMETRIC REPRESENTATION AND TRIGONOMETRIC FORM

It is implicit in our construction of the complex numbers that the mapping $a + bi \to (a, b)$ is a one-one mapping of the set C of all complex numbers onto the set of all ordered pairs of real numbers. Now in ordinary plane analytic geometry we represent points in the plane by their coordinates, that is, by ordered pairs of

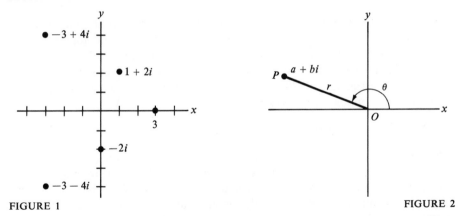

FIGURE 1   FIGURE 2

real numbers. Accordingly, we may represent a point in the plane by a single complex number. In other words, we shall sometimes find it convenient to associate with the complex number $a + bi$ the point with rectangular coordinates (a, b), and to say that this point has *coordinate* $a + bi$. A number of examples are given in Figure 1. It will be observed that a real number, that is, a complex number of the form $a + 0i$, is the coordinate of a point on the x-axis. A number of the form $0 + bi$, sometimes called a *pure imaginary*, is the coordinate of a point on the y-axis. We may also observe that a complex number $a + bi$ and its conjugate $a - bi$ are coordinates of points that are symmetrically located with respect to the x-axis.

Instead of specifying points in a plane by means of rectangular coordinates, we may of course use polar coordinates. If P is the point with nonzero coordinate $a + bi$, the distance of P from the origin O of coordinates is the positive real number $r = \sqrt{a^2 + b^2}$. If θ is an angle in standard position with terminal side OP, as in Figure 2, then by the definition of the trigonometric functions we have

$$a = r \cos \theta, \quad b = r \sin \theta.$$

It follows that the complex number $a + bi$ can be expressed in the form

7.13 $a + bi = r(\cos \theta + i \sin \theta).$

We have been assuming that $a + bi \neq 0$. If $a + bi = 0$, then $r = 0$ in 7.13, and θ may be a completely arbitrary angle.

We now introduce some appropriate terms in the following definition.

7.14 DEFINITION The expression on the right side of 7.13 is called the *trigonometric form* of the complex number $a + bi$. The non-negative real number $r = \sqrt{a^2 + b^2}$ is called the *absolute value* of the complex number $a + bi$, and may be designated by $|a + bi|$. The angle θ occurring in 7.13 is called *an angle* of $a + bi$.

Clearly, the nonnegative real number r occurring in the trigonometric form of $a + bi$ is uniquely determined. However, the angle θ is not unique, but if $r \neq 0$ and θ_1 and θ_2 are any two possible angles of $a + bi$, then elementary properties of the sine and cosine functions show that $\theta_1 = \theta_2 + n \cdot 360°$ for some integer n.

As a consequence of these observations, let us point out that if r and s are positive real numbers and we know that

$$r(\cos \theta + i \sin \theta) = s(\cos \phi + i \sin \phi),$$

then necessarily $r = s$ and $\theta = \phi + n \cdot 360°$ for some integer n.

We have previously defined (4.7) absolute values for an ordered integral domain, and we know that the field of complex numbers is not ordered. However, the present definition of absolute value is an extension of the concept for real numbers. For if a is a real number, we may consider it to be the complex number $a + 0i$ and, by 7.14, we have $|a| = \sqrt{a^2}$. But if c is a positive real number, by $\sqrt{c}$ we mean the *positive* square root of c. It follows that $\sqrt{a^2} = a$ if $a \geq 0$, whereas $\sqrt{a^2} = -a$ if $a < 0$. Hence, for a *real* number a, the present meaning of $|a|$ coincides with its meaning according to Definition 4.7.

Let us now illustrate the trigonometric form of a complex number by some examples. First, let us consider the number $-2 + 2i$. As indicated in Figure 3, $|-2 + 2i| = 2\sqrt{2}$, and an angle of $-2 + 2i$ is 135°. Hence, 7.13 takes the form

$$-2 + 2i = 2\sqrt{2}(\cos 135° + i \sin 135°),$$

which is easily verified by direct calculation. Other examples, which the reader may check, are the following:

$$1 + \sqrt{3}i = 2(\cos 60° + i \sin 60°),$$
$$4 = 4(\cos 0° + i \sin 0°),$$
$$-i = 1(\cos 270° + i \sin 270°),$$
$$-2(\cos 40° + i \sin 40°) = 2(\cos 220° + i \sin 220°).$$

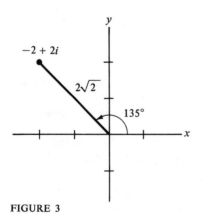

FIGURE 3

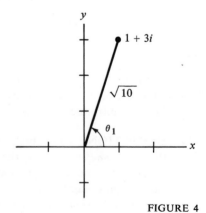

FIGURE 4

It is only in special cases that we can find in degrees an angle of a given complex number. Naturally, an approximation may be obtained by use of trigonometric tables, or an angle may be merely indicated as in the following example. Let us attempt to express $1 + 3i$ in trigonometric form. Clearly, $|1 + 3i| = \sqrt{10}$, but we cannot exactly express its angle in degrees. However, if θ_1 is the positive acute angle such that $\tan \theta_1 = 3$, as indicated in Figure 4, we may write

$$1 + 3i = \sqrt{10}\,(\cos \theta_1 + i \sin \theta_1)$$

as the trigonometric form of $1 + 3i$.

The fact that every complex number can be expressed in trigonometric form is of considerable significance largely because of the following remarkable theorem.

7.15   THEOREM   *If u and v are complex numbers such that in trigonometric form,*

$$u = r(\cos \theta + i \sin \theta),$$

and

$$v = s(\cos \phi + i \sin \phi),$$

then the trigonometric form of uv is given by

7.16 $$uv = rs(\cos (\theta + \phi) + i \sin (\theta + \phi)).$$

Otherwise expressed, $|uv| = |u| \cdot |v|$, and an angle of uv is the sum of an angle of u and an angle of v.

Proof: To establish this result we need only multiply together the trigonometric forms of u and v and then use the simple addition formulas of trigonometry. Thus we have

$$\begin{aligned}
uv &= rs(\cos \theta + i \sin \theta)(\cos \phi + i \sin \phi) \\
&= rs[(\cos \theta \cos \phi - \sin \theta \sin \phi) + i(\cos \theta \sin \phi + \sin \theta \cos \phi)] \\
&= rs[\cos (\theta + \phi) + i \sin (\theta + \phi)],
\end{aligned}$$

and the desired result follows immediately.

The special case of the preceding theorem in which $u = v$ shows at once that

$$u^2 = r^2(\cos 2\theta + i \sin 2\theta).$$

The following generalization of this result is of great importance.

7.17   DE MOIVRE'S THEOREM *If n is an arbitrary positive integer and*

$$u = r(\cos \theta + i \sin \theta),$$

then

7.18 $$u^n = r^n(\cos n\theta + i \sin n\theta).$$

For $n = 3$, we use Theorem 7.15 and the case in which $n = 2$ which has just been established as follows:

$$u^3 = u^2 \cdot u = [r^2(\cos 2\theta + i \sin 2\theta)][r(\cos \theta + i \sin \theta)]$$
$$= r^3(\cos 3\theta + i \sin 3\theta).$$

A general proof along these lines can be easily given by induction.

Example   One application of this theorem will be given in the next section. However, let us point out here how certain trigonometric identities can be obtained in an easy way by use of this theorem. By letting $r = 1$ and, as an example, taking $n = 3$, we see that

$$(\cos \theta + i \sin \theta)^3 = \cos 3\theta + i \sin 3\theta.$$

However, by actually multiplying out the left side, we find that

$$(\cos \theta + i \sin \theta)^3 = \cos^3 \theta - 3 \cos \theta \sin^2 \theta$$
$$+ i(3 \cos^2 \theta \sin \theta - \sin^3 \theta),$$

and it follows that

$$\cos 3\theta + i \sin 3\theta = \cos^3 \theta - 3 \cos \theta \sin^2 \theta$$
$$+ i(3 \cos^2 \theta \sin \theta - \sin^3 \theta).$$

From this equation we get at once the two following trigonometric identities:

$$\cos 3\theta = \cos^3 \theta - 3 \cos \theta \sin^2 \theta,$$

and

$$\sin 3\theta = 3 \cos^2 \theta \sin \theta - \sin^3 \theta.$$

EXERCISES

1. Express each of the following complex numbers in trigonometric form and indicate the points in a coordinate plane that have these numbers as coordinates:

   (a) $-1 - i$,
   (b) $-\sqrt{3} + i$,
   (c) $\sqrt{3} + i$,
   (d) $-1 + \sqrt{3}i$,

   (e) -4,
   (f) $3 - 2i$,
   (g) $2 - 2i$,
   (h) $\cos 26° - i \sin 26°$.

2. Express each of the following complex numbers in the form $a + bi$:

   (a) $4(\cos 45° + i \sin 45°)$,
   (b) $2(\cos 120° + i \sin 120°)$,
   (c) $3(\cos 180° + i \sin 180°)$,
   (d) $3(\cos 270° + i \sin 270°)$,
   (e) $(1/2)(\cos 300° + i \sin 300°)$,
   (f) $12(\cos 0° + i \sin 0°)$,
   (g) $11(\cos 90° + i \sin 90°)$,
   (h) $(\cos 117° + i \sin 117°)(\cos 123° + i \sin 123°)$.

3. Use De Moivre's Theorem to compute each of the following, and then express your answers in algebraic form by evaluating the necessary trigonometric functions:

   (a) $(-1 - i)^5$,
   (b) $(\sqrt{3} - i)^8$,
   (c) $(-i)^{12}$,
   (d) $\left(\dfrac{1}{\sqrt{2}} + \dfrac{i}{\sqrt{2}}\right)^{100}$,

   (e) $\left(-\dfrac{1}{2} - \dfrac{\sqrt{3}i}{2}\right)^6$,
   (f) $(-1 + i)^{10}$,
   (g) $(1 - \sqrt{3}i)^{11}$,
   (h) $(\cos 18° + i \sin 18°)^{10}$.

4. Verify that the points with coordinates

$$(\cos 60° + i \sin 60°)^n, \qquad (n = 1, 2, 3, 4, 5, 6),$$

are the vertices of a regular hexagon inscribed in a circle of radius 1.

5. If u^* is the conjugate of the complex number u, verify each of the following:

   (a) $|u^*| = |u|$,

   (b) $uu^* = |u|^2$,

   (c) $u^{-1} = \dfrac{u^*}{|u|^2}$, if $u \neq 0$.

6. Show that if $u \neq 0$, De Moivre's Theorem also holds for every *negative* integer n.

7. Let $u, v \in \mathbf{C}$, and let P and Q be the points in a coordinate plane having respective coordinates u and v. Let R be the point with coordinate $u + v$. If O is the origin, show that OR is a diagonal of the parallelogram having OP and OQ as adjacent sides.

8. Show that if $u, v \in \mathbf{C}$, then $|u + v| \leq |u| + |v|$.

9. Use De Moivre's Theorem to find trigonometric identities for $\cos 4\theta$ and $\sin 4\theta$.

10. If $u = a + bi$, we have defined $|u| = \sqrt{a^2 + b^2}$. Use this definition to prove directly that if $u, v \in \mathbf{C}$, then $|uv| = |u| \cdot |v|$.

11. If $u, v \in \mathbf{C}$ with $v \neq 0$, prove that $\left|\dfrac{u}{v}\right| = \dfrac{|u|}{|v|}$.

7.6 THE nTH ROOTS OF A COMPLEX NUMBER

In this section we give an important application of the use of the trigonometric form of a complex number. First, we give the following familiar definition.

7.19 DEFINITION Let n be a positive integer greater than 1. If $u, v \in \mathbf{C}$ such that $v^n = u$, we say that v is an *n*th *root* of u.

We shall now prove the following theorem.

7.20 THEOREM *If n is a positive integer greater than* 1, *and*

$$u = r(\cos \theta + i \sin \theta)$$

is a nonzero complex number in trigonometric form, there exist exactly *n* nth roots of u, namely,

7.21　　$r^{1/n}\left(\cos\dfrac{\theta + k\cdot360°}{n} + i\sin\dfrac{\theta + k\cdot360°}{n}\right),$　　$(k = 0, 1, \ldots, n-1).$

Here $r^{1/n}$ represents the principal *n*th root of the positive real number *r*; that is, the positive real *n*th root of *r* whose existence is asserted in Theorem 7.6.

Proof: Suppose that $v = s(\cos\phi + i\sin\phi)$ is an *n*th root of *u*. Then $v^n = u$ and De Moivre's Theorem assures us that

$$s^n(\cos n\phi + i\sin n\phi) = r(\cos\theta + i\sin\theta).$$

It follows that the absolute values of the two members of this equation are equal, and an angle of one must be equal to an angle of the other. Hence, $s^n = r$, so that $s = r^{1/n}$. Moreover, $n\phi = \theta + k\cdot360°$ for some integer *k*, and it follows that $\phi = (\theta + k\cdot360°)/n$. We have therefore shown that every *n*th root *v* of *u* must be of the form

7.22　　　　$v = r^{1/n}\left(\cos\dfrac{\theta + k\cdot360°}{n} + i\sin\dfrac{\theta + k\cdot360°}{n}\right)$

for some integer *k*. Conversely, it is readily verified by De Moivre's Theorem that if *v* is given by 7.22, then $v^n = u$ for *every* choice of the integer *k*. The number of distinct *n*th roots of *u* is therefore the number of ways in which the integer *k* can be chosen in 7.22 so as to give distinct values of *v*. The angles obtained by letting *k* take the values, $0, 1, \ldots, n-1$ have distinct terminal sides, and this fact makes it almost obvious that these *n* values of *k* yield distinct values of *v*. Moreover, if *t* is an arbitrary integer, the Division Algorithm asserts that there exist integers *q* and *r* with $0 \leq r < n$ such that $t = qn + r$, and therefore

$$\frac{\theta + t\cdot360°}{n} = \frac{\theta + r\cdot360°}{n} + q\cdot360°.$$

It is then clear that the angle $(\theta + t\cdot360°)/n$ has the same terminal side as the angle $(\theta + r\cdot360°)/n$. Since $0 \leq r < n$, we see that all possible different values of *v* are obtained if in 7.22 we let *k* take the values $0, 1, \ldots, n-1$. This completes the proof of the theorem.

Example As an example of the use of this theorem, let us find the fifth roots of the complex number $-2 + 2i$. First, we express this number in trigonometric form as follows:

$$-2 + 2i = 2^{3/2}(\cos 135° + i \sin 135°).$$

In the notation of the theorem, we have $r = 2^{3/2}$, $\theta = 135°$, and $n = 5$. Accordingly, the fifth roots of $-2 + 2i$ are the following:

$$2^{3/10}(\cos 27° + i \sin 27°),$$
$$2^{3/10}(\cos 99° + i \sin 99°),$$
$$2^{3/10}(\cos 171° + i \sin 171°),$$
$$2^{3/10}(\cos 243° + i \sin 243°),$$
$$2^{3/10}(\cos 315° + i \sin 315°).$$

An interesting special case of Theorem 7.20 arises if we choose $u = 1$; hence $r = 1$ and $\theta = 0°$. We state this case as follows.

7.23 COROLLARY *The distinct nth roots of* 1 *are the complex numbers*

7.24
$$\cos \frac{k \cdot 360°}{n} + i \sin \frac{k \cdot 360°}{n}, \qquad (k = 0, 1, \ldots, n - 1).$$

By De Moivre's Theorem, we have

$$\left(\cos \frac{360°}{n} + i \sin \frac{360°}{n} \right)^k = \cos \frac{k \cdot 360°}{n} + i \sin \frac{k \cdot 360°}{n}.$$

Hence, the n distinct nth roots of 1, as given in 7.24, may all be expressed as powers of a certain nth root of 1. We have then the following alternate form of the preceding corollary.

7.25 COROLLARY *Let us set*

7.26
$$w = \cos \frac{360°}{n} + i \sin \frac{360°}{n},$$

so that w is the nth root of 1 *having the smallest positive angle. Then the nth roots of* 1 *are the numbers*

7.27
$$w, w^2, w^3, \ldots, w^n = 1.$$

Since all nth roots of 1 have absolute value 1, they are coordinates of points on the circle with radius 1 and center the origin. Moreover, it is clear from 7.24 that they are the vertices of a regular polygon of n sides inscribed in this circle, with one vertex at the real number 1. This fact is of considerable importance in the study of the constructability of regular polygons with ruler and compass.

═══════════════════════════════════

EXERCISES

1. Find the cube roots of 1 and express the answers in algebraic form. Draw a figure showing that these numbers are the coordinates of the vertices of a regular polygon of three sides (equilateral triangle).

2. Do the corresponding thing for the fourth roots of 1.

3. Do the corresponding thing for the eighth roots of 1.

4. Show that the sixth roots of 1 are the cube roots of 1 and their negatives.

5. Find the required roots and express the answers in algebraic form:

    (a) The cube roots of $-2 + 2i$.

    (b) The cube roots of $-8i$.

    (c) The fourth roots of -4.

    (d) The sixth roots of $-i$.

    (e) The fourth roots of $-1 - \sqrt{3}i$.

    (f) The square roots of $-1 + \sqrt{3}i$.

6. In each of the following, express the required roots in trigonometric form:

    (a) The fifth roots of 1.

    (b) The fourth roots of $-1 + i$.

    (c) The fourth roots of $\dfrac{1}{2} + \dfrac{\sqrt{3}i}{2}$.

    (d) The sixth roots of $1 - i$.

    (e) The square roots of $1 + 2i$.

    (f) The fourth roots of $16(\cos 12° + i \sin 12°)$.

7. Show that if v is any one of the nth roots of the nonzero complex number u, and w is given by 7.26, then $v, wv, w^2v, \ldots, w^{n-1}v$ are all the nth roots of u.

8. Show that the multiplicative inverse of an nth root of 1 is also an nth root of 1.

9. If $t \in \mathbf{C}$ such that $t^n = 1$ but $t^m \neq 1$ for $0 < m < n$, t is called a *primitive nth root* of 1. Show each of the following:

    (i) The number w, defined in 7.26, is a primitive nth root of 1.

    (ii) If t is a primitive nth root of 1, then $1, t, t^2, \ldots, t^{n-1}$ are distinct and are all of the nth roots of 1.

    (iii) If t is a primitive nth root of 1, then t^l is also a primitive nth root of 1 if and only if l and n are relatively prime.

 COMMENTARY

1 The approach to the real and complex number systems used in this text is an axiomatic one akin to that used by Hilbert (1862–1943). Also, in optional sections, an approach is indicated starting with the axioms of Peano (1858–1932) for the natural numbers. Hilbert's approach puts into axioms much of what must be developed carefully by the method of Peano. For example, in his book on the real number system, Landau [124] carefully develops the properties of the real numbers from Peano's viewpoint. The approach of Hilbert is to take the major theorems in Landau's book and use them as axioms. Russell (1872–1970) wryly remarked that Hilbert's approach has the usual advantages of theft over honest work.

    The remarkable point of all this is that such arguments could occur as late as 1900 over the number system on which analysis had already been building for almost 300 years. The founders of the calculus were Leibniz (1646–1716) and Newton (1642–1727). A very large body of theorems concerning functions of a real variable had evolved by 1800, and yet, even in 1831, a mathematician as eminent as De Morgan (1806–1871) tells us that negative numbers arise from problems too narrowly stated, but that imaginary numbers are void of meaning, self-contradictory, and absurd. (He felt that all these numbers were nothing more than symbols which obeyed certain manipulative rules.) In other words, until the middle of the nineteenth century, very little effort was spent on the foundations of mathematics. Weierstrass (1815–1897) seems to be the very first to realize that in order to prove theorems about continuous functions, one must have firmly developed properties of the real number system. Dedekind (1831–1916) and Cantor (1845–1918) provided this firm foundation for the real numbers. They, however, assumed properties of the integers and rationals to be understood. So, between 1850 and 1900 the various rigorous approaches to our number systems we now use first evolved. This stands in marked contrast to the thorough treatment given to the same ideas by Euclid (c. 300 B.C.) as they arose in geometry more than 2000 years earlier. In the treatment of incommensurable ratios in metric geometry, Euclid rigorously develops the properties of what we would now call the real number line, using a method created by Eudoxus (c. 400 B.C.) almost a century earlier [139,144].

2   The evolution of our number system is quite interesting. At various times, people had some very unusual notions about what constituted "honest" numbers and what constituted "unreal" numbers. As remarked in the commentary of Chapter 4, numbers as we think of them started arriving in Western culture after 1200. Before this time, Western mathematicians were concerned with "magnitude," i.e., the length of line segments. Essentially, two line segments are called *commensurable* if the ratio of their lengths is a rational number. Of course, the Greeks did not think about rational numbers as we do (as completed entities); instead of $x = (\frac{2}{3})y$, they were forced to say $3x = 2y$, and they defined this latter relation in terms of constructions with straightedge and compass. At any rate, the early Greeks and their precursors worked out, in a geometric way, a satisfactory theory about ratios of whole numbers (rationals).

But now consider a square whose edge is one unit long. What "ratio" is the diagonal to an edge? Today we would say the diagonal has length $\sqrt{2}$. We saw in Section 5.5 that $\sqrt{2}$ is not a ratio of whole numbers; it is not a rational number. It is irrational. The choice of words is interesting; $\sqrt{2}$ is not "unrational" or "nonrational" but "irrational." The words "rational" and "irrational" also are used for kinds of behavior and thinking. The etymology of these words appears to be attached to geometry. Euclid's geometry is eminently rational. Pythagoras (c. 540 B.C.) knew that the length of the diagonal of a square is irrational when the side has length 1, and that bothered him.

What the Pythagoreans showed, then, is that incommensurable ratios exist, i.e., irrational numbers exist. To their thinking, since all ideas of magnitude were based upon ratios of whole numbers, the irrationals presented real problems. One can realize the magnitude of the problem when one knows that these people were not just mathematicians. The business of whole numbers was important to the Pythagoreans, since it was wrapped up in mysticism, philosophy, and religion. The irrationality of $\sqrt{2}$ was serious business.

But Eudoxus (c. 400 B.C.) came along a century later and saved the day. In crude terms, $\sqrt{2}$ (diagonal of our square) is uniquely determined by the "honest ratios" smaller than it. Today, we would put this in words by saying that $\sqrt{2}$ is the least upper bound of the set of all positive rational numbers r such that $r^2 < 2$. You can find it all spelled out in Euclid's (c. 300 B.C.) *Elements*, Book X.

There were a few other numbers which puzzled the Greeks too. One was π, the ratio of the circumference of a circle to its diameter. One of the famous Delian problems posed (and not solved) by the Greeks was: Using straightedge and compass, construct a square whose area is equal to that of a given circle. Even today, there are still those trying to "square the circle." But they are uninformed, for the problem was settled in 1882 by Lindemann (1852–1939) when he showed that π is transcendental. (By straightedge and compass, the only numbers that can be "constructed" are certain roots of polynomial equations with rational coefficients.) Since this problem puzzled thinkers before 450 B.C., and was still puzzling mathematicians until 1882, you can appreciate why the Greeks might have found it tough. Others since have found the solution difficult too.

In any case, π is a transcendental number; it is not a root of a polynomial equation with rational coefficients. Legendre (1752–1833) conjectured that this might be so. In other words, as Euler (1707–1783) pointed out, π may "transcend" algebraic methods and so should be called "transcendental." At that time, of course, no one knew that π (or any other number) is transcendental [133].

From 1200 on, mathematicians in the West worried about those numbers which are irrational but still are roots of rational polynomial equations. Such numbers (e.g., $\sqrt{2}$) are called algebraic irrationals. By using radical signs, many roots of equations could be written down. For example, there are quadratic, cubic, and quartic formulas for roots of polynomial equations using radical signs (e.g., $\sqrt{2}$, $\sqrt[4]{7}$, $\sqrt[3]{-5}$). Very early on these numbers were treated as somewhat strange . . . existing only for purposes of algebraic manipulation. For example, Euler calculated $-1 = \sqrt{-1} \cdot \sqrt{-1} = \sqrt{(-1)(-1)} = \sqrt{1} = 1$ from the formula $\sqrt{a} \cdot \sqrt{b} = \sqrt{ab}$. But eventually, the rules for manipulating radicals (surds) were all worked out. For a long time it was assumed that all algebraic irrationals were expressible in terms of radicals. It came as a surprise when Abel (1802–1829) showed there were algebraic irrationals which do not involve radicals.

Summing up, the whole sequence of discovery of various types of numbers seems to have proceeded along the following path. First, in antiquity, came the whole numbers; then somewhat later, the positive rational numbers. Next came the real algebraic irrationals which can be written using radicals ($\sqrt{2}$ at the time of the Greeks). While these last numbers were forcing themselves on the scene, negative rationals and imaginaries kept appearing (1000–1500). They were annoying because they were useful yet not very believable. Next the algebraic irrationals which cannot be written as radicals marched into the picture (1826). Finally, the transcendental numbers made themselves known (1844). It was only after this long sequence of events that our number system was put on as firm a foundation as the incommensurables are in Euclid's *Elements* [136,139,153].

3   It can honestly be said that the real and complex number systems lie at the very heart of mathematics. There is very little mathematics which does not use these numbers in at least one essential way. This chapter began with a statement of the least upper bound property of the real number system. This property is obviously crucial. In an appendix, Olmsted [129] develops a large number of statements equivalent to the least upper bound property. Landau [124] uses the method set down by Dedekind, while Dubisch [8] uses Cantor's method to develop the real number system. Knopp [112] carefully develops many of the geometric and analytical elementary properties of complex numbers.

chapter eight

MORE GROUP THEORY

In Chapter 2 we introduced some of the most basic ideas of group theory. We return now to the study of groups, and, in particular, to finite groups. In 1872 the Norwegian mathematician, L. Sylow proved some rather remarkable theorems about the existence of certain kinds of subgroups in finite groups. These theorems, called the *Sylow Theorems*, are among the class of theorems called *Anzahl Theorems* (German for number). We shall only discuss the Sylow Theorems, but the Anzahl Theorems describe certain connections between the divisors of a group order and properties of the group.

Recall Lagrange's Theorem, 2.52: the order of a subgroup must divide the order of the group. If a group G has order m and n divides m, is there a subgroup of G of order n? In Exercise 10 of Section 2.7 we have seen that the group of the tetrahedron is of order 12 but has no subgroup of order 6. There is no strong statement about existence of subgroups for arbitrary divisors of the group order. On the other hand, if we restrict ourselves to divisors of the group order which are prime powers, then not only do subgroups of such orders always exist but we can even get some idea as to how many such subgroups there are. This is the basic content of Sylow's fundamentally important theorems.

The proof we shall give here is not at all like Sylow's original proof. In 1915 G. A. Miller published a proof vaguely similar to the one given here. Apparently, no one noticed his paper since the proof was forgotten and only resurfaced again in 1960 in a much more elegant form as something new. The methods used in the proof depend upon certain number theoretic facts, and upon certain properties of permutation groups.

In Section 3 of this chapter we shall prove the existence of certain simple groups. As we shall see, simple groups are the basic building blocks of group theory. Therefore, a description of all simple groups would greatly increase our understanding of group theory. It is for this reason that a major amount of effort over the past two decades has been expended upon attempts to describe the finite simple groups. Progress in this branch of mathematics has been remarkable in this period of time. We shall prove an old result that the alternating groups A_n are all simple if $n \geq 5$.

In Section 4 we study the groups of units in the ring $\mathbf{Z}_n$ for an integer $n > 1$. We will be able to give quite a complete description of this very important abelian group.

In the final section, we sketch, through a sequence of exercises, a more classical proof approximately as given by Sylow for his theorems. We shall use the same language and ideas used in the modern proof; but shall follow the pattern set by Sylow.

8.1   ACTIONS AND ORBITS

Suppose that A is a finite set of elements. In Section 2.8 we defined the symmetric group S upon a set A consisting of all permutations of elements of A. Cayley's Theorem, 2.62, told us how, if we took the set A to be the elements of G, to construct an isomorphism of G into S. We should like to study this type of situation; that is, some set A is given with its symmetric group S and, further, we have a group G and a homomorphism $\phi: G \to S$ which maps G into S. Let us put all this into a definition and give it a name.

8.1   DEFINITION   Assume that A is a set and G is a group. If $\phi: G \to S$ is a homomorphism of G into the symmetric group S upon A then we call ϕ an *action* of G upon A. Often we shall sum all this up by saying that G *acts* upon the set A. Naturally, we shall say the action of G is *trivial* if ϕ maps G onto the identity subgroup of S.

It turns out, as with many words in mathematics, that this definition is not quite the right way to think about actions of groups upon sets. The following theorem gives a more appropriate point of view. Before stating the theorem, let us introduce some unusual but very useful notation. Suppose that $\theta: B \to C$ is a mapping of a set B into a set C. If $b \in B$, we usually denote the image of b under θ by $b\theta$. If we have a long string of mappings and elements, it is difficult to distinguish one from the other since both mappings and elements appear on the same line. To eliminate this difficulty, we shall move the mapping up to the position of an exponent; that is, we set $b\theta = b^\theta$. This way we can always distinguish the mappings from the elements. This form of writing mappings is widely used by algebraists and, for obvious reasons, is called *exponential notation*.

As an example, consider the permutation $\alpha = (12) \in S_3$. In our new exponential notation we would write equations defining α as follows: $1^\alpha = 2$, $2^\alpha = 1$; and $3^\alpha = 3$. If $\beta = (23) \in S_3$ then $\gamma = \alpha\beta = (132)$. We now have $(1^\alpha)^\beta = 2^\beta = 3 = 1^\gamma = 1^{(\alpha\beta)}$. In particular, if several mappings are written in the exponent, they may be composed to form a composite mapping. This composite mapping then appears in the exponential position. We shall use this notation in the following theorem.

8.2   THEOREM   *Suppose that A is a set and G is a group with identity e. The group G acts upon the set A if and only if with each element $a \in G$ we have associated a mapping (also denoted by a) $a: A \to A$ of the set A into itself such that*

(i) $x^e = x$ for all $x \in A$.

(ii) $(x^a)^b = x^{(ab)}$ for all $a, b \in G$ and $x \in A$.

A few remarks are in order before we prove this theorem. If $a, b \in G$, we have mappings $a: A \to A$ and $b: A \to A$. Even though the two group elements a and b may be different elements of G, the two mappings of A given by a and b may be the same mapping. That is, viewed as mappings, a and b may be different names for the same mapping. If we called a the group element and f_a [or $(a)f$ or any other notation], the associated mapping for a we would soon be in a notational nightmare, and that is why we adopt this rather strange use of notation. We should also note the meaning of equation (ii): $(x^a)^b$ means to apply the mapping b to the image of x under the mapping a; and $x^{(ab)}$ means to take the group product $ab = c$ and then apply that mapping to x, $x^{(ab)} = x^c$. Furthermore, if addition is the operation for G then (ii) becomes $(x^a)^b = x^{(a+b)}$.

Let us illustrate the ideas discussed here. Consider the cyclic group Z_9^+ and the homomorphism (Verify!) of Z_9^+ into S_3 given by $[n]\phi = (123)^n$. Then $[2]\phi = (123)^2 = (132)$, and $[5]\phi = (123)^5 = (123)^3 \cdot (123)^2 = (132)$. Let $a = [2]$ and $b = [5]$, and observe that ϕ is an action of Z_9^+ upon $\{1, 2, 3\}$. Certainly $a \neq b$ in Z_9^+; however, $a\phi = b\phi = (132)$ so that $x^a = x(a\phi) = x(b\phi) = x^b$ for all $x \in \{1, 2, 3\}$. For example, $1^a = 3 = 1^b$. Here then is an example of an action where two distinct group elements give rise to the same mapping. We may also illustrate the idea of composition by letting $c = [4]$ so that $c\phi = (123)^4 = (123)^3 \cdot (123) = (123)$ and so that $d = a + c = [2] + [4] = [6]$. Since $d\phi = (123)^6 = \epsilon$, the identity of S_3, with $x = 1$ we have the following equations: $1^{a+c} = 1^d = 1\epsilon = 1$; and $(1^a)^c = [1(132)](123) = 3(123) = 1$. In particular, $x^{a+c} = (x^a)^c$.

Proof of the Theorem: Assume that G acts upon A and that $\phi: G \to S$ is the action. That is, ϕ is a homomorphism of G into the symmetric group S upon A. Fix $a \in G$. Then $a\phi$ is a permutation of A. Let us denote this mapping $a\phi$ by using exponential notation and setting $x(a\phi) = x^a$ for $x \in A$. We now have associated with $a \in G$ a mapping $a: A \to A$, namely, the permutation $a\phi$. Since ϕ is a homomorphism, $e\phi$ is the identity permutation, $x^e = x$, and property (i) holds. If $b \in G$, then by our convention on exponents, $(x^a)^b = (x(a\phi))^b = (x(a\phi))(b\phi) = x[(a\phi)(b\phi)] = x[(ab)\phi] = x^{(ab)}$, proving (ii). We have proven one implication of the theorem.

Assume now that the conclusion of 8.2 holds. For each group element $a \in G$ let us rename the mapping $a: A \to A$ as $a\phi$, i.e., $x^a = x(a\phi)$. One might say that we are undoing the exponential convention here. Then ϕ is a mapping of G into the set of all mappings from A into A since it sends the group element a into the mapping $a\phi$. Let us show that ϕ is a mapping into S, the symmetric group upon A; i.e., if $a \in G$, then $a\phi$ is a permutation of A. To do this, we need only show that each mapping $a\phi$ for $a \in G$ has an inverse mapping. By property (ii) for $x \in A$ we have

$$(x(a\phi))(a^{-1}\phi) = (x^a)^{a^{-1}} = x^{(aa^{-1})} = x^e.$$

But $x^e = x$ by (i) so that the composite mapping $(a\phi)(a^{-1}\phi)$ is the identity mapping of A. A similar argument shows that $(a^{-1}\phi)(a\phi)$ is the identity mapping. We have shown that $a^{-1}\phi$ is a mapping inverse to $a\phi$; therefore, each mapping $a\phi$ is actually one-one and onto, that is, $a\phi$ is a permutation of A.

Finally, let us prove that $\phi: G \to S$ is a homomorphism of G into the symmetric group S upon A. For $a, b \in G$ and any $x \in A$ we have by (ii) that

$$x[(a\phi)(b\phi)] = (x^a)^b = x^{(ab)} = x[(ab)\phi].$$

Since this holds for all $x \in A$, we have

$$(a\phi)(b\phi) = (ab)\phi$$

and $\phi: G \to S$ is a homomorphism. We conclude that ϕ is an action, and G acts upon A. The proof of the theorem is complete.

What good is this theorem? As we shall see in the examples below, it is often easier to verify conditions (i) and (ii) of the theorem than it is to directly prove the existence of an action. We now look at these examples.

Example 1 Let $A = H$ be the set of elements of a finite group H; and suppose that G is a subgroup of H. For each $a \in G$ we define a mapping $a: A \to A$ by setting $x^a = xa$ for all $x \in A$. Then (i) $x^e = xe = x$ and (ii) $(x^a)^b = (xa)^b = (xa)b = x(ab) = x^{(ab)}$ for all $a, b \in G$ and $x \in A$. Therefore, G acts upon the set A by Theorem

8.2. If we take $G = H$, then the action $\alpha: G \to S$ of G into the symmetric group S upon A is precisely the isomorphism used to prove Cayley's Theorem, 2.62; in particular, the action α is an isomorphism of G into a symmetric group.

As an illustration, let us consider the group $G = H = \mathbf{Z}_2^+ = \{[0], [1]\}$, and $A = \mathbf{Z}_2^+$ so that $[0] = e$ is the identity. But $a = [1]$ is not the identity; in particular, $[0]^a = [0] + a = a = [1]$ and $[1]^a = [1] + a = [1] + [1] = [2] = [0]$. That is, the exponential a is the permutation which interchanges $[1]$ and $[0]$ in A.

Example 2 Let C be the group of the cube. Recall Figure 12 of Section 2.5 which is reproduced here as Figure 1 where the x,y,z-axis system was embedded at the center of the cube. Let A_0 be the set consisting of the x-axis, the y-axis, and the z-axis. (We have

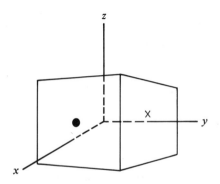

FIGURE 1

used A_0 for this set in order not to confuse it with the group A of symmetries of the x,y,z-axes). If $\sigma \in C$ is a symmetry of the cube, then σ will cause a permutation of the x-, y-, and z-axes determined by the change it effects upon Figure 1. So to each $\sigma \in C$ is associated a permutation $\sigma\phi$ of the set A_0. Now ϕ is an action of C upon A_0. Notice that ϕ gives the desired homomorphism of C onto the symmetric group on three elements of Exercise 7 in Section 2.8.

Example 3 Suppose that H is a finite group with a subgroup G; and fix a positive integer m less than or equal to the order of H. Form the collection A of all subsets containing exactly m elements of H. That is, $X \in A$ is a subset of m elements of H. For $X \in A$ and $a \in G$ set

$$X^a = Xa = \{xa \mid x \in X\}.$$

If $xa, ya \in Xa \subseteq H$ and $xa = ya$, then $x = y$ by cancellation, so that X and X^a have the same number m of elements; and therefore $X^a \in A$. In particular, exponential a is a mapping of the set A into itself. Since $xe = x$ for all $x \in H$, $X^e = X$; further, for $a, b \in H$ we have

$$(X^a)^b = \{yb \mid y \in X^a\} = \{(xa)b \mid x \in X\} = \{x(ab) \mid x \in X\} = X^{(ab)}.$$

We conclude by (i) and (ii) of Theorem 8.2 that G acts upon the set A.

As an illustration, consider the group $\mathbf{Z}_3{}^+ = G = H$ and the two ($m = 2$) element subsets A of $\mathbf{Z}_3{}^+$. The element $a = [2]$ of $\mathbf{Z}_3{}^+$ will act upon this collection. For example, $\{[0], [1]\}^a = \{[0] + a, [1] + a\} = \{[2], [0]\}$. Similarly $\{[1], [2]\}^a = \{[1] + a, [2] + a\} = \{[0], [1]\}$. With $a = [1]$ and $X = \{[0], [2]\}$, what is X^a?

Example 4 Let A be the collection of all subgroups of a finite group G and let $H \in A$. In Exercise 1 of Section 2.9 we saw that $a^{-1}Ha = \{a^{-1}ha \mid h \in H\}$ is also a subgroup of G of the same order as H, so that $a^{-1}Ha \in A$. If $H \in A$, then set $H^a = a^{-1}Ha$ for $a \in G$. For the identity e of G we have

(i) $H^e = e^{-1}He = H$;

and for a, b in G we have

(ii) $(H^a)^b = (a^{-1}Ha)^b = b^{-1}(a^{-1}Ha)b$
$= (b^{-1}a^{-1})H(ab) = (ab)^{-1}H(ab) = H^{(ab)}.$

Therefore, (i) and (ii) of Theorem 8.2 hold and G acts upon this set A.

As an illustration of this example, let us take $G = S_3$ and $H = \{\epsilon, (12)\}$, the subgroup of order 2 in G generated by the transposition (12). Let $\alpha = (123)$ so that

$$H^\alpha = \alpha^{-1}H\alpha = (132)H(123)$$
$$= \{(132)\epsilon(123), (132)(12)(123)\} = \{\epsilon, (23)\}$$

is the subgroup of order 2 in G generated by the transposition (23). If $K = \{\epsilon, (123), (132)\}$ is the subgroup A_3 and $\beta = (12)$, then

$$K^\beta = \beta^{-1}K\beta = (12)K(12)$$
$$= \{(12)\epsilon(12), (12)(123)(12), (12)(132)(12)\}$$
$$= \{\epsilon, (132), (123)\} = K.$$

In this case, $K^\beta = K$ so that exponential β maps the subgroup K to itself.

Example 5   Let $A = H$ be the set of elements of the finite group H, and let G be a subgroup of H. If $x \in A$ and $a \in G$, then set $x^a = a^{-1}xa$. For $x \in A$ and $a, b \in G$ we have

$$\text{(i)} \qquad x^e = e^{-1}xe = x$$

for the identity e of G; and

$$\text{(ii)} \quad (x^a)^b = (a^{-1}xa)^b = b^{-1}(a^{-1}xa)b$$
$$= (b^{-1}a^{-1})x(ab) = (ab)^{-1}x(ab) = x^{(ab)}.$$

Therefore, by Theorem 8.2, G acts upon the set A.

In the specific instance that H is an abelian group, $x^a = a^{-1}xa = x$ for all $a \in G$. In particular, each $a \in G$ gives rise to the trivial permutation of A. In this case the mappings induced by all elements of G are the same, namely, the trivial or identity mapping. This action of G upon A maps G onto the trivial or identity subgroup of a symmetric group.

Consider the case where $A = S_3$ and $G = S_3$. With $x = (12)$ and $\alpha = (123)$ we have $x^\alpha = \alpha^{-1}x\alpha = (132)(12)(123) = (23)$. If $\beta = (23)$, then $x^\beta = \beta^{-1}x\beta = (23)(12)(23) = (13)$.

Example 6   Let G be the cyclic subgroup of S_5 generated by the permutation $\alpha = (123)(45)$ so that

$$G = \{\epsilon, (123)(45), (132), (45), (123), (132)(45)\}$$

and G has order 6. Let $\phi: G \to S_5$ be the identity mapping so that ϕ is an action of G upon the set $A = \{1, 2, 3, 4, 5\}$. In this case we know directly from Definition 8.1 that ϕ is an action upon A.

We shall see these examples again; they will play a significant role in our proofs. It would be worth rechecking each one to make certain that it is clear why an action is given in each case. The finiteness conditions imposed in the examples are mostly unnecessary but have been given to make the arguments and ideas less subtle. Further, we shall apply the ideas here only to finite groups and finite sets.

If we are given an action for a group G upon a set A, there are two ways in which we may obtain new actions. We describe these now.

Method 1   Suppose that G is a finite group which acts upon a set A; and $\phi: G \to S$ is the action mapping G into the symmetric group S

upon the set A. If H is a subgroup of G, then clearly $\phi\colon H \to S$ is also a homomorphism, and therefore, an action of H upon A. In this case, the set has not changed, but the group has. We may obtain new actions upon old sets by restricting the action to subgroups.

To illustrate this method, consider the identity mapping $\phi\colon S_3 \to S_3$ so that ϕ is an action of S_3 upon $A = \{1, 2, 3\}$. Let $H = A_3$ be the alternating group. Since $\phi\colon A_3 \to S_3$ is a homomorphism, A_3 also acts upon the set A.

Method 2   Next suppose that B is a subset of A. Assume that K is a subgroup of S, where the action $\phi\colon G \to S$ maps G into K, i.e., $\phi\colon G \to K$, and that every permutation of K viewed as a mapping of B carries B into itself. That is, the restriction map ρ restricting the domain of an element $\alpha \in K$ to the set B carries α into $\alpha\rho$, a permutation upon B. It is straightforward to prove that $\rho\colon K \to S^*$ is a homomorphism from K into S^*, the symmetric group upon B (Exercise 10). For example, if $x \in B$ and $\alpha, \beta \in K$, then $x\alpha = x'$, $x'\beta = x''$, where $x', x'' \in B$. But $\alpha\rho$ is just α viewed as a mapping of B so that $x(\alpha\rho) = x'$ and, similarly, $x'(\beta\rho) = x''$; therefore, $(\alpha\beta)\rho = (\alpha\rho)(\beta\rho)$. Composing the two homomorphisms, $\phi\colon G \to K$ and $\rho\colon K \to S^*$, we obtain a homomorphism $\phi\rho\colon G \to S^*$. In particular, G also acts upon B. In this case the group remains the same, but the set A and the symmetric group upon it both change. We may obtain a new action from an old one by judiciously passing from A to a subset B of A which is mapped into itself by all elements of G.

This particular construction is far more subtle than the first. Let G be the group and $A = \{1, 2, 3, 4, 5\}$, the set of Example 6. Let us consider the subgroup $K = G$ of S_5, the element α, and the subset $B = \{4, 5\}$ of A of Example 6. Since $4^\alpha = 5$ and $5^\alpha = 4$, all permutations of K map B into itself. The symmetric group upon B is S^* which is isomorphic to S_2. The restriction mapping ρ restricts the domain of α to the subset B so that $\alpha\rho = (45)$. That is, ρ maps the group K homomorphically onto the subgroup $K\rho = \{\epsilon, (45)\}$ of S^*. The composite action $\phi\rho$ now maps α to $\alpha\phi\rho = (45)$. This homomorphism (or action) sends G onto a subgroup of S^* of order 2. The original action ϕ sent G onto the subgroup K of order 6 in S_5. The actions ϕ and $\phi\rho$ are clearly very different.

These two methods allow us to construct a great many actions from the examples just given.

Let us now emphasize a certain point of view. If a group G acts upon a set A, we may visualize that the elements of G "push" the elements of A around. For example, $a \in G$ "pushes" $x \in A$ into x^a. This viewpoint is useful for the following reason. Choose an x from A, and form the set B of all elements of A obtained by "pushing" over and over by elements of G. The set B has the property that it is somehow "generated" by x, and that all elements of B are "pushed" again into B by elements of G. This second point may be restated by saying that G "orbits" ("pushes" around ?) the elements of B among themselves by "pushing." For example, consider again the group G and the set $A = \{1, 2, 3, 4, 5\}$ of Example 6. If we start with $4 \in A$, by "pushing" we obtain $4^\alpha = 5$, $5^\alpha = 4$ so that $B = \{4, 5\}$ is a subset of A obtained by this "pushing" process. If we start with 3 and "push," we obtain $3^\alpha = 1$, $1^\alpha = 2$, $2^\alpha = 3$ so that $B' = \{1, 2, 3\}$ is another subset of A obtained by "pushing." This notion is contained in the following definition.

8.3   **DEFINITION** Suppose that a group G acts upon a set A. A subset B of A is called an *orbit* of G upon A or a *G-orbit* if:

(1) $x^a \in B$ for all $x \in B$ and $a \in G$.

(2) For any $x, y \in B$ there is an $a \in G$ so that $x^a = y$.

This definition says that elements of G "push" the elements of B among themselves, and that it is possible, for any two elements of B, to find an element of G which "pushes" one into the other. For example, if we consider $\mathbf{Z}_4{}^+ = H$ and $G = \{[0], [2]\}$ with the action of Example 1, then $\{[1], [3]\}$ is a G-orbit since $[1]^{[2]} = [3]$ and $[3]^{[2]} = [1]$.

Similarly, the sets $B = \{4, 5\}$ and $B' = \{1, 2, 3\}$ are orbits of the group G of Example 6 with its action upon the set $A = \{1, 2, 3, 4, 5\}$. In this case, the set A is a disjoint union of the two orbits B and B'. The first property of orbits which we shall discuss is described by the following theorem.

8.4   **THEOREM** *Assume that a group G acts upon a set A. The orbits of G upon A form a partition of the elements of A into disjoint subsets. If B is an orbit and $x \in B$, then $B = \{x^a \mid a \in G\}$.*

Proof: To prove this proposition, let us define an equivalence relation. If $x, y \in A$, let us write $x \sim y$ if there exists an $a \in G$ so that $x^a = y$. We will show that $\sim$ is an equivalence relation and that the orbits are the equivalence classes of $\sim$. From this it follows immediately that the orbits partition A.

We must check the properties of an equivalence relation. Observe that $x^e = x$ so that $x \sim x$; in particular, $\sim$ is reflexive. Suppose that $x \sim y$, that is, there is an $a \in G$ so that $x^a = y$. Then $y^{a^{-1}} = (x^a)^{a^{-1}} = x^{(aa^{-1})} = x^e = x$ so that $y \sim x$; therefore $\sim$ is symmetric. Assume that $x \sim y$ and $y \sim z$ so that there are elements $a, b \in G$ such that $x^a = y$ and $y^b = z$. We then have $x^{(ab)} = (x^a)^b = y^b = z$ so that $x \sim z$. We conclude that $\sim$ is transitive and is an equivalence relation upon A.

The equivalence classes of the relation $\sim$ partition the set A. Next we prove that an equivalence class is a G-orbit; accordingly, let B be any equivalence class for $\sim$. Suppose that $x \in B$, $a \in G$, and define $y = x^a$ so that $x \sim y$. Since B is an equivalence class, $y = x^a \in B$; therefore property (1) of Definition 8.3 holds. In view of the fact that B is an equivalence class, for any $x, y \in B$ we must have $x \sim y$; in other words, there is an $a \in G$ such that $x^a = y$. This verifies property (2) of Definition 8.3; thus B is an orbit of G upon A. As a further consequence, for any fixed $x \in B$ and any $y \in B$, $y = x^a$ for some $a \in G$, and therefore $B = \{x^a \mid a \in G\}$ for any $x \in B$. This concludes the proof.

In Example 2 the x-, y-, and z-axes are permuted cyclically among themselves by the rotations around a diagonal of the cube. Therefore the whole set A_0 of that example is an orbit of C upon A_0. On the other hand, in Example 5 if we consider the set $B = \{e\}$ consisting of the identity of G alone, then $e^a = a^{-1}ea = e$ for all $a \in G$. Thus B is an orbit containing one lonely element. If H has order greater than one, then B is not all of $A = H$ so that A is definitely a disjoint union of more than one orbit. In other words, the set on which a group acts may involve one or several orbits.

This theorem tells us that for each action of a group G upon a set A there is associated a collection of orbits. A great deal can be discovered about the action of G upon A by studying the orbits. Our main interest will be in the *size* of orbits; that is, how many elements are in an orbit. Since the orbits partition the set A, the sum of the sizes of the various orbits of G upon A adds up to the size of A. This fact will give us a group theoretic way to count the number of elements in A. Recall the cyclic group G of Example 6 and its action upon $A = \{1, 2, 3, 4, 5\}$. The orbits of G are $B = \{4, 5\}$ and $B' = \{1, 2, 3\}$. These two sets partition A into two disjoint subsets. The size of B is 2 and the size of B' is 3 so that A has size $3 + 2 = 5$. Before making such a count, we must introduce one more concept.

8.5   DEFINITION   Assume that a group G acts upon a set A and that $x \in A$. The set $\{a \mid a \in G, x^a = x\}$ of all elements a of G which fix x (i.e., $x^a = x$) is called the *stabilizer* of x in G.

Recall Example 6, where G is cyclic of order 6 and $A = \{1, 2, 3, 4, 5\}$. The orbits of G upon A are $B = \{4, 5\}$ and $B' = \{1, 2, 3\}$. Let us determine the stabilizer of $4 \in B$. That is, we want the largest subset H of

$$G = \{\epsilon, (123)(45), (132), (45), (123), (132)(45)\}$$

such that $4^\beta = 4$ for all $\beta \in H$. Clearly $H = \{\epsilon, (132), (123)\}$ which is a subgroup of G of order 3 and index 2. But 2 is precisely the size of $B = \{4, 5\}$. That is, the stabilizer of 4 is a subgroup, and the index of this subgroup is the size of B.

The important facts about stabilizers are contained in the following theorem.

8.6   **THEOREM**   *Suppose that a group G acts upon a set A and $B \subseteq A$ is an orbit of G upon A. Then the following conditions hold.*

(*i*) *If $x \in A$ and H is the stabilizer in G of x, then H is a subgroup of G.*

(*ii*) *If G is a finite group, $x \in B$, H is the stabilizer in G of x, and H has index n in G, then B contains exactly n elements (that is, B has size n).*

Proof: We argue first that H is a subgroup of G. For $a, b \in H$ we must verify that $a^{-1}, ab \in H$. To do this, we calculate directly. For any $a, b \in H$, $x^a = x^b = x$ so that

$$x^{a^{-1}} = (x^a)^{a^{-1}} = x^{(aa^{-1})} = x^e = x,$$

and $x^{(ab)} = (x^a)^b = x^b = x$, so that both a^{-1} and ab lie in H. Now we see that H is a subgroup by Theorem 2.30 of Section 2.4.

The second part of the theorem is slightly more difficult. By the previous theorem we know that for $x \in B$, $B = \{x^a \mid a \in G\}$. There are two consequences of this fact. First, since G has finite order m, there are no more than m elements in B; i.e., B is a finite set. Second, we may choose elements $a_1, a_2, \ldots, a_n$ from G so that $x^{a_1}, x^{a_2}, \ldots, x^{a_n}$ are precisely all the distinct elements of B; in particular, B contains exactly n elements. We argue now that the coset Ha_i is precisely the set of all elements b of G such that $x^b = x^{a_i}$, i.e., b "pushes" x into x^{a_i}. If $ha_i \in Ha_i$ for $h \in H$, then $x^{(ha_i)} = (x^h)^{a_i} = x^{a_i}$; and therefore, the elements of Ha_i do "push" x into x^{a_i}. Suppose $x^b = x^{a_i}$ for some $b \in G$ so that $x^{(ba_i^{-1})} = (x^b)^{a_i^{-1}} = (x^{a_i})^{a_i^{-1}} = x^{(a_i a_i^{-1})} = x^e = x$; then $ba_i^{-1} \in H$ since it stabilizes x. We conclude that $b \in Ha_i$, and, therefore, Ha_i is the set of all elements of G "pushing" x into x^{a_i}.

We have shown that each element x^{a_i} is attached to a coset Ha_i determined uniquely as the set of all elements which "push" x into x^{a_i}. The elements $x^{a_1}, x^{a_2}, \ldots, x^{a_n}$ are all distinct so that the cosets Ha_i are all distinct. Suppose Hb is a coset of H in G; then $x^b = x^{a_i}$ for some i so that $b \in Ha_i$, and $Ha_i = Hb$. We now have shown that every coset of H in G corresponds to some x^{a_i}. Therefore, $Ha_1, Ha_2, \ldots, Ha_n$ are all the cosets of H in G; i.e., the index of H in G is n, the number of elements in B. This concludes the proof.

Let us look once again at Example 2. Consider the x-axis, $X \in A_0$. It is not difficult to verify that the stabilizer H of X in C contains the following elements: (1) the identity; (2) the three nontrivial rotations about the x-axis; (3) the nontrivial rotations of 180° about the following four axes: the y-axis, the z-axis, and the two axes obtained by connecting the midpoints of opposite vertical edges of the cube as pictured in Figure 1. Therefore H contains at least eight elements. Since A_0 contains exactly three elements and is a C-orbit, the proposition tells us that H has order $\frac{24}{3} = 8$, where 24 is the order of C. Our description of H must be complete.

In the example of Section 2.4 we chose a geometric configuration connected with the cube and looked at the set of all symmetries of the cube which left this configuration invariant; so, this set formed a subgroup of C. We may verify this principle in general: i.e., if we take a geometric symmetry group G and some configuration X, then the set of symmetries from G leaving the configuration X invariant is a subgroup of G. To make our argument more concrete, we look at the case of the cube. Imagine that space is rigidly connected to our cube; therefore, any symmetry from C will cause a corresponding rigid symmetry of all space. Any geometric configuration may be viewed as collections of points, lines, planes, etc. Any symmetry of C will move the given configuration X of points, lines, etc., into another position giving us another configuration Y. Let A be the set of all configurations obtained in this way. Clearly C acts upon the set A. The given configuration X is an element of A; and the set H of all symmetries from C which fix X is the stabilizer of X in C. Therefore H is a subgroup of C.

In the case of Example 2 discussed above, we applied this analysis to the case where our given configuration was the x-axis as pictured in Figure 1. Other possible given configurations might be: (1) the top face of the cube, (2) a single pair of opposite edges of the cube, (3) the bottom four vertices of the cube, or (4) a particular diagonal of the cube which has been oriented by putting O at the center of the cube and introducing coordinates upon the diagonal. Each one of these configurations has a stabilizer which is a subgroup of C. Can you determine these subgroups and their orders?

In this section we have introduced a very powerful concept. The words introduced to describe various aspects of this concept are action, orbit, and stabilizer. In our proof of the Sylow Theorems we shall make very heavy use of this concept. It is important, therefore, to understand the examples and theorems of this section.

EXERCISES

1. Show that the group of symmetries of the x,y,z-axes acts upon the axes.

2. Show that the group S_n acts upon the set $A = \{1, 2, \ldots, n\}$. If $\alpha \in S_n$, then discuss the orbits upon A of the cyclic group C generated by α. Write α as

a product of disjoint cycles (see Theorem 2.66 of Section 2.8). Is there any connection between the disjoint cycles of α and the orbits of C upon A?

3. Let K be the intersection of all subgroups of S_5 containing $\alpha = (12)$, $\beta = (24)$, $\gamma = (134)$. (i) Show that K acts upon $A = \{1, 2, 3, 4, 5\}$. *Hint:* Try Method 1. (ii) Without finding the actual subgroup K, find the orbits of K upon A.

4. Let H be the smallest subgroup of S_9 containing $\alpha = (135)$, $\beta = (17)$, $\gamma = (2946)$, $\delta = (385)$. (i) Show that H acts upon $A = \{1, 2, 3, 4, 5, 6, 7, 8, 9\}$. (ii) Find the orbits of H upon A without finding H.

5. Consider the smallest subgroup H of S_5 which contains $\alpha = (13)$, $\beta = (24)$, $\gamma = (25)$. (i) Show that H acts upon $A = \{1, 2, 3, 4, 5\}$ and that $B = \{2, 4, 5\}$ is an orbit of H. (ii) Let ρ be the mapping which restricts the domain of elements of H to B. Show that ρ has a nontrivial kernel in H. (iii) Show that the image $H\rho$ of H under ρ is the symmetric group S^* upon $\{2, 4, 5\}$.

6. Suppose that G acts upon the two sets A and B. Show that for $a \in A$, $b \in B$, and $x \in G$, $(a, b)^x = (a^x, b^x)$ defines an action of G upon the Cartesian product $A \times B$ of A and B. If H is the stabilizer in G of $a \in A$ and K is the stabilizer in G of $b \in B$, then what is the stabilizer in G of $(a, b) \in A \times B$?

7. Suppose that K is a subgroup of the symmetric group upon a set A. If $B \subseteq A$ is a subset such that every element α of K maps B into itself then ρ, the restriction map given by restricting the domain of $\alpha \in K$ to B is a homomorphism of K into the symmetric group S^* upon the set B. Give an example of sets A and B and a group K to show that ρ need not be an isomorphism.

8. If $a, b \in G$ where G is a finite group and for some $x \in G$, $a = x^{-1}bx$, then we say that *a is conjugate to b in G*. (See Section 2.11 where "conjugate" is also defined.) (a) Show that "is conjugate to" is an equivalence relation upon G. (b) An equivalence set C is called a *conjugacy class* of G. If a conjugacy class contains exactly m elements, show that m is a divisor of the order of G. *Hint:* Try Example 5.

9. Let S be the group of symmetries of a square.

    (a) Show that S acts upon the vertices of the square.

    (b) Let H be any subgroup of S. Show that there is a set A (vertices, diagonals, etc.) and an element $X \in A$ so that S acts upon A and H is the stabilizer in S of X.

10. Suppose that G is a subgroup of S_n and $A = \{1, 2, \ldots, n\}$, $n > 2$. We call G a Frobenius group if the following conditions hold: (i) A is a single G-orbit;

(ii) if H is the stabilizer of $n \in A$, then $A \setminus \{n\}$ is a single H-orbit; and (iii) if K is the stabilizer of $n - 1 \in A$, then $H \cap K = \{e\}$.

(a) Show that S_3 is a Frobenius group.

(b) Show that the set T of all elements of a Frobenius group which stabilize no element of A contains exactly $n - 1$ elements.

(c) Show that G has order $n(n - 1)$. *Hint:* The order of H is $n - 1$ and A has size n.

Actually, the set $T \cup \{e\}$ is a normal subgroup of G of order n called the Frobenius kernel of G. This fact was first proved by G. Frobenius in 1901. The proof of this fact is surprisingly deep.

11. Show that every nonabelian group G of order 6 is isomorphic to S_3. *Hint:* Show that G must contain subgroups of order 2. Can $G \setminus \{e\}$ be "just" elements of order 3? Consider the action of Example 3 upon the cosets of a subgroup of order 2 in G.

12. Let G be a group. An *automorphism* ϕ of G is an isomorphism $\phi \colon G \to G$ of G *onto* G.

(a) Show that the set Aut (G) of automorphisms of G is a group of permutations upon the elements of G.

(b) Suppose that H is a subgroup of Aut (G) of order p^n and G is of order p^m for a prime p and integers m and n. Show that there is an element $a \in G$, $a \neq e$, which is a fixed point for H, i.e., $a\phi = a$ for all $\phi \in H$. *Hint:* H acts upon G and $\{a\}$, $\{e\}$ are orbits of size one.

13. Assume that G is a finite group with action $\phi \colon G \to S_n$ upon the set $A = \{1, 2, \ldots, n\}$. Let B be an orbit of G upon A. Let $G\phi = K$ be the image of G in S_n and ρ be the mapping which restricts the domain from A to B. By Exercise 7, $\phi\rho$ is an action of G upon B. (Recall Method 2.) We call B a *faithful* orbit if $\phi\rho$ is an isomorphism, i.e., ker $\phi\rho = \{e\}$.

(i) If H is a subgroup of G such that $\cap \, a^{-1}Ha = \{e\}$, where the intersection is over all $a \in G$ and H is the stabilizer of x in an orbit B, then B is a faithful orbit.

(ii) If G is abelian and B is a faithful orbit, then the size of B equals the order of G.

(iii) Give an example of a group G, an action upon a set A, and an orbit $B \subseteq A$, so that B is faithful but the size of B is not equal to the order of G.

14. Prove that if G is a group of order p^m for a prime p and a positive integer m, then there is an element a, not the identity, of G which commutes with every element of G. (*Hint:* Let $A = G = H$ and use the action of G upon A given by Example 5. The sum of the orbit sizes is p^m, the size of A. Observe that the identity e belongs to some orbit. Compare modulo p the sum of the orbit sizes with p^m. Is there more than one orbit of size 1?) Show that a group of order p^2 must be abelian.

15. Assume that G acts upon a finite set A where the action is an isomorphism of G. If $B \subseteq A$ is an orbit of G upon A, and H, the stabilizer in G of $x \in B$, is maximal in G (in the sense that H is the only proper subgroup of G containing H), then $B = \{x^a \mid a \in M\}$ for every normal subgroup $M \neq \{e\}$ of G.

8.2 THE SYLOW THEOREMS

The major idea of the proof used here depends upon a very useful technique: If A is a finite set, then count the number of elements in A in two distinct ways. By comparing the two counts, we may obtain information we did not have before. This idea is the underlying method used in our major proof of this section. Actually, our two counts of the size of A are carried out in identical ways, and our counting procedure is based upon actions and orbits. In fact, at one point we shall weave actions within actions. In order to keep this particular situation clear, we shall prove the necessary lemma now.

8.7   **LEMMA**   *Assume that H is a subgroup of a finite group G and that X is a nonempty subset of G such that $xa \in X$ for all $x \in X$ and $a \in H$.*

   (i) If we define $x^a = xa$ for $x \in X$ and $a \in H$, then this defines an action of H upon X.

   (ii) If $x \in X$, then the stabilizer in H of x is the trivial subgroup $\{e\}$ of H.

   (iii) If H has order h, then every orbit of H upon X has size h. In particular, X has size sh for some positive integer s.

Proof: The fact that for $x \in X$ and $a \in H$, $x^a = xa$ defines an action of H upon X follows from an argument identical to that used in Example 1 of Section 8.1. Therefore (i) holds, and we may now consider (ii). Fix $x \in X$, and let K be the stabilizer in H of x. If $a \in K$, then $xa = x^a = x = xe$ so that by cancellation of $x \in X \subseteq G$

we obtain $e = a$. We conclude that $K = \{e\}$ contains one element. We have proven (ii) holds and now consider (iii). Every coset Ka of K in H contains one element, $Ka = \{a\}$, so that the index of K in H is the order h of H. In other words, by Theorem 8.6, the orbit to which x belongs has size h. But $x \in X$ was arbitrary so that all orbits of H upon X have size h. If there are s different orbits of H upon X then X has size sh since by Theorem 8.4, the orbits partition X into s disjoint subsets each containing exactly h elements. The proof of (iii) and the lemma are complete.

We are now in a position to state and prove the most difficult of Sylow's Theorems.

8.8   **THEOREM** (*Sylow*)   *Suppose that G is a group of order mp^b, where p is a prime and m, b are positive integers. The number l of subgroups of G having order p^b satisfies the congruence*

$$l \equiv 1 \ (\mathrm{mod}\ p).$$

In particular, l is not zero and G contains subgroups of order p^b.

Although Sylow did show that G contains subgroups of order p^b, he did not show that $l \equiv 1 \ (\mathrm{mod}\ p)$ except in the special case where p^b is the highest power of p dividing the order of G [i.e., in the case where $(m,p) = 1$]. In this sense our theorem is slightly more general than that of Sylow.

The proof of this theorem is long and somewhat involved. Example 1, appearing after the proof, works out the specific details of the proof for the case where $G = S_3$, $p^b = 2$, and $m = 3$. It might help to compare the various steps in the proof with parts of Example 1. We will obtain two different counts for the size of a set by showing our method applies in two different situations to give the same answer. To keep the proof clear, we shall first introduce the necessary action, and then we shall prove the theorem in a sequence of lemmas.

Form the collection A of all subsets of elements of G which contain exactly p^b elements; that is, if $X \in A$, then X is a subset of G containing precisely p^b elements. Let us denote the size of this set A by n. We wish to find two formulas for n. The two formulas are obtained by the following observation. The size of A is dependent *only* upon the fact that G is a set of mp^b elements. In particular, if we replace G by a cyclic group G' of order mp^b, then the corresponding set A' defined for G' will also have size n. We shall use the superscript prime (') to indicate computations in A' for the cyclic group G'. We should keep in mind that to each computation in A for G there corresponds a similar computation in A' for G'. For this reason we shall ignore A' and G' until much later in the proof. The reason we bring G' into the picture at all is that we know (Exercise 7 of Section 2.6) that G' contains precisely one subgroup of size p^b.

Now G has an action upon A; in fact, the action we desire is described in Example 3 of Section 8.1. That is, for $a \in G$ and $X \in A$ we set

$$X^a = Xa = \{xa \mid x \in X\}.$$

We designate the orbits of G upon A as $B_1, B_2, \ldots, B_t$, and suppose that the orbit B_i contains precisely k_i elements.

In order to show the direction the proof will take, let us establish a connection between the subgroups of G of order p^b and the orbits of size m of G upon A. Of course, if G contains no subgroups of size p^b, then this lemma says there are no orbits of size m.

8.9   LEMMA   *There is a natural one-one correspondence between subgroups of G of order p^b and orbits of G upon A of size m.*

Proof: Suppose that H is a subgroup of G of order p^b. Then the set $B_0 = \{Ha \mid a \in G\}$ contains $m = mp^b/p^b$ cosets since H has index m in G. We shall prove that this collection B_0 of m elements is the orbit in A which corresponds to H. (If you look in Example 1 at the various orbits, you will see that the ones of size 3 for S_3 are the coset decompositions for the various subgroups of order 2 in S_3.)

We prove now that B_0 is an orbit in A. Since every coset of H contains p^b elements, $B_0 \subseteq A$. If $Ha \in B_0$ and $a_1 \in G$, then $(Ha)^{a_1} = (Ha)a_1 = H(aa_1)$ is a coset of H and lies in B_0. If Ha and Ha_1 are two cosets, then $(Ha_1)^{(a_1^{-1}a)} = Ha_1(a_1^{-1}a) = Ha$ so that properties (1) and (2) of Definition 8.3 hold, proving that B_0 is an orbit of G upon A. In other words, there is a unique orbit B_0 in A of size m corresponding to the subgroup H of order p^b in G.

To complete the proof of the lemma, we must show that if B is an orbit in A of size m, then $B = \{Ya \mid a \in G\}$ is the coset decomposition for a subgroup Y of G of order p^b. To do this, let us fix an orbit B of size m and choose $X \in B$. Since X is a subset of G of size p^b, we may choose $a \in X$, some element of G from X. Notice that Y must be the unique coset containing e in a coset decomposition; therefore, let us set $Y = X^{a^{-1}}$ since $e = aa^{-1} \in Xa^{-1} = X^{a^{-1}} = Y$.

We now show that Y is a group. Let H be the stabilizer in G of Y (recall that $Y \in A$). Since B contains m elements, by Theorem 8.6, the index of H in G is m; that is, the order of H is $mp^b/m = p^b$. We shall prove that $H = Y$.

Notice that $Y^a = Ya = Y$ for all $a \in H$ so that since $e \in Y$, $e^a = ea = a \in Y$ for all $a \in H$. We conclude that $H \subseteq Y$. But both H and Y contain p^b elements; consequently, we must have $H = Y$. Therefore, Y is a subgroup of G of order p^b.

If $a \in G$, then $Y^a = Ya \in B$ is a coset of Y in G. By Theorem 8.4 again, $B = \{Ya \mid a \in G\}$, so that B is the set of cosets of Y in G. This correspondence between subgroups of size p^b and their coset decompositions is therefore the one-one correspondence described by the lemma.

It is now clear that in order to prove the theorem, we must show that the number l of orbits of G upon A of size m satisfies the congruence $l \equiv 1'(\bmod\ p)$. To do this, we shall compute the size of the collection A. Recall that the orbit B_i contains k_i elements. By Theorem 8.4, the orbits of G upon A partition the set A so that A contains $k_1 + \cdots + k_t$ elements. Since n is the number of elements in A, we have the following equation

8.10
$$n = k_1 + k_2 + \cdots + k_t.$$

Next we concentrate upon the integers k_i.

8.11 **LEMMA** *Assume that B is an orbit of G upon A containing exactly k elements. Then $k = mp^{b-c}$ for some nonnegative integer $c \leq b$.*

Proof: From the orbit B choose an element X, and let H be the stabilizer of X in G. Since for any $a \in H$, $X = X^a = Xa = \{xa \mid x \in X\}$, we must have $xa \in X$ for all $x \in X$ and $a \in H$. Lemma 8.7 applies here to H and X, and, in particular, if H has order h, then by part (iii) of Lemma 8.7, the size p^b of X is sh for some integer s. Using Unique Factorization of Integers, we conclude that $h = p^c$ for some $c \leq b$ since $p^b = sh$.

But now by Theorem 8.6, the size k of B is the index of the stabilizer H of X in G; that is, $k = mp^b/p^c = mp^{b-c}$, where $c \leq b$, completing the proof of the lemma.

Using this lemma, we may introduce the notation (for $i = 1, 2, \ldots, t$),

$$k_i = mp^{b-c_i}, \qquad c_i \leq b.$$

In particular, from 8.10 we obtain

$$n = mp^{b-c_1} + \cdots + mp^{b-c_t}$$
$$= m(p^{b-c_1} + \cdots + p^{b-c_t}).$$

Observe that m divides n yielding the following equations.

8.12

(a) $n = mn_0.$

(b) $n_0 = p^{b-c_1} + \cdots + p^{b-c_t}.$

We are now in a position to prove Theorem 8.8 by combining Lemma 8.9 with the following lemma.

8.13   **LEMMA**   *If l is the number of orbits of G upon A of size m, then*

$$l \equiv 1 \pmod{p}.$$

Proof: We will use the equations of 8.12 for G and for a cyclic group G' of order mp^b. First let us number the orbits $B_1, B_2, \ldots, B_t$ so that $B_1, B_2, \ldots, B_l$ have size m and so that the remaining orbits have size unequal to m. In the case where all B_i have size unequal to m, we take $l = 0$ in what follows. If $i = 1, 2, \ldots, l$, then

$$m = k_i = mp^{b-c_i}, \qquad c_i \leq b,$$

so that $c_i = b$ and

$$p^{b-c_i} = p^0 = 1.$$

If $i = l + 1, \ldots, t$, then

$$m \neq k_i = mp^{b-c_i}, \qquad c_i \leq b,$$

so that $c_i < b$ and

$$p^{b-c_i} \equiv 0 \pmod{p}.$$

In particular, equation 8.12(b) becomes

8.14 $$n_0 \equiv l \pmod{p}.$$

What happens when we replace G by the cyclic group G' of order mp^b? Then A' and A have the same size n since each is the collection of all subsets of size p^b in a set of size mp^b. Further, the orbits $B_1', B_2', \ldots, B_{t'}'$ (where almost certainly $t' \neq t$) give rise to numbers k_i', c_i', and l'. The number l' is the number of G'-orbits of size m in A'. By Lemma 8.9, l' is the number of subgroups of G' of order p^b. By Exercise 7 of Section 2.6, G' contains exactly one subgroup of order p^b so that $l' = 1$. Using 8.12 for both G and G', we obtain

$$l \equiv n_0 \equiv l' = 1 \pmod{p}$$

completing the proof of the lemma.

It was the two groups G and G' which gave us the two ways in which to compute the size n of A (or A'). The proof of Theorem 8.8 is now complete.

Remark The size n of A is the number of subsets of size p^b contained in a set of size mp^b. We usually denote n by

$$n = \binom{mp^b}{p^b}$$

and call it a *binomial coefficient* or the number of *combinations* of mp^b things taken p^b at a time. In this notation,

8.15
$$n_0 = m^{-1}\binom{mp^b}{p^b} = \binom{mp^b - 1}{p^b - 1}$$

is also a binomial coefficient, and therefore is an integer. It is possible to prove directly that the binomial coefficient n_0 of 8.15 satisfies $n_0 \equiv 1 \pmod{p}$, eliminating the need for introducing the group G' into the proof. However, this direct proof that $n_0 \equiv 1 \pmod{p}$ via 8.15 is a little messy, though not difficult.

In order to clarify the proof just given, let us look at an example where $G = S_3$, $p^b = 2$, and $m = 3$.

Example 1 We write permutations from $G = S_3$ in cycle notation. The set A consists of all subsets of S_3 of size 2. We have organized the subsets into orbits, the orbits of size 3 occurring first in the list. Notice that the first subset in each orbit of size 3 is a subgroup of order 2. The "extra" orbit is of size 6, and this number is divisible by 2. We have written the values of m, p, b, n, t, k_i, and c_i below.

$$m = 3, \quad p = 2, \quad b = 1, \quad n = 15, \quad t = 4.$$

B_1: $\quad k_1 = 3, c_1 = 1$
$\{e, (12)\}, \quad \{(123), (13)\}, \quad \{(132), (23)\}.$

B_2: $\quad k_2 = 3, c_2 = 1$
$\{e, (23)\}, \quad \{(123), (12)\}, \quad \{(132), (13)\}.$

B_3: $\quad k_3 = 3, c_3 = 1$
$\{e, (13)\}, \quad \{(123), (23)\}, \quad \{(132), (12)\}.$

B_4: $\quad k_4 = 6, c_4 = 0$
$\{e, (123)\}, \{(123), (132)\}, \{(132), e\}, \{(12), (23)\}, \{(23), (13)\},$
$\{(13), (12)\}.$

The first three orbits are the coset decompositions of the subgroups of order 2 in S_3. The last orbit is obtained by multiplying $\{e, (123)\}$ on the right by elements of S_3.

Let us now describe the group G' and the set A' in this same case. That is, we take $G' = \mathbf{Z}_6{}^+$, a cyclic group of order 6. The information about $\mathbf{Z}_6{}^+$ is listed in the same fashion as above.

$$m = 3, \quad p = 2, \quad b = 1, \quad n = 15, \quad t' = 3.$$

B_1': $k_1' = 3, c_1' = 1$
$\{[0], [3]\}, \quad \{[1], [4]\}, \quad \{[2], [5]\}.$

B_2': $k_2' = 6, c_2' = 0$
$\{[0], [1]\}, \quad \{[1], [2]\}, \quad \{[2], [3]\}, \quad \{[3], [4]\}, \quad \{[4], [5]\}, \quad \{[5], [0]\}.$

B_3': $k_3' = 6, c_3' = 0$
$\{[0], [2]\}, \quad \{[1], [3]\}, \quad \{[2], [4]\}, \quad \{[3], [5]\}, \quad \{[4], [0]\}, \quad \{[5], [1]\}.$

Verifying the various numbers and equations of the proof for this particular example should help in illuminating the various steps of the proof of the theorem. We turn now to an example which applies this theorem.

Example 2 Recall the group C of symmetries of the cube. How many subgroups of order 2 are contained in C? Each subgroup of order 2 contains a unique element of order 2; therefore, the number of subgroups of order 2 is equal to the number of elements of order 2. Since each element of C can be described as a rotation about some axis (see Exercise 7 of Section 2.1), each element of order 2 is given by a 180° rotation (either clockwise or counterclockwise) about some axis. (Refer to Figure 1 of Section 8.1 or your cube.) Every axis of symmetry of the cube has a symmetry given by a 180° rotation except for the diagonals of the cube. There are the x,y,z-axes as in Figure 1; further, there are the six axes of symmetry given by connecting the midpoints of opposite edges of the cube. Therefore, there are nine elements (or subgroups) of C of order 2. In this case $9 \equiv 1 \pmod 2$ just as required by the theorem.

The theorem tells us that a finite group contains subgroups of order p^r, where p^r is the highest power of a prime p which divides the order of the group. These particular subgroups are singled out as being quite important.

8.16   DEFINITION Suppose that G is a group of order n and p is a prime which divides n. If the highest power of p which divides n is p^r, a subgroup of G of order p^r is said to be a *Sylow p-subgroup* of G (or just a *Sylow subgroup* of G).

We now state and prove the main theorem of Sylow.

8.17 **THEOREM** (*Sylow*) *Suppose that G is a finite group whose order is divisible by a prime p.*

(i) *If l is the number of Sylow p-subgroups of G, then l divides the order of G and $l \equiv 1 \pmod{p}$.*

(ii) *If H is any subgroup of G whose order is a power of p and P is a fixed Sylow p-subgroup of G, then there is an element a of G such that P contains aHa^{-1}.*

We shall again use actions in the proof of this theorem. Notice that in (i) the congruence $l \equiv 1 \pmod{p}$ follows immediately from Theorem 8.8 applied to the case $b = r$. We shall leave the first part of (i) until the end; so consider part (ii).

Proof of (ii): Let $B = \{Pa \,|\, a \in G\}$ be the collection of cosets of P in G. Since P is a Sylow p-subgroup of G, the number m of cosets in B is relatively prime to p. That is,

8.18
$$m \not\equiv 0 \pmod{p}.$$

If $X \in B$, we set $X^a = Xa$. Recalling the proof of Lemma 8.9 and Example 3 of Section 8.1, it is easy to see that B is a single orbit of G upon the collection A defined in the proof of Theorem 8.8; that is, G acts upon B. This can also be seen by applying Method 2 of Section 8.1. By Theorem 8.2 we can easily show that any subgroup of G also acts upon B (this is Method 1 of Section 8.1). We are interested in the action of H upon B. In passing to a subgroup, we do not have as many group elements to "push" the elements of B around; so it is reasonable that H may have different orbits upon B than G does. In particular, B may not be a single H-orbit. Accordingly, we let $T_1, \ldots, T_s$ be the s orbits of H upon B. For each T_i we choose $X_i \in T_i$ and let H_i be the stabilizer in H of X_i. Since the order of H is a power of p, and since the index of H_i in H divides the order of H, this index is a power p^{c_i} of p. By Theorem 8.6, T_i has size p^{c_i}, the index of the stabilizer H_i in H. Since B has size m, and since the T_i's partition B, we have $m = p^{c_1} + \cdots + p^{c_s}$. From 8.18 we obtain:

8.19
$$p^{c_1} + \cdots + p^{c_s} \not\equiv 0 \pmod{p}.$$

Those terms for which $c_i > 0$ satisfy $p^{c_i} \equiv 0 \pmod{p}$ and, consequently, may be omitted from 8.19. We cannot omit all terms since the sum is not congruent to zero, and therefore some $c_i = 0$.

Let $T = T_i$ and $X = X_i$ be chosen for an i such that $c_i = 0$. The index of H_i in H is $p^{c_i} = p^0 = 1$ so that $H = H_i$ and, further, $T = \{Pa\}$ is a single coset of P in G. Now H stabilizes the coset Pa so that $(Pa)^h = Pah = Pa$ for all $h \in H$. Another way to say this is $Paha^{-1} = P$ for all $h \in H$; or $aha^{-1} \in P$ for all $h \in H$. We conclude that $aHa^{-1} \subseteq P$ for this element $a \in G$. The proof of (ii) is complete.

Remark If H is a Sylow p-subgroup of G, then aHa^{-1} and P have the same order so that $P = aHa^{-1} = b^{-1}Hb$, where $b = a^{-1}$. We shall need this fact.

Proof of (i): To complete the proof of the theorem, we need to show that the number of Sylow p-subgroups divides the order of G. Let A be the set of Sylow p-subgroups of G. If $X \in A$ and $a \in G$, then define

$$X^a = a^{-1}Xa;$$

so that exactly as in Example 4 of Section 8.1 we may show that the above equation defines an action of G upon A. By the Remark, A is a single G-orbit. Let $X \in A$ and let N be the stabilizer of X in G. The number of Sylow p-subgroups is the size of the orbit A which, in turn, is the index of N in G by Theorem 8.6. Since an index of a subgroup divides the order of the group, the number of Sylow p-subgroups of G divides the order of G. The proof of Theorem 8.17 is now complete.

Example 3 Look again at the group C of the cube. It has order 24 so that a Sylow 2-subgroup of C will have order 8. Recall Example 2 of Section 8.1. Following Theorem 8.6 we showed that the stabilizer H of the x-axis of that example was a subgroup of order 8 in C. That is, one Sylow 2-subgroup of C is described as the set of all symmetries of C which carry the x-axis of Figure 1 into itself.

Suppose that P is another Sylow 2-subgroup of C. By Theorem 8.17(ii), we may choose $\sigma \in C$ so that $\sigma P \sigma^{-1} = H$ (H and P both have order 8). Recall that A_0 in Example 2 of Section 8.1 was the set consisting of the x-, y-, and z-axes. Suppose $X \in A_0$ is the x-axis; then X^σ is some other axis. If $\tau \in P$, then $X^{(\sigma\tau)} = X^{(\sigma\tau\sigma^{-1})\sigma}$. Since $\sigma\tau\sigma^{-1} \in H$, the stabilizer of X, we conclude that $(X^\sigma)^\tau = X^{(\sigma\tau\sigma^{-1})\sigma} = X^\sigma$. The group P stabilizes the axis X^σ. If P^* is the stabilizer of the axis X^σ, then a computation like the preceding one shows that $\sigma P^* \sigma^{-1}$ stabilizes X; therefore, $\sigma P^* \sigma^{-1} \subseteq \sigma P \sigma^{-1} = H$, and $P^* \subseteq P$, so that $P^* = P$ is the stabilizer of the axis X^σ.

The above argument tells us that there is a Sylow 2-subgroup connected with each of the axes, and, therefore, there are 3 Sylow 2-subgroups in C. Now $3 \equiv 1 \pmod 2$ as required by Theorem 8.17(i), and 3 divides the order of C which is 24.

Theorem 8.17 holds generally only for Sylow subgroups. Example 2 shows that C contains 9 subgroups of order 2. Certainly 9 does not divide 24; further, it is not true that for any subgroups H and K of order 2 in C that there always is an $\alpha \in C$ such that $\alpha H \alpha^{-1} = K$.

A Sylow 3-subgroup of C will have order 3. The rotations about a diagonal of the cube generate a subgroup of order 3 in C. That is, a Sylow 3-subgroup is given by the set of all symmetries of the cube which stabilize a given oriented diagonal of the cube. Arguing as we did for Sylow 2-subgroups, we find that there are exactly as many Sylow 3-subgroups as there are diagonals of the cube. There are 4 diagonals, 4 divides 24, and $4 \equiv 1 \pmod 3$. Again Theorem 8.17(i) is verified in this case.

More examples of this kind will be covered in the exercises.

EXERCISES

1. Describe all groups of order 15 up to isomorphism.

2. Find the possible numbers of Sylow subgroups in a group of order 1225.

3. Consider the group D of the regular dodecahedron. How many Sylow p-subgroups are there for each prime p dividing the order of D?

4. Show that there are groups of order $2^3 \cdot 3$ which have neither a normal Sylow 2- nor a normal Sylow 3-subgroup.

5. Prove: A group of order $3^3 \cdot 5^2$ must have a normal Sylow 5-subgroup.

6. How many Sylow 5-subgroups are there in A_5? In S_6?

7. Assume that $p > q$, where p and q are primes. (i) If G is a group of order pq, then a Sylow p-subgroup of G is normal in G. (ii) If $q \nmid p - 1$, then G is cyclic.

8. Up to isomorphism, find all groups of order 12.

9. Let $n > 2$ be a positive integer, and U be the group of units of Z_n. Form the set $U \times Z_n^+$ and for $(a, b), (a_1, b_1) \in U \times Z_n^+$ define a product: $(a, b)(a_1, b_1) = (aa_1, ba_1 + b_1)$. Show that $U \times Z_n$ is a group. Show that the set $\{[1]\} \times Z_n$ is a subgroup isomorphic to Z_n^+ and that the set $U \times \{[0]\}$ is a subgroup isomorphic to U. If p divides the order of U, then how many Sylow p-subgroups are there in $U \times Z_n^+$? By taking $n = p^2$, show that for any prime p there is a nonabelian group of order p^3.

10. Suppose that G is a finite group with a Sylow p-subgroup P for a prime p and a normal subgroup K of G whose order is divisible by p. Show that $P \cap K$ is a Sylow p-subgroup of K. Let $N = \{a \mid a \in G, a(P \cap K)a^{-1} \subseteq P \cap K\}$. Show that N is a subgroup such that $NK = \{ab \mid a \in N, b \in K\} = G$.

11. Suppose that P is a group of order p^m for a prime p and a positive integer m, and P_0 is a subgroup of index p in P. Prove that P_0 is normal in P. (*Hint:* Let $A = \{P_0a \mid a \in P\}$ and for $X \in A$ and $a \in P$ set $X^a = Xa$. Compute the kernel of this action by determining the order of a Sylow p-subgroup of S, the symmetric group upon A.) Give an example to show that there are groups of order p^3 for some prime p which contain subgroups of index p^2 which are not normal. (*Hint:* See Exercise 9.)

12. Assume that G is a finite group with a Sylow p-subgroup P for a prime p. Set $N = \{x \mid x \in G, xPx^{-1} \subseteq P\}$. If $a, b \in P$ are chosen so that (i) a and b commute with every element of P and (ii) $cac^{-1} = b$ for some $c \in G$, then there is an element $d \in N$ so that $dad^{-1} = b$. (*Hint:* Consider the group $C = \{x \mid x \in G, xbx^{-1} = b\}$.) Show that P and cPc^{-1} are Sylow p-subgroups of C.

8.3 SOME SIMPLE GROUPS (OPTIONAL)

In Section 2.6 we classified all cyclic groups by showing that any cyclic group is isomorphic to a "well-known" group (namely, Z^+ or Z_n^+). Is it possible to prove a classification theorem for all finite groups? With any reasonable meaning attached to "well known" there is no such theorem now. But there is an obvious "method" available which looks useful for proving such a theorem. The "method" is based upon mathematical induction and has several gigantic flaws. It would go something like the following. Suppose that all groups whose orders were less than or equal to an integer n were isomorphic to certain "well-known" groups. Now consider a group G whose order is $n + 1$. Assume that G has a proper normal subgroup M, that is, M itself has order less than or equal to n and, therefore, is isomorphic to a "well-known" group. Further, the quotient group G/M defined upon the cosets of M in G has order less than or equal to n and is also isomorphic to a "well-known" group. In other words, we may assume that G/M and M are known to us. We know how the cosets G/M multiply, and we know how the elements of M multiply. We need only solve the problem of multiplication of elements within cosets of M in G in order to "know" G.

A rather difficult mathematical machinery has been invented to solve this "patching problem," i.e., "patching" M and G/M together to get G. This machinery is called the "cohomology of groups." Not everything is known about group cohomology; so our "method" has a major flaw right here.

But matters are much worse than this. We assumed that G contained a proper normal subgroup. Suppose it does not—then what? Our "method" using mathematical induction breaks down. In order for this "method" to have any hope of success, we must know all of these basic "building block" groups, that is, all finite groups which contain no nontrivial normal subgroups.

8.20 DEFINITION   A group G is called *simple* if it contains no proper normal subgroups.

The simple groups are interesting in their own right, without reference to the fact that they are the basic building blocks of group theory. Recall that the kernel of a homomorphism of a group G is a normal subgroup of G (Theorem 2.81). Further, if the kernel is G, then every element of G is mapped to the identity, i.e., the homomorphism is trivial; and if the kernel is $\{e\}$, then the homomorphism is an isomorphism (Theorem 2.37). That is, simple groups are precisely those groups for which *every* nontrivial homomorphism *must be an isomorphism*; this is remarkable.

A great deal of recent research has gone into classifying the class of finite simple groups (in some meaningful way). The problem is still far from being solved, but not even the most enthusiastic researcher of 1960 would have predicted that we would know even a little of what is known today. We shall not be able to say very much about simple groups, but we shall say enough to show there really is a problem here. From now on we shall assume that every group mentioned in this section is a finite group. We shall first dispose of the abelian simple groups.

8.21 THEOREM   *A group $G \neq \{e\}$ is a finite abelian simple group if and only if G is cyclic of prime order p.*

Proof: If G is of prime order p, then every nonidentity element a of G is a generator of G by Theorem 2.54; that is, G has no nontrivial subgroups, let alone nontrivial normal subgroups. We conclude that a group of prime order is an abelian simple group.

Next assume that G is a finite abelian simple group of order n. Fix a prime p which divides n; so that by Theorem 8.8, G contains a subgroup H of order p. Now every subgroup of an abelian group is normal, so that H, being normal, cannot be a proper subgroup. From this we conclude that G has order $n = p$ and is cyclic.

Incidentally, the theorem remains true if we omit the condition that G must be finite. This is left to the exercises.

Now that we know all finite abelian simple groups (they are all isomorphic to $Z_p{}^+$ for some prime p by Theorem 2.41)—are there any nonabelian finite simple groups? There are; in fact, we already know of quite a few. Namely, A_n for $n \geq 5$, the alternating groups on five or more symbols, are all simple groups. In the rest of this section we shall prove this fact.

8.22   THEOREM   *The alternating group A_n for $n \geq 5$ is a simple group.*

Our proof will involve computations with 3-cycles. First we shall prove that A_n is generated by 3-cycles, and second, we shall prove that if $N \neq \{e\}$ is a normal subgroup of A_n, then N contains all 3-cycles of S_n. We will conclude that $N = A_n$.

8.23   LEMMA   *Suppose that α is a product of two transpositions in S_n for $n \geq 3$. Then α is the product of one or two 3-cycles in S_n.*

We prove this by enumerating all possibilities below (i, j, k, and l are all different):

$$\alpha = (ij)(ij) = (ijk)(ikj);$$
$$\alpha = (ij)(jk) = (ikj);$$
$$\alpha = (ij)(kl) = (ijk)(ilk).$$

8.24   COROLLARY   *If $n \geq 3$, then every element of A_n is a product of 3-cycles. Further, A_n contains every 3-cycle of S_n.*

By Theorem 2.74, every element α of A_n is a product of an even number of transpositions. Using the lemma, we may write α as a product of 3-cycles. Clearly, every 3-cycle is even and lies in A_n.

Next we suppose that N is a nontrivial normal subgroup of A_n for $n \geq 5$. If $\alpha \in A_n$ and $\beta \in N$, then $\alpha^{-1}\beta\alpha \in N$ since N is normal. Consequently, $\beta^{-1}(\alpha^{-1}\beta\alpha) \in N$ for all $\alpha \in A_n$ and $\beta \in N$. These expressions will be very useful to us. We shall now prove a lemma making it easy to calculate $\alpha^{-1}\beta\alpha$.

8.25 LEMMA *Assume that*

$$\beta = (a_1 a_2 \cdots a_s)(a_{s+1} \cdots a_t) \cdots (a_{r+1} \cdots a_n)$$

is an element of S_n written as a disjoint product of cycles. Assume that

$$\alpha = \begin{pmatrix} a_1 & a_2 & a_3 & \cdots & a_n \\ b_1 & b_2 & b_3 & \cdots & b_n \end{pmatrix}$$

is an element of S_n written as above where the first row $a_1 a_2 a_3 \cdots a_n$ is $123 \cdots n$ written in the same order as appears in β. Then

$$\alpha^{-1}\beta\alpha = (b_1 b_2 \cdots b_s)(b_{s+1} \cdots b_t) \cdots (b_{r+1} \cdots b_n)$$

where the b_i's replace the a_i's in the cycles of β.

This lemma is more difficult to state than it is to prove or use. First let us work an example. Let

$$\beta = (135)(24)(6)$$

and

$$\alpha = \begin{pmatrix} 1 & 3 & 5 & 2 & 4 & 6 \\ 6 & 5 & 4 & 3 & 2 & 1 \end{pmatrix}.$$

Then we obtain the inverse of α by flipping it upside down,

$$\alpha^{-1} = \begin{pmatrix} 6 & 5 & 4 & 3 & 2 & 1 \\ 1 & 3 & 5 & 2 & 4 & 6 \end{pmatrix}.$$

From here it is easy to show that

$$\alpha^{-1}\beta\alpha = (654)(32)(1).$$

For example, $(4)\alpha^{-1}\beta\alpha = (5)\beta\alpha = 1\alpha = 6$ and $(3)\alpha^{-1}\beta\alpha = (2)\beta\alpha = 4\alpha = 2$.

Proof: The proof parallels the method used in the example:

$$\alpha^{-1} = \begin{pmatrix} b_1 & b_2 & \cdots & b_n \\ a_1 & a_2 & \cdots & a_n \end{pmatrix}$$

and

$$(b_j)\alpha^{-1}\beta\alpha = (a_j)\beta\alpha.$$

If a_j is not at the "end" of a cycle of β, then

$$(a_j)\beta\alpha = a_{j+1}\alpha = b_{j+1}.$$

If a_j is at the "end" of a cycle of β which "begins" with a_i, then

$$(a_j)\beta\alpha = a_i\alpha = b_i$$

so that b_j is at the "end" of a cycle of $\alpha^{-1}\beta\alpha$ which "begins" with b_i. The proof is complete (and is shorter than the lemma!).

Next we prove that N contains a 3-cycle.

8.26 LEMMA *If $N \neq \{e\}$ is a normal subgroup of A_n for $n \geq 5$ then N contains a 3-cycle.*

Case 1: If N contains an element

$$\beta = (a_1 a_2 \cdots a_s)(a_{s+1} \cdots a_t) \cdots$$

with a cycle of length $s \geq 4$, then N contains a 3-cycle.

Since β is a product of disjoint commuting cycles, we may write the product with the cycle of length s occurring first. Let $\alpha =$
$$\begin{pmatrix} a_1 & a_2 & a_3 & a_4 & \cdots & a_n \\ a_2 & a_3 & a_1 & a_4 & \cdots & a_n \end{pmatrix} = (a_1 a_2 a_3) \text{ so that}$$

$$\alpha^{-1}\beta\alpha = (a_2 a_3 a_1 a_4 \cdots a_s)(a_{s+1} \cdots a_t) \cdots.$$

Next $\beta^{-1} = (a_s a_{s-1} \cdots a_2 a_1)(a_t \cdots a_{s+1}) \cdots$ so that

$$
\begin{aligned}
\beta^{-1}(\alpha^{-1}\beta\alpha) &= [(a_s a_{s-1} \cdots a_2 a_1)(a_t \cdots a_{s+1}) \cdots] \\
&\quad \cdot [(a_2 a_3 a_1 a_4 \cdots a_s)(a_{s+1} \cdots a_t) \cdots] \\
&= [(a_s a_{s-1} \cdots a_2 a_1)(a_2 a_3 a_1 a_4 \cdots a_s)] \\
&\quad \cdot [(a_t \cdots a_{s+1})^{-1}(a_t \cdots a_{s+1})] \cdots \\
&= (a_1 a_2 a_4).
\end{aligned}
$$

Therefore, $(a_1 a_2 a_4) = \beta^{-1}(\alpha^{-1}\beta\alpha) \in N$.

Case 1 allows us to assume that every element of N is a product of disjoint 2- and 3-cycles.

Case 2: If $\beta \in N$ is a disjoint product of 2-cycles and at least one 3-cycle, then β^2 is a product of disjoint 3-cycles.

We may write $\beta = \tau_1\tau_2 \cdots \tau_s\sigma_1\sigma_2 \cdots \sigma_t$ where the τ_i are transpositions, the σ_j are 3-cycles, and all these cycles are disjoint. Then

$$\beta^2 = \tau_1{}^2\tau_2{}^2 \cdots \tau_s{}^2\sigma_1{}^2\sigma_2{}^2 \cdots \sigma_t{}^2 = \sigma_1{}^2\sigma_2{}^2 \cdots \sigma_t{}^2.$$

Since the square of a 3-cycle is again a 3-cycle, β^2 is a product of disjoint 3-cycles.

Let us next consider the case of a product of 3-cycles.

Case 3: If $\beta \in N$ is a disjoint product of 3-cycles, then N contains a 3-cycle.

If β is a 3-cycle, we are through; therefore, we may assume that β involves at least two 3-cycles:

$$\beta = (a_1a_2a_3)(a_4a_5a_6) \cdots.$$

Let

$$\alpha = \begin{pmatrix} a_1 & a_2 & a_3 & a_4 & a_5 & a_6 & a_7 & \cdots & a_n \\ a_1 & a_2 & a_4 & a_5 & a_3 & a_6 & a_7 & \cdots & a_n \end{pmatrix} = (a_3a_4a_5).$$

Then $\alpha^{-1}\beta\alpha = (a_1a_2a_4)(a_5a_3a_6) \cdots$ so that

$$\begin{aligned} \beta^{-1}(\alpha^{-1}\beta\alpha) &= [(a_3a_2a_1)(a_6a_5a_4) \cdots] \cdot [(a_1a_2a_4)(a_5a_3a_6) \cdots] \\ &= (a_1a_6a_3a_4a_5) \in N. \end{aligned}$$

By Case 1 we conclude that N contains a 3-cycle.

We may now assume that nonidentity elements of N may be written as disjoint products of transpositions. Since the elements of N are even, the number of transpositions must be two, four, or more.

Case 4: If $\beta \in N$ is a product of four or more disjoint transpositions, then N contains a 3-cycle.

Assume that $\beta = (a_1a_2)(a_3a_4)(a_5a_6)(a_7a_8) \cdots$. Set

$$\alpha = \begin{pmatrix} a_1 & a_2 & a_3 & a_4 & a_5 & a_6 & \cdots & a_n \\ a_3 & a_2 & a_5 & a_4 & a_1 & a_6 & \cdots & a_n \end{pmatrix} = (a_1a_3a_5)$$

so that

$$\alpha^{-1}\beta\alpha = (a_3a_2)(a_5a_4)(a_1a_6)(a_7a_8) \cdots.$$

Since $\beta = \beta^{-1}$, we have

$$
\begin{aligned}
\beta^{-1}(\alpha^{-1}\beta\alpha) &= [(a_1a_2)(a_3a_4)(a_5a_6)(a_7a_8)\cdots] \\
&\quad \cdot [(a_3a_2)(a_5a_4)(a_1a_6)(a_7a_8)\cdots] \\
&= (a_1a_3a_5)(a_2a_6a_4) \in N.
\end{aligned}
$$

We now apply Case 3 to complete this case.

We may now assume that any nonidentity element of N is a disjoint product of two transpositions. Up to this point, we have not had to invoke the hypothesis that $n \geq 5$. We must do this now.

Case 5: If $\beta = (a_1a_2)(a_3a_4) \in N$, then N contains a 3-cycle.

Since $n \geq 5$, $\{1, 2, \ldots, n\} = \{a_1, a_2, a_3, a_4, a_5, \ldots\}$, where a_5 is different from a_1, a_2, a_3, a_4. Let

$$
\alpha = \begin{pmatrix} a_1 & a_2 & a_3 & a_4 & a_5 & a_6 & \cdots & a_n \\ a_3 & a_2 & a_5 & a_4 & a_1 & a_6 & \cdots & a_n \end{pmatrix} = (a_1a_3a_5)
$$

so that

$$
\alpha^{-1}\beta\alpha = (a_3a_2)(a_5a_4).
$$

Since $\beta^{-1} = \beta$, we have

$$
\begin{aligned}
\beta^{-1}(\alpha^{-1}\beta\alpha) &= [(a_1a_2)(a_3a_4)] \cdot [(a_3a_2)(a_5a_4)] \\
&= (a_1a_3a_5a_4a_2) \in N.
\end{aligned}
$$

Now Case 1 applies. Observe that Cases 1 to 5 complete the proof of Lemma 8.26.

8.27   LEMMA *If $N \neq \{e\}$ is a normal subgroup of A_n for $n \geq 5$, then $N = A_n$. Theorem 8.22 is true.*

Proof: By the previous lemma, N contains a 3-cycle $(a_1a_2a_3)$. We prove that N contains all 3-cycles. Let $b_1, b_2, b_3 \in \{1, 2, \ldots, n\}$ be distinct, and $b_4, b_5, \ldots, b_n$ the remaining symbols in some order. Set

$$
\alpha = \begin{pmatrix} a_1 & a_2 & a_3 & a_4 & a_5 & \cdots & a_n \\ b_1 & b_2 & b_3 & b_4 & b_5 & \cdots & b_n \end{pmatrix}
$$

so that $\gamma = \alpha^{-1}\beta\alpha = (b_1b_2b_3)$. We cannot conclude that $\gamma \in N$ since we only know that N is normal in A_n and α may not be in A_n. Let $\tau = (b_4b_5)$ so that

$$
\tau^{-1}\gamma\tau = \gamma.
$$

If $\alpha \in A_n$, then $\gamma \in N$. If $\alpha \notin A_n$, then $\alpha\tau \in A_n$ and $(\alpha\tau)^{-1}\beta(\alpha\tau) = \tau^{-1}\gamma\tau = \gamma \in N$. For any $b_1, b_2, b_3 \in \{1, 2, \ldots, n\}$ we conclude that $(b_1 b_2 b_3) \in N$. By Corollary 8.24, $N \supseteq A_n$ so that $N = A_n$ and the only normal subgroups of A_n are $\{e\}$ and A_n, proving that A_n is simple if $n \geq 5$.

Example Let $A = \{1, 2, 3, 4\}$. The group

$$N = \{e, (12)(34), (13)(24), (14)(23)\}$$

is a normal subgroup of A_4. If we number the diagonals of a cube 1, 2, 3, 4 so that the group C of symmetries of the cube act as S_4 upon $A = \{1, 2, 3, 4\}$, then N is the kernel of the homomorphism in Example 1 of Section 2.5. In other words, the proof of Theorem 8.22 works for A_4 right up to Case 5 in the proof of Lemma 8.26. The magic element α of that case does not exist for A_4 because a fifth symbol beyond those in $\{1, 2, 3, 4\}$ is needed to define α.

The alternating groups by no means exhaust all the known nonabelian simple groups. In fact, they are just the beginning. There are an infinite number of simple groups A_n. But there are also an infinite number of other simple groups which are known but not mentioned here. We have barely scratched the surface in our discussion of simple groups.

EXERCISES

1. Prove that there are no infinite abelian simple groups.

2. Prove that the groups S_4, S_3, S_2, A_4, and A_3 are not simple.

3. Prove that no group of order p^m for a prime p and an integer $m > 1$ is simple.

4. Show that a group of order pq for primes p and q is not simple.

5. Prove that for $n \geq 5$, A_n is the only nontrivial normal subgroup of S_n.

6. Prove that every finite group G has a chain of subgroups $H_1, \ldots, H_t$ so that $H_1 = \{e\}$, $H_t = G$, H_i is normal in H_{i+1}, and H_{i+1}/H_i is a simple group. Such a chain is called a *composition series* for G.

7. Show that a group of order 200 is not simple.

8. Up to isomorphism find all groups of order $3 \cdot 5 \cdot 17$.

9. Suppose that in Exercise 6 each of the quotient groups H_{i+1}/H_i is abelian. Show that if $K \supseteq L$ are subgroups of G such that L is normal in K and K/L is simple then K/L is abelian.

10. In Exercise 14 of Section 2.8 it was shown that the group D of symmetries of the regular dodecahedron is isomorphic to A_5. (i) Using the dodeca-hedron and the action of Example 5 in Section 8.1, show that the orbits of D upon $A = D$ consist of B_1, the identity; B_2, the elements of order 2; B_3, the elements of order 3; and B_5, B_6, two orbits containing the elements of order 5. (ii) If N is a normal subgroup of D and B is an orbit as in (i) for which $B \cap N \neq \varnothing$, then show that $B \subseteq N$. (iii) Prove that N is a disjoint union of orbits. (iv) Compute the sizes of the orbits and from (iii) conclude that $N = D$ or $\{e\}$; that is, prove that D is a simple group.

11. Show that of the regular polyhedra, only the groups of the dodecahedron and icosahedron are simple. These two groups are isomorphic to A_5.

12. (This problem could be considered a major project.) Show that there are no nonabelian simple groups of order $n < 60$. If G is simple of order 60, then $G \cong A_5$.

8.4 THE GROUP OF UNITS OF Z_n (OPTIONAL)

For this section we will fix a positive integer $n > 1$. We wish to determine the structure of the group U of units in the ring Z_n. At certain points we shall invoke the Chinese Remainder Theorem, 5.44, of Section 5.7; but it is possible to follow the proofs just by knowing the statement of this theorem. The group U is actually quite interesting. It will allow us to determine the set of all isomorphisms of a cyclic group with itself (Exercises 6, 7). Further, if G is some finite abelian group, then for some choice of n, G is isomorphic to a subgroup of U. These facts alone make the group U interesting. But even more, the group U keeps appearing in various places in algebra; for example, it arises naturally in the study of fields generated over Q, the rational field, by roots of 1 in C, the complex field.

   We shall start with the case $n = p$, a prime; so that, Z_p is a field by Theorem 6.5. In this case we know that the group U of units of the field Z_p is just the set $Z_p \setminus \{[0]\}$; in which case U is a group of order $p - 1$. We shall prove that U

is a cyclic group. The proof depends upon a fact which will be proved in Chapter 10. We shall not be overly rigorous and careful here; Chapter 10 will take care of that problem, but we shall state things clearly. A *polynomial* $f(x)$ over a field F is an expression of the form

8.28 $$f(x) = a_0 + a_1 x + \cdots + a_n x^n,$$

where the coefficients a_j of the powers of x come from some fixed field F. The fact that polynomials do exist in the form of 8.28 will be covered later. If $a_n \neq 0$, we say that $f(x)$ has degree n; as we shall see later, a degree does exist and is uniquely determined by the polynomial $f(x)$. An element a of the field F is called a *root* of $f(x)$ if

$$f(a) = a_0 + a_1 a + \cdots + a_n a^n = 0.$$

The ideas here are similar to ones we know from elementary algebra. In fact, the following theorem (Corollary 10.25), which is proved in Chapter 10, may even be familiar.

8.29 THEOREM *A polynomial $f(x)$ of degree n over a field F cannot have more than n distinct roots in F.*

We shall be interested in the polynomial $f(x) = (-1) + 0 \cdot x + \cdots + 0 \cdot x^{m-1} + 1 \cdot x^m = (-1) + x^m$ of degree m. For a root $a \in F$, $0 = (-1) + a^m$ or $a^m = 1$. Our theorem tells us the following.

8.30 COROLLARY *If F is a field and $m > 0$ is any integer, then there are at most m elements a in F such that $a^m = 1$.*

This is the form in which we shall use the theorem; and in this form, the result is so strong that it forces all finite multiplicative groups in F to be cyclic. To prove this fact, we must study some properties of finite abelian groups. We wish to prove the following theorem.

8.31 THEOREM *If G is a noncyclic finite abelian group of order p^m for a prime p and an integer $m > 0$ with identity e, then there exists an integer $k > 0$ such that there are $p^k + 1$ elements a of G for which $a^{p^k} = e$.*

Proof: Since G is finite, we may sort through the elements of G and find one, b, of largest possible order p^k. Let B be the cyclic group of order p^k generated by the element b. Since G is not cyclic, there is some element c of G not contained in B. The order of c is p^t for some positive integer t; and $c^{p^t} = e \in B$. By our choice of b, $p^t \le p^k$, and therefore $c^{p^k} = e$. The powers of b along with c constitute $p^k + 1$ elements, a, such that $a^{p^k} = e$, completing the proof of the theorem.

8.32   **THEOREM**   *If G is an abelian group of order n, and a Sylow p-subgroup P of G is cyclic for each prime p dividing n, then G is a cyclic group.*

Proof: Every subgroup of an abelian group is normal so that by Theorem 8.17(ii), for each prime p_i dividing n, there is only one Sylow p_i-subgroup of G and it is cyclic of order $p_i^{e_i}$, where $n = p_1^{e_1} p_2^{e_2} \cdots p_t^{e_t}$ is a prime factorization of n. Let $P_1, P_2, \ldots, P_t$ be, respectively, the Sylow p_i-subgroups of G for $i = 1, \ldots, t$; and choose $a_i \in P_i$ to be a generator for the cyclic group P_i. Let $a = a_1 a_2 \cdots a_t$. We shall prove that a has order n.

Let s be the smallest positive integer such that $a^s = e$; so by Theorem 2.45, s is the order of a. By Corollary 2.55 $s \le n$ since $a^n = e$. We shall show that $p_i^{e_i}$ divides s for each i; so we assume that $p = p_i$ and $c = e_i$. Set $m = n/p^c$ so that $p_j^{e_j}$ divides m for each $j \ne i$; and therefore $a^m = a_1^m a_2^m \cdots a_t^m = a_1^m$. Now m and p are relatively prime; hence by Euclid's Lemma, 5.18, there are integers u and v such that

$$up^c + vm = 1.$$

Therefore, $a^{mv} = a_1^{mv} = ea_1^{mv} = a_1^{up^c} a_1^{mv} = a_1^{up^c + mv} = a_1^1 = a_1$. Using the fact that $a^s = e$, we have

$$e = a^{smv} = (a^{mv})^s = a_1^s.$$

But a_1 has order p^c, so that we may conclude by 2.45 that p^c divides $s = s - 0$ because $a_1^s = a_1^0 = e$. Since s is divisible by $p_i^{e_i}$ for each $i = 1, \ldots, t$, we must have $n = p_1^{e_1} \cdots p_t^{e_t}$ dividing s. But $s \le n$ so that $s = n$; and a has order n. In other words, every element of G, which is a group of order n, is a power of a; and so G is a cyclic group, completing the proof.

The previous two theorems have a very nice corollary.

8.33 COROLLARY *Assume that G is a finite abelian group in which, for any positive integer n, there are at most n elements a of G such that $a^n = e$ for e, the identity of G. Then G is a cyclic group.*

Proof: Let P be a Sylow p-subgroup of G for some prime p. There are at most p^k elements a in P such that $a^{p^k} = e$ for every positive integer k. By Theorem 8.31, P must be cyclic; so that for every p dividing the order of G, there is a cyclic Sylow p-subgroup of G. By Theorem 8.32 we conclude that G is cyclic, and the corollary is valid.

We may combine this corollary with Corollary 8.30 to obtain the following useful theorem.

8.34 THEOREM *If F is any field and G is any finite subgroup of the group of units of F, then G is a cyclic group. In particular, the group of units of any finite field is cyclic.*

We have completed our first task; the group U of units of the field $\mathbf{Z}_p$ is cyclic of order $p - 1$.

Next we move to the group U of units of the ring $\mathbf{Z}_{p^a}$ for an integer $a > 1$ and a prime p. If p is odd, we shall show that U is cyclic. If $p = 2$, things go wrong; for example, in $\mathbf{Z}_8$, $[3]^2 = [1]$, $[5]^2 = [1]$, $[1]^2 = [1]$. There are at least three elements a in U such that $a^2 = [1]$, and since $U = \{[1], [3], [5], [7]\}$ has order $2^2 = 4$, Theorem 8.31 tells us that U is not cyclic. We shall examine what goes wrong in the case $p = 2$. We start by computing with congruences modulo p^a. In particular, we raise $(1 + p^t)$ to the p^s-power.

8.35 THEOREM *If p is a prime and s and t are positive integers for which $p > 2$ or $t > 1$, then there is a positive integer b such that*

$$(1 + p^t)^{p^s} = 1 + p^{s+t} + p^{s+t+1}b.$$

Proof: We shall prove this theorem by applying the Binomial Theorem (Exercise 20); but first we need to make an observation concerning binomial coefficients. If $1 \le j \le p - 1$, then the binomial coefficient $\binom{p}{j}$ is divisible by p. To see this, observe that $\binom{p}{j} = \dfrac{p!}{(p - j)!j!}$ is an integer. The denominator $(p - j)!j!$ is a product of integers less than p, i.e., a product of integers relatively prime to p; so that $(p - 1)!j!$ is relatively prime to p. On the other hand, the numerator $p! = p(p - 1)!$ has p as a factor. Since the denominator is relatively prime to p (Exercise 4 of Section 5.5) and the denominator divides the numerator, we now know that the denominator divides $(p - 1)!$. Therefore, as we set out to show, p is a factor of $\binom{p}{j}$ for $1 \leq j \leq p - 1$.

We shall now prove the truth of the following formula by induction upon n.

8.36
$$(1 + p^t c)^{p^n} = 1 + p^{t+n}(c + pc'),$$

where c, n, c' are positive integers. Starting with the case $n = 1$, and using the Binomial Theorem (Exercise 20), we have

$$(1 + p^t c)^p = 1 + \binom{p}{1} p^t c + \cdots + \binom{p}{p-1} p^{(p-1)t} c^{(p-1)} + p^{pt} c^p$$
$$= 1 + p^{t+1} c + p a_2 p^{2t} c^2 + \cdots$$
$$+ p a_{p-1} p^{(p-1)t} c^{(p-1)} + p^{pt} c^p$$

where we have written $p a_j = \binom{p}{j}$ for $j = 2, \ldots, p - 1$ (observe

that these terms are totally absent from the sum if $p = 2$). If $p > 2$ or $p = 2$ and $t > 1$, then $pt \geq t + 2$; also if $p > 2$, then $jt \geq t + 1$ for $j = 2, 3, \ldots, p - 1$ so that

$$(1 + p^t c)^p = 1 + p^{t+1} c + p^{t+2}(a_2 p^{t-1} c^2 + \cdots$$
$$+ a_{p-1} p^{(p-2)t-1} c^{(p-1)} + p^{(p-1)t-2} c^p)$$
$$= 1 + p^{t+1} c + p^{t+2} c'$$
$$= 1 + p^{t+1}(c + pc').$$

We have shown that the formula 8.3 is correct for $n = 1$. Suppose now that $(1 + p^t c)^{p^{n-1}} = 1 + p^{t+n-1}(c + pc')$ for the exponent p^{n-1}. Taking $c'' = c + pc'$ and applying the case for $n = 1$, we have

$$(1 + p^t c)^{p^n} = (1 + p^{t+n-1} c'')^p = 1 + p^{t+n}(c'' + pc''')$$
$$= 1 + p^{t+n}(c + pc' + pc''')$$
$$= 1 + p^{t+n}(c + p(c' + c''')).$$

Therefore, if formula 8.36 is correct for $n - 1$, it is correct for n. We conclude by mathematical induction that 8.36 is correct for all values of $n = 1, 2, 3, \ldots$. Taking $c = 1$, $n = s$, and letting $b = c'$

in 8.36, we have $(1 + p^t)^{p^s} = 1 + p^{t+s}(1 + pb) = 1 + p^{t+s} + p^{t+s+1}b$, completing the proof of the theorem.

Next we compute the order of the group of units of $\mathbf{Z}_n$.

8.37 **THEOREM** *If $n > 1$ is an integer and U is the group of units of the ring $\mathbf{Z}_n$, then the order of U is equal to the number of positive integers j such that $j < n$ and j is relatively prime to n. If $n = p^a$ for a prime p and a positive integer a, then U has order $p^{a-1}(p - 1)$.*

Proof: Recall that $\mathbf{Z}_n = \{[j] \mid j \in \mathbf{Z} \text{ and } 0 \le j < n\}$. If $0 \le j < n$ and $(j, n) = d$ is the greatest common divisor of j and n, then we may factor n and j so that $n = dn'$ and $j = dj'$. If, in addition, $d > 1$, then in $\mathbf{Z}_n$, $[n'] \ne [0]$, so that $[j][n'] = [jn'] = [j'dn'] = [j'n] = [0]$ and $[j]$ is a zero divisor, i.e., $[j]$ is not a unit. On the other hand, if $d = 1$, then by Euclid's Lemma, 5.18, we may choose integers x and y so that

$$jx + ny = 1.$$

We now have

$$[j][x] = [jx] + [0] = [jx] + [ny] = [jx + ny] = [1],$$

so that $[j]$ is a unit. We have shown that $\mathbf{Z}_n$ is the disjoint union of the group U of units and the set of zero divisors; and the set $U = \{[j] \mid j \in \mathbf{Z}, 0 \le j \le n - 1, (j, n) = 1\}$. From this, the first part of the theorem follows immediately.

Applying this fact in the second part, if $0 \le j \le p^a - 1$, then $[j]$ is a zero divisor in $\mathbf{Z}_{p^a}$ if and only if p divides j. Any j such that $0 \le j \le p^a - 1$ and such that $[j]$ is a zero divisor in $\mathbf{Z}_{p^a}$ must factor as $j = pk$; and any number that has such a factorization is a zero divisor. Since there are p^{a-1} choices for k such that $0 \le pk \le p^a - 1$, we conclude that there are p^{a-1} zero divisors in $\mathbf{Z}_{p^a}$. By our previous argument there are $p^a - p^{a-1} = p^{a-1}(p - 1)$ units in $\mathbf{Z}_{p^a}$ since all elements which are not zero divisors are units in $\mathbf{Z}_{p^a}$. The theorem is proved.

We may now prove a theorem giving the structure of the group of units of $\mathbf{Z}_{p^a}$ for $a > 1$.

8.38 **THEOREM** *Let p be a prime and $a > 1$ a positive integer; and suppose that U is the group of units in $\mathbf{Z}_{p^a}$.*

(i) *If $p > 2$, then U is cyclic of order $p^{a-1}(p - 1)$, and $[1 + p]$ has order p^{a-1} in U.*

(ii) If $p^a = 4$, then U is cyclic of order 2.

(iii) If $p = 2$ and $a > 2$, then $[-1]$ generates a cyclic group C of order 2 and $[5]$ generates a cyclic group D of order 2^{a-2} such that

$$U \cong C \times D;$$

and, in particular, $U \cong Z_2^+ \oplus Z_{2^{a-2}}^+$.

Proof: For an integer r we shall use $[r]$ to denote the coset $r + (p^a)$ in Z_{p^a} and $\langle r \rangle$ to denote the coset $r + (p)$ in Z_p. By Exercise 21 of Section 3.2 by setting $[r]\phi = \langle r \rangle$, we obtain a homomorphism of Z_{p^a} onto Z_p. It is easy to show that if $[r]$ is a zero divisor in Z_{p^a}, then $[r]\phi = \langle r \rangle$ is a zero divisor in Z_p. Similarly, if $[r]$ is a unit in Z_{p^a} with inverse $[s]$, then $\langle 1 \rangle = [1]\phi = ([r][s])\phi = ([r]\phi)([s]\phi)$ so that $[r]\phi$ is a unit in Z_p. As shown in the previous proof, each of Z_{p^a} and Z_p is the disjoint union of its group of units and its set of zero divisors. Consequently, ϕ must map the group U of units of Z_{p^a} onto the group $Z_p \backslash \{\langle 0 \rangle\}$ of units of Z_p. Therefore, restricting ϕ to U and considering only multiplication as an operation, $\phi \colon U \rightarrow Z_p \backslash \{\langle 0 \rangle\}$ is a homomorphism of the group of units of Z_{p^a} onto the group of units of Z_p. The image of ϕ upon U is $Z_p \backslash \{\langle 0 \rangle\}$ which has order $p - 1$; and U has order $p^{a-1}(p - 1)$ so that the kernel U_0 of ϕ, viewed as a group homomorphism, has order $p^{a-1}(p - 1)/(p - 1) = p^{a-1}$.

Now the only zero divisor in Z_p is $\langle 0 \rangle$ so that $[1 + p^s]\phi = ([1] + [p^s])\phi = [1]\phi + [p^s]\phi = \langle 1 \rangle + \langle 0 \rangle = \langle 1 \rangle$ for any positive integer s since $[p^s]$ is a zero divisor in Z_{p^a}. We have shown that $[1 + p^s]$ lies in U_0 so that by Theorem 2.53 the order of $[1 + p^s]$ must divide the order p^{a-1} of U_0. In particular, the order p^t of $[1 + p^s]$ will be the smallest positive integer of the form p^t such that $[1 + p^s]^{p^t} = [1]$ by Theorem 2.45.

We now invoke Theorem 8.35 and complete the proof of (i). If p is odd then $[1 + p]^{p^t} = [1 + p^{t+1} + p^{t+2}b]$. When $t = a - 1$ we obtain $[1 + p]^{p^{a-1}} = [1]$; and when $t = a - 2$ we obtain $[1 + p]^{p^{a-2}} = [1 + p^{a-1}] \neq [1]$. We conclude that $[1 + p]$ has order p^{a-1} when p is an odd prime. This proves that a Sylow p-subgroup of U is cyclic.

Let r be a prime divisor of the order of U different from p; and let R be a Sylow r-subgroup of U. Consider ϕ as a homomorphism of R into the group of units of Z_p. Viewed in this way, the kernel of ϕ upon R will be $U_0 \cap R$, where the order of U_0 is a power of p and the order of R is a power of r. So the order of $U_0 \cap R$ is a power of both p and r. Since the only such power is $p^0 = r^0 = 1$, we conclude that $U_0 \cap R = \{[1]\}$. By Theorem 2.37, ϕ upon R is an isomorphism of R into $Z_p \backslash \{\langle 0 \rangle\}$, the group of units of Z_p. By Theorem 8.34 the group of units of Z_p is cyclic; and therefore, a subgroup like $R\phi = \{[s]\phi \mid [s] \in R\}$ must also be cyclic (Theorem 2.47). But ϕ is an isomorphism of R so that R is cyclic. We conclude that for all prime divisors r of the order of U, U has a cyclic Sylow r-subgroup. Since Theorem 8.32 now applies to tell us that U is cyclic, the proof of (i) is complete.

The proof of (ii) is given by the observation that the unit group of Z_4 is $\{[1], [3]\}$, so that we may concentrate our attention upon part (iii), i.e., $p = 2, a > 2$.

Observe that Theorem 8.35 does not tell us about the powers of $3 = 1 + 2$ but does tell us about the powers of $5 = 1 + 2^2$. By that proposition,

8.39
$$[5]^{2^t} = [1 + 2^2]^{2^t} = [1 + 2^{t+2} + 2^{t+3}b].$$

The order of $[5]$ must divide the order of U which is $2^{a-1}(2-1) = 2^{a-1}$; consequently, taking $t = a - 3$ and $a - 2$, we see that $[5]$ has order 2^{a-2} in U. It is trivial to show that $[-1]$ has order 2.

Form the direct product $C \times D$ and define the mapping

8.40
$$([r], [s])\theta = [r][s]$$

of $C \times D$ into U. It is quite straightforward (Exercise 15) to verify that θ is a homomorphism of $C \times D$ into U. We wish to prove that θ is actually an isomorphism. An element $([r], [s])$ will be in the kernel of θ if and only if $[r][s] = [1]$. There are two choices for $[r] \in C$, namely, $[r] = [1]$ and $[r] = [-1]$. But if $[r] = [1]$, then $[1] = [r][s] = [1][s] = [s]$ and to be in the kernel of θ, $([r], [s]) = ([1], [1])$. Suppose now that $[r] = [-1]$ and $([r], [s])$ is in the kernel of θ, in which case $[1] = [-1][s]$, or $[-1] = [s]$, so that $([-1], [s])$ is in the kernel of θ if and only if $[s] = [-1]$. Since $[s]$ is in D and $[5]$ generates D, $[s] = [-1]$ if and only if $[-1]$ is a power of $[5]$. We have shown that θ is an isomorphism (i.e., the kernel of θ is $\{([1], [1])\}$) if and only if $[-1]$ is not a power of $[5]$.

By Exercise 7 of Section 2.6, D contains a unique subgroup of order 2, and in this subgroup $[5]^{2^{a-3}}$ must be the element of order 2. That is, $[5]^{2^{a-3}}$ is the unique element of order 2 in D. By 8.35, with $t = a - 3$, we obtain $[5]^{2^{a-3}} = [1 + 2^{a-1}] \neq [-1]$ so that $[-1]$ does not lie in D. We have shown that the homomorphism θ of $C \times D$ is an isomorphism. The order of $C \times D$ is $2 \cdot 2^{a-2} = 2^{a-1}$ so that the image of $C \times D$ under θ has order 2^{a-1}, which is the order of U. We conclude that the image of $C \times D$ under θ is all of U so that $U \cong C \times D$. The rest of part (iii) now follows from Theorem 2.41. We have completed the proof of Theorem 8.38, and with it, the most difficult part of this section.

We now complete our work on the group of units of Z_n by proving the following theorem.

8.41 THEOREM *For each positive integer $n > 1$ let $U(n)$ denote the group of units in Z_n. If n is a positive integer with a prime factorization $n = p_1{}^{e_1} p_2{}^{e_2} \cdots p_t{}^{e_t}$, then*

$$U(n) \cong U(p_1{}^{e_1}) \times \cdots \times U(p_t{}^{e_t})$$

where each unit group $U(p_i{}^{e_i})$ is described by Theorem 8.38.

Proof: By the Generalized Chinese Remainder Theorem, 5.44, there is a ring isomorphism $\theta : Z_n \to Z_{p_1{}^{e_1}} \oplus \cdots \oplus Z_{p_t{}^{e_t}}$ of Z_n onto the direct sum of the rings $Z_{p_i{}^{e_i}}$ for $i = 1, \ldots, t$. By a slight generalization of Theorem 3.14 in Section 3.2,

$$U' = U(p_1{}^{e_1}) \times \cdots \times U(p_t{}^{e_t})$$

is the group of units in the direct sum of the rings $Z_{p_i{}^{e_i}}$ for $i = 1, \ldots, t$. Since θ is a ring isomorphism of Z_n onto the direct sum, when restricted to the group of units, it is an isomorphism of $U(n)$ onto U'. This fact, along with Theorem 8.38, completes the proof.

The groups $U(n)$ play an important role in many parts of algebra. We shall indicate here one interesting property of the groups $U(n)$. We shall quote two theorems to illustrate our point. The first is proven in Chapter 9. The second is not proven in this text. We shall illustrate the following fact.

8.42 THEOREM *If G is a finite abelian group, then there is an integer n so that G is isomorphic to a subgroup of the group of units of Z_n. In fact, n may be so chosen that its prime factorization is $n = p_1 p_2 \cdots p_t$ where all the p_i's are different odd primes.*

To prove this theorem, we need the following, which is Theorem 9.10 of Chapter 9.

8.43 THEOREM *If G is a finite abelian group, then G is isomorphic to a direct product of cyclic groups of finite order.*

This theorem tells us that, by some isomorphism, we may replace G by a direct product $C_1 \times \cdots \times C_t$ such that each group C_i is cyclic. If we could produce distinct odd primes p_i such that C_i is isomorphic to a subgroup of $U(p_i)$ the group of units in $\mathbf{Z}_{p_i}$ for $i = 1, \ldots, t$, then $C_1 \times \cdots \times C_t$ would be isomorphic to a subgroup of $U(p_1) \times \cdots \times U(p_t) \cong U(n)$, where $n = p_1 \cdots p_t$. Each group C_i is cyclic of some order c_i; and the group of units of $\mathbf{Z}_p$ for an odd prime p is cyclic. Consequently, the proof of Theorem 8.42 depends upon the following lemma.

8.44 LEMMA *If $c_1, c_2, \ldots, c_t$ are positive integers, then there are distinct odd primes $p_1, p_2, \ldots, p_t$ so that for each $i = 1, \ldots, t$, c_i divides $p_i - 1$.*

This lemma may be proved using the following beautiful theorem of Dirichlet (1805–1859) concerning primes in arithmetic sequences.

8.45 DIRICHLET'S THEOREM *If k and l are fixed relatively prime positive integers, then there are infinitely many primes p of the form $p = kn + l$, where n is a positive integer.*

We have purposely only indicated the steps in the proof of Theorem 8.42. A rather nice exposition of Dirichlet's Theorem is contained in Ref. [84]. If this book cannot be found, there is a proof in Ref. [79].

====

EXERCISES

The first few exercises involve the proof of Theorem 8.42.

1. Prove Lemma 8.44 using Dirichlet's Theorem.

2. Using Exercise 1 and Theorem 8.43, supply the missing details in the proof of Theorem 8.42.

The next few exercises illustrate another fact about the unit group of $\mathbf{Z}_n$. An *automorphism* of a group G is an isomorphism of G onto G. The set of all automorphisms of G is denoted by $\text{Aut}(G)$. The set $\text{Aut}(G)$ is endowed with the binary operation given by composition of mappings.

3. Prove that for a group G, $\text{Aut}(G)$ is a group.

4. If $\theta : G \to H$ is an isomorphism of a group G onto a group H, then show that we may define an isomorphism $\Theta : \text{Aut}(G) \to \text{Aut}(H)$ of $\text{Aut}(G)$ onto $\text{Aut}(H)$ as follows: If $\phi \in \text{Aut}(G)$, then we let $\phi\Theta$ be the mapping of H given by $a(\phi\Theta) = (a'\phi)\theta$ for any $a \in H$ and $a' \in G$ such that $a = a'\theta$.

5. Fix an integer $n > 1$, and let U be the group of units of Z_n. For each $u \in U$ define a mapping $\phi_u : Z_n{}^+ \to Z_n{}^+$ by

$$a\phi_u = au$$

for $a \in Z_n{}^+$. Let $U^* = \{\phi_u \mid u \in U\}$. Show that $U^* = \text{Aut}(Z_n{}^+)$ and $U \cong U^*$.

6. Suppose that G is a cyclic group of order $n > 1$ and U is the group of units in Z_n. Prove that $\text{Aut}(G) \cong U$.

7. Prove that if G is cyclic of infinite order then $\text{Aut}(G)$ is cyclic of order 2. In particular, prove that the nonidentity element of $\text{Aut}(G)$ takes every element of G into its inverse.

The next exercises concern ideas about quadratic residues. We call an integer a a *quadratic residue modulo n* for a positive integer n if (i) $[a]$ is a unit in Z_n, and (ii) there is an integer x so that $[x]^2 = [a]$ in Z_n.

8. Let U be an abelian group and define a mapping $\phi : U \to U$ by setting $a\phi = a^2$ for $a \in U$. Prove that ϕ is a homomorphism.

9. Let U be the group of units in Z_n for an integer $n > 1$. Let U_2 be the image of U under the mapping ϕ of Exercise 8. Show that a is a quadratic residue modulo n if and only if $[a] \in U_2$.

10. If $n = p_1{}^{e_1} \cdots p_t{}^{e_t}$ is a prime factorization of n, then compute the order of the group U_2 in Exercise 9. Find the number of quadratic residues a modulo n such that $0 \le a \le n - 1$.

11. Let U_1 be the kernel of the homomorphism ϕ of Exercise 8. Compute the order of U_1 if n is as in Exercise 10. Show that if a is a quadratic residue modulo n and U_1 has order m, then there are precisely m integers x such that $0 \le x \le n - 1$ and $[x]^2 = [a]$ in Z_n.

There is a very famous theorem of Gauss (1777–1855), called the Law of Quadratic Reciprocity, which allows us to determine for any two distinct primes p and q whether or not q is a quadratic residue modulo p. Proofs of the Reciprocity Law may be found in Refs. [74, 76, 85].

12. Prove that if p is an odd prime and U is the group of units of Z_p then there is a unique homomorphism $\psi : U \to \{1, -1\}$ of U *onto* the multiplicative group $\{1, -1\}$. Further, an integer m is a quadratic residue modulo p if and only if $[m]\psi = 1$.

Legendre's symbol is defined for integers m relatively prime to p by setting $(m \mid p) = [m]\psi$. The Reciprocity Law says that for two different odd primes p and q that $(p \mid q)(q \mid p) = -1$ if and only if both p and q are congruent to 3 modulo 4. Further, for an odd prime p, $(2 \mid p) = -1$ if and only if $p \equiv \pm 3 \pmod 8$.

13. Use Quadratic Reciprocity to state whether or not the following are quadratic residues.

    (i) 7 modulo 13.

    (ii) 91 modulo 97.

    (iii) 168 modulo 547.

The following exercises are of a mixed character.

14. If $a > 2$, then what is the order of [3] in $\mathbf{Z}_{2^a}$?

15. Prove that 8.40 does define a homomorphism θ.

16. In each case find the smallest positive integer m such that $[m]$ is a generator of the group of units of $\mathbf{Z}_p$ for the primes $p = 3, 5, 7, 11, 13, 17, 19$.

17. In each case find an integer m such that $[m]$ is a generator of the unit group of $\mathbf{Z}_{p^a}$, where $p^a = 3^5, 11^{10}, 17^3, 19^4$.

18. Suppose that $n > 1$ is a positive integer and k, m are relatively prime positive integral divisors of n. Let a, b be any two integers such that $(a, k) = 1$ and $(b, m) = 1$. Show that there is an integer j such that (i) $(j, n) = 1$, (ii) $j \equiv a \pmod{k}$, and (iii) $j \equiv b \pmod{m}$. Now assume that G is an abelian group of order n and x, y are elements of G whose orders are k, m, respectively. Show that there is an isomorphism $\phi: G \to G$ of G onto G such that $x\phi = x^a$ and $y\phi = y^b$.

19. If G is a cyclic group and a and b are generators of G, then there is an isomorphism $\phi: G \to G$ of G onto G such that $a\phi = b$.

20. The coefficient of $x^i y^j$, where $i + j = n$, a positive integer, in an expansion of $(x + y)^n$ is denoted $\binom{n}{i}$ and is called a *binomial coefficient*. Prove the following:

    (i) $\binom{n}{0} = \binom{n}{n} = 1$.

    (ii) $\binom{n}{i} + \binom{n}{i+1} = \binom{n+1}{i+1}$, $i = 0, 1, 2, \ldots, n - 1$.

    (iii) $\binom{n}{i} = \dfrac{n!}{i!(n-i)!}$ is an integer.

    (iv) A set of size n has exactly $\binom{n}{i}$ subsets of size i for $i = 0, 1, 2, \ldots, n$.

8.5   A PROOF OF SYLOW'S THEOREM (AN EXERCISE)

In this section, by means of a sequence of exercises, we shall prove the following theorem of Sylow.

8.46   THEOREM   *If G is a finite group of order n and p is a prime divisor of n, then G contains l Sylow p-subgroups, where $l \equiv 1 \pmod{p}$.*

This theorem is a special case of Theorem 8.18. The proof that we sketch here, however, is quite different from the one in Section 8.2. The object of the proof here is to build up piece by piece until we obtain a Sylow *p*-subgroup. The first step is to prove that there are elements of order *p* in groups whose orders are divisible by *p*. Even more primitive, we start with abelian groups.

8.47   LEMMA   *If G is an abelian group of order n and p is a prime divisor of n, then G contains an element of order p.*

The proof is by mathematical induction upon *n*, the order of *G*.

1. Prove that the lemma is true if *G* is cyclic.

2. Choose $b \in G$, $b \neq e$, and let *B* be the cyclic group generated by *b*. Show that if *p* divides the order of *B* then the theorem is true.

3. If *p* does not divide the order of *B* then prove that *G/B* contains a coset *aB* of order *p*.

4. Show that the cyclic subgroup generated by *a* contains an element of order *p*.

Use these hints to complete the induction step and prove the lemma.

We wish to prove that a finite group whose order is divisible by a prime *p* contains elements of order *p*. The previous lemma helps us if our group is abelian. We need to find a way to use this lemma when our group is nonabelian; and so we wish to find an abelian subgroup of our group whose order is divisible by *p*.

8.48   LEMMA   *Let G be a group and $\mathbf{Z}(G)$ be the set $\{a \mid a \in G, ab = ba$ for all $b \in G\}$ of elements a of G which commute with every element of G. Then $\mathbf{Z}(G)$ is an abelian subgroup of G, all of whose subgroups are normal subgroups of G.*

The group $\mathbf{Z}(G)$ is called the *center* of *G*, and the notation $\mathbf{Z}(G)$ comes from the German word, "Zentrum," meaning center. Actually, the proof of this lemma should require no hints.

This result gives us the abelian subgroup we want, but why should p divide its order? The fact is, in general, p need not divide the order of $\mathbf{Z}(G)$, the center of a finite group G. But there are circumstances when this does happen.

8.49   LEMMA   *Suppose that G is a finite group whose order n is divisible by a prime p, and assume that the index of every proper subgroup of G is divisible by p. Then the order of $\mathbf{Z}(G)$ is divisible by p.*

This is the major lemma which makes Lemma 8.47 of some value when applied to nonabelian finite groups. We may prove this lemma by means of actions and orbits. Let $A = G$ be the set of elements of G, and show that for $a \in G$ and $x \in A$, $x^a = a^{-1}xa$ defines an action of G upon A.

1. Show that $\mathbf{Z}(G)$ is the union of all orbits of G upon A of size one, i.e., all orbits containing exactly one element.

2. The order n of G satisfies $n \equiv 0 \pmod{p}$. Working modulo p, by summing the various sizes, show that the number f of orbits of size one satisfies $f \equiv 0 \pmod{p}$, and $f \neq 0$.

Using these hints, write a complete proof of Lemma 8.49.

We have reached the point where we may prove one of the theorems of Sylow.

8.50   THEOREM   *If G is a finite group of order n, p is a prime, t is a positive integer, and p^t is a divisor of n, then G contains a subgroup of order p^t.*

This theorem proves both that G contains Sylow p-subgroups and that G contains elements of order p. The latter fact was first proved by Cauchy (1789–1857) and is often called Cauchy's Theorem. The proof of this theorem uses mathematical induction upon n. The following hints may be useful in constructing a proof of this theorem.

1. If H is any proper subgroup of G whose index is relatively prime to p, then H contains a subgroup of order p^t.

2. Suppose that the index of every proper subgroup of G is divisible by p. Then using Lemmas 8.47, 8.48, and 8.49, we may show that G contains a normal subgroup K of order p.

3. Prove that the quotient group of cosets G/K contains a subgroup of the form L/K of order p^{t-1}.

4. The group L has order p^t.

Complete the induction steps and the proof.

Among other things, this theorem tells us that G does contain Sylow p-subgroups, where p is a prime divisor of the order of G. We are now in a position to prove the main theorem of this section, Theorem 8.46. The following steps are suggested.

1. Let A be the set of Sylow p-subgroups of G. Why is the set A not empty?

2. If $P \in A$, then we may prove that P acts upon A by the action $P^a = a^{-1}Pa$ for $a \in P$.

3. If $Q \in A$, then let $B = \{Q^a \mid a \in P\}$. Show that the set B is a P-orbit.

4. Assume that P stabilizes Q. Show that $PQ = \{ab \mid a \in P, b \in Q\}$ is a group with normal subgroup Q. What is the order of PQ? (Maybe Theorem 2.94 applies here.)

5. Show that every orbit of P upon A has a size divisible by P except for one. Which is the exceptional orbit?

6. Count the size of A in two ways and compare your answers modulo p.

Complete the proof of Theorem 8.46.

Except for introducing the ideas of orbits and actions, this proof is very much like the one given by Sylow. He did not do it all alone; various bits and pieces had been proven before he came on the scene. He was the first one to put it all together for a "grand slam."

EXERCISES

The following exercises are intended to clarify certain steps in the proof.

1. What does Lemma 8.49 say about a group of order p^k for a prime p and a positive integer k?

2. Prove that any finite group which satisfies the hypotheses of Lemma 8.49 must have order p^k. Which group orders is Lemma 8.49 designed to cope with in disguised form?

3. Find a group for which $Z(G) = \{e\}$.

4. Why is Theorem 8.50 needed in the proof of Theorem 8.46?

5. Can you "improve" the congruence of Theorem 8.46? For example, if $P \cap Q = \{e\}$ for any two unequal Sylow p-subgroups of G, then is it true that $l \equiv 1 \pmod{p^2}$?

6. At step 4 in the proof of Theorem 8.46, exactly which subgroup of P is the stabilizer in P of Q? Knowing this, what is the best answer possible in Exercise 5?

 # COMMENTARY

1 After Galois, finite group theory slowly evolved as an independent branch of algebra. We have already remarked that Cauchy (1789–1857) and Sylow (1832–1918) were involved in giving conditions for the existence of prime power order subgroups of a group. During this time very few defined precisely what they meant by a group. Cayley (1821–1895), Dedekind (1831–1916), and Kronecker (1823–1891) gave general definitions which skirt the one used by us. The real difficulty is that everyone worked "in context." That is, a group theorist considered permutation groups, or transformation groups, but not abstract groups. Thus it was usually only necessary to state the closure property of group multiplication. All other properties followed from the context. By 1900, however, many sets of defining properties for abstract groups had been given. These definitions came at about the same time as axiomatic formulations did in other parts of mathematics. It was between 1880 and 1900 that the postulational development of mathematics became supreme.

But even in context, many general properties of groups were discovered. It was Jordan (1838–1922) in 1878 who showed the connection of groups of matrices with permutation groups, and in the hands of Schur (1875–1941) and Frobenius (1849–1917) "representation of groups by matrices" became a powerful tool. At the turn of the century, William Burnside (1852–1927) began his probing investigations of group theory. His work delineated a number of the major problems of group theory, some of which have been solved in the past decade.

Certainly any finite group G is generated by a finite number of elements all of finite order. Is the converse true? That is, if a group G is generated by elements $a_1, a_2, \ldots, a_n$ such that $x^m = e$ for any $x \in G$ and a positive integer m (dependent upon x), then must G have finite order? This conjecture comes under the general heading of the "Burnside problem," and only recently has it been shown that the answer to the question is "no."

Burnside proved that if a finite group G has composite order $p^a q^b$ for primes p and q and integers a and b, then G cannot be a simple group. His proof used the representation theory of Frobenius and Schur, and from his deliberations, he felt that if a finite group G has composite odd order then it cannot be a simple group. This "odd order" conjecture and the Burnside problem have provided a major impetus to current activity in finite group theory.

Dickson (1874–1954) in 1900 published a list of the then-known simple groups. Besides the alternating groups A_n for $n \geq 5$, he listed several other infinite "families" of finite simple groups along with five exceptional groups, called the Mathieu (1835–1890) groups, which fit into no known infinite family. It is interesting that all groups in his list have even order. The evolution of tools to tackle the "odd order" problem evolved slowly. Starting in the late 1930s, Richard Brauer (1901–   ) developed a representation theory over finite fields parallel to, but much more involved than, that of Frobenius and Schur. Phillip Hall gave criteria involving Sylow subgroups for a group not to be simple. Blichfeldt (1873–1945), Suzuki, and Brauer carried out deliberate studies of complex matrix groups of small degree.

The development of the representation theories of Schur, Frobenius, and Brauer came to fruition in the 1950s. At the same time, the methods of abstract group theory were reaching a peak of power. P. Hall and G. Higman in 1956 made several major reductions in the Burnside problem. About the same time, Suzuki was able to show that a large class of composite odd order groups could not be simple. It was in this atmosphere that John Thompson began his deliberations upon finite group theory. These centered around the nonsimplicity of composite odd order groups.

Through the efforts of M. Hall, Jr., W. Feit, and J. Thompson, Suzuki's results of the middle 1950s were greatly extended. Then in 1963, Feit and Thompson published a proof of Burnside's conjecture that composite odd order groups are not solvable (their paper runs over 250 pages; the result is considered so difficult that it probably could not be taught in a one-year course at the graduate level) [186]. A year later, two Russian mathematicians, E. S. Golod and I. R. Shafarevich, published a negative solution to the Burnside problem [45]. As is true of most difficult problems of mathematics, these solutions did not end matters, but rather, made much deeper questions meaningful. What are the finite simple groups? What further conditions in the Burnside problem will guarantee finiteness? If a group G satisfying the conditions of the problem is finite, what can we say about the order of G? Since 1964 both the "simple group classification problem" and questions related to the Burnside problem have received major attention, and these areas of group theory are among the most active [29,31,127].

2   Figure 2 shows a rather common type of puzzle. Fifteen numbered 1×1 square tiles are placed in a 4×4 box so that they slide easily. The object is to bring the tiles into the arrangement of Figure 3 by sliding tiles into the empty space. It is

2	3	4	1
6	7	5	9
10	11	12	8
14	15	13	

1	2	3	4
5	6	7	8
9	10	11	12
13	14	15	

FIGURE 2   FIGURE 3

natural to ask: "For which starting positions is it possible to bring the puzzle into the position of Figure 3?" In this commentary we will answer this question.

Number the positions as in Figure 3 giving the empty lower right-hand position the number 16. We may label any *alteration* of the positions by a permutation $\alpha \in S_{16}$, where $i\alpha = j$ means that the tile in position i is relocated into position j (naturally, the "missing tile," number 16, is treated like any other tile). If $\beta \in S_{16}$ is another alteration, then the alteration $\alpha\beta$ corresponds to the change effected by first performing α and then β. For any particular alteration α, the easiest way to effect the change would be to remove all tiles and replace them in their appropriate positions as given by α. The collection of alterations is just the group S_{16}.

The act of sliding a single tile into the empty space will be called a *move*. A sequence $\mathscr{A}$ of moves will start with the empty space in a position i and finish with it at j, and as such, $\mathscr{A}$ causes some alteration α. A sequence $\mathscr{B}$ of moves which starts the empty position at j and moves it to k will cause some alteration β. The combined sequence $\mathscr{A}\mathscr{B}$ will cause the alteration $\alpha\beta$. For any alteration resulting from a sequence of moves, the initial position of the empty space is important. For any product of alterations resulting from appended sequences of moves, e.g., $\alpha\beta$, the position of the empty space after α must coincide with the initial position before β. Therefore, in the following sequences of moves, we must watch the position of the empty space.

Suppose the position of the empty space is at i. Perform a sequence $\mathscr{C}$ of moves in which the empty space circulates but finally comes back to finish at i. The space can slide right-left and up-down. Since it starts and finishes at i, we must have as many up moves as down moves, and as many right moves as left moves (direction being the motion of the empty space). In other words, the sequence $\mathscr{C}$ must involve an even number $2n$ of moves. Each move is a transposition (st), where the empty space is at t and s is adjacent. Thus the alteration γ given by $\mathscr{C}$ is a product of $2n$ transpositions. Therefore, γ is even. We have proven the following.

THEOREM 1 *If the empty space is at position i and a sequence $\mathscr{C}$ of moves gives rise to an alteration γ such that $i\gamma = i$, then γ is an even permutation.*

The alteration from Figure 2 to Figure 3 is

$$\gamma = \begin{pmatrix} 1 & 2 & 3 & 4 & 5 & 6 & 7 & 8 & 9 & 10 & 11 & 12 & 13 & 14 & 15 & 16 \\ 2 & 3 & 4 & 1 & 6 & 7 & 5 & 9 & 10 & 11 & 12 & 8 & 14 & 15 & 13 & 16 \end{pmatrix}$$
$$= (1 \quad 2 \quad 3 \quad 4)(5 \quad 6 \quad 7)(8 \quad 9 \quad 10 \quad 11 \quad 12)(13 \quad 14 \quad 15).$$

This is an odd permutation such that $16\gamma = 16$. Therefore, *no sequence of moves will bring Figure 2 into Figure 3.* The fact that there are starting positions from which it is impossible to reach a desired end position created quite a stir among puzzlers when it was discovered during the nineteenth century.

Usually the empty space starts and finishes at position 16. The alterations with this property are the set of all permutations of S_{15}. The theorem tells us that the alterations which fix 16 and arise as sequences of moves all lie in A_{15}. In fact, they form a subgroup G of A_{15} since any sequence of moves which results in the empty space staying at 16 may be followed by another such sequence of moves. We shall explain now why $G = A_{15}$. In plain language, we will explain why, according to the rules of the puzzle, an alteration is "possible" if and only if it involves an even permutation of the numbers 1, 2, ..., 15.

Choose any thirteen $a_1, a_2, ..., a_{11}, a_{13}, a_{14}$ (the a's are the starting positions of the chosen tiles, and a_{12} has been omitted to simplify notation) of the fifteen tiles. We shall argue that through a sequence of moves we may place a_1 in position 1, a_2 in position 2, ..., a_i in position i, ... Observe that we may place a_1 adjacent to a_2 via a sequence of moves that start and finish with the empty space in position 16. Keeping these two adjacent, we may then jockey them into positions 1 and 2 through a similar sequence of moves. Nail down tiles a_1 and a_2 in positions 1 and 2. Repeat the process for a_3 and a_4 for positions 3 and 4 then nail them down. Continue this process for a_5, a_6, then a_7, a_8. As we nail tiles down, we lose a certain amount of maneuverability. Consequently, we must now change our strategy and use the pairs a_9, a_{13}, moving them into positions 9 and 13, followed by a_{10}, and a_{14}. The remaining "unchosen" tiles a_{12}, a_{15} will be in positions 12 and 15, and there will be no choice as to which is which, as we shall see below. This method of arranging the tiles works and is easy to execute, but is rather complicated to describe.

The heuristic argument outlined above shows that G contains a permutation

$$\sigma = \begin{pmatrix} 1 & 2 & 3 & 4 & 5 & 6 & 7 & 8 & 9 & 10 & 11 & 12 & 13 & 14 & 15 & 16 \\ a_1 & a_2 & a_3 & a_4 & a_5 & a_6 & a_7 & a_8 & a_9 & a_{10} & a_{11} & x & a_{13} & a_{14} & x & 16 \end{pmatrix}.$$

There are two possible ways to insert a_{12}, a_{15}, the remaining "unchosen" tile values, in for the x's. One gives a permutation σ; the other gives the permutation $\sigma' = \sigma(a_{12}a_{15})$. Exactly one of σ, σ' is even; the other is odd. If σ is even, it lies in G; otherwise not.

THEOREM 2 *If $\gamma \in A_{15}$ and the empty space is at 16, then the alteration γ may be affected by some sequence $\mathscr{C}$ of moves.*

These two theorems tell the whole story. In fact, the situation is the same for a puzzle of size $m \times n$, where $m \geq 2$ and $n \geq 2$. Any alteration which lies in A_{mn-1} and no others can be achieved via a sequence of moves.

3 In the Commentary to Chapter 2, we listed quite a few references on group theory. We shall single out some of these for their coverage of the material in this chapter. Herstein [12] gives three different proofs of the Sylow theorems. Proofs of the simplicity of A_n are given in Refs. [24,31,38]. Finally, the group of units U of Z_n is discussed in Refs. [25,79].

A table of the number of nonisomorphic groups of order n for $n \leq 20$ is given on page 52 of Hall [31]. The largest number, namely 14, occurs for $n = 16 = 2^4$. It is expected that there should be a large number of nonisomorphic groups of any given prime power order. On page 96, Rotman [38] lists all nonisomorphic groups of order $n \leq 15$. These lists can be used to gain some insight into the meaning of the Sylow theorems. More general theorems along this line are given in MacDonald [36].

A list (complete up to 1968) of all known finite simple groups is given on page 491 of Gorenstein [29]. One must be familiar with current notation in order to use the table.

The group U of units of Z_n occurs as the Galois group of the cyclotomic extension field of Q by a primitive nth root of unity. For definitions of the words just used and for an exposition of this fact, see Refs. [9,24,51].

chapter nine

FINITE ABELIAN GROUPS

The problem of determining all finite groups is a difficult and, in fact, an unsolved problem. However, in a sense to be made precise later, it is possible to determine all finite *abelian* groups. The purpose of this chapter is to prove the fundamental results in the theory of such groups. We shall make no use of results in Chapter 8.

Let G be a finite abelian group. Throughout this chapter we shall use addition as the operation in G. Of course, everything could just as well be stated in terms of multiplication as the operation. Unless otherwise explicitly stated, we shall always assume that the group G under discussion is a nonzero group, that is, that it does not consist of the identity alone.

Let us recall the following essential facts which will be used frequently in the sequel. Since G has finite order, each element a of G has finite order. If a has order n, Theorem 2.45 shows that n is the least positive integer such that $na = 0$. The order of the zero element is one, all other elements have order greater than one. If a has order n and $k \in \mathbf{Z}$, then $ka = 0$ if and only if $n \mid k$.

9.1 DIRECT SUMS OF SUBGROUPS

If $G_1, G_2, \ldots, G_r$ are subgroups of the abelian group G, we define the *sum*

$$G_1 + G_2 + \cdots + G_r$$

of these subgroups to be the set of all elements of G which can be expressed in the form

$$a_1 + a_2 + \cdots + a_r, \qquad a_i \in G_i (i = 1, 2, \ldots, r).$$

This set is seen to be a subgroup of G, and each G_i is contained in this subgroup since the identity 0 of G is an element of each G_i. Actually, this sum is the smallest subgroup of G which contains all the subgroups G_i.

We now make the following definition.

9.1   **DEFINITION**   If $G_i (i = 1, 2, \ldots, r)$ are subgroups of the abelian group G, the sum $G_1 + G_2 + \cdots + G_r$ is said to be a *direct sum* if and only if the following condition is satisfied:

(i) If $a_i \in G_i (i = 1, 2, \ldots, r)$ such that

$$a_1 + a_2 + \cdots + a_r = 0,$$

then each $a_i = 0$.

We shall indicate that a sum $G_1 + G_2 + \cdots + G_r$ is a direct sum by writing it in the form

$$G_1 \oplus G_2 \oplus \cdots \oplus G_r.$$

It is worth pointing out that condition (i) is equivalent to the following condition:

(ii) If $a_i, b_i \in G(i = 1, 2, \ldots, r)$ such that

$$a_1 + a_2 + \cdots + a_r = b_1 + b_2 + \cdots + b_r,$$

then $a_i = b_i (i = 1, 2, \ldots, r)$.

The equivalence of (i) and (ii) follows readily from the fact that the equation $a_1 + a_2 + \cdots + a_r = b_1 + b_2 + \cdots + b_r$ may be written in the form

$$(a_1 - b_1) + (a_2 - b_2) + \cdots + (a_r - b_r) = 0.$$

We leave the details of the proof of the equivalence of conditions (i) and (ii) as an exercise.

The condition (ii) for a sum $G_1 + G_2 + \cdots + G_r$ to be a direct sum is often expressed by saying that the sum is direct if and only if each element of the sum is *uniquely* expressible in the form

$$a_1 + a_2 + \cdots + a_r, \qquad a_i \in G_i (i = 1, 2, \ldots, r).$$

If G_i has order n_i, we see that in a sum of the type just written there are n_i choices for a_i; and the uniqueness property just mentioned shows that the order of a direct sum $G_1 \oplus G_2 \oplus \cdots \oplus G_r$ is the product $n_1 n_2 \cdots n_r$ of the orders of the respective subgroups G_i.

Since addition is a commutative operation in G, it is clear, for example, that $G_1 \oplus G_2 = G_2 \oplus G_1$. In general, the order in which the subgroups G_i are written in the symbol for their direct sum is immaterial.

As a simple illustration of a general property, suppose that G_1 and G_2 are subgroups of G such that $G = G_1 \oplus G_2$. Now if $G_1 = H_1 \oplus H_2$ and $G_2 = K_1 \oplus K_2$, where H_1 and H_2 are subgroups of G_1, and K_1 and K_2 are subgroups of G_2, then all of H_1, H_2, K_1, K_2 are subgroups of G and

$$G = H_1 \oplus H_2 \oplus K_1 \oplus K_2.$$

It will be clear that a similar result holds for any number of summands. (See Exercise 3 below.) This fact will be useful later on.

Remark To avoid any possible confusion, let us point out that if $G_1, G_2, \ldots, G_r$ are *any* additively written abelian groups (not necessarily subgroups of a given group), according to Section 2.4 the direct sum of these groups would consist of all ordered r-tuples

$$(a_1, a_2, \ldots, a_r),$$

with $a_i \in G_i$ for $i = 1, 2, \ldots, r$, and with addition defined as follows:

$$(a_1, a_2, \ldots, a_r) + (b_1, b_2, \ldots, b_r) = (a_1 + b_1, a_2 + b_2, \ldots, a_r + b_r).$$

Now if the G_i are subgroups of a group G and their sum is direct as defined in 9.1, it is not difficult to prove that the mapping

$$(a_1, a_2, \ldots, a_r) \rightarrow a_1 + a_2 + \cdots + a_r$$

is an isomorphism of the direct sum as defined in Section 2.4 onto the direct sum as defined in this section. Prove it! This fact justi-

fies the use of the term *direct sum* in two different situations. Sometimes, the direct sum as defined in 9.1 is called an *internal* direct
sum (since all groups G_i are subgroups of a given group G and therefore their direct sum is a subgroup of G), and the direct sum of
Section 2.4 is called an *external* direct sum (since the G_i are arbitrary
groups, not given as subgroups of some given group).

Now let G be an abelian group of order $n = p_1{}^{e_1}p_2{}^{e_2} \cdots p_k{}^{e_k}$, where the
p's are distinct primes, $k \geq 1$, and each $e_i \geq 1$. Thus $p_1, p_2, \ldots, p_k$ are the distinct prime divisors of n. Let $G(p_i)$ be the set of all elements of G having order
a power of p_i. The order of the identity 0 of G is $1 = p_i{}^0$ and hence $0 \in G(p_i)$.
Actually, $G(p_i)$ is a subgroup of G since one can see as follows that $G(p_i)$ is closed
under addition. If $a, b \in G(p_i)$, suppose that a has order $p_i{}^m$ and b has order $p_i{}^n$.
If t is the larger of n and m, then $p_i{}^t(a + b) = 0$, and the order of $a + b$ is a divisor
of $p_i{}^t$; hence is a power of p_i. It follows that $a + b \in G(p_i)$, and by Theorem
2.30 we see that $G(p_i)$ is a subgroup of G. Our next goal is to prove the following
theorem.

9.2   THEOREM   *Let G be an abelian group of order n, and let $p_1, p_2, \ldots,$
p_k be the distinct prime divisors of n. If $G(p_i)$ is the subgroup of G
consisting of all elements having order a power of p_i, then*

9.3 $$G = G(p_1) \oplus G(p_2) \oplus \cdots \oplus G(p_k).$$

Much later in this chapter we shall prove that no one of the subgroups
$G(p_i)$ consists of the zero alone. In fact, if $p_i{}^{e_i}$ is the highest power of p_i which
divides n, we shall show that $G(p_i)$ has order $p_i{}^{e_i}$.

Before proving Theorem 9.2, let us introduce some lemmas that will be
helpful in carrying out the proof. It should be noted that in these lemmas n is
temporarily being used to denote the order of an element, not the order of the
group.

9.4   LEMMA   *Suppose that the element a of an abelian group G has
order n. If m is an integer such that $(m, n) = 1$, then $ma = 0$ implies
that $a = 0$.*

Proof: Since $(m, n) = 1$, there exist integers x and y such that
$1 = xm + yn$. Hence $a = xma + yna$. We are assuming that
$ma = 0$, and $na = 0$ since a has order n. It follows that $a = 0$, as
we wished to show.

9.5   LEMMA   *If the element a of the abelian group G has order $n = kl$
with $(k, l) = 1$, then there exist elements b and c of G such that $a =
b + c$, with b and c having respective orders k and l.*

Proof: Since $(k, l) = 1$, there exist integers s and t such that $1 = sk + tl$. Thus we have $a = ska + tla$. Let us show that ska has order l. Clearly, $lska = sna = 0$. Moreover, if $z \in \mathbf{Z}$ such that $zska = 0$, then $n \mid zsk$. But $n = kl$, so we conclude that $kl \mid zsk$ or $l \mid zs$. Now the equation $1 = sk + tl$ implies that $(s, l) = 1$ and therefore $l \mid z$. Accordingly, we conclude that ska has order l. Similarly, tla has order k and if we set $b = tla$ and $c = ska$, we have $a = b + c$, b of order k and c of order l. This completes the proof.

We leave as an exercise the proof by induction of the following generalization of the preceding lemma.

9.6   **LEMMA**   *If the element a of the abelian group G has order $n = n_1 n_2 \cdots n_k$, where $(n_i, n_j) = 1$ for $i \neq j$, then a can be expressed in the form*

$$a = b_1 + b_2 + \cdots + b_k,$$

where b_i has order n_i $(i = 1, 2, \ldots, k)$.

Let us now return to the proof of Theorem 9.2 and show first that the sum $G(p_1) + G(p_2) + \cdots + G(p_k)$ is a direct sum. To this end, suppose that

9.7 $a_1 + a_2 + \cdots + a_k = 0$ $a_i \in G(p_i)$.

By definition of direct sum, we need to prove that each $a_i = 0$. For convenience of notation, let us concentrate on proving that $a_1 = 0$. Each a_i has order a power of p_i, so let us assume that a_i has order $p_i^{t_i}$ $(i = 1, 2, \ldots, k)$. From Equation 9.7, it now follows that

$$p_2^{t_2} p_3^{t_3} \cdots p_k^{t_k} a_1 = 0.$$

Since this coefficient of a_1 is relatively prime to the order of a_1, it follows from Lemma 9.4 that $a_1 = 0$. Similarly, each $a_i = 0$, and this proves that the sum is direct.

Clearly, $G(p_1) \oplus G(p_2) \oplus \cdots \oplus G(p_k) \subseteq G$, so we only need to obtain inclusion the other way. By Corollary 2.53, every element a of G has order a divisor of the order n of G, and therefore the order of a has no prime divisors except for some or all of the $p_i (i = 1, 2, \ldots, k)$. For convenience of notation only, suppose that the order of a contains only the prime factors $p_1, p_2, \ldots, p_u, u \leq k$. By Lemma 9.6, a is expressible as a sum of elements of $G(p_i)$, $i = 1, 2, \ldots, u$. In particular, every element of G is a sum of elements of some or all of the $G(p_i)$, $i = 1, 2, \ldots, k$. We therefore conclude that $G \subseteq G(p_1) \oplus G(p_2) \oplus \cdots \oplus G(p_k)$, and this completes the proof of the theorem.

9.2 CYCLIC SUBGROUPS AND BASES

We shall continue to let G be a finite abelian group. If $a \in G$, let us denote by (a) the cyclic subgroup generated by a. If a has order n, then

$$(a) = \{0, a, 2a, \ldots, (n-1)a\}.$$

9.8   **DEFINITION**   If $a_1, a_2, \ldots, a_k$ are nonzero elements of G such that the sum $(a_1) + (a_2) + \cdots + (a_k)$ is direct, we say that the elements $a_1, a_2, \ldots, a_k$ are *independent* or form an *independent set*.

Suppose that a_i has order $n_i (i = 1, 2, \ldots, k)$. Then, by definition of direct sum, the $a_i (i = 1, 2, \ldots, k)$ are independent if for integers z_i,

$$z_1 a_1 + z_2 a_2 + \cdots + z_k a_k = 0$$

if and only if $z_i a_i = 0$, that is, if and only if $n_i \,|\, z_i \ (i = 1, 2, \ldots, k)$.

Observe that a single element a of G is independent if and only if $a \neq 0$. Clearly, any nonempty subset of an independent set is also independent.

9.9   **DEFINITION**   The set $\{a_1, a_2, \ldots, a_r\}$ forms a *basis* of the abelian group G if and only if the elements of this set are independent and

$$G = (a_1) \oplus (a_2) \oplus \cdots \oplus (a_r).$$

Otherwise expressed, *the group G has a basis if and only if it is expressible as a direct sum of a finite number of cyclic subgroups.*

We may remark that if $G = H_1 \oplus H_2$, where H_1 is a subgroup of G having basis $\{b_1, b_2, \ldots, b_s\}$ and H_2 is a subgroup of G having basis $\{c_1, c_2, \ldots, c_t\}$; then G has a basis $\{b_1, b_2, \ldots, b_s, c_1, c_2, \ldots, c_t\}$. (See Exercise 6 below.)

One of the principal theorems which we shall eventually prove is the following.

9.10   **THEOREM**   *Every finite abelian group has a basis, each element of which has order a power of a prime.*

In view of our definitions, an equivalent formulation of this theorem would be the assertion that *every finite abelian group can be expressed as the direct sum of cyclic subgroups, each of which has order a power of a prime.*

By a generalization of the remark made above, the result of Theorem 9.2 shows that Theorem 9.10 will be true in general when we have established it for each of the groups $G(p_i)$. In the next section we study in some detail a class of groups which will include those of the form $G(p_i)$ as defined in Theorem 9.2.

EXERCISES

1. Prove the equivalence of conditions (i) and (ii) in connection with Definition 9.1.

2. Suppose that $G_i(i = 1, 2, \ldots, r)$ are subgroups of the abelian group G such that the sum $G_1 + G_2 + \cdots + G_r$ is direct. If, for each i, H_i is a subgroup of G_i, prove that the sum $H_1 + H_2 + \cdots + H_r$ is direct.

3. Suppose that $G = G_1 \oplus G_2$, where $G_1 = H_1 \oplus H_2 \oplus H_3$ and $G_2 = K_1 \oplus K_2$. Prove that $G = H_1 \oplus H_2 \oplus H_3 \oplus K_1 \oplus K_2$. Choose an appropriate notation and generalize to an arbitrary finite number of summands.

4. Let G be the additive group of the ring $\mathbf{Z}_{24}$. In the notation of Theorem 9.2, determine the elements of $G(2)$ and of $G(3)$. Verify that these are subgroups of G and that $G = G(2) \oplus G(3)$, thus directly verifying Theorem 9.2 for this particular group.

5. Give an example to show that in a finite nonabelian group the elements which have order a power of some fixed prime need not be a subgroup.

6. Suppose that G_1 and G_2 are subgroups of the abelian group G such that $G = G_1 \oplus G_2$. If $\{a_1, a_2, \ldots, a_r\}$ is a basis of G_1 and $\{b_1, b_2, \ldots, b_s\}$ is a basis of G_2, prove that $\{a_1, a_2, \ldots, a_r, b_1, b_2, \ldots, b_s\}$ is a basis of G. Generalize to direct sums of an arbitrary finite number of subgroups.

7. Illustrate Theorem 9.10 by verifying that for the group G which is the additive group of the ring $\mathbf{Z}_{24}$, a basis of the required kind is $\{3, 8\}$.

8. Prove Lemma 9.6.

9. Prove that a cyclic group of order p^k, where p is a prime and $k \geq 1$, cannot be expressed as a direct sum of two nonzero subgroups. *Hint:* Consider the maximal order that an element can have.

10. If (a) is a cyclic group of order kl with $(k, l) = 1$, prove that there exist elements b and c of (a) of respective orders k and l, such that $(a) = (b) \oplus (c)$.

11. If b and c are elements of an abelian group G with orders k and l respectively, and if $(k, l) = 1$, prove that the sum $(b) + (c)$ is direct and that $(b) \oplus (c)$ is a cyclic subgroup of G of order kl.

9.3 FINITE ABELIAN p-GROUPS

Let us begin with the following definition.

9.11 DEFINITION Let p be a fixed prime. A group is said to be a *p-group* if the order of each of its elements is a power of p.

We may observe that the identity element (the zero) has order p^0. Every other element of a p-group has order p^m for some positive integer m. Thus a nonzero element a of a p-group has order p^m if and only if $p^m a = 0$, $p^{m-1}a \neq 0$. Moreover, if a has order p^m and the order of an element b of the p-group is less than or equal to the order of a, then $p^m b = 0$.

The main goal of this section is to prove the following special case of Theorem 9.10.

9.12 LEMMA *A finite abelian p-group has a basis.*

Throughout this section, let p be a fixed prime and G a finite abelian p-group. As a first step in the proof of Lemma 9.12, we collect a few useful facts for easy reference.

Let H be a subgroup of G and suppose that a is an element of G of order p^m. Since $p^m a = 0 \in H$, there exists a smallest positive integer z (necessarily less than or equal to p^m) such that $za \in H$. Throughout this section it will be convenient to have a distinctive name for this positive integer z. We shall call it the *degree of a relative to H*. Thus, the order of a is the degree of a relative to the zero subgroup.

(A) (i) *If z is the degree of a relative to H and $n \in \mathbf{Z}$, then $na \in H$ if and only if $z \mid n$. In particular, if z is the degree of a relative to H and $za = 0$, then z is the order of a.*

   (ii) *If a has order p^m, and z is the degree of a relative to H, then $z \mid p^m$ and therefore the degree of each element of G relative to any subgroup of G is a power of p.*

Proof: To prove (i), let us use the Division Algorithm to write $n = qz + r$, where $0 \leq r < z$. Thus $na = q(za) + ra$. Now $za \in H$, so if $na \in H$, it follows that $ra \in H$. Since z is the smallest positive integer such that $za \in H$, we conclude that $r = 0$ and therefore $n = qz$. Conversely, if $z \mid n$, it is trivial that $na \in H$. The last statement of (i) follows from the observation that the given conditions imply that z and the order of a divide each other.

Part (ii) follows from (i) by observing that $p^m a = 0 \in H$, and therefore $z \mid p^m$ and z must therefore be a power of p.

(B) *Suppose that H is a subgroup of G and that $a \notin H$. If the order p^m of a is equal to the degree of a relative to H, then the sum $H + (a)$ is direct.*

Proof: To see the truth of this statement, suppose that $h + xa = 0$, where $h \in H$ and $x \in \mathbf{Z}$, and let us prove that $h = 0$ and that $xa = 0$. Now $xa \in H$ and since a has degree p^m relative to H, part (i) of (A) shows that $p^m \mid x$. But this implies that $xa = 0$ since p^m is also the order of a. The equation $h + xa = 0$ then shows that $h = 0$. The sum $H + (a)$ is therefore a direct sum, as we wished to show.

The proof of Lemma 9.12 is carried out by the process of induction. We illustrate the approach by a fairly detailed account of the first two steps in this procedure, and then pass on to the general situation.

Proof: Let a_1 be an element of G of maximum order, say p^{m_1}. If it happens that $G = (a_1)$, we have found a basis $\{a_1\}$ of G consisting of the single element a_1. Suppose, then, that $G \neq (a_1)$. Since a_1 has maximum order among the elements of G, we see that $p^{m_1} c = 0$ for *every* element c of G.

Since $G \neq (a_1)$, we proceed to seek an element a_2 of G such that a_1, a_2 are independent and therefore $(a_1) \oplus (a_2) \subseteq G$.

Let b be an element of G of maximum degree, say p^{m_2}, relative to (a_1). Since $p^{m_1} b = 0 \in (a_1)$, it follows that $p^{m_2} \leq p^{m_1}$, i.e., $m_2 \leq m_1$. Since b has maximum degree relative to (a_1), we see that $p^{m_2} c \in (a_1)$ for every element c of G. Thus if $c \in G$, there exists $y \in \mathbf{Z}$ such that

9.13
$$p^{m_2} c = y a_1.$$

We proceed to show that $p^{m_2} \mid y$. Multiplying the preceding equation by $p^{m_1 - m_2}$, we find that $p^{m_1} c = p^{m_1 - m_2} y a_1$. But $p^{m_1} c = 0$, and since a_1 has order p^{m_1}, we conclude that $p^{m_1} \mid p^{m_1 - m_2} y$, i.e., that $p^{m_2} \mid y$. Thus $y = u p^{m_2}$, where $u \in \mathbf{Z}$. Now applying what we have just proved to the special case in which c is an element b of maximum degree relative to (a_1), we find from 9.13 that there exists $u \in \mathbf{Z}$ such that

9.14
$$p^{m_2} b = p^{m_2} u a_1.$$

We now set

9.15
$$a_2 = b - u a_1,$$

and observe that if $z \in \mathbf{Z}$, then $za_2 \in (a_1)$ if and only if $zb \in (a_1)$. Thus the degree of a_2 relative to (a_1) is p^{m_2}. Now 9.14 and 9.15 imply that $p^{m_2}a_2 = 0$, and it follows from (A) (i) that the order of a_2 is p^{m_2}. We now know from (B) that the sum $(a_1) + (a_2)$ is a direct sum, and we have $G \supseteq (a_1) \oplus (a_2)$. If $G = (a_1) \oplus (a_2)$, we have exhibited a basis $\{a_1, a_2\}$. Otherwise, we can continue this process. We have just completed the case $k = 2$ of the following induction procedure.

Assume that we have found elements $a_1, a_2, \ldots, a_k$ of G of respective orders $p^{m_1}, p^{m_2}, \ldots, p^{m_k}$ such that all of the following are true:

(i) $m_1 \geq m_2 \geq \cdots \geq m_k$.

(ii) *If $c \in G$, there exist integers $y_1, \ldots, y_{k-1}$ such that $p^{m_k}c = y_1 a_1 + \cdots + y_{k-1}a_{k-1}$, and $p^{m_k} | y_i (i = 1, 2, \ldots, k-1)$.*

(iii) *$a_1, a_2, \ldots, a_k$ are independent.*

For convenience, let us set $G_{k-1} = (a_1) \oplus (a_2) \oplus \cdots \oplus (a_{k-1})$ and $G_k = (a_1) \oplus (a_2) \oplus \cdots \oplus (a_k)$. If $G \neq G_k$, we propose to find another element a_{k+1} of G such that all of the above three properties are true with k replaced by $k + 1$.

We may point out that (ii) above states not only that if $c \in G$, then $p^{m_k}c \in G_{k-1}$, but even gives some additional information in that the integers y_i are all divisible by p^{m_k}.

Let b be an element of G of maximum degree, say $p^{m_{k+1}}$, relative to G_k. Thus, if $c \in G$, then $p^{m_{k+1}}c \in G_k$. Observe that since $p^{m_k}b \in G_{k-1} \subseteq G_k$, $p^{m_{k+1}} \leq p^{m_k}$ and $m_k \geq m_{k+1}$, so (i) holds for $k + 1$.

If $c \in G$, since $p^{m_{k+1}}c \in G_k$, then there exist integers $z_1, \ldots, z_k$ such that

9.16 $$p^{m_{k+1}}c = z_1 a_1 + z_2 a_2 + \cdots + z_k a_k.$$

We propose to show that $p^{m_{k+1}} | z_i$ for $i = 1, 2, \ldots, k$. By (ii) of our induction hypothesis, there exist integers $y_1, \ldots, y_{k-1}$ such that

9.17 $$p^{m_k}c = y_1 a_1 + \cdots + y_{k-1}a_{k-1}, \qquad p^{m_k} | y_i \ (i = 1, 2, \ldots, k).$$

If we multiply 9.16 by $p^{m_k - m_{k+1}}$, we obtain $p^{m_k}c = p^{m_k - m_{k+1}}z_1 a_1 + \cdots + p^{m_k - m_{k+1}}z_k a_k$. By equating the right sides of the two preceding equations, we obtain

$$(p^{m_k - m_{k+1}}z_1 - y_1)a_1 + \cdots + (p^{m_k - m_{k+1}}z_{k-1} - y_{k-1})a_{k-1}$$
$$+ p^{m_k - m_{k+1}}z_k a_k = 0.$$

Since, by assumption, $a_1, a_2, \ldots, a_k$ are independent, for each i the coefficient of a_i in this equation must be divisible by the order of a_i. In particular, $p^{m_k} | p^{m_k - m_{k+1}} z_k$ and this implies that $p^{m_{k+1}} | z_k$. Now for $1 \le i < k$, we have $p^{m_i} | (p^{m_k - m_{k+1}} z_i - y_i)$. But $i < k$ and so $m_i \ge m_k$, and thus

$$p^{m_k} | (p^{m_k - m_{k+1}} z_i - y_i).$$

But, as indicated in 9.17, we know that $p^{m_k} | y_i$. It follows that $p^{m_k} | p^{m_k - m_{k+1}} z_i$ and hence that $p^{m_{k+1}} | z_i$. Since this is true for $1 \le i < k$, and we have already shown that $p^{m_{k+1}} | z_k$, we conclude that $p^{m_{k+1}}$ divides every z_i in 9.16. This establishes part (ii) of our induction statement for the case in which k is replaced by $k + 1$.

Now let us apply what we have just proved to the special case in which the element c in 9.16 is chosen to be a particular element b of maximal degree $p^{m_{k+1}}$ relative to G_k. Thus, since $p^{m_{k+1}} | z_i$, there exist integers $u_1, \ldots, u_k$ such that

9.18 $$p^{m_{k+1}} b = p^{m_{k+1}} u_1 a_1 + p^{m_{k+1}} u_2 a_2 + \cdots + p^{m_{k+1}} u_k a_k.$$

We next define

9.19 $$a_{k+1} = b - u_1 a_1 - \cdots - u_k a_k,$$

and observe from this equation that a_{k+1} has the same degree relative to G_k as does b, namely, $p^{m_{k+1}}$. The last two equations show that $p^{m_{k+1}} a_{k+1} = 0$ and it follows from (A)(i) that the order of a_{k+1} is $p^{m_{k+1}}$. By (B), the sum $G_k + (a_{k+1})$ is direct and it follows that $a_1, a_2, \ldots, a_{k+1}$ are independent. We have shown that if $a_1, a_2, \ldots, a_k$ satisfy the induction hypotheses (i), (ii), and (iii), and if $G \ne (a_1) \oplus \cdots \oplus (a_k)$, there exists an element a_{k+1} such that $a_1, \ldots, a_{k+1}$ satisfy (i), (ii), and (iii), with k replaced by $k + 1$. In particular, $G \supseteq (a_1) \oplus \cdots \oplus (a_{k+1})$. Since G is assumed to be a *finite* p-group, these steps must come to an end, and thus for some positive integer r, there exist elements $a_1, a_2, \ldots, a_r$ such that

9.20 $$G = (a_1) \oplus (a_2) \oplus \cdots \oplus (a_r).$$

This completes the proof of Lemma 9.12 which states that every finite abelian p-group has a basis.

In proving 9.20, we have obtained the basis elements $a_1, a_2, \ldots, a_r$ such that their orders are, respectively, $p^{m_1}, p^{m_2}, \ldots, p^{m_r}$ with $m_1 \ge m_2 \ge \cdots \ge m_r \ge 1$.

It follows from 9.20 that the order of G is the product of the orders of the cyclic groups (a_i), namely p^t, where $t = m_1 + m_2 + \cdots + m_r$. In particular, this shows that *the order of a finite abelian p-group is a power of p.*

Before returning to the study of arbitrary abelian groups, let us discuss the question of the uniqueness of a basis for a p-group. Clearly, our construction of a basis indicates that a basis is *not* unique since, as a simple example, a_1 might have been chosen to be any element of maximal order. However, we shall prove the following result.

9.21   THEOREM   *Any two bases of a finite abelian p-group have the same number of elements. Moreover, the orders of the elements of one basis coincide, in some arrangement, with the orders of the elements of any other basis.*

Proof: In proving this theorem, we shall assume that the p-group G has a basis $\{a_1, a_2, \ldots, a_r\}$, with a_i having order p^{m_i}; and that G also has a basis $\{b_1, b_2, \ldots, b_s\}$, with b_i having order p^{n_i}. Moreover, we assume that the notation is chosen so that $m_1 \geq m_2 \geq \cdots \geq m_r \geq 1$ and $n_1 \geq n_2 \geq \cdots \geq n_s \geq 1$. We proceed to prove that $r = s$ and that $m_i = n_i (i = 1, 2, \ldots, r)$.

It will be convenient to consider subgroups pG and G_p of G, defined as follows:

$$pG = \{px \mid x \in G\}$$

and

$$G_p = \{x \mid x \in G, px = 0\}.$$

Thus $pG = \{0\}$ if and only if $G_p = G$.

Making use of the basis $\{a_1, a_2, \ldots, a_r\}$ of G, we leave it to the reader to verify that

9.22
$$\{p^{m_1-1}a_1, p^{m_2-1}a_2, \ldots, p^{m_r-1}a_r\}$$

is a basis of the p-group G_p. Since each of these basis elements has order p and G_p is the direct sum of the cyclic groups generated by these elements, we conclude that G_p has order p^r. In exactly the same way, using the basis $\{b_1, b_2, \ldots, b_s\}$ of G, we see that G_p has order p^s. Hence $p^r = p^s$, and $r = s$. This completes the proof of the first statement of the theorem.

The proof of the second statement is by induction on the order of G, and we therefore assume as an induction hypothesis that the statement is true for all

p-groups with order less than the order of G.   We now make two cases, in the first of which we do not really need this induction hypothesis.

Case 1 $pG = \{0\}$. In this case, every nonzero element (in particular, every basis element) of G has order p.   Hence, $m_i = n_i = 1$ ($i = 1, 2, \ldots, r$).

Case 2 $pG \neq \{0\}$.   In this case, pG is a nonzero subgroup of G. Moreover, the order of pG is less than the order of G, since G necessarily has some elements of order p.   Why?   Using the notation in which the order of a_i is p^{m_i}, not all m_i can equal 1.   Suppose that u is a positive integer so chosen that $m_1 \geq m_2 \geq \cdots \geq m_u > m_{u+1} = \cdots = m_r = 1$.   It may now be verified that pG has a basis

9.23 $\{pa_1, pa_2, \ldots, pa_u\}.$

In like manner, making use of the other given basis $\{b_1, b_2, \ldots, b_r\}$ of G, if v is the positive integer so chosen that $n_1 \geq n_2 \geq \cdots \geq n_v > n_{v+1} = \cdots = n_r = 1$, we see that pG has a basis

9.24 $\{pb_1, pb_2, \ldots, pb_v\}.$

Thus the p-group pG has bases 9.23 and 9.24.   By the first statement of the theorem, already proved, we conclude that $u = v$.   Using the fact that the order of pa_i is p^{m_i-1} and the order of pb_i is p^{n_i-1}, the induction hypothesis as applied to the group pG shows that $m_i - 1 = n_i - 1$ for $i = 1, 2, \ldots, u$.   Since all other m_i and n_i are equal to 1, we have that $m_i = n_i (i = 1, 2, \ldots, r)$.   This concludes the proof of the theorem.

9.4   THE PRINCIPAL THEOREMS FOR FINITE ABELIAN GROUPS

Henceforth, we shall let G be an arbitrary finite abelian group.   Let us assume that G has order n with distinct prime factors $p_1, p_2, \ldots, p_k$.   Thus

9.25 $n = p_1^{e_1} p_2^{e_2} \cdots p_k^{e_k},$

where $e_i > 0$ for all i.   If $G(p_i)$ denotes the subgroup of G consisting of all elements of order a power of p_i, we have proved in Theorem 9.2 that

9.26 $G = G(p_1) \oplus \cdots \oplus G(p_k).$

Now $G(p_i)$ is a p_i-group and as indicated shortly before the statement of Theorem 9.21, its order is a power of p_i. Moreover, from Equation 9.26, the order n of G must be the product of the orders of the groups $G(p_i)$. In view of the unique factorization of n into a product of primes, we conclude from 9.25 that the order of $G(p_i)$ must be $p_i^{e_i} (i = 1, 2, \ldots, k)$. We have therefore proved the first statement of the following lemma.

9.27 LEMMA *Let G be an abelian group of order n.*

> (i) *If p is a prime divisor of n, let p^e be the highest power of p which divides n. Then the subgroup $G(p)$ of G, which consists of all elements with order a power of p, has order p^e. In particular, $G(p) \neq \{0\}$.*

> (ii) *If p is a prime divisor of n, then G contains an element of order p.*

The proof of part (ii) follows at once from the observation that if a is a nonzero element of $G(p)$, then a has order p^t for some positive integer t. Hence $p^{t-1}a$ has order p.

The results of the preceding section show that each subgroup $G(p_i)$ occurring in 9.26 has a basis, say, $\{a_{i1}, a_{i2}, \ldots, a_{ir_i}\}$, and clearly each element of this basis has order a power of p_i. Using this information, Equation 9.26 shows that G has a basis

9.28 $\{a_{11}, a_{12}, \ldots, a_{1r_1}; a_{21}, a_{22}, \ldots, a_{2r_2}; a_{k1}, a_{k2}, \ldots, a_{kr_k}\},$

and each element of this basis has order a power of a prime. This result was stated as Theorem 9.10, one of our principal goals. We have therefore proved part (i) of the following fundamental theorem.

9.29 FUNDAMENTAL THEOREM ON FINITE ABELIAN GROUPS
(i) *Every finite abelian group G has a basis, each element of which has order a power of a prime.*

(ii) *Suppose we have any two bases of a finite abelian group G, with each basis element having a power of a prime. Then the two bases have the same number of elements and the orders of the elements of one basis are, in some arrangement, the same as the orders of the elements of the other basis.*

Proof: To prove part (ii) of this theorem, suppose that one basis of G is given by 9.28. If G has order n, given by 9.25, the order of every element of G is a divisor of n, and hence the only possible primes a power of which can occur as the order of any (basis) element are $p_1, p_2, \ldots, p_k$. Suppose that in a second basis of G, the elements $b_1, \ldots, b_t$ are those whose orders are a power of p_1. Then the elements of G whose elements are a power of p_1 are precisely the elements of the direct sum

9.30
$$(b_1) \oplus \cdots \oplus (b_t).$$

It follows that the direct sum 9.30 is equal to $G(p_1)$. Since from 9.28, we also have

9.31
$$G(p_1) = (a_{11}) \oplus \cdots \oplus (a_{1r_1}),$$

we may apply Theorem 9.21 to the p_1-group $G(p_1)$, and conclude that $t = r_1$, and that the orders of $b_1, \ldots, b_t$ coincide, in some arrangement, with the orders of $a_{11}, \ldots, a_{1r_1}$. Thus the number of basis elements having order a power of p_1 is the same in the two bases, as are also the orders of the elements of the two bases. The same argument applies equally well to each prime p_i, and this completes the proof of the theorem.

Let us next make the following definition.

9.32   DEFINITION   Let G be a finite abelian group. The orders of the elements of a basis (repetitions being allowed), in which each basis element is required to have order a power of a prime, are called the *invariants* (or *elementary divisors*) of G.

Thus, for example, if we say that G has invariants 3, 2^2, 2, 2, it means that G is expressible as a direct sum of cyclic groups of these respective orders. Thus for this group G, we have

$$G \cong C_3 \oplus C_{2^2} \oplus C_2 \oplus C_2,$$

where C_n represents a cyclic group of order n.

The concept of the invariants of an abelian group is important because of the following theorem.

9.33   THEOREM   *Two finite abelian groups are isomorphic if and only if they have the same invariants.*

Proof: One part of this result follows fairly easily from results obtained above. Suppose that $\theta: G \to G'$ is an isomorphism of the finite abelian group G onto the finite abelian group G'. If $\{a_1, a_2, \ldots, a_n\}$ is a basis of G, with each a_i having order a power of a prime, the orders of $a_1, a_2, \ldots, a_n$ are then the invariants of G.

Now, $\{a_1\theta, a_2\theta, \ldots, a_n\theta\}$ is a basis of G' (see Exercise 5 below), and under the isomorphism θ, a_i and $a_i\theta$ have the same order. Hence G' has the same invariants as G.

Conversely, suppose that G and G' have the same invariants. This means that

$$G \cong D_1 \oplus D_2 \oplus \cdots \oplus D_n$$

and

$$G' \cong E_1 \oplus E_2 \oplus \cdots \oplus E_n,$$

where D_i and E_i are cyclic groups of the same order (a power of a prime). Now, by Theorem 2.41, two cyclic groups of the same order are isomorphic. Let $\theta_i : D_i \to E_i$ be an isomorphism of D_i onto E_i. Then, it may be verified that the mapping $\theta : G \to G'$, defined by

9.34
$$(d_1 + d_2 + \cdots + d_n)\theta = d_1\theta_1 + d_2\theta_2 + \cdots + d_n\theta_n,$$

where $d_i \in D_i$, is an isomorphism of G onto G'. (See Exercise 4 below.)

As a simple application of this theorem, let us determine all nonisomorphic abelian groups of order 24. Since the product of the invariants must be 24, we find the following possible systems of invariants: 3, 2^3; 3, 2^2, 2; 3, 2, 2, 2. Thus there are three nonisomorphic abelian groups of order 24. If, as above, we let C_n denote a cyclic group of order n, these three nonisomorphic abelian groups of order 24 are respectively isomorphic to

$$C_3 \oplus C_{2^3}, \qquad C_3 \oplus C_{2^2} \oplus C_2, \qquad C_3 \oplus C_2 \oplus C_2 \oplus C_2.$$

EXERCISES

1. If an abelian group G has invariants 2^2, 5, 5, verify that there exist elements of G of order 20 and none of higher order. Determine the number of elements of order 20.

2. In the notation used in the proof of Theorem 9.21, prove that the set 9.22 is a basis of G_p.

3. In the notation of the same proof, prove that the set 9.23 is a basis of pG.

4. Verify that the mapping θ defined in 9.34 is an isomorphism of G onto G'.

5. Prove: If $\{a_1, a_2, \ldots, a_n\}$ is a basis of an abelian group G and if $\theta \colon G \to G'$ is an isomorphism of G onto G', then $\{a_1\theta, a_2\theta, \ldots, a_n\theta\}$ is a basis of G', and G and G' have the same invariants.

6. If $p_1, p_2, \ldots, p_k$ are distinct primes, show that any two abelian groups of order $p_1 p_2 \cdots p_k$ are isomorphic (and therefore isomorphic to the cyclic group of this order).

7. Suppose that an abelian group G has order $n = p_1^{e_1} p_2^{e_2} \cdots p_k^{e_k}$, where the p's are distinct primes and each $e_i \geq 1$. If among the invariants of G the highest powers of these primes which occur are $p_1^{t_1}, p_2^{t_2}, \ldots, p_k^{t_k}$, prove that there exists an element of G of order $p_1^{t_1} p_2^{t_2} \cdots p_k^{t_k}$ and no element of higher order. *Hint:* Compare Exercise 11 at the end of Section 2.6. Observe also that Exercise 1 above involves a verification of a special case of this result.

8. Verify that there are exactly four nonisomorphic abelian groups of order 100. For each of these groups, determine the maximal order of an element.

9. Show that if a cyclic group G has order p^m (p a prime) and if $t \in \mathbf{Z}$ such that $0 \leq t \leq m$, then G has a subgroup of order p^t.

10. Prove that if p and t are as in the preceding exercise, any abelian group G of order p^m has a subgroup of order p^t. *Hint:* Consider the invariants of G, the result of the preceding exercise, and Exercise 2 of the preceding set.

11. Use Theorem 9.2, Lemma 9.27 and the results of the two preceding exercises to prove the following general result: If an abelian group G has order n and $k \mid n$, then G has a subgroup of order k.

 COMMENTARY

1 The theorem of this chapter completes in a fairly natural way the structure theory of finite abelian groups. What happens if we drop the assumption of finiteness? An example should suffice to show that not every group is a direct product of cyclic groups. If we consider $\mathbf{Q}^+$, the additive group of rationals, then for any two rational numbers p/q and r/s, where p, q, r, s are integers, both p/q and r/s are in the cyclic group generated by $1/qs$. (That is, if $\mathbf{Q}^+$ is a direct sum of cyclic groups, it is actually cyclic.) Since

$\mathbf{Q}^+$ is itself not a cyclic group, it cannot be written as a direct sum of cyclic groups. Further, it is not sufficient to ask that all elements be of finite order. Let $G = \mathbf{Q}^+/\mathbf{Z}^+$ be the quotient group modulo the additive group of integers. Then $q \cdot (p/q + \mathbf{Z}^+) = 0 + \mathbf{Z}^+$ so that $p/q + \mathbf{Z}^+$ has finite order. But again, $p/q + \mathbf{Z}^+$ and $r/s + \mathbf{Z}^+$ both lie in the cyclic group generated by $1/qs + \mathbf{Z}^+$. Consequently, for the same reason that $\mathbf{Q}^+$ is not, G is not a direct sum of cyclic groups. However, *if H is an abelian group and there is a positive integer n such that every element has order less than n, then H is a direct sum of cyclic groups.*

We shall attempt to describe the problem with generalizations to this latter theorem. Let p be a fixed prime integer and consider the smallest subgroup K of the above group G which contains all the elements $1/p^n + \mathbf{Z}^+$ for positive integers n. This group is usually denoted $\mathbf{Z}_{p^\infty}{}^+$. Since K may be hard to envision, let us describe another isomorphic version of K. Let K' be the set of all complex numbers of the form $\cos(2\pi k/p^n) + i \sin(2\pi k/p^n)$ for integers k and positive integers n. That is, K' is the set of all p^nth roots of unity for all n. Then K' is a group under multiplication of complex numbers, and this group is isomorphic to K.

This group K has two interesting properties: first, any proper subgroup of K is finite and cyclic; and second, any nontrivial homomorphic image of K is again isomorphic to K. For any elements $a, b \in K$, either a is in the subgroup generated by b or vice versa, and yet, K is not cyclic. Much of the story with regard to cyclic decomposition is summed up in the following observation. If L is an abelian group such that every element of L has finite order divisible by p, and if L cannot be decomposed into the direct sum of two proper subgroups, then L is isomorphic to $\mathbf{Z}_{p^n}{}^+$ for some n, or L is isomorphic to $\mathbf{Z}_{p^\infty}{}^+$. This observation is a theorem from the theory of infinite abelian groups [28, 33].

2   Algebraic geometry is one of the major branches of contemporary algebra. It has links with the theory of polynomials, commutative rings, group theory, differential geometry, and geometry. Because of the vast prerequisites and the breadth of the subject, we can only take a glimpse at a very pretty example of the algebraic geometry of plane curves. In particular, we shall describe the abelian group defined upon the points of a cubic curve.

Consider the equation

(*)
$$y^2 = x^3 - x + 6,$$

the graph of which is drawn in Figure 1. We shall consider this cubic (*) over the complex field $\mathbf{C}$, but the real graph of Figure 1 will be useful to describe what we mean. A point P on the curve (*) is an ordered pair (a, b) of complex numbers such that $b^2 = a^3 - a + 6$. We shall take G to be the set of all points on the cubic (*) plus one other point which we shall call the point at infinity (∞). To avoid a long winded explanation, we shall ignore the point (∞) and only later give a hint of its presence.

For two points P and Q we wish to define a sum $P + Q$ so that $P + Q = R$ is again a point upon the cubic. We first fix a point $O \in G$. It is not obvious, but different choices for O give isomorphic groups. In order to keep things concrete, let us assume that O is the point as marked upon Figure 1. Suppose that P and Q are also marked. Whether or not the coordinates of P and Q are as marked or are

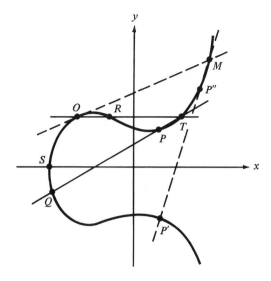

FIGURE 1

complex, using slope intercept formulas it is possible to derive an equation for the line $\overline{PQ}$ passing through P and Q. If we have $P = Q$, we must use some calculus, taking the line $\overline{PP}$ to be the tangent to the cubic at P. We obtain an equation for this line in the form $y = mx + b$. Substituting for y in (*), we obtain a cubic equation with three (possibly repeated) roots. Two of these are accounted for by the x-coordinates of P and Q. The third is the x-coordinate of another point T upon the cubic. Now draw the line $\overline{OT}$. The third point at which this line meets the cubic is R and we take $P + Q = R$. Evidently $\overline{PQ}$ and $\overline{QP}$ are the same line, so that $P + Q = Q + P$. Further, $O + P = P$ so that O is an identity. We shall not complete the verification, but G is an abelian group with this addition.

We may indicate how the point (∞) arises now. Not all equations for lines are of the form $y = mx + b$. There are also the vertical lines $x = a$. These arise when we try to add $S + S$ or $P + P'$ of Figure 1. Making the substitution, for example, in the case $P + P'$, where $\overline{PP'}$ is the line $x = a$, we obtain $y^2 = a^3 - a + 6$ so that $\pm \sqrt{a^3 - a + 6}$ are the y-coordinates of the points P and P'. The third missing point is at infinity and can only be found by observing that our cubic should not really be taken in the "affine" plane but rather in the "projective" plane. Beyond making these comments, we will not attempt to define any of this jargon.

Coming back to finite matters, let us consider the line $\overline{OO}$. It meets the cubic curve at a point M. Using this point, we may find the inverse $-P'$ for P'. In particular, $-P' = P''$, the third intersection of $\overline{P'M}$ with the cubic.

We must not forget that there are complex points on the cubic and complex lines which we cannot sketch upon our graph. But these points may be treated algebraically in the same fashion as the real points. There are points of order n for every positive integer n; further, there are points of infinite order. This example shows a nice connection between ideas of geometry, algebraic equations, and group theory [37, 103].

3   Our theorems on finite abelian groups are also proved in Refs. [5, 6, 21]. There is a connection between these theorems and a theorem about matrices with integer entries. This connection is given in [21, 24]. Further, our theorems may be generalized in various ways.

If there exists a finite number of elements $a_1, a_2, \ldots, a_m$ of an abelian group G such that all elements of G are expressible in the form

$$x_1 a_1 + x_2 a_2 + \cdots + x_m a_m,$$

the x's being integers, the group is said to be *finitely generated*. Of course, every finite abelian group is finitely generated, but $\mathbf{Z}^+$ is a finitely generated infinite group. As the simplest example, the additive group of the integers is generated by the single integer 1. Our Fundamental Theorem (9.29) can be suitably generalized to arbitrary finitely generated abelian groups by allowing the possibility that some of the basis elements have infinite order. The number of basis elements of infinite order is unique, as are the orders of the basis elements of finite order (these being required to have order a power of a prime) [31, 35, 36, 38].

chapter ten

POLYNOMIALS

In elementary algebra an important role is played by polynomials in a symbol "x" with coefficients that are real or complex numbers. In the next section we shall introduce polynomials with coefficients in a commutative ring S with unity, and show that under suitable definitions of addition and multiplication the set of all such polynomials is a ring. Actually, we could just as well start by letting S be an entirely arbitrary ring, but in most of the chapter it is essential that it be commutative and so we simplify matters by making this assumption from the beginning. The restriction that S have a unity is not very important but it does serve to simplify the notation somewhat.

The purpose of this chapter is to introduce polynomials with coefficients in a commutative ring S with unity, and to establish a number of properties of such polynomials. We shall frequently find it necessary or desirable to make additional restrictions on the ring S. In particular, we shall sometimes require that it be a field or a specified one of the fields that have already been studied in detail in previous chapters.

It will be found that a ring of polynomials with coefficients in a *field* has a considerable number of properties in common with the ring $\mathbf{Z}$ of integers. Accordingly, several of the sections of this chapter will closely parallel corresponding material of Chapter 5.

10.1 POLYNOMIAL RINGS

Let S be a commutative ring with unity. Heretofore we have used letters to denote sets or elements of sets, but we now use the letter x in a different way. It is not an element of S, but is just a symbol which we shall use in an entirely formal way. It is customary to call such a symbol an *indeterminate*. It is our purpose in this section to construct a ring which contains S and also has x as an element. This goal will motivate the definitions which we proceed to give.

Let x be an indeterminate and let us consider expressions of the form

10.1
$$a_0 x^0 + a_1 x^1 + a_2 x^2 + \cdots + a_n x^n,$$

where n is some nonnegative integer and $a_i \in S$ $(i = 0, 1, \ldots, n)$. Such an expression is called "a *polynomial* in x with coefficients in S" or simply, "a polynomial in x over S." If i is an integer such that $0 \le i \le n$, we say that a_i is the *coefficient* of x^i in the polynomial 10.1; also we say that $a_i x^i$ is a *term* of the polynomial 10.1 with coefficient a_i.

At this stage we are to think of 10.1 as a purely formal expression. That is, the $+$ signs are not to be considered as representing addition in a ring, and neither is x^i to be considered as a product $x \cdot x \cdots x$ with i factors. Later on, after we have proved the existence of a ring which contains S as well as x, we shall see that in this larger ring we can make these familiar interpretations and thus justify the notation we are using. At the present time, we could logically use some such symbol as $(a_0, a_1, a_2, \ldots, a_n)$ to designate the polynomial 10.1, but the definitions of addition and multiplication of polynomials to be given below will seem more natural with the familiar notation used in 10.1.

For the moment, let S be the ring $\mathbf{Z}$ of integers. Then the following are examples of polynomials in x over $\mathbf{Z}$:

(i) $2x^0 + (-3)x^1 + 4x^2$,   (iii) $0x^0 + 0x^1 + 4x^2$,

(ii) $3x^0$,   (iv) $0x^0 + 2x^1 + (-1)x^2 + 0x^3$.

In order to avoid writing so many terms with zero coefficients, we could agree in the third of these examples to write merely $4x^2$ with the understanding that x^0 and x^1 are assumed to have zero coefficients. Also, it would certainly agree with usual practice if we omitted the terms with zero coefficients in the fourth example and wrote $2x^1 + (-1)x^2$ to designate this polynomial. These simplifications will be possible under general agreements which we now make.

Let us designate the polynomial 10.1 over S by the symbol $f(x)$, and let $g(x)$ be the following polynomial over S:

10.2
$$b_0 x^0 + b_1 x^1 + \cdots + b_m x^m,$$

where $m \geq 0$ and $b_i \in S$ $(i = 0, 1, \ldots, m)$. By the *equality* of $f(x)$ and $g(x)$, written in the usual way as $f(x) = g(x)$, we shall mean that the expressions 10.1 and 10.2 are identical except for terms with zero coefficients. We therefore consider a polynomial as being unchanged by the insertion, or omission, of any number of terms with zero coefficients. In particular, with reference to the above examples, we may write

$$0x^0 + 0x^1 + 4x^2 = 4x^2$$

and

$$0x^0 + 2x^1 + (-1)x^2 + 0x^3 = 2x^1 + (-1)x^2.$$

Also, if we wish, we could write

$$3x^0 = 3x^0 + 0x^1 + 0x^2 + 0x^3$$

and so on.

With this understanding about zero coefficients, if $f(x)$ is a polynomial over S and i is an *arbitrary* nonnegative integer, we may speak of the coefficient of x^i in $f(x)$. For example, in the polynomial $1x^0 + 2x^1 + 3x^2$ over $\mathbf{Z}$, the coefficient of x^{10} is zero. This language often helps to simplify statements about polynomials. As an illustration, we may state again our definition of equality of two polynomials as follows: If $f(x)$ and $g(x)$ are polynomials over S, by $f(x) = g(x)$ we mean that for *every* nonnegative integer i, the coefficients of x^i in $f(x)$ and in $g(x)$ are equal elements of S.

Using the familiar sigma notation for sums, the polynomial 10.1 can be formally written as follows:

$$\sum_{i=0}^{n} a_i x^i.$$

Moreover, in view of our agreement about zero coefficients, we can write an arbitrary polynomial in x over S in the form

$$\sum_{i=0} a_i x^i,$$

with the tacit understanding that all coefficients are zero from some point on, so that this sum may be considered to be a finite sum with an unspecified number of terms.

Now let $S[x]$ denote the set of all polynomials in the indeterminate x over S. We proceed to define operations of addition and multiplication on the set $S[x]$. Of course, these definitions are suggested by the way that one adds and multiplies polynomials in elementary algebra. Let

10.3 $$f(x) = \sum_{i=0} a_i x^i$$

and

10.4
$$g(x) = \sum_{i=0} b_i x^i$$

be elements of $S[x]$. We define addition as follows:

10.5
$$f(x) + g(x) = \sum_{i=0} (a_i + b_i)x^i,$$

that is, for every nonnegative integer i, the coefficient of x^i in $f(x) + g(x)$ is the sum of the coefficients of x^i in $f(x)$ and in $g(x)$. Multiplication in $S[x]$ is defined as follows:

10.6
$$f(x)g(x) = \sum_{i=0} \left(\sum_{k=0}^{i} a_k b_{i-k} \right) x^i.$$

Another way of stating this definition of the product of $f(x)$ and $g(x)$ is to say that for each nonnegative integer i, the coefficient of x^i in the product is the sum (in the ring S) of all products of the form $a_r b_s$, where r and s are nonnegative integers such that $r + s = i$. The first few terms in the product given in 10.6 are as follows:

$$(a_0 b_0)x^0 + (a_0 b_1 + a_1 b_0)x^1 + (a_0 b_2 + a_1 b_1 + a_2 b_0)x^2 + \cdots.$$

We are now ready to state the following theorem.

10.7   THEOREM   *Let $S[x]$ be the set of all polynomials in the indeterminate x over the commutative ring S with unity. If operations of addition and multiplication are defined on $S[x]$ by 10.5 and 10.6, respectively, then*

(i) $S[x]$ is a commutative ring with unity.

(ii) $S[x]$ contains a subring isomorphic to S.

(iii) $S[x]$ is an integral domain if and only if S is an integral domain.

Proof: The commutative and associative laws for addition in $S[x]$ follow from 10.5 since these laws hold in the ring S. Moreover, the polynomial $0x^0$ (which is equal to the polynomial with *all* coefficients zero) is the zero of $S[x]$ since, by 10.5, for each polynomial $f(x)$ we have

$$f(x) + 0x^0 = f(x).$$

Moreover, our definition of addition also shows that

$$\sum_{i=0} a_i x^i + \sum_{i=0} (-a_i)x^i = \sum_{i=0} [a_i + (-a_i)]x^i = 0x^0,$$

and each element of $S[x]$ has an additive inverse in $S[x]$.

To establish that multiplication is commutative, we observe that if $f(x)$ and $g(x)$ are given by 10.3 and 10.4, respectively, then the coefficient of x^i in $g(x)f(x)$ is

$$b_0 a_i + b_1 a_{i-1} + \cdots + b_i a_0,$$

and since S is assumed to be commutative, this is equal to the co-efficient

$$a_0 b_i + a_1 b_{i-1} + \cdots + a_i b_0$$

of x^i in $f(x)g(x)$. Inasmuch as this statement is true for every non-negative integer i, it follows that $f(x)g(x) = g(x)f(x)$, and hence that multiplication in $S[x]$ is commutative.

If $f(x)$ and $g(x)$ are given by 10.3 and 10.4, respectively, and

10.8 $$h(x) = c_0 x^0 + c_1 x^1 + \cdots + c_p x^p$$

is also an element of $S[x]$, the coefficient of x^i in the product $[f(x)g(x)]h(x)$ is found to be the sum of all products of the form $(a_r b_s)c_t$, where r, s, and t are nonnegative integers such that $r + s + t = i$. Similarly, the coefficient of x^i in the product $f(x)[g(x)h(x)]$ is the sum of all products of the form $a_r(b_s c_t)$, with the same restriction on r, s, and t. However, since $(a_r b_s)c_t = a_r(b_s c_t)$ by the associative law of multiplication in S, it follows that

$$[f(x)g(x)]h(x) = f(x)[g(x)h(x)],$$

that is, that multiplication is associative in $S[x]$.

We leave as exercises the proof of the distributive laws, and that if 1 is the unity of S, then $1x^0$ is the unity of $S[x]$. It follows then that $S[x]$ is a commutative ring with unity.

To establish part (ii) of the theorem, let S' denote the set of elements of $S[x]$ of the form ax^0, $a \in S$. It is easy to verify that S' is a subring of $S[x]$. Now the mapping $\theta: S' \to S$ defined by $(ax^0)\theta = a$, $a \in S$, is a one-one mapping of S' onto S. Moreover,

$$(ax^0 + bx^0)\theta = [(a + b)x^0]\theta = a + b = (ax^0)\theta + (bx^0)\theta$$

and

$$[(ax^0)(bx^0)]\theta = [(ab)x^0]\theta = ab = [(ax^0)\theta][(bx^0)\theta],$$

and it follows that θ is an isomorphism of S' onto S. Part (ii) of the theorem is therefore established.

Before proceeding to the proof of part (iii) of the theorem, let us introduce some simplifications of our notation as follows. We shall henceforth identify S' with S, and therefore write simply a in place of ax^0; that is, we shall omit x^0 in writing polynomials. In particular, the zero polynomial will then be designated by the familiar symbol 0. We shall also write x in place of x^1, x^i in place of $1x^i$, and $-ax^i$ in place of $(-a)x^i$. We may now observe that x is itself an element of the ring $S[x]$. If $a \in S$, then also $a \in S[x]$, and ax^i can be interpreted as the product (in the ring $S[x]$) of a times x to the power i. Also, since each individual term of a polynomial 10.1 is itself equal to a polynomial, the $+$ signs occurring in 10.1 can be correctly interpreted as addition in the ring $S[x]$. In other words, we have finally justified the use of the notation appearing in 10.1. Of course, addition is commutative in $S[x]$ and we can write the polynomial 10.1 with the terms in any order. For example, we could just as well write the polynomial 10.1 in the form

$$a_n x^n + a_{n-1} x^{n-1} + \cdots + a_1 x + a_0.$$

In this case, it is customary to say that it is written in *decreasing* powers of x. As given in 10.1, it is written in *increasing* powers of x.

The following familiar concepts are of such great importance that we give a formal definition.

10.9   DEFINITION   Let $f(x)$ be a nonzero element of the ring $S[x]$. If n is the largest nonnegative integer such that x^n has a nonzero coefficient in $f(x)$, we say that $f(x)$ has *degree n*. If $f(x)$ has degree n, the nonzero coefficient of x^n is sometimes called the *leading coefficient* of $f(x)$. The zero polynomial has no degree and therefore also no leading coefficient. The coefficient of x^0 in a polynomial—that is, as now written, the term that does not involve x—is sometimes referred to as the *constant term* of the polynomial. If a polynomial is zero or of degree zero (just a constant term), we call it a *constant polynomial*.

It will be observed that the nonzero elements of S, considered as elements of $S[x]$, are just the polynomials of degree zero. The degree of a polynomial $f(x)$ may be conveniently designated by $\deg f(x)$.

If S is the ring $\mathbf{Z}$ of integers, the polynomials $2 + 3x - x^2$, $4x$, 3, and $x^4 - 2x$ have respective degrees 2, 1, 0, and 4; and respective leading coefficients -1, 4, 3, and 1. The constant terms are, respectively, 2, 0, 3, and 0.

The third part of Theorem 10.7 will follow immediately from the following lemma.

10.10 LEMMA *Let S be an integral domain, and let f(x) and g(x) be non-zero elements of S[x]. Then*

10.11 $\deg(f(x)g(x)) = \deg f(x) + \deg g(x).$

Proof: Since $f(x)$ and $g(x)$ are not zero, they have degrees, and let us suppose that $\deg f(x) = n$ and $\deg g(x) = m$. Then $f(x)$ can be written in the form 10.3 with $a_n \neq 0$, and $g(x)$ in the form 10.4 with $b_m \neq 0$. It now follows by the definition of multiplication (10.6) that $f(x)g(x)$ cannot have degree greater than $n + m$. Moreover, since S is an integral domain and we know that $a_n \neq 0$ and $b_m \neq 0$, it follows that the coefficient $a_n b_m$ of x^{n+m} is not zero, and 10.11 follows at once.

Proof of 10.7(iii): Lemma 10.10 assures us that if $f(x)$ and $g(x)$ are nonzero elements of $S[x]$, with S an integral domain, then the element $f(x)g(x)$ of $S[x]$ has a degree and therefore is not zero. Hence, $S[x]$ is also an integral domain. Since $S \subset S[x]$, it is trivial that if $S[x]$ is an integral domain, then S must be an integral domain. We have thus completed the proof of Theorem 10.7.

The familiar property 10.11 is not necessarily true if S is not an integral domain since, in the above proof, $a_n b_m$ might be zero without either factor being zero. For example, let $S = \mathbf{Z}_6$. If $f(x) = [1] + [2]x$, and $g(x) = [2] + [4]x + [3]x^2$, then $\deg f(x) = 1$ and $\deg g(x) = 2$. However, $f(x)g(x) = [2] + [2]x + [5]x^2$, and $\deg f(x)g(x) = 2$. In this case,

$$\deg(f(x)g(x)) < \deg f(x) + \deg g(x).$$

In this section we have introduced polynomials in *one* indeterminate x. However, this procedure can easily be generalized as follows. If S is a commutative ring with unity, then the polynomial ring $S[x]$ is a commutative ring with unity. If now y is another indeterminate, we may as above construct a ring $(S[x])[y]$ consisting of polynomials in y with coefficients in the ring $S[x]$. It is easy to verify that the elements of this new ring can also be expressed as polynomials in x with coefficients in the ring $S[y]$; in other words, the rings $(S[x])[y]$ and $(S[y])[x]$ are identical. Accordingly, we may denote this ring by $S[x, y]$ and call its elements polynomials in the indeterminates x and y. A double application of Theorem 10.7(iii) then assures us that $S[x, y]$ is an integral domain if and only if S is an integral domain. These statements may be extended in an obvious way to polynomials in any finite number of indeterminates. However, for the most part we shall study polynomials in just one indeterminate.

10.2 THE SUBSTITUTION PROCESS

In defining the polynomial ring $S[x]$, where S is a commutative ring with unity, we have emphasized that x is not to be considered as an element of S. However, if $f(x) = a_0 + a_1 x + \cdots + a_n x^n$ is an element of $S[x]$ and $s \in S$, let us define

10.12
$$f(s) = a_0 + a_1 s + \cdots + a_n s^n.$$

It follows that $f(s)$ is a uniquely determined element of S associated with the polynomial $f(x)$ and the element s of S. Now the importance of this "substitution process" stems from the fact that our definitions of addition and multiplication in $S[x]$ have the same form as though x were an element of S. Let us state this fact more precisely in terms of the mapping $\theta: S[x] \to S$ defined by

10.13
$$f(x)\theta = f(s), \qquad f(x) \in S[x].$$

We may emphasize that in this mapping we are thinking of s as being a fixed element of S. Different elements s of S would, of course, lead to different mappings of $S[x]$ into S. When we said above that addition and multiplication of polynomials were defined "as though x were an element of S," what we really meant was that the operations of addition and multiplication are preserved under the mapping θ, that is, that θ is a homomorphism. Actually, the mapping θ is a homomorphism of $S[x]$ *onto* S (no matter what element s of S is used) since if $a_0 \in S$ and $f(x) = a_0$, clearly $f(x)\theta = a_0$.

We are frequently interested in considering elements r of S such that $f(r) = 0$, and so we make the following definition.

10.14 DEFINITION If $f(x) \in S[x]$ and $r \in S$ such that $f(r) = 0$, we say that r is a *root* of the polynomial $f(x)$.†

In later sections we shall obtain various results about roots of polynomials. However, in order to obtain results of a familiar nature, we shall find it necessary to make some additional restrictions on the ring S. In particular, we shall frequently assume that S is a *field*. As an example to show what may happen if we do not restrict the ring of coefficients, let T be the ring of all subsets of a given set (Example 8 of Section 3.2), and $T[x]$ the ring of polynomials in the indeterminate x with coefficients in T. Since $a^2 = a$ for every element a of T, it is clear that the polynomial $x^2 - x$ of $T[x]$ has as a root *every* element of T. We thus have an example of a polynomial of degree 2 that has more than two roots (if the given

---

† In elementary algebra, *r* is usually said to be a root of the *equation* $f(x) = 0$, in which case x is thought of as an unknown number. However, this is not consistent with the definitions of the preceding section, and we shall continue to write $f(x) = 0$ to mean that $f(x)$ is the zero polynomial.

set has more than one element). In the next section we shall see that this cannot happen in case the ring of coefficients is restricted to be a field.

In places where we have met polynomials before, we usually have viewed them as polynomial functions or polynomial mappings. That is, we have mostly thought of $y = f(x) = 1 + x^2 + x^3$ as a function of x, where x is a variable, taking its values, for example, in the real numbers. It is perfectly legitimate to ask: Why don't we introduce polynomials in $S[x]$ as polynomial functions, where x is a variable upon S? Isn't that essentially what the substitution process is all about; i.e., the substitution process just turns x into a variable which can take the value $x = s$? There are two very good reasons for not doing this; that is, the long exposition of Section 10.1 is really necessary.

If S is a field of characteristic zero, the following objection to defining a polynomial as a polynomial function is not totally valid. But, as we have seen, there are a great many interesting rings which might be chosen for S which do not have characteristic zero. For example, consider the field $\mathbf{Z}_p$ for an odd prime p. Our exposition up to this point shows clearly that the polynomials $f(x) = x^p - [2]x$ and $g(x) = -x$ are unequal polynomials of $\mathbf{Z}_p[x]$. Let us now view $f(x)$ and $g(x)$ as polynomial mappings of $\mathbf{Z}_p$, that is, for an integer a, $f(x)$ is the mapping which sends $[a]$ to $f([a])$ where we are considering here x to be a "variable upon $\mathbf{Z}_p$." Two mappings $f(x): \mathbf{Z}_p \to \mathbf{Z}_p$ and $g(x): \mathbf{Z}_p \to \mathbf{Z}_p$ are equal if $f([a]) = g([a])$ for all $[a] \in \mathbf{Z}_p$; this fundamental point appears in Section 1.2.

We shall now calculate the values of $f(x)$ and $g(x)$ at $x = [a]$ for any $[a] \in \mathbf{Z}_p$. If $[a] = [0]$, then it is easy to see that $f([0]) = [0] = -[0] = g([0])$. On the other hand, if $[a] \neq [0]$, then $[a]$ is a unit in $\mathbf{Z}_p$, so that the order of $[a]$ divides the order $p - 1$ of the group of units of $\mathbf{Z}_p$; consequently, $[a]^{p-1} = [1]$ or $[a]^p = [a]$ by Theorem 2.55. We now have $f([a]) = [a]^p - [2][a] = [a] - [2][a] = -[a] = g([a])$. Therefore $f([a]) = g([a])$ for all values of $[a]$ in $\mathbf{Z}_p$. In other words, even though $f(x)$ and $g(x)$ are different polynomials, they are *not different polynomial mappings*; in fact, they are the *same polynomial mapping* (or function). This is a major drawback to viewing polynomials as we have viewed them in the past, namely, as polynomial mappings or functions.

The second objection may even be conceptually more important. A polynomial *mapping* $f(x)$ of a ring S is a fixed entity into which we may substitute values of the variable x from S. In other words, we usually think of $f(x)$ as being fixed and $x = s \in S$ as the thing which "varies." The substitution process completely reverses this relationship; we shall think of the element $s \in S$ of 10.13 as being fixed and the polynomials $f(x) \in S[x]$ as being the things which "vary"; so that θ of 10.13 is a mapping determined by $s \in S$ which is evaluated at polynomials from $S[x]$. Our traditional view of polynomials as polynomial mappings must be set aside (1) because it can cause "different" polynomials to be the "same" polynomial mapping, and (2) because it is backwards with respect to "what is variable."

The mapping θ defined in 10.13 sends a polynomial $f(x)$ to $f(s)$. As such, it is a homomorphism of $S[x]$ onto S. Now suppose that T is a subring of S containing the unity of S. Clearly then $T[x]$ is a subring of $S[x]$. Restricting θ to $T[x]$ gives a homomorphism of $T[x]$ into S. The image of $T[x]$ under the mapping θ is a subring of S. If $s \notin T$, then the image of $T[x]$ will contain both T and s. In fact, the image will be the smallest subring of S containing both T and s. We shall denote the image of $T[x]$ under θ as $T[s]$ and call it the *ring of polynomials in s over T*.

Let us now turn to some applications of this idea. From this point of view, with $S = \mathbf{C}$, the complex field, $\mathbf{Z} = T$, the ring of integers, and $s = \sqrt{2}$, the set $\{a + b\sqrt{2} \,|\, a, b \in \mathbf{Z}\}$ of Example 1 in Section 3.2 is obviously a ring since it is equal to $\mathbf{Z}[x]\theta$. Similarly, with T the rational field and $s = \sqrt[3]{4}$, we easily see that the set of Example 2 in Section 3.2 is also a ring.

EXERCISES

1. Prove the distributive laws in $S[x]$.

2. If S is a commutative ring with unity, verify that the set of all polynomials of $S[x]$ with zero constant terms is a subring of $S[x]$.

3. Verify that the set of all polynomials of $S[x]$ with the property that all odd powers of x have zero coefficients is a subring of $S[x]$. Is the same true if the word "odd" is replaced by the word "even"?

4. If $\mathbf{Z}$ is the ring of integers and x an indeterminate, let $(\mathbf{Z}[x])^p$ be the subset of $\mathbf{Z}[x]$ consisting of those nonzero polynomials which have as leading coefficient a *positive* integer. Show that the set $(\mathbf{Z}[x])^p$ has all the properties required in 4.4, and hence that $\mathbf{Z}[x]$ is an ordered integral domain.

5. Generalize the preceding exercise by showing that if D is an ordered integral domain, then the polynomial ring $D[x]$ is also an ordered integral domain.

6. Let $h(x)$ be the element $5x^2 - 3x + 4$ of $\mathbf{Z}_6[x]$. (Here we are writing 5, -3, and 4 in place of the more cumbersome [5], [−3], and [4].) By simply trying all the elements of $\mathbf{Z}_6$, find all roots of $h(x)$ in $\mathbf{Z}_6$. How many roots are there? What is the degree of $h(x)$?

7. If $g(x)$ is the element $x^7 - x$ of $\mathbf{Z}_7[x]$, verify that all elements of $\mathbf{Z}_7$ are roots of $g(x)$.

8. If m is a positive integer, how many polynomials are there of degree m over the ring $\mathbf{Z}_n$ of integers modulo n?

9. Let S and T be commutative rings, each with a unity, and suppose that $\theta: S \to T$ is a given homomorphism of S onto T. If a mapping $\phi: S[x] \to T[x]$ is defined by

$$(a_0 + a_1 x + \cdots + a_n x^n)\phi = a_0\theta + (a_1\theta)x + \cdots + (a_n\theta)x^n,$$

prove that ϕ is a homomorphism of $S[x]$ onto $T[x]$.

10. If $f(x) = g(x)h(x)$, where these are elements of $\mathbf{Z}[x]$, and every coefficient of $f(x)$ is divisible by the prime p, prove that every coefficient of $g(x)$ is divisible by p or every coefficient of $h(x)$ is divisible by p. *Hint:* Use the preceding exercise with $S = \mathbf{Z}$, $T = \mathbf{Z}_p$, and $\theta: \mathbf{Z} \to \mathbf{Z}_p$ as defined in 3.15 of Section 3.3. Then consider what $[f(x)]\phi = 0$ implies about the polynomial $f(x)$.

11. If

$$f(x) = \sum_{i=0} a_i x^i$$

is a polynomial over a commutative ring S, let us define the *derivative* $f'(x)$ of $f(x)$ as follows:

$$f'(x) = \sum_{i=1} i a_i x^{i-1}.$$

Prove that

$$[f(x) + g(x)]' = f'(x) + g'(x),$$

and that

$$[f(x)g(x)]' = f(x)g'(x) + f'(x)g(x).$$

12. Prove in all detail that the mapping of 10.13 is a homomorphism of $S[x]$ onto S.

10.3 DIVISORS AND THE DIVISION ALGORITHM

In this and the next two sections we shall study polynomials with coefficients in an arbitrary *field F*. We then know, by Theorem 10.7, that $F[x]$ is necessarily an integral domain. The following definition is essentially a restatement of Definition 5.1 as applied to the integral domain $F[x]$ instead of an arbitrary Euclidean domain.

10.15 DEFINITION Let $F[x]$ be the ring of polynomials in the indeterminate x over an arbitrary field F. If $f(x), g(x) \in F[x]$, $g(x)$ is said to be a *divisor* (or *factor*) of $f(x)$ if there exists $h(x) \in F[x]$ such that $f(x) = g(x)h(x)$. If $g(x)$ is a divisor of $f(x)$, we say also that $f(x)$ is *divisible* by $g(x)$ or that $f(x)$ is a *multiple* of $g(x)$.

It follows immediately from this definition that if c is a nonzero element of F (that is, a polynomial of $F[x]$ of degree zero), then c is a divisor of every element $f(x)$ of $F[x]$. For, since c has a multiplicative inverse c^{-1} in F, we can write $f(x) = c(c^{-1}f(x))$, and this shows that c is a divisor of $f(x)$.

It is also important to observe that if $f(x) = g(x)h(x)$, then also $f(x) = (cg(x))(c^{-1}h(x))$, where c is any nonzero element of F. That is, if $g(x)$ is a divisor of $f(x)$, then $cg(x)$ is also a divisor of $f(x)$ for every nonzero element c of F.

The following result plays just as important a role in the study of divisibility in $F[x]$ as the corresponding result (5.5) does in establishing divisibility properties of the integers; in fact, it is the keystone in the argument that $F[x]$ is a Euclidean domain.

10.16 DIVISION ALGORITHM If $f(x), g(x) \in F[x]$ with $g(x) \neq 0$, there exist unique elements $q(x)$ and $r(x)$ of $F[x]$ such that

10.17 $$f(x) = q(x)g(x) + r(x), \qquad r(x) = 0 \text{ or } \deg r(x) < \deg g(x).$$

We may recall that the zero polynomial has no degree and this fact explains the form of the condition which $r(x)$ is required to satisfy.

Proof: If $f(x)$ and $g(x)$ are given polynomials, the polynomials $q(x)$ and $r(x)$ can easily be computed by the usual process of long division. The existence of such polynomials therefore seems almost obvious. However, we shall give a detailed proof of their existence, and for the moment leave aside the question of their uniqueness. Let us first dispose of two easy cases as follows.

(A) If $f(x) = 0$ or $\deg f(x) < \deg g(x)$, then 10.17 is trivially satisfied with $q(x) = 0$ and $r(x) = f(x)$.

(B) If $\deg g(x) = 0$, so that $g(x) = c$ with c a nonzero element of F, then $f(x) = [c^{-1}f(x)]c$ and 10.17 holds with $q(x) = c^{-1}f(x)$ and $r(x) = 0$.

We are now ready to complete the proof by induction on the degree of $f(x)$. In this case, we shall use the form of the Induction Principle given in Exercise 9 of Section 4.3. If n is a positive integer, let S_n be the statement, "For every polynomial $f(x)$ of degree n and every nonzero polynomial $g(x)$, there exist polynomials $q(x)$ and

$r(x)$ satisfying equation 10.17." Let us now consider the statement S_1. By (A) and (B), we need only consider the case in which $\deg g(x) = 1$; that is, $g(x) = cx + d$, $c \neq 0$. Since $f(x) = ax + b$, $a \neq 0$, we can easily see that

10.18 $f(x) = ac^{-1}(cx + d) + b - ac^{-1}d,$

and 10.17 is satisfied with $q(x) = ac^{-1}$ and $r(x) = b - ac^{-1}d$. Hence, S_1 is true. Now suppose that k is a positive integer with the property that S_i is true for every positive integer $i \leq k$, and let us prove that S_{k+1} is true. Let $f(x) = ax^{k+1} + \cdots$, where $a \neq 0$, be a polynomial of degree $k + 1$ and let $g(x)$ be an entirely arbitrary polynomial. Cases (A) and (B) show that we may assume that $0 < \deg g(x) \leq k + 1$ since otherwise the existence of $q(x)$ and $r(x)$ satisfying 10.17 follows immediately. Suppose that $\deg g(x) = m$, hence that $g(x) = bx^m + \cdots$, with $b \neq 0$ and $0 < m \leq k + 1$. Now it is easily verified that

10.19 $f(x) = b^{-1}ax^{k+1-m}g(x) + [f(x) - b^{-1}ax^{k+1-m}g(x)].$

Perhaps we should point out that this equation is merely the result of taking one step in the usual long-division process of dividing $f(x)$ by $g(x)$. If we set $t(x) = f(x) - b^{-1}ax^{k+1-m}g(x)$, it is easy to see that the coefficient of x^{k+1} in $t(x)$ is zero; hence that $t(x) = 0$ or $\deg t(x) < k + 1$. By (A), or by the assumption that S_i is true for every positive integer $i \leq k$, we know that there exist polynomials $s(x)$ and $r(x)$, with $r(x) = 0$ or $\deg r(x) < \deg g(x)$, such that $t(x) = s(x)g(x) + r(x)$. Substituting in 10.19, we see that

$$f(x) = [b^{-1}ax^{k+1-m} + s(x)]g(x) + r(x),$$

and 10.17 is satisfied. Hence S_{k+1} is true, and it follows that S_n is true for every positive integer n. This completes the proof of the *existence* part of the Division Algorithm. The proof of the fact that $q(x)$ and $r(x)$ are *unique* will be left as an exercise.

It is customary to call $q(x)$ and $r(x)$ satisfying 10.17 the *quotient* and the *remainder*, respectively, in the division of $f(x)$ by $g(x)$. Clearly, $f(x)$ is divisible by $g(x)$ if and only if the remainder in the division of $f(x)$ by $g(x)$ is zero.

A special case of the Division Algorithm of importance is that in which the divisor $g(x)$ is of the special form $x - c$, $c \in F$. In this case the remainder must be zero or have degree zero, that is, it is an element of F. We can thus write

$$f(x) = q(x)(x - c) + r, \qquad r \in F.$$

From this equation it is clear that $f(c) = r$, and hence that

$$f(x) = q(x)(x - c) + f(c).$$

The next two theorems then follow immediately.

10.20 REMAINDER THEOREM *If $f(x) \in F[x]$ and $c \in F$, the remainder in the division of $f(x)$ by $x - c$ is $f(c)$.*

10.21 FACTOR THEOREM *If $f(x) \in F[x]$ and $c \in F$, $f(x)$ is divisible by $x - c$ if and only if $f(c) = 0$, that is, if and only if c is a root of the polynomial $f(x)$.*

We shall now make use of the Factor Theorem to prove the following result.

10.22 THEOREM *Let F be a field and $f(x)$ an element of $F[x]$ of positive degree n and with leading coefficient a. If $c_1, c_2, \ldots, c_n$ are distinct elements of F, all of which are roots of $f(x)$, then*

10.23
$$f(x) = a(x - c_1)(x - c_2) \cdots (x - c_n).$$

Proof: The proof of this theorem is by induction on the degree n of $f(x)$. If n is a positive integer, let S_n be the statement, "The statement of the theorem is true for every polynomial of degree n." We then wish to prove that S_n is true for every positive integer n. The truth of the statement S_1 follows quite easily. If $f(x)$ is of degree 1 and has leading coefficient a, then $f(x) = ax + b$, $a \neq 0$. If c_1 is a root of $f(x)$, we have $f(c_1) = 0$ or $ac_1 + b = 0$. Then $b = -ac_1$ and hence $f(x) = a(x - c_1)$, which is the desired form 10.23 in case $n = 1$.

Now let k be a positive integer such that S_k is true, and consider S_{k+1}. Accordingly, we let $f(x)$ be a polynomial of degree $k + 1$ with leading coefficient a, and let $c_1, c_2, \ldots, c_{k+1}$ be distinct roots of $f(x)$. Since c_1 is a root of $f(x)$, we have $f(c_1) = 0$ and by the Factor Theorem it follows that

10.24
$$f(x) = q(x)(x - c_1).$$

Now it is clear that $\deg q(x) = k$, and the leading coefficient of $q(x)$ is a since a is the coefficient of x^{k+1} in $f(x)$. If c_i ($i \neq 1$) is any other of the given roots of $f(x)$, it follows, using 10.24 and the fact that $f(c_i) = 0$, that

$$q(c_i)(c_i - c_1) = 0.$$

Since the c's are distinct, $c_i - c_1 \neq 0$ and therefore $q(c_i) = 0$. We have therefore shown that the polynomial $q(x)$ in 10.24 is of degree k, has leading coefficient a, and has $c_2, c_3, \ldots, c_{k+1}$ as distinct roots. Since S_k is assumed to be true, it follows that

$$q(x) = a(x - c_2)(x - c_3)\cdots(x - c_{k+1}).$$

Substituting this expression for $q(x)$ in 10.24, we get

$$f(x) = a(x - c_1)(x - c_2)\cdots(x - c_{k+1}).$$

Hence, S_{k+1} is true, and the Induction Principle assures us that S_n is true for every positive integer n. This completes the proof of the theorem.

We next establish the following corollary.

10.25 COROLLARY   *A polynomial $f(x)$ of degree n over a field F cannot have more than n distinct roots in F.*

Since polynomials of degree zero have no roots, in verifying this corollary we may assume that $n \geq 1$. If $c_1, c_2, \ldots, c_n$ are distinct roots of $f(x)$, then $f(x)$ can be written in the form 10.23. Now let c be an arbitrary root of $f(x)$. Since $f(c) = 0$, it follows at once from 10.23 that

$$a(c - c_1)(c - c_2)\cdots(c - c_n) = 0.$$

Since $a \neq 0$, some one of the other factors must be zero, that is, $c = c_i$ for some i. Hence, $c_1, c_2, \ldots, c_n$ are the *only* roots of $f(x)$, and $f(x)$ cannot have more than n distinct roots.

The next corollary is now a simple consequence of this one.

10.26 COROLLARY   *Let $g(x)$ and $h(x)$ be polynomials over a field F with the property that $g(s) = h(s)$ for every element s of F. If the number of elements in F exceeds the degrees of both $g(x)$ and $h(x)$, then necessarily $g(x) = h(x)$.*

Let us set $f(x) = g(x) - h(x)$, and we then have that $f(s) = 0$ for every element s of F. If $f(x) \neq 0$, its degree can certainly not exceed the degrees of both $g(x)$ and $h(x)$, and hence $f(x)$ would have more distinct roots than its degree. Since, by the preceding corollary, this is impossible, we must have $f(x) = 0$. Hence, $g(x) = h(x)$, as required.

The polynomials $f(x)$ and $g(x)$ in $\mathbf{Z}_p[x]$ given at the end of Section 10.2

show that the corollary is false without the restriction upon the number of elements of F. See also Exercise 7 of the preceding set.

10.4 GREATEST COMMON DIVISOR

In this section we shall show that if F is a field, then the ring $F[x]$ of polynomials over F is a Euclidean domain. That is, to repeat Definition 5.8 of Section 5.1, $F[x]$ has the properties of the definition below.

10.27 DEFINITION An integral domain D is called a *Euclidean domain* if there is a mapping $\delta: D \to \mathbf{Z}$ of D into the set of nonnegative integers which satisfies the following two properties:

  (i) If a and b are nonzero elements of D, then $(ab)\delta \geq a\delta$.

  (ii) If a and b are nonzero elements of D, then there exist elements q and r of D so that $a = bq + r$ and $r\delta < b\delta$.

We shall consider D to be $F[x]$, which is an integral domain by Theorem 10.7(iii). We must define the mapping δ. An obvious choice would be $f(x)\delta = \deg f(x)$ for $f(x) \in F[x]$; but this choice does not satisfy property (ii) of the definition since 0 has no degree while 0δ must be an integer. The correct definition of δ is not completely determined, but one easy choice is the following:

  (a) $(f(x))\delta = 1 + \deg f(x)$ if $f(x) \neq 0$.

  (b) $0\delta = 0$.

This somewhat artificial method of coping with the zero polynomial gives us a mapping which by Lemma 10.10 and Theorem 10.16, satisfies the properties of the definition of a Euclidean domain. (Why?) Therefore, we have the following:

10.28 THEOREM *If F is a field, then the ring $F[x]$ of polynomials in an indeterminate x is a Euclidean domain.*

The mapping δ has been constructed to satisfy the definition of a Euclidean domain; obviously, the information given by δ is already known as soon as we know the degree of a polynomial. As a consequence, we shall adhere to common practice and use the mapping "deg" in place of δ, remembering that, strictly speaking, δ is the correct mapping. In other words, δ has served its function so that we can now forget it.

Having proved that $F[x]$ is a Euclidean domain, we may carry over immediately the results of Chapter 5 obtained for Euclidean domains. This list includes a rather large number of definitions and theorems. In order to keep things clear, let us work our way slowly through this list stating the various results for the polynomial ring $F[x]$. It might be a good idea to keep a thumb in Chapter 5 while we translate. First on our agenda is Theorem 5.9; it gives us information about units in $F[x]$. We shall only use part (iv) of that theorem.

10.29 THEOREM *If F is a field and $F[x]$ the polynomial ring over F, then a polynomial $f(x) \in F[x]$ is a unit if and only if $\deg f(x) = 0$, i.e., $f(x)$ is a nonzero constant polynomial.*

This theorem tells us that the units of $F[x]$ are precisely the nonzero constant polynomials in $F[x]$. We may illustrate this result by answering a special case of an interesting question. If S is a commutative ring with unity and $\phi : S[x] \to S[x]$ is an isomorphism of the polynomial ring over S onto itself, then for which rings S does ϕ map S onto S, i.e., $S\phi = S$? Recall that by Theorem 10.7, S is a subring of $S[x]$. We may show that $S\phi = S$ if S is a field F. That is, if $\phi : F[x] \to F[x]$ is an isomorphism of $F[x]$ onto itself, then the restriction of ϕ to the constant polynomials maps F onto itself. By Theorem 10.29, $F\backslash\{0\}$ is the group of units of $F[x]$. Any isomorphism of $F[x]$ onto $F[x]$ must map the group of units onto itself. Therefore ϕ maps $F\backslash\{0\}$ onto $F\backslash\{0\}$. Clearly $0\phi = 0$ so that $F\phi = F$ and ϕ maps F onto itself. To show that the question is not trivial, let us now consider $S = F[y]$ to be the polynomial ring in an indeterminate y so that $S[x] = F[y, x]$ is the polynomial ring in two indeterminates x and y over the field F. Define a mapping ψ of $S[x]$ so that $a\psi = a$ for all $a \in F$, $x\psi = y$, and $y\psi = x$. These conditions define an isomorphism $\psi : S[x] \to S[x]$ of $S[x]$ onto itself which interchanges the indeterminates x and y. When we restrict ψ to S, we obtain $S\psi = F[y] \neq F[x] = S$; in particular, $S\psi \neq S$. The general answer to this question is not known, but the special case for $S = F$, a field, is an immediate consequence of Theorem 10.29.

We shall next translate the results of Section 5.3. The important result of that section is Theorem 5.15 which tells us that every ideal in a Euclidean domain is principal; i.e., it consists of all the multiples of a single fixed element of the Euclidean domain. In other words, if $f(x)$ and $g(x)$ are polynomials in $F[x]$, for a field F, then $g(x)$ divides $f(x)$ if and only if $f(x)$ is a multiple of $g(x)$, or equivalently, $f(x)$ lies in the ideal generated by $g(x)$. Recall that divisors can be controlled "up to units"; and in the case of integers, by requiring that divisors be nonnegative, the "up to units" clause can be eliminated. By a careful choice of units, we may also eliminate the "up to units" clause for polynomials.

Consider a nonzero element c of the field F and a polynomial $f(x)$ from $F[x]$. By choosing c as the multiplicative inverse of the leading coefficient of $g(x)$, the polynomial $cg(x)$ will have the unity 1 of F as its leading coefficient. That is, among all polynomials equal to $f(x)$ up to units, there is a unique polynomial whose leading coefficient is unity. The following definition makes it easy to refer to such polynomials.

10.30 DEFINITION A nonzero element of $F[x]$ is said to be a *monic* polynomial if its leading coefficient is the unity 1 of F.

The existence of monic polynomials greatly simplifies the statements of definitions and theorems for the ring $F[x]$. For example, we need not talk about "a greatest common divisor"; in translating Definition 5.17 to $F[x]$, we may define "*the* greatest common divisor" of two elements of $F[x]$ as follows.

10.31 DEFINITION The monic polynomial $d(x)$ of $F[x]$ is said to be the *greatest common divisor* (g.c.d.) of the nonzero polynomials $f(x)$ and $g(x)$ of $F[x]$ if the following conditions are satisfied:

(i) $d(x)$ is a divisor of both $f(x)$ and $g(x)$.

(ii) Every divisor of both $f(x)$ and $g(x)$ is a divisor of $d(x)$.

As in the case of integers, it is quite easy to verify that two polynomials cannot have more than one g.c.d. We may now translate the information in Section 5.3 concerning ideals and greatest common divisors.

Recall that an *ideal* of $F[x]$ is a subring I of $F[x]$ such that for any polynomial $g(x) \in F[x]$ and any polynomial $h(x) \in I$, the product polynomial $g(x)h(x) \in I$, lies in I. There are a few definitions which will help us in describing an ideal I of $F[x]$. Suppose that $f_1(x), f_2(x), \ldots, f_t(x)$ are fixed polynomials from $F[x]$ for some integer $t > 0$; then a *linear combination* of these polynomials would be an expression of the form

$$g_1(x)f_1(x) + g_2(x)f_2(x) + \cdots + g_t(x)f_t(x)$$

where the coefficients $g_i(x)$, $i = 1, 2, \ldots, t$ are themselves polynomials from $F[x]$. As in Section 5.3 we shall let $(f_1(x), f_2(x), \ldots, f_t(x))$ denote the set of all linear combinations of the polynomials $f_i(x)$, $i = 1, 2, \ldots, t$; and we call the polynomials $f_i(x)$, $i = 1, 2, \ldots, t$ the *generators* of the set $(f_1(x), f_2(x), \ldots, f_t(x))$. When $t = 1$, we call the set $(f_1(x))$ of all multiples of the fixed polynomial $f_1(x)$ *principal*. As we have seen in Theorem 5.14, this notation is useful, since the set $(f_1(x), f_2(x), \ldots, f_t(x))$ of all linear combinations of the $f_i(x)$, $i = 1, 2, \ldots, t$, is an ideal of $F[x]$. This is by no means the entire story. Theorem 5.15 and Corollary 5.16 join together to describe the ideals of $F[x]$ completely.

10.32 THEOREM *If F is a field and I is an ideal in the polynomial ring $F[x]$, then I is principal; and, in particular, if $I \neq (0)$, then there is a unique monic polynomial $f(x)$ in $F[x]$ such that $I = (f(x))$.*

Theorem 5.15 tells us that I is principal, and we may choose $f(x)$ as a polynomial of smallest degree from I. Corollary 5.16 says that any two generators

$f(x)$ and $g(x)$ of I differ by a unit multiple, which by Theorem 10.29 is a constant polynomial. That is, we may choose a unit and multiply by it so that $f(x)$ is monic. Clearly, $f(x)$ is the unique monic generator since multiplying $f(x)$ by any unit other than unity will change $f(x)$ into a polynomial which is not monic.

The important point of this discussion is that if $g(x)$ and $h(x)$ are nonzero polynomials of $F[x]$ and $(g(x), h(x)) = (f(x))$, where $f(x)$ is the unique monic generator of the ideal $(g(x), h(x))$, then $f(x)$ is the greatest common divisor of the polynomials $g(x)$ and $h(x)$. So as a corollary to Theorem 10.32 we obtain Euclid's Lemma, 5.18.

10.33 EUCLID'S LEMMA *If F is a field and $g(x), h(x) \in F[x]$ are non-zero polynomials, then a monic polynomial $f(x) \in F[x]$ is the g.c.d. of $g(x)$ and $h(x)$ if and only if the ideals $(g(x), h(x))$ and $(f(x))$ are equal. From this we conclude:*

  (i) *The g.c.d. of $g(x)$ and $h(x)$ exists and is unique.*

 (ii) *There exist polynomials $r(x)$ and $s(x)$ from $F[x]$ such that*

$$r(x)g(x) + s(x)h(x) = f(x).$$

(iii) *$\deg f(x)$ is the smallest possible value of $\deg (r(x)g(x) + s(x)h(x))$ for any polynomials $r(x)$ and $s(x)$ of $F[x]$ such that $r(x)g(x) + s(x)h(x)$ is nonzero.*

As an example of part (i), suppose that F is the rational field **Q**, $g(x) = x^2 - 5$, and $h(x) = x + 1$ so that

$$\begin{aligned} f(x) = 1 &= -\tfrac{1}{4}(x^2 - 5) + \tfrac{1}{4}(x - 1)(x + 1) \\ &= -\tfrac{1}{4}g(x) + \tfrac{1}{4}(x - 1)h(x). \end{aligned}$$

The g.c.d. of $g(x)$ and $h(x)$ is 1; and we conclude that the ideal $(g(x), h(x))$ is $(1) = F[x]$.

In order to *compute* the g.c.d. of two nonzero polynomials $f(x)$ and $g(x)$ of $F[x]$, we use the Euclidean Algorithm as in the case of integers. By repeated use of the Division Algorithm we obtain the following sequence of equations, it being understood that $r_k(x)$ is the last nonzero remainder [and $r_k(x) = g(x)$ if $r(x) = 0$]:

$f(x) = q(x)g(x) + r(x)$	$\deg r(x) < \deg g(x),$
$g(x) = q_1(x)r(x) + r_1(x)$	$\deg r_1(x) < \deg r(x),$
$r(x) = q_2(x)r_1(x) + r_2(x)$	$\deg r_2(x) < \deg r_1(x),$
$\cdots\cdots\cdots\cdots\cdots\cdots\cdots$	$\cdots\cdots\cdots\cdots\cdots\cdots$
$r_{k-2}(x) = q_k(x)r_{k-1}(x) + r_k(x)$	$\deg r_k(x) < \deg r_{k-1}(x),$
$r_{k-1}(x) = q_{k+1}(x)r_k(x).$	

10.34

Now from these equations it follows that $r_k(x)$ is a divisor of both $f(x)$ and $g(x)$; also that any divisor of both $f(x)$ and $g(x)$ is a divisor of $r_k(x)$. If c is the leading coefficient of $r_k(x)$, then $c^{-1}r_k(x)$ also has these same properties and, moreover, it is a *monic* polynomial. We have therefore outlined a proof of the following result, which is the Euclidean Algorithm as described in Section 5.3.

10.35 THEOREM *Let $r_k(x)$ be the last nonzero remainder in the Euclidean Algorithm as applied to the nonzero polynomials $f(x)$ and $g(x)$ of $F[x]$. If c is the leading coefficient of $r_k(x)$, then $c^{-1}r_k(x)$ is the g.c.d. of $f(x)$ and $g(x)$.*

In a numerical case, the actual calculations may often be simplified by the following observation. If $d(x)$ is the g.c.d. of $f(x)$ and $g(x)$, then also $d(x)$ is the g.c.d. of $af(x)$ and $bg(x)$, where a and b are nonzero elements of F. Hence, instead of the first of Equations 10.34, we might use the similar equation obtained by dividing $af(x)$ by $bg(x)$. In like manner, instead of the second equation we might work with $dg(x)$ and $er(x)$, where d and e are nonzero elements of F; and so on for the other equations. This modification will not affect the validity of the arguments used to show that $c^{-1}r_k(x)$ is the g.c.d. of $f(x)$ and $g(x)$, and may greatly simplify the work involved. Let us give an illustration by finding the g.c.d. of the polynomials

$$f(x) = x^3 + \tfrac{1}{2}x^2 + \tfrac{1}{3}x + \tfrac{1}{6}$$

and

$$g(x) = x^2 - \tfrac{1}{2}x - \tfrac{1}{2}$$

over the field **Q** of rational numbers. In order to avoid fractions, we divide $6f(x)$ by $2g(x)$, obtaining

$$6f(x) = (3x + 3)[2g(x)] + 8x + 4,$$

so that $r(x) = 8x + 4$. If we now divide $2g(x)$ by $r(x)/4$, we see that

$$2g(x) = (x - 1)[r(x)/4].$$

Since $r_1(x) = 0$, the g.c.d. of $f(x)$ and $g(x)$ is obtained from the last nonzero remainder, namely, $8x + 4$ by multiplying it by the multiplicative inverse of its leading coefficient. Hence the g.c.d. of $f(x)$ and $g(x)$ is $x + \tfrac{1}{2}$.

It is sometimes convenient to use the following terminology, which is suggested by the corresponding definition for the integers, and is given as Definition 5.22 in Section 5.3.

10.36 DEFINITION Two nonzero elements $f(x)$ and $g(x)$ of $F[x]$ are said to be *relatively prime* if and only if their g.c.d. is 1.

EXERCISES

1. Complete the proof of the Division Algorithm by showing that the quotient and the remainder are unique.

2. If $f(x) \in F[x]$, show that $f(x)$ has as a factor a polynomial of $F[x]$ of degree 1 if and only if $f(x)$ has a root in F.

3. If F is the field $\mathbf{Z}_7$, use the result of Exercise 7 of the preceding set to show, without calculation, that in $F[x]$ we have

$$x^7 - x = x(x - 1)(x - 2)(x - 3)(x - 4)(x - 5)(x - 6).$$

4. State and prove a corresponding result for the field $\mathbf{Z}_p$, where p is an arbitrary prime.   *Hint:* Consider the multiplicative group of $\mathbf{Z}_p$.

5. Find all roots of $x^3 + [4]x^2 + [4]x + [3]$ in $\mathbf{Z}_5$.

6. Find the g.c.d. of each of the following pairs of polynomials over the field $\mathbf{Q}$ of rational numbers, and express it as a linear combination of the two polynomials:

   (i) $2x^3 - 4x^2 + x - 2$ and $x^3 - x^2 - x - 2$,

   (ii) $x^4 + x^3 + x^2 + x + 1$ and $x^3 - 1$,

   (iii) $x^5 + x^4 + 2x^3 - x^2 - x - 2$ and $x^4 + 2x^3 + 5x^2 + 4x + 4$,

   (iv) $x^3 - 2x^2 + x + 4$ and $x^2 + x + 1$.

7. Find the g.c.d. of each of the following pairs of polynomials over the indicated field, and express it as a linear combination of the two polynomials:

   (i) $x^3 + 2x^2 + 3x + 2$ and $x^2 + 4$; field $\mathbf{Z}_5$,

   (ii) $x^3 + (2i + 1)x^2 + ix + i + 1$ and $x^2 + (i - 1)x - 2i - 2$; field $\mathbf{C}$ of complex numbers,

   (iii) $x^2 + (1 - \sqrt{2})x - \sqrt{2}$ and $x^2 - 2$; field $\mathbf{R}$ of real numbers,

   (iv) $x^4 + x + 1$ and $x^2 + x + 1$; field $\mathbf{Z}_2$.

8. Let $f(x)$ and $g(x)$ be nonzero elements of $F[x]$, where F is a field. If the field F' is an extension of the field F, (i.e., F' is a field containing the subfield F), then $F[x] \subseteq F'[x]$ and we may also consider $f(x)$ and $g(x)$ to be elements of $F'[x]$. Show that the quotient and the remainder in the division of $f(x)$ by

$g(x)$ are the same whether these polynomials are considered as elements of $F[x]$ or of $F'[x]$. In particular, conclude that if there exists an element $h(x)$ of $F'[x]$ such that $f(x) = g(x)h(x)$, then $h(x) \in F[x]$.

9. Verify that the Division Algorithm (10.16) remains true if the field F is replaced by an arbitrary commutative ring S with unity, provided only that $g(x)$ is required to have as leading coefficient a unit of S.

10. By using the result of the preceding exercise, verify that the Factor Theorem and the Remainder Theorem are true if the field F is replaced by a commutative ring S with unity.

11. Give an example to show that Theorem 10.22 is not necessarily true if the field F is replaced by an arbitrary commutative ring S with unity. Where does the proof break down? Verify that the proof of this theorem will remain valid if F is replaced by an integral domain.

10.5   UNIQUE FACTORIZATION IN $F[x]$

We turn our attention now to the results of Section 5.4; but first we restate Definition 5.4 of Section 5.1.

10.37   DEFINITION   A polynomial $p(x)$ of positive degree over a field F is said to be a *prime* (or *irreducible*) polynomial over F if it cannot be expressed as the product of two polynomials of positive degree over F.

If c is a nonzero element of the field F and $f(x) \in F[x]$, then we always have $f(x) = c^{-1}(cf(x))$, so that every polynomial of the form $cf(x)$ is a divisor of $f(x)$. It is easy to verify that a polynomial $f(x)$ of positive degree over F is a prime polynomial over F if and only if the *only* elements of $F[x]$ of positive degree that are divisors of $f(x)$ are of the form $cf(x)$, $c \neq 0$, so that our current definition is a translation of Definition 5.4.

Since the degree of the product of two polynomials over F is the sum of the degrees of the factors, it follows at once from Definition 10.37 that *every element of $F[x]$ of the first degree is necessarily prime over F*.

We may emphasize that the possible divisors of $p(x)$ that are being considered in Definition 10.37 are those which are elements of $F[x]$; that is, they must have coefficients in F. For example, consider the polynomial $x^2 - 2$ over the field $\mathbf{Q}$ of rational numbers.   Now $x^2 - 2$ cannot be factored into the product of two polynomials of the first degree in $\mathbf{Q}[x]$, and hence $x^2 - 2$ is a prime polynomial over $\mathbf{Q}$.   However, if we should consider the same polynomial as a polynomial over the field $\mathbf{R}$ of real numbers, we find that it is *not* prime over $\mathbf{R}$ since we have the factorization $x^2 - 2 = (x - \sqrt{2})(x + \sqrt{2})$ with these factors of the first degree having coefficients in $\mathbf{R}$.   As this example shows, the concept of a polynomial being a prime polynomial is relative to a specified field which contains the coefficients of the given polynomial.

In later sections we shall discuss prime polynomials over each of the familiar fields of elementary algebra.   As for the finite fields of the form $\mathbf{Z}_p$, where p is a prime integer, we may here state without proof the following fact.   For each prime p and each positive integer n, there exists at least one polynomial of degree n over the field $\mathbf{Z}_p$, which is prime over $\mathbf{Z}_p$.

The prime polynomials play essentially the same role in the factorization of an element of $F[x]$ as do the prime integers in the factorization of an integer. We therefore state the following lemma and theorem, which are translations of 5.23 and 5.25, respectively.

10.38   LEMMA   *If $f(x)$ and $g(x)$ are nonzero polynomials over the field F such that $f(x)g(x)$ is divisible by the prime polynomial $p(x)$ over F, then $f(x)$ is divisible by $p(x)$ or $g(x)$ is divisible by $p(x)$.*

10.39   THEOREM   *If $f(x)$ is a polynomial of positive degree over the field F and a is its leading coefficient, then there exist distinct monic prime polynomials $p_1(x), \ldots, p_k(x)$ $(k \geq 1)$ over F such that*

10.40 $$f(x) = a[p_1(x)]^{n_1}[p_2(x)]^{n_2} \cdots [p_k(x)]^{n_k},$$

where the n's are positive integers.   Moreover, such a factorization is unique except for the orders of the factors.

We may emphasize that the prime polynomials in 10.40 are restricted to be monic polynomials, and hence the leading coefficient of the right side is just a, which is given as the leading coefficient of $f(x)$.   Although Theorem 10.39 certainly has some theoretical significance, it is not so very useful from a computational point of view.   For example, from the factorizations of two polynomials in the form 10.40 it is easy to write down their g.c.d. just as in the case of two integers.   However, it is often very difficult to *find* the prime factors of a given polynomial and hence to write it in the form 10.40.   Accordingly, it will usually be very much easier to apply the method of Section 10.4 to find the g.c.d. of two polynomials than to make use of Theorem 10.39.

As an important special case, one or more of the monic prime polynomials occurring in a factorization 10.40 of $f(x)$ may be of the first degree.   In particular, the Factor Theorem (10.21) assures us that $x - c$, $c \in F$, is a factor of $f(x)$ if and only if $f(c) = 0$; that is, if and only if c is a root of the polynomial $f(x)$. The following definition introduces a terminology which is sometimes convenient.

10.41 DEFINITION The element c of F is said to be a root of *multiplicity* $m \geq 1$ of the polynomial $f(x)$ over F if $f(x)$ is divisible by $(x - c)^m$ but not by $(x - c)^{m+1}$. A root of multiplicity two is called a *double root*.

It follows that c is a root of $f(x)$ of multiplicity m if and only if in the factorization 10.40 of $f(x)$ one of the prime factors occurring is $x - c$ and, furthermore, it occurs with the exponent m.

EXERCISES

1. (a) Prove that a polynomial $f(x)$ of degree 2 or 3 over a field F is a prime polynomial over F if and only if the polynomial $f(x)$ has no root in F.

   (b) Show, by means of an example, that a corresponding statement does not hold for polynomials of degree 4.

2. Determine whether or not each of the following polynomials is prime over each of the given fields. If it is not prime, factor it into a product of prime factors over each given field. As usual, $\mathbf{Q}$ is the field of rational numbers, $\mathbf{R}$ the field of real numbers, and $\mathbf{C}$ the field of complex numbers.

   (a) $x^2 + x + 1$ over $\mathbf{Q}$, $\mathbf{R}$, and $\mathbf{C}$;

   (b) $x^2 + 2x - 1$ over $\mathbf{Q}$, $\mathbf{R}$, and $\mathbf{C}$;

   (c) $x^2 + 3x - 4$ over $\mathbf{Q}$, $\mathbf{R}$, and $\mathbf{C}$;

   (d) $x^3 + 2$ over $\mathbf{Q}$, $\mathbf{R}$, and $\mathbf{C}$;

   (e) $x^2 + x + 1$ over $\mathbf{Z}_2$, $\mathbf{Z}_3$, and $\mathbf{Z}_5$;

   (f) $x^3 + x + 1$ over $\mathbf{Z}_2$, $\mathbf{Z}_5$, and $\mathbf{Z}_{11}$;

   (g) $x^4 - 1$ over $\mathbf{Z}_{17}$;

   (h) $x^3 + x^2 + 1$ over $\mathbf{Z}_{11}$;

   (i) $x^2 + 15$ over $\mathbf{R}$ and $\mathbf{C}$.

3. Find all prime polynomials of degree not more than 5 over the field $\mathbf{Z}_2$.

4. In each case the polynomial over the given field has as a root the specified element of the field. Find the multiplicity of this root and complete the factorization of the polynomial into prime factors over the given field.

    (a) $x^4 + x^3 - 3x^2 - 5x - 2$ over **Q**, root -1;

    (b) $x^4 + 1$ over $\mathbf{Z}_2$, root 1;

    (c) $x^4 + 2x^2 + 1$ over **C**, root i;

    (d) $x^4 + 6x^3 + 3x^2 + 6x + 2$ over $\mathbf{Z}_7$, root 4.

5. Prove that there are $(p^2 - p)/2$ monic quadratic polynomials which are prime over the field $\mathbf{Z}_p$.

6. Suppose that $f(x)$ is a polynomial over the field F and let $f'(x)$ be the derivative of $f(x)$, as defined in Exercise 11 of Section 10.2. Prove each of the following:

    (i) If an element c of F is a root of $f(x)$ of multiplicity greater than one, then c is also a root of the polynomial $f'(x)$.

    (ii) If an element c of F is a root of $f(x)$ of multiplicity one, then c is not a root of $f'(x)$.

    (iii) If $f(x)$ can be expressed as a product of elements of $F[x]$ of the first degree, then $f(x)$ and $f'(x)$ are relatively prime if and only if $f(x)$ has no root of multiplicity greater than one.

10.6 RATIONAL ROOTS OF A POLYNOMIAL OVER THE RATIONAL FIELD

If $f(x)$ is a polynomial of degree $n > 0$ over a field F, then clearly $f(x)$ and $cf(x)$ have the same roots for any nonzero element c of F. If, in particular, $f(x)$ has coefficients in the field **Q** of rational numbers and we choose c as the l.c.m. of the denominators of the coefficients of $f(x)$, $cf(x)$ will have coefficients that are integers. In studying the roots of a polynomial with rational coefficients there is therefore no loss of generality in restricting attention to polynomials that have integral coefficients. We shall now prove the following theorem.

10.42 THEOREM *Let*

$$f(x) = a_n x^n + a_{n-1} x^{n-1} + \cdots + a_0, \qquad (a_n \neq 0),$$

be a polynomial of positive degree n with coefficients that are integers. If r/s is a rational number, in lowest terms, which is a root of the polynomial $f(x)$, then r is a divisor of a_0 and s is a divisor of a_n.

We may recall that by saying that r/s is in lowest terms we mean that r and s are relatively prime integers and $s > 0$. However, the requirement that s be positive plays no role in the proof of this theorem.

Proof: Since r/s is assumed to be a root of $f(x)$, we have that

$$a_n\left(\frac{r}{s}\right)^n + a_{n-1}\left(\frac{r}{s}\right)^{n-1} + \cdots + a_0 = 0.$$

If we multiply throughout by the nonzero integer s^n, we obtain

10.43 $$a_n r^n + a_{n-1}r^{n-1}s + \cdots + a_1 r s^{n-1} + a_0 s^n = 0.$$

By transposing the last term to the right side, this equation can be written in the form

$$(a_n r^{n-1} + a_{n-1}r^{n-2}s + \cdots + a_1 s^{n-1})r = -a_0 s^n.$$

Since all letters here represent integers, we see that the integer $a_0 s^n$ is divisible by the integer r. But we are given that r and s are relatively prime, and it therefore follows that a_0 is divisible by r.

By a similar argument, if in 10.43 we transpose $a_n r^n$ to the other side, we can see that a_n is divisible by s.

As an example of the use of this theorem, let us find all rational roots of the polynomial

$$g(x) = 4x^5 + x^3 + x^2 - 3x + 1.$$

If r/s is a rational number, in lowest terms, which is a root of this polynomial, then r must be a divisor of 1 and s a positive divisor of 4. It follows that $r = \pm 1$, $s = 1, 2,$ or 4; and we see that the only possible rational roots are the following: $1, \frac{1}{2}, \frac{1}{4}, -1, -\frac{1}{2}, -\frac{1}{4}$. It is easy to verify by direct calculation that $g(1) \neq 0$, $g(\frac{1}{2}) = 0$, $g(\frac{1}{4}) \neq 0$, $g(-1) = 0$, $g(-\frac{1}{2}) \neq 0$, and $g(-\frac{1}{4}) \neq 0$. Hence, $\frac{1}{2}$ and -1 are the *only* rational roots. If we divide $g(x)$ by $x - \frac{1}{2}$ and then divide the quotient by $x + 1$, we find that

$$g(x) = (x - \tfrac{1}{2})(x + 1)(4x^3 - 2x^2 + 4x - 2).$$

Any root of this third-degree factor is naturally a root of $g(x)$; so its only possible rational roots are therefore $\frac{1}{2}$ and -1. It is easy to verify that $\frac{1}{2}$ is a root and if we again divide by $x - \frac{1}{2}$, we can express $g(x)$ in the form

$$g(x) = (x - \tfrac{1}{2})^2(x + 1)(4x^2 + 4)$$

or

10.44 $$g(x) = 4(x - \tfrac{1}{2})^2(x + 1)(x^2 + 1).$$

We see therefore that $\frac{1}{2}$ is a double root of $g(x)$. Since the quadratic polynomial $x^2 + 1$ has no rational root, it is a prime polynomial over $\mathbf{Q}$ and hence in 10.44 we have $g(x)$ expressed as a product of prime polynomials over $\mathbf{Q}$. For that matter, the polynomial $x^2 + 1$ is prime over the field $\mathbf{R}$ of real numbers and so 10.44 also gives the factorization of $g(x)$ into prime polynomials over $\mathbf{R}$.

EXERCISES

1. Complete the proof of Theorem 10.42 by showing that s is a divisor of a_n.

2. Prove the following corollary of Theorem 10.42. A rational root of a *monic* polynomial with coefficients that are integers is necessarily an integer which is a divisor of the constant term of the polynomial.

3. Find the factorization of the polynomial $g(x)$ of the example given above into prime factors over the field $\mathbf{C}$ of complex numbers.

4. Find all rational roots of each of the following polynomials over the rational field $\mathbf{Q}$:

    (a) $3x^3 + 5x^2 + 5x + 2$,

    (b) $2x^4 - 11x^3 + 17x^2 - 11x + 15$,

    (c) $x^5 - x^4 - x^3 - x^2 - x - 2$,

    (d) $x^3 + x^2 - 2x - 3$,

    (e) $6x^3 - 7x^2 - 35x + 6$,

    (f) $x^5 + 5x^4 + 13x^3 + 19x^2 + 18x + 8$.

    (g) $x^3 - \tfrac{1}{5}x^2 - 4x + \tfrac{4}{5}$,

    (h) $x^7 + x^6 + x^5 + x^4 + x^3 + x^2 + x + 1$.

5. Find all rational roots of each of the following polynomials over the rational field **Q**, and factor each polynomial into a product of prime polynomials over **Q**:

(a) $9x^4 + 6x^3 + 19x^2 + 12x + 2$,

(b) $x^5 - x^4 - 3x^3 + 6x^2 - 4x + 1$,

(c) $4x^4 + 20x^3 + 33x^2 + 20x + 4$,

(d) $2x^4 + 3x^3 + 4x + 6$.

6. Show that each of the following polynomials over **Q** has no rational root:

(a) $x^{1000} - x^{500} + x^{100} + x + 1$,

(b) $x^{12} - x^9 + x^6 - x^3 + 1$,

(c) $x^m + 2x^{m-1} - 2$, m a positive integer ≥ 2.

10.7 PRIME POLYNOMIALS OVER THE RATIONAL FIELD (OPTIONAL)

It was pointed out in Section 10.5 that every polynomial of the first degree over a field F is necessarily a prime polynomial over F. Also, the first exercise at the end of that section asserts that a polynomial of degree 2 or 3 over F is a prime polynomial over F if and only if it has no root in F. If the field F is now taken to be the field **Q** of rational numbers, it is easy to apply Theorem 10.42 to find whether or not a polynomial of degree at most 3 is prime over **Q**. For a polynomial of higher degree it may be exceedingly difficult to determine whether or not it is prime. For example, a polynomial of degree 4 over **Q** may not have a rational root, and therefore may have no factor of the first degree over **Q**, but may be a product of two prime polynomials of degree 2. In this section we shall give some rather special results, which will enable us to show that for *every* positive integer n, there exist polynomials of degree n that are prime over **Q**.

We shall begin by proving two lemmas, the first of which is the following. As usual, $\mathbf{Z}[x]$ is the ring of polynomials in the indeterminate x with coefficients in the ring $\mathbf{Z}$ of integers.

10.45 LEMMA   *Let $f(x)$, $g(x)$, and $h(x)$ be elements of the ring $\mathbf{Z}[x]$ such that $f(x) = g(x)h(x)$. If p is a prime integer which is a divisor of every coefficient of $f(x)$, then p is a divisor of every coefficient of $g(x)$ or a divisor of every coefficient of $h(x)$.*

Proof: The proof of this lemma was stated as Exercise 10 at the end of Section 10.2, and one method of proof was suggested in a hint given there. We here indicate a more elementary proof which, however, does involve a little more calculation. Let us set

$$f(x) = a_0 + a_1 x + \cdots + a_n x^n,$$
$$g(x) = b_0 + b_1 x + \cdots + b_m x^m,$$

and

$$h(x) = c_0 + c_1 x + \cdots + c_k x^k.$$

We are given that each coefficient a_i $(i = 0, 1, \ldots, n)$ is divisible by the prime p. Suppose now that $g(x)$ has at least one coefficient which is not divisible by p, and also that $h(x)$ has at least one coefficient which is not divisible by p, and let us seek a contradiction. To be more precise, let b_s be the *first* coefficient of $g(x)$, when $g(x)$ is written in increasing powers of x, that is not divisible by p; and let c_t be the *first* coefficient of $h(x)$ that is not divisible by p. Since $f(x) = g(x)h(x)$, by considering the coefficients of x^{s+t} on both sides of this equation, we find that

$$a_{s+t} = \cdots + b_{s-1}c_{t+1} + b_s c_t + b_{s+1}c_{t-1} + \cdots.$$

Now, by our choice of s and t, p is seen to be a divisor of every term on the right except the term $b_s c_t$. Since also p is a divisor of a_{s+t}, it follows that p is a divisor of $b_s c_t$. In view of the fact that p is a prime, this implies that p must be a divisor of b_s or a divisor of c_t. We have therefore obtained the desired contradiction. It follows that either $g(x)$ or $h(x)$ must have all coefficients divisible by p, and the proof is complete.

The following lemma, whose proof will be based on the preceding lemma, shows that a polynomial with *integral* coefficients is prime over the field $\mathbf{Q}$ if and only if it cannot be factored into a product of two polynomials of positive degree with *integral* coefficients. It will then be possible to prove that certain polynomials are prime over $\mathbf{Q}$ by making use of special properties of the integers.

10.46 LEMMA   *Let $f(x)$ be an element of $\mathbf{Z}[x]$ such that $f(x) = g(x)h(x)$, where $g(x)$, $h(x) \in \mathbf{Q}[x]$. Then there exist polynomials $g'(x)$, $h'(x)$ of $\mathbf{Z}[x]$ having the same degrees as $g(x)$ and $h(x)$, respectively, such that $f(x) = g'(x)h'(x)$.*

Proof: Let k be the l.c.m. of the denominators of the coefficients of $g(x)$, so that $kg(x)$ has integral coefficients. Similarly, let l be an integer such that $lh(x)$ has integral coefficients. Since $f(x) = g(x)h(x)$, it follows that

10.47
$$klf(x) = g_1(x)h_1(x),$$

where $g_1(x)$ and $h_1(x)$ have integral coefficients. We may then apply the preceding lemma as follows. If p is a prime divisor of kl, it must be a divisor of all coefficients of $g_1(x)$ or of $h_1(x)$; hence p can be divided from both sides of the equation 10.47, and we still have polynomials with integral coefficients. By a repetition of this process, we can divide out every prime factor of kl and finally get $f(x) = g'(x)h'(x)$, where $g'(x)$ and $h'(x)$ have integral coefficients. It is almost trivial that $g'(x)$ has the same degree as $g(x)$, and also that $h'(x)$ has the same degree as $h(x)$. The proof is therefore complete.

We are now ready to prove the *Eisenstein Irreducibility Criterion.*

10.48 THEOREM   *Let $f(x) = a_0 + a_1x + \cdots + a_nx^n$ be a polynomial of positive degree n over the ring $\mathbf{Z}$ of integers, and p a prime integer such that $a_i \equiv 0 \pmod{p}$ for $i = 0, 1, \ldots, n - 1$; $a_n \not\equiv 0 \pmod{p}$, and $a_0 \not\equiv 0 \pmod{p^2}$. Then $f(x)$ is a prime polynomial over $\mathbf{Q}$.*

Proof: The preceding lemma shows that we need only prove that $f(x)$ cannot be factored into a product of two factors of positive degree over $\mathbf{Z}$. Let us assume that

10.49 $a_0 + a_1x + \cdots + a_nx^n$
$$= (b_0 + b_1x + \cdots + b_mx^m)(c_0 + c_1x + \cdots + c_kx^k),$$

where all these coefficients are integers, and clearly $m + k = n$. Since $a_0 = b_0c_0$, the fact that $a_0 \equiv 0 \pmod{p}$ but $a_0 \not\equiv 0 \pmod{p^2}$ shows that exactly one of the integers b_0 and c_0 is divisible by p. Suppose, for convenience of notation, that $c_0 \equiv 0 \pmod{p}$ and that $b_0 \not\equiv 0 \pmod{p}$. Now $a_n = b_mc_k$ and $a_n \not\equiv 0 \pmod{p}$; so $c_k \not\equiv 0 \pmod{p}$. Let s be chosen as the smallest positive integer such that $c_s \not\equiv 0 \pmod{p}$. From what we have just shown we know that there exists such an integer s and that $0 < s \le k$. Now by a consideration of the coefficients of x^s on both sides of 10.49, we see that

$$a_s = b_0c_s + b_1c_{s-1} + \cdots,$$

and, in view of our choice of s, every term on the right with the single exception of b_0c_s is divisible by p. Moreover, $b_0 \not\equiv 0 \pmod{p}$, and $c_s \not\equiv 0 \pmod{p}$; so $a_s \not\equiv 0 \pmod{p}$. However, by our assumptions, the only coefficient of $f(x)$ that is not divisible by p is the leading coefficient a_n. Hence $s = n$, and therefore we must have $k = n$. This shows that in any factorization of $f(x)$ into a product of polynomials with integral coefficients, one of the factors must have degree n. It follows that $f(x)$ is necessarily a prime polynomial over **Q**.

10.50 COROLLARY   *If n is an arbitrary positive integer, there exist polynomials of degree n over* **Q** *that are prime over* **Q**.

This result is easily established by examples. As an illustration, the polynomial $x^n - 2$ over **Q** satisfies all the conditions of the preceding theorem with $p = 2$. Hence, $x^n - 2$ is a prime polynomial over **Q** for each positive integer n. In like manner, each of the following polynomials of degree n over **Q** is prime over **Q**: $x^n + 2$, $x^n + 3$, $3x^n + 2x^{n-1} + 2x^{n-2} + \cdots + 2x + 2$, $x^n + 9x + 3$ $(n > 1)$. The reader will have no difficulty in constructing other examples.

Perhaps we should emphasize that we have not presented a general method for determining whether or not a given polynomial over **Q** is prime over **Q**. This is a difficult problem, and we shall not discuss it further in this book.

10.8 POLYNOMIALS OVER THE REAL OR COMPLEX NUMBERS

In this section we shall discuss some properties of polynomials over the field **R** of real numbers or the field **C** of complex numbers. We begin with a few remarks, essentially established in elementary algebra, about quadratic polynomials; that is, polynomials of degree 2.

Let

$$g(x) = ax^2 + bx + c, \qquad a \neq 0,$$

be a quadratic polynomial with coefficients in the field **C**. Then it is well-known that the polynomial $g(x)$ has roots r_1 and r_2, where

10.51 $$r_1 = \frac{-b + \sqrt{b^2 - 4ac}}{2a}, \qquad r_2 = \frac{-b - \sqrt{b^2 - 4ac}}{2a}.$$

We may point out that, by a special case of Theorem 7.20, every nonzero complex number has two square roots. Hence r_1 and r_2, given by 10.51, are complex numbers and it is easy to verify by direct calculation that

10.52 $$g(x) = a(x - r_1)(x - r_2).$$

Since these first-degree factors have coefficients in **C**, it is apparent that no quadratic polynomial over **C** is a prime polynomial over **C**.

It is customary to call $b^2 - 4ac$ the *discriminant* of the quadratic polynomial $ax^2 + bx + c$. For convenience, let us designate this discriminant by D.

From 10.51 it follows that $r_1 = r_2$ if and only if $D = 0$. However, the factorization 10.52 holds in any case; so $D = 0$ is a necessary and sufficient condition that the polynomial $g(x)$ have a double root.

Now let us assume that the quadratic polynomial $g(x)$ has *real* coefficients. Then the roots r_1 and r_2 will also be real if and only if $D \geq 0$, for only in this case will D have real square roots. The factorization 10.52 of $g(x)$ into factors of the first degree is therefore a factorization over **R** if and only if $D \geq 0$. If $D < 0$, $g(x)$ has no real root and $g(x)$ is therefore prime over **R**.

Let us summarize some of these observations in the following theorem.

10.53 THEOREM *No quadratic polynomial over the field* **C** *of complex numbers is prime over* **C**. *A quadratic polynomial over the field* **R** *of real numbers is prime over* **R** *if and only if its discriminant is negative.*

We have referred above to Theorem 7.20, where it was proved by use of the trigonometric form of a complex number that every nonzero complex number has n nth roots. It may be worth pointing out that the *square* roots of a complex number may also be computed by an algebraic process. As an illustration, let us seek the roots of the polynomial $x^2 + x - (1 + 3i)$ over **C**. By 10.51, these roots can immediately be written down in the form

10.54
$$\frac{-1 \pm \sqrt{5 + 12i}}{2}.$$

Now in order to express these roots in the usual form of complex numbers, we need to compute the square roots of $5 + 12i$. To do so, suppose that s and t are unknown real numbers such that $s + ti$ is a square root of $5 + 12i$. Thus we have

$$(s + ti)^2 = 5 + 12i,$$

or

$$s^2 - t^2 + 2sti = 5 + 12i.$$

In turn, this implies both of the following equations involving the real numbers s and t:

$$s^2 - t^2 = 5, \qquad 2st = 12.$$

If we solve these two simultaneous equations by elementary methods and remember that s and t are real (so that $s^2 \geq 0$ and $t^2 \geq 0$), we find the solutions to be $s = 3$, $t = 2$ and $s = -3$, $t = -2$. Hence, the square roots of $5 + 12i$ are $\pm(3 + 2i)$. Substituting in 10.54, we find that the roots of the polynomial $x^2 + x - (1 + 3i)$ are $1 + i$ and $-(2 + i)$.

We have shown above that no quadratic polynomial is prime over **C**. Another special case of some interest is the following. Let us consider a polynomial of the form $ax^n + b$, where a and b are nonzero complex numbers and n is an arbitrary positive integer greater than 1. Since, by Theorem 7.20, the complex number $-b/a$ has n distinct nth roots and these are obviously roots of the polynomial $ax^n + b$, Theorem 10.22 asserts that this polynomial can be factored over **C** into a product of factors of the first degree. In particular, such a polynomial can never be prime over **C**.

The general theorem which we shall next state is partially suggested by the special cases already discussed. The theorem was first proved by the famous German mathematician Carl Friedrich Gauss (1777–1855), and is of such importance that it has often been called "The Fundamental Theorem of Algebra."†

10.55   THEOREM   *If $f(x)$ is an element of* **C**$[x]$ *of positive degree, there exists an element of* **C** *which is a root of the polynomial $f(x)$.*

If r is a complex number which is a root of the polynomial $f(x)$ of degree n over **C**, then in **C**$[x]$ we can use the Factor Theorem and write

$$f(x) = (x - r)f_1(x),$$

where $f_1(x)$ is of degree $n - 1$. It is then apparent from this observation and Theorem 10.39 that the preceding theorem can be expressed in either of the following alternate forms.

10.56   THEOREM   *The only prime polynomials of* **C**$[x]$ *are the polynomials of the first degree.*

10.57   THEOREM   *If $f(x)$ is an element of* **C**$[x]$ *of positive degree, then $f(x)$ is itself of the first degree or it can be factored in* **C**$[x]$ *into a product of polynomials of the first degree.*

We next consider the question of which polynomials over the real field **R** are prime over **R**. Of course, the polynomials of the first degree are always prime, and we have shown in Theorem 10.53 that the quadratic polynomials over **R** that are prime over **R** are those with negative discriminant. A little later we shall prove that these are the only prime polynomials over **R**. First, however, we need a preliminary result, which is of some interest in itself.

Let

$$f(x) = a_n x^n + \cdots + a_1 x + a_0$$

---

† A proof of this theorem is given in an appendix.

be a polynomial of positive degree with real coefficients. Since $\mathbf{R} \subset \mathbf{C}$, then also $f(x) \in \mathbf{C}[x]$ and Theorem 10.55 states that there exists an element r of $\mathbf{C}$ such that $f(r) = 0$. We now want to make use of the concept of the conjugate of a complex number, introduced in Section 7.4. We recall that if $u = a + bi$ is a complex number, then the conjugate u^* of u is defined by $u^* = a - bi$. It was shown that the mapping $u \rightarrow u^*$ is a one-one mapping of $\mathbf{C}$ onto $\mathbf{C}$, which preserves the operations of addition and multiplication. Now since $f(x)$ is assumed to have real coefficients and a real number is equal to its conjugate, it is not difficult to verify that

$$[f(r)]^* = a_n(r^*)^n + \cdots + a_1 r^* + a_0 = f(r^*).$$

But, since $f(r) = 0$, it follows that $[f(r)]^* = 0$ and therefore $f(r^*) = 0$. That is, r^* is also a root of the polynomial $f(x)$. This result we state as the following theorem.

10.58 THEOREM *If r is a complex number which is a root of the polynomial $f(x)$ with real coefficients, then the conjugate r^* of r is also a root of $f(x)$.*

If it happens that r is a real number, then $r^* = r$, and this theorem has no content. However, if r is not real, then r^* and r are distinct roots of $f(x)$. It follows that in $\mathbf{C}[x]$ we have

$$f(x) = (x - r)(x - r^*)f_1(x),$$

with the degree of $f_1(x)$ two less than the degree of $f(x)$. If $r = a + bi$, then $r^* = a - bi$, and a simple calculation shows that

$$(x - r)(x - r^*) = x^2 - 2ax + a^2 + b^2.$$

We can therefore write

10.59 $$f(x) = (x^2 - 2ax + a^2 + b^2)f_1(x),$$

and the quadratic factor on the right clearly has *real* coefficients. Since also $f(x)$ has real coefficients, it is easy to verify that $f_1(x)$ must have real coefficients (cf. Exercise 8, Section 10.4). It follows that 10.59 gives a factorization of $f(x)$ in $\mathbf{R}[x]$. If $\deg f(x) > 2$, $f(x)$ can therefore not be a prime polynomial over $\mathbf{R}$. This result, combined with Theorem 10.53, completes the proof of the following theorem.

10.60 THEOREM *The only polynomials of $\mathbf{R}[x]$ that are prime over $\mathbf{R}$ are the polynomials of the first degree and the quadratic polynomials with negative discriminant.*

Now a polynomial of *odd* degree clearly cannot be factored into a product of quadratic polynomials. Therefore, if a polynomial $f(x)$ of $\mathbf{R}[x]$ of odd degree is expressed as a product of prime polynomials over $\mathbf{R}$, at least one of these prime polynomials (in fact, an odd number of them) must be of the first degree. This implies that $f(x)$ has at least one real root, and the following is therefore an almost immediate consequence of the preceding theorem.

10.61 COROLLARY   *A polynomial with real coefficients and of odd degree necessarily has a real root.*

Except for quadratic polynomials, and polynomials of the special form $ax^n + b$, we have not given any indication as to how one might actually *find* the real or complex roots of a given polynomial. This is a difficult problem but some information can be found in texts on the "theory of equations." In particular, there do exist algebraic formulas for the roots of polynomials of degrees 3 or 4 with real or complex coefficients. Although these formulas are of great theoretical interest, they are not convenient to use in a numerical case. It is, however, not too difficult to develop methods of approximating the roots to any desired accuracy, and this is what is usually done in practical applications.

EXERCISES

1. Find the roots of each of the following polynomials and express each root in the standard form $a + bi$ of a complex number:

   (a) $x^2 - (3i - 2)x - 5 - i$,   (d) $x^2 + x + 4$,

   (b) $x^2 + ix + 1$,   (e) $x^2 - x + 2 + \sqrt{2}i$,

   (c) $x^2 - (2 + i)x - 1 + 7i$,   (f) $x^2 + 2x + i$.

2. Factor each of the following polynomials of $\mathbf{R}[x]$ into a product of prime polynomials over $\mathbf{R}$:

   (a) $x^3 - 2x - 4$,   (d) $x^4 + 2x^2 - 8$,

   (b) $x^3 - x^2 - 3x + 6$,   (e) $x^4 + x^3 + 2x^2 + x + 1$,

   (c) $x^4 + 1$,   (f) $x^5 + 1$.

10.9 PARTIAL FRACTIONS (OPTIONAL)

In calculus, partial fractions are used to carry out the integration of rational functions. In this section we shall briefly outline a proof of the existence of the required partial fraction decompositions.

It was pointed out in Section 6.4 that, starting with an integral domain D, it is possible to construct a field of quotients of D whose elements are the formal quotients a/b, where $a, b \in D$ and $b \neq 0$. Now if F is a given field, we know that the polynomial ring $F[x]$ is an integral domain, and we are now interested in the field of quotients of this integral domain. An element of this field of quotients is therefore expressible as $f(x)/g(x)$, where $f(x), g(x) \in F[x]$, $g(x) \neq 0$. Such an element is called a *rational form* over F, and the field whose elements are these rational forms is often called the field of rational forms (in the indeterminate x) over F. This field is usually denoted by $F(x)$. It should be observed that, by the same conventions we made in constructing the rational numbers from the integers, a polynomial is a special case of a rational form, and therefore $F[x] \subset F(x)$. However, we are not now primarily interested in properties of the field $F(x)$, but only in individual elements of this field.

If $g(x)$ is a polynomial over F of positive degree, we know by Theorem 10.39 that $g(x)$ can be expressed uniquely as a product of its leading coefficient times a product of powers of distinct monic polynomials that are prime over F. This fact is implicitly used in the proof of the following theorem.

10.62 THEOREM *A rational form $f(x)/g(x)$ over F is expressible as a polynomial over F plus a sum of rational forms over F of the special type $r(x)/[p(x)]^k$, where $p(x)$ is a prime polynomial over F, $[p(x)]^k$ is a divisor of $g(x)$, and $\deg r(x) < \deg p(x)$.*

Expressing $f(x)/g(x)$ as described in this theorem is said to be "expressing $f(x)/g(x)$ as a sum of partial fractions."

As an illustration of the theorem, it may be verified that over the field of real numbers,

10.63
$$\frac{x^2 + x - 1}{x^3(x^2 + 1)} = \frac{2}{x} + \frac{1}{x^2} - \frac{1}{x^3} - \frac{2x + 1}{x^2 + 1}.$$

In this case, the polynomial mentioned in the theorem is the zero polynomial. Moreover, there are just two prime factors of the denominator, namely, x and $x^2 + 1$, and therefore only these two choices for $p(x)$.

Proof: We prove the theorem in two steps as follows. First, suppose that $g(x) = h(x)k(x)$, where $h(x)$ and $k(x)$ are relatively prime polynomials over F. Then we know that there exist polynomials $s(x)$, $t(x)$ over F such that

$$1 = h(x)s(x) + k(x)t(x),$$

from which it follows that

$$\frac{1}{g(x)} = \frac{s(x)}{k(x)} + \frac{t(x)}{h(x)}.$$

Then, since

$$\frac{f(x)}{g(x)} = \frac{f(x)s(x)}{k(x)} + \frac{f(x)t(x)}{h(x)},$$

we see that $f(x)/g(x)$ is expressible as a sum of rational forms with respective denominators $k(x)$ and $h(x)$. If, say, $h(x)$ is now expressible as a product of two relatively prime polynomials, the same procedure can be applied to the rational form $f(x)t(x)/h(x)$. By a repetition of this process we can finally write $f(x)/g(x)$ as a sum of rational forms having denominators which cannot be expressed as a product of two relatively prime polynomials. Each denominator is then a power of a single prime polynomial, and therefore each of these rational forms is of the type $u(x)/[p(x)]^n$, where $p(x)$ is a prime polynomial over F and $[p(x)]^n$ is a divisor of $g(x)$.

We next consider any one such form of the type $u(x)/[p(x)]^n$. If $\deg u(x) < \deg p(x)$, $u(x)/[p(x)]^n$ is already one of the rational forms described in the statement of the theorem. If $\deg u(x) \geq \deg p(x)$, we use the Division Algorithm and write

10.64 $$u(x) = q_0(x)p(x) + r_0(x),$$

where, as usual, $r_0(x) = 0$ or $\deg r_0(x) < \deg p(x)$. If $\deg q_0(x) \geq \deg p(x)$, we divide $q_0(x)$ by $p(x)$ and obtain

$$q_0(x) = q_1(x)p(x) + r_1(x),$$

and by substitution in 10.64 we obtain

$$u(x) = q_1(x)[p(x)]^2 + r_1(x)p(x) + r_0(x).$$

If $\deg q_1(x) \geq \deg p(x)$, we divide $q_1(x)$ by $p(x)$, and continue this process. We omit the details, but essentially as in the proof of Theorem 5.10 it can be shown that for some positive integer m,

10.65 $u(x) = r_m(x)[p(x)]^m + r_{m-1}(x)[p(x)]^{m-1} + \cdots + r_1(x)p(x) + r_0(x),$

where $r_m(x) \neq 0$ and $r_i(x) = 0$ or $\deg r_i(x) < \deg p(x)$ for $i = 0, 1,$ $\ldots, m$. It follows at once from 10.65 that the rational form $u(x)/[p(x)]^n$ is expressible as a polynomial (possibly zero) plus a sum of forms $r(x)/[p(x)]^k$, with $k \leq n$ and therefore $[p(x)]^k$ a divisor of $g(x)$. Since we have already shown that $f(x)/g(x)$ can be expressed as a sum of terms of the form $u(x)/[p(x)]^n$, the proof is therefore complete.

In a numerical case, the steps of the above proof can be actually carried out in order to express a rational form as a sum of partial fractions. Let us give an illustration by carrying out the calculations involved in establishing the example 10.63 given above. Since x^3 and $x^2 + 1$ are relatively prime, by the usual method involving the Euclidean Algorithm, we find that

$$1 = x^3 \cdot x + (x^2 + 1)(1 - x^2),$$

and therefore

$$\frac{1}{x^3(x^2 + 1)} = \frac{x}{x^2 + 1} + \frac{1 - x^2}{x^3}.$$

If we multiply by the given numerator, $x^2 + x - 1$, we obtain

10.66 $\dfrac{x^2 + x - 1}{x^3(x^2 + 1)} = \dfrac{x^3 + x^2 - x}{x^2 + 1} + \dfrac{-x^4 - x^3 + 2x^2 + x - 1}{x^3}.$

We now consider the first term on the right. By the Division Algorithm we see that

$$x^3 + x^2 - x = (x + 1)(x^2 + 1) - 2x - 1,$$

and therefore

$$\frac{x^3 + x^2 - x}{x^2 + 1} = x + 1 - \frac{2x + 1}{x^2 + 1}.$$

Using this equation, we can write Equation 10.66 in the form

$$\frac{x^2 + x - 1}{x^3(x^2 + 1)} = x + 1 - \frac{2x + 1}{x^2 + 1} - x - 1 + \frac{2}{x} + \frac{1}{x^2} - \frac{1}{x^3},$$

and we conclude that

$$\frac{x^2 + x - 1}{x^3(x^2 + 1)} = \frac{2}{x} + \frac{1}{x^2} - \frac{1}{x^3} - \frac{2x + 1}{x^2 + 1}.$$

It will be observed that in simplifying each of the terms on the right of 10.66 a nonzero polynomial occurred, but no polynomial occurred in the final result. It can be shown that whenever $\deg f(x) < \deg g(x)$, as is true in this example, in expressing $f(x)/g(x)$ as a sum of partial fractions no polynomial occurs. Also, it is proved in more advanced texts that the expression of a rational form as a sum of partial fractions is *unique*. However, we shall not prove these facts here.

Over the real field, which is the case of most importance, the only prime polynomials are of the first or second degree. Hence, in this case, the prime polynomials $p(x)$ in the statement of Theorem 10.62 are of the first or second degree. This fact is important in the proof that every rational function can be integrated in terms of elementary functions.

EXERCISES

Use the method of proof of Theorem 10.62 to express each of the following rational forms over the field of real numbers as a sum of partial fractions.

1. $\dfrac{x^2 - x + 1}{x(x^2 + x + 1)}.$

2. $\dfrac{x^2 + x - 2}{x(x^2 - 1)}.$

3. $\dfrac{x + 1}{x(x^2 + 1)^2}.$

4. $\dfrac{1}{(x - 1)^2(x^2 + 2)}.$

5. $\dfrac{2x^2 + x - 1}{x(x - 1)^3}.$

6. $\dfrac{x^4 + 2x^3 + 4}{(x^2 + 1)^2}.$

10.10   QUOTIENT RINGS $F[x]/(s(x))$

In this last section we shall consider quotient rings of the ring $F[x]$ of all polynomials in an indeterminate x over a field F. Recall from Section 3.4 that a quotient ring $F[x]/S$ is defined whenever we have an ideal S of $F[x]$. Theorem 10.32 tells us that either $S = (0)$ or there is a unique monic polynomial $s(x)$ such that $S = (s(x))$ is the set of all multiples of $s(x)$ by elements of $F[x]$. If $S \neq (0)$ and $\deg s(x) = 0$, then $s(x) = 1$ is necessarily unity so that $S = (1) = F[x]$. In a certain sense the cases $S = (0)$ and $S = F[x]$ are the trivial ones. So for most of our discussion we shall assume that $S = (s(x))$, where $s(x)$ is a polynomial of $F[x]$ of *positive* degree. Our purpose is to study the quotient ring $F[x]/(s(x))$. An element of this ring is a coset of the form

$$f(x) + S, \qquad f(x) \in F[x],$$

where $S = (s(x))$. We start by proving the following theorem.

10.67 THEOREM   *The ring $F[x]/(s(x))$ is a field if and only if $s(x)$ is a prime polynomial over F. In any case, if $s(x)$ has positive degree then the ring $F[x]/(s(x))$ contains a subring isomorphic to the field F.*

Proof: We may invoke Theorem 5.31 in the proof. Since $F[x]$ is a Euclidean domain, this theorem tells us that $F[x]/(s(x))$ is an integral domain if and only if $s(x)$ is a prime polynomial. If $s(x)$ is not prime then $F[x]/(s(x))$ is not even an integral domain, let alone being a field, i.e., $F[x]/(s(x))$ contains nonzero divisors of zero. On the other hand, if $s(x)$ is prime then $F[x]/(s(x))$ is an integral domain in which, by Theorem 5.31, every nonzero element has a multiplicative inverse. That is, $F[x]/(s(x))$ is a field.

Let $S = (s(x))$ and for $f(x) \in F[x]$ define $f(x)\theta = f(x) + S$. By Theorem 3.30 θ is a homomorphism of $F[x]$ onto $F[x]/S$ with kernel the ideal S. Since $s(x)$ has positive degree, every nonzero element of $S = (s(x))$ has positive degree. In particular, $F \cap S = (0)$ contains only the zero polynomial. If we restrict the domain of θ from $F[x]$ to F obtaining a homomorphism $\theta' : F \to F[x]/S$ then the kernel of θ' is $F \cap S = (0)$. By Theorem 3.24 θ' must be an isomorphism of F into $F[x]/S$ completing the proof of the theorem.

In the future we shall often find it convenient to identify F with $F\theta$, its image in $F[x]/S$; that is, we merely change the notation by writing a in place of $a + S$, where $a \in F$, whenever it is clear from the context that we are working in the ring $F[x]/S$.

We proceed to point out a convenient, and somewhat more explicit, way to specify the elements of the ring $F[x]/S$. As soon as we have done this, we shall give a number of examples that may help to clarify the material of this section.

We have assumed that the fixed polynomial $s(x)$ of $F[x]$ has positive degree, say k. If c is a nonzero element of F, then a polynomial is a multiple of $s(x)$ if and only if it is a multiple of $cs(x)$, that is, $(s(x)) = (cs(x))$. Accordingly, there is no loss of generality in assuming that $s(x)$ is a monic polynomial since this would only involve choosing c to be the multiplicative inverse of the leading coefficient of $s(x)$. We shall henceforth assume that $s(x)$ is monic and, for later convenience, we choose the notation

10.68 $$s(x) = x^k - s_{k-1}x^{k-1} - \cdots - s_1x - s_0,$$

the coefficients being elements of F. Now let $f(x) + S$ be an element of $F[x]/S$. By the Division Algorithm, we have $f(x) = q(x)s(x) + r(x)$, where $r(x) = 0$ or $\deg r(x) < k$. Since $q(x)s(x) \in S$, it follows that

$$f(x) + S = r(x) + S.$$

Thus every element of $F[x]/S$ can be expressed in the form

10.69 $$a_0 + a_1x + \cdots + a_{k-1}x^{k-1} + S, \qquad \text{each } a_i \in F.$$

We shall next show that every element is *uniquely* expressible in this form. Let $g(x) = a_0 + a_1x + \cdots + a_{k-1}x^{k-1}$ and $h(x) = b_0 + b_1x + \cdots + b_{k-1}x^{k-1}$, and suppose that $g(x) + S = h(x) + S$. This implies that $g(x) - h(x) \in S$ and since S contains no polynomial of degree less than k, we conclude that $g(x) - h(x) = 0$, or $g(x) = h(x)$. Thus each element of $F[x]/S$ can be expressed in exactly one way in the form 10.69.

We proceed to simplify our notation by introducing a simple symbol for the particular element $x + S$ of $F[x]/S$. Let us agree to set

$$j = x + S.$$

Then $j^2 = (x + S)(x + S) = x^2 + S$ and, more generally, for each positive integer m we have

$$j^m = x^m + S.$$

Now, for example, let us consider an element of the form $(a + bx + cx^2) + S$, where $a, b, c \in F$. Clearly,

$$
\begin{aligned}
(a + bx + cx^2) + S &= (a + S) + (b + S)(x + S) + (c + S)(x^2 + S) \\
&= (a + S) + (b + S)j + (c + S)j^2.
\end{aligned}
$$

We agreed above to write a for $a + S$, $a \in F$, so using this notation we have

$$(a + bx + cx^2) + S = a + bj + cj^2.$$

By generalizing this argument, we see that an element $f(x) + S$ of $F[x]/S$ may be written in the simple form $f(j)$. In particular, from 10.69 it follows that the elements of $F[x]/S$ are uniquely expressible in the form

10.70 $$a_0 + a_1j + \cdots + a_{k-1}j^{k-1}, \qquad \text{each } a_i \in F.$$

Since $s(x) \in S$, $s(x) + S$ is the zero of $F[x]/S$, and we have $s(j) = s(x) + S = 0$. Thus, in particular, the element j is a root of the polynomial $s(x)$ (although j is in the ring $F[x]/S$, not in F). Accordingly, we can operate with the elements 10.70 of $F[x]/S$ by merely considering j to be a symbol such that $s(j) = 0$. The sum of two elements of the form 10.70 is immediately an element of the same form. The product of two elements can be expressed in the form 10.70 by multiplying out in the usual way and then replacing each power of j higher than the $(k-1)$st by the element of the form 10.70 to which it is equal. Since $s(j) = 0$, we have from 10.68 that

10.71
$$j^k = s_0 + s_1 j + \cdots + s_{k-1} j^{k-1},$$

and the right side is of the form 10.70. To compute j^{k+1}, we multiply the preceding equation by j and obtain

$$j^{k+1} = s_0 j + s_1 j^2 + \cdots + s_{k-2} j^{k-1} + s_{k-1} j^k.$$

Now the right side is not of the form 10.70, but we can get it into this form by substituting for j^k from 10.71 and collecting coefficients of the different powers of j. In this way we obtain

10.72
$$j^{k+1} = s_{k-1} s_0 + (s_0 + s_{k-1} s_1) j + \cdots + (s_{k-2} + s_{k-1}^2) j^{k-1}.$$

We could proceed in this way to compute higher powers of j. However, the general formulas are not very useful since it is much easier to apply the method directly in any specific case. The following examples will illustrate how this is done.

> **Example 1** Let F be the field **R** of real numbers, and let $s(x)$ be the polynomial $x^2 + 1$. Since $x^2 + 1$ is prime over **R**, we know by Theorem 10.67 that $\mathbf{R}[x]/(x^2 + 1)$ is a field. We proceed to describe this field in some detail.
>
> As a special case of the general discussion above, we may observe that since $x^2 + 1$ is of the second degree, every element of the field $\mathbf{R}[x]/(x^2 + 1)$ is uniquely expressible in the form
>
> **10.73**
> $$a + bj, \qquad a, b \in \mathbf{R},$$
>
> where $j^2 + 1 = 0$. Let us now characterize the field $\mathbf{R}[x]/(x^2 + 1)$ by specifying the operations of addition and multiplication of the elements 10.73. Addition is trivial since
>
> **10.74**
> $$(a + bj) + (c + dj) = (a + c) + (b + d)j,$$

which is immediately of the form 10.73. As for multiplication, we have

$$(a + bj)(c + dj) = ac + (ad + bc)j + bdj^2,$$

or, replacing j^2 by -1,

10.75 $$(a + bj)(c + dj) = (ac - bd) + (ad + bc)j.$$

We have thus shown how to express the product of two elements of the form 10.73 in the same form. The field $\mathbf{R}[x]/(x^2 + 1)$ can now be simply characterized as the field with elements 10.73 and with addition and multiplication given by 10.74 and 10.75, respectively.

Except for an almost trivial difference in notation, the field we have just constructed coincides with the field $\mathbf{C}$ under the mapping

$$a + bj \rightarrow a + bi, \qquad a, b \in \mathbf{R}.$$

Hence, we have now given another method of constructing the field of complex numbers from the field of real numbers.

Example 2 Let F be the field $\mathbf{Z}_2$ of integers modulo 2, whose two elements we shall now write as 0 and 1; and let $s(x) = x^2 + x + 1$. Since neither of the elements of $\mathbf{Z}_2$ is a root of this quadratic polynomial, it follows that this polynomial is prime over $\mathbf{Z}_2$. Hence the ring $\mathbf{Z}_2[x]/(x^2 + x + 1)$ is a field which, for simplicity, we shall designate by F^*. Since also in this example $s(x)$ is a quadratic polynomial, it follows as in the preceding example that the elements of F^* are uniquely expressible in the form

10.76 $$a + bj, \qquad a, b \in \mathbf{Z}_2.$$

As usual, addition of two of these elements is carried out in an almost trivial way as follows:

10.77 $$(a + bj) + (c + dj) = (a + c) + (b + d)j.$$

As for multiplication, we have as in the previous example

$$(a + bj)(c + dj) = ac + (ad + bc)j + bdj^2.$$

However, since $s(j) = 0$, in this example we have $j^2 + j + 1 = 0$, or $j^2 = -j - 1$. Since our coefficients are from the field $\mathbf{Z}_2$, we can just as well write $j^2 = j + 1$. Replacing j^2 by $j + 1$ in the above expression for $(a + bj)(c + dj)$, we obtain

10.78 $$(a + bj)(c + dj) = (ac + bd) + (ad + bc + bd)j,$$

as the general formula for the product of two elements of F^*. Since there are only two elements of $\mathbf{Z}_2$, and therefore only two choices for a and b in 10.76, we see that F^* has only the four elements 0, 1, j, $1 + j$. Using 10.77 and 10.78 or, better still, simply carrying out the calculations in each case, we can construct the following addition and multiplication tables for F^*.

(+)	0	1	j	$1 + j$
0	0	1	j	$1 + j$
1	1	0	$1 + j$	j
j	j	$1 + j$	0	1
$1 + j$	$1 + j$	j	1	0

(·)	0	1	j	$1 + j$
0	0	0	0	0
1	0	1	j	$1 + j$
j	0	j	$1 + j$	1
$1 + j$	0	$1 + j$	1	j

The reader may verify that this field F^* of four elements is isomorphic to the ring of Example 7 of Section 3.2.

Example 3 Let F be the field $\mathbf{Z}_2$, as in the preceding example, but let $s(x) = x^2$. In this case, $s(x)$ is certainly not prime over $\mathbf{Z}_2$. For convenience, let us denote the ring $\mathbf{Z}_2[x]/(x^2)$ by T. We know then, by Theorem 10.67, that T is not a field. The elements of T are again of the form

$$a + bj, \qquad a, b \in \mathbf{Z}_2,$$

only this time $j^2 = 0$. Using this fact, we can obtain the following addition and multiplication tables for T.

(+)	0	1	j	$1 + j$
0	0	1	j	$1 + j$
1	1	0	$1 + j$	j
j	j	$1 + j$	0	1
$1 + j$	$1 + j$	j	1	0

(·)	0	1	j	$1 + j$
0	0	0	0	0
1	0	1	j	$1 + j$
j	0	j	0	j
$1 + j$	0	$1 + j$	j	1

The addition table coincides with the addition table of Example 2, but the multiplication table is different, as it would have to be since T is not a field.

Example 4 Let $s(x)$ be the polynomial $x^3 + x^2 + 1$ over the field
$\mathbf{Q}$ of rational numbers, and consider the ring $\mathbf{Q}[x]/(x^3 + x^2 + 1)$,
which we shall denote by U. It is easy to verify that $x^3 + x^2 + 1$
has no rational root and, since it is of degree 3, it must therefore be
prime over $\mathbf{Q}$. Hence, U is a field. Since the degree of $s(x)$ is 3 in
this case, it follows, as a special case of 10.70, that the elements of
U are uniquely expressible in the form

10.79 $a + bj + cj^2, \qquad a, b, c \in \mathbf{Q}.$

Let us consider the product of two of these elements as an illustration of the procedure by which we obtained 10.71 and 10.72. Since
$s(j) = 0$, we have that $j^3 + j^2 + 1 = 0$, or

10.80 $j^3 = -1 - j^2.$

If we multiply this equation by j and substitute $-1 - j^2$ for j^3, we
get

$$j^4 = -j - j^3 = -j - (-1 - j^2);$$

that is,

10.81 $j^4 = 1 - j + j^2.$

The product of two elements 10.79 can now be computed by using
the distributive laws and then substituting for j^3 and j^4 from 10.80
and 10.81. If we do so, we finally obtain

10.82 $(a + bj + cj^2)(d + ej + fj^2) = ad - bf - ce + cf$
$$+ (ae + bd - cf)j + (af + be + cd - bf - ce + cf)j^2.$$

The field U can therefore be characterized as the field with elements
10.79, with multiplication given by 10.82, and addition carried out
in the obvious way. It should perhaps be remarked that the formula
10.82 is obviously too complicated to be of much practical use. For
example, it would be extremely difficult to find the multiplicative inverse of a given element of U by using this formula for the product
of two elements. Instead, one would use the method of proof of
Theorem 5.31 in order to carry out such a calculation.

There is one further theorem which is implicit in what we have done, and
which is of sufficient importance to warrant an explicit statement, known as the
"Theorem of Kronecker" (1823–1891) since he was the first to prove and use it.

This particular theorem is so important that it can be made the foundation for a purely algebraic study of fields. The corollary below will illustrate the power of this theorem, by showing that we may always find fields in which polynomials factor into linear factors. We neither have to use only fields of characteristic zero nor must we invoke the Fundamental Theorem of Algebra in order to obtain a field in which a polynomial factors into linear factors.

10.83 THEOREM *If F is a field and $f(x)$ is an arbitrary element of $F[x]$ of positive degree, there exists an extension F' of the field F such that $f(x)$ has a root in F'.*

> **Proof:** If $f(x)$ has a root in F, then the result is trivial with $F' = F$. Otherwise, there exists a factor $s(x)$ of $f(x)$, which is of degree at least two and is prime over F. Then let us set $F' = F[x]/(s(x))$. We know that the field F' is an extension of the field F, and we may therefore also consider $f(x)$ to be a polynomial over F'. Moreover, using the notation in which the element $x + (s(x))$ of F' is denoted by j, we have that $s(j) = 0$. That is, the element j of F' is a root of the polynomial $s(x)$ and therefore also of $f(x)$. This completes the proof.

The following result is an easy consequence of the theorem just established.

10.84 COROLLARY *If F is a field and $f(x)$ is an element of $F[x]$ of positive degree, there exists an extension F^* of the field F with the property that $f(x)$ factors in $F^*[x]$ into factors of the first degree.*

> **Proof:** Using the notation of the proof of the theorem, the Factor Theorem assures us that in $F'[x]$ the polynomial $f(x)$ has $x - j$ as a factor. If $f(x)$ does not factor entirely into factors of the first degree over F'; that is, if $F(x)$ contains a prime factor over F' which is of degree at least two, the process can be repeated by constructing a field F'' which contains F' and in which $f(x)$ has another root. It is clear that by a continuation of this process, there exists a field F^* containing F such that $f(x)$ factors in $F^*[x]$ into factors of the first degree, and the corollary is proved.

In order to illustrate this last corollary, let us construct a field where a particular polynomial factors completely into linear factors.

> **Example 5** Let $s(x)$ be the polynomial $x^3 + x^2 + 1$ over the field **Q** of rational numbers, just as in Example 4. The field U contains the root j of $s(x)$, so that by the Factor Theorem, 10.21, $x - j$ is a factor of $s(x)$ in $U[x]$. Dividing $s(x)$ by $x - j$ we obtain:

10.85 $$x^3 + x^2 + 1 = (x - j)[x^2 + (1 + j)x + (j + j^2)].$$

The second factor on the right side of Equation 10.85 is a prime in $U[x]$, though we do not prove this fact now. In other words, if

$$t(x) = x^2 + (1 + j)x + (j + j^2)$$

then $V = U[x]/(t(x))$ is a field in which $t(x)$ has a root. Normally we would use j to denote this root, but since $j \in U$ we shall replace j by k in the notation of Equation 10.70. In the field V, which contains U as a subfield, the equation analogous to Equation 10.71 is

$$k^2 = -(i + j)k - (j + j^2).$$

In the field V, j and k are roots of $s(x) = x^3 + x^2 + 1$; so that the third root is also in V. That is, for some $w \in V$, w is a root of $t(x)$ hence also of $s(x)$. Using this fact we have

$$\begin{aligned} t(x) &= (x - w)(x - k) \\ &= x^2 - (w + k)x + wk \\ &= x^2 + (1 + j)x + (j + j^2). \end{aligned}$$

Equating the coefficients of x we have

$$-(w + k) = 1 + j$$

so that

$$w = -1 - j - k.$$

By substituting w into $s(x)$ it is possible to show directly that w is a root. An easier way is to show that $wk = j + j^2$. In any case, V is a field in which

$$s(x) = (x - j)(x - k)(x + 1 + j + k).$$

A comment is in order here. It is entirely possible that the field U which contains the root j of $s(x)$ could contain another, and therefore, all roots of $s(x)$. In this case $t(x)$ would not be a prime in $U[x]$. Therefore, the statement that $t(x)$ is a prime in $U[x]$ is not obvious. Let us indicate how one might show that $t(x)$ is irreducible.

Start with the polynomial $s(x)$ and treat it as a function $y = s(x)$ of a real variable. Using elementary calculus and graphing techniques it is possible to prove that $s(x)$ has exactly one root a in the real field **R**. In other words, by the Fundamental Theorem of Algebra, 10.55, $s(x)$ has three unequal roots in the complex field: a, u, and u^* where u^* is the complex conjugate of u.

Recall that the substitution mapping of Section 10.2 is a homomorphism of $Q[x]$ into some ring. We may define two possible substitution homomorphisms, $\theta: Q[x] \rightarrow Q[a]$ such that $x\theta = a$, and $\theta': Q[x] \rightarrow Q[j]$ such that $x\theta' = j$. In each case the kernel is $(s(x))$. By applying the Universal Mapping Property, 3.25, we obtain an isomorphism (since $U = Q[j]$)

$$\psi: U \rightarrow Q[a].$$

In other words, U will contain a root of $s(x)$ other than j if and only if $Q[a]$ contains a root of $s(x)$ other than a. Since $Q[a]$ is a subfield of R, the real field, and since u, u^* are not real numbers, we conclude that a is the only root of $s(x)$ in $Q[a]$. Consequently, j is the only root of $s(x)$ in U. In particular, there are no roots of $t(x)$ in U. Since $t(x)$ is of degree two, it must be a prime in $U[x]$.

This sketch can be filled in to give a proof that $t(x)$ is a prime in $U[x]$ (see Exercise 12). The ideas suggested here are very important and will reappear in Chapter 12.

EXERCISES

1. For the given field F and the given polynomial $s(x)$ over F, construct a multiplication table for the ring $F[x]/(s(x))$. Which of these rings are fields?

   (a) $F = Z_2$, $s(x) = x^2 + 1$;

   (b) $F = Z_3$, $s(x) = x^2 + 1$;

   (c) $F = Z_3$, $s(x) = x^2 + x + 1$;

   (d) $F = Z_2$, $s(x) = x^3 + x + 1$;

   (e) $F = Z_2$, $s(x) = x^3 + x^2 + 1$.

2. Determine all positive integers n with $1 < n \leq 7$ such that $Z_n[x]/(x^2 + x + 1)$ is a field.

3. Discuss the field $Q[x]/(x^2 - 2)$ and verify that it is isomorphic to the field of all real numbers of the form $a + b\sqrt{2}$, where $a, b \in Q$.

4. Discuss the field $Q[x]/(x^3 - 2)$, and describe a field of real numbers to which it is isomorphic.

5. It was stated earlier that for each positive integer n and each positive prime p, there exists a polynomial of degree n with coefficients in the field $\mathbf{Z}_p$ which is prime over this field. Use this fact to show that there exists a field with p^n elements. (See Exercise 11 below.)

6. In each case describe a field in which the given polynomial over the specified field has a root. In particular, give a general formula for the product of two elements of the field you describe.

    (a) $x^3 + x + 1$ over $\mathbf{Q}$,

    (b) $x^3 + x^2 + x + 2$ over $\mathbf{Q}$,

    (c) $x^2 - x + 1$ over $\mathbf{R}$,

    (d) $x^3 + x + 1$ over $\mathbf{Z}_2$.

7. (a) Compute the multiplicative inverse of $1 + j + j^2$ in the field U of Example 4 above, and check by use of Formula 10.82.

    (b) Verify that in $U[x]$,

    $$x^3 + x^2 + 1 = (x - j)(x^2 + (1 + j)x + j + j^2).$$

8. An ideal P in a commutative ring R is said to be a *prime ideal* if whenever $a, b \in R$ such that $ab \in P$, then $a \in P$ or $b \in P$. Determine all prime ideals in the ring $\mathbf{Z}$ and in the ring $F[x]$, where F is a field.

9. Prove: If R is a commutative ring with unity, an ideal $P \neq R$ in R is a prime ideal if and only if R/P is an integral domain.

10. Let p be a positive prime and n a positive integer, and consider the polynomial $f(x) = x^{p^n} - x$ as a polynomial over the field $\mathbf{Z}_p$. By Corollary 10.84, there exists an extension F of $\mathbf{Z}_p$ such that $f(x)$ factors in $F[x]$ into factors of the first degree. Prove that $f(x)$ has p^n distinct roots in F. *Hint:* See Exercise 6 of Section 10.5.

11. In the notation of the preceding exercise, prove that the p^n distinct roots of $f(x)$ in F are the elements of a subfield of F, and hence conclude that there exists a field with p^n elements. *Hint:* See Exercise 8 of Section 6.3.

12. Prove that the polynomial $t(x)$ of Example 5 is a prime in $U[x]$.

COMMENTARY

1 Polynomials have a long and interesting history. From the Ahmes papyrus (c. 1650 B.C.) we know that Egyptians were solving simple linear polynomial equations in certain special cases. Of course, in our notation, polynomials did not exist until about 1700 so that what actually were solved by the Egyptians were problems which *today* we would consider to be ones involving polynomial equations. Certain Indian literature from about 600 B.C. shows that the Hindus knew how to solve quadratic equations. From the time of Hippocrates (c. 460 B.C.) until the time of Diophantus (c. 275) the Greeks solved many problems which we would now state in terms of polynomials. However, up until Diophantus, Greek "algebra" was all geometric. That is, magnitudes were represented by line segments so that the problem of finding the roots of a given quadratic equation really involved giving a straightedge and compass construction for line segments representing the roots.

A real symbolic algebra started to appear in India and Arabia after about A.D. 400. Between A.D. 800 and 1100 the Arabs developed algebra to the point where they were considering certain cubic equations. During all this time, there was a fairly steady infusion of ideas into the Middle East from the Orient. Until about 1200, the unenlightened Europeans did little or nothing for algebra. In a Commentary in Chapter 2, we have discussed the history of polynomial equations from 1200 up to about 1840 which brings our survey up to this date.

Between the time of Diophantus (c. 275) and the time of Descartes (1596–1650) another major change was taking place. The idea slowly evolved that algebraic number and geometric magnitude were just different instances of the same concept. Egyptian surveyors laid out their towns using geometric coordinates. Ancient Hindus worked simple arithmetic problems by manipulating simple counters. The concept of geometric magnitude developed along with coordinate systems: the Greeks used methods of latitude and longitude to plot the stars. During the same period, the concept of algebraic and symbolic number evolved: merchants painstakingly evolved accurate and reliable methods for reckoning and keeping track of numbers. Others worked out a primitive symbolic algebra as a means to state and solve problems about numbers.

The first wedding of these ideas appears in work of the Arab mathematician al-Khowârizmi (c. 825) where, in solving problems, he computed via algebra and used rectangular areas to represent algebraic quantities. But it was not until 1637 when Descartes invented his analytic geometry that the joining of these two concepts of number was fairly complete. At last, the full power of the ideas of algebra and geometry could be brought to bear upon one another. This union was a great one for mathematics. From it sprang the ideas of the calculus which was actively developed from 1650 onward.

Another significant idea came to fruition in 1822 when Poncelet (1788–1867) published his work upon projective geometry, where central projection plays a very prominent role. Poncelet was in prison (without access to books) during 1813–1814, and it was during this period that he worked out many of his ideas on geometry. The concept of a central projection is not hard to describe. Imagine a light located at the origin of three-dimensional space. If a line or circle or other geometric figure is drawn on one fixed plane in space, then the figure will cast a shadow on a second fixed plane in space. This "shadow casting" from one plane to another is called a *central projection* from the first plane onto the second. The light rays correspond the points of the first plane with those of the second (provided, of course, that neither plane contains the light source). The light rays parallel to the second plane hit the first plane in a line, and we imagine the nonexistent "shadow" of this line on the second plane to be cast "just beyond" the second plane. This "just beyond" line of the second plane is called the "line at infinity." Certain light rays will pass through the first plane and continue on away from the second plane. For these rays we may imagine that the first plane is a strange mirror, where the rays are reflected *back through* the light source meeting the second plane on the other (wrong) side of the light source. If our first plane is perpendicular to the z-axis, and our figure is a circle with its center on the z-axis then the family of light rays from the source passing through the circle will form a cone with an equation $x^2 + y^2 = kz^2$, where $k > 0$ depends upon the radius of the circle and the position of the first plane. Evidently, the "shadow" of the circle on the second plane will be a conic section (i.e., circle, ellipse, parabola, or hyperbola). The equations for the shadow are given by the equation for the cone and the equation for the second plane. In general, if we consider curves given by polynomials, central projections will change these curves into other curves given by ratios of polynomials.

Using the light rays, we may identify corresponding points on different planes. This way we may imagine that there is only one plane with its "line at infinity." This one plane is called the "real projective plane" (see the Commentary of Chapter 6). The change that a central projection can cause in passing from the first plane to the second can now be imagined to take place in this single plane. Viewed in this way, central projections may be thought of as taking place in a single plane.

Of great importance, are the properties of a curve which do not change by such projections. These properties are called invariants. Invariant theory evolved about 1841 and thrived until about 1900. The computation of certain invariants allowed algebraists to prove theorems about geometry. Some of the leaders in invariant theory included Cayley (1821–1895), Sylvester (1814–1897), Gordon (1837–1912), and Hilbert (1862–1943).

The central projections are not the only interesting ones. Algebraists also consider all one-one onto continuous mappings θ of space such that both θ and θ^{-1} are given by polynomial mappings. These are called birational transformations. The view being put forward here is certainly a far cry from the original ancient Egyptian works upon polynomials. What properties of curves remain invariant under these new kinds of transformations? Consideration of this and related questions has led to the branch of algebra called algebraic geometry. It is certainly correct to say that of all branches of algebra, this area is today one of the most prominent. It brings together many important ideas from analysis, geometry, topology, and algebra [102, 103, 107, 144].

2   The ring $F[x]$ of polynomials over a field F is a principal ideal domain; every ideal can be generated by a single polynomial. The ideal (x, y) in the ring $F[x, y]$ takes two polynomials (for instance, x, y) to generate it. This suggests a generalization: assume that $x_1, x_2, \ldots, x_n$ are indeterminates and I is an ideal in the polynomial

ring $F[x_1, x_2, \ldots, x_n]$ over the field F. Does I have a finite basis, i.e., are there a finite number of polynomials in I which suffice to generate I? Posed in quite different terms this question was a major one from nineteenth century invariant theory. The solution today is known as the Hilbert Basis Theorem [14 Vol. 1, 24], and the answer is that I does have a finite basis. It was P. Gordon (1837–1912) who first proved this theorem for ideals in $F[x, y]$ in 1868. For the next twenty years, Gordon and others were able to extend these results to certain special kinds of ideals in $F[x_1, x_2, x_3]$ and $F[x_1, x_2, x_3, x_4]$. These proofs were extremely difficult, involving intricate computations with immense polynomials. It seemed apparent that there were limitations to what man could prove because of the vast computational difficulties. Actually, even though they worked with polynomials, the way the question was posed, it did not seem to involve ideals, and it apparently only involved certain special kinds of polynomials.

It was D. Hilbert (1862–1943) who in 1888 tore away the obfuscation and showed that the more general question in the form we have posed it could be settled with a rather short argument. This theorem so astounded the mathematical community that Gordon exclaimed, "This is not mathematics; it is theology"; though upon later reflection he admitted that, "Theology has its uses."

These comments should be sufficient to introduce David Hilbert, one of the most profound mathematical thinkers of modern times. He was not a child prodigy, and even in his university work he was more thorough and systematic than brilliant. His originality as a computer pales beside his feeling for the direction and scope of mathematics. He carried out researches in number theory, set theory, algebra, mathematical physics, and analysis. In each case, his influence was so profound as to redefine part of the subject matter and give it new directions.

His insight into the nature of important mathematics, and his view of major problems was so clear that, in his address to the 1900 International Congress of Mathematicians, he posed twenty-three problems which essentially declared the direction of mathematics in this century almost to the present day [188]. The solution of one of Hilbert's 23 problems is occasion for great rejoicing, and brings much fame to the solver. Matiyasevič's negative solution in 1970 to Hilbert's tenth problem (see the Commentary after Chapter 13) created great astonishment throughout the mathematical community [i]. Recently the American Mathematical Society held a special meeting at which the major topic was the current state with regard to Hilbert's problems.

There is an excellent biography of Hilbert by Constance Reid [150]. Included in the book is the preface to Hilbert's address to the 1900 Congress. The book makes contemporaries of Hilbert (such as the brilliant Minkowski (1864–1909)) come alive, while giving an excellent account of Hilbert's mathematical career.

3   Polynomial rings are among the most useful in all of algebra. In Chapter 12, we will see their connection with fields. In Chapter 15, we will see their connection with linear algebra. Herstein [12], Goldstein [11], Birkhoff and MacLane [3], van der Waerden [24], and McCoy [48] are just a few of the references which give good coverage of the topics in this chapter.

In Theorem 10.39 we proved that for a field F, $F[x]$ is a unique factorization domain. If $x_1, x_2, \ldots, x_n$ are indeterminates, then $F[x_1, x_2, \ldots, x_n]$ is also a unique factorization domain. If $n > 1$ then $F[x_1, x_2, \ldots, x_n]$ is neither a principal ideal domain nor a Euclidean domain. In fact, the ideal (x_1, x_2) is not principal. These points are discussed in [12, 24]. It should also be remarked that $\mathbf{Z}[x]$ is a unique factorization domain that is not a principal ideal domain. Incidentally, in [48, 50] and Chapter 16, it is shown that a principal ideal domain is a unique factorization domain.

A chain of ideals $I_1 \subset I_2 \subset I_3 \subset \cdots$ in a polynomial ring $S = F[x_1, x_2, \ldots, x_n]$ over a field F is called an ascending chain. Any such chain can contain only a finite number of distinct ideals of S. If T is a commutative ring with unity such that any ascending chain of ideals contains only a finite number of distinct ideals, then T is called a *Noetherian* ring. Therefore, S is a Noetherian ring. This important class of rings is discussed in [24, 48, 50].

chapter eleven

VECTOR SPACES

In this chapter we shall introduce and study a new class of algebraic systems called *vector spaces*. The concept of a vector, used here to indicate an element of a vector space, is a generalization and abstraction of the concept of vector as the term is used in physics. We begin with a short discussion of vectors in the latter sense, as a partial motivation of the material to follow. We then proceed to give the abstract definition of a vector space and to establish some of the most important properties of such systems.

11.1 VECTORS IN A PLANE

Physical entities such as displacements, forces, and velocities have both magnitude and direction, and are usually called *vectors*. Geometrically, a vector may be represented by a directed line segment, the length of the segment indicating the magnitude of the vector and the direction of the segment specifying the direction of the vector. We shall here consider vectors in a given coordinate plane and, moreover, shall represent the vectors by directed segments emanating from the origin of coordinates. A vector X, as shown in Figure 1, is then completely determined by the coordinates (r, s) of its terminal point. Accordingly, we may just as well call (r, s) the vector; that is, we may identify the vector X with the ordered pair of real numbers that specifies its terminal point, and write $X = (r, s)$.

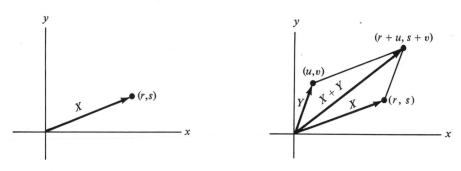

FIGURE 1

FIGURE 2

If X and Y are vectors in the same plane, the so-called parallelogram law states that the *sum* (or resultant) $X + Y$ of these vectors is the vector determined by the diagonal of the parallelogram, two of whose adjacent sides are the line segments representing the vectors X and Y, respectively (Figure 2). If $X = (r, s)$ and $Y = (u, v)$, it is not difficult to show that

11.1 $$X + Y = (r + u, s + v).$$

From our point of view, we propose to take 11.1 as the *definition* of the sum of two vectors. Let us now denote by V_2 the set of all vectors (x, y) where x and y are real numbers. Then, using 11.1 as the definition of addition, we have an operation of addition defined on V_2, and it is easy to verify that V_2 is an abelian group with respect to this operation. The identity $(0, 0)$ of this group we may call the *zero vector*.

When using vectors in physical problems it is customary to consider, for example, that $2X$ is a vector having twice the magnitude of X and the same direction as X. The vector $(-2)X$ is considered to be the vector having twice the magnitude of X and the opposite direction to X. In general, if a is a real number, aX has magnitude $|a|$ times the magnitude of X and has the same or opposite direction to X according as $a > 0$ or $a < 0$. If $X = (r, s)$, simple geometric considerations show that if a is any real number,

11.2 $$aX = a(r, s) = (ar, as).$$

Now real numbers are often called *scalars* to distinguish them from vectors, so 11.2 gives us a multiplication of vectors by scalars. Again, although this scalar multiplication was suggested by a procedure used in physics, we propose to take 11.2 as the *definition* of scalar multiplication.

The set V_2 with addition defined by 11.1 and scalar multiplication defined by 11.2 is an example of a vector space according to the definition to be given presently. Actually, since the scalars are real numbers, it is customary to call V_2 a vector space *over the field* **R** of real numbers and to recognize this fact by the more explicit notation $V_2(\mathbf{R})$, instead of V_2 as used so far.

It is apparent that the above procedure could be generalized in various ways. For example, a consideration of vectors in space instead of in a plane would lead to vectors designated by ordered triples of real numbers, and the set of all such vectors with analogous definitions of addition and scalar multiplication would be a vector space which we might denote by $V_3(\mathbf{R})$. From a purely algebraic point of view, there is no reason why we might not go on and consider vectors to be ordered n-tuples of real numbers, and likewise the field **R** might be replaced by any other field. Actually, we shall not be so much concerned with the nature of the vectors themselves as with certain formal properties involving addition and scalar multiplication of vectors. Of course, the particular properties which are of primary interest are those which will be exhibited in the definition of a vector space we now proceed to give.

11.2 DEFINITION AND SIMPLE PROPERTIES OF A VECTOR SPACE

The following definition will assign a precise meaning to some of the terms which have been used in the preceding section.

11.3   DEFINITION   Let F be a field, and V a nonempty set on which there is defined an operation of addition. The elements of F and of V may be called *scalars* and *vectors*, respectively. We assume that there is also defined on V a scalar multiplication by elements of F; that is, if $a \in F$ and $X \in V$, then aX is a uniquely determined element of V. The set V is then called a *vector space over the field F* if the following conditions are satisfied:

(i) V is an abelian group with respect to addition,

(ii) $a(X + Y) = aX + aY$, $a \in F$; X, $Y \in V$,

(iii) $(a + b)X = aX + bX$, a, $b \in F$; $X \in V$,

(iv) $a(bX) = (ab)X$, a, $b \in F$; $X \in V$,

(v) $1X = X$,   1 the unity of F, $X \in V$.

When it is desirable to exhibit explicitly the particular field F over which V is a vector space, we shall find it convenient to designate the vector space by $V(F)$. However, we shall omit the "F" when the context makes it clear what field is being considered.

Let us now clarify the preceding definition by several examples.

Example 1   Let F be an arbitrary field, n a positive integer, and let $V_n(F)$ be the set of all ordered n-tuples of elements of F. That is, the elements of $V_n(F)$ are of the form $(a_1, a_2, \ldots, a_n)$, where $a_i \in F$ $(i = 1, 2, \ldots, n)$. In $V_n(F)$ we define addition as follows:

11.4 $(a_1, a_2, \ldots, a_n) + (b_1, b_2, \ldots, b_n) = (a_1 + b_1, a_2 + b_2, \ldots, a_n + b_n)$.

Moreover, if $c \in F$, we define scalar multiplication in the following way:

11.5 $c(a_1, a_2, \ldots, a_n) = (ca_1, ca_2, \ldots, ca_n)$.

It is then easy to verify that all properties of a vector space are satisfied, and hence that $V_n(F)$ is a vector space over the field F. Vector spaces of this type are of great importance and will be referred to often in the future. We shall consistently use the notation $V_n(F)$ to designate this vector space.

It is obvious that 11.4 and 11.5 are generalizations of 11.1 and 11.2, so that the use of $V_2(\mathbf{R})$ to designate the vector space discussed in the preceding section is consistent with the more general notation introduced in this example.

Example 2   We now modify the preceding example by using *infinite* sequences of elements of F. Let $W(F)$ be the set of all infinite sequences of elements of F; that is, all expressions of the form

$$(a_1, a_2, a_3, \ldots), \qquad a_i \in F \, (i = 1, 2, 3, \ldots).$$

In $W(F)$ we define addition and scalar multiplication as follows:

$$(a_1, a_2, a_3, \ldots) + (b_1, b_2, b_3, \ldots) = (a_1 + b_1, a_2 + b_2, a_3 + b_3, \ldots),$$

and

$$c(a_1, a_2, a_3, \ldots) = (ca_1, ca_2, ca_3, \ldots).$$

Then $W(F)$ is a vector space over F.

If, instead of using *all* infinite sequences of elements of F, we use just those infinite sequences in which at most a finite number of elements of F are different from zero, we again obtain a vector space. (Cf. Exercise 8 at the end of this section.)

Example 3 Let H be a given field and F a subfield of H. We can consider H to be a vector space over the field F if we use as addition the addition already defined in the field H, and define scalar multiplication in the following obvious way. If $a \in F$ and $c \in H$, let ac be the product of these elements as already defined in the field H. The vector space that we obtain in this way might be designated by $H(F)$. In this special case, Property (i) of Definition 11.3 is satisfied since we are using the additive group of the field H. Properties (ii) and (iii) follow from the distributive laws in H, and Property (iv) is just the associative law in H.

Example 4 Let $F[x]$ be the set of polynomials in an indeterminate x over a field F. We use the usual addition of polynomials, and scalar multiplication is defined in the following way. If $a \in F$ and $f(x) \in F[x]$, then $af(x)$ is the polynomial obtained from $f(x)$ by multiplying all its coefficients by a. Of course, this coincides with the product of two polynomials, as defined previously, where one of the polynomials is just a. It is easily verified that $F[x]$ is then a vector space over the field F.

This example and the preceding one have a common generalization as follows. Let R be a ring which contains a field F as a subring, with R and F having the same unity. Then R is a vector space over F, using addition as already defined in R and scalar multiplication as ring multiplication of elements of R by elements of F. The vector space of Example 3 is obtained by specializing the ring R to be a field, and the vector space of Example 4 by taking R to be the ring $F[x]$ of polynomials in x over F.

Example 5 Let T be the set of all polynomials in an indeterminate x over a field F that have degree at most three, together with the zero polynomial. Then T is a vector space over F if we define addition and scalar multiplication as in the preceding example. We may point out, however, that T is not a ring since it is not closed under multiplication of polynomials.

Now that we have given the definition of a vector space and exhibited some examples, let us prove a few simple properties of vector spaces in general. If X is an element of a vector space V, we shall use the familiar notations of abelian groups with respect to the operation of addition. In particular, we shall denote the additive inverse of X by $-X$. The identity of the abelian group V we shall call the *zero vector* and, for the moment, we shall designate it by O to distinguish it from the zero 0 of the field F. The most fundamental properties of vector spaces are stated in the following theorem.

11.6   THEOREM   *Let V be a vector space over the field F, and let O be the zero vector of V. The following are then true:*

   (i) If $a \in F$, then $aO = O$,

   (ii) If $X \in V$, then $0X = O$,

   (iii) If $a \in F$ and $X \in V$, then $a(-X) = (-a)X = -(aX)$,

   (iv) If $aX = O$, then $a = 0$ or $X = O$.

   Proof: We shall prove (i) and (iii), and leave the proofs of the others as exercises.

   To prove (i), let $a \in F$ and $X \in V$. Then

$$aX = a(X + O) = aX + aO,$$

by Definition 11.3(iii). This shows that aO is the zero vector; that is, that $aO = O$.

   One part of (iii) is a consequence of the following calculation:

$$O = aO = a(X - X) = aX + a(-X).$$

It follows that $a(-X)$ is the additive inverse of aX, and hence that $a(-X) = -(aX)$. In this calculation we have tacitly used 11.3(ii) and 11.6(i). To get the other part of (iii) we use the following calculation in which we assume the truth of 11.6(ii) although the proof has not been written out, and also make use of 11.3(iii):

$$O = 0X = (a - a)X = aX + (-a)X.$$

Hence, $(-a)X = -(aX)$, as we wished to show.

Up to this point we have used different symbols to designate the zero vector and the zero element of the field F. However, in the future we shall not find it necessary to distinguish between these zeros since the context will always make it clear which one is intended. Accordingly, we shall henceforth use the familiar symbol 0 to designate either the zero vector or the zero scalar.

If V is a vector space over a field F, a nonempty subset U of V is naturally called a *subspace* of V if U is itself a vector space over F with respect to the addition and scalar multiplication already defined in V. The following theorem is helpful in identifying subspaces of a given vector space.

11.7   THEOREM   *A nonempty subset U of a vector space V over a field F is a subspace of V if and only if U is closed under addition and scalar multiplication.*

Proof: If U is a subspace of V, it is trivial that U must be closed under addition and scalar multiplication. To show the converse, suppose that U is a nonempty subset of the vector space V, which is closed under addition and scalar multiplication, and let us show that U is indeed a vector space. If $X \in U$ and 1 is the unity of F, we have $(-1)X \in U$. But, by 11.6(iii) and 11.3(v), we see that

$$(-1)X = -(1X) = -X,$$

so that the additive inverse $-X$ of X is in U. Since U is closed under addition, Theorem 2.30 now shows that U is a subgroup of V; that is, that part (i) of Definition 11.3 is satisfied. Properties (ii)–(v) hold in U since they hold in the larger set V. Hence, U is a vector space over F and therefore is a subspace of V.

As an illustration of the use of this theorem, let us show that the set W of all elements of the vector space $V_3(F)$ of the form $(x + 2y, y, -x + 3y)$, where x and y are elements of F, is a subspace of $V_3(F)$. If $(a + 2b, b, -a + 3b)$ and $(c + 2d, d, -c + 3d)$ are elements of W, we see that their sum can be written in the form

$$(a + c + 2(b + d), b + d, -(a + c) + 3(b + d)),$$

and this is seen to be the element of W in which $x = a + c$ and $y = b + d$. This shows that W is closed under addition. Also, if $(a + 2b, b, -a + 3b)$ is an element of W and $r \in F$, it follows that

$$r(a + 2b, b, -a + 3b) = (ra + 2rb, rb, -ra + 3rb),$$

and we have the desired form with $x = ra$ and $y = rb$. Hence, W is also closed under scalar multiplication and, by the preceding theorem, W is therefore a subspace of $V_3(F)$.

EXERCISES

1. In each case, using natural definitions of addition and scalar multiplication which of the following are vector spaces over the indicated field?

(a) The set of all real numbers of the form $a + b\sqrt{2} + c\sqrt[3]{3}$, where a, b, and c are elements of the field $\mathbf{Q}$ of rational numbers; field $\mathbf{Q}$.

(b) The set of all polynomials of degree greater than five over a field F; field F.

(c) The set of all real functions f such that $f(x + 1) = f(x)$; field $\mathbf{R}$ of real numbers.

(d) The set $\{0, x + 2, 2x + 4, 3x + 1, 4x + 3\}$ of polynomials in the indeterminate x over the field $\mathbf{Z}_5$; field $\mathbf{Z}_5$.

(e) The set of all polynomials with zero constant terms over a field F; field F.

2. Prove Theorem 11.6(ii) and (iv).

3. Let V be a vector space over a field F. Prove each of the following "cancellation laws":

(i) If $a, b \in F$ and X is a nonzero element of V such that $aX = bX$, then $a = b$.

(ii) If $X, Y \in V$ and a is a nonzero element of F such that $aX = aY$, then $X = Y$.

4. If F is a field and a_1, a_2, and a_3 are fixed elements of F, show that the set of all ordered triples (x_1, x_2, x_3) of elements of F such that $a_1x_1 + a_2x_2 + a_3x_3 = 0$ is a subspace of $V_3(F)$.

5. Find all subspaces of $V_2(\mathbf{Z}_2)$; of $V_3(\mathbf{Z}_2)$.

6. (a) How many elements are there in the vector space $V_n(\mathbf{Z}_p)$?

(b) Show that the number of elements in any subspace of $V_n(\mathbf{Z}_p)$ is of the form p^k for some nonnegative integer k.

7. Which of the following are subspaces of $V_3(\mathbf{R})$?

   (a) The set of all elements of the form $(x, 2y, 3z)$, where $x, y, z \in \mathbf{R}$.

   (b) The set of all elements of the form (x, y, z), where x, y, and z are rational numbers.

   (c) The set of all elements of the form $(x, 2x, x + 1)$, where $x \in \mathbf{R}$.

   (d) The set of all elements of the form $(x, 0, z)$, where $x, z \in \mathbf{R}$.

   (e) The set of all elements of the form $(x, y, 2)$, where $x, y \in \mathbf{R}$.

   (f) The set of all elements of the form $(x + 2y, x - 3z, 2x + y + z)$, where $x, y, z \in \mathbf{R}$.

8. Verify that the set of all elements of the vector space $W(F)$ of Example 2 in which at most a finite number of elements of F are different from zero is a subspace of $W(F)$.

11.3 LINEAR DEPENDENCE

Throughout this section V will denote a vector space over a field F. Unless otherwise explicitly stated, when we refer to a set $\{X_1, X_2, \ldots, X_m\}$ of vectors, it will be understood that m is a positive integer and the set is therefore a nonempty finite set. The concept which we now define plays a central role in the study of vector spaces.

11.8 DEFINITION A set $\{X_1, X_2, \ldots, X_m\}$ of vectors of a vector space V is said to be a *linearly dependent* set if there exist elements $a_1, a_2, \ldots, a_m$ of F, *not all of which are zero*, such that

11.9 $$a_1 X_1 + a_2 X_2 + \cdots + a_m X_m = 0.$$

If the set $\{X_1, X_2, \ldots, X_m\}$ is not linearly dependent, it is said to be *linearly independent*.

As a matter of language, we shall also sometimes say that the vectors $X_1, X_2, \ldots, X_m$ are linearly dependent or independent according as the set $\{X_1, X_2, \ldots, X_m\}$ is linearly dependent or independent.

Let us emphasize the meaning of the definition by the following remarks. Certainly, a relation of the form 11.9 will always hold if the a's are all equal to zero. If such a relation holds *only* in this case, the set $\{X_1, X_2, \ldots, X_m\}$ is linearly independent. However, if a relation 11.9 holds with at least one of the a's unequal to zero, the set $\{X_1, X_2, \ldots, X_m\}$ is linearly dependent.

Example 1 Let us consider the set $\{X_1, X_2, X_3\}$, where these are the following vectors of $V_3(\mathbf{R})$: $X_1 = (1, 3, 2)$, $X_2 = (1, -7, -8)$, $X_3 = (2, 1, -1)$. It is easily verified that $3X_1 + X_2 - 2X_3 = 0$, and hence that this set is linearly dependent.

Example 2 Find whether the vectors $X_1 = (2, 1, 1, 1)$, $X_2 = (1, 3, 1, -2)$, and $X_3 = (1, 2, -1, 3)$ of $V_4(\mathbf{R})$ are linearly dependent or independent.

In order to solve this problem we need to determine whether there exist real numbers y_1, y_2, y_3, not all of which are zero, such that

11.10
$$y_1 X_1 + y_2 X_2 + y_3 X_3 = 0.$$

Using the definitions of addition and scalar multiplication in $V_4(\mathbf{R})$, Equation 11.10 is equivalent to the following system of simultaneous equations in the unknowns y_1, y_2, and y_3:

$$2y_1 + y_2 + y_3 = 0,$$
$$y_1 + 3y_2 + 2y_3 = 0,$$
$$y_1 + y_2 - y_3 = 0,$$
$$y_1 - 2y_2 + 3y_3 = 0.$$

Systems of equations of this form will be studied in detail in the next chapter, but a systematic use of the methods of elementary algebra is sufficient for our present purposes. First, we proceed as follows to eliminate y_1 from every equation but one. Let us multiply the second equation by 2 and subtract it from the first; then also subtract the second equation from the third and from the fourth. There then results the following system of equations:

$$-5y_2 - 3y_3 = 0,$$
$$y_1 + 3y_2 + 2y_3 = 0,$$
$$-2y_2 - 3y_3 = 0,$$
$$-5y_2 + y_3 = 0.$$

We could, by a similar method, proceed to eliminate y_2 from all the equations of this new system except, say, the first. However, it is not necessary to do so, for if we subtract the first equation from the last, we find that $4y_3 = 0$, and hence we must have $y_3 = 0$. Now, setting

$y_3 = 0$ in the first equation, we see that $y_2 = 0$. Finally, if we set $y_2 = y_3 = 0$ in the second equation, it follows that $y_1 = 0$. We have shown that if y_1, y_2, y_3 are real numbers such that 11.10 holds, then $y_1 = y_2 = y_3 = 0$, and the vectors X_1, X_2, X_3 are therefore linearly independent.

We now state in the following theorem a number of simple, but fundamental, properties of linear dependence or independence.

11.11 THEOREM   *In the following, the indicated vectors are elements of a vector space V over a field F.*

(i) *The set $\{X_1, X_2, \ldots, X_m\}$ is a linearly dependent set if one of the vectors of the set is the zero vector.*

(ii) *The set $\{X\}$, consisting of the one vector X, is linearly independent if and only if $X \neq 0$.*

(iii) *If the set $\{X_1, X_2, \ldots, X_m\}$ is linearly independent, then any nonempty subset of this set is linearly independent.*

(iv) *If the set $\{X_1, X_2, \ldots, X_m\}$ is linearly dependent, then the set $\{X_1, X_2, \ldots, X_m, X\}$ is linearly dependent for each $X \in V$.*

(v) *If $\{X_1, X_2, \ldots, X_m\}$ is a linearly independent set and if b_i, $c_i \in F$ $(i = 1, 2, \ldots, m)$ such that*

$$b_1 X_1 + b_2 X_2 + \cdots + b_m X_m = c_1 X_1 + c_2 X_2 + \cdots + c_m X_m,$$

then $b_i = c_i$ $(i = 1, 2, \ldots, m)$.

(vi) *If $X_i \in V$ $(i = 1, 2, \ldots, m)$ and $r_i \in F$ $(i = 2, 3, \ldots, m)$, are such that the set $\{X_2 + r_2 X_1, X_3 + r_3 X_1, \ldots, X_m + r_m X_1\}$ is a linearly dependent set, then the set $\{X_1, X_2, \ldots, X_m\}$ is a linearly dependent set.*

The reader should try to prove the various parts of this theorem before looking at the proofs below. There should be no difficulty in carrying out the proofs provided the definitions of a vector space and of linear dependence and independence are clearly in mind.

Proof of (i): Suppose, for convenience of notation, that $X_1 = 0$. Then, since $1X_1 = 0$ by 11.3(v), and $0X_i = 0$ by 11.6(ii), it follows that

$$1X_1 + 0X_2 + \cdots + 0X_m = 0.$$

This relation is of the form 11.9 with $a_1 \neq 0$; hence the set $\{X_1, X_2, \ldots, X_m\}$ is linearly dependent.

Proof of (ii): Suppose that $X \neq 0$. If $aX = 0$, it follows from 11.6(iv) that $a = 0$. Hence, the set $\{X\}$ is linearly independent. Of course, in this case a relation of the form 11.9 has just one term on its left side. If $X = 0$, the special case of (i) in which $m = 1$ shows at once that the set $\{X\}$ is linearly dependent.

Proof of (iii): For convenience of notation, let us consider a sub-set of the form $\{X_1, \ldots, X_i\}$ where $1 \leq i < m$, and suppose that $a_1 X_1 + \cdots + a_i X_i = 0$. Then, obviously,

$$a_1 X_1 + \cdots + a_i X_i + 0X_{i+1} + \cdots + 0X_m = 0.$$

Since the set $\{X_1, \ldots, X_m\}$ is linearly independent, it follows that we must have $a_1 = a_2 = \cdots = a_i = 0$. We have therefore shown that $a_1 X_1 + \cdots + a_i X_i = 0$ only if all a's are equal to zero, and hence that the set $\{X_1, \ldots, X_i\}$ is linearly independent.

Proof of (iv): Suppose that $a_1 X_1 + \cdots + a_m X_m = 0$ with $a_j \neq 0$, where j is some integer such that $1 \leq j \leq m$. Then

$$a_1 X_1 + \cdots + a_m X_m + 0X = 0,$$

and since $a_j \neq 0$, we conclude that the set $\{X_1, \ldots, X_m, X\}$ is a linearly dependent set.

Proof of (v): From what is given it follows that

$$b_1 X_1 + b_2 X_2 + \cdots + b_m X_m - (c_1 X_1 + c_2 X_2 + \cdots + c_m X_m) = 0.$$

Then, using 11.6(iii), 11.3(iii), and the fact that V is an abelian group with respect to addition, it is not difficult to show that

$$(b_1 - c_1)X_1 + (b_2 - c_2)X_2 + \cdots + (b_m - c_m)X_m = 0.$$

Since the set $\{X_1, \ldots, X_m\}$ is linearly independent, this equation implies that $b_i - c_i = 0$ and therefore that $b_i = c_i$ ($i = 1, 2, \ldots, m$).

Proof of (vi): By what is given we know that there must exist elements a_i ($i = 2, 3, \ldots, m$) of F, not all of which are zero, such that

$$a_2(X_2 + r_2 X_1) + a_3(X_3 + r_3 X_1) + \cdots + a_m(X_m + r_m X_1) = 0.$$

However, it follows from this equation that

$$(a_2 r_2 + \cdots + a_m r_m)X_1 + a_2 X_2 + \cdots + a_m X_m = 0$$

and, since the a's are not all zero, we see that the set $\{X_1, X_2, \ldots, X_m\}$ is a linearly dependent set. This completes the proof of the theorem.

EXERCISES

1. Determine whether each of the following sets of vectors of $V_3(\mathbf{R})$ is linearly dependent or independent:

  (a) $\{(-1, 2, 1), (3, 1, -2)\}$,
  (b) $\{(1, 3, 2), (2, 1, 0), (0, 5, 4)\}$,
  (c) $\{(2, -1, 1), (1, 2, 3), (0, 1, 2)\}$,
  (d) $\{(1, 0, 0), (0, 1, 0), (0, 0, 1)\}$,
  (e) $\{(1, -2, 1), (0, 1, 2), (1, 1, 1)\}$,
  (f) $\{(1, 0, -1), (2, 1, 3), (-1, 0, 0), (1, 0, 1)\}$.

2. Determine whether each of the following sets of vectors of $V_4(\mathbf{R})$ is linearly dependent or independent:

  (a) $\{(1, -1, 2, 1), (2, 1, 1, 2)\}$,
  (b) $\{(1, 2, 1, 2), (0, 1, 1, 0), (1, 4, 3, 2)\}$,
  (c) $\{(0, 1, 0, 1), (1, 2, 3, -1), (1, 0, 1, 0), (0, 3, 2, 0)\}$,
  (d) $\{(1, 2, -1, 1), (0, 1, -1, 2), (2, 1, 0, 3), (1, 1, 0, 0)\}$.

3. Determine whether each of the following sets of vectors of $V_3(\mathbf{Z}_5)$ is linearly dependent or independent:

  (a) $\{(1, 3, 2), (2, 1, 3)\}$,
  (b) $\{(1, 1, 2), (2, 1, 0), (0, 4, 1)\}$,
  (c) $\{(2, 1, 0), (1, 1, 2), (3, 0, 2)\}$.

4. Prove, giving the reason for each step, that a set consisting of *two* vectors of a vector space is a linearly dependent set if and only if one of these vectors is equal to a scalar times the other.

5. If X_1 and X_2 are vectors of a vector space over the field F, and $a, b \in F$, show that the set $\{X_1, X_2, aX_1 + bX_2\}$ is a linearly dependent set.

6. Let X_1, X_2, and X_3 be vectors of a vector space over the field F, and let a and b be arbitrary elements of F. Show that the set $\{X_1, X_2, X_3\}$ is a linearly dependent set if and only if the set $\{X_1 + aX_2 + bX_3, X_2, X_3\}$ is linearly dependent.

7. Let X_1 and X_2 be linearly independent vectors of a vector space over a field F. If a, b, c, and d are elements of F, prove that the vectors $aX_1 + bX_2$ and $cX_1 + dX_2$ are linearly independent if and only if $ad - bc \neq 0$.

11.4 LINEAR COMBINATIONS AND SUBSPACES

We shall continue to let V be a vector space over a field F.

11.12 DEFINITION   A vector of the form $a_1 X_1 + a_2 X_2 + \cdots + a_m X_m$, where $X_i \in V$ and $a_i \in F$ $(i = 1, 2, \ldots, m)$, is called a *linear combination* (over F) of the vectors $X_1, X_2, \ldots, X_m$.

In the linear combination $a_1 X_1 + a_2 X_2 + \cdots + a_m X_m$, it is sometimes convenient to call a_i the *coefficient* of X_i $(i = 1, 2, \ldots, m)$.

We shall now verify the important fact that *the set U of all linear combinations of given vectors $X_1, X_2, \ldots, X_m$ of V is a subspace of V.* By Theorem 11.7, we only need to show that U is closed under addition and scalar multiplication. Suppose that Y and Z are elements of U. Then,

$$Y = c_1 X_1 + c_2 X_2 + \cdots + c_m X_m$$

and

$$Z = d_1 X_1 + d_2 X_2 + \cdots + d_m X_m,$$

where the c's and d's are elements of F. It follows that

$$Y + Z = (c_1 + d_1)X_1 + (c_2 + d_2)X_2 + \cdots + (c_m + d_m)X_m.$$

Hence, $Y + Z$ is a linear combination of $X_1, X_2, \ldots, X_m$; and therefore an element of U. This shows that U is closed under addition. Now if $r \in F$ and Y is as above, we see that

$$rY = r(c_1 X_1 + c_2 X_2 + \cdots + c_m X_m) = (rc_1)X_1 + (rc_2)X_2 + \cdots + (rc_m)X_m.$$

It follows that $rY \in U$; that is, that U is also closed under scalar multiplication. We have therefore proved that U is a subspace of V.

We now make the following definition.

11.13 DEFINITION If $X_1, X_2, \ldots, X_m$ are elements of a vector space V, the subspace of V which consists of all linear combinations of these vectors will be designated by $[X_1, X_2, \ldots, X_m]$, and called the subspace *generated by* (or *spanned by*) the vectors $X_1, X_2, \ldots, X_m$.

Perhaps we should emphasize the distinction between the sets $\{X_1, X_2, \ldots, X_m\}$ and $[X_1, X_2, \ldots, X_m]$. The former is the set consisting of just the m vectors $X_1, X_2, \ldots, X_m$; whereas the latter consists of all vectors which are expressible as linear combinations of the vectors $X_1, X_2, \ldots, X_m$. Since

$$X_1 = 1X_1 + 0X_2 + \cdots + 0X_m,$$

it is clear that X_1 is a linear combination of the vectors $X_1, X_2, \ldots, X_m$; and hence that $X_1 \in [X_1, X_2, \ldots, X_m]$. In like manner we see that $X_i \in [X_1, X_2, \ldots, X_m]$ for $i = 1, 2, \ldots, m$; and it follows that

$$\{X_1, X_2, \ldots, X_m\} \subseteq [X_1, X_2, \ldots, X_m].$$

Actually, $[X_1, X_2, \ldots, X_m]$ is the smallest subspace of V which contains all the vectors $X_1, X_2, \ldots, X_m$; and this is the reason that we call it the subspace of V *generated by* these vectors.

As a consequence of the fact that $[X_1, X_2, \ldots, X_m]$ is a subspace of V, or by a simple direct calculation, it follows that a linear combination of vectors, each of which is a linear combination of $X_1, X_2, \ldots, X_m$, is itself a linear combination of $X_1, X_2, \ldots, X_m$. Otherwise expressed, if $Y_1, Y_2, \ldots, Y_k$ are elements of the subspace $[X_1, X_2, \ldots, X_m]$, then

$$[Y_1, Y_2, \ldots, Y_k] \subseteq [X_1, X_2, \ldots, X_m].$$

Moreover, we have that

$$[Y_1, Y_2, \ldots, Y_k] = [X_1, X_2, \ldots, X_m]$$

if and only if each of the Y's is a linear combination of the X's, and each of the X's is a linear combination of the Y's. This observation is frequently useful in proving the equality of two subspaces of V.

As an illustration of a subspace generated by given vectors, let us consider the vector space $V_3(\mathbf{R})$ and let $X = (1, 0, -1)$ and $Y = (2, 1, 3)$. Then the subspace $[X, Y]$ of $V_3(\mathbf{R})$ is the set of all vectors of the form $aX + bY$, where $a, b \in \mathbf{R}$. However,

$$aX + bY = a(1, 0, -1) + b(2, 1, 3) = (a + 2b, b, -a + 3b),$$

and hence $[X, Y]$ may be characterized as the set of all vectors of the form $(a + 2b, b, -a + 3b)$, with $a, b \in \mathbf{R}$.

404 · VECTOR SPACES

Now let n be an arbitrary positive integer, and consider the vector space $V_n(F)$, where F is an arbitrary field. Let $E_1, E_2, \ldots, E_n$ be the following vectors of $V_n(F)$:

$$E_1 = (1, 0, 0, \ldots, 0),$$
$$E_2 = (0, 1, 0, \ldots, 0),$$
$$E_3 = (0, 0, 1, \ldots, 0),$$
$$\cdots\cdots\cdots\cdots\cdots$$
$$E_n = (0, 0, 0, \ldots, 1).$$

These are often called the *unit vectors* of $V_n(F)$, and we shall in the future use the notation we have introduced here for these vectors. We observe now that these unit vectors are linearly independent since

$$a_1 E_1 + a_2 E_2 + \cdots + a_n E_n = (a_1, a_2, \ldots, a_n),$$

and this is the zero vector if and only if all a's are equal to zero. It is also almost obvious that the entire space $V_n(F)$ is generated by these unit vectors, that is, that

$$V_n(F) = [E_1, E_2, \ldots, E_n].$$

For if $(c_1, c_2, \ldots, c_n)$ is an arbitrary element of $V_n(F)$, we have

$$(c_1, c_2, \ldots, c_n) = c_1 E_1 + c_2 E_2 + \cdots + c_n E_n,$$

and hence every element of $V_n(F)$ is a linear combination of the unit vectors.

Before stating the next theorem let us observe that in the notation introduced in this section, Theorem 11.11(v) may be stated in the following convenient form.

11.14 COROLLARY *If $X_1, X_2, \ldots, X_m$ are linearly independent vectors of V, each vector of the subspace $[X_1, X_2, \ldots, X_m]$ of V is uniquely expressible as a linear combination of $X_1, X_2, \ldots, X_m$.*

Several important properties are collected in the following theorem.

11.15 THEOREM *Let m be a positive integer and let $X_1, X_2, \ldots, X_m$ be vectors of the vector space V over the field F. The following are then true:*

(i) If $Y = a_1 X_1 + a_2 X_2 + \cdots + a_m X_m$, where the a's are scalars and $a_k \neq 0$ for an integer k such that $1 \leq k \leq m$, then the subspace $[X_1, X_2, \ldots, X_m]$ is unchanged if X_k is replaced by Y; that is,

$$[X_1, \ldots, X_k, \ldots, X_m] = [X_1, \ldots, Y, \ldots, X_m].$$

(ii) *If $X \in V$, then $[X_1, X_2, \ldots, X_m, X] = [X_1, X_2, \ldots, X_m]$ if and only if $X \in [X_1, X_2, \ldots, X_m]$.*

(iii) *If $\{X_1, X_2, \ldots, X_m\}$ is a linearly independent set and if $X \notin [X_1, X_2, \ldots, X_m]$, then $\{X_1, X_2, \ldots, X_m, X\}$ is a linearly independent set.*

Proof: We shall prove parts (i) and (iii) of this theorem, and leave the proof of part (ii) as an exercise.

First, let us consider part (i).   Since Y is a linear combination of the X's, it follows at once that

$$[X_1, \ldots, Y, \ldots, X_m] \subseteq [X_1, \ldots, X_k, \ldots, X_m].$$

To obtain inclusion the other way, it is necessary to show that $X_k \in [X_1, \ldots, Y, \ldots, X_m]$.   However, since it is given that $a_k \neq 0$, we may solve the given relation for X_k as follows:

$$X_k = -a_k^{-1}(-Y + a_1 X_1 + \cdots + a_{k-1} X_{k-1} \\ + a_{k+1} X_{k+1} + \cdots + a_m X_m).$$

Hence $X_k \in [X_1, \ldots, Y, \ldots, X_m]$, and this completes the proof.

To establish part (iii), suppose that $c_1, c_2, \ldots, c_m$ are scalars such that

$$c_1 X_1 + c_2 X_2 + \cdots + c_m X_m + cX = 0,$$

and let us show that all of these scalars must be zero.   If $c \neq 0$, we may write

$$X = -c^{-1}(c_1 X_1 + c_2 X_2 + \cdots + c_m X_m),$$

and it follows that $X \in [X_1, X_2, \ldots, X_m]$, thus violating our hypothesis.   Hence we must have $c = 0$ and it then follows that $c_1 X_1 + c_2 X_2 + \cdots + c_m X_m = 0$.   But since $X_1, X_2, \ldots, X_m$ are linearly independent, we conclude that $c_i = 0$ $(i = 1, 2, \ldots, m)$.   These calculations therefore prove that the vectors $X_1, X_2, \ldots, X_m, X$ are linearly independent, as we wished to show.

Two special cases of the first part of this theorem are of sufficient interest to warrant specific mention.   We therefore state them as the following corollary.

11.16 COROLLARY   *If $X_1, X_2, \ldots, X_m$ are vectors of a vector space V over a field F, then each of the following is true:*

(i) *The subspace* $[X_1, X_2, \ldots, X_m]$ *is unchanged if a vector* X_k $(1 \le k \le m)$ *is replaced by* aX_k, *where* a *is a nonzero scalar; that is,*

$$[X_1, \ldots, X_k, \ldots, X_m] = [X_1, \ldots, aX_k, \ldots, X_m].$$

(ii) *The subspace* $[X_1, X_2, \ldots, X_m]$ *is unchanged if a vector* X_k $(1 \le k \le m)$ *is replaced by* $X_k + bX_l$, *where* $1 \le l \le m, l \ne k,$ *and* b *is any scalar. That is, we have*

$$[X_1, \ldots, X_k, \ldots, X_l, \ldots, X_m]$$
$$= [X_1, \ldots, X_k + bX_l, \ldots, X_l, \ldots, X_m].$$

We conclude this section with the following theorem, which will prove to be extremely useful in the further study of vector spaces.

11.17 THEOREM *If* $X_1, X_2, \ldots, X_m$ *are vectors in the vector space* V, *any* $m + 1$ *vectors in the subspace* $[X_1, X_2, \ldots, X_m]$ *of* V *are linearly dependent.*

Proof: The proof is by induction on m. For the case in which $m = 1$ we need only show that any two vectors of $[X_1]$ are linearly dependent. Let $Y_1 = aX_1$ and $Y_2 = bX_1$ be two vectors of the subspace $[X_1]$. If $a = 0$, then $Y_1 = 0$ and the equation $1\,Y_1 + 0\,Y_2 = 0$ shows that Y_1 and Y_2 are linearly dependent. If $a \ne 0$, the equation $-b\,Y_1 + a\,Y_2 = 0$ shows that Y_1 and Y_2 are linearly dependent. This disposes of the case in which $m = 1$.

To complete the proof by induction, let k be a positive integer such that the statement of the theorem is true for $m = k$, and let us prove it for $m = k + 1$. We shall do so by showing that if $Y_1, Y_2, \ldots, Y_{k+2}$ are arbitrary vectors in the subspace $[X_1, X_2, \ldots, X_{k+1}]$, then these vectors are linearly dependent. Since each Y_i is a linear combination of $X_1, X_2, \ldots, X_{k+1}$, we can write

11.18

$$\begin{aligned}
Y_1 &= a_1 X_1 + a_2 X_2 + \cdots + a_{k+1} X_{k+1}, \\
Y_2 &= b_1 X_1 + b_2 X_2 + \cdots + b_{k+1} X_{k+1}, \\
Y_3 &= c_1 X_1 + c_2 X_2 + \cdots + c_{k+1} X_{k+1}, \\
&\quad \cdot \quad \cdot \quad \cdot \quad \cdot \quad \cdot \quad \cdot \quad \cdot \\
Y_{k+2} &= s_1 X_1 + s_2 X_2 + \cdots + s_{k+1} X_{k+1},
\end{aligned}$$

it being understood that the various coefficients are elements of F. If it happens that in these equations the coefficients of X_1 are all zero, then the Y's are, in fact, linear combinations of the k vectors $X_2, X_3, \ldots, X_{k+1}$. This means that they are elements of the subspace $[X_2, X_3, \ldots, X_{k+1}]$ and, by our assumption that the statement of the theorem is true for $m = k$, we see that any $k + 1$ vectors of the set $\{Y_1, Y_2, \ldots, Y_{k+2}\}$ are linearly dependent. Theorem 11.11(iv) then shows that the set $\{Y_1, Y_2, \ldots, Y_{k+2}\}$ is linearly dependent and we have the desired result.

There remains to dispose of the case in which not all the coefficients of X_1 in Equations 11.18 are zero. Let us assume, for convenience of notation, that $a_1 \neq 0$. Then from the first of Equations 11.18 we have that

$$a_1{}^{-1}Y_1 = X_1 + a_1{}^{-1}a_2 X_2 + \cdots + a_1{}^{-1}a_{k+1} X_{k+1},$$

and from this and the second of Equations 11.18 it follows that the vector $Y_2 - b_1 a_1{}^{-1} Y_1$ is a linear combination of $X_2, \ldots, X_{k+1}$; and is therefore an element of $[X_2, \ldots, X_{k+1}]$. Similarly, using the third of Equations 11.18, we see that $Y_3 - c_1 a_1{}^{-1} Y_1$ is an element of $[X_2, \ldots, X_{k+1}]$. Continuing in this manner, we find that the $k + 1$ vectors

$$Y_2 - b_1 a_1{}^{-1} Y_1, \; Y_3 - c_1 a_1{}^{-1} Y_1, \ldots, \; Y_{k+2} - s_1 a_1{}^{-1} Y_1$$

are elements of $[X_2, \ldots, X_{k+1}]$. But by our assumption that the statement of the theorem is true for $m = k$, these $k + 1$ vectors must be linearly dependent. Theorem 11.11(vi) then shows that the vectors $Y_1, Y_2, \ldots, Y_{k+2}$ are linearly dependent, and the proof of the theorem is complete.

The following simple corollary of this theorem follows immediately from the observation made above that if $E_1, E_2, \ldots, E_n$ are the unit vectors of $V_n(F)$, then $V_n(F) = [E_1, E_2, \ldots, E_n]$.

11.19 COROLLARY *Any $n + 1$ vectors of $V_n(F)$ are linearly dependent.*

EXERCISES

1. Prove that if $\{X_1, X_2, \ldots, X_m\}$ (where $m > 1$) is a linearly dependent set of vectors, then some one of these vectors is a linear combination of the others.

2. Prove Theorem 11.15(ii).

3. Show that if the set $\{X_1, X_2, \ldots, X_m\}$ is linearly independent but the set $\{X_1, X_2, \ldots, X_m, X\}$ is linearly dependent, then X is a linear combination of the vectors $X_1, X_2, \ldots, X_m$.

4. If $m > 1$, $X \in [X_1, X_2, \ldots, X_m]$, and $X \notin [X_1, X_2, \ldots, X_{m-1}]$, prove that $X_m \in [X_1, X_2, \ldots, X_{m-1}, X]$.

5. If $X \in [X_1, X_2, \ldots, X_m]$, show that the set $\{X, X_1, X_2, \ldots, X_m\}$ is a linearly dependent set.

6. Show that if $[X_1, X_2, \ldots, X_m] = [Y_1, Y_2, \ldots, Y_n]$ with $m \neq n$, then at least one of the sets $\{X_1, X_2, \ldots, X_m\}$ and $\{Y_1, Y_2, \ldots, Y_n\}$ is a linearly dependent set.

7. Show that if the nonzero vectors $X_1, X_2, \ldots, X_m$ $(m \geq 2)$ are linearly dependent, there exists an integer k with $2 \leq k \leq m$ such that X_k is a linear combination of $X_1, X_2, \ldots, X_{k-1}$.

11.5 BASIS AND DIMENSION

We now make the following definition.

11.20 DEFINITION The set $\{X_1, X_2, \ldots, X_n\}$ of elements of a vector space V is said to be a *basis* of V if the following two conditions are satisfied:

(i) The set $\{X_1, X_2, \ldots, X_n\}$ is a linearly independent set.

(ii) V is generated by $X_1, X_2, \ldots, X_n$; that is,

$$V = [X_1, X_2, \ldots, X_n].$$

In view of Corollary 11.14, we see that if the set $\{X_1, X_2, \ldots, X_n\}$ is a basis of V, every vector of V is *uniquely* expressible as a linear combination of the vectors $X_1, X_2, \ldots, X_n$.

As an example of the concept of basis of a vector space, we have already observed that the set $\{E_1, E_2, \ldots, E_n\}$ of unit vectors of $V_n(F)$ has both the defining properties and is therefore a basis of $V_n(F)$. In particular, let us now consider the vector space $V_3(\mathbf{R})$. Then, as a special case of the observation just made, the set $\{(1, 0, 0), (0, 1, 0), (0, 0, 1)\}$ is a basis of $V_3(\mathbf{R})$. However, $V_3(\mathbf{R})$ may have other bases as well. For example, let us consider the set $A = \{(1, 2, 1), (2, 1, 0),$

$(1, -1, 2)\}$ of elements of $V_3(\mathbf{R})$. By the method used in Section 11.3 it can be shown that this is a linearly independent set, and we shall not give the details here. Moreover, that each of the unit vectors is a linear combination of the vectors of this set A follows from the following easily verified equations:

$$(1, 0, 0) = -\tfrac{2}{9}(1, 2, 1) + \tfrac{5}{9}(2, 1, 0) + \tfrac{1}{9}(1, -1, 2).$$
$$(0, 1, 0) = \tfrac{4}{9}(1, 2, 1) - \tfrac{1}{9}(2, 1, 0) - \tfrac{2}{9}(1, -1, 2).$$
$$(0, 0, 1) = \tfrac{1}{3}(1, 2, 1) - \tfrac{1}{3}(2, 1, 0) + \tfrac{1}{3}(1, -1, 2).$$

Since $V_3(\mathbf{R})$ is generated by the unit vectors, we see therefore that every element of $V_3(\mathbf{R})$ is a linear combination of the vectors of the set A. Hence, the vectors of this set form a basis of $V_3(\mathbf{R})$.

We have just indicated that a vector space may have more than one basis. However, it is easy to verify that not every vector space has a basis according to our definition.† As an example of a vector space without a basis, let us consider the vector space $F[x]$ of polynomials in an indeterminate x over a field F (Example 4 of Section 11.2). If n is an arbitrary positive integer and $f_1, f_2, \ldots, f_n$ are any n elements of $F[x]$, then no linear combination of these vectors can have a degree exceeding the maximum of the degrees of $f_1, f_2, \ldots, f_n$. Hence, there exist elements of $F[x]$ that are not in the subspace $[f_1, f_2, \ldots, f_n]$, and condition (ii) of Definition 11.20 cannot be satisfied. This shows that the vector space $F[x]$ cannot have a basis. Another example of a vector space without a basis is the vector space $W(F)$ of Example 2 of Section 11.2. However, we shall be primarily concerned with vector spaces that do have bases, and for them the following theorem is fundamental.

11.21 THEOREM *If the vector space V has a basis consisting of n vectors, then every basis of V has exactly n vectors.*

> **Proof:** Suppose that $\{X_1, X_2, \ldots, X_n\}$ and $\{Y_1, Y_2, \ldots, Y_m\}$ are bases of V, and let us show that necessarily $m = n$. We observe that $V = [X_1, X_2, \ldots, X_n] = [Y_1, Y_2, \ldots, Y_m]$ and apply Theorem 11.17 as follows. If $m > n$, any $n + 1$ of the vectors $Y_1, Y_2, \ldots, Y_m$ are linearly dependent, and hence the entire set $\{Y_1, Y_2, \ldots, Y_m\}$ is linearly dependent. However, this contradicts the fact that $\{Y_1, Y_2, \ldots, Y_m\}$ is a basis of V. We conclude therefore that we cannot have $m > n$. By interchanging the roles of the X's and the Y's in this argument, we find also that we cannot have $n > m$. Hence, $m = n$, and the proof is complete.

Since all bases of a vector space have the same number of elements, it is convenient to have a name for the number of such elements.

---

† In more advanced treatises, what we have called a basis is usually called a *finite* basis, and every vector space has either a finite or an infinite basis. See Theorem 16.28 in the last chapter of this book.

11.22 DEFINITION   A vector space V is said to have *dimension n* $(n \geq 1)$ if V has a basis consisting of n elements. The vector space consisting of only the zero vector is said to have *dimension zero*. A vector space is said to have *finite* dimension if it has dimension m for some nonnegative integer m.

We shall write dim $V = n$, to indicate that the vector space V has dimension n.

It is now clear that dim $V_n(F) = n$, since $V_n(F)$ has the basis $\{E_1, E_2, \ldots, E_n\}$. In particular, the vector space $V_2(\mathbf{R})$ discussed in Section 11.1, and which we described geometrically as the set of all vectors in a plane, has dimension 2. This fact should at least help to make our definition of dimension seem a reasonable one.

We shall now prove the following theorem.

11.23 THEOREM   *If* dim $V = n$ *with* $n > 0$, *the following are true:*

(i) *Any* $n + 1$ *vectors of V are linearly dependent.*

(ii) *Any set of n linearly independent vectors of V is a basis of V.*

(iii) *V cannot be generated by fewer than n vectors.*

(iv) *If* $V = [Z_1, Z_2, \ldots, Z_n]$, *then* $\{Z_1, Z_2, \ldots, Z_n\}$ *is necessarily a linearly independent set and therefore a basis of V.*

Proof: If $\{X_1, X_2, \ldots, X_n\}$ is a basis of V, then $V = [X_1, X_2, \ldots, X_n]$ and the first statement of the theorem follows at once from Theorem 11.17.

To prove the second statement, let $\{Y_1, Y_2, \ldots, Y_n\}$ be a set of n linearly independent vectors of V. If this set were not a basis, there would exist an element Y of V such that $Y \notin [Y_1, Y_2, \ldots, Y_n]$. By Theorem 11.15(iii), this would imply that the set $\{Y_1, Y_2, \ldots, Y_n, Y\}$ is a linearly independent set. However, this is impossible by the first part of the present theorem, and we conclude that $\{Y_1, Y_2, \ldots, Y_n\}$ must be a basis of V.

If V were generated by m vectors with $m < n$, Theorem 11.17 would show that V could not contain n linearly independent vectors. However, dim $V = n$ implies that there do exist n linearly independent vectors in V, and we have established part (iii) of the theorem.

To prove part (iv), suppose that $V = [Z_1, Z_2, \ldots, Z_n]$, and let us assume that the set $\{Z_1, Z_2, \ldots, Z_n\}$ is linearly dependent and seek a contradiction. The case in which $n = 1$ is trivial, so we may assume that $n > 1$. Then the linear dependence of the set $\{Z_1, Z_2, \ldots, Z_n\}$ implies that some one of these vectors is a linear combination of the others. (See Exercise 1 of the preceding set.) For convenience of notation, let us suppose that Z_1 is a linear combination of $Z_2, \ldots, Z_n$. By Theorem 11.15(ii), we then see that

$$V = [Z_1, Z_2, \ldots, Z_n] = [Z_2, \ldots, Z_n].$$

However, this violates part (iii) of the present theorem since we now have V generated by fewer than n vectors. This contradiction shows that the set $\{Z_1, Z_2, \ldots, Z_n\}$ is linearly independent, and the proof of the theorem is therefore complete.

The next theorem shows that any set of linearly independent vectors of a vector space of finite dimension is a part of a basis of the space.

11.24 THEOREM *Let V be a vector space of dimension $n > 1$. If $\{X_1, X_2, \ldots, X_r\}$, where $1 \le r < n$, is a set of linearly independent vectors of V, there exist vectors $X_{r+1}, \ldots, X_n$ of V such that $\{X_1, X_2, \ldots, X_n\}$ is a basis of V.*

Proof: This result is established as follows. Since $r < n$, the set $\{X_1, X_2, \ldots, X_r\}$ is not a basis of V and hence there exists a vector X_{r+1} of V such that $X_{r+1} \notin [X_1, X_2, \ldots, X_r]$. By Theorem 11.15(iii), the set $\{X_1, X_2, \ldots, X_r, X_{r+1}\}$ is a linearly independent set. If $r + 1 < n$, we can repeat the argument. Continuing in this way, we must eventually obtain a set of n linearly independent vectors and, by the preceding theorem, this set is a basis of V.

We next consider a few questions about the dimensions of subspaces of a given vector space.

11.25 THEOREM

(i) *If $\dim V = n$ and U is a subspace of V, then U has finite dimension and $\dim U \le n$. Moreover, $U = V$ if and only if $\dim U = n$.*

(ii) *If the nonzero subspace U is generated by the vectors of the set $\{X_1, X_2, \ldots, X_k\}$, there exists a subset of $\{X_1, X_2, \ldots, X_k\}$ which is a basis of U.*

Proof: The first part of this theorem is trivial if $n = 0$; hence we assume that $n > 0$. If U consists only of the zero vector, $\dim U = 0$ and clearly $\dim U \le n$, so there is nothing to prove. Suppose, then, that X_1 is a nonzero vector in U. If $U \ne [X_1]$, let X_2 be an element of U which is not in $[X_1]$. Then X_1 and X_2 are linearly independent. If $U \ne [X_1, X_2]$, let X_3 be an element of U which is not in $[X_1, X_2]$, and again we know that X_1, X_2, and X_3 are linearly independent.

Continuing in this way, we must eventually come to the point at which $U = [X_1, X_2, \ldots, X_r]$, where $r \leq n$ since there can exist at most n linearly independent vectors in V. This shows that dim $U = r \leq n$, and the first statement of part (i) is established. The second statement of this part is an immediate consequence of Theorem 11.23(ii).

Part (ii) can be proved by a simple modification of the above proof. Instead of choosing an arbitrary vector of U not in a specified subspace, one can always choose an element of the generating set $\{X_1, X_2, \ldots, X_k\}$. We leave the details as an exercise. (Cf. also Exercise 6 below.)

If U_1 and U_2 are subspaces of the same vector space V, let us define

11.26 $$U_1 + U_2 = \{X + Y \mid X \in U_1, \, Y \in U_2\}.$$

As will soon be stated in the next theorem, it can be shown that $U_1 + U_2$ is a *subspace* of V. We may remark that since the subspace U_2 contains the zero vector, it follows that if $X \in U_1$, then $X = X + 0$ is also an element of $U_1 + U_2$. That is, $U_1 \subseteq U_1 + U_2$; and, similarly, $U_2 \subseteq U_1 + U_2$. However, in general, $U_1 + U_2$ will contain many vectors other than those in U_1 or U_2. It is quite easy to show that the intersection $U_1 \cap U_2$ of the two subspaces U_1 and U_2 is also a subspace of V. An interesting relationship between the subspaces $U_1, U_2, U_1 + U_2$, and $U_1 \cap U_2$ is given in the second part of the following theorem.

11.27 THEOREM   *Let U_1 and U_2 be subspaces of a vector space V. Then*

(i) *$U_1 \cap U_2$ and $U_1 + U_2$ are subspaces of V,*

(ii) *If V has finite dimension,*

$$\dim (U_1 + U_2) = \dim U_1 + \dim U_2 - \dim (U_1 \cap U_2).$$

Proof: We shall leave as exercises the proof of part (i) of this theorem and the proof of part (ii) for the special case in which dim $(U_1 \cap U_2) = 0$. We then proceed to the proof of part (ii) under the assumption that dim $(U_1 \cap U_2) > 0$. We may remark that since $U_1 \cap U_2$ is a subspace of U_1, it follows that

$$\dim (U_1 \cap U_2) \leq \dim U_1,$$

and, similarly,

$$\dim (U_1 \cap U_2) \leq \dim U_2.$$

First, we dispose of an easy special case as follows. Suppose that

$$\dim (U_1 \cap U_2) = \dim U_1,$$

which implies that $U_1 \cap U_2 = U_1$. It follows that $U_1 \subseteq U_2$, and 11.26 then shows that $U_1 + U_2 = U_2$. Hence, in this case, 11.27(ii) takes the form

$$\dim U_2 = \dim U_1 + \dim U_2 - \dim U_1,$$

which is obviously true. Similar remarks hold if it happens that

$$\dim (U_1 \cap U_2) = \dim U_2.$$

Let us now set $\dim (U_1 \cap U_2) = r$, $\dim U_1 = r + s$, and $\dim U_2 = r + t$. In view of the preceding remarks, we henceforth assume that $r > 0$, $s > 0$, and $t > 0$. The proof will be completed by showing that

$$\dim (U_1 + U_2) = r + s + t.$$

Let $\{X_1, \ldots, X_r\}$ be a basis of $U_1 \cap U_2$. Then, by Theorem 11.24, there exist vectors $Y_1, \ldots, Y_s$ such that $\{X_1, \ldots, X_r, Y_1, \ldots, Y_s\}$ is a basis of U_1; and vectors $Z_1, \ldots, Z_t$ such that $\{X_1, \ldots, X_r, Z_1, \ldots, Z_t\}$ is a basis of U_2. It now follows from 11.26 that every vector of the subspace $U_1 + U_2$ is a linear combination of vectors of the set

11.28 $\{X_1, \ldots, X_r, Y_1, \ldots, Y_s, Z_1, \ldots, Z_t\}.$

We shall show that this is a linearly independent set and hence a basis of $U_1 + U_2$.

Suppose that

11.29 $a_1 X_1 + \cdots + a_r X_r + b_1 Y_1 + \cdots + b_s Y_s + c_1 Z_1 + \cdots + c_t Z_t = 0,$

where all the coefficients are elements of F. Let $Z = c_1 Z_1 + \cdots + c_t Z_t$. It is then clear that $Z \in U_2$, and we see from 11.29 that

$$Z = -(a_1 X_1 + \cdots + a_r X_r + b_1 Y_1 + \cdots + b_s Y_s),$$

and hence also $Z \in U_1$. This shows that $Z \in (U_1 \cap U_2)$, and hence Z is a linear combination of the basis elements $X_1, \ldots, X_r$ of $U_1 \cap U_2$. Thus there exist scalars $e_1, \ldots, e_r$ such that

$$Z = c_1 Z_1 + \cdots + c_t Z_t = e_1 X_1 + \cdots + e_r X_r,$$

and from this it follows that

$$e_1 X_1 + \cdots + e_r X_r - c_1 Z_1 - \cdots - c_t Z_t = 0.$$

But since $\{X_1, \ldots, X_r, Z_1, \ldots, Z_t\}$ is a basis of U_2, this is a linearly independent set and hence all e's and all c's must equal zero. Now, setting all c's equal to zero in 11.29, the linear independence of the set $\{X_1, \ldots, X_r, Y_1, \ldots, Y_s\}$ shows that we must have all a's and all b's equal to zero. Hence, a relation of the form 11.29 holds only if all coefficients are zero. This shows that the set 11.28 is a linearly independent set and therefore a basis of $U_1 + U_2$. Finally, we see that dim $(U_1 + U_2) = r + s + t$ since this is the number of vectors in the basis 11.28. The proof is therefore complete.

EXERCISES

1. Prove Theorem 11.25(ii).

2. Prove Theorem 11.27(i).

3. Prove Theorem 11.27(ii) for the case in which dim $(U_1 \cap U_2) = 0$.

4. Find a basis for $V_3(\mathbf{R})$ which contains the vectors $(1, -1, 0)$ and $(2, 1, 3)$.

5. If $\mathbf{C}(\mathbf{R})$ is the vector space of the field of complex numbers over the field $\mathbf{R}$ of real numbers (cf. Example 3 of Section 11.2),

   (i) Find the dimension of $\mathbf{C}(\mathbf{R})$.

   (ii) Show that $\{a + bi, c + di\}$ is a basis of $\mathbf{C}(\mathbf{R})$ if and only if $ad - bc \neq 0$.

6. Let $A = \{X_1, X_2, \ldots, X_m\}$ be a set of nonzero vectors of a vector space V. A subset $\{Y_1, Y_2, \ldots, Y_k\}$ of A is said to be a *maximal linearly independent subset* of A if (i) the set $\{Y_1, Y_2, \ldots, Y_k\}$ is linearly independent, and (ii) the set $\{Y_1, Y_2, \ldots, Y_k, Y\}$ is a linearly dependent set for each Y in A other than $Y_1, Y_2, \ldots, Y_k$. Show that if $\{Y_1, Y_2, \ldots, Y_k\}$ is a maximal linearly independent subset of A, then $[X_1, X_2, \ldots, X_m] = [Y_1, Y_2, \ldots, Y_k]$ and, in particular,

   $$\dim [X_1, X_2, \ldots, X_m] = k.$$

7. Find the dimension of the subspace

$$[(1, 2, 1, 0), (-1, 1, -4, 3), (2, 3, 3, -1), (0, 1, -1, 1)]$$

of $V_4(\mathbf{R})$.

8. Let U_1 and U_2 be subspaces of $V_4(\mathbf{R})$ as follows: $U_1 = [(1, 2, -1, 0), (2, 0, 1, 1)]$ and $U_2 = [(0, 0, 0, 1), (1, 0, 1, 0), (0, 4, -3, -1)]$. Find dim U_1, dim U_2, dim $(U_1 \cap U_2)$, dim $(U_1 + U_2)$, and verify the truth of Theorem 11.27(ii) in this particular case.

9. Give an example to show that the following statement is false: If $\{X_1, X_2, \ldots, X_n\}$ is a basis of V, and U is a subspace of V, then some subset of $\{X_1, X_2, \ldots, X_n\}$ is a basis of U.

10. Let T be the set of all subspaces of a nonzero vector space V, and let an operation of addition be defined on the set T by 11.26. Verify that T is not a group with respect to this operation of addition.

11. We have observed that a field may be considered to be a vector space over any subfield. Suppose that F_1, F_2, and F_3 are fields such that $F_1 \subset F_2 \subset F_3$. If $\{X_1, X_2, \ldots, X_m\}$ is a basis of F_2 over F_1, and $\{Y_1, Y_2, \ldots, Y_n\}$ is a basis of F_3 over F_2, prove that the mn elements $X_i Y_j$ $(i = 1, 2, \ldots, m; j = 1, 2, \ldots, n)$ are a basis of F_3 over F_1.

12. If U_1 is a subspace of a vector space V of finite dimension, prove that there exists a subspace U_2 of V such that $V = U_1 + U_2$ and $U_1 \cap U_2 = \{0\}$. Exhibit an example to show that there may be more than one such subspace U_2.

13. If U_1, U_2, and U_3 are subspaces of a vector space V, and $U_1 \subseteq U_3$, prove that $U_1 + (U_2 \cap U_3) = (U_1 + U_2) \cap U_3$.

11.6 HOMOMORPHISMS OF VECTOR SPACES

The familiar concept of homomorphism of rings or of groups can easily be modified so as to apply to vector spaces over the same field.

11.30 DEFINITION Let V and W be vector spaces over the same field F. A mapping $\theta: V \to W$ is called a *homomorphism* (or a *linear transformation*) of V into W if the following are true:

(i) $(X + Y)\theta = X\theta + Y\theta,$ $X, Y \in V,$

(ii) $(cX)\theta = c(X\theta),$ $c \in F, X \in V.$

The first of these conditions merely states that addition is preserved under the mapping θ; that is, if we ignore scalar multiplication, θ is a homomorphism of the abelian group V into the abelian group W. The second condition asserts that also scalar multiplication is preserved under the mapping θ.

Just as in the case of rings or of groups, if there exists a homomorphism of V onto W, we may say that V is *homomorphic to* W or that W is a *homomorphic image* of V. Naturally, a homomorphism which is a one-one mapping is called an *isomorphism*, and if there exists an isomorphism of V onto W we say that V is *isomorphic* to W or that W is an *isomorphic image* of V, and may indicate this fact by writing $V \cong W$.

The following are some examples of homomorphisms of vector spaces. We leave it to the reader to verify that they are indeed homomorphisms.

Example 1 Let $\theta: V_3(F) \to V_2(F)$ be defined by

$$(a_1, a_2, a_3)\theta = (a_1, a_3).$$

Then θ is a homomorphism of $V_3(F)$ onto $V_2(F)$.

Example 2 Let $\phi: V_2(F) \to V_3(F)$ be defined by

$$(a_1, a_2)\phi = (a_1 + a_2, a_1, a_2).$$

This mapping ϕ is a homomorphism, but it is not an onto mapping.

Example 3 Let $\gamma: V_2(\mathbf{R}) \to V_2(\mathbf{R})$ be defined by

$$(a_1, a_2)\gamma = (2a_1 - 3a_2, a_1 + 2a_2).$$

Then γ is a homomorphism of $V_2(\mathbf{R})$ into $V_2(\mathbf{R})$. (See Exercise 1 below.)

The following theorem lists a few of the basic facts about homomorphisms of vector spaces.

11.31 THEOREM *Let $\theta: V \to W$ be a homomorphism of the vector space V over the field F into the vector space W over F. Then each of the following is true:*

(i) If 0 is the zero of V, then 0θ is the zero of W.

(ii) *If $X \in V$, then $(-X)\theta = -(X\theta)$.*

(iii) *Let k be an arbitrary positive integer. If $X_i \in V$ and $c_i \in F$ $(i = 1, 2, \ldots, k)$, then*

$$(c_1 X_1 + c_2 X_2 + \cdots + c_k X_k)\theta$$
$$= c_1(X_1 \theta) + c_2(X_2 \theta) + \cdots + c_k(X_k \theta).$$

(iv) *If U is a subspace of V and $U\theta = \{X\theta \mid X \in U\}$, then $U\theta$ is a subspace of W.*

Proof: Since parts (i) and (ii) do not involve scalar multiplication, they are properties of group homomorphisms which have already been established. They also follow easily from 11.30(ii) by letting c be 0 and -1, respectively. The proofs of (iii) and (iv) are straight-forward, and we leave them as an exercise.

We may emphasize that both of the properties (i) and (ii) of 11.30, used to define a homomorphism, are special cases of 11.31(iii), so that we *could* have used property 11.31(iii) as the definition of a homomorphism.

If $\theta : V \to W$ is a homomorphism (always assuming that V and W are vector spaces over the *same* field), let us define the *kernel* of θ, which we shall write as ker θ, as follows:

$$\ker \theta = \{X \mid X \in V, X\theta = 0\}.$$

We may now state the following result.

11.32 THEOREM *If $\theta : V \to W$ is a homomorphism of V into W, then ker θ is a subspace of V. Moreover, θ is an isomorphism if and only if ker $\theta = \{0\}$.*

Proof: We shall leave the first statement as an exercise, and give only the proof of the second.

Since, by Theorem 11.31(i), we know that $0\theta = 0$, it follows that if θ is a one-one mapping, no other element can have 0 as an image; hence ker $\theta = \{0\}$. Conversely, if ker $\theta = \{0\}$ and $X\theta = Y\theta$, it follows from Theorem 11.31 that $(X - Y)\theta = 0$. Thus, $(X - Y) \in$ ker θ, and therefore $X - Y = 0$, or $X = Y$. This shows that θ is a one-one mapping, as we wished to prove.

In the case of vector spaces of finite dimension, we have the following result.

11.33 THEOREM *If V and W are vector spaces of finite dimension over a field F, then $V \cong W$ if and only if $\dim V = \dim W$.*

Proof: We shall give an outline of a proof of this theorem and suggest that the reader supply the details. Suppose, first, that $\theta: V \to W$ is an isomorphism of V onto W. Let us assume that dim $V = n$ and dim $W = m$. The desired result is trivial in case $n = 0$, so we assume that $n > 0$. If $\{X_1, X_2, \ldots, X_n\}$ is a basis of V, it can be verified that $\{X_1\theta, X_2\theta, \ldots, X_n\theta\}$ is a basis of W. Hence dim $W = n$, and therefore $n = m$.

Conversely, suppose that dim $V =$ dim $W = n > 0$. Let $\{X_1, X_2, \ldots, X_n\}$ be a basis of V and $\{Y_1, Y_2, \ldots, Y_n\}$ a basis of W. Each element X of V can be uniquely expressed as a linear combination of the basis elements of V. We leave it to the reader to verify that the mapping θ defined by

$$(a_1 X_1 + a_2 X_2 + \cdots + a_n X_n)\theta = a_1 Y_1 + a_2 Y_2 + \cdots + a_n Y_n,$$

where the a's are arbitrary elements of F, is the desired isomorphism of V onto W.

Since for each $n > 0$, $V_n(F)$ has dimension n, we have the following immediate consequence of the preceding theorem.

11.34 COROLLARY *If the vector space V has dimension $n > 0$ over the field F, then $V \cong V_n(F)$.*

The content of this theorem suggests that when studying vector spaces of finite dimension we could limit ourselves to vector spaces of the form $V_n(F)$. However, it is often simpler not to make this restriction. Moreover, the most important properties of vector spaces are, and should be, independent of notation. This is the reason that we have waited until this point to prove Corollary 11.34.

The next result gives some important information about the *existence* of homomorphisms of V into W, in case V has finite dimension.

11.35 THEOREM *Let V and W be vector spaces over the field F, and suppose that dim $V = n > 0$. If $\{X_1, X_2, \ldots, X_n\}$ is a basis of V, and $Y_1, Y_2, \ldots, Y_n$ are arbitrary elements of W, there exists exactly one homomorphism α of V into W such that $X_i\alpha = Y_i$ $(i = 1, 2, \ldots, n)$.*

Proof: If $X \in V$, there exist unique elements $a_1, a_2, \ldots, a_n$ of F such that

11.36 $$X = a_1 X_1 + a_2 X_2 + \cdots + a_n X_n.$$

We now define a mapping α of V into W as follows:

11.37 $(a_1 X_1 + a_2 X_2 + \cdots + a_n X_n)\alpha = a_1 Y_1 + a_2 Y_2 + \cdots + a_n Y_n.$

Although the proof is straightforward let us verify that α is indeed a homomorphism. If X, given by 11.36, and

11.38 $$X' = b_1 X_1 + b_2 X_2 + \cdots + b_n X_n$$

are any elements of V, then

$$
\begin{aligned}
(X + X')\alpha &= [(a_1 + b_1)X_1 + (a_2 + b_2)X_2 + \cdots + (a_n + b_n)X_n]\alpha \\
&= (a_1 + b_1)Y_1 + (a_2 + b_2)Y_2 + \cdots + (a_n + b_n)Y_n \\
&= (a_1 Y_1 + a_2 Y_2 + \cdots + a_n Y_n) \\
&\quad + (b_1 Y_1 + b_2 Y_2 + \cdots + b_n Y_n) \\
&= X\alpha + X'\alpha,
\end{aligned}
$$

and thus addition is preserved under the mapping α.

Also, if X is given by 11.36 and $c \in F$, then

$$
\begin{aligned}
(cX)\alpha &= (ca_1 X_1 + ca_2 X_2 + \cdots + ca_n X_n)\alpha \\
&= ca_1 Y_1 + ca_2 Y_2 + \cdots + ca_n Y_n \\
&= c(a_1 Y_1 + a_2 Y_2 + \cdots + a_n Y_n) \\
&= c(X\alpha).
\end{aligned}
$$

This shows that scalar multiplication is preserved, and we have that α is a homomorphism. Moreover, if in 11.37 we choose all a's to be zero except $a_i = 1$, we see that $X_i \alpha = Y_i$ for $i = 1, 2, \ldots, n$.

These calculations verify that the mapping α, defined by 11.37, is one homomorphism of V into W satisfying the requirement of the theorem. To show that there is only one such homomorphism, let $\theta : V \to W$ be any homomorphism of V into W with the property that $X_i \theta = Y_i$ for $i = 1, 2, \ldots, n$. Then if $X \in V$, given by 11.36, we have from Theorem 11.31(iii) that

$$
\begin{aligned}
X\theta &= (a_1 X_1 + a_2 X_2 + \cdots + a_n X_n)\theta \\
&= a_1(X_1\theta) + a_2(X_2\theta) + \cdots + a_n(X_n\theta) \\
&= a_1 Y_1 + a_2 Y_2 + \cdots + a_n Y_n.
\end{aligned}
$$

It follows that $X\alpha = X\theta$ for every X in V, and $\alpha = \theta$, as we wished to show.

It may be emphasized that we have just shown that if two homomorphisms of V into W have the same effect upon the elements of a basis of V, then the homomorphisms are equal. This fact is often expressed by saying that *a homomorphism of V into W is completely determined by its effect upon a basis of V.*

In particular, if we wish to show that two homomorphisms β and γ of V into W are equal, we may do so by showing that each element of a basis of V has the same image under β as under γ.

If V and W are vector spaces over the same field F, it is customary to denote the set of all homomorphisms of V into W by $\text{Hom}_F(V, W)$. In the following section we shall give definitions of addition and scalar multiplication in such a way that $\text{Hom}_F(V, W)$ becomes itself a vector space over F.

EXERCISES

1. Prove that the homomorphism γ of Example 3 is an isomorphism of $V_2(\mathbf{R})$ onto $V_2(\mathbf{R})$.

2. If $\theta: V_3(F) \to V_2(F)$ is the homomorphism defined in Example 1, determine ker θ and verify that it is a subspace of $V_3(F)$.

3. Define a mapping $\theta: V_3(\mathbf{Q}) \to V_3(\mathbf{Q})$ by each of the equations below, where $\mathbf{Q}$ is the rational field. In each case, determine if θ is a homomorphism, and if so, find the kernel of θ showing that it is a subspace of $V_3(\mathbf{Q})$.

    (i) $(a_1, a_2, a_3)\theta = (2a_2, a_3, -a_1)$.

    (ii) $(a_1, a_2, a_3)\theta = (a_1a_2 + a_3, 0, 0)$.

    (iii) $(a_1, a_2, a_3)\theta = (a_1 + a_2, 0, a_2)$.

    (iv) $(a_1, a_2, a_3)\theta = (a_1 + a_2 + a_3, a_1, a_2)$.

    (v) $(a_1, a_2, a_3)\theta = (a_1 - a_2, a_2 - a_3, a_3 - a_1)$.

4. Let V be the vector space of all polynomials with real coefficients (cf. Example 4 of Section 11.2), and denote the derivative of $f(x)$ by $f'(x)$. Verify that the mapping $\theta: V \to V$ defined by $f(x)\theta = f'(x)$ is a homomorphism of V into V. What is the kernel of this homomorphism?

5. Prove Theorem 11.31(iii) and (iv).

6. Prove the first statement of Theorem 11.32.

7. Complete the proof of Theorem 11.33.

8. Prove that there is a Universal Mapping Property for vector spaces analogous to Theorem 2.38 for groups and Theorem 3.25 for rings.

11.7 HOM$_F$ (V, W) AS A VECTOR SPACE

Throughout this section, V and W will be vector spaces over the same field F. For simplicity, we shall often denote the set of all homomorphisms of V into W by H instead of using the more explicit, but more cumbersome, Hom$_F$ (V, W). Our present goal is to define addition in H and scalar multiplication of elements of H by elements of F in such a way that H becomes a vector space over F. We may observe that if V has positive dimension and $W \neq \{0\}$, Theorem 11.35 assures us that H has elements other than the trivial homomorphism which maps every element of V into the zero of W.

If $\alpha, \beta \in H$, we propose to define a mapping $\alpha + \beta$ as follows:

11.39
$$X(\alpha + \beta) = X\alpha + X\beta, \quad X \in V.$$

It is clear that, as so defined, $\alpha + \beta$ is a mapping of V into W. The proof that it is a homomorphism follows from the defining properties (11.3) of a vector space and properties of homomorphisms. The details of the proof are as follows, in which $X_1, X_2 \in V$ and $a \in F$:

$$
\begin{aligned}
(X_1 + X_2)(\alpha + \beta) &= (X_1 + X_2)\alpha + (X_1 + X_2)\beta && \textit{(by 11.39)}\\
&= X_1\alpha + X_2\alpha + X_1\beta + X_2\beta && \textit{(by 11.30(i))}\\
&= X_1\alpha + X_1\beta + X_2\alpha + X_2\beta && \textit{(by 11.3(i))}\\
&= X_1(\alpha + \beta) + X_2(\alpha + \beta) && \textit{(by 11.39),}
\end{aligned}
$$

and

$$
\begin{aligned}
(aX_1)(\alpha + \beta) &= (aX_1)\alpha + (aX_1)\beta && \textit{(by 11.39)}\\
&= a(X_1\alpha) + a(X_1\beta) && \textit{(by 11.30(ii))}\\
&= a(X_1\alpha + X_1\beta) && \textit{(by 11.3(ii))}\\
&= a[X_1(\alpha + \beta)] && \textit{(by 11.39).}
\end{aligned}
$$

These calculations show that both addition and scalar multiplication are preserved under the mapping $\alpha + \beta$ of V into W; hence $\alpha + \beta \in H$, and 11.39 defines an addition on the set H.

Next, if $\alpha \in H$ and $c \in F$, we define a mapping $c\alpha$ of V into W as follows:

11.40
$$X(c\alpha) = (cX)\alpha \quad X \in V.$$

Again, we show that $c\alpha \in H$ by verifying that the mapping $c\alpha$ is a homomorphism of V into W. Let $X_1, X_2 \in V$ and $a \in F$. Then

$$
\begin{aligned}
(X_1 + X_2)(c\alpha) &= [c(X_1 + X_2)]\alpha && \textit{(by 11.40)}\\
&= (cX_1 + cX_2)\alpha && \textit{(by 11.3(ii))}\\
&= (cX_1)\alpha + (cX_2)\alpha && \textit{(by 11.30(i))}\\
&= X_1(c\alpha) + X_2(c\alpha) && \textit{(by 11.40),}
\end{aligned}
$$

and

$$
\begin{aligned}
(aX_1)(c\alpha) &= [c(aX_1)]\alpha && \textit{(by 11.40)} \\
&= [(ca)X_1]\alpha && \textit{(by 11.3(iv))} \\
&= [(ac)X_1]\alpha && \textit{(comm. of F)} \\
&= [a(cX_1)]\alpha && \textit{(by 11.3(iv))} \\
&= a[(cX_1)\alpha] && \textit{(by 11.30(ii))} \\
&= a[X_1(c\alpha)] && \textit{(by 11.40)}.
\end{aligned}
$$

Thus, both addition and scalar multiplication are preserved under the mapping $c\alpha$, and therefore $c\alpha \in H$.

We can now state the following significant result.

11.41 THEOREM *Let V and W be vector spaces over the same field F, and let $H = \mathrm{Hom}_F(V, W)$ be the set of all homomorphisms of V into W. If addition and scalar multiplication by elements of F are defined respectively by 11.39 and 11.40, H becomes a vector space over F.*

Proof: In order to establish this result, it is necessary to verify all the defining properties 11.3 of a vector space. We shall sketch a proof that H is an abelian group with respect to addition, and leave the proofs of 11.3(ii) – (v) as an exercise.

If $\alpha, \beta, \gamma \in H$ and $X \in V$, we have

$$
\begin{aligned}
X[(\alpha + \beta) + \gamma] &= X(\alpha + \beta) + X\gamma && \textit{(by 11.39)} \\
&= (X\alpha + X\beta) + X\gamma && \textit{(by 11.39)} \\
&= X\alpha + (X\beta + X\gamma) && \textit{(by 11.3(i))} \\
&= X\alpha + X(\beta + \gamma) && \textit{(by 11.39)} \\
&= X[\alpha + (\beta + \gamma)] && \textit{(by 11.39)}.
\end{aligned}
$$

Since $X[(\alpha + \beta) + \gamma] = X[\alpha + (\beta + \gamma)]$ for every X in V, we have $(\alpha + \beta) + \gamma = \alpha + (\beta + \gamma)$, and addition in H is associative.

Moreover, since for α, β in H and X in V,

$$
\begin{aligned}
X(\alpha + \beta) &= X\alpha + X\beta \\
&= X\beta + X\alpha \\
&= X(\beta + \alpha),
\end{aligned}
$$

we conclude that $\alpha + \beta = \beta + \alpha$, and addition is commutative.

If we define $\zeta: V \to W$ by $X\zeta = 0$ for every X in V, then ζ is the identity for the abelian group H since clearly $X(\alpha + \zeta) = X\alpha + X\zeta = X\alpha$, and therefore $\alpha + \zeta = \alpha$ for $\alpha \in H$.

Finally, if $\alpha \in H$, then the inverse $-\alpha$ of α is the mapping (clearly a homomorphism) defined by

$$X(-\alpha) = -(X\alpha).$$

For then

$$
\begin{aligned}
X(\alpha + (-\alpha)) &= X\alpha + X(-\alpha) \\
&= X\alpha - (X\alpha) = 0 = X\zeta
\end{aligned}
$$

for every X in V, and it follows that $\alpha + (-\alpha) = \zeta$.

We have verified the defining properties of an abelian group, and hence we have shown that H is an abelian group with respect to addition. As stated above, we shall omit the proofs of the other properties of a vector space.

In case both V and W have finite dimension over the field F, the dimension of the vector space $\text{Hom}_F (V, W)$ is given by the following theorem.

11.42 THEOREM *If the vector spaces V and W have respective dimensions $n > 0$ and $m > 0$ over the field F, then the vector space $\text{Hom}_F (V, W)$ has dimension nm over F.*

Proof: The proof consists in exhibiting a basis of the space $H = \text{Hom}_F (V, W)$ with nm elements. Let $\{X_1, X_2, \ldots, X_n\}$ be a basis of V and $\{Y_1, Y_2, \ldots, Y_m\}$ a basis of W. For each $i = 1, 2, \ldots, n$ and $j = 1, 2, \ldots, m$, there exists (by Theorem 11.35) a unique homomorphism α_{ij} of V into W such that

11.43
$$
\begin{aligned}
X_i\alpha_{ij} &= Y_j \\
X_k\alpha_{ij} &= 0 \quad \text{for } k \neq i.
\end{aligned}
$$

Since $1 \leq i \leq n$ and $1 \leq j \leq m$, there are n choices for i and m choices for j; hence we have nm elements α_{ij} of H. We propose to prove that these elements form a basis of the vector space H. First of all, let us show that these α_{ij} are linearly independent. Suppose that

$$\beta = \sum_{j=1}^{m} \sum_{i=1}^{n} c_{ij}\alpha_{ij} = 0,$$

the c_{ij} being elements of F. Let k be an arbitrary, but fixed, element of the set $1, 2, \ldots, n$. Since we are assuming that $\beta = 0$, we have

$$0 = X_k\beta = \sum_{j=1}^{m} \sum_{i=1}^{n} c_{ij}(X_k\alpha_{ij}).$$

But, using the definition 11.43 of the α_{ij}, we see that we have

$$0 = \sum_{j=1}^{m} c_{kj} Y_j.$$

However, the Y's form a basis of W and hence are linearly independent. Accordingly, we conclude that $c_{kj} = 0$ for $j = 1, 2, \ldots, m$. Since k was an arbitrary element of the set $1, 2, \ldots, n$, we have shown that $c_{ij} = 0$ for all choices of i and j; hence the α_{ij} are linearly independent elements of the vector space H.

We must now show that any element of H is a linear combination of the elements α_{ij}. Let γ be an arbitrary element of H. For each $k \in \{1, 2, \ldots, n\}$, $X_k\gamma \in W$ and therefore $X_k\gamma$ can be expressed as a linear combination of the basis elements of W. Thus there exist elements a_{kj} of F such that

11.44
$$X_k\gamma = a_{k1} Y_1 + a_{k2} Y_2 + \cdots + a_{km} Y_m \qquad (k = 1, 2, \ldots, n).$$

We shall show that

$$\gamma = \sum_{j=1}^{m} \sum_{i=1}^{n} a_{ij}\alpha_{ij}$$

by showing that the elements of H occurring on the two sides of this equation have the same effect upon the elements of the basis $\{X_1, X_2, \ldots, X_n\}$ of V. If X_k is any one of these basis elements, we find (using 11.43) that

$$X_k\left(\sum_{j=1}^{m} \sum_{i=1}^{n} a_{ij}\alpha_{ij}\right) = \sum_{j=1}^{m} \sum_{i=1}^{n} a_{ij}(X_k\alpha_{ij})$$

$$= \sum_{j=1}^{m} a_{kj} Y_j$$

$$= X_k\gamma \qquad\qquad (by\ 11.44).$$

It follows that

$$\gamma = \sum_{j=1}^{m} \sum_{i=1}^{n} a_{ij}\alpha_{ij},$$

and we conclude that the α_{ij} do indeed form a basis of the vector space H over F. Since there are nm of the elements α_{ij}, we have shown that the vector space H over F has a basis with nm elements, and therefore has dimension nm. This completes the proof of the theorem.

In this section we have shown that under appropriate definitions of addition and scalar multiplication, $\text{Hom}_F(V, W)$ is a vector space over F. In the next section we shall study in detail the important special case in which $\dim V = n > 0$ and $\dim W = 1$. The special case in which $V = W$ will be the principal subject of Chapter 15.

EXERCISES

1. Complete the proof of Theorem 11.41 by verifying that the defining properties 11.3(ii) – (v) of a vector space hold in H.

2. In the notation of Theorem 11.42, let $V = W = V_2(\mathbf{Q})$ and use the basis of unit vectors in both V and W. Determine the element $2\alpha_{11} + 3\alpha_{21}$ by exhibiting its effect upon the basis of unit vectors.

3. If V is a vector space over a field F and $\alpha, \beta \in \text{Hom}_F(V, V)$, verify that the mapping $\alpha\beta$ is defined and that it is in fact an element of $\text{Hom}_F(V, V)$.

4. If V and W are vector spaces over a field F with $\dim V = n > 0$, and $\dim W = m > 0$, what is the dimension of the vector space $\text{Hom}_F(\text{Hom}_F(V, W), V)$?

11.8 DUAL VECTOR SPACES

Throughout this section, V will denote a vector space of finite dimension over the field F, and W will have dimension 1, that is, W is F considered as a vector space over itself. We shall thus be concerned with the vector space $\text{Hom}_F(V, F)$ of homomorphisms of V into F.

An element of $\text{Hom}_F(V, F)$ is thus a mapping† $\alpha: V \to F$ such that if $X, Y \in V$ and $a \in F$, then

$$(X + Y)\alpha = X\alpha + Y\alpha$$

and

$$(aX)\alpha = a(X\alpha).$$

---

† A mapping of V into F which has these properties is sometimes called a *linear functional* on V. In this language, $\text{Hom}_F(V, F)$ consists of all linear functionals on V.

As a special case of Theorem 11.42, we see that if dim $V = n > 0$, then the dimension of $\text{Hom}_F(V, F)$ is also n.

In order to emphasize that we are considering $\text{Hom}_F(V, F)$ as a vector space, let us now denote it by V', and let us denote elements of V' by X', Y', and so on. Thus if $X' \in V'$, X' is a homomorphism of V into F, so that for each $X \in V$, $XX' \in F$. If $\{X_1, X_2, \ldots, X_n\}$ is a basis of V and we take as a basis of F the unity 1 of F, following the proof of Theorem 11.42 with $m = 1$ (and using the new notation) we see that V' has a basis $\{X_1', X_2', \ldots, X_n'\}$ where for each $k = 1, 2, \ldots, n$, the homomorphism X_k' of V into F is defined by

11.45 $X_k X_k' = 1, \qquad X_i X_k' = 0 \text{ for } i \neq k.$

Let us now make the following definition.

11.46 DEFINITION   The vector space $V' = \text{Hom}_F(V, F)$ is called the *dual* of the vector space V. If $\{X_1, X_2, \ldots, X_n\}$ is a basis of V and the X_k' are defined by 11.45, the basis $\{X_1', X_2', \ldots, X_n'\}$ of V' is called the *dual basis* to $\{X_1, X_2, \ldots, X_n\}$.

Now that we have the vector space V' over F, for the moment we need not try to keep in mind what its elements are, but just consider that we have a vector space V' over F. There is then no reason why we should not consider *its* dual, which we may denote by V''. Thus elements of V'' are homomorphisms of V' into F, that is, $V'' = \text{Hom}_F(V', F)$. If dim $V = n$, we know that dim $V' = n$ and dim $V'' = n$, so that these vector spaces are all isomorphic (by Theorem 11.33). This fact in itself is of no great importance, but it is significant that there exists a *natural* isomorphism of V onto V''. As we proceed to obtain this result, it will become apparent that the roles of V and V' are essentially interchangeable in that the dual of V', although not V itself, is isomorphic to V in a special way.

We observed above, when the present notation was introduced, that if $X \in V$ and $X' \in V'$, then $XX' \in F$. If X' is fixed, the mapping X' of V into F may be indicated by writing

$$X \rightarrow XX', \qquad X \in V.$$

Now if X is fixed, let us consider the mapping

$$X' \rightarrow XX', \qquad X' \in V'$$

of V' into F. For convenience, let us denote this last mapping by T_X, that is, T_X is defined by

11.47 $X' T_X = XX', \qquad X' \in V'.$

Not only is T_X a mapping of V' into F, we shall show that it is actually a homomorphism of V' into F, and is therefore an element of V''. Then we shall prove an important result which asserts that the mapping $X \to T_X$ is actually an isomorphism of V onto V''.

We proceed to verify that addition and scalar multiplication are preserved under the mapping T_X. If $Y', Z' \in V'$, we have

$$\begin{aligned}(Y' + Z')T_X &= X(Y' + Z') & \text{(by 11.47)}\\ &= XY' + XZ' & \text{(by 11.39)}\\ &= Y'T_X + Z'T_X & \text{(by 11.47)}.\end{aligned}$$

Also, if $Y' \in V'$ and $a \in F$,

$$\begin{aligned}(aY')T_X &= X(aY')\\ &= (aX)Y' & \text{(by 11.40)}\\ &= a(XY') & (Y' \text{ is a homomorphism})\\ &= a(Y'T_X) & \text{(by 11.47)}.\end{aligned}$$

Thus, T_X is a homomorphism of V' into F and hence an element of V'', the dual of V'. As a matter of fact, *every* element of V'' is of the form T_X for some $X \in V$. This is one consequence of the following principal theorem about dual vector spaces of finite dimension.

11.48　THEOREM　*Let V be a vector space of finite dimension and let $\theta: V \to V''$ be the mapping defined by $X\theta = T_X$, $X \in V$; where T_X is given by 11.47. Then θ is an isomorphism of V onto V''.*

Proof:　Let us first verify that θ is a homomorphism of V into V''. If $X, Y \in V$ and $X' \in V'$, we have (by use of 11.47 and the fact that X' is a homomorphism of V into F) that

$$\begin{aligned}X'T_{X+Y} &= (X + Y)X'\\ &= XX' + YX'\\ &= X'T_X + X'T_Y\\ &= X'(T_X + T_Y),\end{aligned}$$

by definition of addition in V''. Also, if $X' \in V'$ and $c \in F$, we have

$$\begin{aligned}X'T_{cX} &= (cX)X'\\ &= c(XX')\\ &= c(X'T_X)\\ &= X'(cT_X),\end{aligned}$$

by definition of the element cT_X of V''. These calculations show that $T_{X+Y} = T_X + T_Y$ and that $T_{cX} = cT_X$, that is, that $\theta: V \to V''$ has the properties that

$$(X + Y)\theta = X\theta + Y\theta$$

and

$$(cX)\theta = c(X\theta),$$

and θ is indeed a homomorphism of V into V''. There are various ways of showing that the mapping θ is one-one and onto. One method is as follows.

Let $\{X_1, X_2, \ldots, X_n\}$ be a basis of V and $\{X'_1, X'_2, \ldots, X'_n\}$ the dual basis (11.45) of V'. For each $k = 1, 2, \ldots, n$, the following hold (by 11.47):

$$X'_k T_{X_k} = X_k X'_k = 1,$$
$$X'_i T_{X_k} = X_k X'_i = 0 \qquad \text{for } i \neq k.$$

By the same argument that led to the conclusion that $\{X'_1, \ldots, X'_n\}$ was a basis of V' (dual to the basis $\{X_1, \ldots, X_n\}$ of V), it follows that $\{T_{X_1}, T_{X_2}, \ldots, T_{X_n}\}$ is a basis of V'' dual to the basis $\{X'_1, X'_2, \ldots, X'_n\}$ of V'. Now if $X \in V$ and we write $X = a_1 X_1 + a_2 X_2 + \cdots + a_n X_n$, it is true that

$$\begin{aligned} X\theta &= (a_1 X_1 + a_2 X_2 + \cdots + a_n X_n)\theta \\ &= a_1 T_{X_1} + a_2 T_{X_2} + \cdots + a_n T_{X_n}. \end{aligned}$$

We leave the verification of this fact as an exercise. Now since $\{X_1, X_2, \ldots, X_n\}$ is a basis of V and $\{T_{X_1}, T_{X_2}, \ldots, T_{X_n}\}$ is a basis of V'', it follows easily, as indicated near the end of the sketch of the proof of Theorem 11.33, that θ is an isomorphism of V onto V'', as we wished to show.

11.9 QUOTIENT VECTOR SPACES AND DIRECT SUMS

In this section we shall present very briefly a few additional concepts about vector spaces which will be useful for later reference. Since they are analogous to concepts which have previously been introduced for rings or groups, we shall sometimes state the basic results without proofs or with only an outline of a proof.

First, let us present the concept of a quotient space, which corresponds to the idea of a quotient ring or a quotient group. Let V be a vector space over a field F, and let U be a subspace of V. With respect to addition (ignoring scalar multiplication for the moment), we know that U is a subgroup of V. Accordingly, we may consider the set of all cosets of U in V, such a coset being of the form

$$X + U = \{X + Y \mid Y \in U\},$$

X being a fixed element of V. We know from our study of groups that addition of cosets is well-defined by

11.49 $$(X_1 + U) + (X_2 + U) = (X_1 + X_2) + U.$$

We now propose to define scalar multiplication of cosets by elements of F. If $X + U$ is a coset and $a \in F$, we define

11.50 $$a(X + U) = aX + U,$$

that is, $a(X + U)$ is the coset which contains aX. Let us show that scalar multiplication is well-defined by 11.50. To this end, suppose that $X_1 + U = X_2 + U$, and let us show that $aX_1 + U = aX_2 + U$. Now $X_1 + U = X_2 + U$ implies that $X_1 - X_2 \in U$. Then since U is a subspace of V, we see that $a(X_1 - X_2) \in U$ and it follows that $aX_1 + U = aX_2 + U$, and this completes the proof.

We may now state the following theorem (cf. Theorem 2.79).

11.51 THEOREM *Let U be a subspace of the vector space V over F. With respect to the definition 11.49 of addition and 11.50 of scalar multiplication, the set of all cosets of U in V is a vector space over F, usually called the* quotient space *of V by U and denoted by V/U. Moreover, the mapping $\theta : V \to V/U$ defined by $X\theta = X + U$, $X \in V$, is a homomorphism of V onto V/U with kernel U.*

Proof: We know from our study of groups that V/U is an abelian group with respect to addition and that the zero element is the coset U. To verify that it is in fact a vector space over F, the remaining properties (11.3) of the definition of a vector space must be verified. This will be listed as an exercise below.

If V has finite dimension, the following theorem states how dim V and dim U/V are related.

11.52 THEOREM *Suppose that the vector space V has finite dimension $n > 0$ over the field F, and let U be a subspace of V of dimension m. Then the vector space V/U has dimension $n - m$.*

Proof: We know that $m \leq n$. If $m = 0$, then $U = \{0\}$ and $V/U \cong V$, so this case is trivial. Let us therefore assume that $m > 0$.

Let $\{X_1, X_2, \ldots, X_m\}$ be a basis of U. By Theorem 11.24, this set can be extended to a basis

11.53
$$\{X_1, \ldots, X_m, X_{m+1}, \ldots, X_n\}$$

of V. We shall show that

11.54
$$\{X_{m+1} + U, \ldots, X_n + U\}$$

is a basis of the vector space V/U.

First, let $X + U$ be any element of V/U, and let us write X as a linear combination of the basis elements 11.53 of V,

$$X = a_1 X_1 + \cdots + a_n X_n.$$

Then since $a_1 X_1 + \cdots + a_m X_m$ is an element of U, we see that by 11.49 and 11.50

$$
\begin{aligned}
X + U &= (a_{m+1} X_{m+1} + \cdots + a_n X_n) + U \\
&= a_{m+1}(X_{m+1} + U) + \cdots + a_n(X_n + U).
\end{aligned}
$$

Thus every element of V/U is a linear combination of the elements 11.54.

Next, let us show that the elements 11.54 are linearly independent. Suppose that the b's are elements of F such that

$$b_{m+1}(X_{m+1} + U) + \cdots + b_n(X_n + U) = U,$$

U being the zero of the space V/U. This equation implies that $b_{m+1} X_{m+1} + \cdots + b_n X_n$ is an element of U. However, this fact implies that $b_{m+1} = \cdots = b_n = 0$ since the set 11.53 is linearly independent and each element of U is expressible as a linear combination of the basis elements $X_1, X_2, \ldots, X_m$ of U. Accordingly, we have shown that the elements 11.54 of V/U are linearly independent. We have thus completed the proof that the elements 11.54 are a basis of V/U. Since there are $n - m$ of these basis elements, the dimension of V/U is indeed $n - m$.

We now state the following theorem which is analogous to Theorem 3.31 for rings and Theorem 2.81 for groups. We shall list its proof as an exercise below.

11.55 THEOREM *Let V and W be vector spaces over the same field, and let $\theta: V \to W$ be a homomorphism of V onto W. Then $W \cong V/(\ker \theta)$.*

Next let us introduce the concept of direct sum of subspaces. If $U_1, U_2, \ldots, U_r$ are subspaces of the vector space V over the field F, by 11.26 we see that $U_1 + U_2 + \cdots + U_r$ is the subspace of V consisting of all elements of V expressible in the form

$$X_1 + X_2 + \cdots + X_r, \qquad X_i \in U_i \, (i = 1, 2, \ldots, r).$$

In keeping with the concept of the direct sum of subgroups of a given abelian group, let us make the following definition.

11.56 DEFINITION If $U_i \ (i = 1, 2, \ldots, r)$ are subspaces of the vector space V, the sum $U_1 + U_2 + \cdots + U_r$ is said to be a *direct sum* if and only if each element X of this sum is *uniquely* expressible in the form

$$X = X_1 + X_2 + \cdots + X_r, \qquad X_i \in U_i.$$

As was pointed out when direct sums of abelian groups were introduced, an equivalent way of stating that a sum $U_1 + U_2 + \cdots + U_r$ is direct is to assert that *if $X_i \in U_i \ (i = 1, 2, \ldots, r)$ are such that $X_1 + X_2 + \cdots + X_r = 0$, then every $X_i = 0$.*

We may indicate that a sum $U_1 + U_2 + \cdots + U_r$ is direct by writing it in the form $U_1 \oplus U_2 \oplus \cdots \oplus U_r$.

Let us conclude this section by proving the following simple consequence of previously established facts.

11.57 THEOREM *If U_1 and U_2 are subspaces of the vector space V such that $V = U_1 \oplus U_2$, then $U_1 \cong V/U_2$ and $U_2 \cong V/U_1$.*

Proof: Each element X of V is uniquely expressible in the form $X_1 + X_2$, where $X_1 \in U_1$ and $X_2 \in U_2$. The mapping $\theta: V \to U_1$ defined by $X\theta = X_1$ is a homomorphism of V onto U_1. Moreover, $\ker \theta = U_2$. From Theorem 11.55, it follows that $U_1 \cong V/U_2$. Similarly, $U_2 \cong V/U_1$.

11.10 INNER PRODUCTS IN $V_n(F)$

For the moment, we consider only vector spaces of the form $V_n(F)$. The concept now to be introduced will be particularly useful in the following chapter.

11.58 DEFINITION If $X = (a_1, a_2, \ldots, a_n)$ and $Y = (b_1, b_2, \ldots, b_n)$ are elements of $V_n(F)$, the *inner product* $X \cdot Y$ of X and Y is defined as follows:

$$X \cdot Y = a_1 b_1 + a_2 b_2 + \cdots + a_n b_n.$$

It is clear that the inner product of two vectors is a scalar; that is, it is an element of F. The following theorem gives the most important properties of inner products. Since the proofs are quite simple, we leave them as exercises.

11.59 THEOREM

(i) *If* $X, Y \in V_n(F)$, *then* $X \cdot Y = Y \cdot X$.

(ii) *If* $X, Y, Z \in V_n(F)$, *then* $(X + Y) \cdot Z = X \cdot Z + Y \cdot Z$.

(iii) *If* $X, Y \in V_n(F)$ *and* $a \in F$, *then*
$$(aX) \cdot Y = X \cdot (aY) = a(X \cdot Y).$$

(iv) *If* $Y, X_1, X_2, \ldots, X_m$ *are elements of* $V_n(F)$, *and* $a_1, a_2, \ldots, a_m$ *are elements of* F, *then*
$$(a_1 X_1 + a_2 X_2 + \cdots + a_m X_m) \cdot Y$$
$$= a_1 (X_1 \cdot Y) + a_2 (X_2 \cdot Y) + \cdots + a_m (X_m \cdot Y).$$

EXERCISES

1. Let $\theta: V \to V''$ be the mapping defined in the statement of Theorem 11.48. If $X_1, X_2, \ldots, X_n$ are elements of V, verify in detail that
$$(a_1 X_1 + a_2 X_2 + \cdots + a_n X_n)\theta = a_1 T_{X_1} + a_2 T_{X_2} + \cdots + a_n T_{X_n}.$$

2. If θ is as in the preceding exercise, prove that θ is a one-one mapping by verifying that $\ker \theta = \{0\}$.

3. In the notation of Theorem 11.51 verify that all the defining properties of a vector space are satisfied in V/U.

4. Let U_1 and U_2 be subspaces of the vector space V such that $V = U_1 \oplus U_2$. If $\{X_1, X_2, \ldots, X_h\}$ is a basis of U_1 and $\{Y_1, Y_2, \ldots, Y_k\}$ is a basis of U_2, prove that $\{X_1, X_2, \ldots, X_h, Y_1, Y_2, \ldots, Y_k\}$ is a basis of V. Prove that not every basis of V is the union of bases of U_1 and of U_2.

5. State and prove a generalization of the preceding exercise to the case in which V is the direct sum of any finite number of subspaces.

6. Prove Theorem 11.55.

7. Prove Theorem 11.59.

8. If X and Y are nonzero vectors of $V_2(\mathbf{R})$, show that $X \cdot Y = 0$ if and only if the directed line segments which represent these vectors, in the sense described in Section 11.1, are perpendicular.

9. If $Y \in V_n(F)$, use Theorem 11.59 to show that the set of all vectors X of $V_n(F)$ such that $X \cdot Y = 0$ is a subspace of $V_n(F)$.

10. If S is a nonempty *subset* of $V_n(\mathbf{R})$, let us define $S^\perp = \{X \mid X \in V_n(\mathbf{R}), Y \cdot X = 0$ for every Y in $S\}$.

    (i) Prove that $S^\perp$ is a *subspace* of $V_n(\mathbf{R})$.

    (ii) Prove that $S \cap S^\perp = \{0\}$ or $S \cap S^\perp = \varnothing$.

    (iii) Show, by an example, that part (ii) would be false if we were using $V_n(\mathbf{C})$ in place of $V_n(\mathbf{R})$.

11. Assume that $\theta: V \to W$ is a homomorphism of the vector space V onto the vector space W. If $\{X_1, X_2, \ldots, X_n\}$ is a basis for V then prove that θ is an isomorphism if and only if $\{X_1\theta, X_2\theta, \ldots, X_n\theta\}$ is a basis for W. In particular, an isomorphism onto W takes a basis to a basis.

 COMMENTARY

1 Josiah Willard Gibbs (1839–1903) created a vector algebra that, even in its most elementary form, can yield easy answers to questions that seem very difficult when tackled in other ways. A vector in three dimensions

$$X = a\mathbf{i} + b\mathbf{j} + c\mathbf{k}$$

where $a, b, c \in \mathbf{R}$ may be thought of as a directed line segment running from the origin $\mathbf{O} = (0, 0, 0)$ to a point $\mathbf{P} = (a, b, c)$. In this notation, the length of X is given by

$$|X| = \sqrt{a^2 + b^2 + c^2}.$$

If $Y = a'\mathbf{i} + b'\mathbf{j} + c'\mathbf{k}$ is another vector then the inner or dot product is defined to be

$$X \cdot Y = aa' + bb' + cc'.$$

In another interpretation, we may view $X \cdot Y$ as $|X||Y| \cos \theta$ where θ is the angle between X and Y. In three dimensions there also is a vector or crossproduct defined as

$$X \times Y = (bc' - b'c)\mathbf{i} + (ca' - c'a)\mathbf{j} + (ab' - a'b)\mathbf{k}.$$

The vector $X \times Y$ has length $|X||Y|\sin\theta$ where θ is the angle between X and Y, and $X \times Y$ points in a direction perpendicular to the plane of X and Y.

Let us illustrate this vector algebra by solving two problems for analytic geometry. Suppose $\mathbf{P}$ is a point and π is a plane in three dimensions. What is the distance from $\mathbf{P}$ to π? There is a formula for this distance given in elementary analytic geometry, but by using vector algebra we can give a very easy derivation of this formula. The equation for the plane has the form

$$\pi: ax + by + cz = d$$

and the point

$$\mathbf{P} = (a', b', c').$$

If (x, y, z) and (x_0, y_0, z_0) lie on the plane then

$$a(x - x_0) + b(y - y_0) + c(z - z_0) = 0.$$

If $W = (x - x_0)\mathbf{i} + (y - y_0)\mathbf{j} + (z - z_0)\mathbf{k}$ then this equation says that

$$X \cdot W = 0$$

or that $\cos\theta = 0$ where θ is the angle between X and W. The vector W is parallel to the line from (x_0, y_0, z_0) to (x, y, z), and therefore, parallel to the plane. So $\theta = 90°$ for all such W, and X is perpendicular to the plane. We may make X one unit long by dividing by $|X|$. Thus $X/|X|$ is one unit long and perpendicular to the plane π.

The vector $T = (a' - x_0)\mathbf{i} + (b' - y_0)\mathbf{j} + (c' - z_0)\mathbf{k}$ is parallel to and of the same length as the line from $\mathbf{P}$ to the point (x_0, y_0, z_0) upon the plane π. Furthermore, if ϕ is the angle between T and X, then the shortest distance from $\mathbf{P}$ to π is along a line $\overline{\mathbf{PQ}}$ where $\mathbf{Q}$ is a point on π and $\mathbf{P_1Q_1}(x_0, y_0, z_0)$ is a right triangle. Thus $|T|\cos\phi = T\cdot(X/|X|)$ is the desired distance from $\mathbf{P}$ to the plane. Writing this out we obtain

$$\pm\,\text{distance from }\mathbf{P}\text{ to }\pi = \frac{a'a + b'b + c'c - d}{\sqrt{a^2 + b^2 + c^2}},$$

the familiar formula for this distance where $\mathbf{P} = (a', b', c')$ and $\pi: ax + by + cz = d$.

Let us tackle a more difficult problem. Suppose l_1 and l_2 are two skew straight lines in space. What is the distance from l_1 to l_2? We shall give a description of the vector solution in this case. We find a vector X parallel to l_1 and a vector Y parallel to l_2. We shall assume that l_1 is not parallel to l_2 so that the lines are definitely skewed. Then $X \times Y$ is perpendicular to the direction of both l_1 and l_2. In other words, the shortest distance from l_1 to l_2 is along a line parallel to $X \times Y$. Let $\mathbf{Q} = (x, y, z)$ be on l_1 and $\mathbf{R} = (x_0, y_0, z_0)$ be on l_2. Then $W = (x - x_0)\mathbf{i} + (y - y_0)\mathbf{j} + (z - z_0)\mathbf{k}$ is parallel to the line connecting $\mathbf{Q}$ to $\mathbf{R}$ and of length equal to the distance from $\mathbf{Q}$ to $\mathbf{R}$. The projection of this segment onto the line of $X \times Y$ will be the distance by simple trigonometry. Thus if θ is the angle between $X \times Y$ and W then

$$\pm\,\text{distance from }l_1\text{ to }l_2 = \pm|W|\cos\theta$$
$$= W\cdot(X \times Y)/|X \times Y|.$$

This formula may be worked out to give equations in the components. Many other similar problems can be solved using the algebra of vectors to reason geometrically. One sees why Gibbs' vector algebra is so useful [61, 138].

2   Throughout this chapter we have assumed that vector spaces are always finite dimensional. Let us consider a particular vector space which is not finite dimensional. Let l_2 be the set of all sequences $\{a_n\}$ of real numbers a_n such that $\sum_{n=1}^{\infty} |a_n|^2$ converges. From elementary calculus, we know that if $\sum_{n=1}^{\infty} |a_n|^2$ and $\sum_{n=1}^{\infty} |b_n|^2$ both converge and $c \in \mathbf{R}$ then $\sum_{n=1}^{\infty} |a_n b_n|$, $\sum_{n=1}^{\infty} (|a_n|^2 + |b_n|^2)$, and $\sum_{n=1}^{\infty} |ca_n|^2$ all converge. In particular, $\sum_{n=1}^{\infty} |a_n + b_n|^2$ converges. Using these facts we may now make the following definitions.

(1) If $c \in \mathbf{R}$ and $\{a_n\} \in l_2$ then we set $c \cdot \{a_n\} = \{ca_n\}$.

(2) If $\{a_n\}, \{b_n\} \in l_2$ then we set $\{a_n\} + \{b_n\} = \{a_n + b_n\}$.

These operations make l_2 into a vector space over the real field.

One way to think of this vector space is as "infinite-tuples" $\{a_n\} = (a_1, a_2, \ldots, a_n, \ldots)$ of real numbers subject to the condition that $\sum_{n=1}^{\infty} |a_n|^2$ converges. Certainly the infinite-tuples $\{e_n^{(i)}\}$ where $e_n^{(i)} = 0$ if $n \neq i$ and $e_n^{(i)} = 1$ if $n = i$ lie in l_2 for $i = 1, 2, 3, \ldots$ and form a linearly independent set. We conclude that l_2 is not finite dimensional. The set of sequences $\{e_n^{(i)}\}$ for $i = 1, 2, 3, \ldots$ is not even a basis for l_2 since not every sequence is a *finite* linear combination of the $\{e_n^{(i)}\}$'s; in particular, $\{1/n\}$ is an example of such a sequence. However, l_2 contains the set of "finite sequences," i.e., sequences $\{a_n\}$ such that $a_n = 0$ except for a finite number of values of $n = 1, 2, 3, \ldots$. This set of finite sequences is a subspace of l_2 for which the collection of sequences $\{e_n^{(i)}\}$, $i = 1, 2, 3, \ldots$ does serve as a basis.

In Section 11.10 we introduced an inner or dot product upon $V_n(F)$; we may do the same thing for l_2. If $\{a_n\}, \{b_n\} \in l_2$ then $\sum_{n=1}^{\infty} a_n b_n$ converges so that we may set

(3) $\{a_n\} \cdot \{b_n\} = \sum_{n=1}^{\infty} a_n b_n$.

In fact, this inner product has all the properties of Theorem 11.59.

We may now combine this algebra with some calculus. In three dimensional Euclidean space we defined the distance between a point $\mathbf{P} = (a, b, c)$ and a point $\mathbf{Q} = (x, y, z)$ to be

$$d(\mathbf{P}, \mathbf{Q}) = \sqrt{(a - x)^2 + (b - y)^2 + (c - z)^2}.$$

Working analogously in l_2 we may define a distance between $\{a_n\}, \{b_n\} \in l_2$ by

(4) $d(\{a_n\}, \{b_n\}) = \sqrt{\sum_{n=1}^{\infty} (a_n - b_n)^2}$
$= \sqrt{(\{a_n\} - \{b_n\}) \cdot (\{a_n\} - \{b_n\})}$.

Once we have a measure of distance, we may study many of the properties of l_2 analogous to those studied in the calculus. For example, if $\{A_n\}$ is a sequence of real numbers which satisfies the Cauchy Criterion (i.e., for any $\epsilon > 0$, there is a $N > 0$ such that $n, m \geq N$ imply that $|A_n - A_m| < \epsilon$) then we know that $\{A_n\}$ is a convergent sequence. Similarly, we may consider a sequence $\{B_n\}$ from l_2, that is, $B_n = \{a_s^{(n)}\} \in l_2$, such that for any $\epsilon > 0$ there is an $N > 0$ such that $n, m \geq N$ imply that $d(B_n, B_m) < \epsilon$. Such a sequence in l_2 would naturally be called a Cauchy sequence since it satisfies the Cauchy Criterion. Is there an element $\{b_s\} \in l_2$ such that the sequence $\{B_n\}$ "converges" to $\{b_s\}$? The answer is "yes" and it combines some of the ideas of vector spaces with those of the calculus [109, 116].

The Cauchy Criterion for sequences of real numbers is equivalent to the least upper bound property of the real field [129]. Either one of these properties is taken to mean that the real field is *complete* (i.e., sequences which "ought" to converge do so, functions which "ought" to be continuous are). The space l_2 also shares the Cauchy property, i.e., it is complete. However, there are subtleties. The real numbers have the property that every closed bounded subset is "compact," in the sense that the Heine-Borel Theorem holds for these subsets. On the other hand, there are closed bounded subsets of l_2 which are not compact.

It was Hilbert who realized that the Cauchy Criterion gives the "right" definition of completeness for infinite dimensional spaces. As a consequence of this and many other contributions, complete vector spaces with an inner product related to their distance measures are called (real) *Hilbert spaces*. Such spaces serve as a meeting point of analysis and algebra, using the best of both to describe mathematical structures.

3   The most important theorem we have proved about vector spaces is that n-dimensional vector spaces have bases and any basis consists of n vectors. In [10, 57] this fact is discussed from several different points of view. We have just scratched the surface of a very extensive branch of mathematics called linear algebra. Elementary coverage of this subject can be found in [60, 62, 63, 67].

The commutativity of multiplication in a field F plays essentially no role in obtaining many properties of a vector space over F. With proper care, everything which we have done can be done as well for vector spaces over a "noncommutative field" or, as it is usually called, a "division ring." That is, all field properties are assumed, including the existence of a multiplicative inverse for each nonzero element, except that multiplication is not required to be commutative. Vector spaces over a division ring are studied in detail in vol. 2 of Jacobson [14]. Chapter 9 of this same reference is an exposition of results about infinite dimensional vector spaces, including such concepts as infinite bases of such spaces.

chapter twelve

FIELD EXTENSIONS

The object of this chapter is to discover properties of roots of polynomials with coefficients in a field F. We shall encounter fields which contain one another, so let us recall some terminology. If F and K are fields with $F \subseteq K$, then F is a *subfield* of K and K is an *extension* of F. If $F \subset K$, that is, if F is a subfield of K unequal to K, we may call F a *proper subfield* of K and K a *proper extension* of F.

If F is a subfield of K, we proved in Section 6.6 that F and K have the same unity, which we shall now usually denote by 1. This fact implies that F and K have the same characteristic, which we know is either zero or a prime.

We proved in Theorem 6.28 that a field of characteristic zero has a subfield isomorphic to the field $\mathbf{Q}$ of rational numbers, and in Theorem 6.10 that a field of characteristic the prime p has a subfield isomorphic to $\mathbf{Z}_p$. Another way of stating these facts is to say that every field of characteristic zero is (essentially) an extension of the field $\mathbf{Q}$ of rational numbers, and every field of characteristic p is (essentially) an extension of the field $\mathbf{Z}_p$ of integers modulo p. The word "essentially" as used here is to be interpreted as "if we do not distinguish between isomorphic fields."

In this chapter we shall present a few of the most basic concepts and results about extensions of a given field, and (in the Commentary at the end of the chapter) give references to some further aspects of this very extensive branch of algebra.

In previous chapters we have already obtained some results which are important for our present purposes; we shall refer to them and, in some cases, give a short review of them at appropriate points in the exposition.

12.1 CLASSIFICATION OF EXTENSIONS

Let us start right off by establishing a connection between extensions of a field F and vector spaces over F. Throughout this section we shall let K be a fixed extension of the field F. Recall from Example 3 of Section 11.2 that K may be viewed as a vector space over F in the following way. The field K, with its addition, is an abelian group denoted by K^+. Further, a scalar multiplication of elements of K^+ by elements of F is given by multiplication in the field K since $F \subseteq K$. In this way, K^+ is a vector space over F. When we want to discuss K as a vector space over F, we shall denote the additive abelian group of K by K^+ to call attention to the fact that multiplication is being partially ignored. This view of K^+ as a vector space over F is an important one because we now can discuss dimension.

12.1 · DEFINITION Let K be an extension of the field F. If the dimension of K^+ considered as a vector space over F is finite, we say that K is a *finite extension* of F; otherwise, it is an *infinite extension* of F. If the dimension of K^+ over F is the integer $n \geq 1$, we may say that n is the *degree* of K over F, or the *index* of F in K.

The word *finite* in this definition refers to dimension and not to numbers of elements. In particular, the index of F in K is *not the same* as the index of the abelian group F^+ in the abelian group K^+. If F is a finite extension of F, it is customary to denote the degree of K over F by $[K:F]$.

Using our knowledge of bases in vector spaces we may prove the very important *Product of the Degrees* theorem.

12.2 THEOREM *If the field K is a finite extension of the field L, and the field L is a finite extension of the field F (that is, $K \supseteq L \supseteq F$) then K is a finite extension of the field F and*

$$[K:F] = [K:L][L:F].$$

Proof: Assume that $n = [K:L]$ and $m = [L:F]$ are the degrees of the extensions over their subfields. In particular, K^+ is an n-dimensional vector space over L, and so by Definition 11.22 there is a basis $v_1, v_2, \ldots, v_n$ of vectors from K^+ over L. Similarly there is a basis $u_1, u_2, \ldots, u_m$ of vectors from L^+ over F. Since $K \supseteq L$ all these basis vectors lie in the extension field K of F. Consequently, we may form the nm products

12.3 $u_i v_j \qquad i = 1, 2, \ldots, m; \qquad j = 1, 2, \ldots, n.$

We shall prove that these vectors form a basis for the vector space K^+ viewed over the field F.

In our proof we shall be concerned with sums constructed from elements a_{ij} from F of the form

12.4
$$\sum a_{ij} u_i v_j$$

where it is understood that in this sum i takes values from 1 to m, and j independently takes values from 1 to n. First let us prove that the vectors of 12.3 are linearly independent. Accordingly, assume that the sum 12.4 is equal to 0. Writing the resulting equation in a more explicit way we obtain:

$$\sum_{j=1}^{n} (a_{1j}u_1 + a_{2j}u_2 + \cdots + a_{mj}u_m)v_j = 0.$$

The coefficient $b_j = a_{1j}u_1 + a_{2j}u_2 + \cdots + a_{mj}u_m$ is an element of L. Therefore the equation may be written as

$$b_1 v_1 + b_2 v_2 + \cdots + b_n v_n = 0$$

where the b's lie in L. Since the v's form a basis for K^+ over L we must have $b_j = 0$ for $j = 1, 2, \ldots, n$. It follows that

$$a_{1j}u_1 + a_{2j}u_2 + \cdots + a_{mj}u_m = 0$$

for each $j = 1, 2, \ldots, n$. However, the u's are linearly independent over F and the a's lie in F so that $a_{ij} = 0$ for all i and j. We conclude that the vectors of 12.3 are linearly independent.

To conclude the proof we must show that every element of K^+ is expressible as a linear combination 12.4 over F. Since the v's form a basis of K^+ over L, every element of K^+ may be written in the form

12.5
$$c_1 v_1 + c_2 v_2 + \cdots + c_n v_n$$

the c's being elements of L. But since the u's form a basis of L^+ over F, each c is expressible as a linear combination $c_j = a_{1j}u_1 + a_{2j}u_2 + \cdots + a_{mj}u_m$ where the a's are in F. Substituting these linear expressions for the c's into 12.5 and using the distributive laws we obtain the expression 12.4. That is, every element of K^+ is a linear combination 12.4 where the a's lie in F. We have thus completed the proof that the $nm = [K:L][L:F]$ elements of 12.3 of K^+ form a basis for K^+ over F, and the theorem is established.

Except for examples, our main interest will be in a field K which is a finite extension of a field F. It is in this case where polynomials from $F[x]$ play a very prominent role. Let us now establish this connection between field elements and polynomials.

12.6   DEFINITION   If $F \subseteq K$, an element u of K is said to be *algebraic* over F if there exists some nonzero element $f(x)$ of the polynomial ring $F[x]$ such that $f(u) = 0$. An element of K which is not algebraic over F is said to be *transcendental over F*.

If $u \in K$ lies in F, then u is a root of $x - u \in F[x]$, so that every element of F is algebraic over F. The real number $\sqrt{2}$ is algebraic over $\mathbf{Q}$ since it is a root of the polynomial $x^2 - 2 \in \mathbf{Q}[x]$. It is known, although proofs are somewhat involved, that the real numbers π and e are transcendental over $\mathbf{Q}$.

From the definition, it is clear that if we have extensions K and L of the field F such that $K \supseteq L \supseteq F$, and u is an element of K which is algebraic over F, u is also algebraic over L since $L[x] \supseteq F[x]$. Using the notion *algebraic* we may divide extensions of F into two types.

12.7   DEFINITION   An extension K of a field F is said to be an *algebraic extension* of F if every element of K is algebraic over F. If K contains at least one element which is transcendental over F, K is called a *transcendental extension* of F.

The real field $\mathbf{R}$ is a transcendental extension of the rational field $\mathbf{Q}$ since π is in $\mathbf{R}$ and is transcendental over $\mathbf{Q}$. We shall not give a proof here, but the set of all complex numbers which are algebraic over $\mathbf{Q}$ is a subfield A of the complex field $\mathbf{C}$. Every element of A is algebraic over $\mathbf{Q}$ but A is not a finite extension of $\mathbf{Q}$. On the other hand, any finite extension K of a field F must necessarily be algebraic over F as we now prove.

12.8   THEOREM   *If K is a finite extension of F, then it is an algebraic extension of F.*

Proof: If $n = [K:F]$ then by Theorem 11.23(*i*) any set of $n + 1$ vectors of K^+ is linearly dependent over F since K^+ has dimension n over F. Thus, if $u \in K^+$, the $n + 1$ elements $1, u, u^2, \ldots, u^n$ of K^+ are linearly dependent over F. There exist scalars $a_0, a_1, a_2, \ldots, a_n$ of F, not all zero, such that $a_0 + a_1u + a_2u^2 + \cdots + a_nu^n = 0$. In particular, u is a root of the nonzero polynomial

$$f(x) = a_0 + a_1x + a_2x^2 + \cdots + a_nx^n \in F[x]$$

and must be algebraic. We conclude that K is an algebraic extension of F.

We shall now bring Theorem 10.67 into the picture to show how it relates polynomials to fields. In this way, we will also have a method for determining the degree of certain extension fields. The basic idea uses the substitution process of Section 10.2. In that section we found that if $u \in K$ is fixed then there is a homomorphism $\theta: F[x] \to K$ such that for a polynomial $f(x) \in F[x]$ we have

12.9
$$f(x)\theta = f(u).$$

This homomorphism θ sends the polynomial ring onto a subring $F[u]$ called the *ring of polynomials in u* in the field K. The elements of $F[u]$ are not really polynomials, but are the result of substituting u in for the indeterminate x in polynomials of $F[x]$. Notice that $F[u]$ is the smallest subring of K containing both F and u. We shall exploit the homomorphism θ defined by Equation 12.9 in the following theorem.

12.10 THEOREM *Let K be an extension of the field F, and fix u, an element of K. Define a mapping $\theta: F[x] \to K$ by Equation 12.9. Then θ is a homomorphism of $F[x]$ into K with image $F[u]$.*

(1) *If θ is an isomorphism then u is transcendental over F.*

(2) *If θ is not an isomorphism then u is algebraic over F and there is a unique nonzero monic prime polynomial $s[x] \in F[x]$ of smallest degree such that $s(u) = 0$; in particular, u is algebraic over F. Further;*

    (i) $\ker \theta = (s(x))$;

    (ii) *the equation*
$$(f(x) + (s(x)))\alpha = f(u)$$
    defines an isomorphism of the field $F[x]/(s(x))$ onto the field $F[u]$; and

    (iii) *if $\deg s(x) = n$ then $1, u, u^2, \ldots, u^{n-1}$ is a basis for $F[u]^+$ over F and $n = [F[u]:F]$.*

Proof: The fact that 12.9 defines a homomorphism of $F[x]$ into K follows from Section 10.2. The kernel of θ is the set of all polynomials $f(x) \in F[x]$ such that $f(u) = 0$. In particular, if θ is an isomorphism, $\ker \theta = (0)$ so that u is the root of no polynomial of $F[x]$. Consequently, u is transcendental over F, proving part (1).

If θ is not an isomorphism then ker $\theta \neq (0)$ so that u is the root of some nonzero polynomial of $F[x]$; that is, u is algebraic over F. By Theorem 10.32 there is a unique monic nonzero polynomial $s(x) \in F[x]$ which generates the kernel ideal of θ; so that ker $\theta = (s(x))$. By this same theorem we know that $s(x)$ has to be the monic polynomial of smallest degree in ker θ; in particular, $s(x)$ is the unique monic nonzero polynomial of smallest degree in $F[x]$ such that $s(u) = 0$. By the Fundamental Theorem of Ring Homomorphisms (3.31) we conclude that the mapping of part (2) (ii) definies an isomorphism of $F[x]/(s(x))$ onto $F[u]$. Since K is a field, $F[u]$ is an integral domain. By Theorem 5.31 for Euclidean domains, $s(x)$ must be a prime in $F[x]$. In this case, Theorem 5.31 or Theorem 10.67 applies to tell us that $F[x]/(s(x))$, hence also $F[u]$, is a field.

Only part (2) (iii) of the theorem remains to be proven. Clearly $F[x]/(s(x))$ is a vector space over F by Theorem 10.67. The statement 10.70 says precisely that if deg $s(x) = n$ then

$$1 + (s(x)), \quad x + (s(x)), \ldots, x^{n-1} + (s(x))$$

form a basis for $F[x]/(s(x))$ over F. Since α is a vector space isomorphism, by Exercise 11 of Section 11.10 α takes a basis to a basis. We conclude that $1, u, u^2, \ldots, u^{n-1}$ is a basis for $F[u]^+$ over F and $n = [F[u]:F]$ completing the proof of the theorem.

As we shall see, this theorem is very useful. Suppose we have a nonzero element u of K and want to generate a field containing u and F. Starting naively we would first form the ring $F[u]$ of all polynomials in u. We would now be assured that we have the smallest subring of K containing both F and u. This ring is an integral domain. By a method analogous to that of Section 6.4 we could form all fractions $a/b = ab^{-1}$ where $a, b \in F[u]$, $b \neq 0$, recalling that $b^{-1} \in K$. The result of this would be a field, much as the ring of quotients of the integers is the rational field. We could then be assured that the resulting ring is a field and is the smallest one containing both F and u. In other words, the resulting field L is the intersection of all subfields of K containing both F and u. If u is algebraic over F, we need not form quotients. Part (2)(ii) of the preceding theorem tells us that $L = F[u]$ is already a field. On the other hand, if u is transcendental over F, then part (1) of that theorem tells us that u has no inverse in $F[u]$ and we *must* form quotients in order to obtain the field L.

12.11 DEFINITION Let K be an extension of the field F, and S a set of elements of K. The intersection of all subfields of K which contain F and S is a subfield of K which we denote by $F(S)$ and call the field *generated by S over F*. We also say that $F(S)$ is the field obtained from F by *adjoining* the elements of S. If the set $S = \{u_1, u_2, \ldots, u_n\}$ is finite then we write $F(S) = F(u_1, u_2, \ldots, u_n)$. In the case where $F(S) = F(u)$ is obtained by adjoining the single element u then we call $F(u)$ a *simple extension* of F.

We shall be concerned with the case where S is a finite set. If $S = \{u_1, u_2, \ldots, u_n\}$ then we may obtain fields by successively adjoining to F first u_1, then u_2, then u_3, etc. It seems clear that the resulting field would also be the smallest field containing both F and S. In particular, $F(u_1, u_2, \ldots, u_n)$ is equal to the sequence of simple extensions $F(u_1)(u_2)\ldots(u_n)$. In the case where $u_1, u_2, \ldots, u_n$ are all algebraic over F, part (2)(ii) of the theorem tells us that

$$F(u_1, u_2, \ldots, u_n) = F(u_1)(u_2)\ldots(u_n)$$
$$= F[u_1][u_2]\ldots[u_n]$$
$$= F[u_1, u_2, \ldots, u_n]$$

is just the ring of polynomials in $u_1, u_2, \ldots, u_n$. That is, the homomorphism $\psi\colon F[x_1, x_2, \ldots, x_n] \to K$ from the ring of polynomials in indeterminates $x_1, x_2, \ldots, x_n$ defined by

$$f(x_1, x_2, \ldots, x_n)\psi = f(u_1, u_2, \ldots, u_n)$$

where $f(x_1, x_2, \ldots, x_n) \in F[x_1, x_2, \ldots, x_n]$ actually maps this polynomial ring onto the subfield $F(u_1, u_2, \ldots, u_n)$ of K.

Dropping back to the case where $n = 1$ and u is algebraic over F, we see that the polynomial $s(x)$ of part (2) of the preceding theorem plays a prominent role in considering u. Therefore, we single it out with the following definition.

12.12 DEFINITION If $u \in K$ is algebraic over F, the unique monic polynomial of least degree over F of which u is a root may be called the *minimal polynomial* of u over F.

By Theorem 12.10(2) the minimal polynomial $m(x)$ of u over F is certainly prime in $F[x]$. Further, if $f(x) \in F[x]$ and $f(u) = 0$, then by Theorem 12.10(2)(i), $m(x)$ divides $f(x)$ in $F[x]$. Let us now look at some examples illustrating these ideas.

Example 1 Consider the subfield $Q(\sqrt[3]{2})$ of R. The minimal polynomial of $\sqrt[3]{2}$ over Q is $x^3 - 2$ since $(\sqrt[3]{2})^3 - 2 = 0$ and since $x^3 - 2$ is a prime in $Q[x]$. By Theorem 12.10(2) $Q(\sqrt[3]{2}) = Q[\sqrt[3]{2}]$ and $[Q(\sqrt[3]{2}):Q] = 3$. The elements $1, \sqrt[3]{2}, \sqrt[3]{4}$ form a basis for $Q(\sqrt[3]{2})^+$ over Q so that every element of $Q(\sqrt[3]{2})$ may be written in the form $a + b\sqrt[3]{2} + c\sqrt[3]{4}$ where $a, b, c \in Q$.

Example 2 Since π is transcendental over $\mathbf{Q}$, the subfield $\mathbf{Q}(\pi)$ of $\mathbf{R}$ is not isomorphic to $\mathbf{Q}[\pi]$. It is not a finite extension of $\mathbf{Q}$ and is a transcendental extension of $\mathbf{Q}$. The elements of $\mathbf{Q}(\pi)$ are all of the form $f(\pi)/g(\pi)$ where $f(x), g(x) \in \mathbf{Q}[x]$ are polynomials and $g(x) \neq 0$. The subfields $\mathbf{Q}(\pi^n)$ for integers $n \geq 1$ form an infinite collection of subfields between $\mathbf{Q}(\pi)$ and $\mathbf{Q}$. Each simple extension $\mathbf{Q}(\pi^n)$ is a transcendental extension of $\mathbf{Q}$, but for any $n \geq 1$, $\mathbf{Q}(\pi)$ is an algebraic extension of $\mathbf{Q}(\pi^n)$.

Example 3 Consider the subfield $\mathbf{Q}(\sqrt{2}, \sqrt{3})$ of $\mathbf{R}$. The minimal polynomial of $\sqrt{2}$ over $\mathbf{Q}$ is $x^2 - 2$. Therefore, $[\mathbf{Q}(\sqrt{2}):\mathbf{Q}] = 2$, and $\mathbf{Q}(\sqrt{2})$ is the set of all linear combinations of the form $a + b\sqrt{2}$ where $a, b \in \mathbf{Q}$. By squaring any such combination, we can easily show that $\mathbf{Q}(\sqrt{2})$ contains no roots of $x^2 - 3$; that is, $x^2 - 3$ is a prime in $\mathbf{Q}(\sqrt{2})[x]$. We conclude that $x^2 - 3$ is the minimal polynomial of $\sqrt{3}$ over $\mathbf{Q}(\sqrt{2})$. Since $\mathbf{Q}(\sqrt{2}, \sqrt{3}) = \mathbf{Q}(\sqrt{2})(\sqrt{3})$ and since $[\mathbf{Q}(\sqrt{2}, \sqrt{3}):\mathbf{Q}(\sqrt{2})] = 2$ the Product of the Degrees theorem tells us that $[\mathbf{Q}(\sqrt{2}, \sqrt{3}):\mathbf{Q}] = 4$. Since $1, \sqrt{2}$ is a basis for $\mathbf{Q}(\sqrt{2})^+$ over $\mathbf{Q}$ and since $1, \sqrt{3}$ is a basis for $\mathbf{Q}(\sqrt{2}, \sqrt{3})^+$ over $\mathbf{Q}(\sqrt{2})$ we know, as a special case of what was proved about 12.3, that $1, \sqrt{2}, \sqrt{3}, \sqrt{6}$ is a basis for $\mathbf{Q}(\sqrt{2}, \sqrt{3})^+$ over $\mathbf{Q}$. Every element of $\mathbf{Q}(\sqrt{2}, \sqrt{3})$ can be written in the form $a + b\sqrt{2} + c\sqrt{3} + d\sqrt{6}$ where $a, b, c, d \in \mathbf{Q}$. By multiplying out the expressions 1, $(\sqrt{2} + \sqrt{3})$, $(\sqrt{2} + \sqrt{3})^2$, $(\sqrt{2} + \sqrt{3})^3$ and using vector space arguments over $\mathbf{Q}$ we can prove that these four vectors of $\mathbf{Q}(\sqrt{2}, \sqrt{3})^+$ also form a basis over $\mathbf{Q}$. Therefore, $\mathbf{Q}(\sqrt{2}, \sqrt{3}) = \mathbf{Q}(\sqrt{2} + \sqrt{3})$ is a simple extension of $\mathbf{Q}$ obtained by adjoining $\sqrt{2} + \sqrt{3}$.

EXERCISES

1. Show that $1, (\sqrt{2} + \sqrt{3}), (\sqrt{2} + \sqrt{3})^2, (\sqrt{2} + \sqrt{3})^3$ is a basis for $\mathbf{Q}(\sqrt{2}, \sqrt{3})^+$ over $\mathbf{Q}$. Find the minimal polynomial $s(x)$ of $\sqrt{2} + \sqrt{3}$ over $\mathbf{Q}$. What is the minimal polynomial of $\sqrt{2} + \sqrt{3}$ over $\mathbf{Q}(\sqrt{2})$? Factor $s(x)$ into primes in $\mathbf{Q}(\sqrt{2})[x]$.

2. Prove that if $a, b \in \mathbf{Q}$ then $\mathbf{Q}(\sqrt{a}, \sqrt{b}) = \mathbf{Q}(\sqrt{a} + \sqrt{b})$.

3. Let $w = (-1 + \sqrt{-3})/2$ so that $w^3 = 1$. What is the minimal polynomial of $w\sqrt[3]{2}$ over $\mathbf{Q}$? What is the minimal polynomial of $\sqrt[3]{2}$ over $\mathbf{Q}$? Using

Theorem 12.10, prove that $Q(w\sqrt[3]{2})$ and $Q(\sqrt[3]{2})$ are unequal isomorphic fields. Compute $[Q(\sqrt[3]{2}, w\sqrt[3]{2}):Q]$.

4. Consider the polynomial $s(x) = x^2 + x + 1$. If 3 does not divide $p - 1$ for a prime p then prove that $s(x)$ is a prime in $Z_p[x]$. Construct a field containing p^2 elements where 3 does not divide $p - 1$. Hint: $(x - 1)s(x) = x^3 - 1$.

5. Find the minimal polynomial over Q of the following complex numbers.

   (1) $\sqrt{4 - \sqrt{2}}$ (3) $\sqrt{-1} + \sqrt{7}$

   (2) $\sqrt[3]{5 - \sqrt{3}}$ (4) $\sqrt{2} - \sqrt[3]{5}$

6. If the complex number u is algebraic over Q then prove that there is a positive integer n such that nu is a root of a monic polynomial $f(x) \in Z[x]$.

7. Prove that if K is a finite field (i.e. a field with a finite number of elements) then K contains exactly p^n elements for a power of some prime p.

8. Prove that for each positive integer n, $[Q(\pi):Q(\pi^n)] = n$. You may assume that π is transcendental over Q. Prove that $Q(\pi^n)$, $n \geq 1$, is an infinite collection of different fields. If m and n are positive integers prove that $Q(\pi^n)$ and $Q(\pi^m)$ are isomorphic fields.

9. Prove that $Q(\sqrt{2})$ and $Q(\sqrt{3})$ are not isomorphic fields.

10. Show that for any positive integer n, there is a subfield K of the complex field C such that $[K:Q] = n$.

11. Show that if L is an extension of the real field R lying in the complex field C then $L = C$ or $L = R$.

12.2 GEOMETRIC CONSTRUCTIONS

Probably everyone has studied geometric construction at one time in the past. Using only straightedge and compass we learn how to bisect angles and construct regular pentagons. Our tools are pencil, paper, straightedge (unmarked ruler), and compass. Our compass differs slightly from the hypothetical Greek compass in that our compass will hold a fixed radius when lifted from the paper, whereas theirs would flop closed when lifted. In constructing something, the actual construction we do is not as important as the *proof* that the procedure we use does, in fact, lead to the desired conclusion. Because our tools are mechanical, any real construction is only approximate. In the sense of proof, it is possible to show that any construction with our compass can be done, with a little more work, using a Greek compass. For this reason, we shall assume that a compass may be used as dividers, i.e., it does not flop closed when lifted.

In this section we shall determine *algebraic* necessary and sufficient conditions for *geometric* constructions to exist. This unexpected translation of geometry into algebra will allow us to settle some otherwise very difficult problems. The vehicle for this translation is the Cartesian plane. We shall carefully define what we mean by constructibility, and when we have done this, the usual rules for geometric construction will be recognized in our definitions. Our construction procedure will be based upon constructing points; using the constructed points, we may construct lines and circles. The emphasis upon points also simplifies the translation of our problem into algebra. Our definitions are inductive; that is, geometrically constructible entities are made from previously constructed geometric entities.

We must start somewhere, so we start with two distinct fixed points O and X in the plane.

12.13 DEFINITION

(1) A *constructible point P* is either O or X, or is given by an intersection of one of the following kinds:

    (a) an intersection of two constructible lines;

    (b) an intersection of a constructible line with a constructible circle; or

    (c) an intersection of two constructible circles.

(2) A *constructible line l* is a line which passes through two constructible points.

(3) A *constructible radius* is the distance between two constructible points.

(4) A *constructible circle C* is a circle whose center is at a constructible point and whose radius is constructible.

(5) A *constructible angle A* is an angle formed by two constructible lines.

Sometimes constructions are effected by drawing circles with arbitrary radii. Since the radii cannot be known, within broad limits they are irrelevant. In particular, any unknown radius may be replaced by a radius which is a constructible one. Similar arguments may be made for points, lines, etc. That is, anything which we may lazily construct can also be constructed *using only constructible figures* as defined in 12.13.

We now introduce a Cartesian coordinate system upon the plane in a very special way. Since O and X are constructible points, the line $\overline{OX}$ is a constructible line. We label the point O as $(0, 0)$ and the point X as $(1, 0)$. In our argument we shall require a large number of constructions. These are all fairly easy, and therefore, are stated as exercises within the text.

Exercise 1 Show that the line $\overline{OX}$ and the unit radius from O to X are constructible. Show that given these we may introduce coordinates upon $\overline{OX}$ where all points are labeled $(a, 0)$ for a real number a. Further, prove that the points labeled $(a, 0)$, where a is an *integer*, are all constructible points. Show that any positive integer a is a constructible radius.

Next we erect a perpendicular to $\overline{OX}$ passing through the point O. With center O and radius 1 we describe a counterclockwise arc from X to Y, the first point at which the arc encounters the perpendicular to $\overline{OX}$. The point Y is labeled $(0, 1)$.

Exercise 2 Prove that $\overline{OY}$ is a constructible line and that Y is a constructible point.

Naturally we call $\overline{OX}$ the *x*-axis and $\overline{OY}$ the *y*-axis. We now have the desired Cartesian plane. Each point in the plane is uniquely represented by a real number pair (a, b). We may now define constructible numbers.

12.14 DEFINITION A real number a is called a *constructible number* if there is a real number b such that (a, b) is a constructible point in the plane.

Exercise 3 Prove that the point (a, b) is constructible for $a, b \in \mathbf{R}$ if and only if both $(a, 0)$ and $(0, b)$ are constructible points. Prove that if (a, b) is constructible, then (b, a) is constructible. Also prove that the sum and difference of two constructible numbers are both constructible.

We come now to our first algebraic theorem.

12.15 THEOREM *The set T of all constructible numbers is a subfield of the field $\mathbf{R}$ of real numbers.*

We indicate how, if a and b are constructible numbers with $b \neq 0$, to construct a/b. Let $A = (0, a)$ and $B = (0, b)$. Construct the line $\overline{AX}$ and the line $\overline{BC}$ parallel to $\overline{AX}$ passing through B. The point $C = (c, 0)$ is the intersection of $\overline{BC}$ with $\overline{OX}$. Then $c = a/b$. (See Figure 1).

Exercise 4 Prove Theorem 12.15.

Our main interest is not in the field T, but rather its subfields. From our remarks, it is evident that the set **Z** of integers is a set of constructible numbers. From this we have an immediate corollary to the theorem.

12.16 COROLLARY *All rational numbers are constructible. If K is a field of constructible numbers and u is a constructible number then $K(u)$ is a field of constructible numbers.*

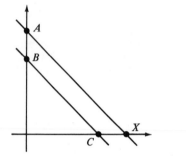

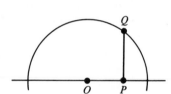

FIGURE 1

FIGURE 2

Let us now introduce the algebraic terminology which we shall use.

12.17 DEFINITION If K is an extension of a field F and $[K:F] = 2$ then K is called a *quadratic extension* of F. We call K a *multiply quadratic extension* of F if there are subfields $L_0, L_1, \ldots, L_n$ of F for some positive integer n such that

12.18 $$K = L_n \supset L_{n-1} \supset \cdots \supset L_1 \supset L_0 = F$$

where $[L_i:L_{i-1}] = 2$ for $i = 1, 2, \ldots, n$.

Our major theorem is the following one.

12.19 THEOREM *A number $a \in \mathbf{R}$ is constructible if and only if a lies in some multiply quadratic extension $K \subseteq \mathbf{R}$ of the rational field $\mathbf{Q}$.*

Let us first assume that K is a multiply quadratic extension of $\mathbf{Q}$. We will argue that every number in K is constructible. First we sketch an argument to show that if a is a positive constructible number then $\sqrt{a}$ is constructible. Start with a circle of radius $(a + 1)/2$ centered at O. Mark the point $P = ((a - 1)/2, 0)$ and erect a perpendicular to $\overline{OX}$ at P. The perpendicular meets the circle at a point Q and the y-coordinate of Q is $\sqrt{a}$. (See Figure 2).

Exercise 5   Prove that if $a > 0$ is a constructible number then $\sqrt{a}$ is constructible.

Because of the chain of 12.18 and the fact that $K \subseteq \mathbf{R}$, the following lemma is sufficient to prove that every number in K is constructible.

12.20 LEMMA   *If K is a quadratic extension of a field F of characteristic 0 then $K = F(\sqrt{a})$ for some $a \in F$ where $\sqrt{a}$ is a root of the polynomial $x^2 - a \in F[x]$.*

Proof: Choose $u \in K \backslash F$. Clearly u is not in the subspace with basis 1 over F so that by Theorem 11.15 1, u are linearly independent over F. But $[K:F] = 2$ so that 1, u form a basis for K^+ over F by Theorem 11.23. By Theorem 11.23 1, u, u^2 are linearly dependent so that there are coefficients $c, d, e \in F$, not all zero, such that $cu^2 + du + e = 0$. Since u is not in F, $c \neq 0$; and therefore, by dividing by c, we may assume that $u^2 + du + e = 0$. In particular, u is a root of the polynomial $f(x) = x^2 + dx + e$. This polynomial is irreducible over F since a root u does not lie in F. Let $a = d^2 - 4e$. We shall prove that $K = F(\sqrt{a})$.
   Let $v = 2u + d$. Then $v^2 = 4u^2 + 4du + d^2 = 4(u^2 + du + e) + (d^2 - 4e) = a$ so that v is a root of $x^2 - a$. That is, we may take $\sqrt{a} = v$. Notice that 1, v is also a basis for K^+ over F since $u = 2^{-1}(v - d)$ so that $K = F(v)$. The lemma is proved.

To prove the converse part of Theorem 12.19 requires a little more work. We isolate the various steps in the lemma below. For purposes of clarity we shall assume that F is some fixed subfield of the real field $\mathbf{R}$.

12.21 LEMMA   (1) *If the coordinates of the points P and Q all lie in F then the coefficients for the equation $y = mx + b$ (or $x = b$) of the line $\overline{PQ}$ all lie in F.*

(2) *If the coordinates of the points P and Q all lie in F then the distance from P to Q lies in an extension K of F where $K \subseteq \mathbf{R}$ and $[K:F] = 1$ or 2.*

(3) *If the coordinates of the point $P = (a, b)$ and the number $r > 0$ all lie in F then the coefficients of the equation $r^2 = (x - a)^2 + (y - b)^2$ for the circle centered at P of radius r all lie in F.*

(4) *If the coefficients of the equations for two nonparallel lines all lie in F then the coordinates of the intersection point lie in F.*

(5) *If the coefficients of the equations for an intersecting line and circle all lie in F then there is an extension $K \subseteq \mathbf{R}$ of F so that $[K:F] = 1$ or 2 and the coordinates of the intersection points all lie in K.*

(6) *If the coefficients of the equations for two intersecting circles all lie in F then there is an extension $K \subseteq \mathbf{R}$ of F so that $[K:F] = 1$ or 2 and the coordinates of the intersection points all lie in K.*

We should remark that we are taking the unconventional view that r, a, b are "coefficients" of the equation $(x - a)^2 + (y - b)^2 = r^2$. Strictly speaking, r^2 is a coefficient but r is not. The proof of this lemma involves some elementary algebra, further, comparing this lemma with Definitions 12.13 and 12.17 shows that the coordinates of any constructible point lie in some multiply quadratic extension $K \subseteq \mathbf{R}$ of $\mathbf{Q}$.

Exercise 6 Prove Lemma 12.21 and Theorem 12.19.

As a corollary to this theorem we obtain the following useful criterion.

12.22 COROLLARY *If a real number a is contructible then its minimal polynomial will have degree 2^m for some integer $m \geq 0$.*

Proof: Assume that a is constructible, so that a lies in some multiply quadratic extension $K \subseteq \mathbf{R}$ of $\mathbf{Q}$. In Definition 12.14 $[L_i:L_{i-1}] = 2$ so that by Theorem 12.2 $[K:\mathbf{Q}] = 2^n$. Since $\mathbf{Q}(a)$ is a subfield of K, by the same theorem, $[\mathbf{Q}(a):\mathbf{Q}] = 2^m$ for some integer $m \geq 0$. By Theorem 12.8, a is a root of some polynomial. Finally, by Theorem 12.10 $\mathbf{Q}(a) = \mathbf{Q}[a]$, and the minimal polynomial of a has degree 2^m.

The following three problems played a prominent role in Greek geometry. If we assume that π is transcendental over $\mathbf{Q}$ then we shall show that all three problems have no solution. These problems have only been settled in the past 350 years. The problems are as follows.

(1) (Duplicating the cube) Given the edge of a cube, construct the edge of a cube whose volume is twice the original volume.

(2) (Trisecting angles) Give a construction procedure which will trisect any angle.

(3) (Squaring the circle) Given a circle and its center, construct a square whose area is equal to that of the given circle.

In the first problem we may assume that O and X are the end points of an edge of the given cube, where the volume of the cube is 1. We wish to construct an edge for a cube of volume 2. Such an edge could run from O to the point $P = (a, 0)$ where a is a real root of the equation $x^3 - 2 \in \mathbf{Q}[x]$. But this polynomial is prime over $\mathbf{Q}$, is the minimal polynomial of a, and has degree 3 which is not a power of 2. The construction in (1) is impossible.

The impossibility of Problem (2), a trisection procedure, is contained in the following lemma.

12.23 LEMMA *60° is a constructible angle. 20° is not a constructible angle.*

Proof: Center angles at O with one side being $\overline{OX}$. Then the co-ordinates of the point P on the unit radius circle and the second side of an angle θ are $(\cos \theta, \sin \theta)$. For 60° the point P is $(1/2, \sqrt{3}/2)$ where $1/2$ and $\sqrt{3}/2$ are clearly contructible numbers. So 60° is a construct-ible angle.

The x-coordinate of the point P for $\theta = 20°$ is $\cos \theta$, but then $\cos 3\theta = 1/2$. Using trigonometric identities we may show that

$$8 \cos^3 \theta - 6 \cos \theta - 1 = 0.$$

Certainly, neither or both of $\cos \theta$ and $2 \cos \theta$ are constructible. But $2 \cos \theta$ is a root of the polynomial

$$f(x) = x^3 - 3x - 1 \in \mathbf{Q}[x].$$

It is straightforward to check that $f(x)$ has no rational roots, and therefore, is irreducible over $\mathbf{Q}$. Since 3 is not a power of 2, we conclude that $\cos \theta$ is not constructible, and therefore, 20° is not a constructible angle.

In Problem (3) we may take the segment from O to X to be a radius and O to be the center of the circle. The area of such a circle is π. The edge of the required square must be $\sqrt{\pi}$. If this number is constructible then certainly $\pi = (\sqrt{\pi})^2$ is contructible. On the other hand, since π is transcendental over $\mathbf{Q}$, it is the root of no polynomial in $\mathbf{Q}[x]$. Therefore π is not constructible, and Problem (3) has no solution.

Exercise 7 Show that it is not possible to construct a regular 7-sided polygon.

Exercise 8 Show, by using algebra, that a regular pentagon is con-structible. Use your algebraic solution to devise a construction procedure for a pentagon. Algebraically analyze the standard construction procedure for a pentagon. [*Hint:* In a construction, we successively obtain new points from old ones. Algebraically analyze the coordinates of the successive points.]

Exercise 9 Give constructions for the following numbers.

$$\sqrt{7}, \quad 1 + \sqrt{7}, \quad \sqrt{1 + \sqrt{7}}.$$

Exercise 10 Use Lemma 12.21 to prove that a regular nine-sided polygon is not constructible. Prove that a regular three-sided polygon is constructible.

Exercise 11 Describe a procedure whereby one could construct a regular polygon of $2^n \cdot 3 \cdot 5$ sides where $n \geq 0$ is an integer.

12.3 ISOMORPHISMS OF SPLITTING FIELDS

We shall now consider carefully the connection between a nonconstant polynomial $f(x) \in F[x]$ over a field F and an extension of F in which $f(x)$ may be written as a product of linear factors. If F is a subfield of the complex field $\mathbf{C}$, then the Fundamental Theorem of Algebra, 10.55 (see 10.57 also), tells us that in $\mathbf{C}[x]$, $f(x)$ may be written as a product of polynomials all of degree one. If F is not a subfield of $\mathbf{C}$, then the methods of Section 10.10 allow us to construct an extension K of F such that $f(x)$ is equal to a product of linear factors in $K[x]$ (see Corollary 10.84). The construction of such a field K was actually carried out in Example 5 of Section 10.10. The critical step in the construction of such an extension K of F over which $f(x)$ is a product of linear factors is the observation that if $s(x)$ is a prime factor of $f(x)$ in $F[x]$, then $L = F[x]/(s(x))$ is a field (10.67) and the coset $j = x + (s(x))$ is a root of $f(x) \in F[x]$ in L. The Factor Theorem, 10.21, supplies the connection between a root j of $f(x)$ and a linear factor $x - j$ of $f(x)$. In other words, any field which contains all the roots of $f(x)$ should be large enough to factor $f(x)$ completely into linear factors.

Suppose that K is an extension field of F which contains all roots of $f(x)$. We may form the intersection L over all subfields of K containing all the roots of $f(x)$ and the subfield F. This subfield L is certainly uniquely determined as a subfield of K, and further, it must be the smallest subfield of K containing F in which $f(x)$ factors into a product of polynomials of degree one. This subfield L will play a special role in our considerations.

12.24 DEFINITION If $f(x) \in F[x]$ is a nonconstant polynomial and K is an extension of F such that $f(x)$ can be expressed as a product of factors of the first degree in $K[x]$, then we say that $f(x)$ *splits in K.* We call K a *splitting field of $f(x)$ over F* if $f(x)$ splits in K, but $f(x)$ does not split in any proper subfield of K containing F.

Example 1   The polynomial $x^2 - 2 \in Q[x]$ splits in the real field **R** since $x^2 - 2 = (x - \sqrt{2})(x + \sqrt{2})$.   A splitting field for $x^2 - 2$ in **R** over **Q** is $Q(\sqrt{2})$.   Observe that $Q(\sqrt{2}) = Q(\sqrt{2}, -\sqrt{2})$ is generated over **Q** by the roots of $x^2 - 2$ in **R**.

12.25 THEOREM   *Let $f(x)$ be a polynomial of degree $n > 1$ of the polynomial ring $F[x]$, and let K be an extension of F in which $f(x)$ splits.   We may write $f(x)$ in the form*

12.26 $$f(x) = a(x - u_1)(x - u_2) \cdots (x - u_n)$$

where $a \in F$ and the u's are in K.   Then the field $L = F(u_1, u_2, \ldots, u_n)$ is the unique splitting field of $f(x)$ over F in K, and L is a finite extension of F.

Proof: The Theorem on Unique Factorization, 10.39, assures us that $f(x)$ is a product of primes in $K[x]$, and by definition, these primes are all of degree one.   The coefficient a is the leading coefficient of $f(x)$, and therefore it lies in F; thus the factorization given in 12.26 exists in $K[x]$.   Since this factorization is unique, it will only exist over extensions L of F in K which contain all the u's.   Clearly then $L = F(u_1, u_2, \ldots, u_n)$ is the unique smallest subfield of K containing F in which $f(x)$ splits.   The elements $u_1, u_2, \ldots, u_n$ are all roots of $f(x)$.   Therefore, by Theorem 12.10(2)(i), the minimal polynomial of u_i over $F_{i-1} = F(u_1, u_2, \ldots, u_{i-1})$ (let $F_0 = F$) is a divisor of $f(x)$ in $F_{i-1}[x]$ for $i = 1, 2, \ldots, n$.   Thus $[F_i : F_{i-1}] \leq \deg f(x) = n$ for $i = 1, 2, \ldots, n$ by Theorem 12.10.   By the Product of the Degrees Theorem, 12.2, we conclude that $[F(u_1, u_2, \ldots, u_n) : F] \leq n^n$ is finite, completing the proof of the theorem.

Another way of stating what we have just proved is to assert that a splitting field of a polynomial $f(x)$ over F is obtained by adjoining to F all the roots of $f(x)$ (which are assumed to be in some field K in which $f(x)$ splits).

It should be noted that in this theorem we started with a field K in which $f(x)$ splits and have then shown that there is a *unique subfield L* of K which is a splitting field of $f(x)$ over F.   Assume that K' is another extension of F in which $f(x)$ splits.   By this same argument there is a unique splitting field L' of $f(x)$ over F in the field K'.   A major object of this section is to prove that L and L' are iso-

morphic fields; in other words, any two splitting fields for $f(x)$ over F are isomorphic. The proof of this fact is somewhat involved, though we have already
discussed the main ideas. To illustrate these ideas let us consider another example.

Example 2 The extension $L = \mathbf{Q}(\sqrt{2})$ of $\mathbf{Q}$ in $\mathbf{R}$ is the splitting field
of $x^2 - 2 \in \mathbf{Q}[x]$ in $\mathbf{R}$. Following the methods described in Section
10.10, $j = x + (x^2 - 2)$ is a root of $x^2 - 2$ in the field $L' = \mathbf{Q}[x]/(x^2 - 2)$. Since $L' = \mathbf{Q}(j)$, we conclude that L' is also a
splitting field of $x^2 - 2$ over $\mathbf{Q}$. By Theorem 12.10(2)(ii) the field L'
is isomorphic to the field L by the mapping

$$(a + bj)\alpha = a + b\sqrt{2}$$

where $a, b \in \mathbf{Q}$. In this particular case, these two splitting fields are
isomorphic by a mapping α which is the identity upon $\mathbf{Q}$, and which
maps the root j of $x^2 - 2$ in L' to the root $\sqrt{2}$ of $x_2 - 2$ in L.

Some of the properties of this isomorphism α are named in the following
definition.

12.27 DEFINITION Let $\beta: R \to S$ be an isomorphism of the ring R onto
the ring S. Suppose that R_1 and S_1 are subrings of R and S respectively, and that $\alpha: R_1 \to S_1$ is an isomorphism of R_1 onto S_1. If
when restricted to R_1, β equals α (that is, $a\beta = a\alpha$ for all $a \in R_1$) then
we call the isomorphism β an *extension* of the isomorphism α; we also
may say that β *extends* α, or that α *can be extended* to an isomorphism
β of R onto S.

In Example 2, α extends the identity mapping of $\mathbf{Q}$ onto $\mathbf{Q}$ to an isomorphism of L' onto L. Let us point out one special case, important for our present
purposes, in which a rather natural extension of an isomorphism does exist.

12.28 THEOREM *Let* $\alpha: F \to F'$ *be an isomorphism of the field F onto the
field F'. Define a mapping* $\alpha_x: F[x] \to F'[x]$ *by*

12.29 $\quad (a_0 + a_1x + \cdots + a_nx^n)\alpha_x = (a_0\alpha) + (a_1\alpha)x + \cdots + (a_n\alpha)x^n$

where $a_i \in F$, $i = 0, 1, 2, \ldots, n$. Then α_x is an extension of α to an
isomorphism of $F[x]$ onto $F'[x]$. A polynomial $p(x) \in F[x]$ is a
prime if and only if $p(x)\alpha_x$ is a prime in $F'[x]$.

Proof: The proof that $\alpha_x: F[x] \to F'[x]$ is an isomorphism of $F[x]$
onto $F'[x]$ is a straightforward calculation (see Exercise 9 of Section

10.2). For a constant polynomial $a \in F[x]$, $a\alpha_x = a\alpha$ so that α_x is an extension of α. If $f(x) = g(x)h(x)$ is a product of two polynomials in $F[x]$ then $f(x)\alpha_x = [g(x)\alpha_x][h(x)\alpha_x]$ so that $p(x) \in F[x]$ is a prime if and only if $p(x)\alpha_x \in F'[x]$ is also a prime.

This theorem allows us to prove a very useful generalization of Theorem 12.10.

12.30 LEMMA *Let $\alpha: F \to F'$ be an isomorphism of the field F onto the field F'. Assume that $p(x) \in F[x]$ is a monic prime polynomial, K is an extension of F containing a root u of $p(x) \in F[x]$, and K' is an extension of F' containing a root u' of $p(x)\alpha_x \in F'[x]$ where α_x is defined by Equation 12.29. Then there is a unique isomorphism β of $F(u)$ onto $F'(u')$ which is an extension of the isomorphism α and has the property that $u\beta = u'$.*

Proof: Let $\theta: F[x] \to K$ be the homomorphism of $F[x]$ into K given by the substitution process, i.e., for $f(x) \in F[x]$

$$f(x)\theta = f(u).$$

Similarly we define a homomorphism $\theta': F'[x] \to K'$ by setting

$$f'(x)\theta' = f'(u')$$

for $f'(x) \in F'[x]$ (see Section 10.2).

By Theorem 12.28 there is an isomorphism $\alpha_x: F[x] \to F'[x]$ of $F[x]$ onto $F'[x]$ which extends α. In addition, the polynomial $p(x)\alpha_x \in F'[x]$ is a monic prime, and the element u' is a root of $p(x)\alpha_x$. Refer now to Figure 3 and notice that we have a composite homomorphism $\alpha_x\theta': F[x] \to F'(u')$ of $F[x]$ into $F'(u')$. We also have a homomorphism $\theta: F[x] \to F(u)$ of $F[x]$ into $F(u)$. This situation should remind us of the Universal Mapping Property, 3.25, for rings. We shall verify that the hypotheses of that theorem hold in this instance.

The polynomial $p(x)$ is a monic prime and u is a root so that $p(x) \in$

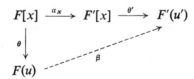

FIGURE 3

$F[x]$ must be the minimal polynomial of u over F. By Theorem 12.10, $F[x]\theta = F[u]$ is a field so that $F[u] = F(u)$, and θ maps $F[x]$ onto $F(u)$. The kernel ideal of θ is $(p(x))$.

This same argument applied to $F'[x]$, θ', and u' shows that θ' maps $F'[x]$ onto $F'(u')$, and the kernel ideal of θ' is $(p(x)\alpha_x)$. Since α_x is an isomorphism of $F[x]$ onto $F'[x]$, we conclude that $\alpha_x\theta'$ maps $F[x]$ onto $F'(u')$, and the kernel ideal of $\alpha_x\theta'$ is $(p(x))$.

Clearly now $\ker \theta = (\ker \alpha_x)\theta'$ so that the Universal Mapping Property, 3.25, applies to tell us that there is a unique homomorphism $\beta: F(u) \rightarrow F'(u')$ such that $\theta\beta = \alpha_x\theta'$. Since $\alpha_x\theta'$ maps $F[x]$ onto $F'(u')$, β maps $F(u)$ onto $F'(u')$. We may argue that β is an isomorphism either by observing that $F(u)$ is a field and has only trivial ideals, or by observing that $\ker \beta = (\ker \alpha_x\theta')\theta = (p(x))\theta = (0)$. If $a \in F$ then $a\beta = a\theta\beta = a\alpha_x\theta' = a\alpha$ so that β extends α. Finally, $x\theta = u$ and $x\theta' = u'$ so that $u\beta = x\theta\beta = x\alpha_x\theta' = x\theta' = u'$ completing the proof of the lemma.

We are now in a position to prove the very important *Isomorphism Extension Theorem*.

12.31 THEOREM   *Let $\alpha: F \rightarrow F'$ be an isomorphism of the field F onto the field F', and let $f(x)$ be an arbitrary nonconstant polynomial in the ring $F[x]$. Set $f'(x) = f(x)\alpha_x \in F'[x]$ where α_x is defined by Equation 12.29. Let K be a splitting field of $f(x)$ over F, and K' a splitting field for $f'(x)$ over F'.*

(i) *There is an isomorphism $\beta: K \rightarrow K'$ of K onto K' which is an extension of α.*

(ii) *If $p(x)$ is a monic prime factor of $f(x)$ in $F[x]$, $u \in K$ is a root of $p(x)$, and $u' \in K'$ is a root of $p(x)\alpha_x$ then the isomorphism β of part (i) may be chosen so that $u\beta = u'$.*

Proof: By Theorem 12.25 K is a finite extension of F. In particular, we may prove the theorem by induction upon the degree $[K:F]$ of K over F. If $[K:F] = 1$ then $K = F$ and $f(x)$ splits in F. Thus $p(x) = x - u$ and $p(x)\alpha_x = x - u'$. In this case we may take $\beta = \alpha$ in parts (i) and (ii). We therefore assume that $[K:F] > 1$ and that the theorem holds for all subfields $F_0 \supseteq F$ of K such that $[K:F_0] < [K:F]$.

As we have just shown above, if $p(x)$ is linear then $p(x) = x - u$ and $u\alpha = u'$. In this particular case, part (ii) follows immediately from part (i). Since $K \neq F$, and $f(x)$ does not split in F, we may assume that $p(x)$ is actually a prime factor of $f(x)$ in $F[x]$ of degree greater than one.

Consider the extensions $F(u)$ of F and $F'(u')$ of F'. By Lemma 12.30 there is an extension $\alpha^*: F(u) \to F'(u')$ of α to an isomorphism mapping $F(u)$ onto $F'(u')$ such that $u\alpha^* = u'$ (see Figure 4). By Theorem 12.10, $F(u) = F[u]$ and $[F(u):F] = \deg p(x) > 1$. In particular, the Product of Degrees Theorem, 12.2, says that

$$[K:F] = [K:F(u)][F(u):F] > [K:F(u)].$$

If $u = u_1, u_2, \ldots, u_n$ are all the roots of $f(x)$ in K then $K = F(u_1, u_2, \ldots, u_n) = F(u)(u_2, u_3, \ldots, u_n)$ is the splitting field for $f(x)$ over $F(u)$. A similar argument shows that K' is the splitting field for $f'(x)$ over $F'(u')$. Our induction hypothesis applies to $F(u)$, $F'(u')$,

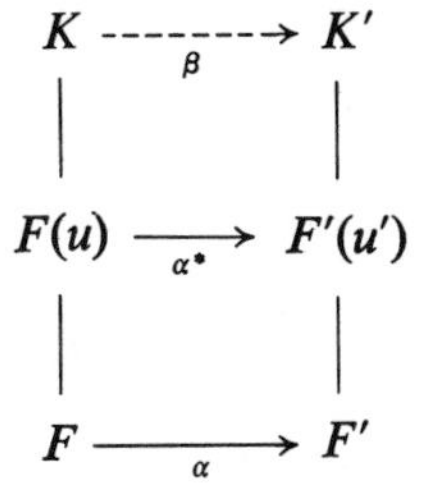

FIGURE 4

α^*, K, and K', therefore, by part (i) of the theorem there is an isomorphism $\beta: K \to K'$ of K onto K' which extends α^*. But α^* extends α so that β also extends α. For part (ii) we notice that $u\beta = u\alpha^* = u'$. This completes the induction and the proof of the theorem.

In the special case where $F = F'$ and α is the identity, we obtain the following very important corollary.

12.32 COROLLARY *If F is a field and $f(x) \in F[x]$ is a nonconstant polynomial then any two splitting fields of $f(x)$ over F are isomorphic.*

We may use this theorem to construct isomorphisms as the following example shows.

Example 3 Recall that $\mathbf{Q}(\sqrt{2})$ is the splitting field in $\mathbf{R}$ of $x^2 - 2 \in \mathbf{Q}[x]$. In the theorem, let us take $\mathbf{Q} = F = F'$, α the identity mapping, $u = \sqrt{2}$, and $u' = -\sqrt{2}$. Then part (ii) of the theorem implies that there is an isomorphism $\beta: \mathbf{Q}(\sqrt{2}) \to \mathbf{Q}(\sqrt{2})$ such that

$$(a + b\sqrt{2})\beta = a - b\sqrt{2}$$

for any $a, b \in \mathbf{Q}$. This mapping is a nonidentity isomorphism of the field $\mathbf{Q}(\sqrt{2})$ onto itself.

EXERCISES

1. Find the splitting field in $\mathbf{C}$ of $x^3 - 5 \in \mathbf{Q}[x]$.

2. Give the splitting field K in $\mathbf{C}$ of $x^2 - 3 \in \mathbf{Q}[x]$. Find a basis for K^+ over $\mathbf{Q}$. Give all isomorphisms of the field K onto itself.

3. Find a splitting field of $x^2 + x + 1 \in \mathbf{Z}_2[x]$.

4. Let K be the splitting field of $x^3 + x^2 + 1 \in \mathbf{Q}[x]$ in $\mathbf{C}$. Prove that K is isomorphic to the field V of Example 5 in Section 10.10.

5. Suppose that K is an extension of a field F and that $f(x) \in F[x]$ is nonconstant. Prove that $u \in K$ is algebraic over F if and only if $f(u)$ is algebraic over F.

6. Prove that if $p(x) = x^3 - \pi \in \mathbf{Q}(\pi)[x]$ and $q(x) = x^3 - e \in \mathbf{Q}(e)[x]$ then a splitting field K of $p(x)$ over $\mathbf{Q}(\pi)$ is isomorphic to a splitting field of $q(x)$ over $\mathbf{Q}(e)$. You may assume that e and π are transcendental over $\mathbf{Q}$.

7. Let K be a subfield of $\mathbf{C}$ such that $[K:\mathbf{Q}] = 2$. Prove that $K = \mathbf{Q}(u)$ where $u \in \mathbf{C}$ is a root of $x^2 - a \in \mathbf{Q}[x]$ for some $a \in \mathbf{Q}$.

8. Let K be an extension of the field F. Prove that the set L of all elements of K which are algebraic over F is a subfield of K. The field L is called the *algebraic closure* of F in K. Prove that if $K = \mathbf{C}$ and $F = \mathbf{Q}$ and $f(x) \in L[x]$ is a nonconstant polynomial then $f(x)$ has a root in L.

9. Prove that if K is an extension of F of degree 2^n for a positive integer n, then a prime polynomial $f(x) \in F[x]$ which is of odd prime degree must also be a prime in $K[x]$.

10. The real number $1 + \sqrt[3]{5}$ is a root of the polynomial $f(x) = x^3 - 3x^2 + 3x - 26 \in \mathbf{Q}[x]$.

>   (i) Knowing this, construct the splitting field K for $f(x)$ in $\mathbf{C}$. Find all isomorphisms of the field K onto itself.
>
>   (ii) Show that this set G of isomorphisms is a group of order $[K:\mathbf{Q}]$ with multiplication given by composition of mappings.
>
>   (iii) Show that G is isomorphic to S_3.
>
>   (iv) For each subgroup H of G define the set $L = \{u \mid u \in K, u\alpha = u$ for all $\alpha \in G\}$. Show that L is a subfield of K.
>
>   (v) In part (iv) show that $[G:H] = [L:\mathbf{Q}]$, where $[G:H]$ is taken to be the number of cosets of H in G.

12.4 FINITE FIELDS

In this section we will apply what we have learned to study fields containing a finite number of elements. If K is a field containing n elements for some positive integer n then we call K a *finite field* and say that K has *order* n. Most of our arguments will depend upon facts which have been proven at various points within the text. In Theorem 6.9 of Section 6.2, for example, we showed that a field has characteristic zero or a prime p. Theorem 6.28 of Section 6.6 showed that a field of characteristic zero contains an isomorphic copy of $\mathbf{Q}$, the rational field. In particular, a finite field K must have characteristic a prime p. Theorem 6.10 of Section 6.2 showed that in a finite field K the set $\{m \cdot 1 \mid m \in \mathbf{Z}\}$ is a subfield isomorphic to $\mathbf{Z}_p$, the integers modulo p where the prime p is the characteristic of K. We shall always identify $\mathbf{Z}_p$ with this isomorphic subfield; in particular, we shall view any field K whose characteristic is the prime p as an extension field of $\mathbf{Z}_p$, the integers modulo p.

A major aim of this section is to prove that, up to isomorphism, there is exactly one finite field K of order p^n for any prime p and any positive integer n. The field K of order p^n is determined as a splitting field for the polynomial $x^{p^n} - x$ over $\mathbf{Z}_p$. Consequently, our study will start with this polynomial. One fact, which was established in Corollary 10.84 of Section 10.10, and of which we will make repeated use, is that given any nonconstant polynomial $f(x) \in F[x]$ for a field F, there is an extension K of F in which $f(x)$ splits. In Theorem 12.25 of Section 12.3 we showed that in K there is a unique finite extension of F generated by the roots of $f(x)$ which is a splitting field of $f(x)$ over F. That is, every nonconstant polynomial $f(x) \in F[x]$ has a splitting field over F which is a finite extension of F.

12.33 THEOREM *Assume that p is a prime, n is a positive integer, and K is an extension of the field $\mathbf{Z}_p$ in which $x^{p^n} - x \in \mathbf{Z}_p[x]$ splits. Then there are p^n distinct roots $u_1, u_2, \ldots, u_{p^n}$ of this polynomial in K, and in $K[x]$*

12.34
$$x^{p^n} - x = (x - u_1)(x - u_2) \cdots (x - u_{p^n}).$$

Proof: By Theorem 12.25 of Section 12.3 $x^t - x$ where $t = p^n$ certainly has a factorization given by 12.34 in $K[x]$. Therefore, we need only prove that the u's are all distinct. Since $x^t - x = x(x^{t-1} - 1)$ and 0 is not a root of $x^{t-1} - 1$ we conclude that 0 is not a repeated root of $x^t - x$. Let $u \in K$ be any nonzero root of $x^t - x$.

By the Factor Theorem, 10.21, we know that

$$x^t - x = (x - u)f(x)$$

for some polynomial $f(x) \in K[x]$. Since $u \neq 0$ and u is a root of $x^t - x$, we must have $u^{t-1} - 1 = 0$ and $u^{t-1} = 1$. A straightforward long-division calculation shows that

$$f(x) = x^{t-1} + ux^{t-2} + u^2x^{t-3} + \cdots + u^{t-2}x$$

so that

$$\begin{aligned}
f(u) &= u^{t-1} + u^{t-1} + u^{t-1} + \cdots + u^{t-1} \qquad (t - 1 \text{ times})\\
&= (t - 1)u^{t-1}\\
&= (t - 1) \cdot 1\\
&= -1
\end{aligned}$$

since $t \cdot 1 = p^n \cdot 1 = 0$ in K. We apply the Factor Theorem, 10.21, of Section 10.3 to tell us that $x - u$ is not a factor of $f(x)$. Since $x^t - x$ is a product of linear factors in $K[x]$ we conclude that $x^t - x$ must have $t = p^n$ distinct (i.e. not repeated) roots.

NOTE: This proof is crude but elementary. To suggest a more sophisticated method, consider a real polynomial mapping $f(x) \in \mathbf{R}[x]$. Now $f(x)$ will have a real repeated root $a \in \mathbf{R}$ if and only if $f(x) = (x - a)^2 g(x)$ for some $g(x) \in \mathbf{R}[x]$. Taking the derivative we see that

$$f'(x) = (x - a)[2g(x) + (x - a)g'(x)].$$

In particular, it can be shown that $f(x)$ has a repeated root if and only if $f(x)$ and $f'(x)$ have a nonconstant g.c.d. By defining "formal derivatives" and developing their properties we may show that $f(x) \in F[x]$ has a repeated root in an extension field K of any field $\mathbf{F}$ if and only if $f(x)$ and the "formal derivative" of $f(x)$ have a nonconstant g.c.d. in $\mathbf{F}[x]$. For example, the "formal derivative" of $x^{p^n} - x$ in $\mathbf{Z}_p[x]$ is $p^n x^{p^n} - 1 = -1$ so that $x^{p^n} - x$ and -1 have g.c.d. 1; therefore, $x^{p^n} - x$ has no repeated roots. Since we have no real need for "formal derivatives," we shall not derive any of their properties. (Exercise 11 of Section 10.2)

This theorem tells us that a splitting field L of $x^{p^n} - x$ over $\mathbf{Z}_p$ will contain at least the p^n roots of this polynomial. We wish to argue that L contains at most p^n elements, and therefore, exactly p^n elements.

12.35 DEFINITION   If K is a field and $\alpha: K \rightarrow K$ is an isomorphism of K onto K then α is called an *automorphism* of K. The set $\mathscr{G}$ of all automorphisms of K is called the *Galois group* of K. If K is a finite field then the mapping $\theta: K \rightarrow K$ defined by

$$a\theta = a^p$$

for $a \in K$ and p, the prime characteristic of K, is called the *Frobenius automorphism* of K.

In order to justify our usage of words in this definition, it is necessary to show that $\mathscr{G}$ is a group (Exercise 12) and that θ is an automorphism of K.

12.36 THEOREM   *Assume that K is a finite field of characteristic the prime p. Then K has order p^n where $[K:\mathbf{Z}_p] = n$. The group of units U of K is cyclic of order $p^n - 1$. The Frobenius automorphism is an automorphism of K of order n.*

Proof: Since K has finite order, certainly K^+ has some finite dimension $n = [K:\mathbf{Z}_p]$ over $\mathbf{Z}_p$, so we let $v_1, v_2, \ldots, v_n$ be a basis for K^+ over $\mathbf{Z}_p$. The elements of K have the form $a_1 v_1 + a_2 v_2 + \cdots + a_n v_n$ where each a_i is one of the p elements in $\mathbf{Z}_p$. Therefore, there are exactly p^n elements in K. This argument reverses to show that if K has order p^n then $[K:\mathbf{Z}_p] = n$.

From Exercise 8 of Section 6.3 we know that $(a + b)\theta = (a\theta) + (b\theta)$ and $(ab)\theta = (a\theta)(b\theta)$ for all $a, b \in K$ where θ is the Frobenius automorphism. The group of units U of K has order $p^n - 1$ where p and $p^n - 1$ are relatively prime numbers. Consequently, $a\theta = 1$ if and only if $a = 1$, and θ is an isomorphism of U. We conclude that θ is an isomorphism of K; and since K is of finite order, it maps K onto K.

The group U is a finite multiplicative group in a field so that by Theorem 8.34 of Section 8.4 U is a cyclic group. A generator a of U has order $p^n - 1$, and thus, $a\theta^n = a^{p^n} = a$. If $1 \leq j < n$, $a\theta^j =$

$a^{p^j} \neq a$ so that θ has order n. This completes the proof of the theorem.

Automorphisms of fields are very useful as we shall see.

12.37 THEOREM If K is a field and $\mathscr{S}$ is a subset of the Galois group $\mathscr{G}$ of K then the set

$$L = \{a \mid a \in K, a\alpha = a \text{ for all } \alpha \in \mathscr{S}\}$$

is a subfield of K. If L is a subfield of K then the set

$$\mathscr{T} = \{\alpha \mid \alpha \in \mathscr{G}, a\alpha = a \text{ for all } a \in L\}$$

is a subgroup of $\mathscr{G}$.

The proof of this theorem is a simple computation and is left to the exercises. We may now prove our big theorem.

12.38 THEOREM *Any finite field K has order p^n for a prime p and a positive integer n. For any prime p and any positive integer n there is a field of order p^n. A field of order p^n is the splitting field of $x^{p^n} - x \in \mathbf{Z}_p[x]$. Any two fields of order p^n are isomorphic.*

Proof: The first assertion of the theorem follows from Theorem 12.36. Let K be a splitting field for $x^{p^n} - x \in \mathbf{Z}_p[x]$, and K_0 be the set of roots of $x^{p^n} - x$ in K. Observe that K is a finite extension of $\mathbf{Z}_p$, and thus, is a finite field.

Turning now to K_0, by Theorem 12.33, it contains exactly p^n elements. Let θ be the Frobenius automorphism of K. Then the set K_0 is precisely the subset of all $a \in K$ such that $a\theta^n = a^{p^n} = a$. Hence, by Theorem 12.37 with $\mathscr{S} = \{\theta^n\}$ we conclude that K_0 is a subfield of K. Certainly $x^{p^n} - x$ splits in K_0 so that $K_0 = K$. We have proven that the splitting field of $x^{p^n} - x$ over $\mathbf{Z}_p$ is a field K of order p^n.

Let K be any field of order p^n. By Theorem 12.36 the group of units U of K is cyclic of order $p^n - 1$. Since for any $a \in U, a^{p^n-1} = 1$, we must have $a^{p^n} - a = 0$, so we conclude that K is a set of p^n distinct roots of $x^{p^n} - x$, and therefore, K is a splitting field of $x^{p^n} - x$ over $\mathbf{Z}_p$.

If K and K' are two fields of order p^n then both are splitting fields of $x^{p^n} - x$ over $\mathbf{Z}_p$. The proof of the theorem is complete since Corollary 12.32 of Section 12.3 tells us that the splitting field of a polynomial over a given field is unique up to isomorphism.

Example 1 Let us construct a field K of order $3^3 = 27$. Theorem 12.10 of Section 12.1 tells us that if $p(x)$ is a prime in $\mathbf{Z}_3[x]$ of degree 3 then $K = \mathbf{Z}_3[x]/(p(x))$ will have order 27. A polynomial of degree 3 will be irreducible in $\mathbf{Z}_3[x]$ if and only if it has no roots in $\mathbf{Z}_3$. Since $x^3 - x$ has every element of $\mathbf{Z}_3$ as a root, $x^3 - x - 1$ has no roots in $\mathbf{Z}_3$ and is a prime in $\mathbf{Z}_3[x]$. Therefore,

$$K = \mathbf{Z}_3[x]/(x^3 - x - 1)$$

is a field of order 27. Let $j = x + (x^3 - x - 1)$ as in Section 10.10.

The group U of units is cyclic of order $27 - 1 = 26$. Next, let us find a generator of this group. Since $(\mathbf{Z}_3 = \{0, +1, -1\})$

$$j^3 = j + 1,$$

j does not have order 1 or 2, therefore j has order 13 or 26. Using the Frobenius automorphism, $a\theta = a^3$, we have

$$j^9 = (j + 1)^3 = j^3 + 1 = j - 1.$$
$$j^{12} = (j + 1)(j - 1) = j^2 - 1.$$
$$j^{13} = j^3 - j = 1.$$

Since j has order 13, $-j$ has order 26 and is a generator of U.

Example 2 Let us examine fields of order $2^3 = 8$. By factoring

$$x^8 - x = x(x - 1)(x^3 + x + 1)(x^3 + x^2 + 1)$$

in $\mathbf{Z}_2[x]$ where $\mathbf{Z}_2 = \{0, 1\}$ we obtain two prime cubic polynomials in $\mathbf{Z}_2[x]$. We set

$$K = \mathbf{Z}_2[x]/(x^3 + x + 1) \qquad \text{where } j = x + (x^3 + x + 1)$$

and

$$K' = \mathbf{Z}_2[x]/(x^3 + x^2 + 1) \qquad \text{where } j' = x + (x^3 + x^2 + 1).$$

By Theorem 12.10 both K and K' are fields of order 8. Both are splitting fields for $x^8 - x$ so that K is isomorphic to K' and $x^3 + x + 1$ splits in both fields. A straightforward computation shows that $j' + 1$ is a root to $x^3 + x + 1$ in K'. Now, by Theorem 12.31, there is an isomorphism $\beta: K \to K'$ which is the identity upon $\mathbf{Z}_2$, and for which $j\beta = j' + 1$. Further, by Theorem 12.10, $K = \mathbf{Z}_2[j]$ and $1, j, j^2$ is a $\mathbf{Z}_2$-basis for K^+ over $\mathbf{Z}_2$. Similarly, $1, (j' + 1), (j' + 1)^2$

is a $\mathbf{Z}_2$-basis for K'^{+} over $\mathbf{Z}_2$. Therefore, for $a, b, c \in \mathbf{Z}_2$ β must be given by the equation

$$(a + bj + cj^2)\beta = a + b(j' + 1) + c(j' + 1)^2.$$

Since a field K of order p^n is unique up to isomorphism, we may speak of *the* field of order p^n. The field of order p^n is denoted by $\mathrm{GF}(p^n)$ and is called the *Galois field* of order p^n. If K is a subfield of order p^m in $\mathrm{GF}(p^n)$ then K is the splitting field for $x^{p^m} - x$ in $\mathrm{GF}(p^n)$ over $\mathbf{Z}_p = \mathrm{GF}(p)$ so that K is the unique subfield of order p^m in $\mathrm{GF}(p^n)$. Consequently, there is no confusion in speaking of $\mathrm{GF}(p^m)$ as a subfield of $\mathrm{GF}(p^n)$.

12.39 THEOREM *If p is a prime and m, n are positive integers then $\mathrm{GF}(p^m)$ is a subfield of $\mathrm{GF}(p^n)$ if and only if m divides n.*

Proof: If $\mathrm{GF}(p^m)$ is a subfield of $\mathrm{GF}(p^n)$ then the Product of the Degrees Theorem, 12.2, says that

$$[\mathrm{GF}(p^n):\mathbf{Z}_p] = [\mathrm{GF}(p^n):\mathrm{GF}(p^m)][\mathrm{GF}(p^m):\mathbf{Z}_p].$$

Since $[\mathrm{GF}(p^n):\mathbf{Z}_p] = n$ and $[\mathrm{GF}(p^m):\mathbf{Z}_p] = m$ we conclude that m divides n.

Assume now that m divides n. Then the group U of units of $\mathrm{GF}(p^n)$ is cyclic of order $p^n - 1$ and

$$p^n - 1 = (p^m - 1)(p^{m(n-1)} + p^{m(n-2)} + \cdots + p^m + 1)$$

so that U contains a unique cyclic subgroup U_0 of order $p^m - 1$. The set $U_0 \cup \{0\}$ is a complete set of roots of $x^{p^m} - x$ in $\mathrm{GF}(p^n)$. Consequently, the splitting field $\mathrm{GF}(p^m)$ of $x^{p^m} - x$ over $\mathbf{Z}_p$ is a subfield of $\mathrm{GF}(p^n)$, completing the proof.

We may now prove a theorem about irreducible polynomials in $\mathbf{Z}_p[x]$.

12.40 THEOREM *A prime polynomial $f(x) \in \mathbf{Z}_p[x]$ for a prime p is a factor of $x^{p^n} - x$ in $\mathbf{Z}_p[x]$ for a positive integer n if and only if the degree t of $f(x)$ divides n.*

Proof: By Theorem 12.10, $K = \mathbf{Z}_p[x]/(f(x))$ is a field of degree t over $\mathbf{Z}_p$, and therefore, may be taken to be $\mathrm{GF}(p^t)$. Now K contains a root u of both $x^{p^t} - x$ and $f(x)$. Thus $x - u$ divides the g.c.d. $g(x)$ of $x^{p^t} - x$ and $f(x)$ in $\mathrm{GF}(p^t)[x]$. Both $x^{p^t} - x$ and $f(x)$ lie in $\mathbf{Z}_p[x]$ so that we may compute the g.c.d. in $\mathbf{Z}_p[x]$. Accordingly, $g(x)$ lies in $\mathbf{Z}_p[x]$ and has positive degree. Since $f(x)$ is a prime in $\mathbf{Z}_p[x]$ we conclude that $g(x)$ differs from $f(x)$ by a unit of $\mathbf{Z}_p[x]$. Therefore $f(x)$ divides $x^{p^t} - x$. The polynomial $x^{p^t} - x$ splits in K, so that $f(x)$ does also. In other words, the splitting fields of $f(x)$ and $x^{p^t} - x$ are the same.

Suppose that t divides n, so that $GF(p^t)$ is a subfield of $GF(p^n)$. Since $f(x)$ is a factor of $x^{p^t} - x$ and since this latter polynomial has no repeated roots, it must be that $f(x)$ has no repeated roots. By Equation 12.34 we conclude that $f(x)$ divides $x^{p^n} - x$.

Finally suppose that $f(x)$ divides $x^{p^n} - x$. Then $f(x)$ splits in $GF(p^n)$ so that $K = GF(p^t)$ is a subfield. The proof is now complete since t divides n by Theorem 12.39.

We shall end our results upon Galois fields with the following theorem.

12.41 THEOREM *The Galois group $\mathcal{G}$ of $GF(p^n)$ is cyclic of order n. Theorem 12.37 gives a unique one-one correspondence between subgroups $\mathcal{S}$ of $\mathcal{G}$ and subfields L of $GF(p^n)$.*

Proof: Since the Frobenius automorphism θ has order n, in order to prove that $\mathcal{G}$ is cyclic, it is sufficient to prove that every element of $\mathcal{G}$ is a power of θ. Let $\alpha \in \mathcal{G}$ and u a generator of the group U of units of $GF(p^n)$. For some positive integer s, $u\alpha = u^s$. In fact, if m is any integer then $(u^m)\alpha = (u\alpha)^m = (u^s)^m = (u^m)^s$. Since $GF(p^n) = U \cup \{0\}$, we conclude that $a\alpha = a^s$ for any $a \in GF(p^n)$. The group U is cyclic of order $p^n - 1$ so that we certainly may assume that $s \leq p^n - 1$. Suppose that $s = tp^m$ where t is relatively prime to p, and m is a nonnegative integer. Taking $\alpha' = \alpha\theta^{-m}$ we have $a\alpha' = a\alpha\theta^{-m} = a^t$ for $a \in GF(p^n)$. We will have shown that $\alpha = \theta^m$ if we prove that $t = 1$.

Consider the polynomial $g(x) = (x + 1)^t - x^t - 1$. Then for $a \in GF(p^n)$, $g(a) = (a + 1)^t - a^t - 1 = (a + 1 - a - 1)\alpha' = 0$. In particular, $x - a$ divides $g(x)$ for every $a \in GF(p^n)$. Thus $x^{p^n} - x$ divides $g(x)$ so that $g(x) = 0$ since otherwise we would have $\deg g(x) \geq p^n > s \geq t \geq \deg g(x)$. Expanding this polynomial in powers of x shows that if $t > 1$ then the coefficient of x in $g(x)$ is $t \cdot 1$. That is, $t \cdot 1 = 0$ in $GF(p^n)$ or p divides t. But t is relatively prime to p so that we must have $t = 1$; therefore $\alpha = \theta^m$.

The subgroups of $\mathcal{G}$ are generated by θ^m where m is a divisor of n. The subfields of $GF(p^n)$ are the fields $GF(p^m)$ where m is a divisor of n. The set of elements fixed by all powers of θ^m where m is a divisor of n is the set of roots to $x^{p^m} - x$ which, in turn, is $GF(p^m)$. On the other hand, θ^t fixes every element of $GF(p^m)$ for a divisor m of n if and only if every element of $GF(p^m)$ is a root to $x^{p^t} - x$. Thus $x^{p^m} - x$ divides $x^{p^t} - x$. If u is a generator of the group of units of $GF(p^m)$ then the minimal polynomial $f(x) \in \mathbf{Z}_p[x]$ of u must be of degree m

since $[\mathrm{GF}(p^m):\mathbf{Z}_p] = m$ and $\mathbf{Z}_p[u] = \mathrm{GF}(p^m)$. Consequently, $f(x)$ divides $x^{p^m} - x$ and also $x^{p^t} - x$. By Theorem 12.40 m divides t and θ^t is in the subgroup generated by θ^m. The proof of the theorem is now complete.

The subfields $\mathrm{GF}(p^m)$ of $\mathrm{GF}(p^n)$ are in one-one correspondence with the subgroups generated by θ^m in $\mathscr{G}$ where m runs over the divisors of n. Let $\mathscr{G}_m$ be the subgroup generated by θ^m, so that the index m of $\mathscr{G}_m$ in $\mathscr{G}$ is equal to $[\mathrm{GF}(p^m):\mathbf{Z}_p]$. We now introduce the notation $[\mathscr{G}:\mathscr{T}] = t$ for the index of a subgroup $\mathscr{T}$ in a group $\mathscr{G}$. This notation is in common usage and gives the following corollary an easily remembered form.

12.42 COROLLARY *If $\mathscr{G}$ is the Galois group of $\mathrm{GF}(p^n)$ and $\mathscr{T} \subseteq \mathscr{G}$, $L \subseteq \mathrm{GF}(p^n)$ are a corresponding subgroup and subfield pair then*

$$[\mathscr{G}:\mathscr{T}] = [L:\mathbf{Z}_p].$$

This theorem and its corollary constitute the Fundamental Theorem of Galois Theory for Finite Fields. There is a corresponding theorem for fields which are not finite; but this theorem is much more difficult to prove. In our final examples, we will use our theorem, 2.41, to help us construct polynomials.

Example 3 Suppose that K is a finite field of order p^n and θ is the Frobenius automorphism of K. If $u \in K$ then what is the minimal polynomial for u in $\mathbf{Z}_p[x]$? Let $g(x)$ be the (monic) minimal polynomial of u over $\mathbf{Z}_p$. The coefficients of $g(x)$ lie in $\mathbf{Z}_p$ and $a\theta = a$ for all $a \in \mathbf{Z}_p$ so that $g(u)\theta^m = g(u\theta^m) = 0$ and the set

$$S = \{u\theta^m \mid m \in \mathbf{Z}\}$$

is a set of roots of $g(x)$. Suppose that

$$S = \{u = u_1, u_2, \ldots, u_t\}.$$

Form the polynomial

$$h(x) = (x - u_1)(x - u_2)\cdots(x - u_t).$$

If we apply θ to the u's, $h(x)$ remains unchanged. That is, θ fixes the coefficients of $h(x)$. Hence the coefficients of $h(x)$ all lie in the field fixed by θ which is $\mathbf{Z}_p$, the splitting field of $x^p - x$, and thus $h(x) \in \mathbf{Z}_p[x]$. Now $h(x)$ divides $g(x)$ and $g(x)$ is prime so that $g(x) = h(x)$. We have found a method for determining $g(x)$.

Example 4 Construct the field GF(3^6). In Example 1 we constructed the field GF(3^3). It is a subfield $L = \mathbf{Z}_p[j]$ of $K = $ GF(3^6). Since $-j$ has order 26 in L, it generates the group U of units of L. The square of any generator of U has order 13 so that $-j$ is not a square in L, hence $x^2 + j$ is irreducible over L. By Theorem 12.10, the splitting field of $x^2 + j$ over L must be GF(3^6). Let u be a root of $x^2 + j$ in K. To find the minimal polynomial of u in $\mathbf{Z}_p[x]$ we could use the preceding example. Instead, let us apply that analysis to $x^2 + j$. In particular, by applying θ to j several times, we obtain the elements $j, j + 1, j - 1$. Since θ does not change the coefficients of

$$f(x) = (x^2 + j)(x^2 + j + 1)(x^2 + j - 1)$$

we conclude that $f(x) \in \mathbf{Z}_3[x]$ is the minimal polynomial of u. Therefore, GF(3^6) $\cong \mathbf{Z}_3[x]/(f(x))$. Explicitly,

$$f(x) = x^6 - x^4 - x^2 + 1.$$

Example 5 How many irreducible polynomials of positive degree n are there in $\mathbf{Z}_p[x]$? Let $K = $ GF(p^n). If $u \in K$ lies in no proper subfield of K then $K = \mathbf{Z}_p[u]$ so that u has a minimal polynomial $f(x)$ in $\mathbf{Z}_p[x]$ of degree n. Since $f(x)$ is a factor of $x^{p^n} - x$, it has no repeated roots. In particular, if S is the set of all elements of K lying in no proper subfield of K then the set S partitions into disjoint subsets of size n; each subset being the set of n roots of a prime polynomial in $\mathbf{Z}_p[x]$ of degree n. If S has size m then there are m/n primes in $\mathbf{Z}_p[x]$ of degree n.

For example, if $K = $ GF(3^6) then the proper subfields of K are GF(3), GF(3^2), and GF(3^3). Therefore, there are $(3^2 - 3) + (3^3 - 3) + 3 = 33$ elements of K lying in proper subfields of K. Of the remaining elements there are $(3^6 - 33)/6 = 116$ subsets of 6 elements each belonging to the irreducible polynomials of degree 6 in $\mathbf{Z}_p[x]$. Therefore, there are 116 prime polynomials of degree 6 in $\mathbf{Z}_3[x]$.

EXERCISES

1. Construct the field GF(8). Find all its subfields.

2. Find a prime polynomial of degree 5 in $\mathbf{Z}_2[x]$ and then construct GF(32).

3. Find the splitting field of $x^3 + 2x + 5$ over (i) $\mathbf{Z}_3$, (ii) $\mathbf{Z}_5$, (iii) $\mathbf{Z}_7$.

4. Compute the number of prime polynomials of degree 10 in $\mathbf{Z}_3[x]$.

5. Prove that the set $\mathscr{G}$ of Definition 12.35 is a group with multiplication given by composition of mappings.

6. Prove Theorem 12.37.

7. Show that the polynomials $f(x) = x^3 + x + 1$ and $g(x) = x^3 + x^2 + 1$ are prime over $\mathbf{Z}_5$. Exhibit an isomorphism from $\mathbf{Z}_5[x]/(f(x))$ onto $\mathbf{Z}_5[x]/(g(x))$.

8. Prove that every finite field K of characteristic the prime p is a simple extension of $\mathbf{Z}_p$.

9. Find a prime polynomial $f(x)$ of degree 6 in $\mathbf{Z}_7[x]$. Using this polynomial construct GF(7^6). Find all subfields of this field.

10. Assume that p is a prime and m is a positive integer not divisible by p. Prove that there is a positive integer n so that the group of units GF(p^n) contains a cyclic group of order m.

11. Suppose that F is a finite field, $f(x) \in F[x]$ is a prime polynomial, and K is an extension of F which contains a root of $f(x)$. Prove that $f(x)$ splits in K.

12. Suppose that a prime $p = 2^m + 1$.

   (i) Show that the group U of units of $K = $ GF(2^{2m}) contains an element u of order p.

   (ii) Show that $K = \mathbf{Z}_2(u)$.

   (iii) Let G be the Galois group of K over $\mathbf{Z}_2$. Show that G has an action upon the subgroup V of U generated by u and that this action is an isomorphism of G.

   (iv) Using (iii) prove that $2m$ divides $p - 1$.

   (v) Prove that m is a power of 2.

   This exercise shows that if $2^m + 1$ is prime then m is a power of 2 and $2^m + 1$ is a Fermat prime.

13. Construct the field $K = $ GF(2^2) and its Galois group G. Form the pairs $H = \{(\alpha, u) \mid u \in K\backslash\{0\}, \alpha \in G\}$.

   (i) Show that H is a group isomorphic to S_3 when we define multiplication via

   $$(\alpha, u)(\alpha', u') = (\alpha\alpha', [u\alpha']u').$$

(ii) If $(\alpha, u) \in H$ and v is in the $\mathbf{Z}_2$-vector space K^+ then show that by setting

$$v(\alpha, u) = (v\alpha)u,$$

(α, u) defines a vector space isomorphism of K^+ onto itself.

(iii) Form the set $L = \{(a, v) \mid a \in H, v \in K^+\}$. Define

$$(a, v)(a', v') = (aa', [va'] + v).$$

Prove that with this multiplication, L is a group isomorphic to S_4.

 # COMMENTARY

1 One of the most beautiful branches of algebra is the Galois theory of equations. In Section 12.4 we alluded to this theory for finite fields. However, from an intuitive point of view, the most interesting field to start with is the rational field. This "starting field" is called the *ground* field. One then studies finite extensions of the ground field. If we consider a polynomial $f(x)$ with rational coefficients then there is a unique subfield K of the complex field $\mathbf{C}$ which is a splitting field of the polynomial. Traditionally, Galois theory focused upon the polynomial $f(x)$ and its roots in K. This viewpoint made Galois theory one of the most abstruse branches of mathematics, but since about 1900 a "modern" viewpoint has evolved which is, at the same time, more abstract and easier to understand. The modern approach focuses upon the field K and the subfields L such that $K \supseteq L \supseteq \mathbf{Q}$. We will start by describing this modern version and later relate it to the polynomial $f(x)$.

An *automorphism* σ of K is an isomorphism of the field K onto itself. An automorphism σ of K is said to *fix* a subfield L of K if $a\sigma = a$ for all elements a of L, that is, σ *fixes* the elements of L. The set F of all elements fixed by σ is a subfield of K called the *fixed field* of σ. We will be considering the set G of all automorphisms of K which fix the ground field $\mathbf{Q}$. The composition of any two automorphisms of G is again an automorphism of G. In fact, with this multiplication, G is a group called the *Galois* group of K over $\mathbf{Q}$.

Because K is the splitting field of a rational polynomial $f(x)$, there are two remarkable facts about K and G which we may now state.

(1) The intersection of all the fixed fields for all elements σ of G (i.e., the fixed field of G) is the ground field $\mathbf{Q}$.

(2) The order n of G is finite and equal to the degree $[K:\mathbf{Q}]$ of K over the ground field $\mathbf{Q}$.

The first property says that if a is an element of K and $a\sigma = a$ for all elements σ of G then a is in $\mathbf{Q}$. The second property gives us the order of the Galois group.

Let us change the situation very slightly by altering the ground field. Let L be a field between K and $\mathbf{Q}$, that is, $K \supseteq L \supseteq \mathbf{Q}$. Since the coefficients of $f(x)$ lie in

Q and $L \supseteq$ **Q**, it is still true that K is the splitting field of $f(x)$ over L; only the ground field has changed. Let H be the group of all automorphisms of K which fix L. Since $L \supseteq$ **Q**, all these automorphisms fix **Q** also. In particular, the Galois group H of K over L is a subgroup of the Galois group G of K over **Q**. In addition, properties (1) and (2) both hold with H in place of G and L in place of **Q**. In particular, if K is the splitting field of some polynomial over some ground field $L \supseteq$ **Q** then the properties (1) and (2) always hold for the Galois group H of K over L. With the exception of one detail, the properties (1) and (2) as applied to K, L, and H contain the essence of Galois theory.

Let us now complicate matters a little bit more. The intermediate field L may itself be a splitting field of some different polynomial $g(x)$ over **Q**. As we shall see, this need not be the case; but when it is, something very interesting occurs. At this point we have G, the Galois group of K over **Q**; H, a subgroup of G which is the Galois group of K over L; and (since L is the splitting field of $g(x)$ over **Q**) a new Galois group M of L over **Q** which is not a subgroup of G. The automorphisms in G all carry the elements of L into itself, and therefore, by restriction of their domain to L, we obtain automorphisms of L. The final missing detail may now be stated.

(3) If $K \supseteq L \supseteq$ **Q** where L is the splitting field of $g(x)$ then the restriction mapping of automorphisms of G to L is a homomorphism of the Galois group G of K over **Q** onto the Galois group M of L over **Q** with kernel H, where H is the Galois group of K over L. In particular, H is normal in G and $G/H \cong M$. Conversely, if H is normal in G then L is the splitting field of some polynomial over **Q**.

Properties (1) to (3) we may call the Fundamental Theorem of Galois Theory. In property (3) we could have taken the ground field to be a field F where $L \supseteq F \supseteq$ **Q** in place of **Q**. In particular, properties (1) to (3) do not change if we alter the ground field.

We have glossed rather lightly over a tremendous amount of algebra. Let us now discuss this in relation to an example. At the same time we will indicate connections with other parts of algebra. Consider the polynomial $f(x) = x^3 - 2 \in$ **Q**$[x]$, and let K be the splitting field of $x^3 - 2$ in **C**, the complex field. Let $\omega = (-1 + \sqrt{-3})/2$ so that $\omega^3 = 1$. Then the roots of $x^3 - 2$ are: $\sqrt[3]{2}$, $\omega\sqrt[3]{2}$, and $\omega^2\sqrt[3]{2}$. We shall label these roots a_1, a_2, a_3 respectively. Then

$$x^3 - 2 = (x - a_1)(x - a_2)(x - a_3)$$

and the coefficients of $x^3 - 2$ are $-(a_1 + a_2 + a_3) = 0$, $a_1a_2 + a_1a_3 + a_2a_3 = 0$, and $-a_1a_2a_3 = -2$.

Let G denote the Galois group of K over **Q**. If $\sigma \in G$ then, since **Q** is fixed by σ, it fixes the coefficients of $x^3 - 2$. That is, σ acts upon the set of roots of $x^3 - 2$. Since $K = $ **Q**(a_1, a_2, a_3), if $a_i\sigma = a_i$ for $i = 1, 2, 3$ then $a\sigma = a$ for all $a \in K$. In this situation, $\sigma = e$, the identity. We have just proven that knowing what σ does to the three roots a_1, a_2, a_3 determines what σ does to every $a \in K$; that is, the action of G upon the set of roots of $x^3 - 2$ is an isomorphism of G into S_3, the symmetric group on $\{1, 2, 3\}$ (that is, σ gives rise to the permutation $\bar\sigma$ where $i\bar\sigma = j$ if and only if $a_i\sigma = a_j$). In general, the isomorphisms of G into the corresponding symmetric groups S_n (here S_3) are "into," not "onto." Not all permutations need occur in this way.

Historically, things came in the reverse order. We shall illustrate the historical development by starting with Lagrange's method for solving polynomial equations. Consider the polynomials in $\mathbf{Q}[x_1, x_2, x_3]$. If $h(x_1, x_2, x_3)$ is such a polynomial and $\alpha \in S_3$, then set $h^\alpha(x_1, x_2, x_3) = h(x_{1\alpha}, x_{2\alpha}, x_{3\alpha})$. If $s_1 = x_1 + x_2 + x_3$, $s_2 = x_1x_2 + x_1x_3 + x_2x_3$, and $s_3 = x_1x_2x_3$ then $s_i\alpha = s_i$ for all $\alpha \in S_3$. Substituting a_i for x_i, $i = 1, 2, 3$, we discover that $h(a_1, a_2, a_3)$ is a rational number if $h^\alpha(x_1, x_2, x_3) = h(x_1, x_2, x_3)$ for all $\alpha \in S_3$. This is certainly the case for the polynomials s_j since they give the coefficients of $x^3 - 2$ upon substituting the a's. Lagrange (1736–1813) discovered this criterion. That is, if h is a polynomial in n variables which remains unchanged by any permutation of S_n and $a_1, a_2, \ldots, a_n$ are the roots of a rational polynomial of degree n, then substituting the a's into h gives a rational number. We understand why this is so: there is an isomorphism of the Galois group of the splitting field over $\mathbf{Q}$ of this polynomial onto $\bar{G}$, a subgroup of S_n, and any polynomial in the a's which is unchanged by all the permutations of $\bar{G}$ must be rational by property (1). Lagrange was a bit heavy-handed in using all the permutations of S_n instead of restricting his attention to $\bar{G}$.

Let us stop at this point to make a few isolated observations. Historically, the symmetric group arose exactly as we have described: as permutations upon the roots of a polynomial. In Lagrange's theory the polynomials s_j play a prominent role. They are called the *elementary symmetric polynomials* of degree three. Such polynomials occur for all degrees and have uses in combinatorics and polynomial algebra. The "normal" situation in Galois theory is to consider the splitting field of a polynomial. Property (3) tells us that the "normal" situation is related to normal subgroups. This is the origin of the word "normal" as applied to groups.

Before proceeding further, let us interpret (1) to (3) in our current situation. Property (1) tells us that if $\bar{G}$ is the image of the Galois group in S_3 then $h(a_1, a_2, a_3)$ is rational if and only if $h^\alpha = h$ for all $\alpha \in \bar{G}$. Property (2) tells us that the order n of $\bar{G}$ is $[K:\mathbf{Q}]$. A straightforward computation shows that $[K:\mathbf{Q}] = 6$ in this instance. In particular, $\bar{G} = S_3$ since $\bar{G}$ and S_3 both have order 6. Property (3) relates the normal subgroups of $\bar{G}$ to the subfields L of K which are splitting fields for polynomials. The only nontrivial normal subgroup of S_3 is A_3, so that there is some subfield L of K such that $[L:\mathbf{Q}]$ equals the order of $M = S_3/A_3$ which is 2. In Section 12.2 we saw that $L = \mathbf{Q}(\sqrt{m})$ for some rational number m. We turn next to a search for this m.

Some subgroup H of the Galois group G maps onto A_3 and L is the fixed field of H. By (1), an element a of K is in L if and only if $a\sigma = a$ for all $\sigma \in H$. From Lagrange's viewpoint, $h(a_1, a_2, a_3)$ is in L if and only if $h^\alpha = h$ for all $\alpha \in A_3$. We want to choose h so that $h(a_1, a_2, a_3)$ tells us $\sqrt{m}$. That is, we do not want $h(a_1, a_2, a_3) \in \mathbf{Q}$. In other words, we want $h^\beta \neq h$ for some $\beta \in S_3$ where $\beta \notin A_3$. Recall that the Vandermonde polynomial $V = V(x_1, x_2, x_3)$ of Section 2.8 has the interesting property that $V^\alpha = \pm V$ for all $\alpha \in S_3$. Thus $V^\alpha = V$ for all $\alpha \in A_3$ and $V^\beta = -V$ for $\beta \in S_3$ and $\beta \notin A_3$. In other words, V^2 is unchanged by elements of S_3. In our case, $V(a_1, a_2, a_3)^2$ is a function of the coefficients of $x^3 - 2$, and has the value

$$V(a_1, a_2, a_3)^2 = -6^2 \cdot 3.$$

Consequently, we have

$$V(a_1, a_2, a_3) = \pm 6\sqrt{-3}.$$

Since $\omega = a_2/a_1$ and $\sqrt{-3} = 2\omega + 1$, we know that $L = \mathbf{Q}(\sqrt{-3})$ is in K. Further, L is the splitting field of $x^2 + 3$.

The Galois group H of K over L maps onto A_3 in S_3 and is cyclic of order 3. In a fashion analogous to the above, Lagrange found a polynomial $h \in L[x_1, x_2, x_3]$ so that $h^\alpha \neq h$ for any α in A_3, but $(h^3)^\alpha = h^3$ for all $\alpha \in A_3$. These polynomials h are called Lagrange resolvents. Then $K = L(h(a_1, a_2, a_3))$.

Lagrange's method for solving the cubic proceeds as follows:

(i) find a polynomial V so that $V^\alpha \neq V$ for some $\alpha \in S_3$ but $(V^2)^\alpha = V^2$ for all $\alpha \in S_3$;

(ii) $V(a_1, a_2, a_3)^2$ may be written as a function of the coefficients of the polynomial so that $\sqrt{m} = V(a_1, a_2, a_3)$ may be computed explicitly;

(iii) find a polynomial $h \in \mathbf{Q}(\sqrt{m})[x_1, x_2, x_3]$ so that $h^\alpha \neq h$ for some $\alpha \in A_3$ but $(h^3)^\alpha = h^3$ for all $\alpha \in A_3$; and

(iv) h^3 may be written as a function of $\sqrt{m}$ and the coefficients of the polynomial so that $h(a_1, a_2, a_3) = \sqrt[3]{b}$ may be computed explicitly.

The field K is then $\mathbf{Q}(\sqrt{m}, \sqrt[3]{b})$ and the roots are rational expressions in $\sqrt{m}$ and $\sqrt[3]{b}$. The first step involves the group S_3/A_3, and the third step involves the group A_3. At each stage, the group involved is finite and cyclic.

We now come to a general theme which emerges out of this development. If we change the degree of the polynomial, then Lagrange's method for finding roots will work so long as the appropriate symmetric group T has a chain of subgroups $T_0 \supseteq T_1 \supseteq T_2 \supseteq \cdots \supseteq T_t$ so that $T_0 = T$, $T_t = \{e\}$ and T_i is a normal subgroup of T_{i-1} with the factor group T_{i-1}/T_i cyclic of finite order, $i = 1, 2, \ldots, t$. This situation occurs for S_2, S_3, S_4 but not for S_5, S_6, S_7, $\ldots$. Any group T which has a chain of T_i's as above is called a *solvable* group because it is related to the necessary conditions for solving equations. Lagrange did not get away from the symmetric groups, and thus did not find general methods for solving equations.

Really, to apply Lagrange's method, the only important group in the symmetric group is $\bar{G}$, the image of the Galois group. In our case, $\bar{G} = S_3$, but it need not be the case that $\bar{G}$ is all of the appropriate symmetric group. It was Galois who observed this point first. He then showed that Lagrange's method works if $\bar{G}$ is a solvable group, and that no algebraic method works if $\bar{G}$ is not a solvable group. Let us set this out in a little more detail.

What is an expression by radicals? The number $\sqrt[3]{7 + \sqrt{2}}/\sqrt[7]{-5}$ is an example of one. This number lies in the field L_3 where $L_1 = \mathbf{Q}(\sqrt[7]{-5})$, $L_2 = L_1(\sqrt{2})$, and $L_3 = L_2(\sqrt[3]{7 + \sqrt{2}})$. That is, such a number is an element of a field L which is obtained from $\mathbf{Q}$ by extending $\mathbf{Q}$ via a sequence of nth roots for various n's.

What can we say about polynomials whose roots are expressible by radicals? The expression of the roots of $p(x) \in \mathbf{Q}[x]$ as radicals means that the splitting field K of $p(x)$ is obtained via a sequence of extensions by nth roots for various n's. The multiply quadratic extensions of Section 12.2 are examples of such fields. The Galois group of K over $\mathbf{Q}$ acts as a subgroup $\bar{G}$ of the symmetric group upon the roots of $p(x)$ in K. In building up K we go from a field L to a field $L(\sqrt[n]{d})$ where $d \in L$. The

Galois group H of K over L contains the Galois group H_0 of K over $L(\sqrt[n]{d})$ as a normal subgroup. The fact that $L(\sqrt[n]{d})$ is the splitting field of $x^n - d \in L[x]$ corresponds to the fact that the factor group H/H_0 is cyclic of order n.

We have the following remarkable theorem. *The roots of $p(x) \in Q[x]$ may be written in radicals if and only if the Galois group G of a splitting field K of $p(x)$ over Q is a solvable group,* that is, G contains a sequence of subgroups $G_0 \supseteq G_1 \supseteq G_2 \supseteq \cdots \supseteq G_t$ such that $G_0 = G$, $G_t = \{e\}$, and G_i is a normal subgroup of G_{i-1} where G_{i-1}/G_i is a finite cyclic group for $i = 1, 2, \ldots, t$. We have seen Lagrange's contribution to this theorem. Ruffini (1762–1822) skirted this result and almost showed that the quintic is not solvable by radicals. Abel (1802–1829) observed a part of this in the case of the quintic and demonstrated, in general, that quintics are not solvable by radicals. Galois (1811–1832) discovered the criterion above and settled the question of solvability by radicals in general.

Since the Galois group is isomorphic to a subgroup of the symmetric group on the roots of a polynomial $p(x) \in Q[x]$, polynomials of degrees 1, 2, 3, and 4 all have splitting fields with solvable Galois groups. In particular, if $\deg p(x) = 1, 2, 3$, or 4 then the roots of $p(x)$ may be written as radicals, a fact contained in Cardan's *Ars Magna* (1545). On the other hand, A_5 is a simple group. The Galois group G for the splitting field K of $p(x) = x^5 - 4x^3 + 2$ over Q permutes the roots just as S_5 does. In particular, G is isomorphic to S_5 and, consequently, is not solvable. Therefore, $p(x)$ has no solution by radicals.

There are two points which we have suppressed throughout this discussion. Both follow from Properties (1) to (3). First, there is a one-one correspondence between the subfields L of K containing Q ($K \supseteq L \supseteq Q$) and the subgroups H of the Galois group G of K over L. The subgroup H of G corresponds to the subfield L of K if H is the set of all automorphisms of K which fix L. Via this correspondence, $\{e\}$ "belongs" to K and G "belongs" to Q. In other words, if we write containment diagrams for subfields of K and subgroups of G, turning the diagram upside-down for G, then the diagrams will look the same.

Set $L_i = Q(a_i)$, $F = Q(\sqrt{-3})$, and $H_i = \{e, (jk)\}$ where $\{j, k\} = \{1, 2, 3\}\backslash\{i\}$, $i = 1, 2, 3$. The diagrams for $x^3 - 2$ are illustrated below. Along each edge, the appropriate index is listed. In these diagrams L_i and H_i correspond, $i = 1, 2, 3$, F and A_3 correspond, K and $\{e\}$ correspond, and Q and S_3 correspond.

Second, a subfield L is the splitting field of a rational polynomial if and only if its corresponding subgroup H is normal in G. In other words, the fields L_i, $i = 1, 2, 3$,

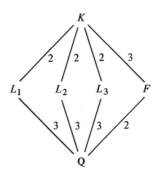

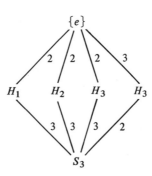

are not splitting fields of polynomials over Q, but the fields K, F, and **Q** are. They are respectively the splitting fields for $x^3 - 2$, $x^2 + 3$, and x.

Galois theory serves as a very powerful tool of algebra. Polynomials pervade much of algebra, and, wherever they occur, Galois theory is also a useful tool. The ideas of this theory have been generalized and extended to many rings (other than fields). Both the generalizations of Galois theory and the older Galois theory we have described here remain fairly active branches of algebra [6, 9, 18, 51, 52, 53, 54, 55, 146].

2   The study of fields is a large branch of algebra with many applications in algebra, analysis, and geometry. The little bit that we have discussed leads into Galois Theory, the theory of polynomial equations, and algebraic number theory. We have already had occasion to mention algebraic number theory as it applies to Fermat's Last Theorem (Commentary of Chapter 3). This application is a special case of the generalization of the notion of integer. We have seen that if $p(x) \in$ **Z**$[x]$ is a polynomial with a rational root m/n then the integer n must divide the leading coefficient and m must divide the constant term of $p(x)$. As a consequence, if $p(x)$ is monic then its rational roots must be integers. This suggests a possible extension: call a complex number a an *algebraic integer* if a is a root of a monic polynomial $p(x) \in$ **Z**$[x]$. A remarkable property, not unlike the same fact for algebraic numbers over **Q**, is that the sum and product of two algebraic integers is again an algebraic integer. We shall not prove this fact, but one consequence is that the set $\mathcal{O}$ of all algebraic integers *in a finite extension field F* of **Q** is an integral domain and F is the field of fractions formed from the elements of $\mathcal{O}$.

For this ring $\mathcal{O}$ we may investigate many of the same properties as hold for the integers **Z**. For example, what is the structure of the group of units U of $\mathcal{O}$? It turns out that this group can be quite involved; the Dirichlet unit theorem tells us that U is isomorphic to the direct sum of a finite cyclic group with a finite direct sum of several copies of **Z**$^+$. The theorem even tells us how many copies of **Z**$^+$ appear as direct summands. This suggests that rings of algebraic integers may have much richer groups of units than the particular examples of domains studied in the text [71, 79].

Passing from the group of units to other elements, we may ask questions similar to those posed in Chapters 5 and 10. For example, when will $\mathcal{O}$ be a unique factorization domain? From our theorems of Chapter 5 we know that $\mathcal{O}$ is such a domain if it is a Euclidean domain. The unfortunate truth is that $\mathcal{O}$ is rarely Euclidean. If we restrict our attention to just those fields $F =$ **Q**$(\sqrt{-d})$ which are quadratic extensions of **Q** for positive integral values of d then $\mathcal{O}$ is a Euclidean domain if and only if $d = 1, 2, 3, 7, 11$ [71]. All other such values of d give domains $\mathcal{O}$ which have no Division Algorithm.

We will prove in Chapter 16 that principal ideal domains are unique factorization domains. It is remarkable that the converse is also true for $\mathcal{O}$. That is, if $\mathcal{O}$ is a unique factorization domain then all of its ideals are principal. In our search for unique factorization domains, we need look no further than the principal ideal domains. There are more rings $\mathcal{O}$ which have this property than have the property of a Division Algorithm. For example, it is true that if $F =$ **Q**$(\sqrt{-19})$ then every ideal of $\mathcal{O}$ is principal, but $\mathcal{O}$ is not a Euclidean domain. Even in this more general setting, the rings $\mathcal{O}$ which are principal ideal domains are almost as scarce as Euclidean ones.

For example, if $F = \mathbf{Q}(\sqrt{-6})$, $F = \mathbf{Q}(\sqrt{-23})$, or $F = \mathbf{Q}(\sqrt{-d})$ for most choices of positive integral nonsquare d, then $\mathcal{O}$ does not have unique factorization into primes (and, therefore, is not a principal ideal domain).

Since most of the rings $\mathcal{O}$ are not unique factorization domains, we would naturally want to search for a slightly more general property which will simplify the arithmetic in $\mathcal{O}$ in a way similar to the simplification given by unique factorization when it is present. All is not lost; instead of focusing upon the factorization of an element a of $\mathcal{O}$, we should concentrate upon the factorization of the ideal (a) into a product of ideals. In order to do this we need to define the *product* of ideals. The most obvious definition is chosen; the product IJ of two ideals I and J is defined to be the ideal generated by all products ab where a is in I and b is in J. It is an important theorem that if I is a nontrivial ideal in $\mathcal{O}$ then

(*)
$$I = I_1 I_2 \cdots I_t$$

where the I_j are nontrivial ideals of $\mathcal{O}$ of a special kind called *prime* ideals. The ideals of (*) are unique except for their ordering. The word prime is appropriately used naming these ideals. An integer $p > 1$ is called a prime if and only if for any two integers a and b such that p divides ab, then p divides a or p divides b (see Lemma 5.23 of Section 5.4). Stated in terms of ideals, ab is in (p) if and only if a is in (p) or b is in (p). Accordingly, we say that an ideal J of $\mathcal{O}$ is a *prime* ideal if for any a and b of $\mathcal{O}$ such that ab is in J then a is in J or b is in J.

To a small degree, the identity (*) retrieves unique factorizations in $\mathcal{O}$. In fact, elementary texts on algebraic number theory almost lead one to believe that (*) is as good as unique factorization. This is definitely not the case when it comes to the arithmetic of elements. The difficulty is that the ideals I_j need not be principal; in particular, the expression for an element a of I written as an element of the product in (*) need not be unique, need not be a product, and may be quite complicated.

Let us look at a specific example of a "factorization into prime ideals." We saw that if $F = \mathbf{Q}(\sqrt{-5})$ so that $\mathcal{O} = \mathbf{Z}[\sqrt{-5}]$ then

$$6 = (1 - \sqrt{-5})(1 + \sqrt{-5})$$

and

$$6 = 2 \cdot 3$$

where the factors on the right are all irreducible. It can be shown that the ideal $(3, 1 + \sqrt{-5})$ is a prime ideal. In fact, the factorization of (6) into a product of prime ideals of $\mathcal{O}$ is

$$(6) = (3, 1 + \sqrt{-5})(3, 1 - \sqrt{-5})(2, 1 + \sqrt{-5})^2.$$

It is easy to see that any element from the product on the right lies in the ideal (6). Letting $\alpha = 1 + \sqrt{-5}$ so that $\alpha^* = 1 - \sqrt{-5}$, we obtain

$$6 = 3^2\alpha(\alpha - 2) + 3^2 \cdot 2^2 - \alpha^*\alpha(\alpha - 2) - \alpha\alpha^*2^2$$

so that 6 (and so any element of (6)) lies in the product on the right. This proves the equality. We obtained the ideal (6) as a product of prime ideals, but the expression for the number 6 in this factorization is very complicated [76, 83].

3   Goldstein [11], Herstein [12], Fraleigh [9], and Dean [6] give good coverage of the topics of this chapter plus much more. We have already mentioned references to Galois Theory in the first commentary. Many other topics are associated with the theory of fields. We have defined transcendental elements. The references [12, 24, 50] all discuss transcendental extensions.

In Theorem 12.33 we showed that $x^{p^n} - x \in \mathbf{Z}_p[x]$ has distinct roots. We have also shown that if $p(x)$ is a prime polynomial in $\mathbf{Z}_p[x]$ then $p(x)$ divides $x^{p^n} - x$ for some integer n. In particular, all prime polynomials in $\mathbf{Z}_p[x]$ have distinct roots in some extension field of $\mathbf{Z}_p$. It is also true that if $p(x)$ is a prime polynomial with rational coefficients then $p(x)$ has no repeated roots in the complex field $\mathbf{C}$. We might guess that if F is any field and $p(x)$ any prime polynomial of $F[x]$ then $p(x)$ has no repeated roots in a splitting field. This is true if F either has characteristic zero or is a finite field. Oddly, there are infinite fields F of prime characteristic, and prime polynomials $f(x)$ of degree greater than one with coefficients in F such that $f(x)$ has only one root in a splitting field; that is, $f(x) = x^n - a$ and in some extension of F, $f(x) = (x - b)^n$. A polynomial with no repeated roots is called *separable*; a polynomial, like $f(x)$, with repeated roots is called *inseparable*. The existence of inseparable prime polynomials in $F[x]$ plays a role in the study of extensions of F and in the Galois theory over F. These ideas are discussed in [14 Vols. 1, 3; 50].

From any integral domain we may form a field of fractions much as we did for $\mathbf{Z}$ in Section 6.4. In particular, we may form the field $F(x)$ associated with the domain $F[x]$ of polynomials over a field F. Similarly, if S is a ring of functions which is a domain, then we may form a function field associated with S. These function fields have been extensively studied [50, 56].

The last chapter of [1] is an excellent reference on finite fields.

chapter thirteen

SYSTEMS
OF LINEAR
EQUATIONS

There are several different ways of approaching the study of systems of linear equations. In this chapter we shall apply the theory of vector spaces as developed in Chapter 11. After determinants have been introduced in the next chapter we shall then indicate how the theory of determinants may be used as an alternate approach to the subject.

Section 13.2 is devoted to a systematic procedure which can be used to determine whether or not a given system of linear equations has a solution and, if it does have a solution, for finding all solutions. However, we shall be primarily interested in obtaining theoretical results such as various conditions under which a system of linear equations will have a solution. Matrices will be introduced in Section 13.3 and a few simple properties will be established. Finally, these properties will be used to obtain additional results about systems of linear equations.

13.1 NOTATION AND SIMPLE RESULTS

If F is a given field and $x_1, x_2, \ldots, x_n$ are indeterminates, it is customary to call a polynomial of the form $a_1x_1 + a_2x_2 + \cdots + a_nx_n$, where the coefficients are elements of F, a *linear form* (over F). Using the sigma notation for sums, such a linear form may be written as

$$\sum_{k=1}^{n} a_k x_k.$$

The linear form in which all coefficients are zero may be called the *zero linear form* and designated simply by 0.

Now let m and n be positive integers and let us consider a system of linear equations which we may write in the following explicit form:

13.1
$$
\begin{aligned}
a_{11}x_1 + a_{12}x_2 + \cdots + a_{1n}x_n &= b_1, \\
a_{21}x_1 + a_{22}x_2 + \cdots + a_{2n}x_n &= b_2, \\
\cdots \cdots \cdots \cdots \cdots \cdots \cdots \cdots & \\
a_{m1}x_1 + a_{m2}x_2 + \cdots + a_{mn}x_n &= b_m.
\end{aligned}
$$

It is understood that all the coefficients a_{ij}, as well as the *constant terms* b_i, are elements of a given field F. If we wish to emphasize the particular field involved, we shall speak of a system of linear equations *over F*.

As a matter of notation, it should be observed that a_{ij} is the coefficient of x_j in the ith equation. That is, the first subscript of a_{ij} specifies the equation and the second one the indeterminate of which it is the coefficient.

Now the left sides of Equation 13.1 are linear forms over F, and the right sides are elements of F. Since a linear form over F is not an element of F, it is clear that the "=" is being used in a different sense than heretofore. However, it is common practice to write Equations 13.1 with the understanding that what we really mean is that we are seeking elements $t_1, t_2, \ldots, t_n$ of F such that if we replace x_i by t_i ($i = 1, 2, \ldots, n$), each equation will yield a true equality of elements of F. Then, if $t_1, t_2, \ldots, t_n$ are such elements, it is customary to say that $x_1 = t_1, x_2 = t_2, \ldots, x_n = t_n$ is a *solution* of the Equations 13.1; also, that in this solution x_i has the *value* t_i ($i = 1, 2, \ldots, n$). Of course, it is quite possible for a given system of linear equations not to have any solution. In fact, we shall eventually obtain various tests for the existence of a solution.

The symbols $x_1, x_2, \ldots, x_n$ are often called *unknowns*, and we may therefore call 13.1 *a system of m linear equations in n unknowns*.

There are various alternative ways of writing the Equations 13.1, each of which is convenient for certain purposes. First, we observe that we may write them in the following form:

13.2
$$\sum_{j=1}^{n} a_{ij}x_j = b_i, \qquad (i = 1, 2, \ldots, m).$$

In this form, each value of i gives one equation, and in each equation j takes values from 1 to n. We next introduce a quite different notation which will be most useful.

Let us define elements of the vector space $V_n(F)$ as follows:

13.3
$$
\begin{aligned}
A_1 &= (a_{11}, a_{12}, \ldots, a_{1n}), \\
A_2 &= (a_{21}, a_{22}, \ldots, a_{2n}), \\
&\ \cdot\ \cdot\ \cdot\ \cdot\ \cdot\ \cdot\ \cdot\ \cdot\ \cdot \\
A_m &= (a_{m1}, a_{m2}, \ldots, a_{mn}).
\end{aligned}
$$

It will be noted that A_i is composed of the coefficients in the ith equation of the system 13.1. Let us also formally write $X = (x_1, x_2, \ldots, x_n)$. Then in terms of the inner product of vectors as defined in the last section of Chapter 11, we may write the system 13.1 in the following form:

13.4
$$A_1 \cdot X = b_1, \ A_2 \cdot X = b_2, \ldots, A_m \cdot X = b_m.$$

A *solution* of this system of equations is then an element $T = (t_1, t_2, \ldots, t_n)$ of $V_n(F)$ such that

$$A_1 \cdot T = b_1, \ A_2 \cdot T = b_2, \ldots, A_m \cdot T = b_m.$$

It is now fairly easy to give one condition which must hold provided this system of linear equations has a solution. First, let us make the following definition.

13.5 DEFINITION The system 13.4 of linear equations is said to be *compatible* if for every choice of elements s_i of F such that $\sum_{i=1}^{m} s_i A_i = 0$, then necessarily $\sum_{i=1}^{m} s_i b_i = 0$ also.

We shall now show that if the system 13.4 of linear equations has a solution, the system is compatible. For if T is a solution and

$$\sum_{i=1}^{m} s_i A_i = 0,$$

it follows from Theorem 11.59 that

$$\sum_{i=1}^{m} s_i b_i = \sum_{i=1}^{m} s_i(A_i \cdot T) = \sum_{i=1}^{m} (s_i A_i) \cdot T = \left(\sum_{i=1}^{m} s_i A_i\right) \cdot T = 0 \cdot T = 0,$$

and the system is therefore compatible.

It should be clear that when we say that a system is compatible, we are only saying in a precise way that if the system has a solution and if there exist elements of F such that when we multiply both sides of the equations by these elements of F and add corresponding members we obtain the zero linear form on the left, then we must also obtain the zero element of F on the right. Later on we shall prove the converse of what was proved above, that is, we shall prove that if a system of linear equations is compatible, the system necessarily has a solution. This is a fairly deep result and in order to prove it we shall have to wait until the proper machinery is available.

There is still another useful way of writing Equations 13.1 (or 13.2 or 13.4). First, we shall introduce elements of $V_m(F)$ consisting of coefficients in our system of equations that lie in a fixed vertical line. Let us set

13.6
$$A^1 = \begin{bmatrix} a_{11} \\ a_{21} \\ \vdots \\ a_{m1} \end{bmatrix}, A^2 = \begin{bmatrix} a_{12} \\ a_{22} \\ \vdots \\ a_{m2} \end{bmatrix}, \ldots, A^n = \begin{bmatrix} a_{1n} \\ a_{2n} \\ \vdots \\ a_{mn} \end{bmatrix}, B^1 = \begin{bmatrix} b_1 \\ b_2 \\ \vdots \\ b_m \end{bmatrix}.$$

The elements 13.3 of $V_n(F)$ are written as *row vectors*, and we may call these elements 13.6 of $V_m(F)$ *column vectors* to indicate that we have used a vertical arrangement instead of a horizontal one. Heretofore we have always used row vectors only because they are simpler to write. When we have occasion to use column vectors we shall use superscripts (not to be confused with exponents) to indicate this fact. Now in terms of the column vectors defined in 13.6 the entire system 13.1 of linear equations can be written in the following vector form:

13.7
$$x_1 A^1 + x_2 A^2 + \cdots + x_n A^n = B^1.$$

In the notation used here we may now observe that *the Equation* 13.7 (or the system 13.1) *has a solution if and only if the vector B^1 is a linear combination of the vectors $A^1, A^2, \ldots, A^n$.* In view of Theorem 11.15(ii) and Theorem 11.25(i), we can restate this result in the following useful form.

13.8   THEOREM   *The Equation 13.7 (or the system 13.1) has a solution if and only if*

$$\dim [A^1, A^2, \ldots, A^n, B^1] = \dim [A^1, A^2, \ldots, A^n].$$

The usual method of finding solutions of a given system of linear equations involves a procedure for simplifying the form of the system. The following concept is a useful one in this connection.

13.9 DEFINITION Two systems of linear equations are said to be *equivalent* if they have exactly the same solutions.

There are simple operations on a system of linear equations that always yield equivalent systems. In order to simplify the wording, we shall speak of multiplying an equation by an element of F to mean the multiplying of both members of the equation by this element. Also, we shall speak of adding two equations to mean the adding of corresponding members of the two equations.

13.10 DEFINITION We shall say that we perform an *elementary operation* on a system of linear equations if we do any one of the following:

Type 1. Interchange two equations.

Type 2. Multiply an equation by a nonzero element of F.

Type 3. Add to one equation d times a different equation, where $d \in F$.

As suggested by what was said above, the importance of these elementary operations stems from the following theorem.

13.11 THEOREM *If a system of linear equations is obtained from a given system by applying in turn a finite number of elementary operations, the two systems are equivalent.*

Proof: It is obvious that we need only show that *one* elementary operation always yields an equivalent system. The desired result is quite trivial for an elementary operation of Type 1 or 2, and so we consider an elementary operation of Type 3. Let us write our given system of equations in the form 13.4 and, for convenience of notation, suppose that the operation consists of multiplying the second equation by $d \in F$ and adding it to the first equation. Using the fact that

$$A_1 \cdot X + d(A_2 \cdot X) = (A_1 + dA_2) \cdot X,$$

the resulting system can be written in the following form:

13.12 $(A_1 + dA_2) \cdot X = b_1 + db_2, \ A_2 \cdot X = b_2, \ldots, A_m \cdot X = b_m.$

Now if T is a solution of 13.4 so that, in fact, $A_i \cdot T = b_i$ for $i = 1, 2, \ldots, m$ it is clear that T is also a solution of 13.12. Conversely, let T be a solution of 13.12, so that, in particular, $(A_1 + dA_2) \cdot T = b_1 + db_2$ and $A_2 \cdot T = b_2$. Since

$$(A_1 + dA_2) \cdot T = A_1 \cdot T + d(A_2 \cdot T) = b_1 + db_2,$$

it follows that $A_1 \cdot T = b_1$. Moreover, the two systems 13.4 and 13.12 are identical except for the first equation, and hence T is a solution of every equation of the system 13.4. We have thus shown that every solution of either of these systems is a solution of the other and they are therefore equivalent.

If one of the equations of a system is the *zero equation*; that is, if all the coefficients and the constant term are zero, the deletion of this equation from the system will clearly yield an equivalent system. We shall not call this an elementary operation, but we shall use this operation without hesitation whenever it is helpful to do so.

Theorem 13.11, together with the operation of deletion of one or more zero equations, may be applied in a systematic way to find all solutions (if any) of a given system. What we do is to apply a sequence of operations until a system is obtained in a simple enough form that the solutions are apparent. These solutions must then be the solutions of the original system. Although it is a very simple procedure, we shall find that it leads to interesting theoretical results as well as to a practical method for actually finding the solutions of a given system. This procedure will be discussed in detail in the following section.

13.2 ECHELON SYSTEMS

We shall begin by discussing two examples which will clarify the ideas involved. First, let us consider the following system of four equations in five unknowns over the field $\mathbf{Q}$ of rational numbers:

13.13
$$\begin{aligned}
x_1 + x_2 + x_3 + 2x_4 + 3x_5 &= 13, \\
-2x_1 - 2x_2 + x_3 + 3x_4 - 4x_5 &= 5, \\
3x_1 + 3x_2 \quad\quad + x_4 + 5x_5 &= 10, \\
x_1 + x_2 + 2x_3 - x_4 + 9x_5 &= 18.
\end{aligned}$$

In order to eliminate x_1 from all equations but the first, we perform, in turn, the following three elementary operations of Type 3 on this system. First, we multiply the first equation by 2 and add it to the second equation, then multiply the first equation by -3 and add it to the third, and finally multiply the first equation by

−1 and add it to the last equation. We thus obtain the following system which, by Theorem 13.11, must be equivalent to the system 13.13:

$$x_1 + x_2 + x_3 + 2x_4 + 3x_5 = 13,$$
$$3x_3 + 7x_4 + 2x_5 = 31,$$
$$-3x_3 - 5x_4 - 4x_5 = -29,$$
$$x_3 - 3x_4 + 6x_5 = 5.$$

We now proceed to eliminate x_3 from all these equations but one. Since in the last equation the coefficient of x_3 is unity, we shall work with this equation and, in order to systematize the procedure, we use an elementary operation of Type 1 to interchange the second and fourth equation in order to get this equation in the second position. We then eliminate x_3 from all the equations except this new second one by multiplying it in turn by the proper elements of $\mathbf{Q}$ and adding to the other equations. In this way we obtain the system

$$x_1 + x_2 + 5x_4 - 3x_5 = 8,$$
$$x_3 - 3x_4 + 6x_5 = 5,$$
$$-14x_4 + 14x_5 = -14,$$
$$16x_4 - 16x_5 = 16.$$

It is now clear that the third equation (as well as the fourth) can be simplified by an elementary operation of Type 2. Accordingly, we divide the third equation by -14 (multiply it by $-1/14$), and then proceed as above to eliminate x_4 from all equations except the third. We thus obtain the following system of equations:

$$x_1 + x_2 + 2x_5 = 3,$$
$$x_3 + 3x_5 = 8,$$
$$x_4 - x_5 = 1,$$
$$0 = 0.$$

Finally, we omit the last equation and rewrite the system in the following form:

13.14
$$x_1 = 3 - x_2 - 2x_5,$$
$$x_3 = 8 \qquad - 3x_5,$$
$$x_4 = 1 \qquad + x_5.$$

This system is equivalent to the given system 13.13, and in the present form its solutions are apparent. If x_2 and x_3 are replaced by arbitrary elements of $\mathbf{Q}$, and the other unknowns are calculated from 13.14, we obtain a solution; and all solutions can be obtained in this way. To express this fact algebraically, let s and t be arbitrary elements of $\mathbf{Q}$. Then

$$x_1 = 3 - s - 2t, \quad x_2 = s, \quad x_3 = 8 - 3t, \quad x_4 = 1 + t, \quad x_5 = t$$

is a solution of the system 13.14, and any solution is of this form for suitable choices of s and t. We have therefore found all solutions of the given system 13.13 since it is equivalent to the system 13.14.

In the example just discussed, we found many solutions of the given system of equations. In other cases there might be exactly one solution or no solution. As a simple illustration of what happens when there is no solution, let us attempt to solve the following system of equations by the same method used above:

13.15
$$\begin{aligned} x_1 + x_2 - x_3 &= 5, \\ x_1 + 2x_2 - 3x_3 &= 8, \\ x_2 - 2x_3 &= 1. \end{aligned}$$

First, we multiply the first equation by -1 and add it to the second equation, obtaining the following system:

$$\begin{aligned} x_1 + x_2 - x_3 &= 5, \\ x_2 - 2x_3 &= 3, \\ x_2 - 2x_3 &= 1. \end{aligned}$$

Of course, it is obvious that this system has no solution, but let us proceed to eliminate x_2 from the first and third equations. After doing this, we have the system

$$\begin{aligned} x_1 + x_3 &= 2, \\ x_2 - 2x_3 &= 3, \\ 0 &= -2. \end{aligned}$$

Finally, in order to standardize the procedure, we divide the third equation by -2 and then multiply the new third equation by -2 and add it to the first; and also multiply the new third equation by -3 and add it to the second. We obtain in this way the following system of equations which is equivalent to the given system 13.15:

13.16
$$\begin{aligned} x_1 + x_3 &= 0, \\ x_2 - 2x_3 &= 0, \\ 0 &= 1. \end{aligned}$$

Now the symbol "0" on the left in the last of these equations stands for the zero linear form $0x_1 + 0x_2 + 0x_3$. It is clear that if the unknowns in the zero linear form are replaced by arbitrary elements of the field $\mathbf{Q}$, we obtain the zero element of $\mathbf{Q}$. Hence the last equation has no solution, and therefore the system 13.16 has no solution. We shall presently see that any system of linear equations which does not have a solution is reducible to a system in which one of the equations takes the form $0 = 1$.

We have considered these two examples in detail as illustrations of the procedure which can be applied to any system of linear equations. Suppose, now, that we wish to solve an arbitrary system 13.1 of m linear equations in n unknowns. We shall assume that not *all* the coefficients a_{ij} are zero since, otherwise, the problem is completely trivial. Moreover, if all the coefficients of x_1 were zero, we could just as well consider only $x_2, \ldots, x_n$ as the unknowns and ignore x_1. However, to use a general notation, let us suppose that x_{k_1} is the *first* one of the unknowns $x_1, x_2, \ldots, x_n$ which has a nonzero coefficient in any of the equations. Normally, of course, we will have $k_1 = 1$. By interchanging equations, if necessary, we can obtain an equivalent system in which the coefficient of x_{k_1} is not zero in the *first* of our equations. Then, by dividing the first equation by the coefficient of x_{k_1} in this equation, we can obtain unity as the coefficient of x_{k_1}. Next, by elementary operations of Type 3, as in the examples, we can obtain a system in which the coefficient of x_{k_1} is zero in every equation other than the first. The system is then of the following type:

$$x_{k_1} + b_{1\,k_1+1}x_{k_1+1} + \cdots + b_{1n}x_n = c_1,$$
$$b_{2\,k_1+1}x_{k_1+1} + \cdots + b_{2n}x_n = c_2,$$
$$\cdots \cdots \cdots \cdots \cdots \cdots$$
$$b_{m\,k_1+1}x_{k_1+1} + \cdots + b_{mn}x_n = c_m.$$

Now suppose that x_{k_2} is the first one of the unknowns $x_{k_1+1}, \ldots, x_n$ which has a nonzero coefficient in any of these equations *except the first*. By interchanging equations, if necessary, we can be sure that the coefficient of x_{k_2} is not zero in the second equation. We can then use elementary operations to make this coefficient unity and the coefficient of x_{k_2} zero in all equations except the second. By continuing this process, we find that for some positive integer r, with $r \leq m$ and $r \leq n$, there exist positive integers $k_1 < k_2 < \cdots < k_r$ such that our system of equations can be reduced by use of elementary operations to a system of the following form:

$$x_{k_1} + \cdots + 0x_{k_2} + \cdots + 0x_{k_r} + \cdots = d_1,$$
$$x_{k_2} + \cdots + 0x_{k_r} + \cdots = d_2,$$
$$\cdots \cdots \cdots \cdots \cdots \cdots$$

13.17
$$x_{k_r} + \cdots = d_r,$$
$$0 = d_{r+1},$$
$$\cdots \cdots \cdots$$
$$0 = d_m.$$

In this system if i is an integer such that $1 \leq i \leq r$, the coefficient of x_{k_i} is different from zero *only* in the ith equation. For example, the coefficients of $x_{k_2}, \ldots, x_{k_r}$ are all zero in the first equation, but if there are any other unknowns, their coefficients need not be zero. In the second equation the coefficient of x_j is zero for all $j < k_2$; also, the coefficients of $x_{k_3}, \ldots, x_{k_r}$ are zero. Corresponding remarks hold for the other equations. If it happens that $r = m$, then there will be no equations with 0 as left member.

Let us now assume that the given system 13.1, and therefore also the equivalent system 13.17, has a solution. It is then apparent that we must have $d_{r+1} = \cdots = d_m = 0$. Accordingly, we conclude that *if* the system 13.1 has a solution, it can be reduced by elementary operations, together with the possible deletion of zero equations, to a system of the form

13.18
$$
\begin{aligned}
x_{k_1} + \cdots + 0x_{k_2} + \cdots + 0x_{k_r} + \cdots &= d_1, \\
x_{k_2} + \cdots + 0x_{k_r} + \cdots &= d_2, \\
\cdot \quad \cdot \quad \cdot \quad \cdot \quad \cdot \quad \cdot \quad \cdot \quad \cdot \quad \cdot \quad \cdot & \\
x_{k_r} + \cdots &= d_r,
\end{aligned}
$$

with the same understanding about zero coefficients as in 13.17. We now observe that it is easy to find all solutions of the system 13.18. If we assign arbitrary values from the field F to the unknowns (if any) other than $x_{k_1}, x_{k_2}, \ldots, x_{k_r}$, the value of x_{k_1} is uniquely determined from the first equation, the value of x_{k_2} is uniquely determined from the second equation, and so on. Not only does this procedure give a method of finding solutions but, since the values of $x_{k_1}, x_{k_2}, \ldots, x_{k_r}$ are uniquely determined by the values of the other unknowns, it follows that *every* solution of the system 13.18 can be obtained by the process just described.

Let us next assume that the given system 13.1, and therefore also the equivalent system 13.17, does not have a solution. If $d_{r+1} = \cdots = d_m = 0$, the argument given above would show that the system has a solution. Accordingly, at least one of the equations in 13.17 must be of the form $0 = d$ with $d \neq 0$. For theoretical purposes it is helpful to continue our simplification procedure a little further as follows. By interchanging equations, if necessary, we can consider the equation $0 = d$ to be the first one of those with 0 as left member. Then, by elementary operations, we can write this equation in the form $0 = 1$ and also reduce the constant terms in all other equations to 0. Finally, after deleting any zero equations, our system takes the following form:

13.19
$$
\begin{aligned}
x_{k_1} + \cdots + 0x_{k_2} + \cdots + 0x_{k_r} + \cdots &= 0, \\
x_{k_2} + \cdots + 0x_{k_r} + \cdots &= 0, \\
\cdot \quad \cdot \quad \cdot \quad \cdot \quad \cdot \quad \cdot \quad \cdot \quad \cdot \quad \cdot \quad \cdot & \\
x_{k_r} + \cdots &= 0, \\
0 &= 1.
\end{aligned}
$$

A system of linear equations either of the form 13.18 or of the form 13.19 is said to be an *echelon* system.

The conclusions we have reached above may be summarized in the following theorem.

13.20 THEOREM *Any system 13.1 of linear equations with coefficients over a field F can be reduced by means of elementary operations, and the possible deletion of zero equations, to an echelon system.*

The given system has a solution if and only if it is reducible to an echelon system of the form 13.18. In this case, the unknowns (if any) other than $x_{k_1}, x_{k_2}, \ldots, x_{k_r}$, can be assigned arbitrary values from F, and the corresponding values of $x_{k_1}, x_{k_2}, \ldots, x_{k_r}$, are then uniquely determined from the Equations 13.18. Moreover, all solutions can be obtained in this way.

The given system does not have a solution if and only if it is reducible to an echelon system of the form 13.19.

If $r = n$, the Equations 13.18 take the form $x_1 = d_1, x_2 = d_2, \ldots, x_n = d_n$; and the system clearly has exactly one solution. If $r < n$, there is at least one unknown which can be assigned arbitrary values; hence there will be more than one solution. The following result is therefore an immediate consequence of the preceding theorem.

13.21 COROLLARY *A system 13.1 of linear equations has a* unique *solution if and only if the number r of equations in the corresponding echelon system 13.18 is equal to the number n of unknowns.*

EXERCISES

In each of Exercises 1 to 6, reduce the given system of linear equations over the field **Q** to an echelon system, and find all solutions.

1. $x_1 - x_2 + 2x_3 = 5,$
$2x_1 + x_2 - x_3 = 2,$
$2x_1 - x_2 - x_3 = 4,$
$x_1 + 3x_2 + 2x_3 = 1.$

2. $2x_1 - 2x_2 + 2x_3 - 3x_4 = 11,$
$x_1 - x_2 + x_3 - 2x_4 = 4,$
$-x_1 + x_2 + 6x_4 = -3,$
$9x_1 - 9x_2 + 7x_3 - 23x_4 = 43.$

3. $2x_1 - x_2 + x_3 = 4,$
$3x_1 + x_2 - 5x_3 = 1,$
$x_1 - 5x_2 + 8x_3 = 5,$
$2x_1 - 2x_2 + x_3 = 3.$

4. $2x_1 - 4x_2 + 3x_3 - 5x_4 = -3,$
$4x_1 + 2x_2 + x_3 \quad\quad = 4,$
$6x_1 - 2x_2 + 4x_3 - 5x_4 = 1.$

5. $2x_1 - x_2 - x_3 + x_4 + 8x_5 = -4,$
$x_1 + x_2 + 4x_3 + x_4 + 4x_5 = 3,$
$-3x_1 + 2x_2 + 3x_3 + x_4 - 5x_5 = 3,$
$2x_1 - 2x_2 - 3x_3 + 2x_4 + 12x_5 = -10.$

6. $3x_1 - x_2 + x_3 - x_4 = 1,$
$2x_1 + 3x_2 - x_3 + 5x_4 = 7,$
$2x_1 + 5x_2 - x_3 + 12x_4 = 8,$
$x_1 - 2x_2 + 2x_3 + x_4 = 5.$

In each of Exercises 7–9, the equations are over the field $\mathbf{Z}_3$. Reduce to an echelon system, find the number of solutions in each case, and actually exhibit all of them.

7. $x_1 + 2x_2 + x_3 = 2,$
$2x_1 + x_2 + 2x_3 = 1,$
$x_1 + x_2 + x_3 = 2.$

8. $x_1 + 2x_2 + x_3 \quad\quad = 0,$
$x_2 + x_3 + x_4 = 0,$
$2x_1 + 2x_2 + x_3 \quad\quad = 2,$
$2x_2 \quad\quad + x_4 = 2.$

9. $2x_1 + x_2 + 2x_3 + x_4 = 1,$
$x_1 + 2x_2 + x_3 + 2x_4 = 2.$

In each of Exercises 10–12, apply Definition 13.5 to show that the given system of linear equations over the field $\mathbf{Q}$ is *not* compatible. Verify also that the system does not have a solution.

10. $x_1 + x_2 + 3x_3 = 4,$
$2x_1 + 2x_2 + 6x_3 = 12.$

11. $2x_1 - x_2 + x_3 = 1,$
$x_1 + x_2 - 3x_3 = 2,$
$5x_1 - x_2 - x_3 = 1.$

12. $x_1 - x_2 + x_3 - x_4 = 2,$
$x_1 + x_2 + x_3 + x_4 = 1,$
$2x_1 \quad\quad + 2x_3 \quad\quad = 3,$
$x_1 + 3x_2 + x_3 + 3x_4 = 2.$

13.3 MATRICES

The concept of a matrix arises naturally in mathematics in several different ways. For example, in carrying out the calculations of the preceding section it is obvious that we could avoid writing down the unknowns, as well as the signs of addition and equality, and just work with the array of coefficients and constant terms. This possibility suggests the introduction of some suitable notation and terminology for discussing rectangular arrays of elements of a field. We proceed to give the necessary definition and to establish those properties which will be particularly useful in the study of systems of linear equations. Matrices will be studied in somewhat more detail in Chapter 15.

Let p and q be positive integers, and let c_{ij} $(i = 1, 2, \ldots, p; j = 1, 2, \ldots, q)$ be elements of a field F. The rectangular array C defined by

13.22
$$C = \begin{bmatrix} c_{11} & c_{12} & \cdots & c_{1q} \\ c_{21} & c_{22} & \cdots & c_{2q} \\ \multicolumn{4}{c}{\cdots\cdots\cdots\cdots} \\ c_{p1} & c_{p2} & \cdots & c_{pq} \end{bmatrix}$$

is called a *matrix* (over F) with p rows and q columns, or simply a $p \times q$ matrix. As suggested by this language, the elements of the matrix C that occur in a horizontal line constitute a *row* of the matrix, and those in a vertical line a *column* of the matrix. It will be observed that an element c_{ij} occurs at the intersection of the ith row and jth column.

The rows of the matrix C may be considered to be elements of the vector space $V_q(F)$. More precisely, we may define the *row vectors* of C to be the following vectors of $V_q(F)$:

$$C_1 = (c_{11}, c_{12}, \ldots, c_{1q}),$$
$$C_2 = (c_{21}, c_{22}, \ldots, c_{2q}),$$
$$\cdots\cdots\cdots\cdots$$
$$C_p = (c_{p1}, c_{p2}, \ldots, c_{pq}).$$

Similarly, the *column vectors* of C are defined to be the following vectors of $V_p(F)$:

$$C^1 = \begin{bmatrix} c_{11} \\ c_{21} \\ \vdots \\ c_{p1} \end{bmatrix}, C^2 = \begin{bmatrix} c_{12} \\ c_{22} \\ \vdots \\ c_{p2} \end{bmatrix}, \ldots, C^q = \begin{bmatrix} c_{1q} \\ c_{2q} \\ \vdots \\ c_{pq} \end{bmatrix}.$$

It will be observed that a row vector of the matrix C may be considered to be a matrix with one row and q columns, and likewise a column vector may be considered to be a matrix with p rows and one column.

The concepts introduced in the following definition play an important role in the sequel.

13.23 DEFINITION   The subspace of $V_q(F)$ generated by the row vectors of C, that is, in the notation of 11.13, the subspace $[C_1, C_2, \ldots, C_p]$ of $V_q(F)$, is called the *row space* of the matrix C.   The dimension of this row space is called the *row rank* of the matrix C.

Similarly, the subspace of $V_p(F)$ generated by the column vectors of C; that is, the subspace $[C^1, C^2, \ldots, C^q]$ of $V_p(F)$ is called the *column space* of C and its dimension is the *column rank* of C.

In order to compute the row rank, or the column rank, of a given matrix, we may use a method suggested by the way in which we simplified a system of linear equations in the preceding section.   Let us first define elementary operations on a matrix as follows (cf. 13.10).

13.24 DEFINITION   Let C be a matrix over a field F.   We shall say that we perform an *elementary row (column) operation* on C if we do any one of the following.

Type 1.   Interchange two rows (columns).

Type 2.   Multiply the elements of one row (column) by a nonzero element of F.

Type 3.   Add to the elements of one row (column) d times the corresponding elements of a different row (column), where $d \in F$.

We thus have three different types of elementary row operations, and three types of elementary column operations.   For simplicity of statement, we now restrict attention to elementary row operations.   From our present point of view, the importance of the elementary row operations is that if a matrix D is obtained from a matrix C by means of an elementary row operation, C and D have the same row spaces and therefore also the same row rank.   This fact is obvious for an elementary row operation of Type 1, and Corollary 11.16 assures us that it is also true for operations of Types 2 and 3.   Since the row rank of a matrix is not changed by application of *one* elementary row operation, the same must also be true for a finite sequence of elementary row operations.   Of course, similar remarks hold for column operations, and we summarize these remarks in the following theorem.

13.25 THEOREM   *If one matrix is obtained from another by a finite sequence of elementary row (column) operations, the two matrices have the same row space (column space), and therefore also the same row rank (column rank).*

In view of this theorem, we may compute the row rank of a given matrix by applying elementary row operations until a matrix is obtained in a form in which its row rank is apparent. This row rank must then also be the row rank of the given matrix. Let us illustrate this procedure by finding the row rank of the following matrix over the field $\mathbf{Q}$ of rational numbers:

13.26
$$\begin{bmatrix} 0 & -1 & 3 & -1 & 0 & 2 \\ -1 & 1 & -2 & -2 & 1 & -3 \\ 2 & -1 & 4 & 4 & -1 & 8 \\ 1 & -2 & 5 & 1 & -1 & 5 \end{bmatrix}.$$

It will be observed that the procedure is essentially that used in the preceding section to reduce a system of linear equations to an echelon system.

First, we get an element 1 in the upper left-hand position by interchanging the first two rows and then multiplying the new first row by -1. We then have the matrix

$$\begin{bmatrix} 1 & -1 & 2 & 2 & -1 & 3 \\ 0 & -1 & 3 & -1 & 0 & 2 \\ 2 & -1 & 4 & 4 & -1 & 8 \\ 1 & -2 & 5 & 1 & -1 & 5 \end{bmatrix}.$$

Next we use elementary row operations of Type 3 to reduce all elements of the first column, except this first element, to zero. More specifically, we multiply the first row by -2 and add it to the third row; also multiply the first row by -1 and add it to the fourth row. This gives us the matrix

$$\begin{bmatrix} 1 & -1 & 2 & 2 & -1 & 3 \\ 0 & -1 & 3 & -1 & 0 & 2 \\ 0 & 1 & 0 & 0 & 1 & 2 \\ 0 & -1 & 3 & -1 & 0 & 2 \end{bmatrix}.$$

Now if we add the third row to each of the other rows, and then interchange the second and third rows, we obtain

$$\begin{bmatrix} 1 & 0 & 2 & 2 & 0 & 5 \\ 0 & 1 & 0 & 0 & 1 & 2 \\ 0 & 0 & 3 & -1 & 1 & 4 \\ 0 & 0 & 3 & -1 & 1 & 4 \end{bmatrix}.$$

We next multiply the third row by -1 and add it to the last row, obtaining a row consisting entirely of zeros. We then multiply the third row by $1/3$ in order to get 1 as its first nonzero element. Finally, we multiply the third row by -2 and add it to the first row. This gives the matrix

$$\begin{bmatrix} 1 & 0 & 0 & 8/3 & -2/3 & 7/3 \\ 0 & 1 & 0 & 0 & 1 & 2 \\ 0 & 0 & 1 & -1/3 & 1/3 & 4/3 \\ 0 & 0 & 0 & 0 & 0 & 0 \end{bmatrix}.$$

Since the unit vectors $(1, 0, 0)$, $(0, 1, 0)$, and $(0, 0, 1)$ of $V_3(\mathbf{Q})$ are linearly independent, it is easy to see that the first three rows of this matrix are linearly independent, and hence that its row rank is 3. Since this matrix has been obtained from the given matrix by use of elementary row operations, it follows that the matrix 13.26 also has row rank 3.

In this example, we have computed the *row* rank of the given matrix. It is obvious that we could similarly apply elementary column operations to determine the column rank of the matrix. However, the next theorem is of considerable theoretical importance and will show us that both of these ranks are known as soon as one of them has been determined.

13.27 THEOREM   *The row rank of any matrix is equal to its column rank.*

Let C be the $p \times q$ matrix 13.22 over a field F, and suppose that the row rank C is r and that its column rank is s. The main part of the proof consists in showing that any $r + 1$ column vectors of C are linearly dependent, and hence that $s \leq r$. If $r = p$, this is clearly true since, by Corollary 11.19, any $p + 1$ vectors of $V_p(F)$ are linearly dependent. Suppose, then, that $r < p$. Since the row rank is r, by Theorem 11.25(ii) there must exist r row vectors that constitute a basis of the row space of C. Let us rearrange the rows (if necessary) so that the first r rows are a basis of the row space. It is not difficult to see that this cannot change the linear dependence or independence of column vectors, and hence it does not change the column rank. As a matter of notation, we may therefore assume, without loss of generality, that the *first* r rows of C form a basis of the row space of C. Since we wish to show that the column rank s of C cannot exceed r, we may certainly assume that $q > r$ since, otherwise, there is nothing to prove. Let us now write out matrix C in a somewhat more explicit form as follows:

$$C = \begin{bmatrix} c_{11} & c_{12} & \cdots & c_{1r} & c_{1\,r+1} & \cdots & c_{1q} \\ c_{21} & c_{22} & \cdots & c_{2r} & c_{2\,r+1} & \cdots & c_{2q} \\ \cdot & \cdot & \cdot & \cdot & \cdot & \cdot & \cdot \\ c_{r1} & c_{r2} & \cdots & c_{rr} & c_{r\,r+1} & \cdots & c_{rq} \\ \cdot & \cdot & \cdot & \cdot & \cdot & \cdot & \cdot \\ c_{p1} & c_{p2} & \cdots & c_{pr} & c_{p\,r+1} & \cdots & c_{pq} \end{bmatrix}.$$

As introduced earlier, let $C_1, C_2, \ldots, C_p$ be the row vectors, and $C^1, C^2, \ldots, C^q$ the column vectors of C. Moreover, let

$$S^1 = \begin{bmatrix} c_{11} \\ c_{21} \\ \vdots \\ c_{r1} \end{bmatrix}, \; S^2 = \begin{bmatrix} c_{12} \\ c_{22} \\ \vdots \\ c_{r2} \end{bmatrix}, \ldots, S^q = \begin{bmatrix} c_{1q} \\ c_{2q} \\ \vdots \\ c_{rq} \end{bmatrix}$$

be elements of $V_r(F)$ consisting of column vectors taken from the first r rows of C. We shall now prove the following lemma.

13.28 LEMMA   *If $t_i \in F$ ($i = 1, 2, \ldots, q$) such that*

$$\sum_{i=1}^{q} t_i S^i = 0, \quad \text{then also} \sum_{i=1}^{q} t_i C^i = 0.$$

Proof: If $T = (t_1, t_2, \ldots, t_q)$, then $\sum_{i=1}^{q} t_i S^i = 0$ can be expressed using inner products of vectors as follows:

13.29 $C_1 \cdot T = 0, \; C_2 \cdot T = 0, \ldots, C_r \cdot T = 0.$

Now if C_k is an arbitrary row vector of C, then C_k is a linear combination of the vectors $C_1, C_2, \ldots, C_r$, since they form a basis of the row space of C. That is, there exist elements a_j ($j = 1, 2, \ldots, r$) of F such that

$$C_k = a_1 C_1 + a_2 C_2 + \cdots + a_r C_r.$$

It follows from Theorem 11.35(iv) that

$$\begin{aligned} C_k \cdot T &= (a_1 C_1 + a_2 C_2 + \cdots + a_r C_r) \cdot T \\ &= a_1 (C_1 \cdot T) + a_2 (C_2 \cdot T) + \cdots + a_r (C_r \cdot T). \end{aligned}$$

Hence, in view of 13.29, we see that $C_k \cdot T = 0$. Again, changing the notation, the fact that $C_k \cdot T = 0$ for $k = 1, 2, \ldots, p$ assures us that

$$\sum_{i=1}^{q} t_i C^i = 0,$$

as we wished to prove. The lemma is therefore established.

It is now easy to complete the proof that any $r + 1$ column vectors of C are linearly dependent. Merely for convenience of notation, let us show that the first $r + 1$ column vectors $C^1, C^2, \ldots, C^{r+1}$ are linearly dependent. We know that the column vectors $S^1, S^2, \ldots, S^{r+1}$ must be linearly dependent since we have $r + 1$ vectors of $V_r(F)$. Hence, there exist elements b_i $(i = 1, 2, \ldots, r + 1)$ of F, not all zero, such that

$$\sum_{i=1}^{r+1} b_i S^i = 0.$$

The lemma then asserts that also

$$\sum_{i=1}^{r+1} b_i C^i = 0,$$

and since the coefficients are not all zero we have shown that the first $r + 1$ column vectors of C are linearly dependent. Similarly, any set of $r + 1$ column vectors of C is a linearly dependent set, and we have therefore shown that $s \leq r$.

We may restate what we have proved up to this point as follows. *The column rank of an arbitrary matrix cannot exceed the row rank of the matrix.* One simple way to complete the proof of the theorem is as follows. If C is the matrix 13.22, let us write a new matrix C', called the *transpose* of C, obtained from C by interchanging rows and columns. That is, C' is defined as follows:

$$C' = \begin{bmatrix} c_{11} & c_{21} & \cdots & c_{p1} \\ c_{12} & c_{22} & \cdots & c_{p2} \\ \cdot & \cdot & \cdot \cdot \cdot \cdot & \cdot \\ c_{1q} & c_{2q} & \cdots & c_{pq} \end{bmatrix}.$$

Now the row space and column space of the matrix C' are, respectively, the column space and the row space of C. Hence, the row rank of the matrix C' is s and its column rank is r. Since we have already proved that the column rank of a matrix can never exceed the row rank, we see that $r \leq s$. Moreover, we already know that $s \leq r$, and we conclude that $s = r$. This completes the proof of the theorem.

In view of this theorem, we need not distinguish between the row rank and the column rank of a matrix. Accordingly, we make the following definition.

13.30 DEFINITION   The common value of the row rank and the column rank of a matrix is called simply the *rank* of the matrix.

We shall sometimes find it convenient to refer to the (row) rank of a matrix by which we shall mean that the truth of the statement we are making is most easily seen by consideration of the *row* rank.

EXERCISES

In each of Exercises 1–5, use elementary row operations to determine the (row) rank of the given matrix over the rational field.

1. $\begin{bmatrix} -1 & 2 & 1 \\ 1 & -1 & 2 \\ 1 & 1 & 4 \end{bmatrix}$.

2. $\begin{bmatrix} 2 & -1 & 1 \\ 3 & 1 & 2 \\ 1 & 0 & -1 \\ 0 & 1 & 0 \end{bmatrix}$.

3. $\begin{bmatrix} 1 & 3 & -1 & 2 \\ 0 & 1 & 2 & -5 \\ 2 & 3 & -8 & 19 \end{bmatrix}$.

4. $\begin{bmatrix} 0 & 1 & -1 & 2 \\ 2 & -1 & 0 & 1 \\ 1 & 1 & 1 & 1 \end{bmatrix}$.

5. $\begin{bmatrix} 1 & 2 & -1 & 3 \\ 4 & 1 & 2 & 1 \\ 3 & -1 & 1 & 2 \\ 1 & 2 & 0 & 1 \end{bmatrix}$.

6. Use elementary row operations to determine the rank of the following matrix over the field $\mathbf{Z}_5$:

$$\begin{bmatrix} 2 & 1 & 1 & 3 & 4 \\ 4 & 1 & 2 & 1 & 2 \\ 1 & 2 & 1 & 1 & 2 \\ 1 & 3 & 0 & 2 & 1 \end{bmatrix}.$$

7. Show that, by use of both elementary row and column operations, the matrix

$$\begin{bmatrix} -1 & -2 & 1 & 3 \\ 1 & 0 & 1 & 1 \\ 3 & 2 & 1 & -1 \\ 1 & -4 & 5 & 9 \end{bmatrix}$$

over the rational field can be reduced to the matrix

$$\begin{bmatrix} 1 & 0 & 0 & 0 \\ 0 & 1 & 0 & 0 \\ 0 & 0 & 0 & 0 \\ 0 & 0 & 0 & 0 \end{bmatrix}.$$

8. Determine the dimension of each of the following subspaces of the appropriate vector space $V_n(\mathbf{Q})$. *Hint:* Form a matrix whose row vectors are the given vectors and find the (row) rank of the matrix.

    (a) $[(1, 2, -1), (3, 1, 2), (1, -3, 4)]$,

    (b) $[(3, -1, 4), (2, 1, 3), (1, 0, 2)]$,

    (c) $[(0, 1, 1, 2), (-2, 1, 0, 1), (3, 1, 5, 2), (1, 0, 3, -1)]$,

    (d) $[(-1, 2, -1, 0), (0, 3, 1, 2), (1, 1, -2, 2), (2, 1, 0, -1)]$.

13.4 APPLICATIONS TO SYSTEMS OF LINEAR EQUATIONS

We shall now apply the results of the preceding section to a further study of systems of linear equations.

Associated with the system 13.1 of m linear equations in n unknowns are two matrices. The $m \times n$ matrix

$$\begin{bmatrix} a_{11} & a_{12} & \cdots & a_{1n} \\ a_{21} & a_{22} & \cdots & a_{2n} \\ \cdot & \cdot & \cdots & \cdot \\ a_{m1} & a_{m2} & \cdots & a_{mn} \end{bmatrix}$$

consisting of the coefficients of the various unknowns, is called the *matrix of the coefficients* of the given system. The $m \times (n + 1)$ matrix

$$\begin{bmatrix} a_{11} & a_{12} & \cdots & a_{1n} & b_{1} \\ a_{21} & a_{22} & \cdots & a_{2n} & b_{2} \\ \cdot & \cdot & \cdots & \cdot & \cdot \\ a_{m1} & a_{m2} & \cdots & a_{mn} & b_{m} \end{bmatrix}$$

is called the *augmented matrix* of the system.   The augmented matrix differs from the matrix of the coefficients only in that it has an additional column which consists of the constant terms of the equation.

In the notation introduced in 13.6, the column vectors of the augmented matrix are $A^1, A^2, \ldots, A^n, B^1$.   Since dim $[A^1, A^2, \ldots, A^n, B^1]$ is the (column) rank of the augmented matrix, we have at once the following consequence of Theorem 13.8.

13.31 THEOREM   *A system 13.1 of linear equations over a field F has a solution if and only if the rank of the matrix of the coefficients of the system is equal to the rank of the augmented matrix of the system.*

We can obtain some further information by again considering the reduction of a given system to an echelon system.   An elementary operation on a system of linear equations, as defined in 13.10, has the same effect on the augmented matrix of the system as the corresponding elementary row operation on this matrix, as defined in 13.24.   Hence, if the given system is reduced to an echelon system of equations by a sequence of elementary operations and, possibly, the deletion of zero equations, we know that the augmented matrix of the echelon system can be obtained from the augmented matrix of the given system by a sequence of elementary row operations and the possible deletion of zero rows.

In view of these observations, we see that the process of reducing a given system of linear equations to an echelon system can be carried out by working with the augmented matrix of the system.   For example, suppose that we wish to solve the following system of linear equations over the rational field:

$$\begin{aligned}
x_1 + 2x_2 + 2x_3 &= 5, \\
x_1 - 3x_2 + 2x_3 &= -5, \\
2x_1 - x_2 + x_3 &= -3.
\end{aligned}$$

The augmented matrix of this system is the matrix

$$\begin{bmatrix} 1 & 2 & 2 & 5 \\ 1 & -3 & 2 & -5 \\ 2 & -1 & 1 & -3 \end{bmatrix}.$$

We omit the details, but by use of elementary row operations this matrix can be reduced to the form

$$\begin{bmatrix} 1 & 0 & 0 & -1 \\ 0 & 1 & 0 & 2 \\ 0 & 0 & 1 & 1 \end{bmatrix}.$$

This is clearly the augmented matrix of the following echelon system of equations, which must be equivalent to the given system:

$$x_1 \qquad\qquad = -1,$$
$$x_2 \qquad = 2,$$
$$x_3 = 1.$$

Accordingly, we conclude that the only solution of the given system is $x_1 = -1$, $x_2 = 2$, $x_3 = 1$.

As indicated by this example, there is some economy of effort in working with the augmented matrix instead of with the equations themselves. However, it is of more importance that we can apply these ideas to obtain some general theoretical results.

Since the presence or absence of a zero row cannot affect the row space of a matrix, we conclude from Theorem 13.25 that the row space of the augmented matrix of a given system of linear equations is equal to the row space of the augmented matrix of the echelon system 13.18 or 13.19 to which the given system is equivalent. Moreover, an elementary row operation on the augmented matrix of a system of linear equations induces an elementary row operation on the matrix of coefficients as well.

If a system 13.1 of linear equations has a solution, we know that it is reducible to an echelon system of the form 13.18. The (row) rank of the matrix of the coefficients, and also of the augmented matrix, of this echelon system is easily seen to be r. Hence, in view of the observations made above, and Theorem 13.25, we know that the ranks of the matrix of the coefficients and of the augmented matrix of the given system 13.1 must both be equal to r. If the given system does not have a solution, it is reducible to an echelon system of the form 13.19. For this system, the (row) rank of the matrix of coefficients is r, whereas the (row) rank of the augmented matrix is $r + 1$.

These observations, together with Theorem 13.20, yield the following result.

13.32 THEOREM *Suppose that the augmented matrix and the matrix of coefficients of a system 13.1 of linear equations over a field F both have rank r. If r is less than the number n of unknowns, certain n − r of the unknowns can be assigned arbitrary values from F and the values of the other r unknowns are then uniquely determined. Moreover, all solutions can be obtained in this way. The system has a unique solution if and only if r = n.*

Let us now return to a consideration of the compatibility condition as defined in 13.5. We have already proved that if a system of linear equations has a solution, it is compatible. We shall now complete the proof of the following theorem.

13.33 THEOREM   *A system 13.1 of linear equations over a field F has a solution if and only if it is compatible.*

Proof: We assume that our system is compatible and shall show that it has a solution.

We continue to denote the row vectors of the matrix of coefficients of the given system by $A_1, A_2, \ldots, A_m$. It will also be convenient to denote the row vectors of the augmented matrix by $(A_1, b_1), (A_2, b_2), \ldots, (A_m, b_m)$.

Suppose that the matrix of coefficients has rank r. We shall prove that the augmented matrix also has rank r by showing that, if $r < m$, any $r + 1$ row vectors of the augmented matrix are linearly dependent. For convenience of notation, let us consider the *first* $r + 1$ row vectors. Since the coefficient matrix has rank r, we know that the $r + 1$ row vectors $A_1, A_2, \ldots, A_{r+1}$ of the coefficient matrix are linearly dependent. Accordingly, there exist elements c_i ($i = 1, 2, \ldots, r + 1$) of F, not all of which are zero, such that

$$\sum_{i=1}^{r+1} c_i A_i = 0.$$

The compatibility condition now assures us that also

$$\sum_{i=1}^{r+1} c_i b_i = 0,$$

and we therefore conclude that

$$\sum_{i=1}^{r+1} c_i (A_i, b_i) = 0.$$

Since the c's are not all zero, we have shown that the first $r + 1$ row vectors of the augmented matrix are linearly dependent. The same argument applies to any set of $r + 1$ row vectors. This proves that the augmented matrix must have rank r, and Theorem 13.31 then shows that the system has a solution. This completes the proof of the theorem.

The next result will show that we may be able to delete some of the equations of a given system without affecting the solutions of the system. Let us now write our system of equations in the form given in 13.4, which we here repeat for convenience of reference:

13.34 $A_1 \cdot X = b_1, A_2 \cdot X = b_2, \ldots, A_m \cdot X = b_m.$

We shall proceed to prove the following theorem.

13.35 THEOREM *If the row vectors* $(A_{l_1}, b_{l_1}), (A_{l_2}, b_{l_2}), \ldots, (A_{l_r}, b_{l_r})$
form a basis of the row space of the augmented matrix of the system
13.34, then the system is equivalent to the system consisting of the
following r equations:

13.36 $A_{l_1} \cdot X = b_{l_1}, A_{l_2} \cdot X = b_{l_2}, \ldots, A_{l_r} \cdot X = b_{l_r}.$

> **Proof:** It is obvious that any solution of the entire system 13.34 is
> also a solution of the subsystem 13.36. To prove the theorem, we
> therefore need only show that any solution T of the system 13.36
> is also a solution of the system 13.34. If $A_j \cdot X = b_j$ is an arbitrary
> one of the Equations 13.34, we shall show that $A_j \cdot T = b_j$. Since
> the row vector (A_j, b_j) is a linear combination of the given basis of
> the row space of the augmented matrix, there exist elements d_i
> $(i = 1, 2, \ldots, r)$ of F such that
>
> $$A_j = \sum_{i=1}^{r} d_i A_{l_i}, \qquad b_j = \sum_{i=1}^{r} d_i b_{l_i}.$$
>
> Using these equations, simple properties of the inner product, and
> the fact that T is a solution of the system 13.36, we find that
>
> $$A_j \cdot T = \left(\sum_{i=1}^{r} d_i A_{l_i} \right) \cdot T = \sum_{i=1}^{r} d_i (A_{l_i} \cdot T) = \sum_{i=1}^{r} d_i b_{l_i} = b_j.$$
>
> Hence, T is, in fact, a solution of each equation of the system 13.34,
> and the proof is complete.

13.5 SYSTEMS OF LINEAR HOMOGENEOUS EQUATIONS

In this section we briefly discuss a special case of some importance. A linear
equation is said to be *homogeneous* if its constant term is zero. A system of m
linear homogeneous equations can then be written in the form 13.34 with all
b's equal to zero. That is, such a system is of the form

13.37 $A_1 \cdot X = 0, A_2 \cdot X = 0, \ldots, A_m \cdot X = 0.$

Since the augmented matrix of this system differs from the matrix of coefficients only by having an additional zero column, the ranks of these matrices
are the same. Theorem 13.31 then asserts that the system must have a solution.

However, this fact is also immediately obvious from the observation that the equations are satisfied if we set $x_1 = 0$, $x_2 = 0, \ldots, x_n = 0$; otherwise expressed, $X = 0$ is a solution of the system. This solution, in which all the unknowns are assigned the value zero, is usually called the *trivial solution* of the system. Any other solution is a nontrivial solution.

The principal theorem about the solutions of a system of linear homogeneous equations is the following.

13.38 THEOREM   *If r is the rank of the matrix of coefficients of a system 13.37 of linear homogeneous equations in n unknowns over a field F, then the solutions of the system form a subspace of dimension $n - r$ of the vector space $V_n(F)$.*

Proof: First, let us verify that the set of all solutions is a subspace of $V_n(F)$. If S and T are solutions of the system 13.37, so that $A_i \cdot S = 0$ and $A_i \cdot T = 0$ for $i = 1, 2, \ldots, m$, then also

$$A_i \cdot (S + T) = A_i \cdot S + A_i \cdot T = 0,$$

so that $S + T$ is a solution. Similarly, if S is a solution and $c \in F$, then $A_i \cdot (cS) = c(A_i \cdot S) = c0 = 0$ for $i = 1, 2, \ldots, m$; and cS is also a solution. We have thus shown that the subset of $V_n(F)$ consisting of all solutions of the system 13.37 is closed under addition and scalar multiplication. This subset is then a subspace of $V_n(F)$ by Theorem 11.7. We now proceed to prove that this subspace has dimension $n - r$.

If $r = n$, we know from Theorem 13.32 that there is a unique solution, and it must therefore be the trivial solution. The subspace of all solutions is then the zero vector space and, by definition, it has dimension zero. Having disposed of this simple case, we shall henceforth assume that $r < n$.

By Theorem 13.20, we know that our system 13.37 can be reduced to an echelon system of the form 13.18 in which all the d's are zero. For convenience of notation, let us suppose that $k_1 = 1$, $k_2 = 2$, $\ldots, k_r = r$. Then this echelon system, which is equivalent to our given system, can be written in the form

13.39
$$\begin{aligned}
x_1 &= c_{1\,r+1}x_{r+1} + \cdots + c_{1n}x_n, \\
x_2 &= c_{2\,r+1}x_{r+1} + \cdots + c_{2n}x_n, \\
&\cdots \cdots \cdots \cdots \cdots \cdots \cdots \\
x_r &= c_{r\,r+1}x_{r+1} + \cdots + c_{rn}x_n.
\end{aligned}$$

where the c's are fixed elements of F. We can get a particular solution by setting $x_{r+1} = 1$, $x_{r+2} = \cdots = x_n = 0$, and solving for the corresponding values of $x_1, x_2, \ldots, x_r$. In this way we obtain the solution

$$T_1 = (c_{1\,r+1}, \ldots, c_{r\,r+1}, 1, 0, \ldots, 0).$$

In like manner, each of the following is a solution:

$$T_2 = (c_{1\,r+2}, \ldots, c_{r\,r+2}, 0, 1, \ldots, 0),$$
$$\cdot \;\; \cdot \;\; \cdot \;\; \cdot \;\; \cdot \;\; \cdot \;\; \cdot \;\; \cdot \;\; \cdot \;\; \cdot \;\; \cdot \;\; \cdot \;\; \cdot$$
$$T_{n-r} = (c_{1n}, \ldots, c_{rn}, 0, 0, \ldots, 1).$$

By a consideration of the last $n - r$ entries in these vectors, it is seen that $T_1, T_2, \ldots, T_{n-r}$ are linearly independent elements of $V_n(F)$. Moreover, if $T = (t_1, t_2, \ldots, t_n)$ is an arbitrary solution of the system 13.39, and therefore of 13.37, a straightforward calculation will show that $T = t_{r+1}T_1 + \cdots + t_nT_{n-r}$. We have therefore shown that the set $\{T_1, T_2, \ldots, T_{n-r}\}$ is a basis of the subspace of $V_n(F)$ consisting of all solutions of the system 13.37. The dimension of this subspace is thus $n - r$, and the proof of the theorem is complete.

Since a vector space of positive dimension must contain nonzero vectors, we have at once the following corollary.

13.40 COROLLARY *A system of linear homogeneous equations has a nontrivial solution if and only if the rank of the matrix of coefficients of the system is less than the number of unknowns.*

The following result is an immediate consequence of the fact that the rank of a matrix cannot exceed the number of rows in the matrix.

13.41 COROLLARY *A system of linear homogeneous equations with fewer equations than unknowns always has a nontrivial solution.*

EXERCISES

In each of Exercises 1–3, solve the given system of linear equations over the rational field by working only with the augmented matrix of the system.

1. $x_1 - 2x_2 + x_3 = -6,$
    $3x_1 + 4x_2 + 2x_3 = 5,$
    $-x_1 + 3x_2 - x_3 = 8.$

2. $2x_1 - x_2 + 3x_3 \qquad = 3,$
    $x_1 + 2x_2 - x_3 - 5x_4 = 4,$
    $x_1 + 3x_2 - 2x_3 - 7x_4 = 5.$

3. $2x_1 - 2x_2 - x_3 + 5x_4 + x_5 = 7,$
    $3x_1 - 3x_2 + 2x_3 + 4x_4 + 2x_5 = 15,$
    $x_1 - x_2 + x_3 + x_4 + x_5 = 6.$

In each of Exercises 4–7, solve the given system of linear homogeneous equations over the rational field and find a basis for the vector space of all solutions.

4. $x_1 - x_2 + 2x_3 = 0,$
    $2x_1 + x_2 - 5x_3 = 0,$
    $x_1 - 4x_2 + 11x_3 = 0.$

5. $x_1 + 3x_2 - 4x_3 - 4x_4 + 9x_5 = 0,$
    $x_1 + x_2 - 2x_3 - x_4 + 2x_5 = 0,$
    $2x_1 - x_2 + 2x_3 + 4x_4 - 8x_5 = 0.$

6. $4x_1 + 2x_2 - x_3 + 3x_4 = 0,$
    $x_1 - 2x_2 + 2x_3 - x_4 = 0,$
    $2x_1 + 6x_2 - 5x_3 + 5x_4 = 0,$
    $3x_1 + 14x_2 - 12x_3 + 11x_4 = 0.$

7. $2x_1 + 3x_2 + x_3 - x_4 = 0,$
    $3x_1 - 2x_2 - x_3 + x_4 = 0.$

In Exercises 8 and 9, solve the given system of linear homogeneous equations over the field Z_5 and find a basis for the vector space of all solutions.

8. $2x_1 + 3x_2 + x_4 = 0,$
    $x_1 + 2x_2 + 2x_4 = 0,$
    $4x_1 + 4x_2 + x_4 = 0.$

9. $x_1 + 2x_2 + 3x_3 + x_4 = 0,$
    $3x_1 + x_2 + x_3 + 2x_4 = 0,$
    $4x_1 + 3x_2 + x_3 + 2x_4 = 0,$
    $x_1 + 2x_2 + 4x_3 + 3x_4 = 0.$

10. Use Corollary 13.41 to show that any $n + 1$ vectors of a vector space $V_n(F)$ are linearly dependent (11.19).

11. Show that if T is one solution of the system 13.34 of linear equations, every solution of the system is of the form $T + S$, where S is a solution of the system 13.37 of linear homogeneous equations.

 COMMENTARY

1 Since this chapter discussed solutions of equations, perhaps it is not too far out of place to discuss Diophantine equations here. Suppose that $x_1, x_2, \ldots, x_n$ are variables which take their values in the set of *integers* **Z**. A *Diophantine equation* is one of the form

(1) $$f(x_1, x_2, \ldots, x_n) = 0$$

where $f(x_1, x_2, \ldots, x_n)$ is a polynomial function with integral coefficients. For example,

(2) $$x^2 + y^2 - z^2 = 0$$

and

(3) $$x^3 + y^3 - z^3 = 0$$

are Diophantine equations. The object is to find integral values for $x_1, x_2, \ldots, x_n$ which satisfy the equation (1). In fact, today we would say that we have solved the equation (1) if we have shown either that (1) has no integral solutions, as is the case with

$$x^2 + 3 = 0$$

or we have given all possible sets of solutions [80].

There are very old tablets of Babylonian and Egyptian origin indicating that very simple Diophantine problems were considered even then. The equations are named after Diophantus (*c.* 250) of Alexandria. In his *Arithmetica* (the same one used for note taking by Fermat (1601–1665)) he posed certain problems which give rise to equations as in (1). Strictly speaking, these were not problems as we view them today since Diophantus allowed rational solutions and required that only one solution (not all) be found. The strict adherence to integral solutions is a clear demand in an Arabic text of Brahmagupta (*c.* 628).

Fermat may be said to be the modern originator of Diophantine problems. From his time to the present, most of the greatest mathematicians have thought about Diophantine equations of one sort or another. Gauss (1777–1855) considered the quadratic forms which represent integers [75]. These give rise to equations of the form

$$ax^2 + bxy + cy^2 + m = 0.$$

A special case of Gauss' forms

$$x^2 - ay^2 - 1 = 0$$

was considered by Brahmagupta. Euler (1707–1783) named this Pell's (1610–1685) equation and worked upon its solutions. Lagrange (1736–1813) gave the complete solution in 1766.

Liouville (1809–1882) injected a basically new idea into the subject by showing that it is difficult to approximate very closely roots of algebraic equations by rational numbers, and that transcendental numbers are much more easily approximated. Using these techniques, he first proved the existence of transcendental numbers in 1844. In the hands of Thue (1863–1922) this method assumed great power. He gave methods by which one can show for certain Diophantine equations that there are at most a finite number of solutions. The initial theorem by Thue has been improved by Siegel and finally by Roth [79]. These methods did not include the possibility for giving upper bounds for the size of solutions. That is, they used proof by contradiction in such a way that, although it was shown that the set of solutions is finite, and, hence, upper bounds *must exist* (just take the largest solution), no way was known to *find* such bounds. More recently, Baker has given methods which allow us to compute upper bounds [69]. These bounds are usually colossal, but hypothetically, one can check all possibilities less than a given bound to see if a solution does exist.

In 1900 Hilbert (1862–1943) asked for an algorithm which could be used to solve any given Diophantine problem. This famous problem 10 of Hilbert was solved in 1970 by Matiyasevič when he showed that no such algorithm could exist [i]. In a certain sense, this theorem says that no mathematician can ever "polish off" Diophantine problems as a branch of number theory.

2   Let us consider an equation which leads to an interesting problem. The additive group $\mathbf{R}^+$ of the real field is a vector space over the rational field $\mathbf{Q}$. A function (or mapping) $f: \mathbf{R} \to \mathbf{R}$ such that

(1) $$f(x + y) = f(x) + f(y)$$

is called a *linear function*. An obvious example of a linear function $g(x)$ is one of the form

(2) $$g(x) = ax$$

where a is a fixed real number. Must all linear functions have the form (2)? This question was studied by Euler (1707–1783) and has fascinated many mathematicians since that time. Let us first show that an f satisfying equation (1) must be a linear transformation of $\mathbf{R}^+$ into $\mathbf{R}^+$; that is, f is a member of $\mathrm{Hom}_{\mathbf{Q}}(\mathbf{R}^+, \mathbf{R}^+)$. For a positive integer $n = 1 + 1 + \cdots + 1$ we easily may show that $nf(x) = f(nx)$. Using other arguments we obtain $f(0) = 0$, $f(-x) = -f(x)$, and finally $f((m/n)x) = (m/n)f(x)$ for integers m, n with $n \neq 0$. In particular,

$$af(x) = f(ax) \qquad \text{all } a \in \mathbf{Q}.$$

This calculation might suggest that $f(ax) = af(x)$ for all values of a, even though we have only given a proof for rational values of a. Such a suggestion is wrong; there are particular "wild" functions $f(x)$ which satisfy (1) but not (2). To build them we must adopt a viewpoint given in Chapter 12. We view $\mathbf{R}^+$ as a vector space over $\mathbf{Q}$. Analogous to our definition of a basis of a finite dimensional vector space, a basis for $\mathbf{R}^+$ over $\mathbf{Q}$ is a set $\mathscr{B}$ of real numbers such that (i) the elements of $\mathscr{B}$ are linearly independent over $\mathbf{Q}$, and (ii) each element $Y \in \mathbf{R}^+$ may be expressed, for some positive integer t, in the form

$$Y = a_1 X_1 + a_2 X_2 + \cdots + a_t X_t$$

where the a's lie in $\mathbf{Q}$ and the X's lie in $\mathscr{B}$. Using the Axiom of Choice, one can prove that such bases $\mathscr{B}$, called *Hamel bases*, exist (see Theorem 16.28). Think of $f(x)$ as a linear transformation of the vector space $\mathbf{R}^+$ over $\mathbf{Q}$. As in Theorem 11.35, if we choose an element $W_j \in \mathbf{R}^+$ for each $X_j \in \mathscr{B}$ then the equations

$$f(X_j) = W_j, \qquad X_j \in \mathscr{B}$$

define a unique linear transformation of $\mathbf{R}^+$. This function $f(x)$ will satisfy (1).

Forget for a moment that $f(x)$ is a linear transformation; view it as a real valued function. If $f(1) = a$ then, in order to satisfy (2), we must have

$$f(X_j) = aX_j = W_j$$

for every $X_j \in \mathscr{B}$. That is, the ratios W_j/X_j must all equal the constant a. Since the choice of W's is arbitrary, we may certainly choose them so that (2) fails. In this way, we obtain a "wild" function $f(x)$ for which (1) holds but (2) fails.

It is interesting to speculate about the appearance of the graph of such a "wild" function $f(x)$. A problem whose solution requires only elementary calculus is to show that if $f(x)$ satisfies (1) and is continuous on any open interval then it satisfies (2). That is, a "wild" $f(x)$ must be very badly noncontinuous. It has been shown that $f(x)$ may be defined so that its graph is dense in the plane. Certain of these "wild" $f(x)$ have been shown to have very exotic behavior [187].

3   Linear equations are now viewed as a special application of linear algebra. As such, they are covered in books on linear algebra [10, 57, 58, 62, 63]. Many problems of applied mathematics reduce to solving systems of linear equations. Efficient methods for doing this are discussed in texts on numerical analysis [113, 115, 117] and applied linear algebra [67].

chapter fourteen

DETERMINANTS

The theory of determinants, to be introduced in this chapter, had its origin in the study of systems of linear equations, but it has other applications as well. We shall establish some of the more fundamental properties of determinants and also give a new characterization of the rank of a matrix. Finally, we shall show how determinants may be used in solving a system of linear equations. A few other applications of the theory of determinants will be found in the following chapter.

14.1 PRELIMINARY REMARKS

The purpose of this section is to motivate the general definition of a determinant to be given in the next section, and to introduce some convenient notation.

The concept of a determinant arises in the attempt to find general *formulas* for the solution of a system of linear equations with the same number of equations as unknowns. Suppose, first, that we have the following system of two equations in two unknowns over a field F:

$$a_{11}x_1 + a_{12}x_2 = b_1,$$
$$a_{21}x_1 + a_{22}x_2 = b_2.$$

If, by the familiar methods of elementary algebra, we eliminate x_2 from these two equations, the resulting equation takes the form

$$(a_{11}a_{22} - a_{12}a_{21})x_1 = b_1a_{22} - b_2a_{12}.$$

Similarly, if we eliminate x_1, we obtain the equation

$$(a_{11}a_{22} - a_{12}a_{21})x_2 = b_2a_{11} - b_1a_{12}.$$

The expression $a_{11}a_{22} - a_{12}a_{21}$, which occurs as the coefficient of x_2 in this last equation, and also as the coefficient of x_1 in the preceding equation, is called the *determinant* of the matrix of coefficients of the given system of equations.

The case of two equations in two unknowns is too simple to furnish much of a hint about the general case, so let us briefly consider a system of three equations in three unknowns. Such a system may be written in the form

14.1
$$a_{11}x_1 + a_{12}x_2 + a_{13}x_3 = b_1,$$
$$a_{21}x_1 + a_{22}x_2 + a_{23}x_3 = b_2,$$
$$a_{31}x_1 + a_{32}x_2 + a_{33}x_3 = b_3.$$

It can be shown by direct calculations, although in this case the calculations would be fairly tedious, that if we eliminate any *two* of the unknowns, the coefficient of the remaining unknown in the resulting equation is as follows:

14.2 $a_{11}a_{22}a_{33} - a_{11}a_{23}a_{32} - a_{12}a_{21}a_{33} + a_{12}a_{23}a_{31} + a_{13}a_{21}a_{32} - a_{13}a_{22}a_{31}.$

This expression is called the *determinant* of the matrix of coefficients of the system 14.1. For convenience, let A be this matrix of coefficients; that is, let

$$A = \begin{bmatrix} a_{11} & a_{12} & a_{13} \\ a_{21} & a_{22} & a_{23} \\ a_{31} & a_{32} & a_{33} \end{bmatrix}.$$

Then the determinant of A, defined by 14.2, may be denoted by $|A|$.

Let us now make a few observations about $|A|$. Clearly, $|A|$ is an element of F associated with the matrix A over F. Moreover, $|A|$ consists of an algebraic sum of six terms, each of which is a product of three elements of the matrix A. In each of these products, the integers 1, 2, 3 occur, in this order, as the first subscripts; also, these three integers occur, in some order, as the second subscripts. That is, each product is a product of elements of A, one from each row and one from each column. Another way of expressing this fact is to say that a typical term of 14.2 is of the form $\pm a_{1i_1}a_{2i_2}a_{3i_3}$, where i_1, i_2, and i_3 are the integers 1, 2, and 3 in some order. Using the notation for permutations, introduced in Chapter 2, let α be the permutation of the set $\{1, 2, 3\}$ defined by $1\alpha = i_1$, $2\alpha = i_2$, $3\alpha = i_3$. Then a typical term of 14.2 can be written in the form

14.3 $$\pm a_{1\ 1\alpha}a_{2\ 2\alpha}a_{3\ 3\alpha}.$$

Now the symmetric group S_3 on three symbols has six elements, and since there are six terms in 14.2, we see that for every α in S_3 a term of the form 14.3 occurs as a summand.

Let us now discuss the choice of sign in a term 14.3. Since, in 14.2, three of the terms are prefixed by a "$+$" sign and the other three by a "$-$" sign, it is clear that the choice of sign must depend in some way on the permutation α. We recall that a permutation is called *even* or *odd* according as it can be expressed as a product of an even or an odd number of transpositions. Moreover, we have proved, in Theorem 2.74, that half of the permutations of any symmetric group S_n are even, and half are odd. As suggested by these remarks, it is true that the choice of sign in a term 14.3 depends only on whether α is an even or an odd permutation. Before stating this fact in a more precise form, let us recall the following convenient definition (2.75).

14.4   DEFINITION   If α is a permutation of a finite set, we define

$$\text{sign } \alpha = \begin{cases} +1 \text{ if } \alpha \text{ is an even permutation,} \\ -1 \text{ if } \alpha \text{ is an odd permutation.} \end{cases}$$

Using this notation, we now assert that

14.5 $$|A| = \sum_{\alpha \epsilon S_3} (\text{sign } \alpha)a_{1\ 1\alpha}a_{2\ 2\alpha}a_{3\ 3\alpha},$$

it being understood that the sum is to be taken over all elements α of S_3. Since there are six elements in S_3, there will be six terms in this sum. To verify that this sum is exactly the expression 14.2, we need only verify that the signs of the terms are correct. For example, let us consider the product $a_{13}a_{22}a_{31}$. For this product, we have $1\alpha = 3$, $2\alpha = 2$, $3\alpha = 1$; and it follows that α is the transposition (13). Hence, this α is odd and therefore sign $\alpha = -1$. Accordingly, in the sum on the right of 14.5 we would have $-a_{13}a_{22}a_{31}$, and this product appears, with this same sign, in the expression 14.2. In like manner all the other terms may be verified. (See Exercise 4 in the next set of exercises.)

In the next section we shall give the definition of the determinant of an $n \times n$ matrix. The definition is a natural generalization of 14.5, and this is the reason that we have here gone to so much trouble to express 14.2 in the simple form 14.5.

14.2 GENERAL DEFINITION OF DETERMINANT

Unless otherwise stated, throughout this chapter the matrices with which we shall be concerned will be *square* matrices, that is, $n \times n$ matrices for some positive integer n. It is often convenient to call an $n \times n$ matrix a matrix of *order n*. Although a considerable part of what we shall do would remain valid if the elements of the matrices were elements of a commutative ring, for simplicity we shall always assume that the elements are from a field F. A matrix A, of order n, over a field F may be written in the following explicit form:

14.6
$$A = \begin{bmatrix} a_{11} & a_{12} & \cdots & a_{1n} \\ a_{21} & a_{22} & \cdots & a_{2n} \\ \cdot & \cdot & \cdots & \cdot \\ a_{n1} & a_{n2} & \cdots & a_{nn} \end{bmatrix},$$

it being understood that $a_{ij} \in F$ $(i, j = 1, 2, \ldots, n)$. If the order n is apparent from the context and therefore does not need to be mentioned explicitly, it is sometimes convenient to indicate the above matrix A by writing simply $A = (a_{ij})$.

The line joining the upper left-hand element of a square matrix and the lower right-hand element is often called the *principal diagonal* of the matrix. The elements $a_{11}, a_{22}, \ldots, a_{nn}$ of the matrix A are the elements on its principal diagonal.

We shall continue to denote by S_n the set (actually a group) of all permutations of the set $\{1, 2, \ldots, n\}$. Then, as suggested by the discussion of the preceding section, we shall make the following definition.

14.7 DEFINITION If A is the matrix 14.6 of order n over a field F, the *determinant* of A, denoted by $|A|$, is defined as follows:

14.8
$$|A| = \sum_{\alpha \in S_n} (\text{sign } \alpha) a_{1\,1\alpha} a_{2\,2\alpha} \cdots a_{n\,n\alpha}.$$

The determinant of A may also be denoted by $|a_{ij}|$, or by

$$\begin{vmatrix} a_{11} & a_{12} & \cdots & a_{1n} \\ a_{21} & a_{22} & \cdots & a_{2n} \\ \cdot & \cdot & \cdots & \cdot \\ a_{n1} & a_{n2} & \cdots & a_{nn} \end{vmatrix}.$$

Since the symmetric group S_n has $n!$ elements, we see that if A is a square matrix of order n, $|A|$ is a sum of $n!$ terms, each of which is a product of n elements, one from each row and one from each column of A. Moreover, half of these terms will have a "$+$" sign and the other half a "$-$" sign prefixed. We may make one other observation as follows. If in 14.8 we consider the term obtained when α is the identity permutation, which is an even permutation, we see that one of the terms is the product of the elements on the principal diagonal of A.

We shall presently develop methods for computing the determinant of a given matrix that will be much simpler than applying the definition directly. Before proceeding to do so, we shall present two simple examples to illustrate some of the ideas presented so far.

Example 1 Suppose that the matrix A is of order 5, and let us find the sign of the term involving the product $a_{13}a_{24}a_{35}a_{42}a_{51}$ in $|A|$. For this term, we have $1\alpha = 3$, $2\alpha = 4$, $3\alpha = 5$, $4\alpha = 2$, $5\alpha = 1$. Using the notation of cycles, we find that α can be expressed in the form $\alpha = (135)(24) = (13)(15)(24)$. Hence α is an odd permutation and sign $\alpha = -1$. Accordingly, one of the terms in $|A|$ is $-a_{13}a_{24}a_{35}a_{42}a_{51}$.

Example 2 Suppose that the matrix A is of order 6 and let us find the sign of the term involving the product $a_{45}a_{31}a_{12}a_{64}a_{23}a_{56}$ in $|A|$. As a matter of convenience, let us first rearrange this product so that the first (row) subscripts appear in their natural order. Of course, we can do so because multiplication is commutative in F. We get in this way the product $a_{12}a_{23}a_{31}a_{45}a_{56}a_{64}$. For this product, we have $1\alpha = 2$, $2\alpha = 3$, $3\alpha = 1$, $4\alpha = 5$, $5\alpha = 6$, $6\alpha = 4$; and it follows that $\alpha = (123)(456)$. This is an even permutation and therefore sign $\alpha = +1$. Accordingly, the given product occurs with a "$+$" sign.

EXERCISES

1. If A is a matrix of order 5, find the sign of the term in $|A|$ which involves each of the following products:

    (a) $a_{15}a_{24}a_{33}a_{42}a_{51}$,

    (b) $a_{13}a_{21}a_{32}a_{45}a_{54}$,

    (c) $a_{11}a_{25}a_{32}a_{43}a_{54}$,

(d) $a_{14}a_{25}a_{33}a_{41}a_{52}$,

(e) $a_{52}a_{41}a_{34}a_{25}a_{13}$,

(f) $a_{24}a_{45}a_{12}a_{53}a_{31}$.

2. If the matrix B is obtained from the matrix A of order n by multiplying all elements of one row by the element c of F, show that $|B| = c|A|$.

3. Apply Definition 14.7 to show that

$$\begin{vmatrix} a & b \\ c & d \end{vmatrix} = ad - bc.$$

4. Apply Definition 14.7 to show that the expression 14.2 is actually the determinant of the matrix $A = (a_{ij})$ of order three.

5. Show that if all the elements of one row, or of one column, of a matrix A are zero, then $|A| = 0$.

6. Suppose that for the matrix A, exhibited in 14.6, all the elements above the principal diagonal are zero. Show that in this case the determinant of A is just the product of the elements on the principal diagonal.

14.3 SOME FUNDAMENTAL PROPERTIES

In the definition of the determinant of a matrix A, the rows play a role somewhat different from that of the columns. That is, we have written each term as a product of elements with the row (first) subscripts in their natural order, and have then determined the sign of the term by a consideration of the column (second) subscripts. However, we shall now show that in this definition we could just as well reverse the roles of the rows and the columns.

In the preceding chapter we have defined the *transpose* C' of any matrix C to be the matrix obtained from C by interchanging rows and columns. We shall now prove the following theorem.

14.9 THEOREM *If A' is the transpose of the square matrix A, then*
$|A'| = |A|$.

14.6   **Proof:** Let $A = (a_{ij})$ be the matrix of order n exhibited in 14.6. It will be helpful to set $b_{ji} = a_{ij}$ $(i, j = 1, 2, \ldots, n)$, so that $A' = (b_{ij})$ or, in more detail,

$$A' = \begin{bmatrix} b_{11} & b_{12} & \cdots & b_{1n} \\ b_{21} & b_{22} & \cdots & b_{2n} \\ \cdot & \cdot & \cdots & \cdot \\ b_{n1} & b_{n2} & \cdots & b_{nn} \end{bmatrix}.$$

Applying Definition 14.8, we see that

14.10
$$|A'| = \sum_{\alpha \in S_n} (\text{sign } \alpha) b_{1\,1\alpha} b_{2\,2\alpha} \cdots b_{n\,n\alpha}.$$

For the moment, let α be a fixed element of S_n and let us consider the following term in this sum:

14.11
$$(\text{sign } \alpha) b_{1\,1\alpha} b_{2\,2\alpha} \cdots b_{n\,n\alpha}.$$

Since multiplication is commutative in F, we can rearrange the order of the factors in this product in any way we wish. That is, if β is an arbitrary permutation of $\{1, 2, \ldots, n\}$, we can write the expression 14.11 in the form

14.12
$$(\text{sign } \alpha) b_{1\beta\,1\beta\alpha} b_{2\beta\,2\beta\alpha} \cdots b_{n\beta\,n\beta\alpha}.$$

In particular, let us choose $\beta = \alpha^{-1}$. Then, since, by a remark preceding the statement of Theorem 2.74, $\text{sign } \beta = \text{sign } \alpha$, this expression 14.12 can be written as follows:

14.13
$$(\text{sign } \beta) b_{1\beta\,1} b_{2\beta\,2} \cdots b_{n\beta\,n}.$$

Now if α_1 and α_2 are elements of the group S_n, we know that $\alpha_1^{-1} = \alpha_2^{-1}$ if and only if $\alpha_1 = \alpha_2$. If follows that *every* element of S_n is uniquely expressible in the form β^{-1} for $\alpha \in S_n$. By the equality of the expressions 14.11 and 14.13, we can therefore rewrite 14.10 as follows:

$$|A'| = \sum_{\beta \in S_n} (\text{sign } \beta) b_{1\beta\,1} b_{2\beta\,2} \cdots b_{n\beta\,n}.$$

However, using the fact that $b_{ji} = a_{ij}$, it then follows that

$$|A'| = \sum_{\beta \in S_n} (\text{sign } \beta) a_{1\,1\beta} a_{2\,2\beta} \cdots a_{n\,n\beta}.$$

The sum on the right is clearly $|A|$, and we have therefore proved that $|A'| = |A|$.

In the next theorem we shall determine the effect on $|A|$ of an elementary operation on the square matrix A, as defined in 13.24.

14.14 THEOREM *The effect of an elementary operation of each of the three types may be described as follows:*

Type 1. If the matrix B is obtained from the matrix A by interchanging two rows (columns), then $|B| = -|A|$.

Type 2. If the matrix C is obtained from the matrix A by multiplying all elements of one row (column) by the nonzero element r of F, then $|C| = r|A|$.

Type 3. If the matrix D is obtained from the matrix A by multiplying all elements of one row (column) by an element of F and adding them to the corresponding elements of a different row (column), then $|D| = |A|$.

First, let us observe that an elementary column operation on a matrix A induces an elementary row operation of the same type on the transpose A' of A. Hence, the preceding theorem assures us that in proving the present theorem we may limit ourselves to elementary *row* operations only.

Proof for type 1 Suppose that the matrix B is obtained from the matrix A by interchanging rows k and l, $k < l$. Then $B = (b_{ij})$, where $b_{ij} = a_{ij}$ if $i \neq k$ and $i \neq l$; $b_{kj} = a_{lj}$, $b_{lj} = a_{kj}$. Then, by the definition of determinant, we have

$$|B| = \sum_{\alpha \in S_n} (\text{sign } \alpha) b_{1\,1\alpha} \cdots b_{k\,k\alpha} \cdots b_{l\,l\alpha} \cdots b_{n\,n\alpha}$$

$$= \sum_{\alpha \in S_n} (\text{sign } \alpha) a_{1\,1\alpha} \cdots a_{l\,k\alpha} \cdots a_{k\,l\alpha} \cdots a_{n\,n\alpha}$$

$$= \sum_{\alpha \in S_n} (\text{sign } \alpha) a_{1\,1\alpha} \cdots a_{k\,l\alpha} \cdots a_{l\,k\alpha} \cdots a_{n\,n\alpha}.$$

In this last sum we have merely changed the order of the factors in each product by interchanging $a_{l\,k\alpha}$ and $a_{k\,l\alpha}$ so that the first subscripts are in their natural order.

Now if $\alpha \in S_n$, let us set $\beta = (kl)\alpha$ so that $i\beta = i\alpha$ if $i \neq k$ and $i \neq l$; $k\beta = l\alpha$, $l\beta = k\alpha$. It is clear that $\text{sign } \beta = -\text{sign } \alpha$ and, moreover, every element of S_n is expressible uniquely in the form $(kl)\alpha$ with $\alpha \in S_n$. Using all these facts, and referring to the last form for $|B|$ given above, we see that

$$|B| = -\sum_{\beta \in S_n} (\text{sign } \beta) a_{1\,1\beta} \cdots a_{k\,k\beta} \cdots a_{l\,l\beta} \cdots a_{n\,n\beta} = -|A|,$$

and the proof is complete.

Before proceeding to the proof of the other two parts of the theorem, we point out the following consequence of what we have just proved.

14.15 COROLLARY *If a matrix A has two rows (columns) that are identical, then* $|A| = 0$.

Suppose, for convenience of statement, that the first two rows of A are alike. If B is obtained from A by interchanging the first two rows, clearly A and B are identical and therefore $|B| = |A|$. However, by the first part of the preceding theorem, we have $|B| = -|A|$. Accordingly, $|A| = -|A|$, and we conclude that $|A| = 0$. In this final argument we have tacitly assumed that the characteristic of the field F is different from 2. The statement of Corollary 14.15 remains true even for this case, but the proof must make direct use of the definition of a determinant. However, this case is not a very important one for our purposes, and we shall omit its proof.

We now return to a consideration of the other parts of Theorem 14.14.

Proof for type 2   Since each term in the sum which defines a determinant contains exactly one element from each row, the desired result follows almost immediately from the definition. Actually, the argument holds equally well even if r happens to be zero although, by the definition of elementary operations on a matrix, this would not be an elementary operation. We can, however, conclude that if all the elements of one row (column) of a matrix are zero, then the determinant of the matrix is zero. (Cf. Exercises 2 and 5 of the preceding set.)

Proof for type 3   The same argument would apply in general but, for simplicity, let us assume that the matrix D is obtained from the matrix A by multiplying the elements of the second row by $s \in F$ and adding them to the corresponding elements of the first row. If $D = (d_{ij})$, then for $j = 1, 2, \ldots, n$ we have $d_{ij} = a_{ij}$ if $i \neq 1$; $d_{1j} = a_{1j} + sa_{2j}$. Accordingly,

$$
\begin{aligned}
|D| &= \sum_{\alpha \in S_n} (\text{sign } \alpha) d_{1\,1\alpha} d_{2\,2\alpha} \cdots d_{n\,n\alpha} \\
&= \sum_{\alpha \in S_n} (\text{sign } \alpha)(a_{1\,1\alpha} + sa_{2\,1\alpha}) a_{2\,2\alpha} \cdots a_{n\,n\alpha} \\
&= \sum_{\alpha \in S_n} (\text{sign } \alpha) a_{1\,1\alpha} a_{2\,2\alpha} \cdots a_{n\,n\alpha} \\
&\quad + s \sum_{\alpha \in S_n} (\text{sign } \alpha) a_{2\,1\alpha} a_{2\,2\alpha} a_{3\,3\alpha} \cdots a_{n\,n\alpha} \\
&= |A| + s \sum_{\alpha \in S_n} (\text{sign } \alpha) a_{2\,1\alpha} a_{2\,2\alpha} a_{3\,3\alpha} \cdots a_{n\,n\alpha}.
\end{aligned}
$$

Now this last sum is the determinant of a matrix obtained from A by replacing the first row by the second row. Since two rows are identical, the determinant of the matrix is zero by Corollary 14.15. The above calculations therefore show that $|D| = |A|$, and the proof of the theorem is complete.

If the row vectors of a matrix A are linearly dependent, a finite sequence of elementary row operations of Type 3 will reduce the matrix to one which has a zero row. The determinant of such a matrix is zero and, since the value of a determinant is not changed by elementary row operations of this type, we conclude that also $|A| = 0$. We have therefore proved the following result.

14.16 COROLLARY *If the row vectors (column vectors) of the square matrix A are linearly dependent, then* $|A| = 0$.

Theorem 14.14 is exceedingly useful in actually computing the value of the determinant of a given matrix. However, this theorem is usually used in conjunction with the principal theorem of the following section, and we shall therefore postpone any further discussion of these matters until that theorem has also been established.

14.4 EXPANSION IN TERMS OF A ROW OR COLUMN

Let us make the following convenient definitions.

14.17 DEFINITION

   (i) A matrix obtained from a given matrix (not necessarily a square matrix) by deleting certain rows or columns, or both, is called a *submatrix* of the given matrix.

   (ii) If A is a square matrix of order n, the square submatrix M_{ij} of order $n - 1$ obtained by deleting the ith row and the jth column of A is called the *minor of the element* a_{ij}.

   (iii) If A is a square matrix, the *cofactor* A_{ij} of a_{ij} in $|A|$ is defined as follows: $A_{ij} = (-1)^{i+j}|M_{ij}|$.

In order to illustrate these concepts, suppose that A is the following matrix of order 4:

14.18
$$A = \begin{bmatrix} a_{11} & a_{12} & a_{13} & a_{14} \\ a_{21} & a_{22} & a_{23} & a_{24} \\ a_{31} & a_{32} & a_{33} & a_{34} \\ a_{41} & a_{42} & a_{43} & a_{44} \end{bmatrix}.$$

Then the minor of the element a_{43} in this matrix is the matrix

14.19
$$M_{43} = \begin{bmatrix} a_{11} & a_{12} & a_{14} \\ a_{21} & a_{22} & a_{24} \\ a_{31} & a_{32} & a_{34} \end{bmatrix},$$

and the cofactor A_{43} of a_{43} in $|A|$ is given by:

$$A_{43} = (-1)^7 |M_{43}| = -|M_{43}|.$$

We may emphasize that a minor is a *matrix*, whereas a cofactor is an element of the underlying field F.

The reason for the name *cofactor* will be suggested by the following result.

14.20   LEMMA   *If $A = (a_{ij})$ is a square matrix of order n, the sum of all the terms in $|A|$ which contain the arbitrary fixed element a_{ij} of A is $a_{ij}A_{ij}$.*

Proof: We shall first prove this lemma for the special case in which $i = 1$ and $j = 1$.   In $|A|$, the sum of all terms which contain a_{11} can be written in the form

$$a_{11} \sum_{\substack{\alpha \in S_n \\ 1\alpha = 1}} (\text{sign } \alpha) a_{2\,2\alpha} a_{3\,3\alpha} \cdots a_{n\,n\alpha},$$

it being understood that the sum is over all permutations α of S_n such that $1\alpha = 1$.   Of course, if we let S_{n-1} be the set of all permutations of the set $\{2, 3, \ldots, n\}$, this sum can be written in the form

$$a_{11} \sum_{\gamma \in S_{n-1}} (\text{sign } \gamma) a_{2\,2\gamma} a_{3\,3\gamma} \cdots a_{n\,n\gamma}.$$

Moreover, although the notation is slightly different from that which we have previously used, the sum occurring here is just the determinant of the matrix

$$\begin{bmatrix} a_{22} & a_{23} & \cdots & a_{2n} \\ a_{32} & a_{33} & \cdots & a_{3n} \\ \cdot & \cdot & \cdot \cdot \cdot \cdot & \cdot \\ a_{n2} & a_{n3} & \cdots & a_{nn} \end{bmatrix}.$$

Now this is the minor M_{11} of the element a_{11} of the matrix A; hence the sum of all the terms in $|A|$ which contain a_{11} is $a_{11}|M_{11}|$.   Since, by Definition 14.17(iii), $A_{11} = (-1)^2 |M_{11}| = |M_{11}|$, the sum of all these terms is $a_{11}A_{11}$.   This proves the lemma for the special case in which $i = 1$ and $j = 1$. For later reference, let us restate what we have proved in the following form. *In the determinant of an arbitrary square matrix, the sum of the terms which contain the element in the upper left-hand corner of the matrix is just this element times the determinant of its minor.* We shall make use of this fact in the proof of the lemma for the case in which i and j are arbitrary. However, let us first illustrate the method to be used by an example.

For the moment, let A be the matrix of order 4 given in 14.18, and let us show that the sum of the terms in $|A|$ which contain the element a_{43} is $a_{43}A_{43}$, where $A_{43} = -|M_{43}|$. We proceed to perform on A a finite sequence of elementary operations of Type 1 to get a matrix with a_{43} in the upper left-hand corner and, moreover, in such a way that the minor of a_{43} in this new matrix is the same as its minor in A. More specifically, we first interchange rows 3 and 4, then rows 2 and 3, and then rows 1 and 2. This gives us the following matrix in which the element a_{43} now occurs in the first row:

$$\begin{bmatrix} a_{41} & a_{42} & a_{43} & a_{44} \\ a_{11} & a_{12} & a_{13} & a_{14} \\ a_{21} & a_{22} & a_{23} & a_{24} \\ a_{31} & a_{32} & a_{33} & a_{34} \end{bmatrix}.$$

We now interchange columns 2 and 3 in this matrix, and finally columns 1 and 2. We then have the matrix B given by

$$B = \begin{bmatrix} a_{43} & a_{41} & a_{42} & a_{44} \\ a_{13} & a_{11} & a_{12} & a_{14} \\ a_{23} & a_{21} & a_{22} & a_{24} \\ a_{33} & a_{31} & a_{32} & a_{34} \end{bmatrix}.$$

Now the minor of the element a_{43} *of this matrix B* is obtained by deleting the row and column which contain a_{43}, that is, the first row and first column. This minor is therefore as follows:

$$\begin{bmatrix} a_{11} & a_{12} & a_{14} \\ a_{21} & a_{22} & a_{24} \\ a_{31} & a_{32} & a_{34} \end{bmatrix}.$$

It will be seen that this matrix is exactly the minor M_{43} of the element a_{43} *in the matrix A*, as given in 14.19. Moreover, the special case of the lemma which we have already proved, as applied to the matrix B, shows that the sum of the terms in $|B|$ which contain a_{43} is $a_{43}|M_{43}|$. To obtain the matrix B from A we applied five elementary operations of Type 1, each of which changed the sign of the determinant. Hence, $|A| = -|B|$ and the sum of all the terms in $|A|$ which contain a_{43} is $-a_{43}|M_{43}| = a_{43}A_{43}$.

Of course, it would have been possible to get from A a matrix with a_{43} in the upper left-hand corner merely by interchanging the first and fourth rows, and the first and third columns. However, had we done so, the minor of a_{43} in the matrix so obtained would not have been M_{43} and we would have had to do some more work before reaching the desired conclusion.

To complete the proof of the lemma, we use the same method as in this illustration. Again, let A be a matrix of order n, and let i and j be fixed integers, distinct or identical, from the set $\{1, 2, \ldots, n\}$. By $i - 1$ successive interchanges of adjacent rows and $j - 1$ successive interchanges of adjacent columns we can obtain a matrix C with the element a_{ij} in the upper left-hand corner and with the further important property that the minor of a_{ij} in the matrix C is exactly the minor M_{ij} of the element a_{ij} in the given matrix A. Clearly, $|C| = (-1)^{i+j-2}|A| = (-1)^{i+j}|A|$. Now, applying the special case of the lemma which has already been proved, we see that the sum of all the terms in $|C|$ which contain the element a_{ij} is $a_{ij}|M_{ij}|$. Hence, the sum of all the terms in $|A|$ which contain the element a_{ij} is $(-1)^{i+j}a_{ij}|M_{ij}| = a_{ij}A_{ij}$. This completes the proof of the lemma.

Since, by the definition of a determinant, every term contains exactly one element from the first row, it is clear that every term in $|A|$ contains exactly one of the elements $a_{11}, a_{12}, \ldots, a_{1n}$. It follows at once from the lemma that

$$|A| = a_{11}A_{11} + a_{12}A_{12} + \cdots + a_{1n}A_{1n}.$$

It is obvious that a similar argument applies to the elements of any fixed row or column, and we therefore have the following important result.

14.21 THEOREM   *Let $A = (a_{ij})$ be a matrix of order n over a field. Then*

14.22
$$|A| = \sum_{j=1}^{n} a_{kj}A_{kj}, \qquad (k = 1, 2, \ldots, n),$$

and also

14.23
$$|A| = \sum_{i=1}^{n} a_{il}A_{il}, \qquad (l = 1, 2, \ldots, n).$$

It is customary to say that 14.22 gives the expansion of $|A|$ in terms of the kth row, and 14.23 the expansion in terms of the lth column. Since the cofactor of an element in $|A|$ is the determinant of a matrix of order $n - 1$, these expansions express the determinant of a matrix of order n in terms of determinants of matrices of order $n - 1$. This fact is of great value in computing the determinant of a given matrix. We shall presently give some examples, but first let us establish another result which follows easily from Theorem 14.21 and is of considerable interest in itself.

14.24 THEOREM *Let $A = (a_{ij})$ be a matrix of order n over a field. Then, if $k \neq l$, we have*

14.25
$$\sum_{j=1}^{n} a_{kj} A_{lj} = 0,$$

and

14.26
$$\sum_{i=1}^{n} a_{ik} A_{il} = 0.$$

Theorem 14.21 states that the sum of the products of the elements of any row (column) of a matrix by their respective cofactors is the determinant of the matrix. This theorem states that the sum of the products of the elements of any row (column) by the cofactors of the corresponding elements of a *different* row (column) is always zero.

> **Proof:** To prove 14.25, let k and l be distinct integers of the set $\{1, 2, \ldots, n\}$, and let D be the matrix obtained from A by deleting its lth row and replacing it by its kth row. Since two rows of D are identical, Corollary 14.15 assures us that $|D| = 0$. Moreover, the cofactor of an element a_{kj} of the lth row of D coincides with A_{lj}, the cofactor of the corresponding element a_{lj} of A. Accordingly, the sum appearing in 14.25 is, by the preceding theorem applied to D, the expansion of $|D|$ in terms of its lth row. Since $|D| = 0$, this proves 14.25. A similar argument, using columns instead of rows, will establish 14.26.

We now proceed to give examples which may help to clarify the theory that has been presented so far. In particular, we shall give illustrations of how certain of our results may be used in actually computing the determinant of a given matrix. It will be understood that the elements are from the field of rational numbers.

Example 1 Find the value of the following determinant:

$$\begin{vmatrix} 1 & 3 & 2 \\ -2 & 1 & -1 \\ 0 & 1 & 4 \end{vmatrix}.$$

We shall compute the value of this determinant in two different ways. First, let us use 14.22 to expand the determinant in terms of its first row as follows:

$$\begin{vmatrix} 1 & 3 & 2 \\ -2 & 1 & -1 \\ 0 & 1 & 4 \end{vmatrix} = 1 \cdot \begin{vmatrix} 1 & -1 \\ 1 & 4 \end{vmatrix} - 3 \cdot \begin{vmatrix} -2 & -1 \\ 0 & 4 \end{vmatrix} + 2 \cdot \begin{vmatrix} -2 & 1 \\ 0 & 1 \end{vmatrix}.$$

Of course, the minus sign in the second term is caused by the fact that the cofactor of an element in the first row and second column is $(-1)^3$ times its minor. Now it was observed in Exercise 3 at the end of Section 14.2 that

$$\begin{vmatrix} a & b \\ c & d \end{vmatrix} = ad - bc,$$

and hence it is easy to find the value of each of our determinants of order two. Doing so, we obtain as the value of the given determinant

$$1 \cdot (4 + 1) - 3(-8) + 2(-2) = 25.$$

Now let us carry out the calculation in a different way using an elementary operation as follows. If we multiply the first row by 2 and add to the second, we know by Theorem 14.14 that the determinant is unchanged. We then expand in terms of the first column. The calculations are as follows:

$$\begin{vmatrix} 1 & 3 & 2 \\ -2 & 1 & -1 \\ 0 & 1 & 4 \end{vmatrix} = \begin{vmatrix} 1 & 3 & 2 \\ 0 & 7 & 3 \\ 0 & 1 & 4 \end{vmatrix} = 1 \cdot \begin{vmatrix} 7 & 3 \\ 1 & 4 \end{vmatrix} = 25.$$

Example 2   Find the value of the following determinant:

$$\begin{vmatrix} 5 & -3 & 12 & 2 \\ 6 & 4 & 8 & 6 \\ 3 & -1 & 8 & -1 \\ 4 & 2 & 12 & 4 \end{vmatrix}.$$

Of course, it would be possible to expand this determinant in terms of some row or column and then proceed to evaluate each of the four determinants of order three that would be involved. However, it is much less work to use elementary operations in such a way as to get all elements but one of some row or column equal to zero, and then to expand in terms of that particular row or column. One possible way to apply this procedure is indicated by the following calculations, which we shall explain briefly below:

$$\begin{vmatrix} 5 & -3 & 12 & 2 \\ 6 & 4 & 8 & 6 \\ 3 & -1 & 8 & -1 \\ 4 & 2 & 12 & 4 \end{vmatrix} = 2 \begin{vmatrix} 5 & -3 & 12 & 2 \\ 3 & 2 & 4 & 3 \\ 3 & -1 & 8 & -1 \\ 4 & 2 & 12 & 4 \end{vmatrix} = 8 \begin{vmatrix} 5 & -3 & 3 & 2 \\ 3 & 2 & 1 & 3 \\ 3 & -1 & 2 & -1 \\ 4 & 2 & 3 & 4 \end{vmatrix}$$

$$= 8 \begin{vmatrix} -4 & -9 & 0 & -7 \\ 3 & 2 & 1 & 3 \\ -3 & -5 & 0 & -7 \\ -5 & -4 & 0 & -5 \end{vmatrix} = -8 \begin{vmatrix} 4 & 9 & 0 & 7 \\ 3 & 2 & 1 & 3 \\ 3 & 5 & 0 & 7 \\ 5 & 4 & 0 & 5 \end{vmatrix}$$

$$= 8 \begin{vmatrix} 4 & 9 & 7 \\ 3 & 5 & 7 \\ 5 & 4 & 5 \end{vmatrix} = 8 \begin{vmatrix} 1 & 4 & 0 \\ 3 & 5 & 7 \\ 5 & 4 & 5 \end{vmatrix} = 8 \begin{vmatrix} 1 & 0 & 0 \\ 3 & -7 & 7 \\ 5 & -16 & 5 \end{vmatrix}$$

$$= 8 \begin{vmatrix} -7 & 7 \\ -16 & 5 \end{vmatrix} = 8(77) = 616.$$

We have first used the second part of Theorem 14.14 to factor 2 from each element of the second row, then have factored 4 from each element of the third column. Next we used elementary operations of Type 3, which did not change the value of the determinant, to get all elements but one of the third column equal to zero. In order to avoid so many minus signs we then multiplied the first, third, and fourth rows by −1. Since each of these operations changed the sign of the determinant, we had to place a minus sign in front. We then expanded in terms of the third column. To evaluate the determinent of order three, we subtracted the second row from the first (multiplied by −1 and added to the first). This was done merely to get 1 in some position. The rest of the calculation should be obvious.

Example 3 Without expanding the determinants, show that

$$\begin{vmatrix} a & b & c \\ d & e & f \\ g & h & i \end{vmatrix} = \begin{vmatrix} c & i & f \\ b & h & e \\ a & g & d \end{vmatrix}.$$

The calculations are as follows, first using Theorem 14.9 and then using the first part of Theorem 14.14 twice:

$$\begin{vmatrix} a & b & c \\ d & e & f \\ g & h & i \end{vmatrix} = \begin{vmatrix} a & d & g \\ b & e & h \\ c & f & i \end{vmatrix} = - \begin{vmatrix} c & f & i \\ b & e & h \\ a & d & g \end{vmatrix} = \begin{vmatrix} c & i & f \\ b & h & e \\ a & g & d \end{vmatrix}.$$

EXERCISES

1. Find the value of each of the following determinants over the rational field:

(a) $\begin{vmatrix} 1 & -3 & 2 \\ -2 & 4 & 3 \\ 3 & 1 & 2 \end{vmatrix}$,

(e) $\begin{vmatrix} 2 & 3 & -2 & -3 \\ 4 & 1 & 2 & 1 \\ 2 & -2 & 3 & 4 \\ 2 & 3 & -1 & 2 \end{vmatrix}$,

(b) $\begin{vmatrix} 1 & 2 & 3 \\ 4 & 5 & 6 \\ 7 & 8 & 9 \end{vmatrix}$,

(f) $\begin{vmatrix} 2 & \frac{1}{2} & -\frac{1}{2} & 1 \\ \frac{2}{3} & \frac{1}{3} & -\frac{2}{3} & -\frac{1}{3} \\ 2 & -2 & 2 & -2 \\ 4 & 6 & 2 & 4 \end{vmatrix}$,

(c) $\begin{vmatrix} 6 & -4 & 8 \\ -2 & 3 & 5 \\ 10 & 4 & 14 \end{vmatrix}$,

(g) $\begin{vmatrix} 2 & -3 & 1 & 2 & 3 \\ 1 & 2 & 2 & 3 & 4 \\ -1 & 1 & 1 & -1 & 1 \\ 2 & 4 & 6 & 4 & 2 \\ 3 & 2 & 1 & -3 & 2 \end{vmatrix}$.

(d) $\begin{vmatrix} \frac{1}{2} & \frac{2}{3} & -\frac{1}{2} \\ -\frac{2}{3} & -\frac{1}{2} & 2 \\ \frac{1}{6} & \frac{1}{2} & \frac{1}{3} \end{vmatrix}$,

2. Find the value of each of the following determinants over the field Z_5:

(a) $\begin{vmatrix} 2 & 3 & 4 \\ 1 & 2 & 3 \\ 3 & 3 & 2 \end{vmatrix}$, (b) $\begin{vmatrix} 3 & 0 & 4 \\ 1 & 2 & 4 \\ 4 & 3 & 2 \end{vmatrix}$, (c) $\begin{vmatrix} 1 & 2 & 3 & 4 \\ 2 & 3 & 4 & 1 \\ 3 & 4 & 1 & 2 \\ 4 & 1 & 2 & 3 \end{vmatrix}$.

3. Without expansion of the determinants involved, verify the following (the elements are from any field):

$$\begin{vmatrix} a_1 & a_2 & a_3 \\ b_1 & b_2 & b_3 \\ c_1 & c_2 & c_3 \end{vmatrix} = \begin{vmatrix} c_1 + 2a_1 & b_1 & a_1 \\ c_3 + 2a_3 & b_3 & a_3 \\ c_2 + 2a_2 & b_2 & a_2 \end{vmatrix}.$$

14.5 THE DETERMINANT RANK OF A MATRIX

In this section we shall consider matrices that are not necessarily square. First we make the following definition.

14.27 DEFINITION An arbitrary matrix C over a field F is said to have *determinant rank r* if there exists a square submatrix of C of order r whose determinant is different from zero, whereas every square submatrix of C of order $r + 1$ has zero determinant. If all elements of C are zero, we define its determinant rank to be zero.

We may notice that if the determinant rank of C is r, not only is the determinant of every square submatrix of order $r + 1$ equal to zero, but also the determinant of every square submatrix of order greater than r is necessarily zero. For example, consider a square submatrix M of order $r + 2$. If $|M|$ is expanded in terms of a row or column, every cofactor is, except possibly for sign, the determinant of a submatrix of C of order $r + 1$, and hence has the value zero. Accordingly, $|M| = 0$; that is, the determinant of every square submatrix of order $r + 2$ has the value zero. By the same kind of argument, the determinant of every square submatrix of order $r + 3$ must now be zero, and so on. Of course, a process of induction is actually involved here.

The following theorem justifies the use of the word *rank* in the above definition.

14.28 THEOREM *The determinant rank of an arbitrary matrix C over a field coincides with its rank as defined in 13.30.*

Proof: Let C be the $p \times q$ matrix given by

$$
C = \begin{bmatrix}
c_{11} & c_{12} & \cdots & c_{1q} \\
c_{21} & c_{22} & \cdots & c_{2q} \\
\cdot & \cdot & \cdots & \cdot \\
c_{p1} & c_{p2} & \cdots & c_{pq}
\end{bmatrix},
$$

and let us assume that C has determinant rank r. If $r = 0$, which means that all elements of C are zero, then the dimension of the row space (or column space) of C is also zero by definition of the dimension of a zero vector space. Hence, also, the rank of C is zero, and this case is easily disposed of. Henceforth, we shall assume that $r > 0$, and shall complete the proof by showing that the row rank of C is r.

It is clear that interchanging rows (or columns) of C cannot affect its row (or column) rank, and also cannot affect its determinant rank since such operations would at most change the *sign* of certain determinants. Accordingly, by making such interchanges we can be sure that the square submatrix of order r in the upper left-hand corner has determinant different from zero. As a matter of notation, let us assume that this is already true for the matrix C; that is, that the determinant of the matrix

14.29

$$\begin{bmatrix} c_{11} & c_{12} & \cdots & c_{1r} \\ c_{21} & c_{22} & \cdots & c_{2r} \\ \cdot & \cdot & \cdot & \cdot \\ c_{r1} & c_{r2} & \cdots & c_{rr} \end{bmatrix}$$

is different from zero. Now, by Corollary 14.16, the row vectors of this matrix are linearly independent; hence the first r row vectors of C must also be linearly independent. If $C_1, C_2, \ldots, C_p$ are the row vectors of C, we therefore know that the set $\{C_1, C_2, \ldots, C_r\}$ is linearly independent and we shall show that it is a basis of the row space of C. This is obviously true if $r = p$, so we henceforth assume that $r < p$. Let s be an arbitrary, but fixed, integer such that $r < s \le p$, and let us show that C_s is a linear combination of $C_1, C_2, \ldots, C_r$. For each integer $t = 1, 2, \ldots, q$, let us consider the matrix $D(t)$ of order $r + 1$ defined as follows:

$$D(t) = \begin{bmatrix} c_{11} & c_{12} & \cdots & c_{1r} & c_{1t} \\ c_{21} & c_{22} & \cdots & c_{2r} & c_{2t} \\ \cdot & \cdot & \cdots & \cdot & \cdot \\ c_{r1} & c_{r2} & \cdots & c_{rr} & c_{rt} \\ c_{s1} & c_{s2} & \cdots & c_{sr} & c_{st} \end{bmatrix}.$$

If $t \le r$, this matrix has two identical columns and hence $|D(t)| = 0$. On the other hand, if $t > r$, $D(t)$ is a square submatrix of C of order $r + 1$, and again $|D(t)| = 0$ since it is given that C has determinant rank r. Accordingly, $|D(t)| = 0$ for $t = 1, 2, \ldots, q$. If $d_1, d_2, \ldots, d_r, d_s$ are the cofactors of the elements of the last column of $D(t)$, it is clear that they do not depend on t, and if we expand $|D(t)|$ in terms of its last column, we find that

$$c_{1t}d_1 + c_{2t}d_2 + \cdots + c_{rt}d_r + c_{st}d_s = 0, \qquad (t = 1, 2, \ldots, q).$$

In terms of row vectors, this equation can be written in the form

$$d_1 C_1 + d_2 C_2 + \cdots + d_r C_r + d_s C_s = 0.$$

Moreover, $d_s \ne 0$ since it is the determinant of the matrix 14.29. It follows that C_s is a linear combination of $C_1, C_2, \ldots, C_r$. Since this is true for each s satisfying $r < s \leq p$, we have proved that $\{C_1, C_2, \ldots, C_r\}$ is indeed a basis of the row space of C, and hence that C has (row) rank r. The proof is therefore complete.

If A is a square matrix of order n, its determinant rank will be less than n if and only if $|A| = 0$. Moreover, the row (column) rank will be less than n if and only if the row vectors (column vectors) are linearly dependent. Accordingly, we have at once the following result, which completes the result of Corollary 14.16.

14.30 COROLLARY *If A is a square matrix over a field, then $|A| = 0$ if and only if the row vectors (column vectors) of A are linearly dependent.*

In view of the equality of all the various ranks of a matrix, in the future we shall usually refer merely to the *rank* of a matrix to mean the row rank, the column rank, or the determinant rank.

14.6 SYSTEMS OF LINEAR EQUATIONS

We now briefly discuss applications of determinants to the problem of finding the solutions of a system of linear equations. We shall first consider a system with the same number of equations as unknowns. Let us therefore consider the following system of linear equations over a field F:

14.31
$$\sum_{j=1}^{n} a_{ij}x_j = b_i, \qquad (i = 1, 2, \ldots, n).$$

We shall denote by A the matrix of the coefficients in this system of equations and by A_{ij} the cofactor of a_{ij} in $|A|$. If $|A| \neq 0$, it follows that the matrix A has rank n and clearly the augmented matrix of the system also has rank n. We already know from Theorem 13.32 that in this case the system of equations will have a *unique* solution. As hinted at in Section 14.1, the theory of determinants gives us an easy way to write down the solution in this case. The procedure is as follows.

In order to find the value of an arbitrary unknown x_l, we multiply the first equation by A_{1l}, the second by $A_{2l}, \ldots$, the nth by A_{nl}, and add. In the resulting equation, the coefficient of x_l is

$$\sum_{i=1}^{n} a_{il}A_{il}$$

which, by Theorem 14.21, is just $|A|$. If $k \neq l$, the coefficient of x_k in this resulting equation is

$$\sum_{i=1}^{n} a_{ik}A_{il},$$

which is zero by 14.26. Accordingly, the equation takes the following form:

14.32 $$|A|x_l = \sum_{i=1}^{n} b_i A_{il}.$$

For convenience, let us define the matrix $B(l)$ to be the matrix of order n obtained from A by replacing the lth column by the column of constant terms in the system 14.31. It follows that the right side of 14.32 is the expansion of $|B(l)|$ in terms of its lth column. Using this fact, and observing that the above argument holds for each choice of l, we find that

14.33 $$|A|x_l = |B(l)|, \qquad (l = 1, 2, \ldots, n).$$

Up to this point the calculations remain valid even if $|A| = 0$, but we are here concerned with the case in which $|A| \neq 0$. In this case, the preceding equations yield at once the unique solution of our given system of equations in the following explicit form:

14.34 $$x_l = \frac{|B(l)|}{|A|}, \qquad (l = 1, 2, \ldots, n).$$

Actually, our calculations here merely show that *if* $|A| \neq 0$ and *if* the given system of equations has a solution, then that solution is given by 14.34. However, from previous results we know that if $|A| \neq 0$, the system does have a solution and it is therefore given by 14.34. It is also fairly easy to verify directly that 14.34 does furnish a solution (see Exercise 12 below). Our results may be summarized as follows.

14.35 CRAMER'S RULE *If A is the matrix of the coefficients of a system 14.31 of n linear equations in n unknowns over a field, and if $|A| \neq 0$, then the system has the unique solution*

$$x_l = \frac{|B(l)|}{|A|}, \qquad (l = 1, 2, \ldots, n),$$

where $B(l)$ is the matrix obtained from A by replacing the lth column by the column of constant terms.

Example As an illustration of the use of Cramer's Rule, let us solve the following system of equations over the rational field:

$$3x_1 + x_2 - x_3 = 2,$$
$$x_1 + 2x_2 + x_3 = 3,$$
$$-x_1 + x_2 + 4x_3 = 9.$$

For this system, using the notation introduced above, we have

$$A = \begin{bmatrix} 3 & 1 & -1 \\ 1 & 2 & 1 \\ -1 & 1 & 4 \end{bmatrix},$$

$$B(1) = \begin{bmatrix} 2 & 1 & -1 \\ 3 & 2 & 1 \\ 9 & 1 & 4 \end{bmatrix},$$

$$B(2) = \begin{bmatrix} 3 & 2 & -1 \\ 1 & 3 & 1 \\ -1 & 9 & 4 \end{bmatrix},$$

and

$$B(3) = \begin{bmatrix} 3 & 1 & 2 \\ 1 & 2 & 3 \\ -1 & 1 & 9 \end{bmatrix}.$$

We omit the details but the values of the determinants of these matrices are: $|A| = 13$, $|B(1)| = 26$, $|B(2)| = -13$, and $|B(3)| = 39$. The solution of the system is therefore $x_1 = 2$, $x_2 = -1$, $x_3 = 3$.

Although Cramer's Rule applies to the solution of a system of equations involving the same number of equations as unknowns, and then only if the determinant of the coefficients is different from zero, it can frequently be used in a somewhat more general situation as follows. Suppose that we have the following system of r linear equations in n unknowns, for which the rank of the matrix of the coefficients is also r:

14.36
$$\sum_{j=1}^{n} a_{ij}x_j = b_i, \quad (i = 1, 2, \ldots, r).$$

If $r = n$, we may apply Cramer's Rule at once, so let us assume that $r < n$. The matrix of the coefficients must have a square submatrix of order r whose determinant is different from zero.   Suppose, for simplicity, that this submatrix is made up of the *first r* columns.   In this case, we rewrite the system 14.36 in the following form:

14.37
$$a_{11}x_1 + a_{12}x_2 + \cdots + a_{1\,r}x_r = b_1 - a_{1\,r+1}x_{r+1} - \cdots - a_{1n}x_n,$$
$$a_{21}x_1 + a_{22}x_2 + \cdots + a_{2\,r}x_r = b_2 - a_{2\,r+1}x_{r+1} - \cdots - a_{2n}x_n,$$
$$\cdot \quad \cdot \quad \cdot \quad \cdot \quad \cdot \quad \cdot \quad \cdot \quad \cdot \quad \cdot \quad \cdot \quad \cdot \quad \cdot \quad \cdot \quad \cdot \quad \cdot \quad \cdot \quad \cdot \quad \cdot$$
$$a_{r1}x_1 + a_{r2}x_2 + \cdots + a_{r\,r}x_r = b_r - a_{r\,r+1}x_{r+1} - \cdots - a_{rn}x_n.$$

We then replace $x_{r+1}, \ldots, x_n$ by arbitrary elements of the underlying field and since the matrix of the coefficients of $x_1, \ldots, x_r$ has nonzero determinant, we can use Cramer's Rule to solve for the corresponding values of $x_1, \ldots, x_r$.   All solutions of the system 14.37, and therefore of the system 14.36, can be obtained in this way.

We may point out that, by Theorem 13.35, any system of equations that has a solution is equivalent to a system of the form 14.36 so that, at least in theory, the present method is always available.

Example   Let us illustrate how to solve a system of the form 14.36 by considering the following system of linear equations over the rational field:

14.38
$$x_1 + 2x_2 - x_3 + x_4 = 4,$$
$$-x_1 + x_2 + 3x_3 + x_4 = -2,$$
$$x_1 + 5x_2 + x_3 + x_4 = 2.$$

It may be verified that the matrix of the coefficients has rank 3; also that the determinant of the submatrix consisting of the first three columns is zero.   However, the matrix consisting of the first, third, and fourth columns has nonzero determinant.   Accordingly, we replace x_2 by the arbitrary rational number s and solve the following system by Cramer's Rule:

$$x_1 - x_3 + x_4 = 4 - 2s,$$
$$-x_1 + 3x_3 + x_4 = -2 - s,$$
$$x_1 + x_3 + x_4 = 2 - 5s.$$

We omit the details but the solution turns out to be as follows:

$$x_1 = \frac{2 - 7s}{2}, \quad x_2 = s, \quad x_3 = -\frac{2 + 3s}{2}, \quad x_4 = 2.$$

Every solution of the given system 14.38 is then of this form.

In this section we have considered applications of the theory of determinants to the problem of solving a given system of linear equations. Of course, there is no reason why the use of determinants may not be combined with the methods of the preceding chapter. In particular, it may be helpful first to simplify the system somewhat by use of elementary operations, and then at some appropriate stage to apply Cramer's Rule.

EXERCISES

In each of Exercises 1–9, apply Cramer's Rule to solve the given system of linear equations over the rational field.

1. $3x_1 - 5x_2 = 25,$
$x_1 + 4x_2 = -3.$

2. $3x_1 + 6x_2 = -15.$
$x_1 + 4x_2 = 1.$

3. $2x_1 - x_2 + x_3 = 0,$
$x_1 + 2x_2 - 2x_3 = 10,$
$3x_1 - 3x_2 - 5x_3 = 2.$

4. $2x_1 - 4x_2 + x_3 = 4,$
$x_1 + 3x_2 - x_3 = 5,$
$4x_1 - 2x_2 + 3x_3 = 6,$

5. $x_1 - 3x_2 + x_3 = 2,$
$3x_1 + x_2 + x_3 = 1,$
$5x_1 + x_2 + 3x_3 = 3.$

6. $2x_1 + x_2 + 3x_3 - x_4 = 1,$
$x_1 - x_2 + x_3 - x_4 = -5,$
$3x_1 + 2x_2 + 2x_3 - 3x_4 = 1,$
$-x_1 + 3x_2 - x_3 + 2x_4 = 14.$

7. $2x_1 - x_2 + 3x_3 = 4,$
$3x_1 + x_2 - 2x_3 = 3.$

8. $x_1 - 2x_2 + x_3 - x_4 = 2,$
$2x_1 + x_2 - x_3 + 2x_4 = 1,$
$x_1 + x_2 + 3x_3 - 3x_4 = 3.$

9. $x_1 - x_2 + 2x_3 - x_4 = 2,$
$2x_1 + x_2 - 3x_3 + 3x_4 = 0,$
$4x_1 - x_2 + x_3 + 2x_4 = 1.$

In each of Exercises 10 and 11, apply Cramer's Rule to solve the given system of linear equations over the field Z_3.

10. $x_1 + x_2 + 2x_3 = 0.$
$x_1 + x_2 + x_3 = 2,$
$2x_1 + 2x_2 + x_3 = 1.$

11. $x_1 + x_2 + x_3 + x_4 = 2,$
$2x_1 + x_2 + 2x_3 = 2.$
$x_1 + x_3 + x_4 = 1.$

12. Verify that 14.34 actually gives a solution of the *first* equation of the system

14.31. *Hint:* By expanding in terms of the first row, show that the determinant

$$\begin{vmatrix} b_1 & a_{11} & a_{12} & \cdots & a_{1n} \\ b_1 & a_{11} & a_{12} & \cdots & a_{1n} \\ b_2 & a_{21} & a_{22} & \cdots & a_{2n} \\ \cdot & \cdot & \cdot & \cdots & \cdot \\ b_n & a_{n1} & a_{n2} & \cdots & a_{nn} \end{vmatrix}$$

has the value

$$b_1|A| - a_{11}|B(1)| - a_{12}|B(2)| - \cdots - a_{1n}|B(n)|.$$

 # COMMENTARY

1 Leibnitz (1646–1716) seems to have been the first explicitly to calculate and use determinants. In 1693 he used them in considering several equations in two unknowns. Cramer (1704–1752) published his rule for solving equations in 1750, and Laplace (1749–1827) gave his general rule for expansion of a determinant by minors in a paper of 1772. In 1776, Vandermonde (1735–1796) published a systematic development of properties of determinants. In 1815 Cauchy (1789–1857) used determinants and wrote the coefficients in the matrix fashion we now use. It was also Cauchy who discovered the product rule for determinants.

Moving into analysis, in 1841, Jacobi (1804–1851) showed how to differentiate a determinant if the entries are differentiable functions. He and Catalan (1814–1894) then worked out the rules for change of variable in a multiple integral which give rise to the determinant now called a Jacobian.

Determinants arise in the theory of invariants. Let us work out a single example. Suppose that we start with a *quadratic form*

$$f(x, y) = ax^2 + 2bxy + cy^2$$

where a, b, c are real numbers. Let us change variables by the linear equations

(*)
$$\begin{aligned} x_0 &= c_{11}x + c_{12}y, \\ y_0 &= c_{21}x + c_{22}y. \end{aligned}$$

After substituting and expanding we obtain a new form

$$F(x_0, y_0) = Ax_0{}^2 + 2Bx_0y_0 + Cy_0{}^2$$

where A, B, C are real numbers determined by $a, b, c, c_{ij}, 1 \le i, j \le 2$. Since the forms $f(x, y)$ and $F(x_0, y_0)$ are quadratic, their discriminants are $b^2 - ac$ and $B^2 - AC$.

In fact, if d is the determinant of the coefficient matrix of c_{ij}, $1 \le i, j \le 2$, for the equations (*) then

$$B^2 - AC = d^2(b^2 - ac).$$

The relationship between the discriminants of the two forms $f(x, y)$ and $F(x_0, y_0)$ is via the square of the determinant of the transforming coefficients. Apparently Boole (1815–1864) observed this point first, and then Cayley (1821–1895) with his friend Sylvester (1814–1897) erected invariant theory upon this observation. If we have a form

$$g(x, y) = a_0 x^n + a_1 x^{n-1} y + \cdots + a_{n-1} xy^{n-1} + a_n y^n$$

and transform the variables by (*) to obtain a form

$$G(x_0, y_0) = A_0 x_0^n + A_1 x_0^{n-1} y_0 + \cdots + A_{n-1} x_0 y_0^{n-1} + A_n y_0^n$$

then an *invariant* $I = I(a_0, a_1, \ldots, a_n)$ is defined to be a polynomial in the coefficients of the forms $g(x, y)$ and $G(x_0, y_0)$ related by the equation

$$I(A_0, A_1, \ldots, A_n) = d^m I(a_0, a_1, \ldots, a_n)$$

for some integer m. In particular, the discriminant of a quadratic form is an invariant.

There are many other invariants, and a good deal of effort has gone into their computation. Invariant theory received its strongest development between 1850 and 1900, and the workers on this topic discovered many new computational properties and applications of determinants [7, 37].

Many of the results about solutions of systems of linear equations were first stated in terms of determinants. In this context, H. J. S. Smith (1826–1883) introduced augmented matrices in 1861. The general results about solutions to m linear equations in n unknowns were set down by Cayley and others, and these results began appearing in textbooks during the last third of the nineteenth century. In 1858, Cayley proved that every $n \times n$ matrix satisfies its characteristic equation (a theorem now called the Cayley-Hamilton theorem). From this point on, matrices and determinants were closely linked, with their properties being developed jointly. Further improvements were made by changing the rings from which coefficients are taken to form determinants. Since a known theorem about determinants for one ring may become false in another ring, many of the most classical theorems on determinants have received different proofs for different rings at different times.

In recent times, we do not usually distinguish a branch of algebra called "Theory of Determinants." Rather, these useful gadgets are used as tools in many branches of algebra [57, 62, 63, 65, 66, 144].

2   Most writers agree that the greatest mathematicians of recorded history were: Archimedes of Syracuse (c. 287–212 B.C.) who foresaw some of the basic ideas of the infinitesimal calculus some 1800 years before its time; Isaac Newton (1642–1727) who, at the same time as Leibniz (1646–1716), created the infinitesimal calculus as well as

applying it to obtain remarkable results in physics; and C. F. Gauss (1777–1855), the creator of systematic number theory. Gauss is probably the most famous of child prodigies. There is an often told story that while his father was settling some accounts and paying off his workers, the young Gauss, then aged three, watched his father's calculations. At one point the boy expostulated that his father had made an error in computing payment. Upon checking calculations, they found the child was correct.

A story which has more authenticity is told of Gauss at age ten. His school master asked the class to sum the integers from 1 to 100, possibly expecting to keep his charges busy while he did other things. Gauss, in a matter of seconds, wrote his answer 5,050 upon his slate and submitted it. His answer was not arrived at by calculational gymnastics, but rather by an understanding of the sum of an arithmetic progression.

Very few prodigies go on to rank among the greatest of thinkers. But Gauss' profound abilities remained with him to the end of his days. At age 17 he gave a straightedge-compass construction for a regular polygon of 17 sides; the only other prior constructions of polygons with a prime p number of sides were those of the Greeks for $p = 3, 5$. Further, he showed that if such a polygon can be constructed with straightedge and compass, then the prime p must be a Fermat prime $p = 2^{2^n} + 1$, and any such polygon can be constructed. Implicit in this work is that there is no general procedure for trisecting angles, and that given a cube of volume one there is no procedure which will allow us to construct the edge of a cube of volume two. In other words, two of the Delian problems (trisecting an angle, duplicating a cube) of Greek mathematics were settled for the first time.

Gauss was the first to observe the real importance of and give a proof of the Fundamental Theorem of Algebra. During his lifetime he gave four distinctly different proofs. He also introduced the notion of congruence for integers and, therefore, can be credited with the creation of the ring $\mathbf{Z}_n$ of integers modulo n.

A binary quadratic form is a polynomial in two indeterminates x, y with integer coefficients A, B, C having the form

$$f(x, y) = Ax^2 + Bxy + Cy^2.$$

If we substitute integers a, b for x, y then $f(a, b) = m$ is an integer. Gauss considered the problem: for which integers m is $m = f(a, b)$ for some choice of a, b? A special case of this are the integers m such that $a^2 + b^2 = m$. For odd primes p, we have $a^2 + b^2 = p$ if and only if $p \equiv 1 \pmod 4$. This special case, the so called "Two Square Problem," was asserted by Fermat (1601–1665) and proved by Euler (1707–1783). Gauss gave a general formulization of the problem of representing integers by integral quadratic forms which has not been significantly improved upon to this day.

Unfortunately, the general representation problem, which Gauss develops in his treatise *Disquisitiones Arithmeticae* [75], does not have an answer which is as simple as that for the "Two Square" problem posed by Fermat. Gauss pushed his solution very far, introducing a whole succession of profound ideas; yet, despite this, a completely "satisfactory" solution eluded him. About a hundred years later, the mathematical world found out why. It follows from the "class field theory," initiated by Hilbert, and developed by Artin, Takagi, and others, that Gauss had

pushed his solution about as far as it will go. The "defects" in his theory are inherent in the complicated nature of mathematical truth.

Gauss also discovered non-Euclidean geometry, put complex numbers onto a
firm intuitive footing, and carried out important researches in probability, celestial
mechanics, and function theory. There are several excellent accounts of Gauss' life
and his mathematics [140, 142].

3   There are many references to determinants. Among the more elementary
treatments are [3, 60, 62, 63, 67]. At first glance, it might seem that determinants are
a very good way to solve linear systems of equations via Cramer's rule. Actually,
the method outlined in Section 13.2 is much better. In working a problem, the number of additions and multiplications required is a measure of efficiency. Cramer's
rule is very inefficient compared to the method of Section 13.2. Therefore, determinants have very little practical application, so that their major value is of a purely
theoretical nature. That is, they are very useful in proving some theorems [113, 115].

chapter fifteen

LINEAR TRANSFORMATIONS AND MATRICES

In Section 11.6 we defined a homomorphism of a vector space V over a field F into a vector space W over the same field. We also introduced the descriptive notation $\mathrm{Hom}_F(V, W)$ for the set of all homomorphisms of V into W. In this chapter we shall be concerned with the important special case in which V and W coincide.

It is customary to call a homomorphism of a vector space V into itself a *linear transformation* of V. We shall adopt this terminology and, for simplicity, we shall find it convenient to denote the set of all linear transformations of a given vector space by L instead of using the notation $\mathrm{Hom}_F(V, V)$.

Let L be the set of all linear transformations of the vector space V over F. As a special case of results of Section 11.7, we know that under suitable definitions of addition and scalar multiplication L is itself a vector space over F. In the present setting, we will have also an operation of multiplication defined on L and it will turn out that L is a ring with respect to these operations of addition and multiplication. Actually, L is an example of an *algebra over F*, according to the definition to be given in Section 15.2.

For our purposes, the most important case is that in which V is a vector space of finite dimension n over F and, in this case, we shall show how to set up a one-one mapping of the set L onto the set F_n of all matrices of order n over F. This leads to natural definitions of addition, multiplication, and scalar multiplication in F_n in such a way that F_n becomes an algebra over F which is isomorphic to the algebra L. The rest of the chapter has to do with certain properties of linear transformations or of matrices and, in particular, with the interplay between these two concepts.

The topics introduced in this chapter play an important role in algebra and have been studied extensively. Our treatment gives merely a brief introduction to some of the basic ideas and methods.

15.1 NOTATION AND PRELIMINARY REMARKS

Let V be a vector space over a field F, and let us denote the unity of F by 1. Heretofore, we have denoted mappings by lower case Greek letters, but we shall henceforth denote linear transformations of a vector space by capital script letters such as $\mathscr{A}$, $\mathscr{B}$, and $\mathscr{C}$.

Using the present notation and terminology, we proceed to review briefly what we already know about linear transformations. First, we recall that a mapping $\mathscr{A}: V \to V$ of the vector space $V(F)$ into itself is, by definition, a linear transformation of V if addition and scalar multiplication are preserved under the mapping $\mathscr{A}$, that is, if the following hold:

15.1   If $X, Y \in V$ then $(X + Y)\mathscr{A} = X\mathscr{A} + Y\mathscr{A}$,

and

15.2   If $X \in V$ and $c \in F$, then $(cX)\mathscr{A} = c(X\mathscr{A})$.

From Theorem 11.31(iii), we see that if k is an arbitrary positive integer, $X_i \in V$ and $c_i \in F$ ($i = 1, 2, \ldots, k$), then

15.3 $(c_1 X_1 + c_2 X_2 + \cdots + c_k X_k)\mathscr{A} = c_1(X_1\mathscr{A}) + c_2(X_2\mathscr{A}) + \cdots + c_k(X_k\mathscr{A})$.

Of course, both 15.1 and 15.2 are special cases of 15.3. Moreover, the special cases of 15.2 in which c is 0 or -1, respectively, show that $0\mathscr{A} = 0$ and that for each X in V, $(-X)\mathscr{A} = -(X\mathscr{A})$.

A very useful property of a linear transformation is that it maps a subspace of V onto a subspace of V. That is, as indicated in Theorem 11.31(iv), if U is a subspace of V, then $U\mathscr{A} = \{X\mathscr{A} \mid X \in U\}$ is a subspace of V. In particular, $V\mathscr{A} = V$ if and only if $\mathscr{A}$ is an onto mapping.

For later reference, let us recall the definitions of addition and scalar multiplication (11.39 and 11.40) on the set L of all linear transformations of the vector space V over F. They are, respectively, as follows, it being understood that $\mathscr{A}$ and $\mathscr{B}$ are elements of L and $c \in F$:

15.4 $X(\mathscr{A} + \mathscr{B}) = X\mathscr{A} + X\mathscr{B}, \qquad X \in V,$

and

15.5 $$X(c\mathscr{A}) = (cX)\mathscr{A}, \qquad X \in V.$$

Theorem 11.41 then shows that if addition and scalar multiplication are so defined, L is a vector space over F.

15.2 ALGEBRA OF LINEAR TRANSFORMATIONS

In order to have a convenient way to state the principal result of this section, we first make the following definition.

15.6 DEFINITION Let S be a nonempty set on which there are defined binary operations of addition and multiplication, and also a scalar multiplication by elements of a field F. We shall call S an *algebra* over the field F if the following conditions are satisfied:

(i) S is a ring with respect to the operations of addition and multiplication,

(ii) S is a vector space over F with respect to the operations of addition and scalar multiplication,

(iii) If $u, v \in S$ and $a \in F$, then $(au)v = u(av) = a(uv)$.

It will be observed that part (iii) of this definition is a condition which involves both multiplication and scalar multiplication.

Throughout this section we shall continue to let L denote the set of all linear transformations of a vector space V over a field F. We already have (15.4 and 15.5) definitions of addition and scalar multiplication on L, and we proceed to introduce an operation of multiplication on L.

Multiplication of mappings has been defined in Section 1.3, where it was also shown that the associative law of multiplication always holds. In accordance with the general definition of multiplication of mappings, we define the product $\mathscr{A}\mathscr{B}$ of elements of L as follows:

15.7 $$X(\mathscr{A}\mathscr{B}) = (X\mathscr{A})\mathscr{B}, \qquad X \in V.$$

Certainly, $\mathscr{A}\mathscr{B}$ is a mapping of V into V, but we must show that it is in fact a linear transformation of V. In this case, the calculations are as follows where, again, $X, Y \in V$ and $c \in F$:

$$\begin{aligned}
(X + Y)(\mathscr{A}\mathscr{B}) &= ((X + Y)\mathscr{A})\mathscr{B} && \textit{(by 15.7)} \\
&= (X\mathscr{A} + Y\mathscr{A})\mathscr{B} && \textit{(by 15.1)} \\
&= (X\mathscr{A})\mathscr{B} + (Y\mathscr{A})\mathscr{B} && \textit{(by 15.1)} \\
&= X(\mathscr{A}\mathscr{B}) + Y(\mathscr{A}\mathscr{B}) && \textit{(by 15.7)},
\end{aligned}$$

and

$$
\begin{aligned}
(cX)(\mathscr{A}\mathscr{B}) &= ((cX)\mathscr{A})\mathscr{B} && \textit{(by 15.7)} \\
&= (c(X\mathscr{A}))\mathscr{B} && \textit{(by 15.2)} \\
&= c((X\mathscr{A})\mathscr{B}) && \textit{(by 15.2)} \\
&= c(X(\mathscr{A}\mathscr{B})) && \textit{(by 15.7)}.
\end{aligned}
$$

We have thus established the two defining properties of a linear transformation, and therefore $\mathscr{A}\mathscr{B} \in L$. Hence, 15.7 actually defines an operation of multiplication on the set L.

We may now state the following important result.

15.8 THEOREM *Using the respective Definitions 15.4, 15.7, and 15.5 of addition, multiplication, and scalar multiplication, the set $L = \text{Hom}_F(V, V)$ of all linear transformations of a vector space V over F is an algebra over F.*

Proof: We have already shown (Theorem 11.41) that L is a vector space over F, so part (ii) of Definition 15.6 holds. The zero of this vector space L is the linear transformation $\mathcal{O}: V \to V$ defined by

15.9 $$X\mathcal{O} = 0, \qquad X \in V,$$

that is, it maps every element of V into the zero.

Since L is a vector space, addition in L has all the properties required of addition in a ring. Moreover, we have the associative law of multiplication since the elements of L are mappings. In order to show that L is a ring there remains only to prove the distributive laws. Suppose that $\mathscr{A}$, $\mathscr{B}$, and $\mathscr{C}$ are elements of L, and let us prove that

15.10 $$\mathscr{A}(\mathscr{B} + \mathscr{C}) = \mathscr{A}\mathscr{B} + \mathscr{A}\mathscr{C}.$$

To establish this result, we shall prove that for every element X of V, X has the same image under the linear transformation $\mathscr{A}(\mathscr{B} + \mathscr{C})$ as under the linear transformation $\mathscr{A}\mathscr{B} + \mathscr{A}\mathscr{C}$. The calculations are as follows:

$$
\begin{aligned}
X[\mathscr{A}(\mathscr{B} + \mathscr{C})] &= (X\mathscr{A})(\mathscr{B} + \mathscr{C}) && \textit{(by 15.7)} \\
&= (X\mathscr{A})\mathscr{B} + (X\mathscr{A})\mathscr{C} && \textit{(by 15.4)} \\
&= X(\mathscr{A}\mathscr{B}) + X(\mathscr{A}\mathscr{C}) && \textit{(by 15.7)} \\
&= X(\mathscr{A}\mathscr{B} + \mathscr{A}\mathscr{C}) && \textit{(by 15.4)}.
\end{aligned}
$$

We leave as exercises the proof of the other distributive law and property (iii) of Definition 15.6.

It is quite easy to verify that the identity mapping $\mathscr{T}$ of V into V, that is, the mapping defined by

15.11 $$X\mathscr{T} = X, \qquad X \in V,$$

is a linear transformation and hence an element of L. Moreover, if $\mathscr{A} \in L$, we have that $\mathscr{A}\mathscr{T} = \mathscr{T}\mathscr{A} = \mathscr{A}$, and thus the algebra L has $\mathscr{T}$ as unity.

It is now natural to consider the question of which elements of L have multiplicative inverses. One characterization of these elements is given in the following theorem.

15.12 THEOREM   *An element $\mathscr{A}$ of $L = \mathrm{Hom}_F(V, V)$ has a multiplicative inverse $\mathscr{A}^{-1}$ in L if and only if $\mathscr{A}$ is a one-one mapping of V onto V.*

Proof: From Theorem 1.11 and remarks following the proof of that theorem, we know that there exists a *mapping $\mathscr{A}^{-1}$* which is an inverse of the *mapping $\mathscr{A}$* if and only if $\mathscr{A}$ is a one-one mapping of V onto V. There remains only to show that if $\mathscr{A}$ is a linear transformation which has an inverse mapping, this inverse mapping is also a linear transformation. Suppose, then, that $\mathscr{A}$ is a linear transformation of V which has an inverse *mapping $\mathscr{A}^{-1}$*. Thus $\mathscr{A}\mathscr{A}^{-1} = \mathscr{A}^{-1}\mathscr{A} = \mathscr{T}$ and $\mathscr{A}$ is a one-one mapping of V onto V. In fact, $\mathscr{A}^{-1}$ is the mapping defined by

15.13 $$(X\mathscr{A})\mathscr{A}^{-1} = X, \qquad X \in V.$$

We complete the proof of the theorem by verifying that the mapping $\mathscr{A}^{-1}$ defined by 15.13 is indeed a linear transformation and therefore an element of L.

Since $\mathscr{A}$ is an onto mapping, we may let $X\mathscr{A}$ and $Y\mathscr{A}$ be any elements of V. Then

$$
\begin{aligned}
(X\mathscr{A} + Y\mathscr{A})\mathscr{A}^{-1} &= [(X + Y)\mathscr{A}]\mathscr{A}^{-1} && (by\ 15.4) \\
&= X + Y && (by\ 15.13) \\
&= (X\mathscr{A})\mathscr{A}^{-1} + (Y\mathscr{A})\mathscr{A}^{-1} && (by\ 15.13).
\end{aligned}
$$

This establishes property 15.1 of a linear transformation. Likewise, if $X\mathscr{A} \in V$ and $c \in F$, we have

$$
\begin{aligned}
[c(X\mathscr{A})]\mathscr{A}^{-1} &= [(cX)\mathscr{A}]\mathscr{A}^{-1} && (by\ 15.2) \\
&= cX && (by\ 15.13) \\
&= c[(X\mathscr{A})\mathscr{A}^{-1}] && (by\ 15.13).
\end{aligned}
$$

Hence property 15.2 is also satisfied, and $\mathscr{A}^{-1}$ is an element of L. This completes the proof of the theorem.

We may pause to point out one consequence of part of the calculations used to prove Theorem 15.8. Let us, for the moment, ignore scalar multiplication in V and consider V to be merely an abelian group with operation addition. Moreover, instead of linear transformations of the vector space V, let us consider group homomorphisms of V into itself, that is, we assume Property 15.1 but not Property 15.2. If we define addition and multiplication of homomorphisms by 15.4 and 15.7, respectively, the parts of the proof of Theorem 15.8 which do not involve scalar multiplication show that the set of all these homomorphisms is a ring with unity. Otherwise expressed, if Hom (G, G) is the set of all homomorphisms of an abelian group G into itself, Hom (G, G) is a ring with respect to natural definitions of addition and multiplication. In Exercises 10 and 11 below it is indicated that *every* ring is isomorphic to a subring of Hom (G, G) for some abelian group G. This fact plays an important role in certain parts of the theory of rings.

EXERCISES

1. Which of the following mappings of $V_2(\mathbf{R})$ into $V_2(\mathbf{R})$ are linear transformations of $V_2(\mathbf{R})$?

    (a) $(x_1, x_2)\mathscr{A} = (0, 0)$,

    (b) $(x_1, x_2)\mathscr{A} = (3x_1 + x_2, x_1 + x_2)$,

    (c) $(x_1, x_2)\mathscr{A} = (x_1 + 1, x_1 + x_2)$,

    (d) $(x_1, x_2)\mathscr{A} = (x_2, x_1)$,

    (e) $(x_1, x_2)\mathscr{A} = (2x_1 - x_2, x_1x_2)$,

    (f) $(x_1, x_2)\mathscr{A} = (x_1 - 3x_2, x_1 - 3x_2)$,

    (g) $(x_1, x_2)\mathscr{A} = (x_1, x_2)$.

2. Let $V(F)$ be a vector space with a basis $\{X_1, X_2, X_3\}$, and let $\mathscr{A}$ be a linear transformation of V such that

$$X_1\mathscr{A} = X_2, \quad X_2\mathscr{A} = X_3 + X_2, \quad X_3\mathscr{A} = X_2.$$

    (i) Determine the vector $(X_1 + 2X_2 - X_3)\mathscr{A}$.

    (ii) Find a basis of the subspace $V\mathscr{A}$ of V.

    (iii) Find a subspace U of V of dimension two such that $U\mathscr{A} = U$.

    (iv) Can you find a subspace U of V of dimension two such that $U\mathscr{A} \subset U$?

3. Let V be the vector space of all polynomials in an indeterminate x over a field F (Example 4, Section 11.2). Verify that each of the following mappings of V into V is a linear transformation of V:

   (a) $f(x)\mathscr{A} = -f(x)$,

   (b) $f(x)\mathscr{A} = 0$,

   (c) $f(x)\mathscr{A} = f(x)$,

   (d) $f(x)\mathscr{A} = f(-x)$,

   (e) $f(x)\mathscr{A} = f(0)$,

   (f) $f(x)\mathscr{A} = f(x^2)$,

   (g) $f(x)\mathscr{A} = f(x) + f(-x)$.

4. If $V(F)$ is the vector space of the preceding exercise, verify that the mapping defined by $f(x)\mathscr{A} = f'(x)$, where $f'(x)$ is the derivative of $f(x)$ (Exercise 11 of Section 10.2), is a linear transformation of V. Determine the subspace $V\mathscr{A}$ of V in case F has characteristic zero. Do the same thing in case F has characteristic the prime p.

5. Let $\mathscr{A}$ and $\mathscr{B}$ be linear transformations of the vector space $V_2(\mathbf{R})$ defined as follows:

$$(x_1, x_2)\mathscr{A} = (2x_1 + x_2, x_1 - x_2),$$
$$(x_1, x_2)\mathscr{B} = (x_1, x_1 + 3x_2).$$

Exhibit in a similar manner each of the following linear transformations of $V_2(\mathbf{R})$: $\mathscr{A} + \mathscr{B}$, $\mathscr{A}\mathscr{B}$, $\mathscr{B}\mathscr{A}$, $5\mathscr{A}$, $-\mathscr{A}$, $\mathscr{A}^2$.

6. If $\mathscr{A}$ is as in the preceding exercise, verify that $\mathscr{A}^2 - \mathscr{A} = 3\mathscr{T}$, where $\mathscr{T}$ is the unity of the algebra of all linear transformations of $V_2(\mathbf{R})$.

7. Find nonzero linear transformations $\mathscr{C}$ and $\mathscr{D}$ of $V_2(\mathbf{R})$ such that $\mathscr{C}\mathscr{D} = \mathscr{O}$.

8. Complete the proof of Theorem 15.8 by proving the other distributive law and Property (iii) of Definition 15.6.

9. If L is the algebra of all linear transformations of a vector space V over F, verify that the set of all elements of L that are one-one mappings of V onto V is a group with respect to the operation of multiplication.

10. Let R be an arbitrary ring with unity, and let G be the additive group of R. If $a \in R$, the mapping $\theta_a: G \to G$ defined by $x\theta_a = xa$, $x \in G$, is a homomorphism of G into G. Show that the mapping $a \to \theta_a$ is an isomorphism of the ring R onto a subring of Hom (G, G). [Cf. the proof of Theorem 2.62.]

11. Show that every ring is isomorphic to a subring of Hom (G, G) for some abelian group G. [*Hint:* Use the result of Exercise 16 of Section 3.3.]

15.3 THE FINITE DIMENSIONAL CASE

Heretofore, we have considered linear transformations of an entirely arbitrary vector space V over a field F. In this section we shall restrict V to have finite dimension $n > 0$. Of course, this restriction assures us that V has a basis consisting of n vectors, and we shall exploit the existence of a basis in obtaining the results to follow.

We begin by stating the following result, which is a special case of Theorem 11.35, since it is important for our present purposes.

15.14 COROLLARY *If* $\{X_1, X_2, \ldots, X_n\}$ *is a basis of* V *and* $Z_1, Z_2, \ldots, Z_n$ *are arbitrary elements of* V, *there exists exactly one linear transformation* $\mathscr{A}$ *of* V *such that*

15.15 $$X_i \mathscr{A} = Z_i, \qquad (i = 1, 2, \ldots, n).$$

As in the preceding section, we shall continue to let L denote the algebra of all linear transformations of the vector space V.

We have already observed that if U is a subspace of the vector space V and $\mathscr{A} \in L$, then $U\mathscr{A}$ is also a subspace of V. We can now say something as follows about the dimensions of these subspaces.

15.16 THEOREM *If* U *is a subspace of the vector space* V *of finite dimension and* $\mathscr{A} \in L$, *then* dim $(U\mathscr{A}) \leq$ dim U. *Moreover, if* $\mathscr{A}$ *has a multiplicative inverse* $\mathscr{A}^{-1}$ *in* L, *then* dim $(U\mathscr{A}) =$ dim U.

Proof: We may remark that the theorem remains true for an arbitrary vector space V provided only that the subspace U has finite dimension. However, our assumption that V has finite dimension assures us that every subspace of V necessarily has finite dimension. The result is trivial if either U or V has dimension zero. Suppose that dim $U = k > 0$ and that $\{X_1, X_2, \ldots, X_k\}$ is a basis of U, so that every element of U is expressible in the form

$$c_1 X_1 + c_2 X_2 + \cdots + c_k X_k, \qquad c_i \in F \ (i = 1, 2, \ldots, k).$$

Since

$$(c_1 X_1 + c_2 X_2 + \cdots + c_k X_k)\mathscr{A}$$
$$= c_1(X_1 \mathscr{A}) + c_2(X_2 \mathscr{A}) + \cdots + c_k(X_k \mathscr{A}),$$

it follows at once that $U\mathscr{A}$ is generated by the k vectors $X_1 \mathscr{A}$, $X_2 \mathscr{A}$, $\ldots$, $X_k \mathscr{A}$. That is, in the notation of Chapter 11, we have

$$U\mathscr{A} = [X_1 \mathscr{A}, X_2 \mathscr{A}, \ldots, X_k \mathscr{A}].$$

Hence dim $(U\mathscr{A}) \leq k$, and since dim $U = k$, we see immediately that dim $(U\mathscr{A}) \leq$ dim U.

Now if $\mathscr{A}$ has a multiplicative inverse $\mathscr{A}^{-1}$ in L, we apply the result just established with U replaced by $U\mathscr{A}$ and $\mathscr{A}$ by $\mathscr{A}^{-1}$. Accordingly, we find that

$$\dim U = \dim ((U\mathscr{A})\mathscr{A}^{-1} \leq \dim (U\mathscr{A})).$$

Since we proved above that always dim $(U\mathscr{A}) \leq$ dim U, we conclude that dim $(U\mathscr{A}) =$ dim U, and the proof is complete.

The dimension of the subspace $V\mathscr{A}$ of V gives some important information about the linear transformation $\mathscr{A}$. For convenience of reference, we therefore make the following definition.

15.17 DEFINITION Suppose that dim $V = n$ and let $\mathscr{A}$ be a linear transformation of V. Then dim $(V\mathscr{A})$ is called the *rank* of the linear transformation $\mathscr{A}$. If the rank of $\mathscr{A}$ is less than n, $\mathscr{A}$ is said to be *singular*; if the rank of $\mathscr{A}$ is n, $\mathscr{A}$ is said to be *nonsingular*.

In the next section we shall show how matrices are related to linear transformations, and justify the use of the word "rank" in terms of the previous definition of rank of a matrix.

We now proceed to prove the following theorem.

15.18 THEOREM *Let V be a vector space of dimension $n > 0$, and let $\{X_1, X_2, \ldots, X_n\}$ be a basis of V. If $\mathscr{A} \in L$, the following are all equivalent:*

  (i) $\mathscr{A}$ is a one-one mapping.

  (ii) If $X \in V$ such that $X\mathscr{A} = 0$, then $X = 0$.

  (iii) $\{X_1 \mathscr{A}, X_2 \mathscr{A}, \ldots, X_n \mathscr{A}\}$ is a basis of V.

  (iv) $\mathscr{A}$ is an onto mapping, that is, $V\mathscr{A} = V$.

  (v) $\mathscr{A}$ is nonsingular.

  (vi) $\mathscr{A}$ has a multiplicative inverse $\mathscr{A}^{-1}$ in L.

By saying that these statements are equivalent, we mean that each one implies all the others.

Proof: It is quite easy to show as follows, without any restriction on V, that (i) and (ii) are equivalent. If (i) holds and $X\mathscr{A} = 0$, then we have $X\mathscr{A} = 0\mathscr{A}$, and thus $X = 0$. Conversely, if (ii) holds and $X\mathscr{A} = Y\mathscr{A}$, then $(X - Y)\mathscr{A} = 0$ and we conclude that $X - Y = 0$, that is, that $X = Y$.

Next, let us prove the equivalence of (ii) and (iii). If (ii) holds and $\sum c_i(X_i\mathscr{A}) = 0$, it follows that $(\sum c_iX_i)\mathscr{A} = 0$ and (ii) implies that $\sum c_iX_i = 0$. However, the X's are linearly independent, so every $c_i = 0$. This shows that $\{X_1\mathscr{A}, X_2, \ldots \mathscr{A}, X_n\mathscr{A}\}$ is a linearly independent set and, by Theorem 11.23(ii), it is a basis of V. Conversely, let us assume (iii) and prove (ii). Suppose that $X\mathscr{A} = 0$. Since we may write $X = \sum c_iX_i$, it follows that

$$0 = X\mathscr{A} = \left(\sum c_iX_i\right)\mathscr{A} = \sum c_i(X_i\mathscr{A}).$$

But, by our assumption, $\{X_1\mathscr{A}, X_2\mathscr{A}, \ldots, X_n\mathscr{A}\}$ is a basis of V and therefore these vectors are linearly independent. We conclude that every $c_i = 0$, and hence that $X = 0$. We have thus proved the equivalence of (ii) and (iii).

Since $V\mathscr{A} = [X_1\mathscr{A}, X_2\mathscr{A}, \ldots, X_n\mathscr{A}]$, clearly (iii) implies (iv). Conversely, if $V\mathscr{A} = V$, we must have $V = [X_1\mathscr{A}, X_2\mathscr{A}, \ldots, X_n\mathscr{A}]$ and these vectors are a basis of V (by Theorem 11.23(iv)). Thus (iii) and (iv) are equivalent.

Since $V\mathscr{A}$ is a subspace of V, $V\mathscr{A} = V$ if and only if $\dim(V\mathscr{A}) = \dim V$, that is, if and only if $\mathscr{A}$ is nonsingular. This shows the equivalence of (iv) and (v).

Finally, we have proved in Theorem 15.12, without any restriction on V, that (vi) is equivalent to (i) and (iv) together. But, with the present restriction that V have finite dimension, what we have already proved above shows that (i) and (iv) are equivalent. This completes the proof of the theorem.

We may observe that since $\mathscr{A}$ is the multiplicative inverse of $\mathscr{A}^{-1}$, the equivalence of (v) and (vi) shows that $\mathscr{A}^{-1}$ *is nonsingular if and only if $\mathscr{A}$ is nonsingular.*

Let us now give an example to indicate one possible way of actually computing the multiplicative inverse of a given nonsingular linear transformation.

Example Compute the multiplicative inverse of the linear transformation of $V_2(\mathbf{R})$ defined by $(x_1, x_2)\mathscr{A} = (x_1 - x_2, x_1 + x_2)$.

Solution: The set $\{(1, 0), (0, 1)\}$ of unit vectors is a basis of $V_2(\mathbf{R})$, and we find that

$$(1, 0)\mathscr{A} = (1, 1), \quad (0, 1)\mathscr{A} = (-1, 1).$$

Now these image vectors $(1, 1)$ and $(-1, 1)$ are linearly independent and form a basis of $V_2(\mathbf{R})$. By definition of $\mathscr{A}^{-1}$, we therefore have

$$(1, 1)\mathscr{A}^{-1} = (1, 0), \quad (-1, 1)\mathscr{A}^{-1} = (0, 1).$$

But an arbitrary element (x_1, x_2) of $V_2(\mathbf{R})$ can be expressed in the form

$$(x_1, x_2) = \frac{x_1 + x_2}{2}(1, 1) + \frac{x_2 - x_1}{2}(-1, 1),$$

and hence we see that $\mathscr{A}^{-1}$ is as follows:

$$(x_1, x_2)\mathscr{A}^{-1} = \frac{x_1 + x_2}{2}(1, 1)\mathscr{A}^{-1} + \frac{x_2 - x_1}{2}(-1, 1)\mathscr{A}^{-1}$$

$$= \frac{x_1 + x_2}{2}(1, 0) + \frac{x_2 - x_1}{2}(0, 1)$$

$$= \left(\frac{x_1 + x_2}{2}, \frac{x_2 - x_1}{2}\right).$$

The reader may check these calculations by verifying that if (x_1, x_2) is an element of $V_2(\mathbf{R})$, then

$$(x_1, x_2)\mathscr{A}\mathscr{A}^{-1} = (x_1, x_2)\mathscr{A}^{-1}\mathscr{A} = (x_1, x_2).$$

The next theorem will give some information about the rank of a product of linear transformations. For convenience, we shall designate the *rank* of the linear transformation $\mathscr{A}$ by "rank $\mathscr{A}$."

15.19 THEOREM

(i) *If $\mathscr{A}, \mathscr{B} \in L$, then rank $(\mathscr{A}\mathscr{B}) \leq$ rank $\mathscr{A}$ and also rank $(\mathscr{A}\mathscr{B}) \leq$ rank $\mathscr{B}$.*

(ii) *If $\mathscr{A}, \mathscr{B} \in L$ and $\mathscr{A}$ is nonsingular, then rank $(\mathscr{A}\mathscr{B}) =$ rank $(\mathscr{B}\mathscr{A}) =$ rank $\mathscr{B}$.*

Proof: Let us first apply Theorem 15.16 with U replaced by $V\mathscr{A}$ and $\mathscr{A}$ by $\mathscr{B}$. We then have dim $((V\mathscr{A})\mathscr{B}) \leq$ dim $(V\mathscr{A})$. But, by the definition of the product of linear transformations, $(V\mathscr{A})\mathscr{B} = V(\mathscr{A}\mathscr{B})$. Accordingly, we see that rank $(\mathscr{A}\mathscr{B}) \leq$ rank $\mathscr{A}$.

Since $V\mathscr{A} \subseteq V$, it follows that $V(\mathscr{A}\mathscr{B}) = (V\mathscr{A})\mathscr{B} \subseteq V\mathscr{B}$. Hence, $\dim (V(\mathscr{A}\mathscr{B})) \leq \dim (V\mathscr{B})$, that is, rank $(\mathscr{A}\mathscr{B}) \leq$ rank $\mathscr{B}$. We have thus established part (i) of the theorem.

To prove the second part, suppose that $\mathscr{A}$ is nonsingular. Then, by the preceding theorem, $\mathscr{A}$ has a multiplicative inverse $\mathscr{A}^{-1}$, and we can write $\mathscr{B} = \mathscr{A}^{-1}(\mathscr{A}\mathscr{B})$. Now the part of the theorem already proved assures us that the rank of a product does not exceed the rank of either factor. Hence, rank $\mathscr{B} \leq$ rank $(\mathscr{A}\mathscr{B})$. On the other hand, we know from part (i) of the theorem that rank $(\mathscr{A}\mathscr{B}) \leq$ rank $\mathscr{B}$, and we conclude that rank $(\mathscr{A}\mathscr{B}) =$ rank $\mathscr{B}$. To show that also rank $(\mathscr{B}\mathscr{A}) =$ rank $\mathscr{B}$, we need only write $\mathscr{B} = (\mathscr{B}\mathscr{A}) \mathscr{A}^{-1}$, and apply a similar argument.

The following important result is a special case of the second part of the theorem just proved, and is also an easy consequence of Theorem 15.18.

15.20 COROLLARY *The product of two nonsingular linear transformations of a vector space V is itself a nonsingular linear transformation of V.*

If $\mathscr{A}$ is a linear transformation of V, we have previously called the set of all vectors X of V such that $X\mathscr{A} = 0$ (actually a subspace of V by Theorem 11.32) the *kernel* of $\mathscr{A}$. However, in the present setting it is frequently given an alternate name as follows.

15.21 DEFINITION If $\mathscr{A} \in L$, the subspace of V consisting of all vectors X of V such that $X\mathscr{A} = 0$ is called the *null space* of $\mathscr{A}$. The dimension of the null space of $\mathscr{A}$ is called the *nullity* of $\mathscr{A}$.

The equivalence of conditions (ii) and (v) of Theorem 15.18 shows that if $\mathscr{A}$ is nonsingular (that is, has rank n), then the nullity of $\mathscr{A}$ is zero. This is a special case of the following theorem.

15.22 THEOREM *If* $\dim V = n$ *and* $\mathscr{A}$ *is a linear transformation of V of rank r, then* $\mathscr{A}$ *has nullity* $n - r$.

Proof: In view of the preceding remarks we may restrict attention to the case in which the nullity k of $\mathscr{A}$ is positive. Let $\{Y_1, \ldots, Y_k\}$ be a basis of the null space of $\mathscr{A}$, and let us extend this set to a basis

$$\{Y_1, \ldots, Y_k, \ldots, Y_n\}$$

of V. Since $Y_i\mathscr{A} = 0$ $(i = 1, 2, \ldots, k)$, it follows easily that

$$V\mathscr{A} = [Y_{k+1}\mathscr{A}, \ldots, Y_n\mathscr{A}].$$

We shall now show that $\{Y_{k+1}\mathscr{A}, \ldots, Y_n\mathscr{A}\}$ is a linearly independent set and hence a basis of $V\mathscr{A}$. Suppose that

$$c_{k+1}(Y_{k+1}\mathscr{A}) + \cdots + c_n(Y_n\mathscr{A}) = 0,$$

where $c_{k+1}, \ldots, c_n$ are elements of F. It follows that

$$(c_{k+1}Y_{k+1} + \cdots + c_nY_n)\mathscr{A} = 0,$$

and hence that

$$c_{k+1}Y_{k+1} + \cdots + c_nY_n$$

is in the null space of $\mathscr{A}$. Hence, this vector is a linear combination of the basis elements $Y_1, Y_2, \ldots, Y_k$ of this null space. However, since $\{Y_1, \ldots, Y_n\}$ is a linearly independent set, we conclude that $c_{k+1} = 0, \ldots, c_n = 0$. This shows that $\{Y_{k+1}\mathscr{A}, \ldots, Y_n\mathscr{A}\}$ is a linearly independent set, and therefore a basis of $V\mathscr{A}$. Accordingly, $\dim(V\mathscr{A}) = n - k$. Since $r = \dim(V\mathscr{A})$, it follows that $r = n - k$ or $k = n - r$, and the proof of the theorem is complete.

EXERCISES

1. In each of the following, find the rank of the linear transformation $\mathscr{A}$ of $V_3(\mathbf{R})$ and find a basis for the null space of $\mathscr{A}$:

   (a) $(x_1, x_2, x_3)\mathscr{A} = (x_1 + 2x_2 - x_3, 2x_1 + x_2 + x_3, x_2 - x_3)$,

   (b) $(x_1, x_2, x_3)\mathscr{A}$
   $$= (2x_1 - x_2 + x_3, x_1 + 2x_2 - x_3, x_1 + 7x_2 - 4x_3),$$

   (c) $(x_1, x_2, x_3)\mathscr{A} = (x_1 + x_2, x_1 + x_2, x_1 + x_2)$.

2. In each of the following, find the multiplicative inverse of the given linear transformation of $V_2(\mathbf{R})$:

   (a) $(x_1, x_2)\mathscr{A} = (2x_1 - x_2, x_1 + x_2)$,

   (b) $(x_1, x_2)\mathscr{A} = (x_2, -x_1)$,

   (c) $(x_1, x_2)\mathscr{A} = (x_1 + 2x_2, 2x_1 + x_2)$.

3. Find the multiplicative inverse of the linear transformation $\mathscr{A}$ of $V_3(\mathbf{R})$ defined by $(x_1, x_2, x_3)\mathscr{A} = (x_1 + x_2 + x_3, x_2 + x_3, x_3)$.

4. Let V be the vector space consisting of all polynomials of degree not greater than two, together with the zero polynomial, over the real field **R**. Now let $\mathscr{A}$ be the linear transformation of V defined by $f(x)\mathscr{A} = f'(x)$, where $f'(x)$ is the derivative of $f(x)$. [See Exercise 11 of Section 10.2].

(a) Find the rank and the nullity of each of the following linear transformations of V: $\mathscr{A}, \mathscr{A}^2, \mathscr{A}^3, \mathscr{A} + \mathscr{T}$, where $\mathscr{T}$ is the unity.

(b) Find the multiplicative inverse of the one of these linear transformations which is nonsingular.

In Exercises 5–8 the vector space is assumed to have finite dimension.

5. If c is a nonzero element of F and $\mathscr{A} \in L$, show that the rank of $(c\mathscr{A})$ is equal to the rank of $\mathscr{A}$.

6. Prove that if $\mathscr{A}, \mathscr{B} \in L$ and $\mathscr{A}\mathscr{B}$ is nonsingular, then both $\mathscr{A}$ and $\mathscr{B}$ are nonsingular.

7. Prove that if $\mathscr{A}, \mathscr{B} \in L$ and $\mathscr{A}\mathscr{B} = \mathscr{T}$, then also $\mathscr{B}\mathscr{A} = \mathscr{T}$, and therefore $\mathscr{B} = \mathscr{A}^{-1}$.

8. If $\mathscr{A}, \mathscr{B} \in L$, prove that rank $(\mathscr{A} + \mathscr{B}) \leq$ rank $\mathscr{A}$ + rank $\mathscr{B}$. [*Hint:* Using the Definition 11.26 of the sum of two subspaces, observe that $V(\mathscr{A} + \mathscr{B}) \subseteq V\mathscr{A} + V\mathscr{B}$.]

9. Let $W(F)$ be the vector space of Example 2 of Section 11.2 whose elements are infinite sequences of elements of F. Verify that the mapping $\mathscr{A}: W \to W$ defined by

$$(a_1, a_2, a_3, \ldots)\mathscr{A} = (0, a_1, a_2, \ldots)$$

is a linear transformation of W which is a one-one mapping but not an onto mapping. Similarly, verify that the mapping $\mathscr{B}: W \to W$ defined by

$$(a_1, a_2, a_3, \ldots)\mathscr{B} = (a_2, a_3, \ldots)$$

is an onto mapping but not a one-one mapping. Why do these examples not violate Theorem 15.18?

15.4 ALGEBRA OF MATRICES

We have defined a basis of a vector space to be a set of vectors having certain properties, and the order of writing down these vectors was of no significance.

Now, however, we wish to specify an order for the elements of a basis, and shall then speak of an *ordered basis*. Thus, for example, if $\{X_1, X_2, X_3\}$ is a basis of a vector space V of dimension three, then X_1, X_2, X_3 and X_2, X_1, X_3 would be different ordered bases of V, although the *sets* $\{X_1, X_2, X_3\}$ and $\{X_2, X_1, X_3\}$ are equal.

Throughout this section we shall let V be a vector space of dimension $n > 0$ over a field F, L the algebra of all linear transformations of V, and X_1, $X_2, \ldots, X_n$ a fixed ordered basis of V.

Suppose, now, that $\mathscr{A} \in L$. Then each of the vectors $X_i\mathscr{A}$ is uniquely expressible as a linear combination of the basis elements, so that there exist elements a_{ij} ($i, j = 1, 2, \ldots, n$) of F, *uniquely determined* by $\mathscr{A}$, such that

$$\begin{aligned}
X_1\mathscr{A} &= a_{11}X_1 + a_{12}X_2 + \cdots + a_{1n}X_n, \\
X_2\mathscr{A} &= a_{21}X_1 + a_{22}X_2 + \cdots + a_{2n}X_n, \\
&\cdots \cdots \cdots \cdots \cdots \cdots \cdots \cdots \cdots \cdots \\
X_n\mathscr{A} &= a_{n1}X_n + a_{n2}X_2 + \cdots + a_{nn}X_n.
\end{aligned}$$

Of course, we may also write these equations in the following condensed form:

15.23
$$X_i\mathscr{A} = \sum_{j=1}^{n} a_{ij}X_j, \qquad (i = 1, 2, \ldots, n).$$

Let us restate what we have just observed in the following way. Each linear transformation $\mathscr{A}$ of V has associated with it, by Equations 15.23, a unique matrix $A = (a_{ij})$ of order n over F. Conversely, if $A = (a_{ij})$ is a given matrix of order n over F, Corollary 15.14 shows that there exists a unique linear transformation $\mathscr{A}$ of V such that $\mathscr{A}$ and A are related as in Equations 15.23.

Let us henceforth denote by F_n the set of all matrices of order n over F. To avoid any possible confusion, perhaps we should state that two elements of F_n are considered as equal only if they are identical. That is, if (a_{ij}) and (b_{ij}) are elements of F_n, $(a_{ij}) = (b_{ij})$ means that $a_{ij} = b_{ij}$ for all $i, j = 1, 2, \ldots, n$.

We can now state in the following precise way what we have observed above. The mapping

15.24
$$\mathscr{A} \to A = (a_{ij}), \qquad \mathscr{A} \in L,$$

defined by Equations 15.23, is a one-one mapping of L onto F_n.

Inasmuch as L is an algebra over F, it is almost obvious that we can use this one-one mapping of L onto F_n to define operations of addition, multiplication, and scalar multiplication of F_n in such a way that F_n will be an algebra over F, which is isomorphic to L. We proceed to consider each of these operations in turn.

First, let us consider addition, and let $A = (a_{ij})$ and $B = (b_{ij})$ be elements of F_n. Suppose, further, that under the mapping 15.24, $\mathscr{A} \to A$ and $\mathscr{B} \to B$. Then

$$X_i\mathscr{A} = \sum_{j=1}^{n} a_{ij}X_j, \qquad (i = 1, 2, \ldots, n),$$

and

$$X_i\mathscr{B} = \sum_{j=1}^{n} b_{ij}X_j, \qquad (i = 1, 2, \ldots, n).$$

Now, by the definition of addition of linear transformations, it follows that

$$X_i(\mathscr{A} + \mathscr{B}) = X_i\mathscr{A} + X_i\mathscr{B} = \sum_{j=1}^{n} (a_{ij} + b_{ij})X_j, \qquad (i = 1, 2, \ldots, n).$$

Accordingly, we see that under the mapping 15.24,

$$\mathscr{A} + \mathscr{B} \rightarrow (a_{ij} + b_{ij}).$$

This leads us to *define* addition in F_n as follows:

15.25 $$(a_{ij}) + (b_{ij}) = (a_{ij} + b_{ij}).$$

That is, the element of the matrix $A + B$ in any fixed position is obtained by adding the elements of A and of B that are in that position. Taking $n = 2$, and F to be the field of rational numbers, we have as an illustration:

$$\begin{bmatrix} 4 & 0 \\ -2 & 3 \end{bmatrix} + \begin{bmatrix} -1 & 2 \\ 1 & 2 \end{bmatrix} = \begin{bmatrix} 3 & 2 \\ -1 & 5 \end{bmatrix}.$$

We next consider multiplication, and let A and B be as above. Then, using first the definition of a product of linear transformations, we have the following:

$$X_i(\mathscr{A}\mathscr{B}) = (X_i\mathscr{A})\mathscr{B} = \left(\sum_{k=1}^{n} a_{ik}X_k\right)\mathscr{B}$$
$$= \sum_{k=1}^{n} a_{ik}(X_k\mathscr{B})$$
$$= \sum_{k=1}^{n} a_{ik}\left(\sum_{j=1}^{n} b_{kj}X_j\right), \qquad (i = 1, 2, \ldots, n).$$

By rearranging the order of summation, this can be written in the form

$$X_i(\mathscr{A}\mathscr{B}) = \sum_{j=1}^{n} \left(\sum_{k=1}^{n} a_{ik}b_{kj}\right)X_j, \qquad (i = 1, 2, \ldots, n).$$

Hence, under the mapping 15.24,

$$\mathscr{A}\mathscr{B} \rightarrow \left(\sum_{k=1}^{n} a_{ik}b_{kj}\right).$$

Accordingly, we *define* multiplication in F_n as follows:

15.26
$$(a_{ij})(b_{ij}) = \left(\sum_{k=1}^{n} a_{ik}b_{kj} \right).$$

This definition may be stated in words as follows. The element in the ith row and jth column of the product AB is the sum of the products of the elements of the ith row of A by the corresponding elements of the jth column of B. This can be expressed in another way as follows. Let $A_1, A_2, \ldots, A_n$ be the row vectors of A; and let $B^1, B^2, \ldots, B^n$ be the column vectors of B. In terms of inner products of vectors, we may then write 15.26 in the following alternate form:

$$AB = (A_i \cdot B^j).$$

Example As a simple example of multiplication of matrices, using the same matrices as were used above to illustrate addition, we have

$$\begin{bmatrix} 4 & 0 \\ -2 & 3 \end{bmatrix} \begin{bmatrix} -1 & 2 \\ 1 & 2 \end{bmatrix}$$

$$= \begin{bmatrix} 4(-1) + 0(1) & 4(2) + 0(2) \\ -2(-1) + 3(1) & -2(2) + 3(2) \end{bmatrix} = \begin{bmatrix} -4 & 8 \\ 5 & 2 \end{bmatrix}.$$

On the other hand, the reader may verify that

$$\begin{bmatrix} -1 & 2 \\ 1 & 2 \end{bmatrix} \begin{bmatrix} 4 & 0 \\ -2 & 3 \end{bmatrix} = \begin{bmatrix} -8 & 6 \\ 0 & 6 \end{bmatrix},$$

and clearly the commutative law of multiplication does not hold in F_n.

Finally, we consider scalar multiplication. If $c \in F$ and $\mathscr{A} \in L$, by the definition of scalar multiplication in L, we have

$$X_i(c\mathscr{A}) = c(X_i \mathscr{A}) = c \sum_{j=1}^{n} a_{ij}X_j = \sum_{j=1}^{n} (ca_{ij})X_j, \qquad (i = 1, 2, \ldots, n).$$

This suggests that we *define* scalar multiplication in F_n as follows:

15.27
$$c(a_{ij}) = (ca_{ij}).$$

Otherwise expressed, if $A \in F_n$, cA is the matrix obtained by multiplying every element of A by c. As a simple example, we have

$$2 \begin{bmatrix} 4 & 0 \\ -2 & 3 \end{bmatrix} = \begin{bmatrix} 8 & 0 \\ -4 & 6 \end{bmatrix}.$$

We have now defined addition, multiplication, and scalar multiplication on F_n in such a way that all of these operations are preserved under the mapping 15.24. That is, if under this mapping $\mathscr{A} \to A$ and $\mathscr{B} \to B$, then $\mathscr{A} + \mathscr{B} \to A + \mathscr{B}$, $\mathscr{A}\mathscr{B} \to AB$, and $c\mathscr{A} \to cA$ for $c \in F$. We have therefore established the following result, it being understood that an isomorphism of two algebras over F means an isomorphism as rings and also as vector spaces.

15.28 THEOREM *With addition, multiplication, and scalar multiplication defined respectively by 15.25, 15.26, and 15.27, the set F_n of all matrices of order n over F is an algebra over F. Moreover this algebra is isomorphic to the algebra $L = \mathrm{Hom}_F(V,V)$ of all linear transformations of a vector space V of dimension n over F.*

It follows easily from 15.25 that the zero element of the algebra F_n is the matrix of order *n* all of whose elements are zero. We shall usually designate this zero matrix by the familiar symbol 0. Of course, this matrix is the image of the zero linear transformation under the mapping 15.24.

If $\mathscr{I}$ is the unity of L, we have $X_i\mathscr{I} = X_i$ ($i = 1, 2, \ldots, n$), and the image of $\mathscr{I}$ under the mapping 15.24 is the matrix with 1's on the principal diagonal and zeros elsewhere. This matrix must then be the unity of F_n, as can also be verified by use of 15.26. The unity of F_n will henceforth be denoted by I. For example, if $n = 3$, we have

$$I = \begin{bmatrix} 1 & 0 & 0 \\ 0 & 1 & 0 \\ 0 & 0 & 1 \end{bmatrix}.$$

Perhaps we should emphasize that the isomorphism 15.24 of L onto F_n depends upon the ordered basis of L which is used. A different ordered basis would lead to a different isomorphism, so that there are many different isomorphisms of L onto F_n.

The matrix A which corresponds to the linear transformation $\mathscr{A}$ under the mapping 15.24 may be referred to as *the matrix of $\mathscr{A}$ relative to the ordered basis* $X_1, X_2, \ldots, X_n$ of V. In a later section we shall determine the relationship which exists between the matrices of a linear transformation of a vector space V relative to two different ordered bases of V.

The following theorem will justify the use of some of our previous terminology.

15.29 THEOREM *The rank of a linear transformation $\mathscr{A}$ of a vector space V is equal to the rank of the matrix of $\mathscr{A}$ relative to any ordered basis of V.*

We use the above notation according to which $\mathscr{A}$ and A are related by Equations 15.23. Since $V\mathscr{A} = [X_1\mathscr{A}, X_2\mathscr{A}, \ldots, X_n\mathscr{A}]$, Theorem 11.25(ii) shows that the rank of $\mathscr{A}$ is the maximal number of linear independent vectors in the set $\{X_1\mathscr{A}, X_2\mathscr{A}, \ldots, X_n\mathscr{A}\}$. Similarly, the (row) rank of the matrix A is the maximum number of linearly independent vectors in the set $\{A_1, A_2, \ldots, A_n\}$ of row vectors of A. The main part of the proof that these numbers are the same is in establishing the following lemma.

15.30 LEMMA

If c_i $(i = 1, 2, \ldots, n)$ are elements of F, then

$$\sum_{i=1}^{n} c_i(X_i\mathscr{A}) = 0 \quad \text{if and only if} \quad \sum_{i=1}^{n} c_i A_i = 0.$$

Proof: To prove this lemma, we first observe that Equations 15.23 show that

$$\sum_{i=1}^{n} c_i(X_i\mathscr{A}) = \sum_{i=1}^{n} c_i\left(\sum_{j=1}^{n} a_{ij}X_j\right) = \sum_{j=1}^{n} \left(\sum_{i=1}^{n} c_i a_{ij}\right)X_j.$$

Since the X's are linearly independent, it follows that

$$\sum_{i=1}^{n} c_i(X_i\mathscr{A}) = 0 \quad \text{if and only if} \quad \sum_{i=1}^{n} c_i a_{ij} = 0 \quad \text{for} \quad j = 1, 2, \ldots, n.$$

However, this last set of equations can be written in the vector form $\sum_{i=1}^{n} c_i A_i = 0$, and the lemma is established.

We shall leave as an exercise the application of this lemma to complete the proof of the theorem.

Just as for linear transformations, it is customary to call a matrix of order n *singular* or *nonsingular* according as its rank is less than n or equal to n.

We conclude this section with some examples illustrating how one computes the matrix of a linear transformation relative to a given ordered basis.

Example 1 Find the matrix of the linear transformation $\mathscr{A}$ of $V_2(\mathbf{R})$ defined by $(x_1, x_2)\mathscr{A} = (3x_1 + x_2, x_1 - 2x_2)$ relative to the ordered basis $(1, 0), (0, 1)$ of $V_2(\mathbf{R})$. Do the same thing relative to the ordered basis $(1, 1), (-1, 1)$ of $V_2(\mathbf{R})$.

Solution: Equations 15.23 in this case become the following:

$$(1, 0)\mathscr{A} = (3, 1) = 3(1, 0) + (0, 1),$$
$$(0, 1)\mathscr{A} = (1, -2) = (1, 0) - 2(0, 1).$$

Thus the matrix of $\mathscr{A}$ relative to the ordered basis $(1, 0)$, $(0, 1)$ is as follows:

$$\begin{bmatrix} 3 & 1 \\ 1 & -2 \end{bmatrix}.$$

Let us now compute the matrix of $\mathscr{A}$ relative to the ordered basis $(1, 1)$, $(-1, 1)$ of $V_2(\mathbf{R})$. Some of the calculations in the example of the preceding section can be used to show that in this case we have

$$(1, 1)\mathscr{A} = (4, -1) = \tfrac{3}{2}(1, 1) - \tfrac{5}{2}(-1, 1),$$
$$(-1, 1)\mathscr{A} = (-2, -3) = -\tfrac{5}{2}(1, 1) - \tfrac{1}{2}(-1, 1).$$

Accordingly, the matrix

$$\begin{bmatrix} \tfrac{3}{2} & -\tfrac{5}{2} \\ -\tfrac{5}{2} & -\tfrac{1}{2} \end{bmatrix}$$

is the matrix of $\mathscr{A}$ relative to the ordered basis $(1, 1)$, $(-1, 1)$ of $V_2(\mathbf{R})$.

Example 2 Let V be the vector space over $\mathbf{R}$ consisting of all polynomials in an indeterminate x of degree no more than two, together with the zero polynomial. Let $\mathscr{A}$ be the linear transformation of V defined by $f(x)\mathscr{A} = f'(x)$, where $f'(x)$ is derivative of $f(x)$.

(i) Find the matrix of $\mathscr{A}$ relative to the ordered basis x^2, x, 1 of V.

(ii) Find the matrix of $\mathscr{A}$ relative to the ordered basis $x + 1$, 1, $x^2 + x$ of V.

Solution: (i) The following equations

$$x^2\mathscr{A} = 0 \cdot x^2 + 2x + 0,$$
$$x\mathscr{A} = 0 \cdot x^2 + 0 \cdot x + 1,$$
$$1\mathscr{A} = 0 \cdot x^2 + 0 \cdot x + 0 \cdot 1,$$

show that the matrix of $\mathscr{A}$ relative to the ordered basis x^2, x, 1 of V is as follows:

$$\begin{bmatrix} 0 & 2 & 0 \\ 0 & 0 & 1 \\ 0 & 0 & 0 \end{bmatrix}.$$

(ii) It is easy to verify that

$$(x + 1)\mathscr{A} = 0 \cdot (x + 1) + 1 \cdot 1 + 0 \cdot (x^2 + x),$$
$$1\mathscr{A} = 0 \cdot (x + 1) + 0 \cdot 1 + 0 \cdot (x^2 + x),$$
$$(x^2 + x)\mathscr{A} = 2 \cdot (x + 1) - 1 \cdot 1 + 0 \cdot (x^2 + x).$$

Thus the matrix of $\mathscr{A}$ relative to the ordered basis $x + 1,\ 1,\ x^2 + x$ of V is the following:

$$\begin{bmatrix} 0 & 1 & 0 \\ 0 & 0 & 0 \\ 2 & -1 & 0 \end{bmatrix}.$$

EXERCISES

1. Let A and B be the following matrices of order 3 over $\mathbf{Q}$:

$$A = \begin{bmatrix} 1 & -1 & 2 \\ 0 & 1 & 3 \\ 2 & 1 & -2 \end{bmatrix}, \qquad B = \begin{bmatrix} 1 & -2 & 3 \\ 2 & 1 & -1 \\ 1 & 0 & 1 \end{bmatrix}.$$

Compute each of the following: AB, BA, A^2, B^2, $(A + B)^2$.

2. If A is the matrix of the preceding exercise, verify that $A^3 - 10A + 15I = 0$.

3. If A is the same matrix as above, and

$$C = -\tfrac{1}{15}\begin{bmatrix} -5 & 0 & -5 \\ 6 & -6 & -3 \\ -2 & -3 & 1 \end{bmatrix},$$

verify that $AC = CA = I$, and hence that $C = A^{-1}$.

4. If B is an element of F_n which has an inverse B^{-1} in F_n, verify that the mapping $A \to B^{-1}AB$ $(A \in F_n)$ is an isomorphism of the algebra F_n onto itself.

5. Complete the proof of Theorem 15.29.

6. Find the matrix of each linear transformation of Exercise 1 of the preceding section relative to the ordered basis $(1, 0, 0)$, $(0, 1, 0)$, $(0, 0, 1)$ of $V_3(\mathbf{R})$.

7. Let V be the vector space of Example 2 above, and let $\mathscr{C}$ and $\mathscr{D}$ be linear transformations of V defined as follows: $f(x)\mathscr{C} = f(x + 1), f(x)\mathscr{D} = f(x - 1)$. Find the matrix of $\mathscr{C}$ and of $\mathscr{D}$ relative to the ordered basis x^2, x, 1 of V, and verify that these matrices are multiplicative inverses of each other.

8. Let V be the vector space of all polynomials of degree not greater than three, together with the zero polynomial, over a field F. Find the matrix of each of

556 · LINEAR TRANSFORMATIONS AND MATRICES

the following linear transformations of V relative to the ordered basis 1, x, x^2, x^3 of V:

(a) $f(x)\mathscr{A} = 2f(x)$,

(b) $f(x)\mathscr{A} = f(-x)$,

(c) $f(x)\mathscr{A} = f(x) + f(-x)$.

9. If A is the matrix of a linear transformation $\mathscr{A}$ relative to an ordered basis $X_1, X_2, \ldots, X_n$ of a vector space V, how would you describe the matrix of $\mathscr{A}$ relative to an ordered basis obtained from the given one by interchanging X_i and X_j $(i \neq j)$?

10. Verify directly that F_n has dimension n^2 as a vector space over F. (Cf. Theorems 11.42 and 15.28.)

11. Show that the subset of F_n consisting of those matrices all of whose elements below the principal diagonal are zero is a subalgebra of the algebra F_n.

12. If D is any element of F_n, let D' be the *transpose* of D. Prove that if A, $B \in F_n$, then $(A + B)' = A' + B'$ and $(AB)' = B'A'$. Prove also that if A has a multiplicative inverse, then A' has a multiplicative inverse and that $(A^{-1})' = (A')^{-1}$.

15.5 LINEAR TRANSFORMATIONS OF $V_n(F)$

As the reader may have already noticed in some of the examples and exercises, the results of the preceding section take a particularly simple form if we restrict V to be a vector space $V_n(F)$ and use the unit vectors $E_1, E_2, \ldots, E_n$ (as defined in Section 11.4) as our ordered basis. The isomorphism 15.24 of L onto F_n is now given by $\mathscr{A} \to A = (a_{ij})$, where

15.31 $$E_i\mathscr{A} = \sum_{j=1}^{n} a_{ij}E_j = (a_{i1}, a_{i2}, \ldots, a_{in}), \qquad (i = 1, 2, \ldots, n).$$

That is, $E_i\mathscr{A}$ is just the ith row vector A_i of the matrix A, and we can write 15.31 in the simpler form

15.32 $$E_i\mathscr{A} = A_i, \qquad (i = 1, 2, \ldots, n).$$

An arbitrary element X of $V_n(F)$ can be written in the form

15.33 $$X = (x_1, x_2, \ldots, x_n) = \sum_{i=1}^{n} x_i E_i,$$

and it follows from 15.32 that

$$X\mathscr{A} = \left(\sum_{i=1}^{n} x_i E_i\right)\mathscr{A} = \sum_{i=1}^{n} x_i(E_i\mathscr{A}) = \sum_{i=1}^{n} x_i A_i.$$

Thus $(V_n(F))\mathscr{A}$ is simply the row space of the matrix A, and we observe that in this special case it is particularly evident that the rank of $\mathscr{A}$ is equal to the (row) rank of the corresponding matrix A.

There is still another way of writing $X\mathscr{A}$. It is in terms of the inner product of X by the column vectors of A, as follows:

15.34 $$X\mathscr{A} = (X \cdot A^1, X \cdot A^2, \ldots, X \cdot A^n).$$

So far we have carefully distinguished between linear transformations and matrices. However, since L and F_n are isomorphic algebras it is possible to use an identical notation for these concepts. Whenever we wish to do so we shall henceforth consider that a matrix A of order n *is* the corresponding linear transformation $\mathscr{A}$ of $V_n(F)$ relative to the unit vectors. That is, in view of 15.32, the linear transformation A of $V_n(F)$ is the linear transformation which maps the ith unit vector E_i onto the ith row vector A_i of A. We may then write 15.34 as follows:

15.35 $$XA = (X \cdot A^1, X \cdot A^2, \ldots, X \cdot A^n).$$

It will be observed that XA can be computed by using a "row by column" multiplication of the *vector* X of $V_n(F)$ by the *matrix* A of F_n.

Example Let us illustrate the use of this notation. Suppose that

$$A = \begin{bmatrix} 1 & -1 & 2 \\ 0 & 1 & -1 \\ 3 & 2 & 1 \end{bmatrix}$$

is considered as a linear transformation of the vector space $V_3(\mathbf{Q})$. Then the image of the vector $(2, -1, 3)$ under the mapping A, as given by 15.35, is computed as follows:

$$(2, -1, 3)\begin{bmatrix} 1 & -1 & 2 \\ 0 & 1 & -1 \\ 3 & 2 & 1 \end{bmatrix}$$
$$= (2\cdot 1 + (-1)0 + 3\cdot 3, 2(-1) + (-1)1 + 3\cdot 2,$$
$$2\cdot 2 + (-1)(-1) + 3\cdot 1) = (11, 3, 8).$$

15.6 ADJOINT AND INVERSE OF A MATRIX

If we think of a matrix A of order n as being a linear transformation of $V_n(F)$, it is clear from Theorem 15.18 that A has a multiplicative inverse A^{-1} in F_n if and only if it is nonsingular. Moreover, by Theorem 14.28, A is nonsingular if and only if $|A| \neq 0$. We now proceed to show how determinants may be used to compute the multiplicative inverse of a nonsingular matrix.

If $A = (a_{ij}) \in F_n$, we use the notation of the preceding chapter and let A_{ij} denote the cofactor of the element a_{ij} in $|A|$. That is, $A_{ij} = (-1)^{i+j} |M_{ij}|$, where M_{ij} is the minor of a_{ij} in the matrix A. We now consider a certain matrix whose elements are cofactors of elements of A. It will be convenient to make the following definition.

15.36 DEFINITION If $A \in F_n$, the *adjoint* of A (which we shall write as adj A) is the transpose of the matrix (A_{ij}) of F_n; that is,

$$\text{adj } A = \begin{bmatrix} A_{11} & A_{21} & \cdots & A_{n1} \\ A_{12} & A_{22} & \cdots & A_{n2} \\ \cdot & \cdot & \cdot \cdot \cdot & \cdot \\ A_{1n} & A_{2n} & \cdots & A_{nn} \end{bmatrix}.$$

We shall now prove the following result.

15.37 THEOREM *If $A \in F_n$, then*

$$A(\text{adj } A) = (\text{adj } A)A = |A| \cdot I.$$

Moreover, if A is nonsingular, then

$$A^{-1} = |A|^{-1} \text{ adj } A.$$

Proof: By the definition of the product of two matrices, we see that the element in the pth row and the qth column of $A(\text{adj } A)$ is $\sum_{k=1}^{n} a_{pk} A_{qk}$. By 14.25 and 14.22, this element has the value zero if $p \neq q$, and is just $|A|$ if $p = q$. Hence, each element of the principal diagonal of the matrix $A(\text{adj } A)$ is $|A|$, and all other elements are zero. That is,

$$A(\text{adj } A) = |A| \cdot I,$$

where I is the unity of F_n. A similar argument, using 14.23 and 14.26, will show that also

$$(\text{adj } A)A = |A| \cdot I,$$

and the first statement of the theorem is established.   Using this result and the Definition 15.27 of scalar multiplication of matrices, we now see that

$$A(|A|^{-1} \text{ adj } A) = (|A|^{-1} \text{ adj } A)A = I,$$

and hence that $A^{-1} = |A|^{-1} \text{ adj } A$. This completes the proof of the theorem.

Example   As an illustration of this theorem, let C be the matrix

$$\begin{bmatrix} 1 & -1 & 2 \\ 0 & 1 & 2 \\ 1 & -3 & -4 \end{bmatrix}$$

of order three over **Q**.   Then a calculation shows that

$$\text{adj } C = \begin{bmatrix} 2 & -10 & -4 \\ 2 & -6 & -2 \\ -1 & 2 & 1 \end{bmatrix}.$$

The reader may now verify that

$$C(\text{adj } C) = (\text{adj } C)C = \begin{bmatrix} -2 & 0 & 0 \\ 0 & -2 & 0 \\ 0 & 0 & -2 \end{bmatrix} = -2I.$$

Accordingly, we have that

$$C^{-1} = -\tfrac{1}{2} \text{ adj } C = \begin{bmatrix} -1 & 5 & 2 \\ -1 & 3 & 1 \\ \tfrac{1}{2} & -1 & -\tfrac{1}{2} \end{bmatrix}.$$

Now that we have available the concept of the inverse of a matrix, it may be of interest to give a brief indication of how matrix methods may be used, in place of Cramer's Rule, to solve a system of n linear equations in n unknowns over a field F.   Suppose that we have the following system of equations:

$$\sum_{j=1}^{n} a_{ij}x_j = b_i, \qquad (i = 1, 2, \dots, n).$$

Let $A = (a_{ij})$ be the matrix of coefficients in this system of equations.   Moreover, let us set $B = (b_1, b_2, \dots, b_n)$ and $X = (x_1, x_2, \dots, x_n)$, where we may now consider $x_1, x_2, \ldots, x_n$ as unknown elements of F. Then it may be verified that the above system can be written in the following simple form

15.38 $$XA' = B,$$

it being understood that A' is the transpose of A. Let us now assume that A is nonsingular. Hence, also, A' is nonsingular (why?), and if we multiply the preceding equation on the right by the multiplicative inverse of A', we obtain

$$(XA')(A')^{-1} = B(A')^{-1}.$$

However,

$$(XA')(A')^{-1} = X(A'(A')^{-1}) = X,$$

and therefore

15.39 $$X = B(A')^{-1}.$$

This, then, is the solution of the system 15.38. Of course, it is the same solution as would be obtained by use of Cramer's Rule (14.35).

> **Example** As as example of the use of this notation, suppose that we have the following system of three linear equations in three unknowns over the field **Q** of rational numbers:
>
> $$\begin{array}{rcrcrcr} x_1 & & & + & x_3 & = & 1, \\ -x_1 & + & x_2 & - & 3x_3 & = & -2, \\ 2x_1 & + & 2x_2 & - & 4x_3 & = & 3. \end{array}$$
>
> This system can be written in the form 15.38 as follows:
>
> $$(x_1, x_2, x_3)\begin{bmatrix} 1 & -1 & 2 \\ 0 & 1 & 2 \\ 1 & -3 & -4 \end{bmatrix} = (1, -2, 3).$$
>
> The matrix appearing here is the matrix C whose inverse was computed above, and the solution 15.39 is therefore obtained by the following calculation:
>
> $$(x_1, x_2, x_3) = (1, -2, 3)\begin{bmatrix} -1 & 5 & 2 \\ -1 & 3 & 1 \\ \frac{1}{2} & -1 & -\frac{1}{2} \end{bmatrix} = (\tfrac{5}{2}, -4, -\tfrac{3}{2}).$$
>
> The unique solution is therefore $x_1 = \tfrac{5}{2}, x_2 = -4, x_3 = -\tfrac{3}{2}$.

EXERCISES

1. Let the matrix

$$A = \begin{bmatrix} 1 & 2 & 1 \\ -1 & 1 & -4 \\ -1 & 4 & -7 \end{bmatrix}$$

over Q be considered as a linear transformation of $V_3(Q)$. Verify that under this linear transformation both of the vectors $(2, 1, 3)$ and $(1, -1, 4)$ map into the vector $(-2, 17, -23)$. What does this fact tell us about the matrix A? Find a basis of the null space of A.

2. Find the adjoint and, if the matrix is nonsingular, the multiplicative inverse of each of the following matrices over Q:

(a) $\begin{bmatrix} 1 & 2 \\ 2 & -3 \end{bmatrix}$,

(e) $\begin{bmatrix} 2 & -1 & 3 \\ 1 & 2 & -1 \\ 1 & -8 & 9 \end{bmatrix}$.

(b) $\begin{bmatrix} -2 & 1 \\ 0 & 2 \end{bmatrix}$,

(f) $\begin{bmatrix} 0 & 0 & 2 \\ 1 & 0 & 3 \\ 3 & 4 & 2 \end{bmatrix}$,

(c) $\begin{bmatrix} 1 & -1 & 1 \\ -1 & 1 & 1 \\ 1 & 1 & -1 \end{bmatrix}$.

(g) $\begin{bmatrix} 1 & 0 & -1 & 0 \\ 0 & 2 & 0 & -3 \\ 2 & 0 & 0 & 1 \\ 1 & 0 & 1 & 2 \end{bmatrix}$.

(d) $\begin{bmatrix} 2 & -1 & 0 \\ 1 & 3 & -2 \\ 2 & 1 & 1 \end{bmatrix}$,

3. Use the method illustrated above to solve each of the following systems of linear equations over Q:

(a) $\begin{aligned} 2x_1 - x_2 + x_3 &= 2, \\ 3x_1 + x_2 - 2x_3 &= -1, \\ x_1 + 2x_2 + 3x_3 &= 3. \end{aligned}$

(b) $\begin{aligned} x_1 + x_2 - x_3 &= -3, \\ 2x_1 - x_2 - x_3 &= 8, \\ x_1 - 3x_2 + x_3 &= 17. \end{aligned}$

4. Of the sixteen matrices of order two over the field Z_2 verify that ten are singular and six are nonsingular.

5. Show that the group of all nonsingular matrices of order two over $\mathbf{Z}_2$ with operation multiplication is isomorphic to the symmetric group on three symbols.

15.7 EQUIVALENCE OF MATRICES

All matrices will be of order n over a field F, that is, elements of F_n. We now study again the elementary row and column operations, as defined in 13.24. There are three types, and we begin by introducing some notation that will be helpful in referring to specific elementary operations.

Let $\mathscr{R}_{ij}$ stand for the operation of interchanging the ith and jth rows, let $\mathscr{R}_i(c)$ stand for the operation of multiplying the ith row by the nonzero element c of F, and let $\mathscr{R}_{ij}(d)$ stand for the operation of adding to the jth row d times the ith row, where $d \in F$.

In like manner, let $\mathscr{C}_{ij}$, $\mathscr{C}_i(c)$, and $\mathscr{C}_{ij}(d)$ represent the corresponding *column* operations.

When we speak of an elementary operation, we shall mean either an elementary row operation or an elementary column operation.

It is an important fact that the effect of an elementary operation can always be canceled by an elementary operation. Suppose, first, that matrix B is obtained from matrix A by applying an elementary operation $\mathscr{R}_{ij}$. Then, it is clear that the same operation $\mathscr{R}_{ij}$ applied to B will yield A again. Likewise, if the operation $\mathscr{R}_i(c)$ applied to A gives B, the operation $\mathscr{R}_i(c^{-1})$ applied to B gives A. In like manner, if $\mathscr{R}_{ij}(d)$ applied to A yields B, then $\mathscr{R}_{ij}(-d)$ applied to B yields A. Of course, similar statements apply to the column operations as well.

We now make the following definition.

15.40 DEFINITION If A, $B \in F_n$, we say that A is *equivalent* to B, and write $A \sim B$, if it is possible to pass from A to B by a finite sequence of elementary operations.

From the observations just made it follows easily that if $A \sim B$, then $B \sim A$. This is one of the defining properties (1.12) of an equivalence relation. The other two properties are obviously satisfied, and hence $\sim$ is an equivalence relation defined on F_n.

We may remark that other equivalence relations are often defined on F_n, and we shall briefly consider another one in Section 15.9. However, in the theory of matrices it is customary to use the words "equivalence" and "equivalent" with reference to the particular equivalence relation now being considered.

Now that we have an equivalence relation defined on F_n, we may consider the equivalence sets relative to this equivalence relation. The main part of this section will be devoted to the determination of these equivalence sets. Otherwise expressed, we shall find conditions under which two elements of F_n will be equivalent. For convenience of reference, let us first state the following fact, which is a consequence of Theorem 13.25 and Definition 13.30.

15.41 LEMMA   *If $A \sim B$, then rank A = rank B.*

We shall presently prove the converse of this lemma, from which it will follow that the elements of an equivalence set are just those matrices with some specified rank. Before proving this converse, we shall establish one more lemma.

If r is an integer ($0 \leq r \leq n$), let us denote by I_r the element of F_n which has a 1 in the first r places of the principal diagonal and zeros elsewhere. Clearly, I_0 is the zero and I_n the unity I of F_n. As an illustration of this notation, if $n = 3$, we have

$$I_1 = \begin{bmatrix} 1 & 0 & 0 \\ 0 & 0 & 0 \\ 0 & 0 & 0 \end{bmatrix} \quad \text{and} \quad I_2 = \begin{bmatrix} 1 & 0 & 0 \\ 0 & 1 & 0 \\ 0 & 0 & 0 \end{bmatrix}.$$

We are now in a position to state the following result.

15.42 LEMMA   *If $A \in F_n$ and rank $A = r$, then $A \sim I_r$.*

The method of proof of this lemma is suggested by the procedure used in Chapter 13 to reduce a system of linear equations to an echelon system. However, we can here carry the simplification somewhat further since we may use column operations as well as row operations.

Before proceeding, we give an example to illustrate the method of proof and also to clarify the notation and terminology introduced so far. Let us consider the following matrix over the rational field $\mathbf{Q}$:

$$D = \begin{bmatrix} 1 & 2 & -1 \\ 3 & 1 & 2 \\ 2 & -1 & 3 \end{bmatrix}.$$

First, we perform the operations $\mathscr{R}_{12}(-3)$ and $\mathscr{R}_{13}(-2)$. That is, we multiply the first row by -3 and add it to the second row, then multiply the first row by -2 and add it to the third row. Next, we perform the column operations $\mathscr{C}_{12}(-2)$ and $\mathscr{C}_{13}(1)$. At this stage, we have a 1 in the upper left-hand corner and zeros in all other positions of the first row and first column. Actually, we have the matrix

$$\begin{bmatrix} 1 & 0 & 0 \\ 0 & -5 & 5 \\ 0 & -5 & 5 \end{bmatrix}.$$

We now proceed to perform elementary operations that do not involve the first row or first column. In particular, the operation $\mathscr{R}_{23}(-1)$ makes the last row zero, and then the operation $\mathscr{R}_2(-1/5)$ places a 1 in the second row and second column. Finally, the column operation $\mathscr{C}_{23}(1)$ gives us the matrix I_2. We have therefore showed that $D \sim I_2$. Moreover, the elementary operations that we performed were as follows, and in this order:

15.43 $\mathscr{R}_{12}(-3),\ \mathscr{R}_{13}(-2),\ \mathscr{C}_{12}(-2),\ \mathscr{C}_{13}(1),\ \mathscr{R}_{23}(-1),\ \mathscr{R}_2(-\tfrac{1}{5}),\ \mathscr{C}_{23}(1).$

As a matter of fact, we could have obtained the same result by first performing all the specified row operations in the order in which they appear above, *followed* by the column operations in their specified order (or vice versa). However, we are here only concerned with the fact that there is at least one sequence of elementary operations by which we can pass from D to I_2.

> **Proof:** Let us return to the proof of the lemma and let $A = (a_{ij})$ be an element of F_n. If $r = 0$, then $A = 0$, and the result is trivial; hence we assume that $r > 0$. Then A has at least one nonzero element. If necessary, we can use suitable operations $\mathscr{R}_{ij}$ and $\mathscr{C}_{ij}$ (interchanging rows and interchanging columns) in order to get a nonzero element in the upper left-hand corner. For convenience, let us assume that a_{11} itself is different from zero. Then a matrix of the form
>
> $$B = \begin{bmatrix} 1 & 0 & 0 & \cdots & 0 \\ 0 & b_{22} & b_{23} & \cdots & b_{2n} \\ 0 & b_{32} & b_{33} & \cdots & b_{3n} \\ \cdot & \cdot & \cdot & \cdot & \cdot \\ 0 & b_{n2} & b_{n3} & \cdots & b_{nn} \end{bmatrix}$$
>
> can be obtained from A by the following sequence of elementary operations: $\mathscr{R}_1(a_{11}{}^{-1}),\ \mathscr{R}_{12}(-a_{21}),\ \mathscr{R}_{13}(-a_{31}),\ \ldots, \mathscr{R}_{1n}(-a_{n1}),$ $\mathscr{C}_{12}(-a_{11}{}^{-1}a_{12}),\ \mathscr{C}_{13}(-a_{11}{}^{-1}a_{13}),\ \ldots, \mathscr{C}_{1n}(-a_{11}{}^{-1}a_{1n}).$
>
> If some element b_{ij} of B ($i \le 2 \le n, j \le 2 \le n$) is different from zero, we can get such a nonzero element in the second row and second column by suitable interchange of rows and of columns. Then, proceeding as above, working only with rows and columns other than the first, we can apply elementary operations to B and get a matrix of the form
>
> $$C = \begin{bmatrix} 1 & 0 & 0 & \cdots & 0 \\ 0 & 1 & 0 & \cdots & 0 \\ 0 & 0 & c_{33} & \cdots & c_{3n} \\ 0 & 0 & c_{43} & \cdots & c_{4n} \\ \cdot & \cdot & \cdot & \cdot & \cdot \\ 0 & 0 & c_{n3} & \cdots & c_{nn} \end{bmatrix}.$$

If some c_{ij} is different from zero, this process can be repeated. We thus finally obtain a matrix of the form I_s with $s \leq n$. Now it was given that the rank of A is r, and it is obvious that the rank of I_s is s. Since we have $A \sim I_s$, Lemma 15.41 assures us that $r = s$, and the proof is complete.

We are now ready to prove the following result.

15.44 THEOREM *If $A, B \in F_n$, then $A \sim B$ if and only if rank $A =$ rank B.*

Proof: Of course, one part of this theorem is merely Lemma 15.41. To prove the other part, suppose that rank $A =$ rank $B = r$. Then, by the preceding lemma, $A \sim I_r$ and $B \sim I_r$. By the symmetric and transitive properties of the equivalence relation $\sim$, it follows at once that $A \sim B$.

We have now obtained one characterization of the equivalence sets relative to the equivalence relation $\sim$. The elements of an equivalence set $[A]$ are precisely those elements of F_n that have the same rank as A. A little later we shall obtain another characterization of these equivalence sets.

It is a fact of considerable importance in the theory of matrices that elementary operations can be effected by matrix multiplication. We shall briefly indicate how this is done, and then give a few simple consequences of this fact.

We begin by defining certain matrices of F_n as follows. Let E_{ij}, $E_i(c)$, and $E_{ij}(d)$ be the matrices obtained by applying the respective elementary operations $\mathscr{R}_{ij}$, $\mathscr{R}_i(c)$, and $\mathscr{R}_{ij}(d)$ to the matrix I. As examples, if $n = 3$, we have the following:

$$E_{12} = \begin{bmatrix} 0 & 1 & 0 \\ 1 & 0 & 0 \\ 0 & 0 & 1 \end{bmatrix}, \quad E_2(c) = \begin{bmatrix} 1 & 0 & 0 \\ 0 & c & 0 \\ 0 & 0 & 1 \end{bmatrix}, \quad E_{12}(d) = \begin{bmatrix} 1 & 0 & 0 \\ d & 1 & 0 \\ 0 & 0 & 1 \end{bmatrix}.$$

We now make the following definition.

15.45 DEFINITION A matrix of the form E_{ij}, $E_i(c)$, or $E_{ij}(d)$ is called an *elementary matrix.* It is understood that i and j are distinct integers from the set $\{1, 2, \ldots, n\}$, that c is a nonzero element of F, and that $d \in F$.

We may remark on the fact that we have defined elementary matrices in terms of *row* operations on I. However, the elementary column operations on I would yield the same set of matrices. In fact, if we apply the operations $\mathscr{C}_{ij}$, $\mathscr{C}_i(c)$, and $\mathscr{C}_{ij}(d)$ to I, it may be verified that we obtain E_{ij}, $E_i(c)$, and $E_{ji}(d)$, respectively. Note that although, by definition, the application of $\mathscr{R}_{ij}(d)$ to I gives $E_{ij}(d)$, the application of $\mathscr{C}_{ij}(d)$ to I yields $E_{ji}(d)$ with subscripts interchanged.

We now assert that an elementary row (column) operation on a matrix A can be achieved by multiplying A on the left (right) by an elementary matrix. More precisely, we have the following theorem.

15.46 THEOREM *The result of applying an elementary row operation $\mathscr{R}_{ij}$, $\mathscr{R}_i(c)$, or $\mathscr{R}_{ij}(d)$ to a matrix A is to obtain the matrix $E_{ij}A$, $E_i(c)A$, or $E_{ij}(d)A$, respectively. The result of applying an elementary column operation $\mathscr{C}_{ij}$, $\mathscr{C}_i(c)$ or $\mathscr{C}_{ij}(d)$ to a matrix A is to obtain the matrix AE_{ij}, $AE_i(c)$, or $AE_{ji}(d)$, respectively.*

If $A = (a_{ij})$ and $n = 3$, we may illustrate certain parts of this theorem by the following calculations. In each case, the matrix product is obviously the matrix obtained from A by the corresponding elementary operation:

$$E_{12}A = \begin{bmatrix} 0 & 1 & 0 \\ 1 & 0 & 0 \\ 0 & 0 & 1 \end{bmatrix} \begin{bmatrix} a_{11} & a_{12} & a_{13} \\ a_{21} & a_{22} & a_{23} \\ a_{31} & a_{32} & a_{33} \end{bmatrix} = \begin{bmatrix} a_{21} & a_{22} & a_{23} \\ a_{11} & a_{12} & a_{13} \\ a_{31} & a_{32} & a_{33} \end{bmatrix},$$

$$E_{12}(d)A = \begin{bmatrix} 1 & 0 & 0 \\ d & 1 & 0 \\ 0 & 0 & 1 \end{bmatrix} \begin{bmatrix} a_{11} & a_{12} & a_{13} \\ a_{21} & a_{22} & a_{23} \\ a_{31} & a_{32} & a_{33} \end{bmatrix}$$

$$= \begin{bmatrix} a_{11} & a_{12} & a_{13} \\ a_{21} + da_{11} & a_{22} + da_{12} & a_{23} + da_{13} \\ a_{31} & a_{32} & a_{33} \end{bmatrix},$$

$$AE_{12} = \begin{bmatrix} a_{11} & a_{12} & a_{13} \\ a_{21} & a_{22} & a_{23} \\ a_{31} & a_{32} & a_{33} \end{bmatrix} \begin{bmatrix} 0 & 1 & 0 \\ 1 & 0 & 0 \\ 0 & 0 & 1 \end{bmatrix} = \begin{bmatrix} a_{12} & a_{11} & a_{13} \\ a_{22} & a_{21} & a_{23} \\ a_{32} & a_{31} & a_{33} \end{bmatrix},$$

$$AE_{12}(d) = \begin{bmatrix} a_{11} & a_{12} & a_{13} \\ a_{21} & a_{22} & a_{23} \\ a_{31} & a_{32} & a_{33} \end{bmatrix} \begin{bmatrix} 1 & 0 & 0 \\ d & 1 & 0 \\ 0 & 0 & 1 \end{bmatrix} = \begin{bmatrix} a_{11} + da_{12} & a_{12} & a_{13} \\ a_{21} + da_{22} & a_{22} & a_{23} \\ a_{31} + da_{32} & a_{32} & a_{33} \end{bmatrix}.$$

By a separate consideration of each of the cases involved, the reader should have no difficulty in convincing himself of the truth of Theorem 15.46, or even in supplying a formal proof. Accordingly, we shall omit the proof.

Suppose, now, that we can pass from a matrix A to a matrix B by a finite sequence of elementary operations. The preceding theorem says that B can be obtained from A by successive multiplications by elementary matrices. As an example, consider the matrix D which is transformed into I_2 by the elementary operations 15.43. After applying $\mathscr{R}_{12}(-3)$ to D we have the matrix $E_{12}(-3)D$, after applying $\mathscr{R}_{13}(-2)$ to this matrix we have $E_{13}(-2)E_{12}(-3)D$, after applying the column operation $\mathscr{C}_{12}(-2)$ to this matrix we have $E_{13}(-2)E_{12}(-3)DE_{21}(-2)$, and so on. We finally obtain in this way that

15.47 $\qquad E_2(-\tfrac{1}{5})E_{23}(-1)E_{13}(-2)E_{12}(-3)DE_{21}(-2)E_{31}(1)E_{32}(1) = I_2.$

Now each elementary matrix is nonsingular since it is obtained from the non-singular matrix I by an elementary operation. Moreover, we know by Corollary 15.20 that a product of nonsingular matrices is nonsingular. Hence, if we set

$$S = E_2(-\tfrac{1}{5})E_{23}(-1)E_{13}(-2)E_{12}(-3)$$

and

$$T = E_{21}(-2)E_{31}(1)E_{32}(1),$$

S and T are nonsingular matrices and Equation 15.47 can be written in the form

15.48 $\qquad\qquad\qquad\qquad\qquad SDT = I_2.$

This is an illustration of one part of the following general theorem.

15.49 THEOREM   *If A, $B \in F_n$, then $A \sim B$ if and only if there exist non-singular matrices P and Q such that $B = PAQ$.*

    Proof: If $B = PAQ$, where P and Q are nonsingular, it follows from Theorem 15.19 (ii) that A and B have the same rank, and Theorem 15.44 then shows that $A \sim B$.

      Conversely, let us assume that $A \sim B$. In view of Theorem 15.46, this implies that there exist elementary matrices $P_1, P_2, \ldots, P_k$ and $Q_1, Q_2, \ldots, Q_l$ such that

$$B = P_k \cdots P_2 P_1 A Q_1 Q_2 \cdots Q_l.$$

If we set $P = P_k \cdots P_2 P_1$ and $Q = Q_1 Q_2 \cdots Q_l$, P and Q are non-singular since they are products of nonsingular matrices, and $B = PAQ$ as required. The proof of the theorem is therefore complete.

    This theorem shows that the equivalence set $[A]$ consists of all those elements of F_n of the form PAQ, where P and Q are nonsingular elements of F_n.

    An important special case of some of the preceding results is that in which A is taken to be the unity I of F_n. Since I is nonsingular, Theorem 15.44 asserts that $I \sim B$ if and only if B is nonsingular. Moreover, Theorem 15.46, as applied in the proof of the preceding theorem, shows that $I \sim B$ if and only if B is expressible as a product of elementary matrices. We have then the following result.

15.50 COROLLARY   *A matrix is expressible as a product of elementary matrices if and only if it is nonsingular.*

15.8 THE DETERMINANT OF A PRODUCT

The results of the preceding section enable us to prove the following theorem about determinants.

15.51 THEOREM *If A, $B \in F_n$, then $|AB| = |A| \cdot |B|$.*

First, we dispose of the case in which A is singular. In view of Theorem 14.28, this means that $|A| = 0$. Moreover, by Theorem 15.19, we see that rank $(AB) < n$, so that also $|AB| = 0$. Hence, the theorem is true in this case.

We next show that the desired result is true if A is an elementary matrix. Let us state this special case as follows.

15.52 LEMMA *If E is an elementary matrix, then $|EB| = |E| \cdot |B|$.*

Proof: To prove this lemma, we consider, in turn, each of the three types of elementary matrices. Since each such matrix is obtained from I by an elementary operation, and $|I| = 1$, it follows from Theorem 14.14 that $|E_{ij}| = -1$, $|E_i(c)| = c$, and $|E_{ij}(d)| = 1$. Then, by again applying the same theorem and Theorem 15.46, we can verify each of the following:

$$|E_{ij}B| = -|B| = |E_{ij}| \cdot |B|,$$
$$|E_i(c)B| = c|B| = |E_i(c)| \cdot |B|,$$
$$|E_{ij}(d)B| = |B| = |E_{ij}(d)| \cdot |B|,$$

and the lemma is established.

Suppose now that A is an arbitrary nonsingular matrix. Corollary 15.50 then assures us that it can be expressed as a product of elementary matrices. The desired result is now easily completed by induction. Let S_m be the statement "For every matrix B of F_n, and for every matrix A of F_n which can be expressed as a product of m elementary matrices, we have $|AB| = |A| \cdot |B|$." Then S_1 is true by the preceding lemma. Let us now assume that S_k is true and prove that S_{k+1} is true. Suppose, then, that $A = E_1 E_2 \cdots E_{k+1}$, where these are elementary matrices. It follows that

$$
\begin{aligned}
|AB| = |E_1(E_2 E_3 \cdots E_{k+1}B)| &= |E_1| \cdot |E_2 E_3 \cdots E_{k+1}B| && \text{(by } S_1\text{)} \\
&= |E_1| \cdot |E_2 E_3 \cdots E_{k+1}| \cdot |B| && \text{(by } S_k\text{)} \\
&= |E_1 E_2 \cdots E_{k+1}| \cdot |B| && \text{(by } S_1\text{)} \\
&= |A| \cdot |B|.
\end{aligned}
$$

Hence, S_{k+1} is true, and it follows that S_m is true for every positive integer m. The theorem is therefore established.

We may observe that an alternate way of stating the result just proved is to say that multiplication is preserved under the mapping $\theta : F_n \rightarrow F$ defined by $A\theta = |A|$, $A \in F_n$. Clearly, $I\theta = 1$, where 1 is the unity of F and, moreover, we know that an element A of F_n has a multiplicative inverse in F_n if and only if $|A|$ has a multiplicative inverse in F (that is, is not zero). Finally, we state the following corollary of the preceding theorem.

15.53 COROLLARY   *If A is a nonsingular element of F_n, then $|A^{-1}| = |A|^{-1}$.*

EXERCISES

1. For each of the following matrices over $\mathbf{Q}$, write down a sequence of elementary operations that will reduce it to the form I_r:

(a) $\begin{bmatrix} 1 & -1 \\ 2 & 3 \end{bmatrix}$,

(b) $\begin{bmatrix} 2 & -1 \\ 4 & -2 \end{bmatrix}$,

(c) $\begin{bmatrix} 0 & -1 & 2 \\ 1 & 2 & -1 \\ 1 & -1 & 5 \end{bmatrix}$,

(d) $\begin{bmatrix} 1 & -2 & 3 \\ 2 & 1 & 4 \\ -2 & 1 & 2 \end{bmatrix}$.

2. Show that the multiplicative inverse of each elementary matrix is also an elementary matrix.

3. Let $E_1 E_2 \cdots E_s$ be a product of elementary matrices. Since $E_1 E_2 \cdots E_s = E_1 E_2 \cdots E_s I$, Theorem 15.46 says that we can compute this product by first applying the elementary row operation corresponding to E_s to the matrix I, then the elementary row operation corresponding to E_{s-1} to this new matrix, and so on. Use this method to compute the matrices S and T occurring in 15.48.

4. For each matrix A of Exercise 1, find nonsingular matrices P and Q such that $PAQ = I_r$.

5. Express each of the nonsingular matrices of Exercise 1 as a product of elementary matrices.

6. By actual calculation of the determinants involved, verify Theorem 15.51 for the following matrices:

$$A = \begin{bmatrix} 3 & -1 & 0 \\ 1 & 0 & 3 \\ -2 & 1 & 1 \end{bmatrix}, \quad B = \begin{bmatrix} 2 & 1 & -3 \\ 0 & 1 & 2 \\ 1 & 3 & 2 \end{bmatrix}.$$

7. Let us write $A \equiv B$ if there exist nonsingular matrices P and Q such that $B = PAQ$. Use this definition to verify that $\equiv$ is an equivalence relation on F_n. (Of course, Theorem 15.49 shows indirectly that $\equiv$ coincides with the equivalence relation $\sim$.)

8. Let us define $A \approx B$ to mean that it is possible to pass from A to B by a finite sequence of elementary *row* operations. Prove each of the following:

   (a) $\approx$ is an equivalence relation defined on F_n,

   (b) $A \approx B$ if and only if there exists a nonsingular matrix P such that $B = PA$,

   (c) If A is nonsingular, then $A \approx I$.

9. Show that if A can be reduced to I by a sequence of elementary *row* operations, then A^{-1} is the matrix obtained by starting with the matrix I and applying in turn the same sequence of elementary row operations.

10. Use the method of the previous problem to compute the multiplicative inverse of the following matrix over $\mathbf{Q}$:

$$\begin{bmatrix} 3 & 1 & -2 \\ 2 & 0 & 3 \\ -1 & 1 & -6 \end{bmatrix}.$$

15.9 SIMILARITY OF MATRICES

We now consider the problem, suggested by the results of Section 15.4, of determining the relationship which holds between the matrices of a linear transformation of a vector space relative to two different ordered bases of the space.

Throughout this section $X_1, X_2, \ldots, X_n$ will denote a fixed ordered basis of a vector space V of dimension $n > 0$ over a field F, and L will denote the algebra of all linear transformations of V. The word *matrix* will mean an element of F_n.

If $\mathscr{A} \in L$ and

15.54 $$X_i\mathscr{A} = \sum_{j=1}^{n} a_{ij}X_j, \qquad (i = 1, 2, \ldots, n).$$

we showed in Section 15.4 that the mapping $\mathscr{A} \to A = (a_{ij})$ is an isomorphism of L onto F_n. Let us now denote this isomorphism by ϕ, that is, $\phi : L \to F_n$ is the mapping defined by $\mathscr{A}\phi = A$, where $\mathscr{A}$ and A are related by 15.54. We have called A the matrix of $\mathscr{A}$ relative to the ordered basis $X_1, X_2, \ldots, X_n$ of V.

Now suppose that $Y_1, Y_2, \ldots, Y_n$ is another ordered basis of V and that

$$Y_i\mathscr{A} = \sum_{j=1}^{n} a_{ij}{}^* Y_j, \qquad (i = 1, 2, \ldots, n).$$

Then $A^* = (a_{ij}{}^*)$ is the matrix of the linear transformation $\mathscr{A}$ relative to the ordered basis $Y_1, Y_2, \ldots, Y_n$. We propose to determine how the matrices A and A^* are related.

By Corollary 15.14, there exists a unique linear transformation $\mathscr{B}$ of V such that

15.55 $$X_i\mathscr{B} = Y_i, \qquad (i = 1, 2, \ldots, n).$$

Then, if the c_i are elements of F, we have that

$$\left(\sum c_i X_i\right)\mathscr{B} = \sum c_i(X_i\mathscr{B}) = \sum c_i Y_i$$

and, since the Y's also are a basis of V, we see that $V\mathscr{B} = V$. Thus $\mathscr{B}$ is nonsingular and, by Theorem 15.18, it has a multiplicative inverse $\mathscr{B}^{-1}$ in L. Clearly,

$$Y_i\mathscr{B}^{-1} = X_i, \qquad (i = 1, 2, \ldots, n),$$

and we have

$$X_i(\mathscr{B}\mathscr{A}\mathscr{B}^{-1}) = (X_i\mathscr{B})(\mathscr{A}\mathscr{B}^{-1}) = Y_i(\mathscr{A}\mathscr{B}^{-1}) = (Y_i\mathscr{A})\mathscr{B}^{-1}$$

$$= \left(\sum_{j=1}^{n} a_{ij}{}^* Y_j\right)\mathscr{B}^{-1} = \sum_{j=1}^{n} a_{ij}{}^* X_j \qquad (i = 1, 2, \ldots, n).$$

This shows that the matrix of the linear transformation $\mathscr{B}\mathscr{A}\mathscr{B}^{-1}$ relative to the ordered basis $X_1, X_2, \ldots, X_n$ is precisely A^*. In the notation introduced above we therefore have

$$(\mathscr{B}\mathscr{A}\mathscr{B}^{-1})\phi = A^*.$$

But since ϕ is an isomorphism of L onto F_n, it follows that

$$(\mathscr{B}\mathscr{A}\mathscr{B}^{-1})\phi = (\mathscr{B}\phi)(\mathscr{A}\phi)(\mathscr{B}\phi)^{-1} = A^*.$$

If we denote $\mathscr{A}\phi$ by A and $\mathscr{B}\phi$ by B, this equation shows that

15.56 $$BAB^{-1} = A^*.$$

We have therefore shown that if A and A^* are matrices of the same linear transformation relative to two different ordered bases, then there exists a nonsingular matrix B such that 15.56 holds.

Conversely, suppose we are given matrices A, A^*, and a nonsingular matrix B such that Equation 15.56 holds. Suppose that $\mathscr{A}$ and $\mathscr{B}$ are the linear transformations of V such that $\mathscr{A}\phi = A$ and $\mathscr{B}\phi = B$. If we *define* the Y's by

$$Y_i = X_i\mathscr{B}, \qquad (i = 1, 2, \ldots, n),$$

Theorem 15.18 shows that $Y_1, Y_2, \ldots, Y_n$ is an ordered basis of V. Moreover, our calculations above show that BAB^{-1} is the matrix of $\mathscr{A}$ relative to the ordered basis $Y_1, Y_2, \ldots, Y_n$ of V. Since we are now assuming Equation 15.56, we have completed the proof of the following theorem.

15.57 THEOREM

    (*i*) *If A is the matrix of a linear transformation $\mathscr{A}$ relative to an ordered basis $X_1, X_2, \ldots, X_n$ of V, and A^* is the matrix of the same linear transformation relative to an ordered basis $Y_1, Y_2, \ldots, Y_n$ of V, then there exists a nonsingular matrix B such that $BAB^{-1} = A^*$. In fact, B is the matrix of the nonsingular linear transformation, relative to the ordered basis $X_1, X_2, \ldots, X_n$ of V, defined by $X_i\mathscr{B} = Y_i, (i = 1, 2, \ldots, n)$.*

    (*ii*) *Conversely, if matrices A and A^* are given and there exists a nonsingular matrix B such that $BAB^{-1} = A^*$, then A and A^* are matrices of the same linear transformation relative to properly chosen ordered bases of V.*

Example Let us illustrate the above calculations by returning to Example 1 of Section 15.4. Here, using the present notation, we have $n = 2$, $V = V_2(\mathbf{R})$, $X_1 = (1, 0)$, $X_2 = (0, 1)$, $Y_1 = (1, 1)$, and $Y_2 = (-1, 1)$. Moreover,

$$A = \begin{bmatrix} 3 & 1 \\ 1 & -2 \end{bmatrix}, \quad A^* = \begin{bmatrix} \frac{3}{2} & -\frac{5}{2} \\ -\frac{5}{2} & -\frac{1}{2} \end{bmatrix}.$$

We proceed to compute the nonsingular matrix B such that $BAB^{-1} = A^*$. Using 15.55 to define the linear transformation $\mathscr{B}$, we obtain

$$(1, 0)\mathscr{B} = (1, 1) = (1, 0) + (0, 1),$$
$$(0, 1)\mathscr{B} = (-1, 1) = -(1, 0) + (0, 1).$$

Thus the matrix B of $\mathscr{B}$ relative to the ordered basis $(1, 0)$, $(0, 1)$ of $V_2(\mathbf{R})$ is as follows:

$$B = \begin{bmatrix} 1 & 1 \\ -1 & 1 \end{bmatrix}.$$

This is the desired matrix B. The reader may verify that

$$B^{-1} = \begin{bmatrix} \frac{1}{2} & -\frac{1}{2} \\ \frac{1}{2} & \frac{1}{2} \end{bmatrix}$$

and then that $BAB^{-1} = A^*$.

We still have not justified the title of this section. Let us therefore make the following definition.

15.58 DEFINITION If C, $D \in F_n$, we say that C is *similar* to D if there exists a nonsingular element T of F_n such that $C = TDT^{-1}$.

We leave as an exercise the proof that similarity is an equivalence relation defined on F_n. By Theorem 15.57, the matrices in an equivalence set are precisely the matrices of a single linear transformation relative to all possible ordered bases of V. In particular, two elements of F_n are similar if and only if they are the matrices of the same linear transformation relative to two ordered bases of V. However, it is a fairly difficult problem to develop a constructive method of determining whether or not two given matrices are similar. A few partial results will be obtained in the remainder of this chapter, but we shall refer to texts on linear algebra for a full treatment of the subject.

One simple fact which follows readily from Theorem 15.57 and Corollary 15.53 is the following.

15.59 COROLLARY *If C and D are similar elements of F_n, then $|C| = |D|$.*

EXERCISES

1. Prove that similarity of matrices, as defined in 15.58, satisfies the three defining properties (1.12) of an equivalence relation.

2. In Example 2 of Section 15.4 we found matrices of a given linear transformation relative to two different ordered bases of a certain vector space. If we now call these matrices A and A^*, find a nonsingular matrix B such that $BAB^{-1} = A^*$.

3. Show, by an example, that two matrices having the same determinant need not be similar. [*Hint:* Take one of the matrices to be *I*.]

4. Let V be the vector space of all polynomials in an indeterminate x over $\mathbf{Q}$ of degree no more than two, together with the zero polynomial. Let $\mathscr{A}: V \to V$ be the linear transformation of V defined by $f(x)\mathscr{A} = f(x + 1)$. (i) Find the matrix A of $\mathscr{A}$ relative to the ordered basis 1, x, x^2 of V. (ii) Find the matrix A^* of $\mathscr{A}$ relative to the ordered basis x^2, x, 1 of V. (iii) Find a nonsingular matrix B such that $BAB^{-1} = A^*$.

5. Let $A = (a_{ij})$ be the matrix of a linear transformation $\mathscr{A}$ of $V_n(F)$ relative to the ordered basis of unit vectors $E_1, E_2, \ldots, E_n$. If $B = (b_{ij})$ is a nonsingular element of F_n, verify that the matrix of $\mathscr{A}$ relative to the ordered basis $B_1, B_2, \ldots, B_n$ consisting of the row vectors of B is BAB^{-1}.

15.10 INVARIANT SUBSPACES

The following concept is an important one in the further study of linear transformations.

15.60 DEFINITION Let $\mathscr{A}$ be a linear transformation of the vector space V over the field F. A subspace V_1 of V with the property that $V_1\mathscr{A} \subseteq V_1$ is said to be *invariant under* $\mathscr{A}$.

   If the subspace V_1 of V is invariant under the linear transformation $\mathscr{A}$ of V, it is clear that $\mathscr{A}$ induces a linear transformation $\mathscr{A}_1$ of V_1 defined by $X\mathscr{A}_1 = X\mathscr{A}$ for each X in V_1. We shall often refer to this linear transformation $\mathscr{A}_1$ as "$\mathscr{A}$ restricted to V_1" or as "$\mathscr{A}$ acting on V_1."
   There are two trivial subspaces of V, namely $\{0\}$ and V itself, both of which are clearly invariant under every linear transformation of V. Let us now assume that V has finite dimension $n > 0$ and that V_1 is a subspace of dimension m which is invariant under the linear transformation $\mathscr{A}$ of V. We shall assume that $0 < m < n$ since $m = 0$ and $m = n$ give the trivial subspaces just mentioned, and there is not much more to be said about them. By Theorem 11.24, there exists a basis $\{X_1, X_2, \ldots, X_n\}$ of V with the property that $\{X_1, X_2, \ldots, X_m\}$ is a basis of V_1. Now since V_1 is invariant under $\mathscr{A}$, it follows that each of $X_1\mathscr{A}, X_2\mathscr{A}, \ldots, X_m\mathscr{A}$ is expressible as a linear combination of $X_1, X_2, \ldots, X_m$. In this case, Equations

15.23 which define the matrix A of the linear transformation $\mathscr{A}$ relative to the given basis of V take the form

15.61

$$
\begin{aligned}
X_1\mathscr{A} &= a_{11}X_1 + a_{12}X_2 + \cdots + a_{1m}X_m, \\
X_2\mathscr{A} &= a_{21}X_1 + a_{22}X_2 + \cdots + a_{2m}X_m, \\
&\;\cdot\;\cdot\;\cdot\;\cdot\;\cdot\;\cdot\;\cdot\;\cdot\;\cdot\;\cdot\;\cdot\;\cdot\;\cdot\;\cdot \\
X_m\mathscr{A} &= a_{m1}X_1 + a_{m2}X_2 + \cdots + a_{mm}X_m, \\
X_{m+1}\mathscr{A} &= a_{m+11}X_1 + a_{m+12}X_2 + \cdots\cdots + a_{m+1n}X_n, \\
&\;\cdot\;\cdot\;\cdot\;\cdot\;\cdot\;\cdot\;\cdot\;\cdot\;\cdot\;\cdot\;\cdot\;\cdot\;\cdot\;\cdot \\
X_n\mathscr{A} &= a_{n1}X_1 + a_{n2}X_2 + \cdots\cdots + a_{nn}X_n.
\end{aligned}
$$

Thus, if we set $A = (a_{ij})$, we see that $a_{ij} = 0$ for $i = 1, 2, \ldots, m$ provided $j > m$. A convenient way of indicating the form of this matrix A is to write

15.62

$$
A = \begin{bmatrix} A_1 & O \\ B & C \end{bmatrix}
$$

where

15.63

$$
A_1 = \begin{bmatrix}
a_{11} & a_{12} & \cdots & a_{1m} \\
a_{21} & a_{22} & \cdots & a_{2m} \\
\cdot & \cdot & \cdots & \cdot \\
a_{m1} & a_{m2} & \cdots & a_{mm}
\end{bmatrix},
$$

B is a matrix with $n - m$ rows and m columns, C is a matrix with $n - m$ rows and $n - m$ columns, and O is a matrix with m rows and $n - m$ columns, all of whose elements are zero.

It is clear from Equations 15.61 that the matrix A_1 appearing in the upper left corner of A is the matrix of $\mathscr{A}$ restricted to V_1, relative to the ordered basis $X_1, X_2, \ldots, X_m$ of V_1.

Example As a simple illustration of these concepts, let $\mathscr{A}$ be the linear transformation of $V_3(F)$ defined by

15.64

$$
(x_1, x_2, x_3)\mathscr{A} = (x_1 + 2x_2 + x_3, x_1 + x_2 + x_3, x_3),
$$

and let V_1 be the subspace of $V_3(F)$ consisting of all elements of the form $(x_1, x_2, 0)$. Since, by 15.64, we see that

$$
(x_1, x_2, 0)\mathscr{A} = (x_1 + 2x_2, x_1 + x_2, 0),
$$

it is clear that V_1 is invariant under $\mathscr{A}$. In this simple case, the unit vectors E_1, E_2, E_3 of $V_3(F)$ are such that E_1, E_2 is a basis of V_1, and Equations 15.61 become in this case:

$$\begin{aligned} E_1\mathscr{A} &= E_1 + E_2, \\ E_2\mathscr{A} &= 2E_1 + E_2, \\ E_3\mathscr{A} &= E_1 + E_2 + E_3. \end{aligned}$$

Thus the matrix A of $\mathscr{A}$ relative to the ordered basis E_1, E_2, E_3 of $V_3(F)$ is as follows:

$$A = \begin{bmatrix} 1 & 1 & 0 \\ 2 & 1 & 0 \\ 1 & 1 & 1 \end{bmatrix}.$$

This is the desired form 15.62 with

$$A_1 = \begin{bmatrix} 1 & 1 \\ 2 & 1 \end{bmatrix}, \quad B = [1 \quad 1], \quad C = [1], \quad O = \begin{bmatrix} 0 \\ 0 \end{bmatrix}.$$

Moreover, as is to be expected from the general discussion above, A_1 is the matrix of $\mathscr{A}$ restricted to V_1, relative to the ordered basis E_1, E_2 of V_1.

A situation of special interest is that in which V is the direct sum of two invariant subspaces. Suppose that $V = V_1 \oplus V_2$, with both V_1 and V_2 being invariant under the linear transformation $\mathscr{A}$. If dim $V_1 = m$, then dim $V_2 = n - m$. Let $X_1, X_2, \ldots, X_m$ be an ordered basis of V_1 and $X_{m+1}, X_{m+2}, \ldots, X_n$ an ordered basis of V_2. Then (see Exercise 4 at end of Section 11.10)

15.65 $$X_1, X_2, \ldots, X_n$$

is an ordered basis of V, and Equations 15.61 determining the matrix A of $\mathscr{A}$ relative to the ordered basis 15.65 of V are such that also $a_{ij} = 0$ for $i = m + 1$, $\ldots, n$ provided $j = 1, 2, \ldots, m$. In this case, the matrix A is of the form

15.66 $$A = \begin{bmatrix} A_1 & O \\ O & A_2 \end{bmatrix},$$

where A_i is the matrix of $\mathscr{A}$ restricted to $V_i(i = 1, 2)$. Of course, A_i is a square matrix of order equal to the dimension of V_i.

A convenient way of writing the matrix A given by 15.66 is as follows:

$$A = \text{diag}(A_1, A_2).$$

The type of argument used above will yield the following general result whose proof we omit.

15.67 THEOREM   *Let $\mathcal{A}$ be a linear transformation of the vector space V of finite dimension. If*

$$V = V_1 \oplus V_2 \oplus \cdots \oplus V_r,$$

where each V_i is invariant under $\mathcal{A}$ then the matrix A of $\mathcal{A}$ relative to a suitably chosen ordered basis of V is of the form

$$A = \mathrm{diag}\,(A_1, A_2, \ldots, A_r),$$

where A_i is the matrix of $\mathcal{A}$ restricted to V_i (relative to an ordered basis of V_i which is a part of the chosen basis of V).

If a linear transformation $\mathcal{A}$ of a vector space V is given, we have as yet no information as to how one might try to find subspaces of V which are invariant under $\mathcal{A}$. After developing suitable machinery, we shall make a little progress in this direction in the following section.

15.11 POLYNOMIALS IN A LINEAR TRANSFORMATION

Throughout this section we shall continue to let $\mathcal{A}$ be a given linear transformation of a vector space V of finite dimension $n > 0$ over a field F, and let L be the algebra of all linear transformations of V.

Let $f(x)$ be an element of the polynomial ring $F[x]$ in an indeterminate x over F. If

$$f(x) = a_0 + a_1 x + \cdots + a_k x^k,$$

we shall denote by $f(\mathcal{A})$ the linear transformation

$$a_0 \mathcal{T} + a_1 \mathcal{A} + \cdots + a_k \mathcal{A}^k$$

of V, and call $f(\mathcal{A})$ a *polynomial in $\mathcal{A}$ over F.* Let us denote by $F[\mathcal{A}]$ the set of all polynomials in $\mathcal{A}$ over F. Then $F[\mathcal{A}]$ is an algebra over F, actually a subalgebra of L. Moreover, $F[\mathcal{A}]$ is a *commutative* subalgebra of L. That is, multiplication is commutative in $F[\mathcal{A}]$ although it is not generally commutative in L.

The mapping $\theta: F[x] \to F[\mathcal{A}]$ defined by $f(x)\theta = f(\mathcal{A})$ is a ring homomorphism† of $F[x]$ onto $F[\mathcal{A}]$. The Fundamental Theorem on Ring Homo-

---

† Since scalar multiplication is also preserved under the mapping θ, we might more precisely refer to θ as a homomorphism of the *algebra* $F[x]$ onto the *algebra* $F[A]$.

morphisms then asserts that $F[\mathscr{A}] \cong F[x]/\ker \theta$. Now the ideal $\ker \theta$ in $F[x]$ consists of those polynomials $f(x)$ of $F[x]$ such that $f(\mathscr{A}) = 0$. We first prove that $\ker \theta \neq \{0\}$.

15.68 THEOREM *There exists a nonzero polynomial $f(x)$ of $F[x]$ such that $f(\mathscr{A}) = 0$.*

Proof: By Theorem 11.42, the algebra L (considered as a vector space over F) has dimension n^2. Accordingly, the subalgebra $F[\mathscr{A}]$ of L has dimension no greater than n^2. It follows that the $n^2 + 1$ elements $\mathscr{T}$, $\mathscr{A}, \mathscr{A}^2, \ldots, \mathscr{A}^{n^2}$ of $F[\mathscr{A}]$ are linearly dependent over F. Thus there exists a nonzero polynomial $f(x)$ of degree no greater than n^2 such that $f(\mathscr{A}) = 0$. This $f(x)$ is in $\ker \theta$ and therefore $\ker \theta \neq \{0\}$.

Since every ideal in $F[x]$ is a principal ideal, we may write $\ker \theta = (m(x))$, where $m(x)$ is a polynomial of least degree such that $m(A) = 0$. Clearly, $m(\mathscr{A}) = 0$ implies that $cm(\mathscr{A}) = 0$ for every element c of F, so there is no loss of generality in assuming that $m(x)$ is a *monic* polynomial.

15.69 DEFINITION The unique monic polynomial $m(x)$ over F of least degree such that $f(\mathscr{A}) = 0$ is called the *minimal polynomial of $\mathscr{A}$.*

We have only proved that $m(x)$ exists and that $\deg m(x) \leq n^2$. It is true that always $\deg m(x) \leq n$, but this is not important for our present purposes.

We shall frequently use without specific reference the following fact which is a consequence of the above observations: If $m(x)$ is the minimal polynomial of $\mathscr{A}$ and $g(x) \in F[x]$, then $g(\mathscr{A}) = 0$ if and only if $m(x)$ divides $g(x)$ (which we shall write $m(x) \mid g(x)$, a notation previously used in the case of integers).

We shall next prove the following result.

15.70 THEOREM *Let $m(x)$ be the minimal polynomial of the linear transformation $\mathscr{A}$. Then $\mathscr{A}$ is nonsingular if and only if the constant term in $m(x)$ is different from zero.*

Proof: Let

$$m(x) = c_0 + c_1 x + \cdots + c_{k-1} x^{k-1} + x^k$$

be the minimal polynomial of $\mathscr{A}$, and let us first assume that $c_0 \neq 0$. If we set

$$\mathscr{B} = -[c_0^{-1} c_1 \mathscr{T} + c_0^{-1} c_2 \mathscr{A} + \cdots + c_0^{-1} \mathscr{A}^{k-1}],$$

then using the fact that $m(\mathscr{A}) = 0$, it follows by a straightforward calculation that $\mathscr{A}\mathscr{B} = \mathscr{B}\mathscr{A} = \mathscr{T}$. Thus $\mathscr{A}$ has a multiplicative inverse and is therefore nonsingular.

Next, suppose that $c_0 = 0$ and let us show that $\mathscr{A}$ is singular. Since $m(\mathscr{A}) = 0$, we now have that

$$\mathscr{A}(c_1 \mathscr{T} + c_2 \mathscr{A} + \cdots + c_{k-1}\mathscr{A}^{k-2} + \mathscr{A}^{k-1}) = 0.$$

The second factor on the left side of this equation cannot be zero, for otherwise we would have a polynomial $g(x)$ of degree $k - 1$ such that $g(\mathscr{A}) = 0$, and this is impossible in view of the fact that $m(x)$ is the minimal polynomial of $\mathscr{A}$. Accordingly, $\mathscr{A}$ is a divisor of zero in L and therefore cannot have a multiplicative inverse in L. This shows that $\mathscr{A}$ is singular and completes the proof of the theorem.

We may remark that although we have been considering the minimal polynomial of a *linear transformation*, in view of the previously established isomorphism of the algebra of linear transformations of a vector space of order n over F and the algebra F_n of all matrices of order n over F, we could just as well speak of the minimal polynomial of a *matrix*. In fact, in a numerical case, it is often simpler to work with a matrix than with the corresponding linear transformation.

The following lemma will be useful in obtaining some relationships between the concept of minimal polynomial and that of invariant subspace.

15.71   LEMMA   *Let $\mathscr{A}$ be a linear transformation of the vector space V of finite dimension, and suppose that $V = V_1 \oplus V_2$, with both V_1 and V_2 being invariant under $\mathscr{A}$. Let $\mathscr{A}_i$ $(i = 1, 2)$ be the linear transformation $\mathscr{A}$ restricted to V_i, and suppose that $m_i(x)$ is the minimal polynomial of $\mathscr{A}_i$. Then the minimal polynomial of $\mathscr{A}$ is the least common multiple of $m_1(x)$ and $m_2(x)$.*

Proof: Let $m(x)$ be the minimal polynomial of $\mathscr{A}$. Then, $m(\mathscr{A}) = 0$ implies that $Xm(\mathscr{A}) = 0$ for each X in V. In particular, $Xm(\mathscr{A}) = 0$ for each X in V_1, that is, $m(\mathscr{A}_1) = 0$ and it therefore follows that $m_1(x) \mid m(x)$. Similarly, we have that $m_2(x) \mid m(x)$.

Now suppose that $f(x) \in F[x]$ such that $m_1(x) \mid f(x)$ and $m_2(x) \mid f(x)$. Since $m_i(\mathscr{A}_i) = 0$, it follows that $Xf(\mathscr{A}_i) = 0$ for every X in V_i $(i = 1, 2)$. Now if $Y \in V$, we have $Y = Y_1 + Y_2$, with $Y_i \in V_i$. Hence

$$Yf(\mathscr{A}) = Y_1 f(\mathscr{A}) + Y_2 f(\mathscr{A}) = Y_1 f(\mathscr{A}_1) + Y_2 f(\mathscr{A}_2) = 0.$$

This implies that $f(\mathscr{A}) = 0$ and it follows that $m(x) \mid f(x)$. By definition of the least common multiple, these calculations show that $m(x)$ is indeed the least common multiple of $m_1(x)$ and $m_2(x)$, completing the proof.

We next prove another lemma as follows.

15.72 LEMMA *Let $\mathscr{A}$ be a linear transformation of the vector space V of finite dimension over a field F, and let $m(x)$ be the minimal polynomial of $\mathscr{A}$. If $m(x) = f(x)g(x)$, where $f(x)$ and $g(x)$ are monic polynomials of positive degrees with $(f(x), g(x)) = 1$, then there exist subspaces V_1 and V_2 of V such that the following are true:*

(i) *V_1 and V_2 are invariant under $\mathscr{A}$.*

(ii) *$V = V_1 \oplus V_2$.*

(iii) *The minimal polynomial of $\mathscr{A}$ acting on V_1 is $f(x)$ and the minimal polynomial of $\mathscr{A}$ acting on V_2 is $g(x)$.*

Proof: We define subspaces V_1 and V_2 of V as follows and show that they have the specified properties:

$$V_1 = \{X \mid X \in V, \quad Xf(\mathscr{A}) = 0\},$$
$$V_2 = \{X \mid X \in V, \quad Xg(\mathscr{A}) = 0\}.$$

Otherwise expressed, V_1 and V_2 are, respectively, the null spaces of the linear transformations $f(\mathscr{A})$ and $g(\mathscr{A})$.

To prove (i), we observe that if $X \in V_1$, then $(X\mathscr{A})f(\mathscr{A}) = (Xf(\mathscr{A}))\mathscr{A} = 0\mathscr{A} = 0$, and $X\mathscr{A} \in V_1$. Similarly, V_2 is invariant under $\mathscr{A}$.

We next pass to the proof of (ii). Since $(f(x), g(x)) = 1$, there exist polynomials $r(x)$ and $s(x)$ of $F[x]$ such that $1 = r(x)f(x) + s(x)g(x)$, and it follows from this equation that $\mathscr{I} = r(\mathscr{A})f(\mathscr{A}) + s(\mathscr{A})g(\mathscr{A})$. Hence, if $X \in V$, we have

15.73 $$X = X\mathscr{I} = Xr(\mathscr{A})f(\mathscr{A}) + Xs(\mathscr{A})g(\mathscr{A}).$$

But since the minimal polynomial $m(x)$ of $\mathscr{A}$ is equal to $f(x)g(x)$, it follows that

$$Xr(\mathscr{A})f(\mathscr{A}) \in V_2 \quad \text{and} \quad Xs(\mathscr{A})g(\mathscr{A}) \in V_1;$$

hence the preceding equation shows that $V = V_1 + V_2$. To show that this sum is direct, suppose that $X_1 + X_2 = 0$, with $X_1 \in V_1$ and $X_2 \in V_2$. Thus $X_1 f(\mathscr{A}) = 0$ and $X_2 g(\mathscr{A}) = 0$. From Equation 15.73 it follows that $X_1 = X_1 s(\mathscr{A})g(\mathscr{A})$ and that $X_2 = X_2 r(\mathscr{A})f(\mathscr{A})$. We therefore have

15.74 $$X_1 s(\mathscr{A})g(\mathscr{A}) + X_2 r(\mathscr{A})f(\mathscr{A}) = 0.$$

SECTION 15.11   POLYNOMIALS IN A LINEAR TRANSFORMATION   · 581

Multiplying by $s(\mathscr{A})g(\mathscr{A})$ on the right, we obtain

$$X_1[s(\mathscr{A})g(\mathscr{A})]^2 = 0.$$

But

$$X_1 = X_1 s(\mathscr{A})g(\mathscr{A}) = X_1[s(\mathscr{A})g(\mathscr{A})]^2,$$

and we conclude that $X_1 = 0$. Similarly, we may show that $X_2 = 0$, and the sum $V_1 + V_2$ is a direct sum, as we wished to show.

To prove (iii), let us denote by $\mathscr{A}_i$ the linear transformation $\mathscr{A}$ restricted to V_i, and let $m_i(x)$ be the minimal polynomial of $\mathscr{A}_i$. By definition of V_1, $Xf(\mathscr{A}) = 0$ for every X in V_1, and this fact simply asserts that $f(\mathscr{A}_1)$ is the zero linear transformation on V_1, that is, that $f(\mathscr{A}_1) = 0$. This then implies that $m_1(x) \mid f(x)$. In like manner, it may be verified that $m_2(x) \mid g(x)$.

Now if $Y \in V$, we may write $Y = Y_1 + Y_2$, where $Y_i \in V_i$. Then

$$
\begin{aligned}
Ym_1(\mathscr{A})m_2(\mathscr{A}) &= Y_1 m_1(\mathscr{A})m_2(\mathscr{A}) + Y_2 m_1(\mathscr{A})m_2(\mathscr{A}) \\
&= Y_1 m_1(\mathscr{A}_1)m_2(\mathscr{A}_1) + Y_2 m_1(\mathscr{A}_2)m_2(\mathscr{A}_2) \\
&= 0,
\end{aligned}
$$

since $m_i(x)$ is the minimal polynomial of $\mathscr{A}_i$ $(i = 1, 2)$. Hence $m_1(\mathscr{A})m_2(\mathscr{A}) = 0$ and, since the minimal polynomial of A is $f(x)g(x)$, it follows that $f(x)g(x) \mid m_1(x)m_2(x)$. But we have already proved that $m_1(x) \mid f(x)$ and that $m_2(x) \mid g(x)$. Since all these polynomials are monic, it may be shown that $f(x) = m_1(x)$ and $g(x) = m_2(x)$. We leave the verification of this fact as an exercise. This completes the the proof of the lemma.

We now briefly apply the results just obtained to a more general situation as follows. Let $\mathscr{A}$ be a linear transformation of the vector space V of finite dimension over the field F, and let $m(x)$ be the minimal polynomial of $\mathscr{A}$. Then, by Theorem 10.39, we may write

15.75 $$m(x) = [p_1(x)]^{n_1}[p_2(x)]^{n_2} \cdots [p_k(x)]^{n_k},$$

where the n_i and k are positive integers and the $p_i(x)$ are the distinct monic divisors of $m(x)$ which are prime over F. We may now apply the preceding lemma to establish the following result.

15.76 THEOREM   *If the linear transformation $\mathscr{A}$ of V has minimal polynomial $m(x)$ which can be factored as in 15.75, then there exist subspaces V_i $(i = 1, 2, \ldots, k)$ of V, all of which are invariant under $\mathscr{A}$, such that*

$$V = V_1 \oplus V_2 \oplus \cdots \oplus V_k.$$

Moreover, if $\mathscr{A}_i$ denotes the linear transformation $\mathscr{A}$ restricted to V_i, the minimal polynomial of $\mathscr{A}_i$ is $[p_i(x)]^{n_i}$ for $i = 1, 2, \ldots, k$.

Proof: If the number k of distinct monic prime factors of $m(x)$ is one, there is nothing to prove, and the case in which $k = 2$ is disposed of by the preceding lemma. We now complete the proof by induction on k. Suppose, as an induction hypothesis, that the stated result is true for every linear transformation (of any vector space of finite dimension) whose minimal polynomial has $k - 1$ distinct prime factors, and let $\mathscr{A}$ be a linear transformation whose minimal polynomial has k distinct prime factors (given in 15.75). Now, for the moment, let $f(x) = [p_1(x)]^{n_1}$ and

$$g(x) = [p_2(x)]^{n_2} \cdots [p_k(x)]^{n_k}.$$

Then $(f(x), g(x)) = 1$ and we apply the preceding lemma to obtain subspaces V_1 and V_1' of V, both invariant under $\mathscr{A}$, and such that

$$V = V_1 \oplus V_1'.$$

By the lemma, the minimal polynomial of $\mathscr{A}$ restricted to V_1 is $f(x)$ and the minimal polynomial of $\mathscr{A}$ restricted to V_1' is $g(x)$. Let $\mathscr{A}'$ denote the linear transformation $\mathscr{A}$ restricted to V_1'. Since the minimal polynomial of $\mathscr{A}'$ has $k - 1$ distinct monic prime factors, the induction hypothesis assures us that there exist subspaces $V_2, \ldots, V_k$ of V_1', all of them invariant under $\mathscr{A}'$ and such that both of the following are true:

(i) $V_1' = V_2 \oplus \cdots \oplus V_k$,

(ii) The minimal polynomial of $\mathscr{A}'$ restricted to V_i is $[p_i(x)]^{n_i}$ for $i = 2, 3, \ldots, k$.

Now $\mathscr{A}'$ is itself the restriction of $\mathscr{A}$ to V_1', so $\mathscr{A}$ restricted to V_i is identical with $\mathscr{A}'$ restricted to V_i for $i = 2, 3, \ldots, k$. We leave as an exercise the verification that since $V = V_1 \oplus V_1'$, it follows from (i) above that

$$V = V_1 \oplus V_2 \oplus \cdots \oplus V_k.$$

This completes the proof of the theorem.

We may remark that, in view of Theorem 15.67, if the hypotheses of Theorem 15.76 are satisfied, the matrix of $\mathscr{A}$ relative to a properly chosen basis of V is of the form diag $(A_1, A_2, \ldots, A_k)$, where A_i is the matrix of $\mathscr{A}$ restricted to V_i.

Our results show that in any further analysis of linear transformations, we could restrict attention to the special case in which the minimal polynomial of the linear transformation is a power of a prime polynomial. However, we shall not go further into these questions. Actually, the procedure of the present section is not yet very helpful in a numerical situation since we have not developed a method for calculating the minimal polynomial—we have only proved its existence. A few simple special cases in which the calculations can be carried out directly will appear in the following list of exercises.

EXERCISES

1. If $A = \begin{bmatrix} a & b \\ c & d \end{bmatrix}$ is a matrix of order two over the field $\mathbf{Q}$, do each of the following:

  (i) Show that the minimal polynomial of A is of the first degree if and only if A is a *scalar matrix*, that is, of the form rI for some $r \in \mathbf{Q}$.

  (ii) If $h(x) = x^2 - (a + d)x + ad - bc$, verify that $h(A) = 0$.

  (iii) If A is not a scalar matrix, prove that the minimal polynomial of A is $h(x)$.

2. Use the result of the preceding exercise to write down the minimal polynomial of each of the following matrices over $\mathbf{Q}$:

$$\begin{bmatrix} 1 & 2 \\ 1 & 0 \end{bmatrix}, \quad \begin{bmatrix} 1 & 1 \\ -1 & 2 \end{bmatrix}.$$

3. Let $E_1 = (1, 0)$ and $E_2 = (0, 1)$ be the unit vectors of $V_2(\mathbf{Q})$, and let $\mathscr{A}$ be the linear transformation of $V_2(\mathbf{Q})$ defined by

$$(1, 0)\mathscr{A} = (1, 2),$$
$$(0, 1)\mathscr{A} = (1, 0).$$

By the preceding exercise, the minimal polynomial of $\mathscr{A}$ is $x^2 - x - 2 = (x - 2)(x + 1)$. Find bases for the nullspaces V_1 and V_2 of $\mathscr{A} - 2\mathscr{T}$ and $\mathscr{A} + \mathscr{T}$, respectively, and verify (Lemma 15.72) that $V = V_1 \oplus V_2$, that each V_i is invariant under $\mathscr{A}$, and that the minimal polynomial of $\mathscr{A}$ restricted to V_1 or to V_2 is, respectively, $x - 2$ and $x + 1$.

4. Let E_1, E_2, E_3, E_4 be the unit vectors of $V_4(\mathbf{Q})$ and let $\mathscr{A}$ be the linear transformation of $V_4(\mathbf{Q})$ defined by

$$
\begin{aligned}
E_1\mathscr{A} &= E_1 + 2E_3,\\
E_2\mathscr{A} &= E_2 + E_4,\\
E_3\mathscr{A} &= E_1,\\
E_4\mathscr{A} &= -E_2 + 2E_4.
\end{aligned}
$$

If $V_1 = [E_1, E_3]$ and $V_2 = [E_2, E_4]$, verify that both V_1 and V_2 are invariant under $\mathscr{A}$, and that $V_4(\mathbf{Q}) = V_1 \oplus V_2$. Then apply Lemma 15.71 and the results of Exercise 3 to show that the minimal polynomial of $\mathscr{A}$ is $(x - 2)(x + 1) \times (x^2 - 3x + 3)$. Verify that the matrix of $\mathscr{A}$ relative to the ordered basis E_1, E_3, E_2, E_4 of $V_4(\mathbf{Q})$ is of the form

$$
\text{diag}\left(\begin{bmatrix} 1 & 2 \\ 1 & 0 \end{bmatrix}, \begin{bmatrix} 1 & 1 \\ -1 & 2 \end{bmatrix}\right).
$$

5. Find an ordered basis of $V_4(\mathbf{Q})$ relative to which the matrix of the linear transformation A of the preceding exercise is of the form

$$
\text{diag}\left([2], [-1], \begin{bmatrix} 1 & 1 \\ -1 & 2 \end{bmatrix}\right).
$$

6. Prove that $F[\mathscr{A}]$ has a subalgebra which is isomorphic to F.

7. Complete the proof of part (iii) of Lemma 15.72 by proving in detail that $f(x) = m_1(x)$ and $g(x) = m_2(x)$.

8. Fill in the details which were omitted near the end of the proof of Theorem 15.76.

15.12 CHARACTERISTIC VECTORS AND CHARACTERISTIC ROOTS

Let V continue to be a vector space of dimension $n > 0$ over a field F, and $\mathscr{A}$ a linear transformation of V. Under the mapping $\mathscr{A}: V \to V$, the zero vector certainly maps into itself but, in general, there may very well be no other vector with this property. However, we shall here be concerned with vectors satisfying the weaker condition that they "almost" map into themselves in the sense that they map into scalar multiples of themselves. Let us give the following definition.

15.77 DEFINITION If $c \in F$ and X is a nonzero element of V such that

15.78
$$X\mathscr{A} = cX,$$

then X is called a *characteristic vector* of $\mathscr{A}$ corresponding to the *characteristic root* c of $\mathscr{A}$.

Let us emphasize that a nonzero vector X is a characteristic vector of $\mathscr{A}$ if and only if there exists an element c of F such that 15.78 holds. Conversely, an element c of F is a characteristic root of $\mathscr{A}$ if and only if there exists a nonzero vector X of V such that 15.78 holds.

In the literature, a great number of adjectives, such as *latent*, *proper*, and *eigen*, have been used in place of *characteristic* in Definition 15.77.

We may observe from the definition that if X is a characteristic vector corresponding to a characteristic root c, any nonzero scalar multiple of X is also a characteristic vector corresponding to the same characteristic root. Moreover, in view of the definition of a subspace being invariant under a linear transformation, we see that the nonzero vector X is a characteristic vector of $\mathscr{A}$ if and only if the one-dimensional subspace $[X]$ of V is invariant under $\mathscr{A}$.

Let us now consider the problem of determining the characteristic roots of a linear transformation $\mathscr{A}$. If $\mathscr{T}$ is the unity linear transformation, Equation 15.78 may be written in the form

15.79
$$X(\mathscr{A} - c\mathscr{T}) = 0.$$

Since X is required to be a nonzero vector, Theorem 15.18 shows that the linear transformation $\mathscr{A} - c\mathscr{T}$ must be singular. Suppose that A is the matrix of $\mathscr{A}$ relative to some ordered basis of V. Then, by Theorem 15.29, the matrix $A - cI$ is singular, and this implies that $|A - cI| = 0$. Conversely, if this determinant is zero, the linear transformation $\mathscr{A} - c\mathscr{T}$ must be singular and there will exist a nonzero vector X satisfying Equation 15.79.

We may point out that it did not matter which matrix of $\mathscr{A}$ we selected above. As a matter of fact, if we had used a different ordered basis, the results of the preceding section show that the matrix of $\mathscr{A}$ would be of the form BAB^{-1} for some nonsingular matrix B. However,

$$|BAB^{-1} - cI| = |B(A - cI)B^{-1}| = |A - cI|,$$

by Theorem 15.51 and Corollary 15.53.

Let us, for the moment, change our point of view as follows. If λ is an indeterminate (a customary notation in this particular setting), then

$$|A - \lambda I| = \begin{vmatrix} a_{11} - \lambda & a_{12} & \cdots & a_{1n} \\ a_{21} & a_{22} - \lambda & \cdots & a_{2n} \\ \cdot & \cdot \cdot \cdot \cdot \cdot \cdot \cdot \cdot \cdot \cdot & & \cdot \\ a_{n1} & a_{n2} & \cdots & a_{nn} - \lambda \end{vmatrix},$$

and it is not difficult to verify that this is a polynomial in λ of degree exactly n. In fact, the coefficient of λ^n is $(-1)^n$. We now make another definition as follows.

15.80 DEFINITION If $A \in F_n$ and λ is an indeterminate, the polynomial $|A - \lambda I|$ is called the *characteristic polynomial* † of A, and the roots in F of this polynomial are called the *characteristic roots* of A.

We may state the results of our observations above as the following theorem, which will also justify the two uses of the concept of characteristic root.

15.81 THEOREM *The characteristic roots of a linear transformation $\mathscr{A}$ of V are precisely the characteristic roots of the matrix of $\mathscr{A}$ relative to any ordered basis of V.*

Since the characteristic polynomial is of degree n, a linear transformation clearly cannot have more than n characteristic roots. Of course, it may very well have fewer than n of them.

Let us now give some examples to illustrate these concepts.

Example 1 Let our vector space be $V_3(\mathbf{Q})$, and let $\mathscr{A}$ be the linear transformation whose matrix relative to the usual ordered unit vectors is

$$A = \begin{bmatrix} 2 & 0 & 0 \\ 0 & 0 & -1 \\ 0 & -1 & 0 \end{bmatrix}.$$

Find the characteristic roots and characteristic vectors of $\mathscr{A}$.

Solution: We have

$$|A - \lambda I| = \begin{vmatrix} 2 - \lambda & 0 & 0 \\ 0 & -\lambda & -1 \\ 0 & -1 & -\lambda \end{vmatrix} = -(\lambda - 2)(\lambda - 1)(\lambda + 1).$$

---

† It can be proved that if $f(\lambda)$ is the characteristic polynomial of A, then $f(A) = 0$. This is the well-known Cayley-Hamilton Theorem. In particular, it follows that the minimal polynomial of A is a divisor of its characteristic polynomial. See, e.g., Birkhoff and MacLane [3] or Herstein [12] for a proof of the Cayley-Hamilton Theorem.

Thus the characteristic roots of A (and of $\mathscr{A}$) are 2, 1, and -1. To find a characteristic vector of $\mathscr{A}$ corresponding to the root 2, we use the notation introduced in Section 16.6 and seek a nonzero vector $X = (x_1, x_2, x_3)$ such that $X(A - 2I) = 0$. In more detail, this equation is the following:

$$(x_1, x_2, x_3)\begin{bmatrix} 0 & 0 & 0 \\ 0 & -2 & -1 \\ 0 & -1 & -2 \end{bmatrix} = (0, -2x_2 - x_3, -x_2 - 2x_3) = 0.$$

Accordingly, we must have $-2x_2 - x_3 = 0$ and $-x_2 - 2x_3 = 0$, and these equations imply that $x_2 = x_3 = 0$. Thus $(x_1, 0, 0)$ is a characteristic vector of A for each $x_1 \neq 0$. In particular, $(1, 0, 0)$ is a characteristic vector of A corresponding to the characteristic root 2.

We omit the calculations but the reader may verify that $(0, 1, -1)$ and $(0, 1, 1)$ are characteristic vectors corresponding, respectively, to the characteristic roots 1 and -1. Again, these vectors are uniquely determined except for a nonzero scalar multiplier. In this example, there were three distinct characteristic roots and we remark that the three characteristic vectors corresponding to these three roots are linearly independent. This fact illustrates a theorem to be proved presently.

Example 2 Let $\mathscr{A}$ be the linear transformation of $V_3(\mathbf{Q})$ whose matrix relative to the ordered unit vectors is

$$A = \begin{bmatrix} 0 & 1 & -1 \\ 0 & 3 & 0 \\ 1 & 2 & 0 \end{bmatrix}.$$

Find characteristic roots and characteristic vectors of $\mathscr{A}$.

Solution: In this case, we find that

$$|A - \lambda I| = \begin{vmatrix} -\lambda & 1 & -1 \\ 0 & 3 - \lambda & 0 \\ 1 & 2 & -\lambda \end{vmatrix} = -(\lambda - 3)(\lambda^2 + 1).$$

This polynomial has only the root 3 in $\mathbf{Q}$, so there is only one characteristic root. Solving the equation

$$(x_1, x_2, x_3)\begin{bmatrix} -3 & 1 & -1 \\ 0 & 0 & 0 \\ 1 & 2 & -3 \end{bmatrix} = 0,$$

we find that the only characteristic vectors corresponding to the characteristic root 3 are those of the form $(0, x_2, 0)$, where $x_2 \neq 0$. In particular, $(0, 1, 0)$ is such a characteristic vector.

The same calculations and conclusions apply if we replace the field **Q** by **R**. However, if we had assumed that A was the matrix of a linear transformation of $V_3(\mathbf{C})$, we would have found three characteristic roots since, in **C**, $\lambda^2 + 1 = (\lambda + i)(\lambda - i)$.

The subject which we have here just barely introduced is an extensive one, and full expositions can be found in any text on "Linear Algebra." However, we shall now prove one more theorem which is of a rather special nature as follows.

15.82 THEOREM *Suppose that a linear transformation $\mathscr{A}$ of a vector space V of dimension $n > 0$ over a field F has n distinct characteristic roots $\lambda_1, \lambda_2, \ldots, \lambda_n$ in F. If $X_1, X_2, \ldots, X_n$ are characteristic vectors corresponding, respectively, to these characteristic roots, then $\{X_1, X_2, \ldots, X_n\}$ is a linearly independent set of vectors.*

Proof: The proof is an inductive one as follows. First, since characteristic vectors are different from zero, a set consisting of just one of these vectors is linearly independent. Let us assume that a set of any k of these vectors is linearly independent and prove that if $k < n$, the same is true for any set of $k + 1$ vectors. For convenience of notation, let us prove that $\{X_1, X_2, \ldots, X_{k+1}\}$ is a linearly independent set. Suppose that c_i $(i = 1, 2, \ldots, k + 1)$ are elements of F such that

15.83
$$c_1 X_1 + c_2 X_2 + \cdots + c_{k+1} X_{k+1} = 0.$$

If $c_1 \neq 0$, we proceed as follows. From 15.83, we must have

$$\begin{aligned}0 &= (c_1 X_1 + c_2 X_2 + \cdots + c_{k+1} X_{k+1})\mathscr{A} \\ &= c_1(X_1\mathscr{A}) + c_2(X_2\mathscr{A}) + \cdots + c_{k+1}(X_{k+1}\mathscr{A}) \\ &= c_1\lambda_1 X_1 + c_2\lambda_2 X_2 + \cdots + c_{k+1}\lambda_{k+1} X_{k+1}.\end{aligned}$$

If we multiply 15.83 by λ_1 and subtract from this equation, we obtain

$$c_2(\lambda_2 - \lambda_1)X_2 + \cdots + c_{k+1}(\lambda_{k+1} - \lambda_1)X_{k+1} = 0.$$

Since, by our hypothesis, $\{X_2, \ldots, X_{k+1}\}$ is a linearly independent set of k vectors, we must have $c_2(\lambda_2 - \lambda_1) = \cdots = c_{k+1}(\lambda_{k+1} - \lambda_1) = 0$. However, the λ's are *distinct*, and it follows that $c_2 = \cdots = c_{k+1} = 0$.

Equation 15.83 then shows that $c_1 X_1 = 0$ and we conclude that $c_1 = 0$, thus contradicting our assumption that $c_1 \neq 0$. Accordingly, we must have $c_1 = 0$. Substituting this value of c_1 in 15.83, we again use the linear independence of $\{X_2, \ldots, X_{k+1}\}$ to conclude that *all* c's are zero. Therefore, any set of $k + 1$ of our vectors is linearly independent, and this completes the proof of the theorem.

Let us draw one further conclusion from this theorem. Since $\{X_1, X_2, \ldots X_n\}$ is a linearly independent set of vectors, it is a basis of V. The matrix of $\mathscr{A}$ relative to the ordered basis $X_1, X_2, \ldots, X_n$ of V is determined from the following equations:

$$
\begin{aligned}
X_1 \mathscr{A} &= \lambda_1 X_1, \\
X_2 \mathscr{A} &= \qquad \lambda_2 X_2, \\
\cdots \cdots & \cdots \cdots \cdots \cdots \\
X_n \mathscr{A} &= \qquad\qquad\qquad \lambda_n X_n.
\end{aligned}
$$

That is, the matrix of $\mathscr{A}$ relative to this basis has $\lambda_1, \lambda_2, \ldots, \lambda_n$ down the principal diagonal and zeros elsewhere. We thus have the following corollary of the theorem just established.

15.84 COROLLARY *If a linear transformation $\mathscr{A}$ of a vector space V over F has n distinct characteristic roots in F, relative to a suitably chosen ordered basis of V the matrix of $\mathscr{A}$ has these characteristic roots on the principal diagonal and zeros elsewhere.*

Finally, in view of Theorem 15.57 and Definition 15.58, this result can be expressed in an alternate form as follows.

15.85 COROLLARY *If a matrix A of F_n has n distinct characteristic roots in F, then A is similar to a matrix with these characteristic roots on the principal diagonal and zeros elsewhere.*

EXERCISES

1. If the matrix A of Example 2 above is considered to be the matrix of a linear transformation $\mathscr{A}$ of $V_3(\mathbf{C})$, find all characteristic vectors of $\mathscr{A}$.

2. Let $\mathscr{A}$ be the linear transformation of $V_2(\mathbf{Q})$ whose matrix relative to the ordered unit vectors is

$$\begin{bmatrix} 2 & 0 & 1 \\ 0 & 1 & 0 \\ 0 & 0 & 2 \end{bmatrix}.$$

Find all characteristic roots and all characteristic vectors of $\mathscr{A}$.

3. Show that the set of all characteristic vectors of a linear transformation $\mathscr{A}$ of a vector space V which correspond to a fixed characteristic root of $\mathscr{A}$ is a subspace of V.

4. If A is a matrix *of order two*, use the results of Exercise 1 of the preceding set to prove that the characteristic polynomial of A coincides with the minimum polynomial of A if and only if A is not a scalar matrix. [Note: We have used different symbols for the indeterminates in the two cases, but that fact is unimportant.]

5. Let

$$A = \begin{bmatrix} 2 & -2 & 2 \\ 0 & 1 & 1 \\ -4 & 8 & 3 \end{bmatrix}$$

be the matrix of a linear transformation $\mathscr{A}$ of $V_3(\mathbf{Q})$ relative to the usual ordered basis of unit vectors.

(i) Verify that $\mathscr{A}$ has three distinct characteristic roots and find a characteristic vector corresponding to each characteristic root.

(ii) Find an ordered basis of $V_3(\mathbf{Q})$ relative to which the matrix of $\mathscr{A}$ has these characteristic roots on the principal diagonal and zeros elsewhere.

(iii) Find a nonsingular matrix B such that BAB^{-1} is of the form described in part (ii).

 COMMENTARY

1 We have defined an algebra in this chapter and viewed the algebra of matrices. Cayley (1821–1895) was the inventor of this algebra, but this is not where the story begins. With our axiomatic view of mathematics, it is easy to drop this or that axiom

and ask, "What do we have now?" In algebra, however, this approach dates from the late nineteenth century. Before this time new systems were discovered by their utility or as organic extensions of old systems. The basis for all mathematics lies in the number systems. The field of complex numbers is an algebra over the real field. In fact, any extension K of a field F is an algebra over F. Even though this view of things is useful, as we have seen in Chapter 12, it is a cheap shot, that is, viewing K as an algebra over F is nothing more than renaming a familiar object

The mathematics of number was essentially known by the time of Gauss (1777–1855); for then, even though strong foundations were lacking, all of the most useful number systems (rational, real, and complex) had a firm intuitive footing. All that remained was to lay down strong foundations by giving axioms for these number systems. Of course, this was no small task and took more than a century. Our habit now is to study first those mathematical systems with the fewest axioms. From the point of view of evolution, the systems with more axioms (fields) were developed first, so that branching out to other systems was very difficult. Recognition that the groups of permutations were analogous to the additive and multiplicative groups in fields is hardly a century old. It seems natural then that the first noncommutative rings recognized as such should not only carry fields with them, but should look almost like fields, the missing ingredient being the commutative law. A ring which satisfies all the properties of a field except for the commutative law is called a division ring. If this ring is also an algebra, it is called a division algebra.

Hamilton (1805–1865) in 1843 was the first to discover an honest division algebra, the real quaternion algebra. The quaternion group has the elements $\pm \mathbf{i}$, $\pm \mathbf{j}$, $\pm \mathbf{k}$, ± 1 subject to the rules $\mathbf{i}^2 = \mathbf{j}^2 = \mathbf{k}^2 = -1$, $\mathbf{ij} = \mathbf{k} = -\mathbf{ji}$, $\mathbf{jk} = \mathbf{i} = -\mathbf{kj}$, and $\mathbf{ki} = \mathbf{j} = -\mathbf{ik}$. We may take 1, $\mathbf{i}$, $\mathbf{j}$, $\mathbf{k}$ to be a basis for a vector space over $\mathbf{R}$. By using the distributive law and the rules for multiplication in the quaternion group we obtain the real quaternion algebra. (For further discussion see Section 16.1.) The following year Grassman (1809–1877) published his *Ausdehnungslehre* in which he formulates the most general algebra of which the quaternions are a special case. Suppose that F is a field and $v_1, v_2, \ldots, v_n$ are a basis for an n-dimensional vector space V. Let $v_i v_j = \sum_k c_{ijk} v_k$ where $c_{ijk} \in F$, the indices all taking values from 1 to n. Using this law of composition and the distributive law for multiplication of vectors of V we obtain certain "Grassman algebras." For most choices of coefficients $c_{ijk} \in F$ the "algebra" V does not obey the associative law. Casting aside all these choices, we may show that any algebra A which is finite dimensional over F is isomorphic to a "Grassman algebra" for some choice of n and some choice of $c_{ijk} \in F$. Grassman was little appreciated as a mathematician. On the other hand, Hamilton was recognized as a genius at a young age and felt that his quaternions were his greatest discovery. Grassman's algebras subsumed the quaternions as a special case. Where was everyone?

Even today the quaternion algebra plays a special role among algebras. Hamilton had great hopes for them, and developed a calculus dependent upon them for use in physics. Mirrored in ordinary quaternion multiplication are the inner or dot products and the vector or cross products. Further, using ordinary quaternion multiplication, one can obtain the most general rotations of four-, three-, and two-dimensional real space. It is little wonder that Hamilton had high hopes for his quaternions. It is almost a shame that the American physicist Gibbs (1839–1903) invented a more natural vector algebra in which to cast physical theories. Vestiges of quaternions remain as the unit vectors **i**, **j**, **k** of physics.

Grassman's creation was unusable by its very generality. For which choices of $c_{ijk} \in F$ do we obtain isomorphic algebras? Which choices of $c_{ijk} \in F$ give intrinsically interesting algebras? It turned out, at that time, that the most fruitful choice was made by Cayley (1821–1895): he created our matrix algebra in 1857. In fact, he created it to serve the same function which it serves today, that is, to classify the linear transformations of n-dimensional space. It should be remarked that n-dimensional space was a new idea on the mathematical scene also, receiving major development by Cayley [144].

Once the door was opened, the flood began. Algebras found ready application in many branches of mathematics. Next to their creation, the most significant steps in their development came in the early 1900's at the hands of J. H. M. Wedderburn (1882–1948). He showed that a finite dimensional division algebra over a finite field must be a finite field [6, 51]. If you recall the Commentary after Chapter 6 on projective geometry, then you can see the connections this might have with finite projective planes. The theorem of Desargues (1593–1662) and that of Pappus (c. 300) are two celebrated results of projective geometry [99, 106]. Wedderburn's theorem shows the theorem of Pappus is implied by the theorem of Desargues in a finite projective plane!

Wedderburn did not stop here. He also proved that if A is a finite dimensional "semisimple" algebra over a field F then A is isomorphic to a direct sum of full matrix algebras where each matrix algebra has its entries coming from a finite dimensional division algebra over F. A special case of this theorem occurs when $F = \mathbf{C}$, the complex field, A has no nontrivial ideals, and $ab \neq 0$ for some $a, b \in A$. In that case, A is isomorphic to the algebra of $n \times n$ matrices with entries from $\mathbf{C}$ [24, 42, 121].

As remarked in an earlier commentary (Chapter 3), commutative ring theory evolved from considerations about numbers and about polynomials. Noncommutative ring theory had its beginnings with Hamilton, Grassman, and Cayley. From rather dissimilar origins has evolved that branch of algebra called ring theory. It is little wonder that commutative and noncommutative ring theory are still quite independent of each other.

2   Cayley's invention of matrix algebra in 1857 was of the utmost importance to all of mathematics. Nowadays, we view the algebra in one of its isomorphic disguises, i.e. as the algebra $\text{Hom}_F (V, V)$ of linear transformations of a vector space V over a field F. In this form we may drop all conditions that V be finite dimensional. Let us examine a few of the applications of this most versatile of algebras.

Let us start with a group G. Take the elements $X \in G$ to be basis vectors of a vector space V over a field F. If $Y = c_1 X_1 + c_2 X_2 + \cdots + c_t X_t \in V$ and $a \in G$ then we may set $a\theta \in \text{Hom}_F (V, V)$ equal to that linear transformation of V given by

$$Y(a\theta) = c_1(X_1 a) + c_2(X_2 a) + \cdots + c_t(X_t a)$$

where $X_i a$ is the product of X_i and a in G. The mapping θ is an isomorphism of G into the group of units of the algebra $\text{Hom}_F (V, V)$.

Suppose next we consider a ring S with unity e. It is not necessary to do so, but to keep things concrete let us assume that S is an algebra over a field F. Thus S^+ is a vector space over F. Let $X \in S^+$ be a vector and for $a \in S$ define the mapping $a\theta \in \mathrm{Hom}_F(S^+, S^+)$ given by

$$X(a\theta) = Xa$$

where multiplication takes place in S. The mapping θ again is an isomorphism of S into the algebra $\mathrm{Hom}_F(S^+, S^+)$.

To fix ideas even more, let us assume that the field F is the real field $\mathbf{R}$ and the vector spaces in question are n-dimensional. If we fix a particular basis for our vector space then the isomorphism θ replaces G (or S) by a multiplicative group of matrices (or a ring of matrices), moving from the abstract algebraic object G (or S) into a concrete object involving honest numbers. Once we have matrices, there are many methods available to us which were not before. For example, the fact that for matrices A and B, the determinant of the product $|AB|$ is the product of the determinants $|A| \cdot |B|$ gives us a homomorphism of our group G into the group of units of $\mathbf{R}$. We can ask what happens to our group or ring of matrices under change of basis. We can study characteristic vectors and characteristic values of elements. The tools available for studying G (or S) have suddenly multiplied. This idea has been greatly exploited in algebra and analysis. The subject area is called representation theory and the homomorphism θ is called a representation.

Let us look at a very specific example, the complex field $\mathbf{C}$ as an algebra over the real field $\mathbf{R}$. Consider the basis $1, i$ of $\mathbf{C}^+$ as a vector space over $\mathbf{R}$. If $a + bi$ is a complex number for $a, b \in \mathbf{R}$ then we have

$$1(a + bi)\theta = a \cdot 1 + b \cdot i,$$
$$i(a + bi)\theta = -b \cdot 1 + a \cdot i.$$

In particular, we may view θ as mapping $a + bi$ to the matrix $A = \begin{bmatrix} a & -b \\ b & a \end{bmatrix}$. The set of all matrices of this form with $a, b \in \mathbf{R}$ is a field isomorphic to $\mathbf{C}$. The determinant maps A to $a^2 + b^2 = (a + bi)(a - bi) = |a + bi|^2$. The characteristic polynomial of A is $x^2 - 2ax + a^2 + b^2$ which factors into $(x - (a + bi))(x - (a - bi))$. These equations show some of the connections between facts which we know about complex numbers and natural equations arising from linear algebra [40, 46, 114, 121].

3   The references [57, 60, 62, 63, 67] are a few of the many good texts upon linear algebra. At a much more advanced level are [58, 59]. Section 15.12 is just an introduction to the theorems on canonical forms. These are discussed in more detail in [57, 62, 63]. We have barely touched the subject of inner products. More coverage can be found in [57, 62, 63, 64]. Some of the many connections between linear algebra and geometry are discussed in [64, 100]. Turning more specifically to algebras and algebras of matrices, we can mention [40, 121]. Finally, there is a subject called "Matrix Theory" which is concerned with the computational aspects of matrices. Topics on this subject are discussed in [58, 68].

chapter sixteen

SOME ADDITIONAL TOPICS

The purpose of this chapter is to present a brief account of several different topics of considerable algebraic interest which have not been introduced in preceding chapters. The degree of independence of these topics will be indicated by the fact that the reader may start with Section 16.1, 16.2, 16.3 or 16.5 as desired. References for further reading on these and related topics will be found in the commentaries at the end of the chapter.

16.1 QUATERNIONS

Before introducing the topic of this section, let us formalize as follows a definition which was casually mentioned in the Commentary of Chapter 15.

16.1 **DEFINITION** A ring R with more than one element and having a unity is said to be a *division ring* if every nonzero element of R has a multiplicative inverse in R.

It will be seen that this differs from Definition 6.1 of a field only in that the word "commutative" has been omitted. Thus, a division ring is a field if and only if it is a commutative ring. In particular, every field is a division ring.

We now wish to introduce an example of a division ring which is not a field. This division ring, introduced by Sir William R. Hamilton in 1843, is of special historical interest in that it was the first time that a noncommutative ring was studied, although the name *ring* itself was not introduced until much later. (See the Commentary of Chapter 15.)

The approach which we shall use, suggested by that of Hamilton but given a more general setting, is to consider first how one might construct an algebra over a field F, starting only with a given vector space V over F. In other words, given the vector space V, how can one define a multiplication of elements of V which will make V into an algebra over F? As a simple illustration of the ideas involved, let V be a vector space of dimension two over F, and let $\{X_1, X_2\}$ be a basis of V. The elements of V are therefore uniquely expressible in the form

16.2
$$a_1 X_1 + a_2 X_2, \qquad a_1, a_2 \in F.$$

In order to have a multiplication of such elements defined in such a way as to obtain an algebra, it would have to be true that

16.3 $(a_1 X_1 + a_2 X_2)(b_1 X_1 + b_2 X_2)$
$$= a_1 b_1 X_1{}^2 + a_1 b_2 X_1 X_2 + a_2 b_1 X_2 X_1 + a_2 b_2 X_2{}^2.$$

Thus, *if* we knew the products $X_1{}^2$, $X_1 X_2$, $X_2 X_1$, $X_2{}^2$ of the basis elements, we could use 16.3 to *define* the product of any two elements of V.

Now let us look further at this approach, and suppose that we define the products $X_1{}^2$, $X_1 X_2$, $X_2 X_1$, $X_2{}^2$ to be arbitrarily chosen elements of the form 16.2 and then use 16.3 to define arbitrary products of elements of V. It may be shown that all the properties of a ring hold, with the possible exception of the associative law of multiplication. Furthermore, multiplication of arbitrary elements will be associative provided multiplication of basis elements is associative.

As an almost trivial example of an algebra presented as sketched above, let $\{1, i\}$ be a basis of a vector space of dimension two over the field $\mathbf{R}$ of real numbers. The "1" is the unity of $\mathbf{R}$; and we want it to be the unity of our algebra, so we want to have $1 \cdot 1 = 1$, $1 \cdot i = i \cdot 1 = i$. We also propose to define $i^2 = -1$, so we have a multiplication defined for the elements of a basis. The elements of our algebra are of the form $a + bi$, where a and b are elements of $\mathbf{R}$, and the product of any two elements of the algebra is defined (in accordance with 16.3) to be as follows:

$$(a + bi)(c + di) = ac + (ad + bc)i + bdi^2$$
$$= ac - bd + (ad + bc)i.$$

Of course, it turns out that multiplication is associative and we have the algebra of complex numbers over the reals. This algebra is itself a field, the field of complex numbers.

Hamilton tried very hard, but unsuccessfully, to do something to obtain an algebra of dimension three, which would also be a field over the reals. Although he was not able to carry out this program for dimension three, he did succeed (by a stroke of genius) in doing a comparable thing for dimension four— only it turned out that multiplication was not commutative, so he obtained a division ring which was not a field. It is this example which we now wish to present.

Let $\{1, i, j, k\}$ be a basis of a vector space V of dimension 4 over $\mathbf{R}$. We propose to make V into an algebra Q over $\mathbf{R}$ by defining multiplication of these basis elements and then using the analogue of 16.3 to define multiplication of arbitrary elements of V. The unity 1 of $\mathbf{R}$ is to be the unity of Q, so that we need not further specify how to multiply by 1. Following Hamilton, products of the other basis elements are defined as follows:

16.4

$$i^2 = j^2 = k^2 = -1,$$
$$ij = k,$$
$$ji = -k,$$
$$jk = i,$$
$$kj = -i,$$
$$ki = j,$$
$$ik = -j.$$

An easy way to remember the products of different basis elements is by use of the scheme shown in Figure 1. In this figure, the product of an element by the adjacent one in the direction of the arrows is the remaining basis element. The product in the other order is changed in sign. Thus, for example, $ki = j$ since passing from k to i is in the direction of the arrows, similarly $ik = -j$ since passing from i to k is against the direction of the arrows.

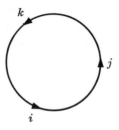

FIGURE 1

The elements of the algebra Q are the linear combinations of the basis elements, that is, they are of the form

16.5 $a + bi + cj + dk,$ $a, b, c, d \in \mathbf{R}.$

The product of any two of these elements is defined by using 16.4 and the other properties that must hold in an algebra (as in 16.3). A detailed calculation will show that this product turns out to be as follows:

16.6 $(a_1 + b_1i + c_1j + d_1k)(a_2 + b_2i + c_2j + d_2k)$
$$= a_1a_2 - b_1b_2 - c_1c_2 - d_1d_2 + (a_1b_2 + b_1a_2 + c_1d_2 - d_1c_2)i$$
$$+ (a_1c_2 + c_1a_2 + d_1b_2 - b_1d_2)j + (a_1d_2 + d_1a_2 + b_1c_2 - c_1b_2)k.$$

It can be shown that the associative property of multiplication does hold and that Q is indeed an algebra over $\mathbf{R}$. Following Hamilton, an element of Q is usually called a *quaternion* and Q is called the algebra of *real quaternions*.

If q is the element 16.5 of Q, the element

$$q^* = a - bi - cj - dk$$

may be called the *conjugate* of q. It follows easily from 16.6 that

$$q^*q = qq^* = a^2 + b^2 + c^2 + d^2.$$

Thus, if q is a nonzero element of Q, not all of the real numbers a, b, c, d can be zero; hence qq^* is a *positive real* number. It follows from the preceding equation that if $q \neq 0$, then $q^{-1} = (qq^*)^{-1}q^*$ since

$$(qq^*)^{-1}q^*q = q(qq^*)^{-1}q^* = 1.$$

This shows that every nonzero element of Q has a multiplicative inverse in Q, and Q is therefore a division ring. Since multiplication of certain quaternions (for example, i and j) is not commutative, Q is not a field. Some additional properties of this division ring Q of real quaternions will be brought out in the exercises below.

EXERCISES

1. Verify that if b, c, and d are real numbers such that $b^2 + c^2 + d^2 = 1$, the quaternion $q = bi + cj + dk$ has the property that $q^2 = -1$. Hence show that there are infinitely many real quaternions q such that $q^2 = -1$.

2. Using a notation which will distinguish between the complex number i and the quaternion i, use 16.4 and 16.6 to define an algebra over the field $\mathbf{C}$ of complex numbers. Verify that the algebra so obtained is not a division ring.

3. If q is a real quaternion, the real number qq^* is often called the *norm* of q and denoted by $N(q)$. If q_1 and q_2 are real quaternions, prove that $N(q_1q_2) = N(q_1)N(q_2)$.

4. Prove that the algebra of Exercise 2 is isomorphic to the algebra of matrices of order two over the field of complex numbers.

16.2 PRINCIPAL IDEAL DOMAINS

The integral domain $\mathbf{Z}$ and the integral domain $F[x]$, where F is a field and x an indeterminate, have a number of similar properties. In particular, in both of these domains there is (in a certain sense) a unique factorization of elements into a product of primes (Theorems 5.26 and 10.39). Both of these theorems are special cases of a more general theorem (Theorem 5.25) on factorization in Euclidean domains. We proved in Section 5.3 that a Euclidean domain is a *principal ideal domain*, i.e., a domain in which every ideal is principal. We then went on in Section 5.4 to prove that a Euclidean domain is a *unique factorization domain*, i.e., a domain in which every nonzero element which is not a unit is either a prime or has a factorization into a product of primes where (except for ordering) the prime factors are uniquely determined up to units. We did note in Section 5.3 an example of a principal ideal domain which is not a Euclidean domain. We also remarked that any principal ideal domain is a unique factorization domain. The purpose of this section is to close the logical gap we left in Chapter 5, that is, to prove the following theorem.

16.7   THEOREM   *A principal ideal domain is a unique factorization domain.*

For the remainder of this section we shall assume that D is a principal ideal domain with unity e. Any ideal I of D is principal, i.e. there is some $a \in D$ such that a generates I or equivalently $(a) = I$. Recall that

$$(a) = \{ax \mid x \in D\}.$$

It will now be convenient to use an alternate notation for (a), namely aD. This notation will perhaps help to emphasize that the principal ideal generated by a consists of the multiples of a. That is $b \in (a) = aD$ if and only if $a \mid b$.

A proof of unique factorization has naturally two parts: (i) a proof that an element factors into a product of primes; and (ii) a proof that any such factorization has the required uniqueness properties. Therefore, Theorem 16.7 may be stated as follows.

16.8 THEOREM

(i) *Let a be a nonzero element of a principal ideal domain D with unity e. If a is not a unit of D, it can be expressed in the form*

16.9
$$a = p_1 p_2 \cdots p_r,$$

where r is a positive integer and the p's are primes in D.

(ii) *The representation 16.9 is unique in the following sense. If also*

$$a = q_1 q_2 \cdots q_s,$$

where the q's are primes in D, then $s = r$ and by a proper choice of notation, q_i and p_i are equal up to units for $i = 1, 2, \ldots, r$.

The proof of part (i) of this theorem involves methods much different from those used in the case of a Euclidean domain (Theorem 5.25), and we shall give this part of the proof in detail. However, we shall not give a detailed proof of part (ii), but will obtain some lemmas from which it should be quite easy to adapt the inductive proof used to establish the corresponding part of Theorem 5.25.

In order to factor the element a of D we commence with $a = m_1 n_1$ where m_1 and n_1 are factors of a. We continue by factoring $m_1 = m_2 n_2$, hoping to return later and worry about n_1. Continuing to factor only the m's we obtain a chain $a = m_1 n_1, m_1 = m_2 n_2, m_2 = m_3 n_3, \ldots, m_{t-1} = m_t n_t$ where $m_1 \mid a, m_2 \mid m_1, m_3 \mid m_2, \ldots, m_t \mid m_{t-1}$. In a Euclidean domain, if none of the n's are units, $m_i \delta < m_{i-1} \delta$ (Theorem 5.9) for $i = 1, 2, \ldots, t$ (we set $m_0 = a$). Since $m\delta$ is a nonnegative integer, well ordering of the integers guarantees that the factoring of m's must finally stop when m_t is a prime. Without any size measure δ, we must show that in a principal ideal domain the factorization process also terminates when m_t is a prime (after all, there is no obvious reason why we could not continue factoring the m's indefinitely, never arriving at a prime). We do this by translating the sequence of divisions into a statement about containment of ideals. The following lemma will prove useful in this respect.

16.10 LEMMA *If $a, b \in D$ then $a|b$ if and only if $aD \supseteq bD$. Further, $aD = bD$ if and only if a and b are equal up to units.*

> **Proof:** If $a|b$ then $b = ac$ for some $c \in D$ so that $bx = acx \in aD$ for all $x \in D$. We conclude that $aD \supseteq bD$. On the other hand, if $aD \supseteq bD$ then $b = be \in bD \subseteq aD$ so that $b = ac$ for some $c \in D$ proving that $a|b$. The second part of the lemma may be proved in the same way as Corollary 5.16 (See Exercise 1).

Recalling the sequence of m's once again we now see that the chain of divisions $m_1|a, m_2|m_1, m_3|m_2, \ldots, m_{t-1}|m_t$ may be written equivalently in the form

$$aD \subseteq m_1 D \subseteq m_2 D \subseteq m_3 D \subseteq \cdots \subseteq m_t D.$$

In order to guarantee that we cannot continue factoring the m's indefinitely, this chain must stop when m_t is a prime. We formalize these ideas in the following lemma.

16.11 LEMMA *In a principal ideal domain D there cannot exist an infinite sequence of ideals $a_1 D, a_2 D, a_3 D, \ldots$, such that each is properly contained in the following, that is, we cannot have*

16.12
$$a_1 D \subset a_2 D \subset a_3 D \subset \cdots.$$

> **Proof:** Suppose, on the contrary, that there is an infinite sequence satisfying 16.12, and let us seek a contradiction.
>
> We assert that the union A of all the ideals $a_i D$ in 16.12 is itself an ideal in D. For if $a, b \in A$, we know that $a \in a_i D$ for some positive integer i and that $b \in a_j D$ for some positive integer j. Suppose, for example, that $i \leq j$. Then $a_i D \subset a_j D$ and hence, in particular, both a and b are in $a_j D$ and therefore $a + b \in a_j D$. But $a_j D \subseteq A$, and we conclude that $a + b \in A$. Clearly, if $a \in A$, then $-a \in A$. Hence A is an additive subgroup of D. Moreover, if $a \in A$ with $a \in a_i D$ and $c \in D$, then $ac \in a_i D \subseteq A$, and A is indeed an ideal in D. Since every ideal in D is a principal ideal, there exists an element a of A such that $A = aD$. Since A is the union of the ideals occurring in 16.12, this implies that $a \in a_i D$ for some positive integer i. Hence

$A = aD \subseteq a_i D$ and, in particular, $a_j D \subseteq a_i D$ for each $j > i$. This gives the desired contradiction and completes the proof.

The fact that 16.12 cannot hold in a principal ideal domain is often expressed by saying that the *ascending chain condition* holds in a principal ideal domain.

Another concept which we shall find useful is the following.

16.13 DEFINITION An ideal A in an arbitrary ring R is said to be a *maximal ideal* if $A \neq R$ and there exists no ideal B in R such that $A \subset B \subset R$.

The relation of this concept to the subject under discussion is given in the following lemma.

16.14 LEMMA *A nonzero ideal aD in a principal ideal domain D is a maximal ideal if and only if a is a prime of D.*

Proof: Suppose, first, that a is prime. Let's assume that aD is not a maximal ideal, and seek a contradiction. Thus, by our assumption, there exists an ideal bD of D such that $aD \subset bD \subset D$. The fact that $aD \subset bD$ implies that $b \mid a$ and that a and b are not equal up to units since, otherwise, we would have $aD = bD$. Moreover, $bD \subset D$ implies that b is not a unit. Thus b is a divisor of a and is neither a unit nor equal to a up to units, contradicting the assumption that a is a prime. We have thus shown that if a is prime, aD is a maximal ideal.

Conversely, suppose that aD is a maximal ideal in D. If a were not prime, it would have a divisor c, neither a unit nor equal to a up to units. But this would imply that $aD \subset cD \subset D$, violating our assumption that aD is a maximal ideal. We conclude that a must be prime, and the proof is complete.

We are now ready to prove the following important first step in the proof of Theorem 16.8(i).

16.15 LEMMA *If the nonzero element a of the principal ideal domain D is not a unit, it has a prime factor.*

Proof: If a is itself prime, there is nothing to prove. If a is not a prime, the preceding lemma shows that aD is not a maximal ideal. Thus there exists an ideal $a_1 D$ such that $aD \subset a_1 D \subset D$. It follows that $a_1 \mid a$ and a_1 is not a unit. Applying the same argument to a_1

we see that if a_1 is not a prime, there exists an element a_2 of R such that $a_2 | a_1$ and a_2 is not a unit. In particular, we would have $aD \subset a_1 D \subset a_2 D \subset D$. Continuing, if a_2 is not a prime, it has a divisor a_3 such that $aD \subset a_1 D \subset a_2 D \subset a_3 D \subset D$. By Lemma 16.11, this process must come to an end after a finite number of steps. Hence there exists a divisor, say a_n, of a such that a_n is a prime. This completes the proof of the lemma.

Using these results, let us now give a proof of part (i) of Theorem 16.8.

Let a be a nonzero element of D, not a unit. By the preceding lemma, a has a prime factor p_1 and we have $a = p_1 c_1$ for some element c_1 of D. If c_1 is not a unit, we therefore have $aD \subset c_1 D$. Applying the same argument to c_1, we see that there exists a prime divisor p_2 of c_1. If we write $c_1 = p_2 c_2$, then $a = p_1 p_2 c_2$ and if c_2 is not a unit, we have $aD \subset c_1 D \subset c_2 D$, and also c_2 has a prime divisor p_3. At this stage, we have $a = p_1 p_2 p_3 c_3$. This process can be continued as long as c_i is not a unit. By Lemma 16.11, there must exist a positive integer r such that c_r is a unit since, otherwise, we would have an infinite sequence $c_i D$ of ideals in D such that

$$aD \subset c_1 D \subset c_2 D \subset \cdots .$$

We conclude that

$$a = p_1 p_2 \cdots p_r c_r,$$

where c_r is a unit. But $p_r c_r$ is a prime and is equal up to units to p_r. Let us change the notation and call it p_r. We thus have an expression 16.9, as we wished to prove. That is, in a principal ideal domain D every nonzero element which is not a unit is expressible as a product of a finite number of primes.

We conclude this section by presenting some results from which the reader may obtain the proof of the second part of the theorem by a simple adaptation of the proof of Theorem 5.25 for Euclidean domains.

If a and b are nonzero elements of a principal ideal domain D, the element d of D is said to be a *greatest common divisor* (g.c.d.) of a and b if the following two conditions are satisfied (see Definition 5.17):

(i) $d|a$ and $d|b$.

(ii) If $c \in D$ such that $c|a$ and $c|b$, then $c|d$.

It may be observed that if d is a g.c.d. of a and b, so also is any element equal to d up to units. In the case of the integers, we found it convenient to obtain a *unique* g.c.d. by requiring that it be positive. Similarly, for polynomials over a field we obtained a *unique* g.c.d. by requiring that it be a monic polynomial. However, in the general case here under discussion there is no obvious way to get a unique g.c.d., so we simply get along without uniqueness since it makes no essential difference anyway.

If $a, b \in D$, by Theorem 5.14 the set

16.16 $(a, b) = \{ax + by \,|\, x, y \in D\}$

is an ideal in D, and therefore is a principal ideal.†

16.17 LEMMA *If a and b are nonzero elements of the principal ideal domain D, any generator d of the ideal 16.16 is a g.c.d. of a and b. In addition, there are elements x, y $\in$ D such that d = ax + by.*

The proof of this lemma is the same as the proof of Euclid's Lemma, 5.18, and will be left to the exercises.

The next result furnishes the essential step in the proof of our theorem. It corresponds with Lemma 5.23 and has a similar proof.

16.18 LEMMA *If p is a prime in the principal ideal domain D and a, b $\in$ D such that p $|$ (ab), then p $|$ a or p $|$ b.*

A generalization of Lemma 16.18 to the product of any finite number of elements of D (instead of just two elements) follows by induction just as Lemma 5.24 is obtained from Lemma 5.23. We omit this proof as well as the rest of the proof of Theorem 16.8. Instead, we shall list these proofs as exercises below.

In conclusion, it may be of interest to state that there exist integral domains which (i) have elements which can be expressed as a product of primes in more than one way (see the second commentary following Chapter 12), (ii) are unique factorization domains but are not principal ideal domains ($F[x, y]$ where F is a field is such an example), and (iii) have nonzero elements which are neither units nor are expressible as a product of primes. Needless to say that these kinds of domains cannot be principal ideal domains. Further discussion along these lines appears in the commentaries at the end of this chapter.

EXERCISES

1. Complete the proof of Lemma 16.10.

2. Prove Lemma 16.17.

---

† According to the definition given in Exercise 13 at the end of Section 3.4, this ideal is the sum $aD + bD$ of the ideals aD and bD.

3. Prove Lemma 16.18.

4. Prove by induction that if p is a prime in the principal ideal domain D which divides a product of any finite number n of elements of D, it must divide at least one of these elements.

5. Complete the proof of Theorem 16.8(ii).

6. Define a least common multiple (l.c.m.) of two nonzero elements of a principal ideal domain. If $aD \cap bD = mD$, verify that m is a l.c.m. of a and b.

7. Let D be the subring of $\mathbf{C}$ of all numbers of the form $a + b(1 + \sqrt{-19})/2$ where $a, b \in \mathbf{Z}$. Prove that D is not a Euclidean domain. The integral domain D is actually a principal ideal domain. Can you prove this?

16.3 MODULES

In this section we introduce a concept which plays an increasingly important role in many aspects of modern algebra. It will be seen that the definition is obtained from the definition of a vector space except that now the scalars are not restricted to be elements of a field but are elements of an arbitrary ring R. For simplicity, we shall assume that R has a unity although, with proper care, one can get along without this restriction.

The concept which we have in mind has a formal definition as follows.

16.19 DEFINITION Let R be a ring with unity e, and M a nonempty set on which there is defined an operation of addition. We also assume that there is defined an operation of scalar multiplication of M by elements of R (that is, if $x \in M$ and $a \in R$, then ax is a uniquely determined element of M). The set M is called an *R-module* (or a *module over R*) if the following conditions are satisfied:

  (i) M is an abelian group with respect to addition.

  (ii) $a(x + y) = ax + ay$, $a \in R; x, y \in M$,

  (iii) $(a + b)x = ax + bx$, $a, b \in R; x \in M$,

  (iv) $a(bx) = (ab)x$, $a, b \in R; x \in M$,

  (v) $ex = x$, e the unity of $R; x \in M$.

Since we are writing the elements of R to the left of elements of M, the above might more precisely be called a *left* R-module, and a *right* R-module could be defined in an analogous way by writing the elements of R on the right. However, we shall consider only left R-modules and shall call them modules in accordance with the definition just given.†

If the ring R remains fixed in a discussion, we shall sometimes refer simply to a module, it being understood that we mean R-module.

As indicated before the definition, if F is a field, an F-module is just a vector space over F. If R is a division ring, it is also customary to call an R-module a vector space over the division ring.

Let us now give some additional examples of modules.

Example 1 Let G be an abelian group with operation addition. If $n \in \mathbf{Z}$ and $a \in G$, we have a definition (given in Section 2.3) of na. Under this definition, it may be verified that G is a $\mathbf{Z}$-module. Accordingly, any abelian group may be considered to be a $\mathbf{Z}$-module.

Example 2 In this example, we generalize in a natural way the concept of a vector space of the form $V_n(F)$. Let R be an arbitrary ring with unity, n a positive integer, and consider the set M of all n-tuples $(a_1, a_2, \ldots, a_n)$ of elements of R. This set becomes an R-module if we define addition and scalar multiplication as follows:

$$(a_1, a_2, \ldots, a_n) + (b_1, b_2, \ldots, b_n) = (a_1 + b_1, a_2 + b_2, \ldots, a_n + b_n),$$

and for $c \in R$,

$$c(a_1, a_2, \ldots, a_n) = (ca_1, ca_2, \ldots, ca_n).$$

Example 3 Let R be a ring with unity, and consider the polynomial ring‡ $R[x]$, where x is an indeterminate. Then $R[x]$ is an R-module with addition the usual addition in $R[x]$ as a ring, and for $c \in R$ and $f(x) \in R[x]$, define $cf(x)$ to be multiplication of the polynomials c and $f(x)$ in $R[x]$.

Example 4 Let A be an ideal in a ring R with a unity. A is an R-module if addition in A is addition in the subring A of R and for $c \in R$ and $a \in A$, ca is the product of c by a in the ring R. In this case, properties (ii) and (iii) of Definition 16.19 are consequences of the distributive laws in R, and (iv) is implied by the associative law of multiplication in R.

---

† If condition (v) is omitted from the definition then an R-module with this condition is sometimes called a *unital* R-module.

‡ Since, for simplicity, we have only defined polynomial rings over a commutative ring, the reader may here assume that R is commutative or assume the true statement that the theory goes over just as well for noncommutative rings.

Example 5 Again, let A be an ideal in R with a unity. Then the additive group of A is a subgroup of the additive group of R. We now let M be the quotient group of the additive group of R by the subgroup of the additive group of A. The elements of M are therefore cosets of the form

$$r + A, \qquad r \in R,$$

and addition of cosets is well-defined by

16.20
$$(r + A) + (s + A) = (r + s) + A.$$

We already know that M is an abelian group with respect to this definition of addition. We propose to make M into an R-module by defining

16.21
$$a(r + A) = ar + A,$$

where a and r are elements of R. We leave it as an exercise to verify that scalar multiplication is well-defined by 16.21 and that M is indeed an R-module under the definitions 16.20 and 16.21.

Having given several examples of modules, we proceed to a brief presentation of a number of concepts. It will be observed that most of these are suggested by what we have already done in the special case of vector spaces. Proofs of some of the statements will be left as exercises.

A subset of an R-module M will be called a *submodule* of M if it is itself an R-module with respect to the operations of addition and scalar multiplication defined in M. A nonempty subset N of M will be a submodule of M if and only if it is closed under addition and $rx \in N$ for $r \in R$ and $x \in N$.

If $x_1, x_2, \ldots, x_n$ are elements of the R-module M, an element of M of the form

$$a_1 x_1 + a_2 x_2 + \cdots + a_n x_n, \qquad \text{each } a_i \in R,$$

may naturally be called a *linear combination* of $x_1, x_2, \ldots, x_n$. The set of all linear combinations of $x_1, x_2, \ldots, x_n$ is a submodule of M. It is the smallest submodule of M (in the sense of set inclusion) which contains the given elements $x_1, x_2, \ldots, x_n$. We call it the *submodule generated by* $x_1, x_2, \ldots, x_n$, and denote it by $[x_1, x_2, \ldots, x_n]$.

If there is a single element x of M which generates M, that is, if $M = [x]$ for some element x of M, we say that M is a *cyclic module* with generator x.

The notation Rx is frequently used in place of $[x]$ to denote such a cyclic module. This is a suggestive notation inasmuch as $Rx = \{ax \mid a \in R\}$.

The R-module M is said to be *finitely generated* if there exists a finite set of elements $\{x_1, x_2, \ldots, x_n\}$ of M such that $M = [x_1, x_2, \ldots, x_n]$.

If $M_1, M_2, \ldots, M_k$ are submodules of an R-module M, the set of all sums of the form

16.22
$$y_1 + y_2 + \cdots + y_k, \qquad y_i \in M_i,$$

is a submodule of M which we denote by

16.23
$$M_1 + M_2 + \cdots + M_k.$$

If each element of this sum is *uniquely* expressed in the form 16.22, we call the sum a *direct sum* and write it in the form

$$M_1 \oplus M_2 \oplus \cdots \oplus M_k.$$

If, in 16.23, each M_i is cyclic and, say, $M_i = Rx_i$, then 16.23 takes the form

$$Rx_1 + Rx_2 + \cdots + Rx_k.$$

A finite set $\{x_1, x_2, \ldots, x_n\}$ of elements of the R-module M is said to be *independent* if

$$a_1 x_1 + a_2 x_2 + \cdots + a_n x_n = 0 \qquad \text{(all } a_i \in R)$$

implies that $a_i = 0$ $(i = 1, 2, \ldots, n)$. If $\{x_1, x_2, \ldots, x_n\}$ is independent and generates M, then $\{x_1, x_2, \ldots, x_n\}$ is said to be a *basis* of M. The concept of a basis is not as important in the study of arbitrary modules as it is for vector spaces since there are many finitely generated modules which do not have a basis. A module which has a basis is often called a *free* module.

The concept of homomorphism is readily applied to modules over the same ring. If M and N are R-modules, a mapping $\theta \colon M \to N$ is said to be an *R-homomorphism* of M into N if both addition and scalar multiplication are preserved, that is, if

$$(x + y)\theta = x\theta + y\theta \qquad x, y \in M,$$

and

$$(ax)\theta = a(x\theta) \qquad a \in R, x \in M.$$

If there exists an R-homomorphism of M onto N, we may say that N is an *R-homomorphic image* of M. Following our usual pattern, an R-homomorphism which is a one-one mapping is called an *R-isomorphism*. If there exists an *R*-isomorphism of M onto N, we also say that M and N are *R-isomorphic*. The notation $M \cong N$ may be used to indicate that M and N are *R*-isomorphic.

EXERCISES

1. If M is an R-module, prove each of the following:

    (i) $a0 = 0$, $a \in R$, 0 the zero of M.

    (ii) $a(-x) = -(ax)$, $a \in R, x \in M$.

    (iii) $(-a)x = -(ax)$, $a \in R, x \in M$.

    (iv) $0x = 0$, $x \in M$, 0 the zero of R and also of M.

2. If M is an R-module, show that the set $\{a \mid a \in R, ax = 0 \text{ for every } x \in M\}$ is an ideal in R.

3. Prove that a nonempty subset N of an R-module M is a submodule of M if and only if it is closed under addition and scalar multiplication.

4. Prove that scalar multiplication is well-defined by 16.21 and that M of Example 5 is an R-module under Definitions 16.20 and 16.21.

5. If θ is an R-homomorphism of the R-module M onto the R-module N, let us define ker $\theta = \{x \mid x \in M, x\theta = 0\}$. Show that ker θ is a submodule of M.

6. If L is a submodule of the R-module M, in a manner suggested by previous situations, introduce the concept of an R-module M/L whose elements are the cosets of the additive subgroup L of the additive group M.

7. In the notation introduced in the two preceding exercises, prove that $M/\text{ker } \theta$ is R-isomorphic to N.

8. A nonzero R-module M is said to be *irreducible* if its only submodules are $\{0\}$ and M. Prove that an irreducible R-module is cyclic.

9. Determine all irreducible **Z**-modules.

10. Find a basis for the module of Example 2.

11. Take A of Example 4 to be R. In Example 5 let A be a submodule of the R-module R (formerly A) of Example 4. Consider the factor group M as defined in Example 5 for this new A. Repeat Exercise 4 for this new M. A submodule A of the R-module R is known as a *left ideal* of R.

16.4 MODULES OVER A PRINCIPAL IDEAL DOMAIN

Our purpose in this section is to introduce enough concepts to be able to state without proof an important theorem about finitely generated modules over a principal ideal domain. We will also give an indication of the generality of this result by interpreting, again without proof, what it says about finite abelian groups and also about the similarity of linear transformations of a vector space over a field. Reference to proofs will be given in the Commentary at the end of the chapter.

Unless otherwise stated, D will denote a principal ideal domain and M a D-module.

If $x \in M$ and there exists a nonzero element c of D such that $cx = 0$, x is said to be a *torsion element* of M. If x is a torsion element of M, the set of all elements c of D such that $cx = 0$ is an ideal A in D. Since in D every ideal is principal, we must have $A = (a)$ for some nonzero element a of D. This element a (or, equally well, any b equal up to units to a) is called the *order* of the element x. It follows that if a is the order of x, then $bx = 0$ for $b \in D$ if and only if $a \mid b$. If x has order a, we may also call a the *order* of the cyclic module Dx.

If M is a $\mathbf{Z}$-module (an abelian group), a torsion element is what we have called an element of finite order.

A *torsion module* is a module in which each element is a torsion element. If M is a finitely generated torsion module, there exist nonzero elements c of D such that $cM = \{cx \mid x \in M\} = \{0\}$. To see this, suppose that $M = x_1 D + x_2 D + \cdots + x_n D$, and that x_i has order c_i for $i = 1, 2, \ldots, n$. Then, in particular, if $c = c_1 c_2 \cdots c_n$, we know that $c \neq 0$ since D is an integral domain, and clearly $cM = 0$. The set of all elements c of R such that $cM = 0$ is again an ideal in R; it is called the *minimal annihilator* of M.

We are now ready to state the following theorem.

16.24 THEOREM

    (i) If M is a finitely generated torsion module over a principal ideal domain D, there exist a finite number of cyclic submodules M_1, $M_2, \ldots, M_r$ of M, each of which has order a power of a prime, such that

16.25 $$M = M_1 \oplus M_2 \oplus \cdots \oplus M_r.$$

(*ii*) *If, in addition to 16.25, we also have*

$$M = N_1 \oplus N_2 \oplus \cdots \oplus N_s,$$

where the N's are cyclic submodules of M, each of which has order a power of a prime of D, then s = r and by a suitable choice of notation, for each i = 1, 2, . . . , r, the order of N_i is the same as (or, equally well, is equal up to units to) the order of M_i.

It will be observed that the second part of this theorem states that, in a certain sense, the expression of a module of the specified kind as a direct sum of submodules having the indicated properties is *unique*.

The orders of the submodules M_i in 16.25 (repetitions being allowed) are called the *elementary divisors* of the module M.

It can be shown that *two torsion D-modules, each of which is finitely generated, are D-isomorphic if and only if their elementary divisors coincide.*

Again, let us consider the special case of a finitely generated abelian group, each element of which has finite order. This requires the group itself to have finite order. The theorem just stated is thus a direct generalization of Theorem 9.29 for finite abelian groups. Compare, also, Theorem 9.33 in relation to the general statement made above about the D-isomorphisms of finitely generated torsion D-modules.

Let us now briefly introduce another situation in which this theorem is applicable. Let $\mathscr{A}$ be a fixed linear transformation of a vector space V of finite dimension over a field F. If t is an indeterminate, we can use $\mathscr{A}$ to make V into an $F[t]$-module in the following way. For each $f(t) \in F[t]$ and $X \in V$, we define

$$f(t)X = Xf(\mathscr{A}).$$

It may be verified that the module properties are satisfied and we do have an $F(t)$-module. Since $F(t)$ is a principal ideal domain, we may apply the results of this section. The minimal polynomial of the linear transformation $\mathscr{A}$, as introduced in Section 15.11, is the minimal annihilator of the $F(t)$-module in the sense defined in the present section. A D-submodule is the same as a subspace invariant under $\mathscr{A}$, as used in Section 15.10. Although, in Theorem 15.76 the submodules V_i are not necessarily cyclic, we know by Theorem 16.24 that the decomposition can be continued until one obtains *cyclic* submodules of orders a power of a prime polynomial.

The elementary divisors of this $F[t]$-module are called the *elementary divisors* of the matrix of $\mathscr{A}$ relative to any ordered basis of V. The next paragraph will hint at why it makes no difference which ordered basis is used.

We have not quite developed all of the machinery needed to prove the following fact, but its truth is the primary reason why the concept of elementary divisors of a matrix is important.

Two square matrices of the same order over a field are similar if and only if their elementary divisors coincide.

16.5   ZORN'S LEMMA

We introduce in this section an important tool of frequent use in algebra as well as in other branches of mathematics.

Let $\mathfrak{M}$ (German *M*) denote a nonempty collection of subsets of some fixed set *S*. A subset $\mathfrak{C}$ (German *C*) of $\mathfrak{M}$ is said to be a *chain* (or a *completely ordered* subset) if for $A, B \in \mathfrak{C}$, either $A \subseteq B$ or $B \subseteq A$. By the *union of the chain* $\mathfrak{C}$ we mean the union of all subsets of *S* which are the elements of $\mathfrak{C}$. An element *M* of $\mathfrak{M}$ is naturally called a *maximal* element of $\mathfrak{M}$ if it is not properly contained in any element of $\mathfrak{M}$. Although we shall take the following statement as an axiom, it is customary to call it a lemma.

16.26   ZORN'S LEMMA   *Let $\mathfrak{M}$ denote a nonempty set of subsets of a fixed set S. If the union of each chain in $\mathfrak{M}$ is an element of $\mathfrak{M}$, then $\mathfrak{M}$ contains one or more maximal elements.*

References to equivalent formulations of this statement and to its role in the theory of sets will be given at the end of this chapter. We shall here consider it to be an axiom and proceed to give two examples illustrating its use.

As a first illustration, let us prove the following result.

16.27   THEOREM   *If R is a ring with unity e, and C is an ideal in R with $C \neq R$, then C is contained in at least one maximal ideal M in R.*

Proof: First, let us recall the Definition 16.13 of a maximal ideal *M* in *R*. It is an ideal with the property that there exists no ideal *N* in *R* such that $M \subset N \subset R$. In the language introduced before the statement of Zorn's Lemma, a maximal ideal in *R* is an ideal which is maximal in the set of all ideals other than *R* itself.

Accordingly, let $\mathfrak{M}$ be the set of all ideals in *R* which contain *C* but not the unity *e* of *R*. Since $C \neq R$, $e \notin C$ and therefore $C \in \mathfrak{M}$, and $\mathfrak{M}$ is not empty. Now let $\mathfrak{C}$ be a chain in $\mathfrak{M}$ and let us denote by *U* the union of the chain $\mathfrak{C}$. Hence *U* consists of all elements of *R* that are in any element of the chain $\mathfrak{C}$ of subsets of R. We proceed to prove that U is an ideal in R, that $e \notin U$, and therefore $U \in \mathfrak{M}$. If $a, b \in U$, there exist ideals A and B belonging to the chain $\mathfrak{C}$ such that $a \in A$ and $b \in B$. Since $\mathfrak{C}$ is a chain, we must have either $A \subseteq B$ or $B \subseteq A$. In either case, both a and b belong to the same ideal in $\mathfrak{C}$, and hence $a + b$ is an element of one of the ideals of the chain. It follows that $a + b \in U$. It is even easier to verify that if $a \in U$, then $-a \in U$; also that if $a \in U$, then $ar \in U$ and $ra \in U$ for each $r \in R$; hence U is an ideal in R. Clearly, also, $e \notin U$ since, otherwise, e would be an element of one of the ideals in $\mathfrak{C}$. But no ideal of $\mathfrak{M}$ (and therefore no ideal of $\mathfrak{C}$) contains e. Therefore, $e \notin U$, and $U \in \mathfrak{M}$ since U satisfies all conditions imposed on the elements of $\mathfrak{M}$. We have shown that the union of each chain in $\mathfrak{M}$ is an element of $\mathfrak{M}$, and Zorn's Lemma then assures us that $\mathfrak{M}$ has maximal elements. Let M be a maximal element of $\mathfrak{M}$ and N an ideal in R such that $M \subset N$. Since N is not an element of $\mathfrak{M}$, we must have $e \in N$. This implies that $N = R$ and therefore that M is indeed a maximal ideal in R. Clearly $C \subseteq M$, and the proof is complete.

As another illustration, let us prove that every vector space has a basis— in a sense now to be made precise.

Let V be a vector space over a field F (a division ring would do just as well). An arbitrary (possibly infinite) nonempty set of elements of V is said to be *linearly independent* if each finite subset is linearly independent. The subspace of V *generated* by an arbitrary nonempty set T of elements of V is the set of all linear combinations of any finite number of elements of T. A set B of elements of V is a *basis* of V if it is a linearly independent set and generates V.

We shall now use Zorn's Lemma to prove the following generalization of Theorem 11.24 for finite dimensional vector spaces, which asserts that any linear independent set of elements is a part of a basis.

16.28 THEOREM *If T is a nonempty linearly independent set of elements of a vector space V over a field F, then there exists a basis of V which contains T.*

Proof: Let $\mathfrak{M}$ be the set of all linearly independent subsets of V which contain the given set T. Since T is itself such a subset, $\mathfrak{M}$ is not empty. Let U be the union of a chain $\mathfrak{C}$ of elements of $\mathfrak{M}$. If $X_1, X_2 \in U$, then $X_1 \in A$ and $X_2 \in B$ for elements A and B of $\mathfrak{C}$. But since $\mathfrak{C}$ is a chain, either $A \subseteq B$ or $B \subseteq A$, and hence both X_1 and X_2 are elements of some one element of $\mathfrak{C}$. More generally, if for an arbitrary positive integer n, $\{X_1, X_2, \ldots, X_n\} \subseteq U$, then $\{X_1, X_2, \ldots, X_n\}$ is contained in some *one* element of $\mathfrak{C}$, and this set is therefore linearly independent. Since each finite set of elements of U is linearly independent, and clearly $T \subseteq U$, it follows that $U \in \mathfrak{M}$. The conclusion of Zorn's Lemma is therefore applicable, and therefore there exists a maximal element B of $\mathfrak{M}$. Let us show that the elements of B generate V and it will follow that B is the basis which we seek.

If $X \in B$, it is trivial that X is expressible as a linear combination of a finite number (actually one!) of elements of B. If $X \in V$, $X \notin B$, the maximal property of B shows that the union $\{X\} \cup B$ cannot be an element of $\mathfrak{M}$. This fact implies that the set $\{X\} \cup B$ cannot be linearly independent, so there must exist finitely many elements $X_1, X_2, \ldots, X_n$ of B such that $\{X, X_1, X_2, \ldots, X_n\}$ is a linearly dependent set. Since $\{X_1, X_2, \ldots, X_n\}$ is a linearly independent set, it follows that X is expressible as a linear combination of X_1, X_2, $\ldots, X_n$ (cf. Theorem 11.15(iii)). We have therefore shown that every element of V is expressible as a linear combination of a finite number of elements of B. The set B is therefore a basis of V and since $T \subseteq B$, the proof is complete.

If V is a nonzero vector space, and X_1 is a nonzero element of V, the set $\{X_1\}$ is linearly independent; hence, by the theorem just established, there is a basis of V which contains the vector X_1. In particular, *any nonzero vector space has a basis.*

16.6 REPRESENTATIONS OF BOOLEAN RINGS

On several occasions we have referred to the ring S of all subsets of some given set A, first introduced in Example 8 of Section 3.2. Throughout this section, it will be convenient to denote addition and multiplication in S by $\oplus$ and $\odot$, respectively. Thus, if $c, d \in S$, we have $c \odot d = c \cap d$. Let us now use the notation $X \backslash Y$ to denote the set of elements of the set X which are not elements of Y. Using this notation, we recall that our definition of addition in the ring S may be given by $c \oplus d = (c \cup d) \backslash (c \cap d)$. The empty set $\varnothing$ is the zero of the ring S, and each element of S is its own additive inverse since $c \oplus c = \varnothing$ for $c \in S$.

Any subring of S is naturally called *a ring of subsets of A*. Since in S every element c is its own additive inverse, we see that a nonempty set K of elements of S will be a subring of S, and therefore a ring of subsets of A, if and only if K is closed with respect to the operations $\oplus$ and $\odot$ in S.

We have also defined (Exercise 8 at the end of Section 3.1) a *Boolean ring* R to be a ring with the property that $a^2 = a$ for every element a of R. It is known (see this same exercise) that a Boolean ring is necessarily a commutative ring and that $a + a = 0$ for $a \in R$. We shall use these facts without further mention.

By the very definition, it follows that a ring of subsets of any given set is a Boolean ring, since always $c \odot c = c \cap c = c$. The purpose of this section is to prove the interesting fact that there are no other Boolean rings. More precisely, we shall prove the following theorem of Stone.

16.29 THEOREM *Every Boolean ring R is isomorphic to a ring of subsets of some set.*

We leave it to the reader to verify that the theorem is trivially true if R has only one element (the zero). Accordingly, we shall henceforth tacitly assume that R has more than one element.

In order to prove the theorem, we must find a set T, some of whose subsets will turn out to form a ring which is isomorphic to R. Clearly, the set T must be related in some significant way to the ring R. Although it is unlikely that it will be obvious how to choose T, we shall show that T may be chosen to be the set of all maximal ideals in R!

We shall prove several lemmas from which the theorem will follow fairly easily. Henceforth, let R be an arbitrary Boolean ring with more than one element, and let T be the set of all maximal ideals in R. We first prove the following useful result.

16.30 LEMMA *If M is an element of T, that is, a maximal ideal in R, and $c \notin M$, then every element of R is expressible in the form $m + cx$ for some $m \in M$ and $x \in R$.*

Proof: To prove the truth of this statement, we observe that the set $C = \{m + cx \mid m \in M, x \in R\}$ is an ideal in R, as is readily verified. Moreover, $M \subseteq C$ since, as a special case, x can be zero. We now assert that $c \in C$ since we obtain c by letting $m = 0$ and $x = c$. Since it was given that $c \notin M$, it follows that the maximal ideal M is properly contained in the ideal C. By definition of maximal ideal, this implies that $C = R$, completing the proof of the lemma.

16.31 LEMMA *If a is a nonzero element of R, there exists in R a maximal ideal M which does not contain a.*

Proof: We leave it to the reader to verify that the set $B = \{ax + x \mid x \in R\}$ is an ideal in R. We assert that $a \notin B$. Suppose, on the contrary, that $a \in B$. Thus $a = ax + x$ for some x in R. Multiplying by a, we obtain

$$a = a^2 = a^2 x + ax = ax + ax = 0,$$

which contradicts the assumption that $a \neq 0$. Hence we conclude that $a \notin B$.

We proceed to apply Zorn's Lemma as follows. Let $\mathfrak{M}$ be the set of all ideals in R which contain the ideal B and do not contain the element a. We have shown above that $a \notin B$, so $B \in \mathfrak{M}$ and $\mathfrak{M}$ is not empty. As in the proof of 16.27, the union of any chain in $\mathfrak{M}$ is an ideal in R and it clearly does not contain a; hence is an element of $\mathfrak{M}$. By Zorn's Lemma, $\mathfrak{M}$ therefore has maximal elements. Let M be such a maximal element. Hence M is an ideal in R, and we wish to show that it is a maximal ideal in R. Accordingly, let N be an ideal such that $M \subset N$, and let us prove that N must be R itself. Suppose that $c \in N$, $c \notin M$. Then, as in the proof of Lemma 16.30, we see that $D = \{m + cx \mid m \in M, x \in R\}$ is an ideal in R which contains M and also contains c, hence M is properly contained in D and, by the fact that M is a maximal element of $\mathfrak{M}$, we see that we must have $a \in D$. But this implies that $ax \in D$ for every $x \in R$. Then, since $B \subseteq M \subset D$, it follows that $ax + x \in D$ and $ax \in D$ so that $x \in D$ for every $x \in R$, and therefore $D = R$. But $D \subseteq N$ and thus $N = R$. This shows that the only ideal which properly contains M is R itself, and M is therefore a maximal ideal in R. Since $a \notin M$, this is an ideal whose existence we wished to prove.

The form of the statements of the next two lemmas is suggested by the fact that we shall presently be interested in maximal ideals that do *not* contain a given element of R.

16.32 LEMMA *If M is a maximal ideal in R and $a, b \in R$, then $ab \notin M$ if and only if $a \notin M$ and $b \notin M$.*

Proof: What we wish to prove is logically equivalent to the statement that if $a, b \in R$, then $ab \in M$ if and only if $a \in M$ or $b \in M$. Let us prove it in this form. We shall do so by assuming that $ab \in M$, $b \notin M$, and showing that we must have $a \in M$.

Since $b \notin M$, Lemma 16.30 asserts the existence of an element m of M and an element x of R such that $a = bx + m$. Multiplying by a, we obtain

$$a^2 = a = abx + am.$$

But since $ab \in M$ and $am \in M$, this equation implies that $a \in M$. On the other hand, if $a \in M$ or $b \in M$ then $ab \in M$, completing the proof.

16.33 LEMMA *If M is a maximal ideal in R and $a, b \in R$, then $a + b \notin M$ if and only if exactly one of the two elements a and b is an element of M.*

Proof: Again, it is a little simpler to prove the following equivalent formulation of this lemma. If $a, b \in R$, then $a + b \in M$ if and only if both a and b are elements of M or neither of them is an element of M.

In order to prove this last statement, suppose first that $a + b \in M$. Then if $a \in M$, it follows that $b = (a + b) - a$ is also an element of M. Similarly, of course, if $b \in M$, then also $a \in M$. So $a + b \in M$ implies that either $a \in M$ and $b \in M$ or $a \notin M$ and $b \notin M$.

Conversely, if $a \in M$ and $b \in M$, it is trivial that $a + b \in M$. The only thing left to prove is that if $a \notin M$ and $b \notin M$, then $a + b \in M$. This fact we can establish as follows.

Since $b \notin M$, by Lemma 16.30, there exist $x \in R$ and $m \in M$ such that

16.34
$$a = bx + m.$$

We observe that $x \notin M$ since otherwise this equation would imply that $a \in M$. Multiplying Equation 16.34 by x, we obtain $ax = bx + mx$. It follows that

$$(a + b)x = ax + bx = ax - bx = mx,$$

so that $(a + b)x \in M$. Since $x \notin M$, Lemma 16.32 now assures us that $a + b \in M$, completing the proof of the lemma.

We are now ready to complete the proof of the theorem. We recall that T denotes the set of all maximal ideals in R. If $a \in R$, let us now denote by T_a the subset of T consisting of all maximal ideals in R which do *not* contain the element a. Then the two preceding lemmas can be stated in the following simple form:

$$T_{ab} = T_a \cap T_b,$$

and

$$T_{a+b} = (T_a \cup T_b) \setminus (T_a \cap T_b).$$

In terms of multiplication $\odot$ and addition $\oplus$ as defined in a ring of subsets of T, these may be written, respectively, as follows:

16.35
$$T_{ab} = T_a \odot T_b,$$

and

16.36
$$T_{a+b} = T_a \oplus T_b.$$

Now let $S = \{T_a \mid a \in R\}$; hence S is a set of subsets of T. Equations 16.35 and 16.36 show that S is closed under the operations $\odot$ and $\oplus$, and hence S is a ring of subsets of T.

Now let $\theta: R \to S$ be defined by $a\theta = T_a$, $a \in R$. By the very definition of S, this mapping is an onto mapping. Moreover, equations 16.35 and 16.36 show that θ is a homomorphism. Lemma 16.31 shows that $T_a \neq 0$ if $a \neq 0$; hence that the kernel of this homomorphism is $\{0\}$. This proves that θ is indeed an isomorphism of R onto a ring of subsets of T, and the proof of the theorem is complete.

An isomorphism of a Boolean ring R onto a ring of subsets is often referred to as a *representation* of R. We conclude with this explanation of the title of this section.

 COMMENTARY

1 Quaternions were mentioned briefly in a commentary at the end of Chapter 15, but here we shall go into a little more detail and make some related remarks, historical and otherwise. Quaternions were introduced by Sir William Rowan Hamilton (Irish) in 1843. The significance of his achievement lies not so much in the importance of the system itself as in the fact that this was the first algebra in which multiplication was not assumed to be commutative. Heretofore, algebraic systems with two operations (addition and multiplication) were thought of as consisting of numbers of some kind and it was simply considered that multiplication was "obviously" commutative. Thus, the very existence of quaternions with noncommutative multiplication and, in particular, the fact that they had applications in physics as well as in mathematics, served to free algebra from the commutativity of multiplication as a necessary requirement. More and more, algebraists began to think of the axioms of a system as "the rules of the game"—not as something forced upon them. This viewpoint, also encouraged by the known existence of noneuclidean geometries, was essential to the very great advances in algebra, as well as in many other branches of mathematics, during the present century.

In order to have a convenient way to state the results of Hamilton, as well as certain generalizations, let us introduce some helpful terminology as follows. If $A = (a_1, a_2, \ldots, a_n)$ is an ordered n-tuple of real numbers, that is, an element of the vector space $V_n(\mathbf{R})$, we define the *norm* of A, denoted by $N(A)$, as follows:

$$N(A) = A \cdot A = a_1^2 + a_2^2 + \cdots + a_n^2.$$

For example, if $n = 2$ and we think of the number pairs which define a complex number, then the norm of a complex number is the square of its absolute value. Using the definition of multiplication of complex numbers, we have that

$$(a_1, a_2)(b_1, b_2) = (a_1 b_1 - a_2 b_2, a_1 b_2 + a_2 b_1);$$

thus the known fact that the absolute value of a product is the product of the absolute values yields at once the same property for norms, that is

$$N(a_1, a_2)N(b_1, b_2) = N[(a_1, a_2)(b_1, b_2)].$$

In more detail, this yields the interesting identity

$$(a_1{}^2 + a_2{}^2)(b_1{}^2 + b_2{}^2) = (a_1b_1 - a_2b_2)^2 + (a_1b_2 + a_2b_1)^2.$$

In the case of n-tuples of real numbers, for an arbitrary positive integer n, suppose that a multiplication of n-tuples has been defined. Then, if A and B are n-tuples and it happens that $N(AB) = N(A)N(B)$, let us say for convenience that "the norm property holds." We have pointed out above that the norm property holds for complex numbers. Moreover, Exercise 3 of Section 16.1 asks the reader to verify that the norm property holds for quaternions.

Hamilton introduced complex numbers as ordered pairs of real numbers (as in Chapter 7 of this book), but he also thought of them geometrically as vectors in a plane, emanating from the origin. Thus he thought of the complex numbers as furnishing a field (although this term was not yet used) whose elements are planar vectors. In accordance with Definitions 16.1 and 15.6, we can say that the complex numbers form a *division algebra* over the reals in the sense of being a division ring (actually commutative) and also an algebra over the reals.

Since three-dimensional vectors arise naturally in physics, Hamilton long sought a definition of multiplication of these vectors such that the field properties would hold, using as addition the familiar vector addition. Actually, he concentrated on trying to obtain a multiplication of ordered triples of real numbers in such a way that the norm property would hold. For a long time he worked on this problem without success, but by a series of geometric, as well as algebraic, considerations decided that one would have to have ordered *quadruples* in order to have any chance of success. Moreover, even with ordered quadruples he made little progress until he took the step of assuming that multiplication might not be commutative. This was just what he needed, and he soon discovered the formula for multiplication of quaternions (quadruples) in such a way that the norm property holds. Then he easily proved that he had, in fact, a division algebra over the reals. Thus, to obtain the result which he sought, he found it necessary to take the bold and imaginative step of eliminating the requirement that multiplication be commutative.

In 1845, just two years after quaternions were introduced by Hamilton, the English mathematician, Arthur Cayley, discovered an interesting algebra of dimension eight over the reals (8-tuples of real numbers). For this algebra, the norm property holds but neither the commutative nor the associative law for multiplication. Naturally, such an algebra is called a *nonassociative division algebra*. It is interesting that elements of Cayley's algebra are still referred to as Cayley *numbers*, the term perhaps suggesting that originally they were thought of as generalized numbers.

In 1878, the German mathematician, G. Frobenius, proved that the *only* division algebras over the reals are the reals themselves, the complex numbers and quaternions. Thus the only division algebras over the reals that are fields (that is, with commutative multiplication) are the field of real numbers and the field of complex numbers. It was not until 1957 that the question was finally settled in case neither commutativity nor associativity of multiplication is assumed. In that year, topological methods were used by Milnor and Bott (American) and independently by

Kervaire (Swiss) to prove that the *only* such algebras are the reals, complex numbers, quaternions, and Cayley numbers.

As to the question for what values of n there exists a multiplication of n-tuples such that the norm property holds, in 1898 the German mathematician, A. Hurwitz, proved that n must be 1, 2, 4, or 8; and in 1949 A. A. Albert (American) showed that, except for minor variations, the *only* possible cases are those already mentioned, namely, the multiplication of reals, complex numbers, quaternions, and Cayley numbers.

Further interesting details of the history of these problems will be found in an excellent chapter by C. W. Curtis in *Studies in Modern Algebra*, edited by A. A. Albert, Mathematical Association of America, 1963.

2   Division rings play an important role in certain so-called "structure theorems" in the theory of rings. We shall here briefly give the necessary definitions, without indicating how the concepts arise in the study of rings, and then state one of the most famous of these theorems. A special case of this theorem was briefly mentioned in a commentary to Chapter 15.

First, let S be an arbitrary ring and let us define addition and multiplication of $n \times n$ matrices whose entries are from S, using the same definitions that we have used for the case in which S is a field. The set of all such matrices then becomes a ring which we call the *complete matrix ring of order n over S*.

If A is a left ideal† (or right ideal or ideal) in a ring R and if there exists a positive integer n such that $a_1 a_2 \cdots a_n = 0$ for every choice of $a_1, a_2, \ldots, a_n$ in A, then A is said to be *nilpotent*. Of course, the zero ideal is trivially nilpotent.

Just as we have defined the ascending chain condition, we say that the *descending chain condition* holds for left ideals in R if there exists no infinite sequence of left ideals each of which is properly contained in the preceding. That is, if $A_1, A_2, \ldots$ is any infinite sequence of left ideals in R, we cannot have

$$A_1 \supset A_2 \supset A_3 \supset \cdots.$$

It can be proved that if the descending chain condition holds for left ideals in a ring R, there exists in R a maximal nilpotent ideal, that is, a nilpotent ideal which contains every nilpotent right ideal and every nilpotent left ideal. This unique maximal nilpotent ideal is called the *radical* of the ring **R**. Thus, saying that the radical of R is the zero ideal is equivalent to saying that R has no nonzero nilpotent ideal.

We can now state the well-known structure theorem which was referred to above.

Wedderburn-Artin Theorem: *A ring R in which the descending chain condition holds for left ideals has zero radical if and only if R is isomorphic to a direct sum of a finite number of rings, each of which is a complete matrix ring over a division ring.*

In the component rings of the direct sum mentioned in this theorem, the order

---

† Left ideals and right ideals were defined in a footnote near the beginning of Section 3.4.

of the matrices and the particular division rings involved may vary from component to component. We may also observe that "left ideal" may just as well be replaced by "right ideal" in the statement of the theorem and in the remarks preceding the theorem.

The above theorem was first proved in 1908 by J. H. M. Wedderburn for the case of finite-dimensional algebras over a field. In the more general case as stated above, it was proved in 1927 by E. Artin.

The Wedderburn-Artin Theorem is called a "structure theorem" because in a certain sense it enables us to "know" all rings satisfying the hypotheses of the theorem. Although we may not actually know all division rings, and so may not completely know everything we might like to know, we do know the type or structure of rings satisfying the hypotheses of the theorem.

The Wedderburn-Artin Theorem only applies to the case of rings with descending chain condition for left ideals. There have been many efforts, with varying degrees of success, to generalize the concept of radical to arbitrary rings and to obtain generalizations or analogues of the Wedderburn-Artin Theorem. In particular, see Jacobson [46]. See also [48] and [49] and other references there given. A comprehensive study of such matters will be found in a book by N. J. Divinsky, *Rings and Radicals*, Mathematical Expositions no. 14, University of Toronto Press, 1965.

We may mention that Wedderburn proved in 1905 another famous theorem which states that a division ring with a finite number of elements is a field, that is, in the case of a finite number of elements, the commutativity of multiplication is a logical consequence of the other defining properties of a division ring. For a proof of this theorem, see Herstein [12] or Dean [6].

3   Let D be an arbitrary integral domain, not necessarily a principal ideal domain. As suggested in a footnote to Definition 5.4, in this general setting it is customary to use the following definitions which we shall now adopt.

An element a of D is *irreducible* if it is not a unit and its only divisors are units and elements equal to a up to units.

An element p of D is *prime* if it is not a unit and it has the property that if $p \,|\, (bc)$, then $p \,|\, b$ or $p \,|\, c$.

It has already been observed that these concepts coincide for principal ideal domains, so we have previously used the word "prime" for what we are now calling "irreducible."

Actually, in any integral domain, if an element is prime, it is necessarily irreducible. This is not difficult to show and the reader is invited to furnish a proof.

However, as pointed out in a commentary to Chapter 12, there exist integral domains in which an irreducible element need not be prime. As an example mentioned in that commentary, let D' be the integral domain consisting of all complex numbers of the form $a + b\sqrt{-5}$, where $a, b \in \mathbf{Z}$. It is easily verified that in this domain

$$6 = 2 \cdot 3,$$
$$6 = (1 + \sqrt{-5})(1 - \sqrt{-5}).$$

It can be shown (see Hardy and Wright [76], or LeVeque [78], for a similar example) that 2, 3, $1 + \sqrt{-5}$ and $1 - \sqrt{-5}$ are all irreducible. But in view of the above, we have

$$2 \cdot 3 = (1 + \sqrt{-5})(1 - \sqrt{-5}),$$

and this shows that no one of these irreducibles is prime. Hence, in D' an irreducible element need not be prime.

In accordance with our present usage, we say that an integral domain D is a *Unique Factorization Domain* (UFD) if each element which is not a unit can be expressed as a product of a finite number of irreducibles, and the factorization is unique in the sense that the order of the factors may be different but the factors in one such factorization differ from those in any other such factorization only by unit factors. Otherwise expressed, the irreducible factors are unique up to units.

Thus, Theorem 16.8 says that a principal ideal domain is a UFD (since irreducible elements are just the prime elements in such a domain). Now it can be proved (see Fraleigh [9]) that if D is a UFD, then so also is $D[x]$, where x is an indeterminate. Thus, in particular, $\mathbf{Z}[x]$ is a UFD, as also is $F[x, y]$ where F is a field and x and y are indeterminates. But it can be shown that neither $\mathbf{Z}[x]$ nor $F[x, y]$ is a principal ideal domain; so the unique factorization property does not imply that every ideal is principal.

Finally, let us observe that there are domains in which it is impossible to factor certain elements into a product of irreducibles in any way. The following example is listed as an exercise in Volume 1 of [14].

Let A be the set of all formal expressions of the form

$$a_1 x^{r_1} + a_2 x^{r_2} + \cdots + a_n x^{r_n},$$

where n is some positive integer (that is, the sums are finite), x is an indeterminate, the elements a_i are elements of a field F, and the r_i are nonnegative rational numbers. If addition is defined in a reasonably obvious way and multiplication is defined by $x^r x^s = x^{r+s}$ and the distributive laws, it can be verified that A is an integral domain. In this domain, the element x^1 is not a unit and does *not* have a factorization into irreducibles. As a hint of what happens, note that for *any* positive rational number t, $x^t = x^{t/2} \cdot x^{t/2}$.

The reader may want to refer again to the commentary to Chapter 12, where there is a discussion of unique factorization of *ideals* in certain domains which are not unique factorization domains.

4   The study of modules was initiated by several German mathematicians, among them being Emmy Noether and Emil Artin. The classic books of van der Waerden [24] on abstract algebra were based in part on lectures by these two outstanding mathematicians.

In recent years it has become increasingly clear that many properties of a ring R are reflected in some way into properties of R-modules or to the existence of an R-module of some particular kind. The study of modules has become in itself an important branch of abstract algebra and, to a considerable extent, the study of rings has given way to the study of modules. We shall here give a brief outline of one special result which may not be typical but will hint at the type of relationship which may hold between rings and modules.

By an R-module we shall continue to mean a left R-module, as in Definition 16.19, but for generality we do not now assume that R necessarily has a unity and we omit part (v) of that definition.

The definition of *irreducible* module given in Exercise 8 of Section 16.3 should now be modified by replacing the hypothesis that the module be a nonzero module by the hypothesis that there exist $a \in R$ and $x \in M$ such that $ax \neq 0$.

If for each nonzero element b of R there exists an element y of M such that $by \neq 0$, we say that M is a *faithful R-module*.

In case R has a faithful irreducible R-module, R is said to be *primitive* (more precisely, *left* primitive). The following theorem, which we state here without proof, therefore characterizes certain rings R for which there does exist a faithful irreducible R-module.

Theorem. *If the descending chain condition holds for left ideals in the ring R, then R is a primitive ring if and only if it is isomorphic to a complete matrix ring over a division ring.*

In view of this result, it is apparent that the concept of primitive ring is a generalization of that of a complete matrix ring over a division ring, in that the two concepts coincide in case the descending chain condition for left ideals holds in the ring.

The study of primitive rings and of an associated concept of radical are carried out in detail in Jacobson [46]. These concepts, due to Jacobson, have played an important role in the study of rings during the last quarter century.

Modules are presented in some detail in Ames [2], Godement [10], and in MacLane and Birkhoff [20]. A proof of our Theorem 16.24 will be found in MacLane and Birkhoff [20] and in Rotman [38]. Further details about this theorem and similarity of matrices will be found in Herstein [12] as well as in MacLane and Birkhoff.

5   The "Axiom of Choice" was briefly mentioned in a footnote in Section 1.3. What is known as "Zorn's Lemma" is logically equivalent to the Axiom of Choice, and we shall now make a few comments about these matters.

Although the Axiom of Choice is sometimes referred to as Zermelo's Postulate (or Axiom), it was É. Borel who pointed out that Zermelo had tacitly made use of this axiom. The axiom in question may be stated as follows:

Axiom of Choice. *If A is a collection of disjoint nonempty sets S_α, then there exists a set S which has as its elements exactly one element x_α of each set S_α in A.*

Clearly, the set S exists if it is possible to choose one element from each of the given sets—hence the name "Axiom of Choice."

If the collection A of sets consists of only a finite number of sets, there seems to be no question but that one element can be chosen from each set. The interesting problem arises when one is required to choose one element from an infinite number of sets.

An illustration, due to Bertrand Russell, is of interest in this connection. If one has an infinite collection of pairs of shoes, there is no difficulty in selecting one shoe from each pair, for one can specify that each right (or left) shoe is to be selected. However, it is not so clear how one could select one sock from an infinite collection of pairs of socks if it is assumed that the socks of each pair are identical.

The use of the word *axiom* is now known to be an appropriate one since a mathematician has the option of assuming it or not. To clarify this statement somewhat, K. Gödel has proved that if the usual axioms of set theory are consistent (in that no contradiction can possibly arise from their use), then the addition of the Axiom of Choice also gives a consistent set of axioms. In the other direction, P. Cohen proved in 1963 that if the usual axioms of set theory are consistent, so also is the set of axioms obtained by adding a denial of the Axiom of Choice.

This situation leads to varying attitudes toward the Axiom of Choice by mathematicians. It is perhaps generally considered preferable to prove results without its use, although probably most mathematicians will use it if necessary, but will call attention to the fact that it has been used. It is certainly true that a fairly substantial part of what is generally accepted (by most mathematicians) as good mathematics leans quite heavily on the Axiom of Choice or on one of its equivalent formulations.

We have stated above that Zorn's Lemma is logically equivalent to the Axiom of Choice. For many applications, particularly in algebra, Zorn's Lemma can be applied more easily than the other equivalent formulations.

A discussion of the Axiom of Choice and of various equivalent formulations, may be found in Wilder [97] and other references there given.

6   The English mathematician and logician, George Boole (1815–1865), was the first to attempt an algebraic approach to logic. The algebraic system which he inspired is still known as *Boolean algebra*. In 1937, the American mathematician M. H. Stone introduced the concept of *Boolean ring* and showed how a Boolean algebra naturally gives rise to a Boolean ring, and vice versa; and in a series of papers exploited their interrelationship to give a number of applications. In this book we have not defined Boolean algebra, but we have introduced a Boolean ring as one in which for every element a, $a^2 = a$. An excellent introduction to Boolean algebra will be found in Whitesitt [131]. Reference to Stone's work will be found in that book as well as in [48] and [49].

Although the concept of representation of a Boolean ring as a ring of subsets of some set is clearly not applicable to rings in general, there does exist an alternate formulation of the Stone representation theorem in terms of a concept of "subdirect sums" which leads to interesting generalizations to certain other rings. We shall not here define this concept or indicate the nature of the theorems which generalize the representation theory. For details and references, see Chapter 3 of [49], together with the notes on that chapter which appear near the end of the book.

APPENDIX

In this appendix we shall give an elementary proof of the Fundamental Theorem of Algebra, 10.55.

A.1   (FUNDAMENTAL THEOREM OF ALGEBRA) *If $f(x)$ is an element of $\mathbf{C}[x]$ of positive degree, there exists an element of $\mathbf{C}$ which is a root of the polynomial $f(x)$.*

Using the Theorem of Kronecker, 10.83, it is possible to construct many extensions F to the rational field $\mathbf{Q}$ so that if $f(x) \in \mathbf{Q}[x]$ is a polynomial with rational coefficients then one of the F's contains a root of $f(x)$. It is also possible to show that these F's may be chosen so that they generate some giant field L such that if $f(x) \in L[x]$ is a polynomial then L contains a root of $f(x)$. The only information we can really obtain about L is algebraic in character. Certainly, if L were isomorphic to a subfield of $\mathbf{C}$, the complex field, we would have more than just algebraic information about L. We would like to show that L may be taken isomorphic to $\mathbf{C}$. We will not use the algebraic construction of Kronecker; instead, we will rely upon the fact that $f(x) \in \mathbf{C}[x]$ is a continuous function as defined in the calculus. As such, a polynomial $f(x) \in \mathbf{C}[x]$ in an indeterminate x will be viewed (via the substitution process, if you like) as a polynomial mapping or function $f(z) \in \mathbf{C}[z]$ in a complex variable z. In other words, $f(z)$ is one of the "usual" polynomials from calculus. A root of $f(z)$ now is a complex number $z_0 \in \mathbf{C}$ such that $f(z_0) = 0$. The object of our proof is to show that 0 is assumed as a value of the mapping $f(z)$. The beauty of this theorem is that we need not hunt up an "abstract" field L in order to factor $f(x)$; we need look no further than $\mathbf{C}$, the complex field.

Kronecker's method for constructing L starts with $\mathbf{Q}$ and adjoins all roots to all polynomials ending up with some field L. We start the other way; $\mathbf{C}$ is given and we must show that it contains all roots of all polynomials. In our proof we must use the defining properties of $\mathbf{C}$; namely, the least upper bound property of Definition 7.2 as applied to the real field $\mathbf{R}$. This property is not algebraic; in fact, it is the foundation upon which the calculus of one variable is built. It is not surprising then that we shall need some calculus in our proof.

Most proofs of the Fundamental Theorem are an odd mix of algebra and analysis. For this reason, algebraists and analysts consider it each other's duty to give a proof of this theorem. Therefore, proofs of this theorem do not appear in as many textbooks as they should.

There is a difficulty with this proof: To say that a proof is elementary is only to say that the methods are elementary. This does not mean the proof is necessarily easy; and our proof is distinctly not easy. There are many proofs of the Fundamental Theorem; several are due to Gauss, himself. One of the nicest and shortest (after the "big theorems" have been proved) uses a theorem of Liouville about functions of a complex variable (see Churchill [110] pp. 125–126). We shall have no sledge hammers to drive our points home; our methods are elementary. Consequently, the proof is somewhat long and a little tedious.

The proof given here is actually quite intuitive. To describe it we must first recall some points. Any complex number can be written uniquely in the form $u = a + bi$ where a and b are real. The absolute value of u is then $|u| = \sqrt{a^2 + b^2}$. (See Definition 7.14.) The absolute value has the following properties:

A.2   If $u, v \in \mathbf{C}$ then

  (i) $|uv| = |u|\,|v|$,

  (ii) (*Triangle Inequality*) $|u + v| \le |u| + |v|$,

  (iii) $||u| - |v|| \le |u - v|$, and

  (iv) $|u| \ge 0$; $u = 0$ if and only if $|u| = 0$.

Parts (i) and (ii) are in Exercises 8 and 10 of Section 7.5. Part (iii) may be derived from (ii) by setting $u = w - v$ in the Triangle Inequality. Part (iv) is a consequence of the definition of $|u|$.

We have all graphed $y = x^n$ for positive integers n and real variables x and y. For large and increasing x, x^{n+1} increases very rapidly; much more rapidly than x^n. The upshot of this is that for large values of x, the highest power of x in a polynomial function $f(x)$ determines the size of $f(x)$. The same thing is true for polynomial functions of a complex variable z; but in this latter case we know that $|f(z)|$ must be large whenever $|z|$ is very large.

Let us put this idea to work for us in the following geometric way. In Figure 1 we have drawn a complex plane where a complex number $z = x + iy$ may be plotted. At O we have erected a vertical axis called the w-axis. Suppose that $f(z)$ is a complex polynomial of positive degree. For a fixed $z = x + iy$ we may mark a point P at a height of $w = |f(z)|$ above the point $x + iy$ in the plane. Imagine we have plotted all such points; since polynomials are continuous functions, we obtain a nice smooth surface above the complex plane. Since $|f(z)| \ge 0$, the surface is above the complex plane; and since $f(z) = 0$ if and only if $|f(z)| = 0$, the surface touches the complex plane precisely at the roots of $f(z)$.

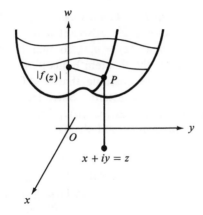

FIGURE 1

Our first theorem will show that our earlier discussion about polynomials is correct, so that for large $|z|$, that is, z far away from O, $|f(z)|$ is very large. In other words, the surface of $f(z)$ pictured in Figure 1 is some kind of giant bowl.

Drop a marble into this giant bowl. It will roll down, eventually coming to rest at the bottom of some depression. Suppose the marble rests over the point $z_0 = x_0 + iy_0$. It is resting at a height of $|f(z_0)|$ above the plane. Since it is resting at the bottom of a depression, we may draw a small circle about the marble so that inside this circle every direction away from the marble is either level or up. In particular, if z' is inside the circle, then $|f(z')| \geq |f(z_0)|$. Let us call this property of z_0 our *depression condition*. Our second theorem is to establish the depression condition, i.e., the marble comes to rest above a point z_0 satisfying the depression condition.

Our final theorem is to show that the bottom of any depression must touch the complex plane. More appropriately, we show that if the point P on the surface given by z and $f(z)$ is above the plane, then no matter how small a circle we draw around P, there is another point Q given by z' and $|f(z')|$ inside that circle which is lower down, that is, $|f(z')| < |f(z)|$. As a consequence, the bottom points on the bowl must touch the complex plane and those values of z are roots of $f(z)$, and that is how the proof works.

Let us now work on the bowl theory; that is, let us show that for large $|z|$, $|f(z)|$ must also be large. First we specify a polynomial function with complex coefficients:

A.3 $$f(z) = a_0 + a_1 z + \cdots + a_n z^n, \quad n > 0, a_n \neq 0.$$

A.4 **THEOREM** *There is a real number $A > 0$ so that if $|z| \geq A$ for $z \in C$, then $|f(z)| \geq |z|^n |a_n|/2$.*

Proof: Let us specify that $z \in \mathbf{C}$ must satisfy $|z| > 1$; this will certainly occur if we take $A \geq 2$. Also set $a = |a_{n-1}| + \cdots + |a_0|$. Notice that if $a = 0$ then $f(z) = a_n z^n$ so that $|f(z)| = |z|^n |a_n| > |z|^n |a_n|/2$. In particular, the theorem holds if $A = 2$, which forces $|z| > 1$. We may therefore assume that $a \neq 0$. The important point is that a is a *positive* constant. Since $|z| > 1$ we must have $|z|^k \geq |z|$ or that $|a_{n-k}|/|z|^k \leq |a_{n-k}|/|z|$ for all $k = 1, 2, \ldots, n$. Using the Triangle Inequality and the fact that $|-a_k| = |a_k|$ we have

A.5
$$|-(a_{n-1}/z + a_{n-2}/z^2 + \cdots + a_0/z^n)| \leq |a_{n-1}|/|z| + \cdots + |a_0|/|z|^n$$
$$\leq (1/|z|)(|a_{n-1}| + \cdots + |a_0|)$$
$$= (1/|z|)a.$$

From A.2(iii) and the above, for $|z| > 1$ we have

$$|f(z)| = |z|^n |a_n - (-(a_{n-1}/z + \cdots + a_0/z^n))|$$
$$\geq |z|^n ||a_n| - |-(a_{n-1}/z + \cdots + a_0/z^n)||.$$

If $|z| \geq 2a/|a_n|$ in addition to $|z| > 1$ then A.5 guarantees that the term $|-(a_{n-1}/z + \cdots + a_n/z^n)|$ is less than or equal to $|a_n|/2$. In other words,

$$|f(z)| \geq |z|^n |a_n| - |a_n|/2| = |z|^n(|a_n|/2).$$

Therefore, we may take A to be the larger of the two numbers 2 and $2a/|a_n|$. This completes the proof of the theorem.

We have established a "global" condition; that is, a condition in the large. We now know that our surface is some kind of giant bowl which ultimately slopes up on the sides. A marble in it cannot roll out by rolling off to infinity. We show next that it must come to rest internally in some depression.

A.6   THEOREM *If $f(z)$ is a polynomial function of degree $n > 0$ with complex coefficients then there exists a complex number z_0 such that for any $z' \in \mathbf{C}$,*

$$|f(z')| \geq |f(z_0)|.$$

Proof: To prove this theorem we shall break it into steps. We shall assume that $f(z)$ is given by A.3. Since $|f(z)| \geq 0$ for all choices of $z \in \mathbf{C}$, if $f(0) = 0$ we may take $z_0 = 0$. Therefore, we assume that $f(0) = a_0 \neq 0$.

By Theorem A.4 we may choose $A > 0$ so that if $|z| \geq A$ then $|f(z)| \geq |z|^n |a_n|/2$. Let B be the largest of 1, A and $2(|a_0| + 1)^2/|a_n|$. Then for $|z| \geq B$ we have

$$f(z) \geq |z|^n |a_n|/2 \geq |z| |a_n|/2$$
$$\geq [2(|a_0| + 1)^2/|a_n|](|a_n|/2) = (|a_0| + 1)^2.$$

We have just completed Step 1 of our proof.

A.7 **(STEP 1)** *There is a constant $B \geq 1$ so that if $|z| \geq B$ then $|f(z)| \geq (|a_0| + 1)^2$.*

To do the next step we recall Definition 7.12. If $u = a + bi$ is a complex number, then we call $u^* = a - bi$ the *complex conjugate* of u. The important facts about conjugates are easy to prove and are listed below. (See Section 7.4.)

A.8 If $u, v \in \mathbf{C}$, then

(i) $(u + v)^* = u^* + v^*$;

(ii) $(uv)^* = u^* v^*$;

(iii) u is a real number if and only if $u = u^*$; and

(iv) $uu^* = |u|^2$.

Write the complex variable z as $z = x + iy$ where x and y are real variables. In $f(z)$, expand the powers of $x + iy$; further, write each coefficient in the form $a_m = c_m + id_m$ where c_m and d_m are real numbers. By expanding and collecting we get

$$f(x + iy) = h(x, y) + ik(x, y)$$

where $h(x, y)$ and $k(x, y)$ are polynomial functions in x and y with real coefficients.

To illustrate this process with $z = x + iy$ notice that for $f(z)$ as below we have:

$$f(z) = (1 + i) + 3z + iz^2$$
$$= (-2xy + 3x + 1) + i(x^2 - y^2 + 3y + 1).$$

Computing now in the general case gives

$$|f(z)|^2 = f(z)f(z)^* = h(x, y)^2 + k(x, y)^2 \geq 0.$$

We have completed another step.

A.9 **(STEP 2)** *If $z = x + iy$ for real variables x and y then*

$$|f(z)|^2 = g(x, y)$$

is a polynomial function in x and y with real coefficients such that $g(x, y) \geq 0$ for all $x, y \in \mathbf{R}$.

The next step involves some calculus and completes the proof of Theorem A.6.

A.10 **(STEP 3)** *There exist real numbers x_0, y_0 so that if $z_0 = x_0 + iy_0$ then $|f(z)| \geq |f(z_0)|$ for all values of $z \in \mathbf{C}$.*

By (Step 1) there is a $B \geq 1$ so that if $|z| \geq B$ then $|f(z)| \geq (|a_0| + 1)^2 \geq 1$. So $|f(z)|^2 \geq |f(z)| \geq (|a_0| + 1)^2$. Writing $z = x + iy$ we have

$$|f(z)|^2 = g(x, y) \geq (|a_0| + 1)^2 > |a_0|^2 = g(0, 0) = |f(0)|^2.$$

That is, on and outside the circle of radius B about O in the complex plane, $|f(z)| > |f(0)|$. So to find out if there are any smaller values of $|f(z)|$ we need only check values such that $|z| < B$.

Consider the closed disc $\{z \,|\, z \in \mathbf{C}, |z| \leq B\}$ centered at O of radius B. Now $g(x, y)$ is a polynomial function on this disc. The following two theorems should sound familiar and do apply to this situation.

A.11 **THEOREM** *If $g(x, y)$ is a polynomial function in real variables x and y with real coefficients then $g(x, y)$ is continuous at all points (x, y) of the plane.*

A.12 **THEOREM** *If $g(x, y)$ is a continuous real valued function on a closed disc in the plane then $g(x, y)$ attains its minimum value on the disc.*

These theorems are contained in Johnson and Kiokemeister's fifth edition. As remarked on page 584 of that text, A.11 is a special case of their Theorem 15.6. Our A.12 is a special case of Theorem 15.28 on page 620 of their text. Rigorous proofs of these theorems can be found in Bartle [108].

These two theorems tell us that there is a $z_0 = x_0 + iy_0$ in a disc of radius B, that is, $|z_0| \leq B$, such that $g(x_0, y_0)$ is minimum on the disc. Now $g(x_0, y_0) \leq g(0, 0) = |f(0)|^2 = |a_0|^2 < |f(z)|^2$ for all $|z| = B$ or $|z| > B$. So $|z_0| \leq B$ implies that $|z_0| < B$. But with $z = x + iy$, $g(x_0, y_0) \leq g(x, y)$ for all $|z| \leq B$, that is, all z in the disc. So

$$|f(z_0)|^2 \leq |f(z)|^2$$

for all values of $z \in \mathbf{C}$, inside or outside the disc. We conclude that

$$|f(z)| \geq |f(z_0)|$$

for all $z \in \mathbf{C}$. This completes the proof of A.10, and with it Theorem A.6.

As our final theorem, we establish a "local" condition; a condition in the small. We shall show that if a point P on the bowl is not resting on the plane then there are points near P which are lower than P itself.

A.13 THEOREM *If $\epsilon > 0$ is given and $z_0 \in \mathbf{C}$ is chosen so that $f(z_0) \neq 0$ then there is a $z \in \mathbf{C}$ so that*

(i) $|z - z_0| < \epsilon$, and

(ii) $|f(z)| < |f(z_0)|$.

The conclusion of this theorem is in direct contradiction with the conclusion of Theorem A.6 unless $|f(z_0)| = 0$, that is, $f(z_0) = 0$, and so z_0 must be a root of $f(z)$. In particular, the Fundamental Theorem A.1 is true. This theorem is then the last step in our proof.

Proof: The proof of this theorem is more computational than the preceding one. We may write $z = u + z_0$ where $u = z - z_0$. Then $f(z) = f(u + z_0)$. Using the binomial theorem we may expand the powers of $u + z_0 = z$ in A.3 and collect like powers of u. The coefficients of the powers of u are polynomials in the a_i's and z_0 so that they are complex numbers. The coefficient of u^n is $b_n = a_n \neq 0$, and the constant term is $b_0 = f(z_0) \neq 0$. Let b_j be the coefficient of u^j in this expansion. It is possible that $b_1 = b_2 = \cdots = b_{m-1} = 0$ counting up from 1 to $m - 1$. Let m be the smallest positive index for which $b_m \neq 0$. Since $b_n = a_n \neq 0$, such an index does exist. We may now write $f(z)$ in the form

A.14 $$f(z) = f(u + z_0) = b_0 + b_m u^m + b_{m+1} u^{m+1} + \cdots + b_n u^n.$$

By Theorem 7.20 we may choose a complex number u_0 such that $u_0{}^m = -b_0/b_m$; in particular, $b_0 + b_m u_0{}^m = 0$. Consider now the polynomial

$$g(h) = h^{-m}(f(hu_0 + z_0) - b_0 - b_m u_0{}^m h^m)$$
$$= b_{m+1} u_0{}^{m+1} h + \cdots + b_n u_0{}^n h^{n-m}$$

in a variable h. By A.9, with $h = x + iy$, $|g(h)|^2 = g_0(x, y)$ is a polynomial in real variables x and y, and $g_0(x, y)$ is always nonnegative. Since $g(0) = 0$, $g_0(0, 0) = 0$. By A.11 $g_0(x, y)$ is continuous at zero so that $\lim_{x \to 0} g_0(x, 0) = 0$. In particular, $\lim_{x \to 0} |g(x)| = \lim_{x \to 0} \sqrt{g_0(x, 0)} = 0$. This limit shows that we may choose a real value for x so small that (a) $0 < x < 1$, (b) $|(xu_0 + z_0) - z_0| = |xu_0| = x|u_0| < \epsilon$, and (c) $|g(x)| < |b_0|$. We prove that if we set $z = xu_0 + z_0$ then z satisfies the conclusion of Theorem A.13.

Condition (i) follows immediately from (b) since $|z - z_0| = |(xu_0 + z_0) - z_0| < \epsilon$. Notice that by our choice of u_0, $b_0 = -b_m u_0{}^m$. Using this fact in the following computation we have:

$$
\begin{aligned}
|f(z_0)| = |b_0| &= x^m |b_0| + (1 - x^m)|b_0| \\
&= x^m |b_0| + |b_0 - b_0 x^m| \\
&= x^m |b_0| + |b_0 + b_m u_0{}^m x^m| \\
&> x^m |g(x)| + |b_0 + b_m u_0{}^m x^m| \\
&= |f(xu_0 + z_0) - b_0 - b_m u_0{}^m x^m| + |b_0 + b_m u_0{}^m x^m| \\
&\geq |f(xu_0 + z_0) - b_0 - b_m u_0{}^m x^m + b_0 + b_m u_0{}^m x^m| \\
&= |f(xu_0 + z_0)| \\
&= |f(z)|.
\end{aligned}
$$

Therefore, Condition (ii) and Theorem A.13 are proven.

Theorem A.6 tells us that our giant bowl has bottom points. This last theorem, on the other hand, tells us that bottom points must be points where the bowl touches the plane. These points of contact are precisely the roots of the polynomial function $f(z)$. Incidentally, our depression condition tells us that the marble may come to rest at a flat place. The exercises below show there are no flat places in the bowl.

EXERCISES

1. Prove that if $z_0 \in \mathbf{C}$ is a root of $f(z)$ (as given in A.3) then there is an $\epsilon > 0$ so that if $0 < |z - z_0| < \epsilon$ then $|f(z)| > 0$.

2. Prove that if $z_0 \in \mathbf{C}$ and $f(z)$ is given by A.3 then if $|f(z)| \geq |f(z_0)|$ for all $z \in \mathbf{C}$ then there is an $\epsilon > 0$ so that for all $z \in \mathbf{C}$ such that $0 < |z - z_0| < \epsilon$ we have $|f(z)| > |f(z_0)|$.

3. Imagine $f(z)$ is given by A.3. What can you say about n, the degree of $f(z)$, if you know that the bowl shaped graph of $f(z)$, as in Figure 1, touches the complex plane at exactly three points? What are the roots of $f(z)$?

BIBLIOGRAPHY

This bibliography is by no means complete. Most of the books mentioned below have been referred to at least once in the notes at the ends of certain chapters. Others are included for a variety of reasons including one or more of the following: they are particularly readable by students with a modest background in abstract algebra, they are standard treatises and give wide coverage, they are modern in spirit, they should be interesting and challenging to the better students.

Several of the listed books have fairly extensive bibliographies which may be used as desired to supplement this one. Recent articles are also referenced in the following:

Annotated Bibliography of Expository Writing in the Mathematical Sciences, Matthew P. Gaffney, Lynn Arthur Steen, ed., Mathematical Association of America, 1976.

GENERAL ALGEBRA

1. Albert, A. Adrian. *Fundamental Concepts of Higher Algebra.* Chicago: University of Chicago Press, 1956.

2. Ames, Dennis B. *An Introduction to Abstract Algebra.* Scranton, Pa.: International, 1969.

3. Birkhoff, Garrett and Saunders MacLane. *A Survey of Modern Algebra* (3rd ed.). New York: Macmillan, 1965.

4. Bourbaki, N. *Algebra, Part 1.* Reading, Mass.: Addison-Wesley, 1973.

5. Burton, David M. *Abstract and Linear Algebra.* Reading, Mass.: Addison-Wesley, 1972.

6. Dean, Richard A. *Elements of Abstract Algebra.* New York: John Wiley, 1967.

7. Dickson, Leonard E. *Algebraic Theories.* New York: Dover, 1959.

8. Dubisch, Roy. *Introduction to Abstract Algebra.* New York: John Wiley, 1965.

9. Fraleigh, John B. *A First Course in Abstract Algebra.* Reading, Mass.: Addison-Wesley, 1967.

10. Godement, Roger. *Algebra.* Paris: Hermann, 1968.

11. Goldstein, Larry J. *Abstract Algebra: A First Course.* Englewood Cliffs, N.J.: Prentice-Hall, 1973.

12. Herstein, I. N. *Topics in Algebra* (2nd ed.). Lexington, Mass.: Xerox Publishing, 1975.

13. Jacobson, Nathan. *Basic Algebra I.* San Francisco: Freeman, 1974.

14. ———. *Lectures in Abstract Algebra.* New York: Van Nostrand, Vol. 1, 1951; Vol. 2, 1953; Vol. 3, 1964.

15. Johnson, Richard E. *University Algebra.* Englewood Cliffs, N.J.: Prentice-Hall, 1966.

16. Kurosh, A. G. *Lectures on General Algebra.* New York: Chelsea, 1963.

17. Lang, Serge. *Algebra.* Reading, Mass.: Addison-Wesley, 1965.

18. ———. *Algebraic Structures.* Reading, Mass.: Addison-Wesley, 1967.

19. Lewis, Donald J. *Introduction to Algebra.* New York: Harper, 1965.

20. MacLane, Saunders and Garrett Birkhoff. *Algebra.* New York: Macmillan, 1967.

21. Mostow, George D., Joseph H. Sampson, and Jean-Pierre Meyer. *Fundamental Structures of Algebra.* New York: Holt, Rinehart, and Winston, 1966.

22. Paley, Hiram and Paul M. Weichsel. *A First Course in Abstract Algebra.* New York: Holt, Rinehart, and Winston, 1966.

23. Perlis, Sam. *Introduction to Algebra.* Waltham, Mass.: Blaisdell, 1966.

24. Van der Waerden, B. L. *Algebra.* Vol. 1, 2 (7th ed.). New York: Ungar, 1970.

25. Weiss, Edwin. *First Course in Algebra and Number Theory.* New York: Academic Press, 1971.

GROUP THEORY

26. Budden, F. J. *The Fascination of Groups.* Cambridge: Cambridge University Press, 1971.

27. Burnside, William. *Theory of Groups of Finite Order* (2nd ed.). New York: Dover, 1955.

28. Fuchs, L. *Infinite Abelian Groups.* Vol. 1. New York: Academic Press, 1970.

29. Gorenstein, Daniel. *Finite Groups.* New York: Harper and Row, 1968.

30. Grossman, I. and W. Magnus. *Groups and their Graphs.* New York: Random House, 1965.

31. Hall, Marshall, Jr. *The Theory of Groups* (Rev. ed.). New York: Macmillan, 1959.

32. Huppert, B. *Endliche Gruppen I.* Berlin: Springer-Verlag, 1967.

33. Kaplansky, Irving. *Infinite Abelian Groups* (Rev. ed.). Ann Arbor: University of Michigan Press, 1968.

34. Kurosh, A. G. *The Theory of Groups.* Vol. I, II. New York: Chelsea, 1960.

35. Lederman, Walter. *Introduction to the Theory of Finite Groups* (4th ed.). New York: Interscience, 1961.

36. Macdonald, Ian D. *The Theory of Groups.* Oxford: Oxford University Press, 1968.

37. Miller, G. A., H. F. Blichfeldt, and L. E. Dickson. *Theory and Applications of Finite Groups.* New York: Dover, 1961.

38. Rotman, Joseph J. *The Theory of Groups* (2nd ed.). Boston: Allyn and Bacon, 1973.

39. Zassenhaus, Hans J. *The Theory of Groups* (2nd ed.). New York: Chelsea, 1958.

RING THEORY

40. Albert, A. Adrian. *Structure of Algebras.* AMS Colloquium Publication 24. Providence R. I.: American Mathematical Society, 1961.

41. Anderson, Frank W. and Kent R. Fuller. *Rings and Categories of Modules.* New York: Springer-Verlag, 1973.

42. Artin, E., C. J. Nesbitt, and R. M. Thrall. *Rings with Minimum Condition.* Ann Arbor: University of Michigan Press, 1944.

43. Burton, David M. *A First Course in Rings and Ideals.* Reading, Mass.: Addison-Wesley, 1970.

44. Gillman, L. and M. Jerison. *Rings of Continuous Functions.* Princeton: Van Nostrand, 1960.

45. Herstein, I. N. *Noncommutative Rings.* Carus Monograph 15. Mathematical Association of America, 1968.

46. Jacobson, Nathan. *Structure of Rings* (Rev. ed.). AMS Colloquium Publication 37. Providence, R.I.: American Mathematical Society, 1964.

47. Kaplansky, Irving. *Commutative Rings*. Boston: Allyn and Bacon, 1970.

48. McCoy, Neal H. *Rings and Ideals*. Carus Monograph 8. Mathematical Association of America, 1948.

49. ———. *The Theory of Rings*. New York: Chelsea, 1973.

50. Zariski, Oscar, and Pierre Samuel. *Commutative Algebra*. Vol. 1, 2. New York: Van Nostrand, 1958.

FIELD THEORY

51. Adamson, Iain T. *Introduction to Field Theory*. London: Oliver and Boyd, 1964.

52. Artin, Emil. *Galois Theory* (2nd ed.). Notre Dame: University of Notre Dame Press, 1944.

53. Dehn, Edgar. *Algebraic Equations*. New York: Dover, 1960.

54. Gaal, L. *Classical Galois Theory with Examples* (2nd ed.). New York: Chelsea, 1973.

55. Postnikov, M. M. *Foundation of Galois Theory*. New York: Macmillan (Pergamon Press), 1962.

56. Winter, David. *The Structure of Fields*. New York: Springer-Verlag, 1974.

LINEAR ALGEBRA

57. Finkbeiner, Daniel T. *Introduction to Matrices and Linear Transformations* (2nd ed.). San Francisco: Freeman, 1966.

58. Gantmacher, F. R. *The Theory of Matrices*. New York: Chelsea, 1959.

59. Greub, W. H. *Linear Algebra*. New York: Springer-Verlag, 1967.

60. Halmos, Paul R. *Finite Dimensional Vector Spaces* (2nd ed.). Princeton: Van Nostrand, 1958.

61. Hay, G. E. *Vector and Tensor Analysis*. New York: Dover, 1953.

62. Hoffman, Kenneth and Ray Kunze. *Linear Algebra* (2nd ed.). Englewood Cliffs, N.J.: Prentice-Hall, 1971.

63. Hohn, Franz E. *Elementary Matrix Algebra* (2nd ed.). New York: Macmillan, 1964.

64. Kaplansky, Irving. *Linear Algebra and Geometry*. Boston: Allyn and Bacon, 1969.

65. Muir, Thomas. *The Theory of Determinants in the Historical Order of Development*. London: Macmillan, Vol. I, 1906; Vol. II, 1911; Vol. III, 1920; Vol. IV, 1923.

66. Muir, Thomas. *A Treatise on the Theory of Determinants*. New York: Dover, 1960.

67. Noble, Ben. *Applied Linear Algebra*. Englewood Cliffs, N.J.: Prentice-Hall, 1969.

68. Wedderburn, J. H. M. *Lectures on Matrices*. New York: Dover, 1964.

NUMBER THEORY

69. Baker, Alan. *Transcendental Number Theory*. Cambridge: Cambridge University Press, 1975.

70. Beiler, Albert H. *Recreations in the Theory of Numbers*. New York: Dover, 1966.

71. Borevich, Z. I. and I. R. Shaferevich. *Number Theory*. New York: Academic Press, 1966.

72. Davenport, H. *Higher Arithmetic: An Introduction to the Theory of Numbers*. New York: Harper, 1960.

73. Dickson, L. E. *History of the Theory of Numbers*. Vol. 1–3. Washington: Carnegie Institution of Washington, 1919.

74. ———. *Introduction to the Theory of Numbers*. New York: Dover, 1957.

75. Gauss, C. F. *Disquisitiones Arithmeticae*. New Haven: Yale University Press, 1966.

76. Hardy, G. H. and E. M. Wright. *An Introduction to the Theory of Numbers* (4th ed.). Oxford: Oxford University Press, 1960.

77. Landau, E. *Elementary Number Theory* (2nd ed.). New York: Chelsea, 1966.

78. Le Veque, William J. *Elementary Theory of Numbers*. Reading, Mass.: Addison-Wesley, 1962.

79. ———. *Topics in Number Theory*. Vol. 1, 2. Reading, Mass.: Addison-Wesley, 1956.

80. Mordell, Louis J. *Diophantine Equations*. New York: Academic Press, 1969.

81. Niven, Ivan and Herbert S. Zuckerman. *An Introduction to the Theory of Numbers*. New York: John Wiley, 1960.

82. Ore, Oystein. *Number Theory and its History*. New York: McGraw-Hill, 1948.

83. Pollard, Harry. *The Theory of Algebraic Numbers*. Carus Monograph 9. Mathematical Association of America, 1961.

84. Rademacher, Hans. *Lectures on Elementary Number Theory*. New York: Blaisdell, 1964.

85. Samuel, Pierre. *Algebraic Theory of Numbers*. Paris: Hermann, 1970.

86. Serre, Jean-Pierre. *A Course in Arithmetic*. New York: Springer-Verlag, 1973.

87. Sierpinski, W. *Elementary Theory of Numbers*. Warsaw: Panstwowe Wydawnictwo Naukowe, 1964.

SET THEORY AND LOGIC

88. Breuer, Joseph. *Introduction to the Theory of Sets.* Englewood Cliffs, N.J.: Prentice-Hall, 1958.

89. Cantor, Georg. *Contributions to the Founding of the Theory of Transfinite Numbers.* New York: Dover, 1963.

90. Frege, G. *The Foundations of Arithmetic.* Oxford: Basil Blackwell, 1950.

91. Halmos, Paul R. *Naive Set Theory.* New York: Van Nostrand-Reinhold, 1960.

92. Kamke, E. *Theory of Sets.* New York: Dover, 1950.

93. Nagel, E. and J. R. Newman. *Godel's Proof.* New York: New York University Press, 1958.

94. Quine, W. V. *The Ways of Paradox and other Essays.* New York: Random House, 1966.

95. Russell, Bertrand. *Principles of Mathematics* (2nd ed.). New York: Norton, 1938.

96. Russell, Bertrand and Alfred North Whitehead. *Principia Mathematica.* Vol. 1–3. Cambridge: Cambridge University Press, 1912.

97. Wilder, Raymond L. *Introduction to the Foundations of Mathematics.* New York: John Wiley, 1952.

98. Wittgenstein, Ludwig. *Remarks on the Foundations of Mathematics.* Boston: Massachusetts Institute of Technology Press, 1967.

GEOMETRY

99. Albert, A. A. and R. Sandler. *An Introduction to Finite Projective Planes.* New York: Holt, Rinehart, and Winston, 1968.

100. Artin, Emil. *Geometric Algebra.* New York: Interscience, 1957.

101. Blumenthal, Leonard M. *A Modern View of Geometry.* San Francisco: Freeman, 1961.

102. Dorwart, Harold L. *The Geometry of Incidence.* Englewood Cliffs, N.J.: Prentice-Hall, 1966.

103. Fulton, William. *Algebraic Curves.* New York: Benjamin, 1969.

104. Heath, Thomas L. (ed.). *The Thirteen Books of Euclid's Elements* (2nd ed.). New York: Dover, 1956.

105. Hilbert, D. and S. Cohn-Vossen. *Geometry and Imagination.* New York: Chelsea, 1952.

106. Hughes, D. R. and F. C. Piper. *Projective Planes.* New York: Springer-Verlag, 1973.

107. Shaferevich, I. R. *Basic Algebraic Geometry.* New York: Springer-Verlag, 1974.

ANALYSIS

108. Bartle, R. G. *The Elements of Real Algebra.* New York: John Wiley, 1964.

109. Berberian, S. K. *Introduction to Hilbert Space.* New York: Oxford University Press, 1961.

110. Churchill, R. V. *Complex Variables and Applications* (2nd ed.). New York: McGraw-Hill, 1960.

111. Johnson, R. E., F. L. Kiokemeister, and E. S. Wolk. *Calculus with Analytic Geometry* (5th ed.). Boston: Allyn and Bacon, 1974.

112. Knopp, Konrad. *Elements of the Theory of Functions.* New York: Dover, 1952.

113. Kunz, Kaiser S. *Numerical Analysis.* New York: McGraw-Hill, 1957.

114. Rickart, Charles E. *General Theory of Banach Algebras.* Princeton: Van Nostrand, 1960.

115. Scarborough, James B. *Numerical Analysis* (5th ed.). Baltimore: Johns Hopkins Press, 1962.

116. Simmons, George F. *Introduction to Topology and Modern Analysis.* New York: McGraw-Hill, 1963.

117. Wendroff, Burton. *First Principles of Numerical Analysis: An Undergraduate Text.* Reading, Mass.: Addison-Wesley, 1969.

GENERAL MATHEMATICS

118. Aleksandrov, A. D., A. N. Kolmogorov, and M. A. Laurent'ev (eds.). *Mathematics: Its Contents, Methods, and Meaning.* Vol. 1–3 (2nd ed.). Cambridge: Massachusetts Institute of Technology Press, 1963.

119. Birkhoff, Garrett. *Lattice Theory* (2nd ed.). AMS Colloquium Publication 25. Providence, R.I.: American Mathematical Society, 1948.

120. Courant, Richard and Herbert Robbins. *What is Mathematics?* New York: Oxford University Press, 1941.

121. Curtis, Charles W. and Irving Reiner. *Representation Theory of Finite Groups and Associative Algebras.* New York: Interscience, 1962.

122. Humphreys, J. E. *Introduction to Lie Algebras and Representation Theory.* New York: Springer-Verlag, 1972.

123. Klein, F., et al. *Famous Problems and other Monographs.* New York: Chelsea, 1962.

124. Landau, Edmund. *Foundations of Analysis.* New York: Chelsea, 1960.

125. Littlewood, D. E. *The Skeleton Key of Mathematics.* London: Hutchinson University Library, 1960.

126. MacLane, Saunders. *Categories for the Working Mathematician.* New York: Springer-Verlag, 1971.

127. Miller, George A. *Collected Works*. Urbana: University of Illinois Press, Vol. 1, 1935; Vol. 2, 1938; Vol. 3, 1946; Vol. 4, 1955; Vol. 5, 1959.

128. Northcott, D. G. *An Introduction to Homological Algebra*. Cambridge: Cambridge University Press, 1960.

129. Olmsted, J. M. H. *The Real Number System*. New York: Appleton-Century-Crofts, 1962.

130. Polya, G. *How to Solve It* (2nd ed.). Garden City, New York: Doubleday, 1957.

131. Whitesitt, J. E. *Boolean Algebra and its Applications*. Reading, Mass.: Addison-Wesley, 1961.

HISTORY

132. Ball, W. W. Rouse. *A Short Account of the History of Mathematics*. New York: Dover, 1960.

133. Beckmann, Petr. *A History of π (pi)*. Boulder: Golem Press, 1971.

134. Bell, Eric Temple. *Development of Mathematics* (2nd ed.). New York: McGraw-Hill, 1945.

135. ———. *Men of Mathematics*. New York: Simon and Schuster, 1962.

136. Boyer, Carl B. *A History of Mathematics*. New York: John Wiley, 1968.

137. Cardan, Jerome. *The Book of My Life*. New York: Dover, 1962.

138. Crowe, Michael J. *A History of Vector Analysis*. Notre Dame: University of Notre Dame Press, 1967.

139. Dantzig, Tobias. *Number: The Language of Science* (4th ed.). Riverside: Free Press, 1967.

140. Dunnington, C. Waldo. *Carl Frederick Gauss: Titan of Science*. New York: Hafner, 1955.

141. Eves, Howard. *An Introduction to the History of Mathematics* (3rd ed.). New York: Holt, Rinehart, and Winston, 1969.

142. Hall, Tord. *Carl Friedrich Gauss*. Cambridge: Massachusetts Institute of Technology Press, 1970.

143. Infeld, Leopold. *Whom the Gods Love*. New York: McGraw-Hill, 1948.

144. Kline, Morris. *Mathematical Thought from Ancient to Modern Times*. New York: Oxford University Press, 1972.

145. ———. *Mathematics in Western Culture*. New York: Oxford University Press, 1966.

146. Kramer, Edna E. *The Nature and Growth of Modern Mathematics*. Vol. I, II. Greenwich: Fawcett Publications, 1970.

147. Ore, Oystein. *Cardano: The Gambling Scholar*. New York: Dover, 1965.

148. ———. *Niels Henrik Abel*. Minneapolis: University of Minnesota Press, 1957.

149. Osen, Lynn M. *Women in Mathematics.* Cambridge: Massachusetts Institute of Technology Press, 1974.

150. Reid, Constance. *Hilbert.* New York: Springer-Verlag, 1970.

151. Smith, David Eugene. *History of Mathematics.* Vol. 1, 2. New York: Dover, 1951.

152. ———. *A Source Book in Mathematics.* Vol. 1, 2. New York: Dover, 1959.

153. Struik, Dirk J. *A Concise History of Mathematics.* New York: Dover, 1948.

RECREATIONAL MATHEMATICS

154. Ball, W. W. Rouse. *Mathematical Recreations and Essays* (Rev. ed.). New York: Macmillan, 1960.

155. Bergamini, David. *Mathematics Life Science Library.* New York: Time-Life Books, 1969.

156. Eves, Howard W. *In Mathematical Circles.* Boston: Prindle, Weber and Schmidt, 1969.

157. ———. *Mathematical Circles Revisited.* Boston: Prindle, Weber and Schmidt, 1971.

158. ———. *Mathematical Circles Squared.* Boston: Prindle, Weber and Schmidt, 1972.

159. Fadiman, Clifton. *Fantasia Mathematica.* New York: Simon and Schuster, 1962.

160. ———. *Mathematical Magpie.* New York: Simon and Schuster, 1964.

161. Gamow, George. *One Two Three . . . Infinity.* New York: New American Library, 1947.

162. Gardner, Martin (ed.). *Martin Gardner's New Mathematical Diversions from Scientific American.* New York: Simon and Schuster, 1966.

163. ——— (ed.). *Martin Gardner's Sixth Book of Mathematical Games from Scientific American.* San Francisco: Freeman, 1971.

164. ——— (ed.). *Mathematical Puzzles of Sam Lloyd.* Vol. 1, 2. New York: Dover, 1959.

165. ——— (ed.). *New Mathematical Diversions from Scientific American.* New York: Simon and Schuster, 1971.

166. ——— (ed.). *Scientific American Book of Puzzles and Diversions.* New York: Simon and Schuster, 1964.

167. ——— (ed.). *Second Scientific American Book of Mathematical Puzzles and Diversions.* New York: Simon and Schuster, 1965.

168. Hadamard, Jacques. *The Psychology of Invention in the Mathematical Field.* New York: Dover, 1945.

169. Hardy, G. H. *A Mathematician's Apology.* Cambridge: Cambridge University Press, 1969.

170. Honsberger, Ross. *Mathematical Gems.* Dolciani Mathematical Expositions 1. Mathematical Association of America, 1973.

171. Kasner, Edward and James Newman. *Mathematics and the Imagination.* New York: Simon and Schuster, 1952.

172. Kline, Morris. *Mathematics in the Modern World: Readings in Scientific American.* San Francisco: Freeman, 1968.

173. Körner, Stephan. *The Philosophy of Mathematics.* New York: Harper, 1960.

174. Moritz, Robert E. *On Mathematics: Collection of Witty, Profound, Amusing Passages about Mathematics and Mathematicians.* New York: Dover, 1942.

175. Newman, James R. (ed.). *The World of Mathematics.* Vol. 1–4. New York: Simon and Schuster, 1956.

176. Phillips, Hubert (Caliban). *My Best Puzzles in Logic and Reasoning.* New York: Dover, 1961.

177. ———. *My Best Puzzles in Mathematics.* New York: Dover, 1961.

178. Rademacher, Hans and Otto Toeplitz. *The Enjoyment of Mathematics.* Princeton: Princeton University Press, 1957.

179. Russell, Bertrand. *Introduction to Mathematical Philosophy.* New York: Simon and Schuster, 1971.

180. Schuh, Fred. *The Master Book of Mathematical Recreations.* New York: Dover, 1968.

181. Steinhaus, H. *Mathematical Snapshots.* New York: Oxford University Press, 1950.

182. Thompson, D'Arcy Wentworth. *On Growth and Form.* John Tyler Bonner (ed.). Abridged ed. New York: Cambridge University Press, 1961.

183. Tietze, Heinrich. *Famous Problems of Mathematics.* New York: Graylock Press, 1965.

184. Waismann, F. *Introduction to Mathematical Thinking.* New York: Harper, 1959.

185. Weyl, Hermann. *Symmetry.* Princeton: Princeton University Press, 1952.

ARTICLES

186. Feit, Walter and John G. Thompson. "Solvability of Groups of Odd Order." *Pacific J. Math.*, **13** (1963), 775–1029.

187. Green, J. W. and W. Gustin. "Quasiconvex Sets." *Can. J. Math.*, **2** (1950), 489–507.

188. Hilbert, David. "Mathematical Problems." *Bulletin Amer. Math. Society*, **8** (1902), 437–479.

189. Matiyasevič, Yuri. "Diophantine Representation of the Set of Prime Numbers." *Soviet Math. Doklady*, **12** (1971), 249–254.

190. Motzkin, I. S. "The Euclidean Algorithm." *Bulletin Amer. Math. Society*, **55** (1949), 1142–1146.

191. Rankin, R. A. "The Difference Between Consecutive Prime Numbers." *J. London Math. Society*, **13** (1938), 242–247.

192. Samuel, P. "About Euclidean Rings." *J. of Algebra*, **19** (1971), 282–301.

193. Weyl, Hermann. "Emmy Noether." *Scripta Mathematica III*, **3** (1935), 201–220.

THE AMERICAN MATHEMATICAL MONTHLY

The Mathematical Association of America publishes a magazine called the *American Mathematical Monthly*, which is at the level of advanced undergraduate and graduate students of mathematics. This remarkable journal has many excellent articles on algebra, a very few of which are included in the following list. An occasional afternoon spent browsing through past issues of the *Monthly* can be very enjoyable.

(a) Anonymous. "Simple Groups." *Amer. Math. Monthly*, **80** (1973), 1128.

(b) Birkhoff, G. "Current Trends in Algebra." *Amer. Math. Monthly*, **80** (1973), 760–782.

(c) Brauer, R. "On a Theorem of Frobenius." *Amer. Math. Monthly*, **76** (1969), 12–15.

(d) Bruckheimer, M., A. C. Bryan, and A. Muir. "Groups which are the Union of Three Subgroups." *Amer. Math. Monthly*, **77** (1970), 52–57.

(e) Cartwright, M. L. "Mathematics and Thinking Mathematically." *Amer. Math. Monthly*, **77** (1970), 20–28.

(f) Cohn, P. M. "Rings of Fractions." *Amer. Math. Monthly*, **78** (1971), 596–615.

(g) ———. "Unique Factorization Domains." *Amer. Math. Monthly*, **80** (1973), 1–18.

(h) Curtis, C. W. "The Classical Groups as a Source of Algebraic Problems." *Amer. Math. Monthly*, **74** (1967), 80–91.

(i) Davis, Martin. "Hilbert's Tenth Problem is Unsolvable." *Amer. Math. Monthly*, **80** (1973), 233–269.

(j) De Meyer, F. "Another Proof of the Fundamental Theorem of Galois Theory." *Amer. Math. Monthly*, **75** (1968), 720.

(k) Dieudonné, J. A. "The Historical Development of Algebraic Geometry." *Amer. Math. Monthly*, **79** (1972), 827–866.

(l) ———. "The Work of Nicholas Bourbaki." *Amer. Math. Monthly*, **77** (1970), 134–145.

(m) Dorwart, H. L. "Irreducibility of Polynomials." *Amer. Math. Monthly*, **42** (1935), 369.

(n)   Dudley, U. "History of a Formula for Primes." *Amer. Math. Monthly*, **76** (1969), 23–28.

(o)   Ellison, W. J. "Waring's Problem." *Amer. Math. Monthly*, **78** (1971), 10–36.

(p)   Gerst, I. and J. Brillhart. "On the Prime Divisors of Polynomials." *Amer. Math. Monthly*, **78** (1971), 250–266.

(q)   Goldstein, L. J. "Density Questions in Algebraic Number Theory." *Amer. Math. Monthly*, **78** (1971), 342–351.

(r)   ———. "A History of the Prime Number Theorem." *Amer. Math. Monthly*, **80** (1973), 599–615.

(s)   Grabiner, J. V. "Is Mathematical Truth Time-Dependent?" *Amer. Math. Monthly*, **81** (1974), 354–365.

(t)   Grost, M. E. "The Smallest Number with a Given Number of Divisors." *Amer. Math. Monthly*, **75** (1968), 725.

(u)   Humphreys, J. E. "Representations of SL(2, p)." *Amer. Math. Monthly*, **82** (1975), 21–39.

(v)   Johnson, C. R. "Positive Definite Matrices." *Amer. Math. Monthly*, **77** (1970), 259–264.

(w)   Kimberling, C. H. "Emmy Noether." *Amer. Math. Monthly*, **79** (1972), 136–149.

(x)   Lefschetz, S. "The Early Development of Algebraic Geometry." *Amer. Math. Monthly*, **76** (1969), 451–460.

(y)   Levinson, N. "Coding Theory: A Counterexample to G. H. Hardy's Conception of Applied Mathematics." *Amer. Math. Monthly*, **77** (1970), 249–258.

(z)   ———. "A Motivated Account of an Elementary Proof of the Prime Number Theorem." *Amer. Math. Monthly*, **76** (1969), 225–245.

(aa) Lloyd, D. B. "Reducibility of Polynomials of Odd Degree." *Amer. Math. Monthly*, **75** (1968), 1081–1084.

(bb) Mahler, K. "How I Became a Mathematician." *Amer. Math. Monthly*, **81** (1974), 981–983.

(cc) Mordell, L. J. "Hardy's 'A Mathematician's Apology.'" *Amer. Math. Monthly*, **77** (1970), 831–836.

(dd) ———. "Reminiscences of an Octogenarian Mathematician." *Amer. Math. Monthly*, **78** (1971), 952–961.

(ee) Polya, G. "Some Mathematicians I Have Known." *Amer. Math. Monthly*, **76** (1969), 746–753.

(ff) Samuel, P. "Unique Factorization." *Amer. Math. Monthly*, **75** (1968), 945–952.

(gg) Shepherdson, J. C. "Weak and Strong Induction." *Amer. Math. Monthly*, **76** (1969), 989–1004.

(hh) Taussky, O. "Sums of Squares." *Amer. Math. Monthly*, **77** (1970), 805–830.

(ii) Vandiver, H. S. "Fermat's Last Theorem: Its History and the Nature of the Known Results Concerning It." *Amer. Math. Monthly*, **53** (1946), 555–578.

(jj) Williams, K. S. "Quadratic Polynomials with the Same Residues." *Amer. Math. Monthly*, **75** (1968), 969–973.

(kk) Wyman, B. F. "What is a Reciprocity Law?" *Amer. Math. Monthly*, **79** (1972), 571–586.

(ll) Young, G. S. "The Linear Functional Equation." *Amer. Math. Monthly*, **65** (1958), 37–38.

(mm) Zassenhaus, H. "What Makes a Loop a Group?" *Amer. Math. Monthly*, **75** (1968), 139–142.

THE WILLIAM LOWELL PUTNAM COMPETITION

The essence of mathematics is problem solving. As such, a particular type of problem has fascinated many mathematicians. We might call such a problem a "Nut." These problems require ingenuity and usually admit fairly short solutions. A special talent is required to crack Nuts quickly, and this talent is worthy of recognition. In each issue, the *American Mathematical Monthly* publishes two sets of Nuts divided as "Elementary" and "Advanced" problems. This division does not mean "Easy" and "Hard." The statement of Elementary problems can be understood with a strong high school mathematical background, whereas understanding the statement of the Advanced problems may require two or more years of college mathematics. The names of those submitting correct solutions to problems are published along with the best solutions.

In addition, each year the Mathematical Association of America sponsors the *Putnam Competition*: a grand Nutcracking competition at various universities and colleges. The high scorers in this competition win scholarship awards and receive recognition for their accomplishments. The Putnam examinations are not "tests for a grade." Rather, they are a chance to pit your wits against "toughies" under time pressure. Anyone who is interested in the Competition can have a sitting arranged at his college or university.

After each competition, the *Monthly* publishes the problems with their solutions and lists the high scoring schools and individuals. See the following recent issues for examples: **81** (1974), p. 1086; **80** (1973) p. 1017; **80** (1973) p. 170; **78** (1971) p. 763.

Cut out, fold, and
tape together

index